Recent Titles in This Series

54 **Robert Greene and S. T. Yau, editors,** Differential geometry (University of California, Los Angeles, July 1990)

53 **James A. Carlson, C. Herbert Clemens, and David R. Morrison, editors,** Complex geometry and Lie theory (Sundance, Utah, May 1989)

52 **Eric Bedford, John P. D'Angelo, Robert E. Greene, and Steven G. Krantz, editors,** Several complex variables and complex geometry (University of California, Santa Cruz, July 1989)

51 **William B. Arveson and Ronald G. Douglas, editors,** Operator theory/operator algebras and applications (University of New Hampshire, July 1988)

50 **James Glimm, John Impagliazzo, and Isadore Singer, editors,** The legacy of John von Neumann (Hofstra University, Hempstead, New York, May/June 1988)

49 **Robert C. Gunning and Leon Ehrenpreis, editors,** Theta functions – Bowdoin 1987 (Bowdoin College, Brunswick, Maine, July 1987)

48 **R. O. Wells, Jr., editor,** The mathematical heritage of Hermann Weyl (Duke University, Durham, May 1987)

47 **Paul Fong, editor,** The Arcata conference on representations of finite groups (Humboldt State University, Arcata, California, July 1986)

46 **Spencer J. Bloch, editor,** Algebraic geometry – Bowdoin 1985 (Bowdoin College, Brunswick, Maine, July 1985)

45 **Felix E. Browder, editor,** Nonlinear functional analysis and its applications (University of California, Berkeley, July 1983)

44 **William K. Allard and Frederick J. Almgren, Jr., editors,** Geometric measure theory and the calculus of variations (Humboldt State University, Arcata, California, July/August 1984)

43 **François Trèves, editor,** Pseudodifferential operators and applications (University of Notre Dame, Notre Dame, Indiana, April 1984)

42 **Anil Nerode and Richard A. Shore, editors,** Recursion theory (Cornell University, Ithaca, New York, June/July 1982)

41 **Yum-Tong Siu, editor,** Complex analysis of several variables (Madison, Wisconsin, April 1982)

40 **Peter Orlik, editor,** Singularities (Humboldt State University, Arcata, California, July/August 1981)

39 **Felix E. Browder, editor,** The mathematical heritage of Henri Poincaré (Indiana University, Bloomington, April 1980)

38 **Richard V. Kadison, editor,** Operator algebras and applications (Queens University, Kingston, Ontario, July/August 1980)

37 **Bruce Cooperstein and Geoffrey Mason, editors,** The Santa Cruz conference on finite groups (University of California, Santa Cruz, June/July 1979)

36 **Robert Osserman and Alan Weinstein, editors,** Geometry of the Laplace operator (University of Hawaii, Honolulu, March 1979)

35 **Guido Weiss and Stephen Wainger, editors,** Harmonic analysis in Euclidean spaces (Williams College, Williamstown, Massachusetts, July 1978)

34 **D. K. Ray-Chaudhuri, editor,** Relations between combinatorics and other parts of mathematics (Ohio State University, Columbus, March 1978)

33 **A Borel and W. Casselman, editors,** Automorphic forms, representations and L-functions (Oregon State University, Corvallis, July/August 1977)

32 **R. James Milgram, editor,** Algebraic and geometric topology (Stanford University, Stanford, California, August 1976)

31 **Joseph L. Doob, editor,** Probability (University of Illinois at Urbana-Champaign, Urbana, March 1976)

(Continued in the back of this publication)

Differential Geometry: Riemannian Geometry

Proceedings of Symposia in
PURE MATHEMATICS

Volume 54, Part 3

Differential Geometry:
Riemannian Geometry

Robert Greene
S. T. Yau
Editors

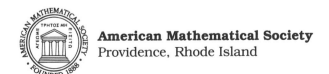

American Mathematical Society
Providence, Rhode Island

PROCEEDINGS OF THE SUMMER RESEARCH INSTITUTE
ON DIFFERENTIAL GEOMETRY
HELD AT THE UNIVERSITY OF CALIFORNIA, LOS ANGELES
LOS ANGELES, CALIFORNIA
JULY 8–28, 1990

with the support from the National Science Foundation
Grant DMS–8913610.

1991 *Mathematics Subject Classification*.
Primary 53A10, 49F20, 58E20, 58E12, 58G11 (Part 1)
81E13, 53C80, 53B50, 32C10, 53C15, 58F05, 83C75 (Part 2)
53C20, 53C40, 58C40, 58G25, 58F07, 58F11, 58F17 (Part 3).

Library of Congress Cataloging-in-Publication Data

Riemannian geometry/Robert Greene and S. T. Yau, editors.

    p.   cm.—(Proceedings of symposia in pure mathematics; v. 54, pt. 3) (Differential geometry; pt. 3)

    "Proceedings of a Summer Research Institute on Differential Geometry, held at the University of California, Los Angeles, July 8–28, 1990"—T.p. verso.

    ISBN 0-8218-1494-X (Part 1)

    ISBN 0-8218-1495-8 (Part 2)

    ISBN 0-8218-1496-6 (Part 3)

    ISBN 0-8218-1493-1 (set) (alk. paper)

    1. Geometry, Riemannian—Congresses. I. Greene, Robert Everist, 1943– . II. Yau, Shing-Tung, 1949– . III. Summer Research Institute on Differential Geometry (1990: University of California, Los Angeles) IV. Series. V. Series: Differential geometry; pt. 3.

QA641.D3833   1993   pt. 3

516.3$'$6 s—dc20   92-34056

[516.3$'$73]   CIP

This publication was typeset using $\mathcal{A}\mathcal{M}\mathcal{S}$-TEX,
the American Mathematical Society's TEX macro system.

10 9 8 7 6 5 4 3 2 1   98 97 96 95 94 93

Dedication.

To Professor S.-S. Chern,

in appreciation of his formative influence on

modern differential geometry.

Contents

(∗ denotes one-hour survey lectures)

Preface   xxi

PART 1

Open Problems in Geometry*
    SHING TUNG YAU   1

Questions and Answers about Area-Minimizing Surfaces and Geometric
    Measure Theory*
    FRED ALMGREN   29

On the Geometrical Origin and the Solutions of a Degenerate
    Monge-Ampère Equation
    DAVID BAO AND TUDOR RATIU   55

Differential Geometry and the Design of Gradient Algorithms*
    ROGER W. BROCKETT   69

Spectral Geometry of V-Manifolds and its Application to Harmonic
    Maps
    YUAN-JEN CHIANG   93

Constructing Harmonic Maps into the Hyperbolic Space
    HYEONG IN CHOI AND ANDREJS TREIBERGS   101

Spherical Minimal Immersions of Spherical Space Forms
    DENNIS DETURCK AND WOLFGANG ZILLER   111

Banach Manifolds of Solutions to Nonlinear Partial Differential Equa-
    tions, and Relations with Finite-Dimensional Manifolds
    JOSEF DORFMEISTER   121

Some New Harmonic Maps
    ROBERT HARDT   141

Existence of Unstable Minimal Surfaces in Manifolds with Homology and Applications to Triply Periodic Minimal Surfaces
JOEL HASS, JON T. PITTS, AND J. H. RUBINSTEIN   147

Closed Minimal Submanifolds in the Spheres*
WU-YI HSIANG   163

Local and Global Behaviour of Hypersurfaces Moving by Mean Curvature*
GERHARD HUISKEN   175

The Level-Set Flow on a Manifold
TOM ILMANEN   193

Unstable Solutions of Two-Dimensional Geometric Variational Problems*
JÜRGEN JOST   205

Harmonic Maps and Superrigidity
JÜRGEN JOST AND SHING TUNG YAU   245

Harmonic Functions of Polynomial Growth on Complete Manifolds
ATSUSHI KASUE   281

The Structure of Constant Mean Curvature Embeddings in Euclidean Three Space
NICK KOREVAAR AND ROB KUSNER   291

Uniformization of Spherical CR Manifolds and the CR Yamabe Problem
ZHONGYUAN LI   299

The Theory of Harmonic Functions and Its Relation to Geometry*
PETER LI   307

Harmonic Maps with Prescribed Singularities
YAN YAN LI AND GANG TIAN   317

Some Recent Results on Harmonic Maps to Spheres*
FANG HUA LIN   327

The Geometry, Topology, and Existence of Periodic Minimal Surfaces*
WILLIAM H. MEEKS III   333

Soap Films and Mathematics
FRANK MORGAN   375

Uniform Boundary Regularity Estimates for Minima of Certain
Quadratic Functionals
LIBIN H. MOU   381

Self-similar Solutions and Asymptotic Behavior of Flows of Nonparametric Surfaces Driven by Gauss or Mean Curvature
VLADIMIR OLIKER   389

A Report on Geometric Quantization
P. L. ROBINSON   403

Motion of Curves by Crystalline Curvature, Including Triple
Junctions and Boundary Points
JEAN E. TAYLOR   417

Recent Progress in Submanifold Geometry*
CHUU-LIAN TERNG   439

Constant Mean Curvature Surfaces in the Heisenberg Group
PER TOMTER   485

Complete Immersions of Constant Mean Curvature*
HENRY C. WENTE   497

Banach Manifolds of Minimal Surfaces in the 4-Sphere
HONGYOU WU   513

On the Isolatedness for the Solutions of Plateau's Problem
STEPHEN ZHENG   541

PART 2

Invariants of Gauss Maps of Theta Divisors
MALCOLM ADAMS, CLINT MCCRORY, TED SHIFRIN,
AND ROBERT VARLEY   1

On Characteristics of Hypersurfaces in Symplectic Manifolds
AUGUSTIN BANYAGA   9

Disprisoning and Pseudoconvex Manifolds
    JOHN K. BEEM   19

On the Holonomy of Lorentzian Manifolds
    L. BERARD BERGERY AND A. IKEMAKHEN   27

Spinors, Dirac Operators, and Changes of Metrics
    JEAN-PIERRE BOURGUIGNON   41

The Hyperkähler Geometry of the ADHM Construction and Quaternionic
    Geometric Invariant Theory
    CHARLES P. BOYER AND BENJAMIN M. MANN   45

Non-Abelian Vortices and a New Yang-Mills-Higgs Energy
    STEVEN B. BRADLOW   85

What Are the Best Almost-Complex Structures on the 6-Sphere?
    EUGENIO CALABI AND HERMAN GLUCK   99

The Failure of Complex and Symplectic Manifolds to be Kählerian
    LUIS A. CORDERO, MARISA FERNÁNDEZ, AND ALFRED GRAY   107

Nonabelian Hodge Theory*
    KEVIN CORLETTE   125

Geometric Invariants and Their Adiabatic Limits
    XIANZHE DAI   145

Geometry of Elementary Particles
    ANDRZEJ DERDZINSKI   157

Existence of Connections with Prescribed Yang-Mills Currents
    DENNIS DETURCK, HUBERT GOLDSCHMIDT, AND
    JANET TALVACCHIA   173

Vector Bundles over Homogeneous Spaces and Complete, Locally
    Symmetric Spaces
    LANCE D. DRAGER AND ROBERT L. FOOTE   183

Margulis Space-Times
    TODD A. DRUMM   191

Incomplete Flat Homogeneous Geometries
    D. DUNCAN AND E. IHRIG   197

Geodesic and Causal Behavior of Gravitational Plane Waves: Astigmatic
   Conjugacy
   PAUL E. EHRLICH AND GERARD G. EMCH   203

Curvature of Singular Spaces via the Normal Cycle
   JOSEPH H. G. FU   211

The Nonintegrable Phase Factor and Gauge Theory
   RONALD O. FULP   223

The Lorentzian Version of the Cheeger-Gromoll Splitting Theorem and
   its Application to General Relativity
   GREGORY J. GALLOWAY   249

Weyl Structures on Self-Dual Conformal Manifolds
   PAUL GAUDUCHON   259

Chiral Anomalies and Dirac Families in Riemannian Foliations
   JAMES F. GLAZEBROOK AND FRANZ W. KAMBER   271

What is the Shape of Space in a Spacetime?
   STEVEN G. HARRIS   287

The Kinematics of the Gravitational Field
   ADAM D. HELFER   297

Support Theorems in Integral Geometry and Their Applications
   SIGURDUR HELGASON   317

Killing Spinors and Eigenvalues of the Dirac Operator
   OUSSAMA HIJAZI   325

Einstein Metrics on Circle Bundles
   GARY R. JENSEN AND MARCO RIGOLI   331

Topological and Differentiable Structures of the Complement of an
   Arrangement of Hyperplanes
   TAN JIANG AND STEPHEN S.-T. YAU   337

The Relationship between the Moduli Spaces of Vector Bundles on K3
   Surfaces and Enriques Surfaces
   HOIL KIM   359

Self-Dual Manifolds with Symmetry
    C. LeBrun and Y. S. Poon   365

Remarks on the Skyrme Model
    Elliott H. Lieb   379

Geometric Settings for Quantum Systems with Isospin
    E. B. Lin   385

Heat Kernels on Covering Spaces and Topological Invariants
    John Lott   391

The Heat Kernels of Symmetric Spaces
    Qi-Keng Lu   401

CR-Geometry and Deformations of Isolated Singularities*
    John J. Millson   411

Generalized Symplectic Geometry on the Frame Bundle of a Manifold
    L. K. Norris   435

Complex Geometry and String Theory*
    D. H. Phong   467

Constructing Non-Self-Dual Yang-Mills Connections on S^4 with
    Arbitrary Chern Number
    Lorenzo Sadun and Jan Segert   529

The Yang-Mills Measure for the Two-Sphere
    Ambar Sengupta   539

Examples of Nonminimal Critical Points in Gauge Theory*
    L. M. Sibner   547

Spectral Invariants of Pseudoconformal Manifolds
    Nancy K. Stanton   551

L_2-Cohomology and Index Theory of Noncompact Manifolds*
    Mark Stern   559

On Calabi-Yau Three-Folds Fibered Over Smooth Complex Surfaces
    Wing-Wah Sung   577

Degeneration of Kähler-Einstein Manifolds. I
    GANG TIAN   595

Flat Connections on Products of Determinant Bundles
    YUE LIN L. TONG   611

Surfaces Riemanniennes à Singularités Simples
    MARC TROYANOV   619

Remarks on Certain Higher-Dimensional Quasi-Fuchsian Domains
    S.-T. YAU AND F. ZHENG   629

L^p-Cohomology: Banach   Spaces   and   Homological   Methods on
    Riemannian Manifolds*
    STEVEN ZUCKER   637

PART 3

Some Concepts and Methods in Riemannian Geometry*
    ROBERT E. GREENE   1

Isometric Immersions of H^n into H^{n+1}
    KINETSU ABE AND ANDREW HAAS   23

The Distance-Geometry of Riemannian Manifolds with Boundary
    STEPHANIE B. ALEXANDER, I. DAVID BERG, AND RICHARD L.
    BISHOP   31

Hypersurfaces and Nonnegative Curvature
    STEPHANIE B. ALEXANDER AND ROBERT J. CURRIER   37

Shortening Space Curves
    STEVEN J. ALTSCHULER   45

Degeneration of Metrics with Bounded Curvature and Applications to
    Critical Metrics of Riemannian Functionals*
    MICHAEL T. ANDERSON   53

The Geometry of Totally Geodesic Hypersurfaces in Hyperbolic Manifolds
    ARA BASMAJIAN   81

Finiteness of Diffeomorphism Types of Isospectral Manifolds
    ROBERT BROOKS, PETER PERRY, AND PETER PETERSEN V   89

Small Eigenvalues of the Laplacian on Negatively Curved Manifolds
  P. Buser, B. Colbois, and J. Dodziuk   95

Geometrically Tame Hyperbolic 3-Manifolds
  Richard D. Canary   99

Isoperimetric Constants and Large Time Heat Diffusion in Riemannian
  Manifolds
  Isaac Chavel and Edgar A. Feldman   111

Large Time Behavior of Solutions of the Heat Equation
  Isaac Chavel and Leon Karp   123

Manifolds with 2-Nonnegative Curvature Operator
  Haiwen Chen   129

The Geometry of Isospectral Deformations
  Dennis DeTurck, Herman Gluck, Carolyn Gordon,
  and David Webb   135

Quadric Representation of a Submanifold and Spectral Geometry
  Ivko Dimitrić   155

Embedded Eigenvalues for Asymptotically Flat Surfaces
  Harold Donnelly   169

Manifolds of Nonpositive Curvature*
  Patrick Eberlein, Ursula Hamenstädt, and
  Viktor Schroeder   179

Topological Rigidity for Compact Nonpositively Curved Manifolds*
  F. T. Farrell and L. E. Jones   229

Almost Nonnegatively Curved Manifolds*
  Kenji Fukaya and Takao Yamaguchi   275

Local Structure of Framed Manifolds
  Patrick Ghanaat   283

Constant Affine Mean Curvature Hypersurfaces of Decomposable Type
  Salvador Gigena   289

Heat Equation Asymptotics
  Peter B. Gilkey   317

Nonnegatively Curved Manifolds Which Are Flat Outside a Compact Set
    R. E. GREENE AND H. WU   327

Spaces of Nonnegative Curvature*
    DETLEF GROMOLL   337

Critical Point Theory for Distance Functions*
    KARSTEN GROVE   357

Spectral Geometry in Higher Ranks: Closed Geodesics and Flat Tori
    DAVID GURARIE   387

Ellipticity of Local Isometric Embeddings
    CHONG-KYU HAN   409

Curve Flows on Surfaces and Intersections of Curves
    JOEL HASS AND PETER SCOTT   415

Curve Shortening, Equivariant Morse Theory, and Closed Geodesics on
    the 2-Sphere
    NANCY HINGSTON   423

Morse Theory for Geodesics
    DIANE KALISH   431

On Some Metrics on $S^2 \times S^{2*}$
    MASATAKE KURANISHI   439

Curve-straightening
    ANDERS LINNÉR   451

Ball Covering Property and Nonnegative Ricci Curvature Outside a
    Compact Set
    ZHONG-DONG LIU   459

General Conjugate Loci are not Closed
    CHRISTOPHE M. MARGERIN   465

Collapsing of Riemannian Manifolds and Their Excesses
    YUKIO OTSU   479

Gromov-Hausdorff Convergence of Metric Spaces*
    PETER PETERSEN V   489

Cartan Connections*
  ERNST A. RUH   505

Eigenvalue Estimates for the Laplacian with Lower Order Terms on
  a Compact Riemannian Manifold
  ALBERTO G. SETTI   521

Positive Ricci Curvature on Compact Simply Connected 4-Manifolds
  JI-PING SHA AND DAGANG YANG   529

Volume Growth and Finite Topological Type
  ZHONGMIN SHEN AND GUOFANG WEI   539

Recent Developments in Sphere Theorems*
  KATSUHIRO SHIOHAMA   551

On the Excess of Open Manifolds
  TAKASHI SHIOYA   577

Local Theory of Affine 2-Spheres
  U. SIMON AND C. P. WANG   585

Riemannian Manifolds with Completely Integrable Geodesic Flows
  R. J. SPATZIER   599

Differentiable Structure on Spheres and Curvature
  YOSHIHIKO SUYAMA   609

Spectral Theory for Operator Families on Riemannian Manifolds*
  Z. I. SZABO   615

Riemannian Foliations and Tautness
  PHILIPPE TONDEUR   667

Eigenvalue Problems for Manifolds with Singularities
  JOHAN TYSK   673

Some Rigidity Aspects of Riemannian Fibrations
  GERARD WALSCHAP   679

Hausdorff Convergence and Sphere Theorems
    JYH-YANG WU   685

Automorphism Groups and Fundamental Groups of Geometric
    Manifolds
    ROBERT J. ZIMMER   693

Preface

The 1990 American Mathematical Society Summer Institute on Differential Geometry took place at University of California, Los Angeles from July 9 to July 27, 1990. This was the largest AMS Summer Institute ever. There were 426 registered participants and 270 lectures. The organizing committee for the Institute consisted of Robert Bryant, Duke University; Eugenio Calabi, University of Pennsylvania; S. Y. Cheng, University of California, Los Angeles; H. Blaine Lawson, State University of New York, Stony Brook; H. Wu, University of California, Berkeley; and as co-chairmen, Robert E. Greene, University of California, Los Angeles, and S. T. Yau, Harvard University.

In the years since the previous AMS Summer Institute on Differential Geometry in 1973, the field has undergone a remarkable expansion, both in the number of people working in the area and in the number and scope of the topics under investigation. Even in the context of the rapid growth of mathematics as a whole during this period, the growth of geometry is striking.

It is our hope that the three volumes of these proceedings, taken as a whole, will provide a broad overview of geometry and its relationship to mathematics in toto, with one obvious exception; the geometry of complex manifolds and the relationship of complex geometry to complex analysis were the subject of one week of the (three-week) 1989 AMS Summer Institute on Several Complex Variables and Complex Geometry. While some topics in complex geometry arose naturally in the 1990 Summer Institute and are covered in these Proceedings, the coverage of this subject in 1989 justified a reduced emphasis in 1990.

Thus the reader seeking a complete view of geometry would do well to add the second volume on complex geometry from the 1989 Proceedings to the present three volumes.

Each week of the 1990 Summer Institute was given a general emphasis as to subject and the Proceedings volumes are organized in the same way. While overlap is natural, and indeed inevitable, the subjects of the volumes are as follows:

I. Partial differential equations on manifolds: harmonic functions and mapping, Monge-Ampre equation, differential systems, minimal submanifolds;

II. Geometry in mathematical physics and related topics: gauge theory, symplectic geometry, complex geometry, L^2 cohomology, Lorentzian geometry;

III. Riemannian geometry: curvature and topology, groups and manifolds, dynamical systems in geometry, spectral theory of Riemannian manifolds.

The articles in these Proceedings are also of several types. We requested broad-ranging surveys from the people who had given one-hour survey lectures at the Institute. These articles are marked with an asterisk in the table of contents. Such surveys were also encouraged from other participants. The remaining articles are either research papers in the usual sense or relatively brief announcements of results. But in these categories as well, we encouraged authors to provide more background information and references to related work than usual. Thus we hope that readers will find the volumes a source of broad perspectives on the rapidly expanding literature of geometry.

The editors themselves essayed two efforts in this direction. Volume I begins with a problem list by S. T. Yau, successor to his 1980 list [Seminar in Differential Geometry, Annals of Math. Studies, no. 102, Princeton University Press]. Volume III begins with an overview by R. E. Greene of some recent trends in Riemannian geometry, in the interests of identifying themes common to the remaining papers in that volume and in outlining certain topics that, as it happened, would not otherwise have been included.

An event such as the 1990 AMS Summer Institute involves the efforts of a great many people. We would like to thank all the participants, both for their participation in the Institute and for their prompt and abundant response to our call for papers for these Proceedings. We are indebted to the staff of the UCLA Mathematics Department for their cooperation during the Summer Institute and to the American Mathematical Society in general and in particular to Mr. Wayne Drady and Ms. Susan Blyth for their invaluable on-the-spot assistance during the Institute. Without the financial support of the National Science Foundation, the Summer Institute could not have had anything like the scope it in fact attained; we are particularly grateful for their willingness to support the participation of advanced graduate students, a willingness that we believe was a valuable investment in the future of geometry. Finally, we are indebted to the publication staff of the American Mathematical Society and particularly to Ms. Alison Buckser, Ms. Donna Harmon, and Ms. Christine Thivierge, whose unfailing patience and helpfulness made the preparation of these volumes a far easier task than it would have been otherwise.

Robert E. Greene
S. T. Yau

Proceedings of Symposia in Pure Mathematics
Volume **54** (1993), Part 3

Some Concepts and Methods in Riemannian Geometry

ROBERT E. GREENE

To introduce the volume of these Proceedings dealing with Riemannian geometry, it seems natural to recall how things stood in Riemannian geometry at the time of the last Summer Institute on Differential Geometry, the Institute at Stanford University in 1973. The decade or so previous to that Institute had seen some signal developments in global Riemannian geometry:

The 1/4-pinched theorem of Berger and Klingenberg had brought to completion the ideas of Rauch on the characterization of (topological) spheres as manifolds admitting metrics of sufficiently pinched positive curvature. The corresponding differentiable sphere theorem had reached essentially its present form soon after ([**SS**] and [**Ru1**]). Berger completed the picture, in even dimensions, by classifying those manifolds which admit exactly 1/4-pinched metrics [**Be1**].

The structure of open manifolds of nonnegative curvature had been elucidated in the work of Cheeger, Gromoll and Meyer [**GrM, CG**]; this work extended to higher dimensions concepts developed by Cohn-Vossen for surfaces and at the same time showed definitively the usefulness of the striking comparison theorem of Toponogov (see, e.g., [**CE**]). These results are global geometric theorems of what one might describe as the classical type: From geometric hypotheses, one deduces that the underlying manifold is a specific manifold topologically (Berger, Klingenberg, et al. in the compact case, Gromoll-Meyer on open manifolds), or one of relatively few possibilities (Berger's 1/4-pinched manifolds) or at least has a certain general structure (Cheeger-Gromoll). Moreover, the proof techniques use classical second-variation techniques (Rauch Comparison) or what amount to integrated versions of second variation (Toponogov Comparison).

As things turned out, the 1973 Summer Institute occurred at the dawn of a new era in geometry, though some of the themes of the new era had already been sounded. One aspect of the new age was the explosive development of

1991 *Mathematics Subject Classification*. Primary 53C23, 53C20; Secondary 53C55.

This paper is in final form and no version of it will be submitted elsewhere.

partial differential equations methods; the scope of this is abundantly clear from the first two volumes of these Proceedings. But in Riemannian geometry in the pure sense, shifts of emphasis and new techniques were also about to appear.

In an earlier survey [G], I traced the evolution of Riemannian geometry towards the study of families of manifolds, rather than characteristics of specific manifolds or structures; that article concentrated on specific results and covered the time up to 1989. Here I want to view the same lines of development not only as they have continued to the present (early 1992) but also more in terms of underlying techniques than just specific theorems. Techniques are ultimately justified by the specific results they produce. But the directions in which a field can move are dictated by the techniques and methods available, barring those occasional burst of originality in which a new method and a profound consequence appear simultaneously.

In the case of Riemannian geometry, at least three basic, new methods have come to the fore in the last two decades. These are

(1) adapting smooth-function concepts to nonsmooth functions,

(2) coordinates in which the metric has maximal regularity,

(3) estimation of coverings, to which is associated metric space compactness results for classes of Riemannian manifolds and, via (2), convergence theorems including the concept of collapsing.

A fourth theme is of a more general sort, namely, the more or less systematic attempt to carry out these and other ideas of Riemannian geometry in the context of Ricci curvature, rather than just sectional curvature.

The author is indebted to Peter Petersen, V for numerous helpful and informative conversations and to G. Kallo for his observations on recent convergence results.

1. Smooth methods for nonsmooth functions

The late fifties and the sixties were the hey-day of Morse theory. The general idea of analyzing a manifold's structure via the critical points of a smooth function on it scored numerous specific triumphs and became part of every geometer or topologist's tool kit. Ironically, this idea, the genesis of which is so clearly geometric intuition, was rather limited in its direct application to Riemannian geometry as such. Its natural geometric application tended to be to the analysis of geodesic behavior; i.e., the technique was applied to spaces of curves, with geodesics as critical points; here the energy function was smooth and amenable to Morse-theoretic analysis. But few (finite-dimensional) Riemannian manifolds come to hand pre-equipped with a smooth function embodying information about their geometry. The functions that arise naturally are generally obtained from the Riemannian distance function, and Riemannian distance has predictable smoothness properties only inside the injectivity radius. Thus the question naturally comes up of how to do Morse-theoretic analysis for nonsmooth functions.

As it happens, this topic has roots in the distant past, distant at least by the standards of mathematical history, in partial differential equations on manifolds. Already in 1959, Matthew Gaffney [**Gaf**] had shown that on a complete Riemannian manifold M, there is a globally Lipschitz continuous C^∞ function $\rho : M \to \mathbb{R}$ with the property that $\rho^{-1}([-\infty, \alpha])$ is compact for each $\alpha \in \mathbb{R}$. One calls a function with such compact sublevel sets an exhaustion function for M. Of course on a complete manifold, the function $y \to \mathrm{dis}_M(y, x)$ is a Lipschitz continuous exhaustion, for each fixed $x \in M$. Gaffney's result amounted to showing that such a distance function—which is generally not smooth—could be smoothed out in such a way as to retain global Lipschitz continuity, or, what is the same, to have a gradient vector field of uniformly bounded length.

Gaffney's result is important in partial differential equations because it enables one to approximate functions by functions of compact support with predictable results as far as derivatives are concerned. In particular, one obtains a graph-norm density statement for certain first-order operators: Let P be a first-order, linear differential operator with coefficients that are uniformly bounded (i.e., there is a uniform bound on the coefficients at each point when the operator is written in a coordinate system orthonormal at that point). Then C^∞ functions of compact support are dense, in the graph norm $\|\varphi\| + \|P\varphi\| + \|P^*\varphi\|$, in the set of all functions in L^2 for which the closure of P and P^* are defined and L^2. This seemingly rather technical result plays an important role in the L^2 $\bar{\partial}$-method for complex manifolds ([**AV**], cf., e.g., [**GW6**])

Partial differential equations requirements led the author and H. Wu to pursue this question of smooth approximation further. In [**GW1, GW2, GW3, GW4**], we showed how smooth approximations could be obtained that preserved certain important geometric properties. In particular, the Riemannian convolution smoothing process developed in [**GW1**] has the following geometrically significant properties:

(a) the C^∞ approximations converge uniformly on compact subsets to the original function;

(b) if the original function is Lipschitz continuous with Lipschitz constant B (i.e., $|f(x) - f(y) \leq B\,\mathrm{dis}(x, y))$, then the C^∞ approximations can be taken to have gradients of Riemannian length $\leq B + \varepsilon$, for any fixed $\varepsilon > 0$.

Property (b) requires only Lipschitz continuity in a neighborhood of a compact set to get the gradient estimate also on that compact set.

In addition, the approximations can be taken to have nowhere vanishing gradient (on a given compact set) provided that the original, nonsmooth function had a direction of strict decrease in a suitable sense. For this, one defines a continuous vector field V (defined in a neighborhood of a compact set K) to be subgradient on a compact set K for a Lipschitz continuous function f if there is a $C > 0$ such that for each $p \in K$.

$$\lim_{t \to 0+} \sup \inf \frac{f(p) - f(\varphi_t(p))}{t} \geq C$$

where φ_t is the "geodesic flow" generated by V; i.e., $\varphi_t(p) = \gamma(t)$ where $\gamma(0) = p$, $\gamma'(0) = V$, and γ is a geodesic. (The definition could equivalently use any other continuous-in-V, smooth-in-t φ with, $\forall p$, $\varphi_0(p) = p$, $\frac{d}{dt}\varphi_t(p) = V(p)$.)

Intuitively, a vector field V is subgradient for f if f decreases at a definite positive rate along V. Now in [GW2], it was shown that if a function f has a subgradient, then the smooth approximations of [GW1] have everywhere a nonvanishing gradient, which indeed can be taken to have length $\geq (1 - \eta)C/\|V\|$, for a given $\eta > 0$ (cf. the argument on p. 278ff, [GW2]). This means that Morse theory applies in effect to continuous functions with subgradient vector fields.

This approach via approximations works particularly well when the original function is convex. In this case, subgradients always exist, away from the minimum set (if any). And the Morse theory via approximation yields a product structure on the complement of the minimum set [GS1, GS2]. Somewhat surprisingly this applies even if the convex function does not have compact level sets [GS1, GS2]. In the case of a strictly convex function with a minimum one can conclude that the manifold is diffeomorphic to $\mathbb{R}^n$; this can be used to provide a proof of the Gromoll-Meyer theorem that is conceptually (and otherwise) simpler than the original proof in [GrM]; this is carried out in [GW4]. The method of [GW1–3] also yields a result for the complex case: a complete Kähler manifold of positive sectional curvature is a Stein manifold [GW4].

While this circle of ideas deals completely effectively with noncritical Morse theory for nonsmooth functions, some conceptual simplifications are possible. The first one is in effect just a technical simplification albeit a significant one conceptually: Note that the existence of a subgradient implies immediately the existence of a one-parameter group of diffeomorphisms along which the function f is strictly decreasing, provided the function f is Lipschitz continuous. One need only note that, if V is subgradient, then so is any sufficiently good C^0 approximation of V so that one can take V to be C^∞ by approximation and use the one-parameter group of V. This simplified picture suffices for many purposes (though not for the full differentiable structure picture, in general, for which the full smoothing picture is needed). This picture is particularly appealing when the function in question is distance as in [GoS] from a fixed point $p \in M$. In this case, it is easy to characterize geometrically when a point is noncritical in the subgradient sense (i.e., when there exists a vector field which is subgradient in a neighborhood of the point). Specifically, this is easily seen to be equivalent at $q \in M$ to the condition that $\exists v \in T_q M$ such that all minimal geodesics from q to

p make an angle strictly less than $\pi/2$ with v.

A second direction of further thought represents an attempt to generalize critical point Morse theory, not just the noncritical theory, to nonsmooth functions. For this, one needs an idea of when a nonsmooth function has a Hessian of index $\geq k$ for some integer q (at a critical point). The natural idea here is to use the concept of a C^∞ support function: a function F supports from above (from below) a function f at p if $F(p) = f(p)$ and $F(x) \geq f(x)$ (resp. $F \leq f$) for all x in a neighborhood of p. Roughly speaking, one would like to define a function f (not necessarily smooth) to have a critical point of index at least k at a point p if it is supported from above at p by a smooth function with a critical point of index q.

Technically, this suffers from certain difficulties; for example, the function on $\mathbb{R}^2$ sending $(x, y) \to |x| - |y|$ should intuitively have index 1 at the origin, but in fact it is not supported above at the origin by any smooth function. Nonetheless, there are circumstances where the support function concept has been of great usefulness. In particular, it defines a natural concept of subharmonicity that occurs, e.g., in the splitting theorem for Ricci curvature [**CG2**]. This method can also be used for cases intermediate between sectional and Ricci curvature [**Wu1, Wu2**]. It also plays a significant role in several complex variables, providing a geometric handle on plurisubharmonicity of the nonsmooth functions arising from distance [**E, GW5**].

2. Preferred coordinates

In Riemannian geometry, it is natural to look for local coordinate systems that are closely related to the Riemannian metric, or even that are "canonical," in some sense of being essentially determined by the metric itself. In its most general form, this idea is virtually as old as Riemannian geometry itself. In fact, in an extended interpretation, the idea is far older than that, since the question of how to make (flat) maps of the round earth goes back many centuries. As far as Riemannian geometry and its predecessor, surface theory, are concerned, coordinates constructed from geodesics dominated the scene for a long time; and in particular, what we now call Riemannian normal coordinates (i.e., the local inverse of the geodesic exponential map) seemed the ideal approach.

In complex geometry, other coordinate systems—isothermal coordinates, and later Bergman representative coordinates—were also used. But these seemed at first to have little impact on the real case. However, isothermal coordinates in particular introduced the idea that especially attractive coordinates could be found by solving partial differential equations.

The obvious advantage of Riemannian normal coordinates is that at one point they provide the greatest possible resemblence to euclidean space, i.e., $g_{ij} = \delta_{ij}$ and $dg_{ij} = 0$. And of course they provide a convenient intuitive visualization of the metric, since, in their polar version, they are intimately related to Jacobi fields. E. Cartan as early as the 1930s already realized that this approach enabled one to compute the metric to infinite order from the curvature and its covariant derivatives of all orders at the center of the coordinates. However, one can see from a derivative count that the estimation of the first derivatives of the g_{ij} in such coordinates on a given neighborhood of the center requires bounds not just on the curvature but also on the covariant derivative of the curvature (on the neighborhood). This was demonstrated explicitly by Kaul [K].

It is again a natural question whether one might not find coordinates in which better estimates of the derivatives of the metric coefficients over the whole coordinate neighborhood were possible if one were prepared to give up the simple and explicit description of the coordinates in terms of geodesics. De Turck and Kazdan [DK] showed that the best possible estimates, best in the sense of (elliptic) regularity theory, were obtained if harmonic coordinates were used. (This question was apparently also investigated, especially for dimension 2, in independent work in the Russian literature, cf., e.g., [N].)

In a remarkable piece of work, Jost and Karcher [JK] showed that such harmonic coordinates could be obtained that would yield optimal estimates on metric coefficients depending only on basic geometric data, i.e., injectivity radius and upper and lower curvature bounds. It is striking that, as a preliminary, coordinates were obtained by geodesic constructions, without elliptic partial differential equations, that already gave better metric coefficient estimates than Riemannian normal coordinates.

Following the pattern long familiar from the Arzela-Ascoli theorem and its applications in complex analysis, one might expect that such estimates would be associated to some convergence properties for (subsequences of) sequences of Riemannian metrics. This idea, of convergence of metrics, is closely related to other ideas of Riemannian manifolds converging as metric spaces. Gromov, pursuing this line of thought, had been led to conjecture [GLP] that, in the presence of uniform sectional curvature bounds and a uniform lower bound on injectivity radius, every sequence of Riemannian metrics (on a fixed manifold) would have a subsequence that converged in the $C^{1,\alpha}$ norm, $0 < \alpha < 1$ (after correction by suitable diffeomorphisms). Rather surprisingly, he seems to have reached this conjecture without the information of [JK], which makes this type of convergence plausible, $C^{k,\alpha}$-norm estimates being usual in elliptic theory. (Perhaps the earlier Russian literature of the Aleksandrov school was suggestive.)

In any case, the conjecture of [GLP] was put on a firm footing by the detailed proof (with the optimal possible regularity) by the present author and H. Wu in [GW8] and independently by S. Peters in [P]. Precisely, it is shown that

THEOREM (Greene-Wu, Peters). *If* $\{g_\lambda : \lambda = 1, 2, 3, \ldots\}$ *is a sequence of* C^∞ *Riemannian metrics on a compact manifold* M *and if* (a) *the injectivity radii* $i(g_\lambda) \geq i_0 > 0$ *and* (b) *the sectional curvatures of the* g_λ *are between* $-\Lambda$

and Λ, where Λ and i_0 are independent of λ, then there is a subsequence $\{g_{\lambda_j} : j = 1, 2, \ldots\}$ and a sequence of C^∞ diffeomorphisms $F_{\lambda_j} : M \to M$ such that the sequence $\{F^*_{\lambda_j} g_{\lambda_j}\}$ converges in the $C^{1,\alpha}$ norms, all $\alpha \in (0, 1)$, to a positive definite Riemannian metric belonging to $C^{1,\alpha}$, all $\alpha \in (0, 1)$.

The development of this result was somewhat convoluted. The result was conjectured in [GLP] with a potential method proof suggested. This method can be carried out but does not yield optimal regularity. The fundamental step toward optimal $C^{1,\alpha}$ regularity was taken by Jost and Karcher on harmonic coordinates in [JK], and the application of their results to convergence was made by Greene-Wu [GW8] and Peters [P2].

Convergence phenomena of this sort are connected with rescaling behavior (cf. [An]). To explain this somewhat explicitly, suppose p is a point of a Riemannian manifold (M, g) and that i_p is the injectivity radius at p and that the absolute value of sectional curvature is $\leq \Lambda$ on the ball around p of radius i_p. According to Jost and Karcher [JK], as already discussed, there are harmonic coordinates on a ball around p of fixed radius $r_p < i_p$ and $C^{1,\alpha}$ estimates of the metric coefficients g_{ij} in these coordinates on, say, the ball of radius $r_p/2$. Here r_p and the estimates depend only on Λ and i_p. In the rescaling viewpoint, this estimation corresponds to the fact that multiplying the metric g by $\lambda > 0$, for increasingly large λ, going to $+\infty$, produces a family of metrics for which injectivity radius (at p) goes to $+\infty$ and sectional curvature goes to 0. The limit metric is thus flat euclidean space, and for flat euclidean space there are global harmonic coordinates, or, in a suitable sense, "the harmonic radius is infinite."

To make the correspondence yet more explicit, one can reason by contradiction: If there were a sequence $\{(M_j, g_j, p_j, i_{p_j}\}$ with $i_{p_j} \geq i_0 > 0$ but with (maximal) "harmonic radius" $r_{p_j} \to 0^+$ as $j \to +\infty$, then $\{M_j, (r_{p_j})^{-2} g_j\}$ would converge in the way indicated to flat $\mathbb{R}^n$. But the $\{M_j, (r_{p_j})^{-2} g_j\}$ all have harmonic radius 1. So one might expect a contradiction of the fact that flat $\mathbb{R}^n$ has harmonic radius $+\infty$. This contradiction of course depends on the limit of the harmonic radii being $\geq$ the harmonic radius of the limit, and this in turn ultimately depends on some type of partial differential equations estimate. But the viewpoint is suggestive and useful.

One might also use this viewpoint to explain why, under the hypotheses of bounded-in-absolute-value sectional curvature, bounded diameter and fixed dimension, a lower bound on volume suffices for convergence. (This is only for philosophical interest: in this case, a lower volume bound and a lower injectivity radius bound are equivalent [Ch2].) If volume is bounded below and with curvature and diameter as indicated, then the Bishop volume comparison result shows that the volume of an r-ball is bounded below by an estimate of the form

$$\lim_{\substack{r \to 0^+ \\ x \in M}} \inf \mathrm{vol}(B(x, r))/\mathrm{vol}_0(r) = 1,$$

where $\mathrm{vol}_0(r)$ = volume of the euclidean r-ball and where the limit occurs uniformly for $x \in M$ and M in the class indicated. (If this uniform estimate failed to hold, the Bishop comparison would yield a contradiction to the assumed global volume bound.) Thus rescaling small balls yields in the limit a manifold with sectional curvature 0 and euclidean volume growth (of balls as a function of radius). Such a manifold is isometric to R^n. As before, one could reach a contradiction if there were a sequence with injectivity radius going to 0, while the other bounds remained satisfied.

For another thing, a (necessarily complete) Riemannian manifold with injectivity radius $i_p = +\infty$ at every point p and with Ricci curvature ≥ 0 is again necessarily flat $\mathbb{R}^n$. This follows from the Cheeger-Gromoll splitting theorem [CG2]: injectivity radius $= +\infty$ implies the existence of lines through p in all directions, so one can successively split off n orthogonal lines. Thus one expects a (finiteness and) convergence result for manifolds with Ricci curvature $\geq -\Lambda$ and injectivity radius $\geq i_0 > 0$. This in fact holds [An], with C^α convergence of metric.

Yet another line of thought suggested by the rescaling viewpoint is that certain types of integral curvature bounds should also imply convergence results. Again the crucial point is that rescaling should result in a limit of zero curvature. Thus one expects convergence for n-dimensional manifolds in the presence of L^p bounds on the curvature tensor's norm if $p > n/2$: if the metric is multiplied by $\lambda > 0$ the pointwise norm of the curvature scales by λ^{-2}, its pth power by λ^{-2p} and the volume form by λ^n, $n =$ dimension. Thus the p-norm integral goes to zero if $-2p + n < 0$ or $p > n/2$. This is put on a precise basis in [Ya]. The scale-invariant case $p = n/2$ is also of special interest [Ga1, Ga2, Ya].

The developments of this type—convergence and finiteness theorem in the presence of some weaker hypotheses than bounded absolute value of sectional curvature—depend ultimately on establishing the required estimates on the metric in harmonic coordinates, and a lower bound for the radius of balls on which harmonic coordinates are defined. The required extensions of the Jost-Karcher estimates began with the fundamental work of Gao [Ga1, Ga2]. This was subsequently extended by Anderson [An] and D. Yang [Ya] and Anderson-Cheeger [AnC]. In the purely geometric context (as opposed to the partial differential equations setting in which the results live most naturally in technical terms) the easiest results to assimilate are those in which L^∞ bounds only are used: Ricci bounded above and below, injectivity radius bounded below, diameter bounded above, implies diffeomorphism finiteness and $C^{1,\alpha}$ precompactness [An]. And Ricci bounded below, injectivity radius bounded below, volume bounded above, implies diffeomorphism finiteness and $C^{1,\alpha}$ precompactness [**An**]. But the other results are also of fundamental significance in the partial differential equations setting. It is also possible to analyze precisely the kinds of singularities which can arise in the absence of a lower bound on injectivity radius, but still assuming a Ricci curvature bound above and below [**AnC**].

3. Packing arguments, metric space convergence, and related topics

One of the most familiar arguments in metric space topology is the construction of a finite ε-net in a (sequentially) compact metric space X, i.e., the construction, given $\varepsilon > 0$, of a finite set $x_1, \ldots, x_k$ of points in X such that $\bigcup_1^k B(x_j, \varepsilon) = X$. The construction, by choosing a maximal set of points with all pairwise distances $\geq \varepsilon$, is, for one thing, the basic step in showing that a sequentially compact metric space is compact; and as such it is encountered by every topology student. Thus it is somewhat surprising that the potential importance of the idea in Riemannian geometry went unobserved for so long.

The potential importance became both observed and actual with the fundamental paper of A. Weinstein [**W**], which contains in embryonic form a surprising amount of what would hatch out later. In particular, [**W**] showed how volume estimates could be used to get covering estimates, here in the case of sectional curvature (positive) pinching. It was later pointed out by Gromov that Ricci curvature bounded below plus diameter bounded above would suffice for the argument. The estimates of this sort proceed as follows: If $x_1, \ldots, x_k$ is an ε-net in a compact Riemannian manifold M constructed by finding a maximal set of points with distances $\geq \varepsilon$, then the open balls $B(x_j, \varepsilon/2)$, $j = 1, \ldots, k$, are pairwise disjoint. Thus

$$\sum_{j=1}^{k} \operatorname{vol}(B(x_j, \varepsilon/2)) \leq \operatorname{vol}(M).$$

Hence a lower bound on $\operatorname{vol}(B(x_j, \varepsilon/2))$ and an upper bound on $\operatorname{vol}(M)$ together estimate from above the number of terms in the sum on the left-hand side, or equivalently the number of points needed for an ε-net.

Lower bounds on the volumes of balls typically require lower bounds on injectivity radius, which are in any case available in [**W**]. But Gromov noted that here such a lower bound on injectivity radius is unnecessary. All one really seeks is an upper bound on

$$\operatorname{vol}(M)/\operatorname{vol} B(x_j, \varepsilon/2).$$

If the diameter $\operatorname{diam}(M) \leq d_0$ and the Ricci curvature $\operatorname{Ric}(M) \geq -(n-1)\Lambda$, $n = \dim M$, then the quotient is bounded by an estimate due to R. Bishop [**BC**]:

$$\operatorname{vol}(M)/\operatorname{vol} B(x_j, \varepsilon/2) = \operatorname{vol} B(x_j, d_0)/\operatorname{vol} B(x_j, \varepsilon/2)$$
$$\leq \operatorname{vol} B_\Lambda(d_0)/\operatorname{vol} B_\Lambda(\varepsilon/2)$$

where $\operatorname{vol} B_\Lambda(r) =$ the volume of the ball of radius r in the complete simply connected n-dimensional space of curvature $-\Lambda$. (Bishop's argument is via Jacobi fields and as such seems to apply only along minimizing geodesics, but the estimate only improves if the ball's radius exceeds the injectivity radius.) Thus one sees that

LEMMA (Gromov). *If M is a compact n-dimensional Riemannian manifold of dimension n, Ricci curvature $\geq -(n-1)\Lambda$, and diameter $\leq d_0$ then M has an ε-net with fewer than or exactly $C(n, \Lambda, d_0, \varepsilon)$ points, where $C(n, \Lambda, d_0, \varepsilon)$ depends only on n, Λ, d_0 and ε.*

This lemma becomes especially interesting if one thinks of the set of all M satisfying the hypotheses as forming a family, call it $\mathscr{C}(n, \Lambda, d_0)$. In some suitable sense, the metric space structures of the members of $\mathscr{C}(n, \Lambda, d_0)$ are determined up to an ε error by $\frac{1}{2}(C(n, \Lambda, d_0, \varepsilon))(C(n, \Lambda, d_0, \varepsilon)-1)$ parameters which are bounded between 0 and d_0, namely the distances between the points of an ε-net. (By repeating points if necessary, one may suppose that the ε-nets have exactly $C(n, \Lambda, d_0, \varepsilon)$ points.) To formulate this kind of controlled behavior, it is convenient to use the notion of distance between compact metric spaces, which arises directly from the classical idea of Hausdorff of the distance between compact sets in a metric space [GLP]. Namely:

(a) A_1, A_2 compact, $A_1 \subset X$, $A_2 \subset X$, X a metric space

$$\operatorname{dis}_X(A_1, A_2) \overset{\text{def}}{=} \inf \left\{ r : A_2 \subset \bigcup_{x \in A_1} B(x, r),\ A_1 \subset \bigcup_{x \in A_2} B(x, r) \right\}.$$

b) X_1, X_2 compact metric spaces

$$\operatorname{dis}(X_1, X_2) = \inf\{\operatorname{dis}_y(\tilde{X}_1, \tilde{X}_2) : \tilde{X}_1 \subset y,\ \tilde{X}_2 \subset Y,$$
$$\tilde{X}_1 \text{ with } Y\text{-metric isometric to } X_1,$$
$$\tilde{X}_2 \text{ with } Y\text{-metric isometric to } X_2\}.$$

This distance, for compact metric spaces, is usually called the Gromov-Hausdorff distance. Roughly speaking, two metric spaces are close together if there is some way to fit them both into a larger metric space so that they are close together as sets. Suppose $p_1, \ldots, p_k$ is an ε-net in X_2 and that

$$|\operatorname{dis}(p_i, p_j) - \operatorname{dis}(q_i, q_j)| < \varepsilon$$

for all $1 \leq i, j \leq k$. Then one can define a metric on $X_1 \cup X_2$ by declaring $\operatorname{dis}(p_i, q_i) = \varepsilon$ and setting, for $p \in X_1$, $q \in X_2$, the distance $\operatorname{dis}(p, q) =$ the inf. of sums $\operatorname{dis}(p, x_1) + \operatorname{dis}(x_1, x_2) + \cdots + \operatorname{dis}(x_{l,q})$ where each consecutive pair (x_i, x_{i+1}) (or (p, x_1) or $(x_{l,q})$) either has both elements in one of X_1 or X_2 or has $x_i = p_m$, $x_{i+1} = q_m$. This construction yields that

$$\operatorname{dis}(X_1, X_2) \leq 2\varepsilon.$$

These observations combined with the estimation for number of points in an ε-net yield easily the following result of Gromov:

PRECOMPACTNESS THEOREM (Gromov). *Let* $\mathscr{C}(n, \Lambda, d_0) =$ *the class of all compact Riemannian manifolds with dimension* n, *Ricci* $\geq -(n-1)\Gamma$, *and diameter* $\leq d_0$. *Then every sequence* $\{(M_i, g_i)\}$ *in* $\mathscr{C}(n, \Gamma, d_0)$ *has a subsequence that is, a Cauchy sequence with respect to the Gromov-Hausdorff distance.*

It is not hard to see that the set of all compact metric spaces is Cauchy complete with respect to Gromov-Hausdorff distance so referring to this result as a precompactness theorem is appropriate.

Ideas of this type have deep roots in the topological study of metric spaces as practiced in the early parts of the century, and to a surprising extent contemporary trends in Riemannian geometry have metric geometry precursors. The reader can pursue this interesting line of thought by reading P. Petersen's survey article on the application metric-topological concepts to geometry in this volume as well as [**Pt**].

Much of the interest of taking limits of sequences of Riemannian manifolds lies in the fact that the limit metric space tends to have local geometry like that of a Riemannian manifold, in some sense at least. Of course, the most favorable case is when the limit metric space is a Riemannian manifold itself, albeit perhaps of less metric regularity; this case has already been in effect discussed in the section on convergence theorems. Roughly speaking, one expects the limit to be a Riemannian manifold only if the members of the sequence have injectivity radius (locally) uniformly bounded from below.

Even in the case of the $C^{1,\alpha}$ convergence results, the limit manifold fails to have curvature in the usual sense, because the curvature tensor involves two derivatives of the metric coefficients. But there is a way in which curvature considerations can nonetheless arise in the $C^{1,\alpha}$ context, or even in situations with for less metric regularity than that, even with *no* regularity in the derivative sense. This has to do with Toponogov's theorem, which in this situation one might think of as an integrated form of the distance behavior for which curvature is the infinitesimal or differentiated form.

By way of illustration, let us consider what should be meant by (sectional) curvature ≥ 0 for this situation. To use Toponogov's theorem, we will suppose at least that we have an inner metric space X with unique minimal geodesics connecting each point p to all sufficiently nearby points q. The usual form of Toponogov's theorem involves angles, but this technicality can be avoided as follows: Suppose $\gamma_1 : [0, 1] \to X$ and $\gamma_2 : [0, 1] \to X$ are (minimal) geodesics with $\gamma_1(0) = \gamma_2(0)$. Fix $t, s \in (0, 1)$ and construct the (plane) euclidean triangle with linear-parameterized lines $L_1 : [0, 1] \to \mathbb{R}^2$ and $L_2 : [0, 1] \to \mathbb{R}^2$, $L_1(0) = L_2(0) = (0, 0)$, $\|L_1(s)\| = \mathrm{dis}(\gamma_1(0), \gamma_1(s))$, $\|L_2(t)\| = \mathrm{dis}(\gamma_1(0), \gamma_2(t))$, $\|L_1(s) - L_2(t)\| = \mathrm{dis}(\gamma_1(s), \gamma_2(t))$. The

condition that X have curvature ≥ 0 is then that $\|L_1(1) - L_2(1)\| \geq$
$\mathrm{dis}(\gamma_1(1), \gamma_2(1))$.

This condition corresponds exactly to the usual intuition that sectional
curvature ≥ 0 implies that geodesics diverge more slowly than do straight
lines in euclidean space. Similar definitions can be made for sectional curvature $\geq K > 0$ and sectional curvature $\geq -K$, $K > 0$, by comparison to
the appropriate simply connected surface of constant Gauss curvature.

The naturalness of these definitions is apparent. Their utility is enhanced
by the fact that they are preserved under passage to the limit (e.g., the limit
of a sequence of Riemannian manifolds of nonnegative curvature is an inner
metric space of nonnegative curvature).

In the case of manifolds that satisfy the hypotheses of the $C^{1,\alpha}$ convergence result of Greene-Wu and Peters, the concept of Gromov-Hausdorff
distance (hereafter GHa convergence) convergence coincides with $C^{1,\alpha}$ convergence. Specifically, let $\{(M_j, g_j)\}$ be a sequence of Riemannian mani-
folds with dimension n, |sectional curvature| $\leq \Gamma$, diameter $\leq d_0$, and
volume $\geq v_0$, where n, Λ, d_0 and $v_0 > 0$ are fixed. For such a sequence
of manifolds, there is an $i_0 > 0$ such that the injectivity radius $i(M_j)$ is
greater than i_0 for all j ([Ch2], see also [HK]). Thus one a priori expects
the GH limit, if it exists, to be a manifold, as already mentioned. Precisely,
the sequence $\{(M_j, g_j)\}$ converges $C^{1,\alpha}$ if and only if it converges GH.
Moreover, in case of convergence, the M_j are, as before, all mutually diffeomorphic for j sufficiently large, and the GH limit space is (isometric to)
this differentiable manifold with the $C^{1,\alpha}$ limit Riemannian metric.

If hypotheses on curvature and so on are not made, the GH convergence af-
fords a wide variety of possibilities. Dimension can drop: for each compact
Riemannian manifold (M, g), $\{(M, g/j) : j = 1, 2, 3, \dots\}$ converges
GH to a point. Dimension can also increase. In particular, Cassorla [Ca]
has shown that any compact inner metric space of finite dimension can be
obtained as the GH limit of compact Riemannian manifolds of dimension
2. Thus one is motivated to consider GH convergence under geometric restrictions that are less restrictive than those of $C^{1,\alpha}$ convergence but are
sufficiently restrictive to avoid completely unpredictable behavior.

Experimentation with two-dimensional examples suggests that bounded
curvature and diameter exert almost complete control over GH convergence.
In particular, these hypotheses prevent GH convergence to a point except in
the case of flat manifolds, e.g., $S^1 \times S^1$, where rescaling by $1/j$ does not
result in unbounded curvature ($0 \times j^2 = 0$ still). At first sight, it would thus
seem that drop in dimension in GH convergence with bounded curvature
could only occur for flat manifolds, or in cases where flat manifolds occur as
factors. It thus came as something of a surprise when Berger pointed out that
S^3 can "collapse" to S^2 with bounded curvature in the following sense: there
is a sequence of metrics $\{g_j\}$ on S^3 with |sectional curvature| uniformly

bounded and with $\{(S^3, g_j)\}$ converging GH to S^2. (The formal concept of GH convergence had not been introduced at the time, but the convergence idea in this case is intuitively clear.) The metrics g_j are obtained by "shrinking" by a multiplicative factor the tangent spaces to the fibers of the Hopf fibration of S^3 over S^2, while leaving the usual S^3 metric unaltered in the directions transversal (orthogonal) to the fibers. With this set-up, $\{(S^3, g_j)\}$ converges GH to S^2 with the metric of constant curvature 4.

The essential aspect that makes Berger's construction possible is the possibility of collapsing the fibers to points without forcing more curvature. For this, the fibers themselves must be collapsible to points with bounded curvature. At this point, it becomes natural to make a definition: A compact manifold M is *almost flat* if there is a sequence of C^∞ Riemannian metrics $\{g_j\}$ on M such that $\{M, g_j\}$ converges GH to a point and such that the metrics g_j have a uniform bound on the absolute value of sectional curvature.

This definition is easily seen, by scaling, to be equivalent to the condition: there is, for each $\varepsilon > 0$, a metric g_ε on M with, for all sectional curvatures of g_ε, and diameter relative to g_ε

$$\text{diameter}^2 \, |\text{sectional curvature}| \le \varepsilon.$$

Gromov ([**Grv1**], cf. [**BK**]) and, later by a different method, Ruh [**Ru2**] produced a topological description of almost flat manifolds: a compact manifold is almost flat if and only if a finite cover is a nilmanifold.

Rather surprisingly, perhaps, there are almost flat manifolds that do not admit any flat metric. This does not contradict other convergence ideas (i.e., $C^{1,\alpha}$ convergence) because the injectivity radius of the metrics g_ε will typically go to 0 as $\varepsilon \to 0^+$ if they are scaled to keep diameter bounded.

A sequence of metrics $\{g_j\}$ on a (compact) manifold M is said to *collapse with bounded curvature* if the sectional curvatures of all the g_j are uniformly bounded in absolute value and if the injectivity radius $i_{p,j}$ of g_j at $p \in M$ converges to zero uniformly in p, i.e.,

$$\lim_{j \to +\infty} \left(\sup_{p \in M} i_p \right) = 0.$$

To simplify terminology, one sometimes says M collapses with bounded curvature (to a limit); e.g., S^3, according to Berger's observation, collapses with bounded curvature to S^2. Roughly, this corresponds to the limit (if it exists) of $\{(M, g_j)\}$ having lower dimension than $n = \text{dimension } M$; a precise form of this relationship will be given momentarily.

Along the lines discussed already, the collapsing of $\{(M, g_j)\}$ would be expected to occur by collapsing to points certain almost flat submanifolds of M, while leaving the metric of M essentially unchanged in directions transversal to the "fibres," i.e., the almost flat submanifolds. This expectation can be given precise substance, but the situation turns out to be more complicated than one sees at first from the easily accessible examples. In particular, the "fibres" need not have constant dimension, in general. Nonetheless, it turns out to be possible to give a detailed description of when and how collapsing can occur. In particular, in [CGv], a necessary and sufficient condition—existence of what is called there an F-structure—is given for a manifold to collapse. Collapse with bounded curvatures gives an F-structure (with fibres which admit locally free torus actions), and if an F-structure exists, collapse with bounded curvature is possible along it. Roughly, an F-structure corresponds to the choice of submanifolds to be collapsed, as above. But the F-structure does not account for all collapsing directions. A more refined version [CFG] does account for all the collapsing. For the rather complicated details, see the references noted.

This might be an appropriate moment to note that E. Ruh, and Min-Oo in many instances and more recently with P. Ghanaat also, have developed much of the material here having to do with almost flat manifolds and collapsing using an entirely different approach. Their alternative to the distance-geometric approach involves construction of connections on the frame bundle via partial differential equations methods. The dearth of discussion of these methods in the present article does not imply lack of belief in their efficacy. Rather, the presence in this same volume of an excellent survey by E. Ruh of this approach makes detailed discussion here redundant. The reader should definitely consult Ruh's survey to round out the picture.

In these general observations on collapsing, no hypothesis is made on diameter. And indeed in some actual cases (e.g., negative curvature, cf. [Grv2]), rescaling to keep diameter bounded may make the uniform bound on sectional curvature impossible. Geometrically, however, it is natural to consider collapse with bounded diameter together with bounded curvature. In this case, it is not hard to see that convergence of $i_{p_0,j}$ to 0 (in our now standardized notation $i_{p_0,j} = $ injectivity radius at p_0 for g_j) for one point $p_0 \in M$ implies that $i_{p,j} \to 0$ uniformly for all $p \in M$. The distance to the first conjugate point is bounded below uniformly in j, so $i_{p_0,j} \to 0$ implies the length of the shortest closed geodesic goes to 0 which implies that $\mathrm{vol}(M) \to 0$ (standard comparison argument, as in [Ch2], cf. [HK]) which implies that $i_{p,j} \to 0$ uniformly. (For that last step, one uses lower bound on volume from upper curvature bound.) This observation and example suggest that collapsing with bounded curvature and bounded diameter is subject to some kind of global uniformity. It is instructive in this regard to consider the 2-dimensional "dumb-bell" example of two unit spheres connected by a long thin tube, which is smoothly joined to the spheres. If the tube is bounded in length (bounded diameter), then it cannot be made extremely thin without introducing large negative curvature. An imaginative and optimistic person might even conjecture from the dumb-bell example that the uniformity is insured by just a lower bound on sectional curvature, with no upper bound needed (even though the proof given seems to need the upper curvature bound, too). These general ideas are put on a precise basis by a result of Grove and Petersen: To state the result, recall that a length space is an inner metric space with the property that the distance between any two points is the infimum of the lengths of rectifiable curves between the two points. The Grove-Petersen result is

If $\{(M, g_j)\}$ is a sequence of Riemannian metrics on a compact connected, n-dimensional manifold M and if diameter $\leq D$ and sectional curvature $\geq k$ for all the metrics g_j, and if $\{(M, g_j)\}$ converges GH to a (compact metric space) limit X then X is a length space with curvature $\geq k$ in the Toponogov sense and there is an integer $m \leq n$ such that X has dimension m at every point.

Note that X may have singular points in the sense of the length space structure not arising from a Riemannian metric on a smooth manifold at and near the point (e.g., $C^{1,\alpha}$ convergence does not hold); and indeed X may not even be a topological manifold. However, Burago, Gromov and Perlman [BGP] have shown that finite-dimensional length spaces with curvature bounded below have a certain degree of regularity. In particular, they show that a finite (Hausdorff) dimensional length space of dimension m has an open dense subset which is an m-dimensional topological manifold. Thus in some sense the whole space is a topological manifold together with a "singular set" which has empty interior.

4. How the methods apply

In the broadest terms, the methods described in the previous sections might well be characterized as a way of connecting the rough phenomena Riemannian distance functions (which typically are only Lipschitz continuous) with the smooth phenomena of differential topology. The theory of noncritical points carries over the basic result of Morse theory to a nonsmooth setting. The preferred coordinate systems give the best possible smoothness to limits of sequences of Riemannian manifolds. And the Gromov-Hausdorff metric (GH) convergence of sequences of manifolds gives a way to analyze limiting processes even when there is effectively no smoothness in the limit.

From one viewpoint, generalized mathematical structures are their own justification. But the classically oriented geometer will seek justification for these newer viewpoints in the contributions they can make to understanding the classical C^∞ situation. And these contributions are abundant enough to justify the new viewpoints with room to spare.

The nonsmooth critical point theory has already proved its utility in a variety of contexts. Its original contexts, of sphere theorems [GoS] and noncompact manifolds of nonnegative curvature and convexity [CG1, GW2, GW4, GS1, GS2] have already been mentioned. In both these general settings, it has continued to be a standard tool. It also played an important role in the estimation of Betti numbers (in [**Grv3**]) for manifolds of nonnegative curvature or, more generally, for manifolds with a lower bound on curvature $\times$ diameter^2. Moreover, the essential point in the proof of the Grove-Petersen finiteness theorem [**GoP1**] (namely, the replacement of Riemannian center-of-mass [**GoK**] by a more general concept) is an application of the same idea of deforming along a vector field of definite decrease for distance. These and other applications of (non)critical point theory are discussed in detail in K. Grove's survey in this volume.

The most natural application of the convergence ideas, including the regularity arising from the preferred coordinates, is to what one might term near-rigidity theorems. One begins with a "rigid" characterization result, say for example Berger's theorem [**Be1**] that an even-dimensional orientable manifold with $1/4 \leq$ sectional curvature ≤ 1 is either homeomorphic to a sphere or isometric to a compact rank one symmetric space. The corresponding near-rigidity result [**Be2**] is: There are positive constants $\varepsilon(2n)$ such that a $2n$-dimensional compact orientable manifold with $+1/4 - \varepsilon(2n) \leq$ sectional curvature ≤ 1 is either homeomorphic to a sphere or diffeomorphic to a compact rank one symmetric space.

There is a clear paradigm for passage from a rigidity result to a near-rigidity one by convergence methods. One supposes, e.g. in the Berger situation, for proof by contradiction that there is a sequence $\{M_j, g_j)\}$ of compact $2n$-dimensional orientable manifolds with $1/4 - 1/j \leq$ sectional curvature $(g_j) \leq 1$. One applies a suitable finiteness theorem (e.g., [**P1**]) to see that it can be supposed without loss of generality that all the M_j have the same topological type and then a suitable convergence theorem (the $C^{1,\alpha}$ convergence of [**GW8**] and [**P2**] in this case) to obtain a limit manifold with $1/4 \leq$ sectional curvature ≤ 1, in the Toponogov sense of curvature. Then the limit manifold is one of the possibilities listed in the rigidity theorem. The last sentence of course presupposes an essential and nonautomatic technical point: the limit manifold a priori does not usually have enough regularity to insure that the classical rigidity theorem actually applies and it must be demonstrated that sufficient regularity occurs. Moreover, there must be a way to go back from the limit to the manifolds near the limit. In this latter regard, Shikata's paper [**Shi**], which is often overlooked, is important. In a sense, this whole scheme serves to isolate the kind of geometric/analytic estimates needed to insure that enough regularity really does hold in the limit. This can be far from easy to prove, cf. the actual argument in [**Be2**].

In retrospect, the theorem that simply connected and pinching-close-to-one implies differentiable sphere can be viewed in this same way: Suppose $\{(M_j, g_j)\}$ is a sequence of n-dimensional manifolds with $1 - 1/j \leq$ sectional curvature ≤ 1. Then according to the injectivity radius estimates of [**Be1**] and [**Kl1**], the injectivity radii $i(g_j) \geq i_0 > 0$, for some fixed i_0. Thus some subsequence of $\{(M_j, g_j)\}$ converges $C^{1,\alpha}$ to a C^∞ n-dimensional manifold with a $C^{1,\alpha}$ Riemannian metric (indeed, one can take the M_j all diffeomorphic to each other and to the limit manifold). This limit Riemannian manifold has constant sectional curvature $+1$ in the Toponogov sense. Thus the result is proved if it is shown (as is true) that a compact $C^{1,\alpha}$ Riemannian manifold of constant Toponogov curvature 1 is a standard sphere. This view does not of course address the question of finding a pinching constant independent of dimension or of estimating the constant explicitly in a fixed dimension. But it does offer a strong intuition as to why the theorem is true.

It is of some interest to trace through the various convergence and collapsing (or lack thereof) ideas in the complex case, or, more specifically, in the Kähler case. In the positive curvature instance, there is only one possible compact Kähler manifold: according to [SiY] (see also [M]), a compact Kähler manifold of positive sectional curvature is biholomorphic to complex projective space $\mathbb{C}P^n$. Complex projective space has of course a standard Kähler metric of constant positive holomorphic sectional curvature $+1$; its sectional curvature varies from $1/4$ to 1 (for $n \geq 2$). No Riemannian metric on $\mathbb{C}P^n$, $n \geq 2$, could be more than quarter-pinched ($\lambda \leq$ sectional curvature ≤ 1, $\lambda > 1/4$) since, by the Berger-Klingenberg theorem, the existence of a more than quarter-pinched metric implies that the universal cover of the manifold is necessarily homeomorphic to a sphere. If $\{g_j\}$ is a sequence of Riemannian metrics on $\mathbb{C}P^n$ ($n \geq 2$) with g_j being λ_j-pinched, and with $\lim \lambda_j = 1/4$, then it is easy to see that the $\{g_j\}$ must satisfy the hypotheses of [GW8], [P2] for $C^{1,\alpha}$ convergence: Since sectional curvature $\geq \lambda_j \geq 1/8$, say, for large j, the diameter in the g_j metric is $\leq$ the diameter of the $(2n\text{-})$ sphere of constant curvature $1/8$ for large j. Also, the volume is less than or equal to the volume of that sphere. On the other hand, the curvature bound, curvature between $1/8$ and 1 for large j, implies an upper bound on the ratio of the generalized Gauss-Bonnet integrand and the volume form. Since the integral of the Gauss-Bonnet integrand is the Euler characteristic of $\mathbb{C}P^n = n + 1 > 0$, one obtains a lower bound on the volume. This in turn implies a lower bound on the injectivity radius. Hence a subsequence of the $\{g_j\}$ converges $C^{1,\alpha}$ (after correction by diffeomorphisms) to a $C^{1,\alpha}$ metric on $\mathbb{C}P^n$ that is $1/4$-pinched in the Toponogov sense. According to Berger [Be2], this limit metric is in fact isometric to the standard Kähler metric. This uniqueness of the possible limit metric implies, incidentally, that the entire sequence $\{g_j\}$ converges to the limit (since every subsequence has a $C^{1,\alpha}$ limit, but only one single limit is possible for all of the convergent subsequences).

The situation for constant negative holomorphic sectional curvature is more complicated. A compact Kähler manifold M^n, $n = \dim_{\mathbb{C}} M$, $n \geq 2$, which admits a sequence of Kähler metrics $\{g_j\}$ that are negatively λ_j

pinched $(-\lambda_j \geq$ sectional curvature $\geq -1)$ with $\lim \lambda_j = 1/4$ also admits
a constant negative holomorphic sectional curvature metric; this was proved
by S. Krantz and the author in [GK] by complex analysis methods. (One can
also prove this by noting that the generalized Schwarz Lemma of [Yau] gives
a uniform upper bound on volume. Then the convergence of the curvature
tensor to one of constant holomorphic sectional curvature implies that the
Chern numbvers satisfy the identities needed to deduce that the canonical
Einstein-Kähler metric has constant negative holomorphic sectional curvature, see [Yau 2] for this last part in detail.) It is natural to try to see this
result in the present convergence framework, at least philosophically.

Now manifolds which have a constant negative holomorphic sectional curvature Kähler metric of course have the ball as universal cover; and their
Kähler metrics of constant negative holomorphic sectional curvature, which
are all constant multiples of each other, are multiples of the "push-down"
of the Bergman metric of the ball via the covering map. Moreover, it is a
standard algebraic fact that a 1/4-pinched Kähler metric (for complex dimension ≥ 2) is always of constant holomorphic sectional curvature. Thus
one might try to see this limit result of [GK] already quoted as an application
of the geometric convergence ideas we have been discussing.

In particular, one might hope that the $\{g_j\}$ sequence converged to the
1/4-pinched Kähler metric on M. That this convergence does happen is
part of the complex analytic argument in [GK]. Our present goal is to see this
geometrically.

Note first that, since (M, g_j), j large, has negative Ricci curvature, it
follows that the (integer) Chern number $c_1^n[M]$ is nonzero, $c_1^n[M]$ being up
to a nonzero constant factor $\int_M \text{Ric} \wedge \cdots \wedge \text{Ric}$ where $\text{Ric} =$ the Ricci form.
Now, again for j large, the Ricci form Ric_j of g_j has eigenvalues bounded
above and below by negative constants. Thus the constancy of the integral
of the nth power of Ric_j as a function of j implies that $\text{vol}(M, g_j)$ is
bounded above and below uniformly for large j.

According to [Grv2], the upper bound on $\text{vol}(M, g_j)$ yields an uniform
upper bound on $\text{diam}_j = \text{diam}(M, g_j)$ (for all large j). In particular, we
now see that the sequence $\{g_j\}$ has a subsequence that converges $C^{1,\alpha}$.
The limit g_0 of such a sequence is in fact again a Kähler metric, in the sense
that the Kähler form of the limit is closed. Now consider the holomorphic
universal cover $\widehat{M}$ (which is diffeomorphic to $\mathbb{R}^{2n}$) of M with the pullback metrics $(\widehat{M}, \hat{g}_j)$. Define maps $(B^{2n}$, Bergman metric) to $(\widehat{M}, \hat{g}_j)$ by
$\exp_j \circ I_x \circ \exp_0^{-1}$ where $\exp_0$ is the exponential map of the Berman metric
at $0 \in B^{2n} \subset \mathbb{C}^n$, I is a complex linear (J-commuting) isometry from
$\mathbb{C}^n$ to the tangent space at $x \in \widehat{M}$, x fixed, and $\exp_j$ is the $\hat{g}_j$-metric
exponential map. By standard Jacobi field arguments, the sequence of maps
thus defined converges to a distance-preserving map from $(B^{2n}$, Bergman

metric) to $(\widehat{M}, g_0)$. The limit map is also holomorphic. Thus, since $\widehat{M}$ holomorphically covers M, one recovers that M is a quotient of the ball.

In the original proof in [**GK**], estimates on the relationships between the metrics g_j were obtained using Yau's generalized Schwarz lemma [**Yau**]. The point of the discussion here is that one can obtain the same picture in the geometric context we have been discussing. More is in fact true about this situation: It is a surprising fact that any $1/4$-pinched *Riemannian* metric on a compact complex quotient of the ball is necessarily (a constant multiple of) the canonical Kähler metric of constant negative holomorphic sectional curvature, obtained by pushing down the metric of the ball ([**H**], see also [**YZ**] for a further generalization; and cf. the survey article of Eberlein, Schroeder and Hamenstadt, this volume).

Perhaps the most natural setting of all for almost-rigidity or quasi-rigidity results is the geometric characterizations of the sphere up to isometry. It has already been discussed how the characterization of the standard sphere as the simply connected manifold of constant sectional curvature 1 corresponds to the differentiable pinching theorem in this general scheme of things. But there are of course other geometrically interesting rigid characterizations of the sphere. Since the survey article of K. Shiohama in this volume covers this topic in detail, no further comments will be offered here except to note how natural the general convergence picture makes it to seek quasi-rigid forms of those characterizations.

Despite the natural inclination of mathematicians to generalize for the sake of generalization, one might well argue that the true test of the significance of a generalization is not only that the generalized situation be attractive mathematically in itself but also that the generalization should provide new insight into the original, more specific situation. On both grounds, the extension of geometry and its methods into the relatively nonsmooth category shows strong signs of being a success. Certainly the possibility of taking limits of sequences of Riemannian manifolds in widely general situations is intrinsically attractive, as is the possibility of doing Morse theory for nonsmooth functions. But even more, these techniques have already produced an impressive body of results of the more classical sort as far as their statement is concerned.

The classicist could argue that theorems about nonsmooth limits are in effect just a notational convenience for expressing estimates about the original smooth objects. And of course this is in effect true. But one could just as well argue that L^2 methods in partial differential equations were just a notation for certain limits of C^∞ functions—which is also true. A new viewpoint, a new terminology, a new notation—past experiences suggest these can transform the face of a subject. To use the words of Hilbert (about Cantor's ideas), the new methods in geometry seem likely to be a "paradise from which we shall never be driven forth".

REFERENCES

[Ab] U. Abresch and D. Gromoll, *On complete manifolds with nonnegative Ricci curvature*, J. Amer. Math. Soc. **3** (1990), 355–374.

[An] M. Anderson, *Convergence and rigidity of manifolds under Ricci curvature bounds*, Invent. Math. **102** (1990), 429–445.

[AnC] M. Anderson and J. Cheeger, *Diffeomorphism finiteness for manifolds with Ricci curvature and $L^{n/2}$ norm of curvature bounded*, Preprint.

[AV] A. Andreotti and E. Vesentini, *Carleman estimates for the Laplace-Beltrami equation on complex manifolds*, Inst. Hautes Études Sci. Publ. Math. **25** (1965), 81–130.

[Be1] M. Berger, *Le varietes riemanniennes 1/4-pincees*, Ann. Scuola Norm. Sup. Pisa **14** (1960), 161–170.

[Be2] ____, *Sur les varietes riemanniennes juste andessons de 1/4*, Ann. Inst. Fourier (Grenoble) **33** (1983), 135–150.

[BC] R. L. Bishop and R. J. Crittenden, *Geometry of manifolds*, Academic Press, New York 1964.

[BGP] Y. Burago, M. Gromov, and G. Perlman, *Aleksandrov's spaces with curvatures bounded from below. I*, Preprint.

[Ch1] J. Cheeger, *Comparison and finiteness theorems for Riemannian manifolds*, Thesis, Princeton Univ., 1967.

[CH2] ____, *Finiteness theorems for Riemannian manifolds*, Amer. J. Math. **96** (1970), 61–74.

[CE] J. Cheeger and D. Ebin, *Comparison theorems in Riemannian geometry*, North Holland Math. Library, vol. 9, North-Holland, 1975.

[CG1] J. Cheeger and D. Gromoll, *On the structure of complete manifolds of nonnegative curvature*, Ann. of Math. (2) **96** (1972), 413–443.

[CG2] ____, *The splitting theorem for manifolds of nonnegative Ricci curvature*, J. Differential Geom. **6** (1971), 119–129.

[CGv] J. Cheeger and M. Gromov, *Collapsing Riemannian manifolds while keeping their curvature bounded. I*, J. Differential Geom. **23** (1983), 309–346; II, J. Differential Geom. **32** (1990), 269–298.

[CFG] J. Cheeger, K. Fukaya, and M. Gromov, *Nilpotent structures and invariant metrics on collapsed manifolds*, Preprint.

[DK] D. De Turck and J. Kazdan, *Some regularity theorems in Riemannian geometry*, Ann. Sci. École Norm. Sup. **14** (1981), 249–260.

[E] G. Elencwajg, *Pseudoconvexite locale dans les varietes Kahleriennes*, Ann. Inst. Fourier (Grenoble) **25** (1975), 295–314.

[F] K. Fukaya, *A boundary of the set of Riemannian manifolds with bounded curvatures and diameters*, J. Differential Geom. **28** (1988), 1–21.

[Gaf] M. Gaffney, *The conservation property of the heat equation on Riemannian manifolds*, Comm. Pure Appl. Math. **12** (1959), 1–11.

[Ga1] L. Z. Gao, $L^{n/2}$*-curvature pinching*, J. Differential Geom. **32** (1990), 713–774.

[Ga2] ____, *Convergence of Riemannian manifolds, Ricci pinching, and $L^{n/2}$-curvature pinching*, J. Differential Geom. **32** (1990), 349–382.

[G] R. E. Greene, *Some recent developments in Riemannian geometry*, Recent Developments in Geometry, Contemp. Math. No. 101, Amer. Math. Soc., Providence, RI, 1989.

[GK] R. E. Greene and S. Krantz, *Deformation of complex structures, estimates for the $\bar{\partial}$-equation, and stability of the Bergman kernel*, Adv. in Math. **41** (1982), 1–86.

[GS1] R. E. Greene and K. Shiohama, *Convex functions on complete noncompact manifolds: differentiable structure*, Ann. Sci. École Norm. Sup. **14** (1981) 357–367.

[GS2] ____, *Convex function on complete noncompact manifolds: topological structure*, Invent. Math. **63** (1981), 129–157.

[GW1] ____, *On the subharmonicity and plurisubharmonicity of geodesically convex functions*, Indiana Univ. Math. J. **22** (1973), 641–653.

[GW2] ____, *Integrals of subharmonic functions on manifolds of nonnegative curvature*, Invent. Math. **27** (1974), 265–298.

[GW3] _____, C^∞ convex functions and manifolds of positive curvature, Acta Math. **137** (1976), 209–245.

[GW4] _____, C^∞ convex functions and manifolds of positive curvature, Acta Math. **137** (1976), 209–245.

[GW5] R. E. Greene and H. Wu, On Kähler manifolds of positive bisectional curvature and a theorem of Hartogs, Ann. Math. Sem. U. Hamburg **47** (1978), 171–185.

[GW6] _____, Function theory on manifolds which possess a pole, Lecture Notes in Math. vol. 699, Springer-Verlag, 1979.

[GW7] _____, Approximation of convex, subharmonic and plurisubharmonic functions, Ann. Sci. École Norm. Sup. **12** (1979), 47–84.

[GW8] _____, Lipschitz convergence of Riemannian manifolds, Pacific J. Math. **131** (1988), 119–141.

[Gm1] D. Gromoll, Differentiation Struckturen und Metriken positiver Krümmung und Sphärenn, Math. Ann. **164** (1966), 353–371.

[GrM] D. Gromoll and W. Meyer, On complete manifolds of positive curvature, Ann. of Math.(2) **90** (1969), 75–90.

[Grv1] M. Gromov, Almost flat manifolds, J. Differential Geom. **13** (1978), 231–241.

[Grv2] _____, Manifolds of negative curvature, J. Differential Geom. **13** (1978), 223–230.

[Grv3] _____, Curvature, diameter, and Betti numbers, Comment. Math. Helv. **56** (1981), 179–195.

[GLP] M. Gromov, J. LaFontaine, and P. Pansu, Structures metriques pour les varietes Riemanniennes, Textes Math. No. 1, CEDIC/Fernant Nathan, Paris, 1981.

[GoK] K. Grove and H. Karcher, How to conjugate C^1 close group actions, Math. Z. **132** (1973), 11–20.

[GoP1] K. Grove and P. Petersen, V, Bounding homotopy types by geometry, Ann. of Math. **128** (1988), 195–206.

[GoP2] _____, Manifolds near the boundary of existence, J. Differential Geom. **33** (1991), 379–394.

[GoS] K. Grove and K. Shiohama, A generalized sphere theorem, Ann. of Math. (2) **106** (1977), 201–211.

[GPW] K. Grove, P. Petersen, V, and J. Y. Wu, Geometric finiteness theorems via controlled topology, Invent. Math. **99** (1990), 205–213.

[H] U. Hamenstadt, Compact manifolds with 1/4-pinched negative curvature, Preprint.

[HK] E. Heintze and H. Karcher, A general comparison theorem with applications to volume estimates for submanifolds, Ann. Sci. École Norm. Sup. **11** (1978), 451–470.

[J] J. Jost, Harmonic mappings between Riemannian manifolds, Proc. Centre Math. Anal. Austral. Nat. Univ., no. 4, Austral. Nat. Univ., Canberra, 1983.

[JK] J. Jost and H. Karcher, Geometrische Methoden zur Gewinnung von a priori Schrander für harmonitsche Abbildungen, Manuscripta Math. **40** (1982), 27–72.

[K] H. Kaul, Schranken für die Christoffelsymbole, Manuscripta Math. **19** (1976), 261–273.

[Kl1] W. Klingenberg, Contributions to Riemannian geometry in the large, Ann. of Math. (2) **69** (1959), 654–666.

[Kl2] _____, Über Riemannsche Mannigfaltigkeiten mit positiver Krümmung, Comment. Math. Helv. **35** (1961) 41–54.

[MR] Min-Oo, E. Ruh, Comparison theorems for compact symmetric spaces, Ann. Sci. École Norm. Sup. **12** (1979), 335–353.

[Mo] S. Mori, Projective manifolds and ample tangent bundles, Ann. of Math. (2) **110** (1979), 593–606.

[N] I. G. Nikolaev, Parallel translation and smoothness of the metric of spaces of bounded curvature, Dokl. Akad. Nauk SSSR **250** (1980), 1056–1058.

[P1] S. Peters, Cheeger's finiteness theorem for diffeomorphism classes of Riemannian manifolds, J. Reine Angew. Math. **394** (1984), 77–82.

[P2] _____, Convergence of Riemannian manifolds, Compositio Math. **62** (1987), 3–16.

[Pt] P. Petersen, V, A finiteness theorem for metric spaces, J. Differential Geom. **31** (1990), 387–396.

[Rh] H. E. Rauch, A contribution to differential geometry in the large, Ann. of Math. (2) **54** (1951), 38–55.

[Ru1] E. Ruh, *Curvature and differentiable structures on spheres*, Comment Math. Helv. **46** (1971), 127–136.

[Ru2] ——, *Almost flat manifolds*, J. Differential Geom. **17** (1982), 1–14.

[SS] I. K. Sabitov and S. Z. Sefel, *Connections between the order of smoothness of a surface and that of its metric*, Sibirsk. Math. Zh. **17** (1976), 687–694.

[ShY] J. P. Sha and D. G. Yang, *Examples of manifolds of positive Ricci curvature*, J. Differential Geom. **29** (1989), 95–103.

[Shi] Y. Shikata, *On the differentiable pinching problem*, Osaka Math. J. **4** (1967), 279–287.

[SHS] K. Shiohama and M. Sugimoto, *On the differentiable pinching problem, with improvement by H. Karcher*, Math. Ann. **195** (1971), 1–16.

[SiY] Y. T. Siu and S. T. Yau, *Compact Kähler manifolds of positive bisectional curvature*, Invent. Math. **59** (1980), 189–204.

[W] A. Weinstein, *On the homotopy type of positively pinched manifolds*, Arch. Math. (Basel) **18** (1967), 523–524.

[Wu1] H. Wu, *An elementary method in the study of nonnegative curvature*, Acta Math. **142** (1979), 57–78.

[Wu2] ——, *On manifolds of partially positive curvature*, Indiana Univ. Math. J. **36** (1987), 525–548.

[Ya] D. Yang, *Convergence of Riemannian manifolds. I*, Ann. Sci. École Norm. Sup. **25** (1992), 77–105; II, Ann. Sci. École Norm. Sup. (to appear).

[Yau] S. T. Yau, *A general Schwarz lemma for Kähler manifolds*, Amer. J. Math. **100** (1978) 197–204.

[Yau2] S. T. Yau, *Calabiś conjecture and some new results in algebraic geometry*, Proc. Nat. Acad. Sci. USA **74** (1977) no. 5, 1798–1799.9

[YZ] S. T. Yau and F. Zheng, *Negatively 1/4-pinched riemannian metric on a compact Kähler manifold*, Invent. Math. **103** (1991), 527–535.

UNIVERSITY OF CALIFORNIA, LOS ANGELES

Proceedings of Symposia in Pure Mathematics
Volume **54** (1993), Part 3

Isometric Immersions of H^n into H^{n+1}

KINETSU ABE AND ANDREW HAAS

Introduction

Let M and $\widetilde{M}$ be the spaceforms of constant curvature c and of dimensions n and $n+1$, respectively. An elementary but fundamental problem in the geometry of submanifolds is to describe the space of all isometric immersions of M into $\widetilde{M}$.

If $c > 0$, the space is very simple. It consists of only the totally geodesic imbeddings [1, 2].

If $c = 0$, the space is considerably richer, but lucidly describable: Each such immersion must be a cylinder built over a plane curve. The result is often referred to for obvious reasons as the "Cylinder Theorem." The theorem is due to Hartman and Nirenberg [4] and Massey [6].

In the hyperbolic case, where $c < 0$, the situation at first glance looks quite different from the others. Isometric immersions seem much more abundant. Indeed, Ferus [3] and Nomizu [7] showed that given a totally geodesic foliation of codimension 1 in M, there is a family of isometric immersions of M into $\widetilde{M}$ for which the relative nullity foliations coincide with the given foliation. Their result completely characterizes the space of nowhere umbilic isometric immersions of M into $\widetilde{M}$.

In this paper, we consider a broader class of isometric immersions of one hyperbolic space into another with codimension 1. We will from now on denote by H^n the hyperbolic space of dimension n. We state our main result as follows.

1991 *Mathematics Subject Classification.* Primary 53C40.

The first author was partially supported by the foreign scholar exchange program at The Science University of Tokyo. He also thanks The Mathematical Institute at Tôhoku University for its hospitality during his stay there. The second author was partially supported by National Science Foundation grant DMS 8702868.

This paper is in final form and no version will be submitted for publication elsewhere.

THEOREM. *Given a differentiable lamination on H^n, there is a family of isometric immersions of H^n into the hyperbolic space H^{n+1} so that the induced relative nullity foliations are completely determined by the lamination.*

While including the result of Ferus and Nomizu, this approach also provides examples of isometric immersions with umbilic sets. In fact, all the possible umbilic sets are realized in this manner. Nevertheless, our method exhibits a shortcoming which may be described as "an overkill." For example, an analytic isometric immersion with umbilic points cannot be obtained using our method of construction.

There are many reasons to believe that the space of isometric immersions of H^n into H^{n+1} is parametrized by a family of properly chosen countable pairs $(c_j(s), k_j(s))$ of real-valued functions on open intervals (a_j, b_j), $j = 1, 2, \ldots$.

Hopefully, more careful analysis of our method, especially involving the above shortcoming, will lead us to a complete description of the space of the isometric immersions.

1. Preliminaries

Let L_1^{n+1} be the Minkowski space of dimension $n+1$. It is R^{n+1} equipped with the indefinite metric induced from the quadratic form:

$$(1) \qquad \langle X, Y \rangle = -x_0 y_0 + \sum_{i=1}^{n} x_i y_i,$$

where $X = (x_0, \ldots, x_n)$ and $Y = (y_0, \ldots, y_n)$ represent points in R^{n+1}.

The locus of the equation $\langle X, X \rangle = -1$ has two connected components. Define H^n to be the connected component with $x_0 > 0$, namely,

$$(2) \qquad H^n = \{X = (x_0, \ldots, x_n) \in L_1^{n+1} : \langle X, X \rangle = -1 \text{ and } x_0 > 0\}.$$

H^n is diffeomorphic to R^n and is a Riemannian manifold of constant curvature -1 with respect to the induced metric $\langle \, , \, \rangle$. We call H^n with the metric $\langle \, , \, \rangle$ the hyperbolic space form of dimension n. If one regards H^n as a submanifold of L_1^{n+1}, H^n is a space-like hypersurface. We will discuss briefly the relevant extrinsic geometry of H^n in L_1^{n+1}.

Denote by $X = (x_0, \ldots, x_n)$ the position vector expressing the point in H^n as a point in L_1^{n+1}. X in turn represents a unit normal vector to H^n at that point in L_1^{n+1}. We also denote by D and ∇ the Levi-Civita connections of L_1^{n+1} and H^n induced from the metric $\langle \, , \, \rangle$, respectively. For any two vector fields X and Y in H^n, we have two well-known equations [5]:

$$(3) \qquad \begin{aligned} D_X Y &= \nabla_X Y + \bar{\alpha}(X, Y), \\ D_X X &= X \quad \text{and} \quad \bar{\alpha}(X, Y) = \langle X, Y \rangle X. \end{aligned}$$

In the above setting, a totally geodesic hypersurface in H^n is given as an open subset of the intersection of H^n and an n-dimensional linear subspace in L_1^{n+1}.

Let U be a connected open subset of H^n. Denote by $\mathscr{F}$ a C^∞-foliation of U by complete totally geodesic hypersurfaces in H^n. Let X denote a unit vector field orthogonal to the leaves of $\mathscr{F}$ everywhere in U. X is a C^∞-vector field.

Define a linear operator $c_Z(x)$ from $\operatorname{Span} X$ into itself at x in U by

$$(4) \qquad\qquad c_Z(x)[X] = -P(\nabla_X Z),$$

where Z is a vector field tangent to the leaf at x and P is the orthogonal projection of TH^n_x onto $\operatorname{Span} X$. c_Z satisfies an ordinary differential equation $dy/dt = y^2 - 1$ along the geodesic through x with the initial vector Z at x. Here, t indicates the arc-length parameter of the geodesic. The only global solutions to the equation are given in the form

$$(5) \qquad\qquad c_Z(t) = -\frac{\sinh t - c_Z(0)\cosh t}{\cosh t - c_Z(0)\sinh t}.$$

Here, $c_Z(0)$ is a constant satisfying $|c_Z(0)| \le 1$. As a consequence, we have that $|c_Z(t)| \le 1$ for all t and $|c_Z(0)| = 1$ if and only if $c_Z(t)$ is identically equal to 1 or -1 for all t. The operator c_Z is often called the conullity operator of the foliation. See [1, 2] for accounts of these facts.

Since U is foliated by complete totally geodesic hypersurfaces, $\overline{U} - U = \partial U$ consists of a disjoint union of either two complete totally geodesic hypersurfaces, one complete totally geodesic hypersurface, or the empty set. Each maximal integral curve of the unit vector field X orthogonal to the foliation intersects with every leaf once and only once. The integral curves futhermore extend continuously to the boundary (see Ferus [3]).

LEMMA 1. *Let $\mathscr{F}_U$ be a totally geodesic foliation in U as described above. Then U admits a global Frobenius coordinate system.*

PROOF. Let X be the unit vector field orthogonal to the leaves of $\mathscr{F}_U$. Choose a maximal integral curve of X and denote it by $\gamma = \gamma(s)$, $s \in (a, b)$, $-\infty \le a < b \le \infty$. We may assume that s is the arc-length parameter. Then, $\gamma_*(d/ds) = \dot\gamma(s) = X(\gamma(s)) = X(s)$, $s \in (a, b)$. Let $\{Z_i(s)\}$, $i = 1, \ldots, n-1$, be a collection of C^∞-vector fields along $\gamma(s)$ that form an orthonormal basis for the tangent space to the leaf passing through $\gamma(s)$ at each s in (a, b). Define a mapping $\Sigma : (a, b) \times R^n \to U$ by

$$(6) \qquad\qquad \Sigma(s, t_1, \ldots, t_{n-1}) = \operatorname{Exp}_{\gamma(s)} \left(\sum_1^{n-1} t_i Z_i(s) \right).$$

Since the curvature of H^n is -1, Σ is a diffeomorphism between $(a, b) \times R^n$ and U such that for a fixed $s \in (a, b)$, $\Sigma(s, t_1, \ldots, t_{n-1})$ is exactly the leaf passing through $\gamma(s)$. Σ gives the desired coordinate system.   Q.E.D.

We will recall a few basic facts about isometric immersions of H^n into H^{n+1}. Let f be such an immersion. For any vector fields X and Y tangent to H^n, we obtain from the Gauss equation that $AX \wedge AY = 0$, where A is the shape operator of the immersion f and $AX \wedge AY$ is a skew symmetric operator acting on a tangent vector Z by $(AX \wedge AY)(Z) = \langle AY, Z \rangle AX - \langle AX, Z \rangle AY$. From $AX \wedge AY = 0$ follows that $\operatorname{rank} A \leq 1$ everywhere. Set $U = \{x \in H^n : \operatorname{rank} A = 1\}$. U is an open subset of H^n and is C^∞-foliated by complete totally geodesic hypersurfaces. This foliation is called the relative nullity foliation of f. See [1, 2] for more details.

U consists of at most countably many open connected components, each of which is C^∞-foliated by complete totally geodesic hypersurfaces. $H^n - U$ is the closed subset where A vanished identically. $H^n - U$ is often called the umbilic subset of f. We call the triple $(U, \mathscr{F}_U, H^n - U)$ a C^∞-*lamination* on H^n associated with the immersion f, where $\mathscr{F}_U$ indicates the relative nullity foliation induced from the immersion f. The lamination, denoted by $\mathscr{L}_f$, is uniquely determined by f.

More generally, let U be an open subset of H^n whose connected components are denoted by U_i, $i = 1, 2, \dots$. Assume that each U_i is endowed with a C^∞-folation $\mathscr{F}_i$ by complete totally geodesic hypersurfaces. We denote by $\mathscr{F}_U$ the disjoint union of $\mathscr{F}_i$'s. Then we call the triple $(U, \mathscr{F}_U, H^n - U)$ a C^∞-*lamination* on H^n. The notion of a geodesic lamination is due to W. Thurston.

2. Construction of isometric immersions of H^2 into H^3

In this section, we restrict our attention to the isometric immersions of H^2 into H^3. Our method of construction can be successfully carried out in higher dimensional cases with a few trivial modifications.

We will begin with a concrete description of the global coordinate system obtained in Lemma 1 in terms of quantities on H^2 regarded as a submanifold in L_1^3.

Let $(U, F_U, H^2 - U)$ be a C^∞-lamination of H^2 with U connected. Denote by $\gamma = \gamma(s)$, $s \in (a, b)$, a maximal integral curve orthogonal to the foliation on U. We will consider γ as a map into $L_1^3 \supset H^2$. Let $Z(s)$ be a unit vector field along $\gamma(s)$ orthogonal to $\dot{\gamma}(s) = \gamma_*(d/ds)$ in $H^2 \subset L_1^3$.

Define a map $\Sigma : (a, b) \times R \to U \subset H^2 \subset L_1^3$ as follows:

$$(7) \qquad \Sigma(s, t) = \gamma(s) \cosh t + Z(s) \sinh t.$$

Then,

$$(8) \quad \begin{aligned} \Sigma_*(\partial/\partial s)(s, t) &= \partial \Sigma/\partial s(s, t) = \dot{\gamma}(s) \cosh t + \dot{Z}(s) \sinh t, \\ \Sigma_*(\partial/\partial t)(s, t) &= \partial \Sigma/\partial t(s, t) = \gamma(s) \sinh t + Z(t) \cosh t. \end{aligned}$$

We have that $\langle \Sigma_*(\partial/\partial s), \Sigma_*(\partial/\partial t) \rangle = 0$, $\Sigma_*(\partial/\partial s) = (\cosh t - c(s) \sinh t) \dot{\gamma}(s)$ and $\langle \Sigma_*(\partial/\partial t), \Sigma_*(\partial/\partial t) \rangle = 1$. Here, $c(s)$ is the curvature of $\gamma(s)$ in H^2, i.e., $\langle d^2 \gamma/ds^2, Z(s) \rangle = c(s) = -\langle \dot{\gamma}(s), \nabla_{\dot{\gamma}(s)} Z(s) \rangle$. As stated in §1, $|c(s)| \leq 1$; hence, $\Sigma_*(\partial/\partial s)$ never vanishes. This implies that Σ is an immersion. It is obvious that Σ is one-to-one and onto; hence Σ is a global diffeomorphism.

LEMMA 2. *Let $\{a_i\}$ and $\{b_i\}$, $i = 1, 2, \ldots,$ be two sequences of real numbers such that $a_1 > a_2 > \cdots$, $b_1 < b_2 < \cdots$ and $a < a_i < b_i < b$, $i = 1, 2, \ldots$. Then there exists a countable family $\{g = g(s),\ g_i = g_i(s)\}$, $i = 1, 2, \ldots,$ of C^∞-real-valued functions defined in R satisfying the following conditions:*

(i) *$g(s)$ is positive if $s \in (a, b)$ and 0 elsewhere;*

(ii) *$g_i(s)$ is positive if $s \in (a_i, b_i)$ and 0 elsewhere.*

(iii) *$d^j g_i / ds^j$ converges to $d^j g / ds^j$ uniformly in any compact subset of R, $i = 1, 2, \ldots, j = 0, 1, 2, \ldots$.*

PROOF. It is well known that there exists a C^∞-function h_i such that $0 < h_i \le 1$ in (a_i, b_i) and $h_i = 0$ outside, $i = 1, 2, \ldots$. Each h_i has compact support $[a_i, b_i]$. Set $m_i = \max\{|d^j h_i / ds^j(s)|,\ s \in [a_i, b_i],\ 0 \le j \le i\}$. Then $0 < m_i < \infty$. Let σ_i be a positive real number such that $m_i \le \sigma_i(1/2)^i$, $i = 1, 2, \ldots$. Define $g_i(s) = \sum_{l=1}^{i} h_l(s)/\sigma_l$, $i = 1, 2, \ldots$. Clearly, $g_i(s)$ is a C^∞-function and satisfies (ii). From the choice of σ_i's, $d^j g_i / ds^j$ converges to $d^j g / ds^j$ uniformly in any compact subset of R, where $g(s) = \lim_{i \to \infty} g_i(s)$. Hence, $g = g(s)$ is a C^∞-function. With this choice of g and g_i's, (i), (ii) and (iii) are satisfied.   Q.E.D.

LEMMA 3. *Let $(U, \mathscr{F}_U, H^2 - U)$ be a C^∞-lamination, where U is connected. Let $c(s)$ be the curvature of an orthogonal trajectory $\gamma(s)$, $s \in (a, b)$, of the folation $\mathscr{F}_U$. Then there is a C^∞-function G in H^2 of the following form:*

(9)
$$G \circ \Sigma(s, t) = \frac{g(s)k(s)}{\cosh t - c(s)\sinh t}, \qquad (s, t) \in (a, b) \times R \ \text{ and } \ G = 0 \text{ in } H^2 - U.$$

Here, (s, t) is the global coordinate system of U described as above and $k(s)$ is a C^∞-function defined in (a, b).

PROOF. Set G_i to be the C^∞-function defined in H^2 by

(10)
$$G_i \circ \Sigma(s, t) = \frac{g_i(s)k(s)}{\cosh t - c(s)\sinh t}, \qquad (s, t) \in (a, b) \times R$$
$$\text{and } G_i = 0 \text{ in } H^2 - U_i.$$

The $g_i(s)$'s are given in Lemma 2 and $U_i = \Sigma((a_i, b_i) \times R)$, $i = 1, 2, \ldots$. Since H^2 is diffeomorphic to R^2, there are global coordinate systems in H^2. Choose one of them and denote it by the pair (u, v), $-\infty < u, v < \infty$. Indeed, we might as well assume that (u, v) is the Cartesian coordinate system in R^2. Then a function in H^2 is C^∞ if all the partials in (u, v) exist and are continuous.

Define a positive real number B_i by

$$B_i = \max\{|\partial^j G_i / \partial u^\alpha \partial v^\beta (\Sigma(s, t))|\alpha + \beta$$
$$= j,\ 0 \le j \le i,\ (s, t) \in (a, b) \times R,\ |t| \le i\}.$$

These B_i's are bounded positive constants. Let τ_i be a positive number such that $B_i \le \tau_i (1/2)^i$, $i = 1, 2, \dots$. Then define G as follows:

$$(11) \qquad G(x) = \sum_{i=1}^{\infty} (1/\tau_i) G_i(x), \qquad x \in H^2.$$

Expressing $g = \sum_{i=1}^{x} (1/\tau_i) g_i$, G has the desired form.   Q.E.D.

Let A_i be a symmetric tensor field of type $(1, 1)$ in H^2 defined as follows:

$$(12) \qquad A_i(\Sigma_*(\partial/\partial t)) = 0 \quad \text{and} \quad A_i(\Sigma_*(\partial/\partial s)) = G_i(s, t)\Sigma_*(\partial/\partial s) \quad \text{in } U,$$
$$A_i = 0 \quad \text{elsewhere.}$$

A_i is obviously a C^∞-symmetric tensor field of type $(1, 1)$.

Let $\{e_i, e_2\}$ be the orthonormal frame field in H^2 obtained from the global coordinate vector fields $\partial/\partial u$ and $\partial/\partial v$ through the Gram-Schmidt orthonormalization process. Then A_i can be expressed as a C^∞-matrix-valued function of H^2 which we write

$$(13) \qquad A_i = \begin{bmatrix} a_{11}(x) & a_{12}(x) \\ a_{21}(x) & a_{22}(x) \end{bmatrix}, \qquad x \in H^2.$$

The components of A_i are C^∞-functions of H^2 and vanish outside U_i. Just as in the proof of Lemma 3, we take the infinite sum of these A_i's with appropriate weights if necessary. The sum is denoted by A, and can be expressed in term of the base $\Sigma_*(\partial/\partial s)$ and $\Sigma_*(\partial/\partial t)$ as follows:

$$(14) \qquad A(\Sigma_*(\partial/\partial s)) = \Psi(x)\Sigma_*(\partial/\partial s) \quad \text{and} \quad A(\Sigma_*(\partial/\partial t)) = 0 \quad \text{in } U,$$
$$A = 0 \quad \text{elsewhere.}$$

Here, Ψ is a C^∞-function in H^2 of the form given in (9) with an appropriate function ψ that replaces g.

LEMMA 4. *A satisfies the Gauss and Codazzi equations in H^2.*

PROOF. We first show that each A_i satisfies these equations. Since rank $A_i \le 1$, the Gauss equation is automatically satisfied. Since A_i is a tensor field, it suffices to show that

$$(15) \qquad (\nabla_{\Sigma_*(\partial/\partial s)} A_i)(\partial/\partial t) = (\nabla_{\Sigma_*(\partial/\partial t)} A_i)(\partial/\partial s) \quad \text{in } U.$$

Note that $A_i = 0$ outside U_i; hence the equations are trivially satisfied there. Since $\Sigma_*(\partial/\partial s)$ and $\Sigma_*(\partial/\partial t)$ form a coordinate frame field, the Lie bracket of these fields is 0. This fact together with $A_i(\Sigma_*(\partial/\partial t)) = 0$ implies that equation (15) is equivalent to the following:

$$(16) \qquad [(\Sigma_*(\partial/\partial t))G_i - G_i c(\Sigma(s, t))]\Sigma_*(\partial/\partial s) = 0.$$

$c(\Sigma(s, t))$ is the linear operator defined in (4) which can be expressed as

$$(17) \qquad c(\Sigma(s, t)) = -\frac{\sinh t - c(\Sigma(\gamma(s)))\cosh t}{\cosh t - c(\Sigma(\gamma(s)))\sinh t},$$

where $\gamma(s)$ is the orthogonal trajectory to the foliation. Equation (16) implies that $dG_i/dt - G_i c(\Sigma(s, t)) = 0$. But $G_i(\Sigma(s, t)) = g_i(s)k(s)/(\cosh t - c(\gamma(s))\sin t)$. Hence,

$$\frac{dG_i}{dt}(\Sigma(s, t)) = -\frac{g_i(s)k(s)(\sinh t - c(\gamma(s))\cosh t)}{(\cosh t - c(\gamma(s))\sinh t)^2}.$$

Substituting this expression in (16), we get the desired equation.

Using the expression (13) for A_i, we see that the Codazzi equation (15) holds for A_i, $i = 1, 2, \ldots$. Since, by the definition of A, the convergence of all partials is uniform in any compact subset of H^2, A also satisfies the Codazzi equation (15). Q.E.D.

PROPOSITION 1. *Given a C^∞-lamination $(U, \mathscr{F}_U, H^2 - U)$, where U is connected, there is an isometric immersion of H^2 so that the associated lamination is precisely $(U, \mathscr{F}_U, H^2 - U)$. In fact, there are infinitely many such immersions, which are parametrized by the space consisting of functions $k(s)$.*

PROOF. Obvious from the proof of the previous lemmas and the fundamental theorem of hypersurfaces in the ambient hyperbolic space (see [5]). Q.E.D.

Now let $(U, \mathscr{F}_U, H^2 - U)$ be any C^∞-lamination on H^2. Then U consists of at most countably many connected components. Denote them by U_i, $i = 1, 2, \ldots$. Each U_i is endowed with a C^∞-foliation $\mathscr{F}_i$. By Lemma 4, the lamination $(U_i, \mathscr{F}_i, H^2 - U_i)$ gives rise to a C^∞-symmetric tensor field A_i of type $(1, 1)$, which satisfies the Gauss and Codazzi equations. Let us express each A_i as a matrix function with respect to the orthogonal frame field e_1 and e_2. For each i, let M_i be the positive real number given by

$$M_i = \text{Max}\left\{\left|\frac{\partial^l a_{pq}}{\partial u^\alpha \partial v^\beta}(u(\Sigma_i(s, t)), v(\Sigma_i(s, t)))\right| : 1 \le p, q \le 2;\right.$$

$$\left. 0 \le l \le i;\ \alpha + \beta = l;\ s \in [\alpha_i, \beta_i];\ |t| \le i\right\},$$

where $\Sigma_i : (\alpha_i, \beta_i) \times R \to U_i$ is the global parameterization in U_i from (7). Set ρ_i to be a positive number such that $M_i \le \rho_i(1/2)^i$. Now define A by the following series:

(18) $$A = \sum_{i=1}^\infty (1/\rho_i)A_i.$$

The series converges uniformly in any compact subset of H^2. The components of A are C^∞-functions. Moreover, it is easy to see that A is nonzero in U and vanishes outside U. Since the A_i's satisfy the Gauss and Codazzi equations, so does A.

THEOREM 1. *Given a C^∞-lamination $(U, \mathscr{F}_U, H^2 - U)$, there exists an isometric immersion for which the associated lamination is precisely the given lamination. Indeed, there are finitely many such immersions.*

PROOF. From (18), we have a C^∞-symmetric tensor field A of type $(1, 1)$ defined in H^2. A satisfies the Gauss and Codazzi equations. By the fundamental theorem of hypersurfaces in H^3, there is an isometric immersion whose shape operator is precisely A. Rank $A = 1$ in U and $A = 0$ elsewhere. Q.E.D.

BIBLIOGRAPHY

1. K. Abe, *Applications of a Riccati type differential equation to Riemannian manifolds with totally geodesic distributions*, Tôhoku Math. J. **25** (1973), 425–444.
2. D. Ferus, *Totally geodesic foliations*, Math. Ann. **188** (1970), 313–316.
3. ____, *On isometric immersions between hyperbolic spaces*, Math. Ann. **205**, (1973), 193–200.
4. P. Hartman and N. Nirenberg, *On spherical image maps whose Jacobians do not change sign*, Amer. J. Math. **81** (1959), 901–920.
5. S. Kobayashi and K. Nomizu, *Foundations of differential geometry*. Vols. I and II, Wiley-Interscience, 1963 and 1969.
6. W. Massey, *Surfaces of Gaussian curvature zero in Euclidean 3-space*, Tôhoku Math. J. **14** (1962), 73–79.
7. K. Nomizu, *Isometric immersions of the hyperbolic plane into the hyperbolic space*, Math. Ann. **205** (1973), 181–192.
8. B. O'Neill and E. Stiel, *Isometric immersions of constant curvature manifolds*, Michigan Math. J. **10** (1963), 335–339.

UNIVERSITY OF CONNECTICUT

Proceedings of Symposia in Pure Mathematics
Volume **54** (1993), Part 3

The Distance-Geometry of Riemannian Manifolds with Boundary

STEPHANIE B. ALEXANDER, I. DAVID BERG,
AND RICHARD L. BISHOP

1. Introduction

One of the consistent goals of modern differential geometry is to draw global conclusions from local hypotheses. For Riemannian manifolds the local hypotheses have usually been restrictions on sectional curvature, but alternative methods, more general because they are expressed directly in terms of distance, were first explored by Alexandrov [Av, **ABN**] and Busemann [**Bu1, Bu2**] 40 years ago. These methods inspired the Toponogov triangle theorem [**T**], and their use has been increasingly exploited and extended in the past decade, most notably in the work of Gromov: since these geometric ideas are expressed in terms of distance, rather than derivatives, they are more suitable for the convergence of spaces [**GLP**]; moreover, by associating polyhedral spaces and orbifolds to discrete groups, it has become sensible to apply the "local-to-global" paradigm to group theory [**Gv1, Gv2**].

Our major goal is to develop a global theory of distance geometry in Riemannian manifolds with boundary. These offer a particularly nice class of *geodesic metric spaces* (metric spaces in which any two points are joined by a distance-realizing curve). Suitable models of real-world phenomena rarely have the homogeneity of manifolds; the optimal way of doing some task is more likely to be modeled by a path subject to boundary constraints. In this note we present several theorems of the local-to-global variety valid in Riemannian manifolds with boundary. They include estimates for conjugate and convexity radii, an extension of the Hadamard-Cartan theorem, and an isoperimetric inequality. Except for the isoperimetric inequality, it is natural and timely to prove these theorems in general geodesic metric spaces. In order to specialize to manifolds with boundary we characterize Alexandrov's

1991 *Mathematics Subject Classification.* Primary 53C20.
This paper is in final form and no version of it will be submitted for publication elsewhere.

upper bound on curvature in terms of sectional curvature of the manifold and its boundary; here, unavoidable pathologies require new techniques of proof. Detailed proofs of these theorems will appear elsewhere [**ABB3**, **AB**]. Generally we assume that the spaces dealt with are connected and complete.

2. Geodesic spaces

The Hadamard-Cartan theorem for locally convex geodesic metric spaces is due to Gromov [**Gv1**, **Gv2**]. We introduce the notion of exponential map and conjugate point and formulate proofs in that setting [**AB**]. Other versions and proofs, independent of our work, may be found in articles by Ballmann [**Ba**] and Plaut [**P**].

For the results on a geodesic metric space M we felt that it was important to introduce a generalization of the exponential map. A *geodesic* is a locally distance-realizing curve parametrized proportionally to arclength by $[0, 1]$. For $m \in M$ let $\mathbf{G}_m$ be the space of geodesics with initial point m having the uniform metric topology. Let $\exp_m : \mathbf{G}_m \to M$ be the endpoint map. A *conjugate point* of m along a geodesic γ is then the endpoint of a subsegment of γ at which $\exp_m$ is not a local homeomorphism onto a neighborhood in M.

A space M is *locally convex* if for nearby geodesics γ and σ, $d(\gamma(t), \sigma(t))$ is a convex function of t. Local convexity is a consequence of nonpositive curvature in Alexandrov's sense, defined below; on the other hand, Minkowski noneuclidean spaces are locally convex but do not have curvature bounded above. In this setting we have the following Theorems 1–3 on geodesic metric spaces [**AB**]. For the first two it is not assumed that M is locally compact.

THEOREM 1 (Hadamard-Cartan 1). *A locally convex space has no conjugate points, and for each m, $\exp_m : \mathbf{G}_m \to M$ is a covering map.*

THEOREM 2. *If M has curvature $\leq K$ in Alexandrov's sense, then the length of a conjugate segment is $\geq \pi / \sqrt{K}$ $(= \infty$ if $K \leq 0$.)*

THEOREM 3 (Hadamard-Cartan 2). *If M is locally compact and $m \in M$ has no conjugate points, then $\exp_m : \mathbf{G}_m \to M$ is a covering map.*

Our remaining local-to-global result is an isoperimetric inequality in Riemannian manifolds with boundary [**ABB3**], due to Yau in the case of Riemannian manifolds [**Y**]. Again the curvature bound is in Alexandrov's sense; for equivalent formulations see the Characterization Theorem below.

THEOREM 4. *Let M be a simply connected n-dimensional Riemannian manifold with boundary having curvature $\leq K < 0$. Then the volume and perimeter of any measurable set E in M satisfy*

$$(n - 1)\sqrt{-K} V(E) \leq P(E).$$

(Following [BZ] the definition of the perimeter $P(E)$ is such that it suffices to prove the inequality for sets E having piecewise smooth boundary.)

3. Curvature bounds in Riemannian manifolds with boundary

The following characterization theorem allows us to interpret all of the preceding results in the setting of Riemannian manifolds with boundary. In our earlier works [ABB1] and [ABB2] we laid a local foundation for the global study of Riemannian manifolds with boundary. Whereas these foundations were concerned with mainly local and regularity properties of geodesics and how they bifurcate, we now have significant curvature estimates, expressed in several ways, and hence have all the consequences given above.

For manifolds with boundary, curvature estimates are necessarily one-sided; the bifurcation of geodesics at the boundary signifies negative infinite curvature. Thus, a main problem is to specify what is meant by an upper bound on curvature. We single out four ways of doing so, and the main purpose of [ABB3] is to specify three of these precisely and establish their equivalence. The fourth way, the extrinsic curvature of spaces isometrically imbedded in Euclidean space, was expressed (crudely) by the radii of external support balls and used as a tool in [ABB1] to obtain local bipoint uniqueness. The other three ways comprise the following

CHARACTERIZATION THEOREM [ABB3]. *In Riemannian manifolds with boundary the following are equivalent conditions on the constant K:*

1. Local triangle comparisons *hold with triangles in a surface S_K of constant curvature K. This gives Alexandrov's notion of a space with curvature $\leq K$. Taking the side-lengths of the compared triangles to be equal, distances between points in the original triangle should be bounded above by the corresponding distances of the comparison triangle in S_K.*

2. *The* Jacobi inequality $f'' \geq -Kf$ *is satisfied by all lengths f of Jacobi fields normal to unit-speed geodesics. (In [ABB1], a reasonable notion of Jacobi field is shown to exist. The differential inequality $f'' \geq -Kf$ is understood here in a weak sense.)*

3. *The* sectional curvatures *of the space at interior points and appropriate sectional curvatures of the boundary are no more than K. Not all sectional curvatures of the boundary have to be bounded by K, only those which are attached to tangent sections all of whose normal curvature vectors point outward.*

4. Remarks on proofs

To show that a locally convex geodesic space has no conjugate points, one needs the local surjectivity of the endpoint map $\exp_m$ on the space $\mathbf{G}_m$ of geodesics from m. Local convexity then implies that $\exp_m$ is a local homeomorphism (i.e., there are no conjugate points) and that the interior metric induced on $\mathbf{G}_m$ by pullback under $\exp_m$ is complete. From this it may be argued that $\exp_m$ is a covering map (Theorem 1 above). Local surjectivity follows from "thin bipoint existence" of geodesics near a given geodesic γ, that is, uniformly close geodesics exist for small variations of both endpoints. Our proof of this proceeds by showing that if thin bipoint existence holds for any subsegment of γ of length at most L, then it holds for any subsegment $\bar{\gamma}$ of length at most $3L/2$. To see this, trisect $\bar{\gamma}$ by points p_0 and q_0. If p and q are within ε of the endpoints of $\bar{\gamma}$, define p_i and q_i recursively by letting p_i be the midpoint of the geodesic pq_{i-1} and q_i be the midpoint of qp_{i-1}; these geodesics exist by the induction hypothesis. Local convexity implies that $d(p_{i-1}, p_i) \leq d(q_{i-2}, q_{i-1})/2$, and similar bounds for q_i. Thus, $\{p_i\}$ and $\{q_i\}$ are Cauchy sequences, by comparison with a geometric series having ratio $1/2$. Then completeness and the induction hypothesis give the required geodesic joining p and q. This "cat's cradle" argument does not use local compactness; instead it fully exploits local convexity.

In a space of curvature $\leq K$ in Alexandrov's sense, the local surjectivity of $\exp_m$ out to distance $\pi/\sqrt{K}$ is obtained by a more delicate cat's cradle argument, still using comparison with a geometric series, but with a ratio obtained from spherical trigonometry. Local injectivity ("thin bipoint uniqueness") follows by a development argument due to Alexandrov. Thus we obtain a lower bound of $\pi/\sqrt{K}$ for conjugate distance (Theorem 2).

Now, instead of a curvature bound, assume that M has no conjugate points and is locally compact. We show $\exp_m$ is a covering map (Theorem 3) by showing that the interior metric on $\mathbf{G}_m$ induced by pullback under $\exp_m$ is complete. For this, we apply a version of the Hopf-Rinow theorem due to Cohn-Vossen [C-V], which states that a locally compact, interior metric space is complete if every half-open minimizing geodesic from a base point extends to a closed interval.

In the Characterization Theorem for Riemannian manifolds with boundary, the implications $1 \to 2 \to 3$ are relatively straightforward because they amount to differentiation and applying known formulas from Riemannian geometry, respectively. For the other two implications, a characteristic difficulty lies in the possibility of unbounded switching behavior, which may, for example, produce Cantor-like coincidence sets between a geodesic and the boundary. Our approach is to develop a calculus of Jacobi fields that is independent of underlying pathology. Indeed, a Jacobi field J may be obtained from a given one-sided geodesic variation with finite endpoint velocities by extracting a subsequence, and $\|J\|$ may be shown to have the regularity properties of a convex function [ABB1]. In particular, $\|J\|$ has right and left derivatives everywhere, and $\|J\|''$ exists almost everywhere and satisfies $\|J\|'' \geq -k\|J\|$ for some sufficiently large positive constant k.

For the implication $3 \to 2$, this convexity constant k must be improved to the sectional curvature bound K. At interior points or points where the geodesic variation is entirely within the boundary this is standard. Where the base geodesic switches between an interior segment and a nontrivial boundary segment, $\|J\|'$ has a nonnegative jump, which just increases convexity. At all the remaining points, which may constitute a set of positive measure, it may be shown that the acceleration of the base geodesic exists and vanishes; so the essential step is to verify K-convexity at such points p. Our method is to extend the manifold with boundary M to a Riemannian manifold N by adding an outward collar along the boundary, and then to express $\|J\|$ as a limit of difference quotients of displacements between geodesics of M, measured through N. The estimates are necessarily delicate, because, as seen by examples, the finitary approximations do not satisfy the desired inequality; the bound K must be increased in a controlled way. Our estimates depend on the first and second variation formulas in N, and the key observation that geodesics sufficiently C^1-close to the base geodesic near p are themselves subject to arbitrary small accelerations from the boundary.

For the implication $2 \to 1$, lengths of Jacobi fields are integrated to show that distances between points of a local geodesic triangle in a Riemannian manifold with boundary M are no greater than the corresponding distances in the comparison space S_K. Following a general construction of Alexandrov, the geodesics in M from a vertex p of a given triangle to its opposite side are "developed" into a variation in S_K that preserves both the lengths of the geodesics and the speed of the curve of endpoints opposite to the vertex. Straightening out the endpoint curve yields the comparison triangle. The corresponding integration argument depends on properties whose verification in Riemannian manifolds with boundary is not immediate, such as the first variation formula for arclength and the Lipschitz continuity of the transverse curves of a geodesic variation.

BIBLIOGRAPHY

[At-Bg] Felix Albrecht and I. David Berg, *Geodesics in Euclidean space with analytic obstacles*, Proc. Amer. Math. Soc. **113** (1991), 201–207.

[AA] Ralph Alexander and S. Alexander, *Geodesics in Riemannian manifolds-with-boundary* Indiana Univ. Math. J. **30** (1981), 481–488.

[ABB1] S. B. Alexander, I. D. Berg, and R. L. Bishop, *The Riemannian obstacle problem*, Illinois J. Math. **31** (1987), 167–184.

[ABB2] ——, *Cauchy uniqueness in the Riemannian obstacle problem*, Differential Geometry, Pensicola 1985 (A. M. Naveira, A. Ferrandez, F. Mascaro, eds.), Lecture Notes in Math., vol. 1209, Springer-Verlag, Berlin-Heidelberg-New York, 1986, pp. 1–8.

[ABB3] ——, *Geometric curvature bounds in Riemannian manifolds with boundary*, Trans. Amer. Math. Soc. (to appear).

[AB] S. B. Alexander and R. L. Bishop, *The Hadamard-Cartan theorem in locally convex metric spaces*, L'Enseign. Math. **36** (1990), 309–320.

[Av] A. D. Alexandrov, *A theorem on triangles in a metric space and some of its applications*, Trudy Mat. Inst. Steklov. **38** (1951), 5–23. (Russian)

[ABN] A. D. Alexandrov, V. N. Berestovskiĭ, and I. G. Nikolaev, *Generalized Riemannian spaces*, Russian Math. Surveys **41** (1986), 1–54.

[Ba] W. Ballmann, *Singular spaces of non-positive curvature*, Sur les Groupes Hyperboliques d'apres Mikhael Gromov (E. Ghys and P. de la Harpe, eds.), Birkhäuser, Boston-Basel-Stuttgart, 1990, chapter 10.

[Bi] R. L. Bishop, *Decomposition of cut loci*, Proc. Amer. Math. Soc. **65** (1977), 133–136.

[BZ] Yu. D. Burago and V. A. Zalgaller, *Geometric inequalities*, Springer-Verlag, Berlin-Heidelberg-New York, 1988.

[Bu1] H. Busemann, *Spaces with non-positive curvature*, Acta Math. **80** (1948), 259–310.

[Bu2] ____, *The geometry of geodesics*, Academic Press, New York-San Francisco-London, 1955.

[C-V] S. Cohn-Vossen, *Existenz Kurzester Wege*, Dokl. Akad. Nauk SSSR **8** (1935), 339–342.

[GLP] M. Gromov, *Structures metriques pour les varietes Riemanniennes*, (J. Lafontaine and P. Pansu, eds.), CEDIC/Fernand Nathan, Paris, 1981.

[G1] ____, *Hyperbolic manifolds, groups and actions*, Riemann Surfaces and Related Topics (Proc., Stony Brook, 1978), (I. Kra and B. Maskit, eds.), Ann. of Math. Stud., No. 97, Princeton Univ. Press Princeton, NJ, 1981, pp. 183–213.

[G2] ____, *Hyperbolic groups*, Essays in Group Theory (S. M. Gersten, ed.), Math. Sci. Res. Inst. Publ., no. 8, Springer-Verlag, New York-Berlin-Heidelberg, 1987, pp. 75–264.

[P] C. Plaut, *A Hadamard-Cartan theorem for metric spaces*, Preprint.

[S] D. Scolozzi, *Un risultato di locale unicita per le geodetiche su varieta con bordo*, Boll. Un. Mat. Ital. B (6) **5** (1986), 309–327.

[T] V. A. Toponogov, *Riemannian spaces having their curvature bounded below by a positive number*, Uspekhi Mat. Nauk **14** (1959), 87–130, English transl., Amer. Math. Soc. Transl., vol. 37, 1964, pp. 291–336.

[W1] F.-E. Wolter, *Distance function and cut loci on a complete Riemannian manifold*, Arch. Math. **32** (1979), 92–96.

[W2] ____, *Cut loci in bordered and unbordered Riemannian manifolds*, Technische Universitat Berlin, FB Mathematik, Dissertation 249S, 1985.

[Y] S.-T. Yau, *Isoperimetric constants and the first eigenvalue of a compact Riemannian manifolds*, Ann. Sci. École Norm. Sup. (4) **8** (1975), 487–507.

UNIVERSITY OF ILLINOIS

Proceedings of Symposia in Pure Mathematics
Volume **54** (1993), Part 3

Hypersurfaces and Nonnegative Curvature

STEPHANIE B. ALEXANDER AND ROBERT J. CURRIER

1. Introduction

This interim report on nonnegative curvature and hypersurfaces of Euclidean and hyperbolic space describes a circle of new and recent results, as well as related work and open questions. Section 2 considers "superconvex" hypersurfaces of hyperbolic space, that is, convex hypersurfaces of nonnegative sectional or Ricci curvature. We describe a connection with the theory of subharmonic functions of finite Riesz mass, from which it may be shown that a superconvex hypersurface with $K \geq 0$ has asymptotic rotational symmetry and is rigid if its asymptotic boundary contains more than one point. M. Gromov has pointed out to us that this rigidity theorem yields an "immersion $\Rightarrow$ imbedding" theorem for hypersurfaces of hyperbolic space (see §4). In §3, we extend the classical theory of nonnegatively curved immersed hypersurfaces of Euclidean space by allowing curvature to be unrestricted on a compact set. From this we obtain conditions guaranteeing the imbeddedness of nonnegatively curved ends of immersed Euclidean hypersurfaces.

The rest of this section describes background and related results. Hypersurfaces are taken to be complete, connected, smooth and of dimension at least two. A convex hypersurface is one that is imbedded as the boundary of a convex body.

The study of nonnegatively curved, immersed hypersurfaces goes back to Hadamard, who showed that a compact, strictly positively curved surface immersed in R^3 is necessarily imbedded as a convex surface [**18**]. Hadamard's theorem was later extended by Stoker [**35**], Chern and Lashof [**14**], and van Heijenoort [**22**]. In particular, van Heijenoort showed that a complete, locally convex immersed hypersurface having a point of strict convexity is a convex hypersurface. Using this, Sacksteder proved a theorem that includes

1991 *Mathematics Subject Classification*. Primary 53C40; Secondary 52A20, 52A55.
This paper is in final form and no version of it will be submitted for publication elsewhere.

all of those just mentioned, namely, that a complete, nonflat, nonnegatively curved immersed hypersurface in Euclidean space is convex [31]. A main point of Sacksteder's intricate proof is to derive local convexity, which does not follow from the curvature hypothesis alone (since locally the outward normal can switch direction across a separating set of flat points), but depends on the global assumption of completeness. The Sacksteder-van Heijenoort theorem completed the classification of nonnegatively curved immersed hypersurfaces of Euclidean space, since a theorem of Hartman and Nirenberg shows that a hypersurface whose curvature is identically zero is immersed as an $(n-1)$-cylinder over a curve [19]. Do Carmo and Lima [9] gave a simpler approach to Sacksteder's theorem for $n > 2$ (see also [29] and the discussion in [15]). A comprehensive treatment of the properties of imbedded convex hypersurfaces and their spherical images is given by Wu in [37]; see also [7].

An important recent application of the theory was made by Smyth and Xavier [33], who showed that the set of principal curvatures of any immersed orientable Euclidean hypersurface either includes both positive and negative values, neither of which are bounded away from 0, or has connected closure. The proof hinges on the fact that the excluded case would yield a parallel hypersurface satisfying the Sacksteder-van Heijenoort theorem. Consequences include higher-dimensional versions of Efimov's theorem [16], and new information about the set of principal curvatures of a convex hypersurface.

All the results that we have mentioned, as well as those discussed below, concern curvature inequalities. Before continuing with this theme, we shall mention the large body of related theorems whose hypothesis involve curvature equalities and whose conclusions specify the image more rigidly. Perhaps the most obvious example is the theorem that a Euclidean hypersurface of constant positive curvature is a round sphere. A deeper result of this sort is Hopf's theorem that a sphere immersed in R^3 with constant mean curvature is imbedded and round [23, pp. 136–141]. A third such example is the characterization by Cheng and Yau [13] of a convex Euclidean hypersurface with constant mean curvature as the imbedded product of a round sphere with an orthogonal Euclidean factor. In hyperbolic space, imbedded hypersurfaces of constant mean curvature are strongly influenced by their asymptotic boundaries [8, 10, 24, 27]. Moreover, constant mean curvature surfaces imbedded in either R^3 [25] or H^3 [26] exhibit asymptotic rotational symmetry. In the next section, similar behavior appears for another class of hypersurfaces in hyperbolic space.

2. Superconvex hypersurfaces in hyperbolic space

In Euclidean space, these four pointwise conditions on an immersed hypersurface are equivalent: (1) semidefiniteness of the second fundamental form, (2) nonnegative Ricci curvature, (3) nonnegative sectional curvature, and (4) infinitesimal support by horospheres. In hyperbolic space, they become four distinct, successively stronger conditions. In terms of the eigenvalues λ_i of the second fundamental form, they correspond to

(1) $\lambda_i \geq 0$;   (2) $\lambda_i \left(\sum_{j=1}^{n} \lambda_j \right) \geq n - 1 + \lambda_i^2$;

(3) $\lambda_i \lambda_j \geq 1$ if $i \neq j$;   (4) $\lambda_i \geq 1$.

(Here the second is written so that it clearly implies the first, up to a choice of a unit normal.) The first condition is quite weak. Even with strict inequality, it need not give an imbedded convex hypersurface [34, p. 124]; moreover, the condition of convexity itself is rather weak, since by a theorem of A. D. Alexandrov, any connected open subset of the 2-sphere is homeomorphic to some complete convex surface in H^3 [1]. By a theorem of the second author [15], condition (4) implies the hypersurface is imbedded and, if noncompact, a horosphere.

Here and in §4, we consider the two intermediate conditions $K \geq 0$ and Ric ≥ 0. Convex hypersurfaces satisfying either of these conditions we regard as "superconvex," and as good candidates to replace convex hypersurfaces in arguments for which convexity alone is not sufficient. Our main tool for investigating the global geometric implications of these partial differential inequalities is the theory of subharmonic functions. By a different approach, using conformal maps and the uniformization theorem, Epstein proved independently that a convex surface in H^3 with $K \geq 0$ that is diffeomorphic to R^2 has a single point in its asymptotic boundary [17].

First we state: *If a convex hypersurface with* Ric ≥ 0 *has more than one component in its asymptotic boundary, then it is an equidistant hypersurface about a geodesic.* This follows from the Cheeger-Gromoll splitting theorem [12], together with a theorem of Volkov and Vladimirova [36] for $n = 2$ and a direct calculation from the Codazzi equation for $n \geq 2$. The theorem of Volkov and Vladimirova states that the only way to immerse the Euclidean plane into H^3 is by imbedding it as a horosphere or by covering an equidistant surface.

While it is well known that nonnegatively curved hypersurfaces of Euclidean space have convex height functions over hyperplanes, it is much less obvious that in hyperbolic space, nonnegative sectional curvature gives rise to height functions that are subharmonic on 2-planes (this is necessary but not sufficient) [4]. Moreover, a nontrivial calculation shows that nonnegative Ricci curvature gives rise to subharmonic height functions. Here the height functions are taken over a horosphere in Busemann coordinates. Using known restrictions on the set where a subharmonic function on the plane can take the value $-\infty$ [21], we obtain the following theorem, to which we shall refer in §4.

ASYMPTOTIC BOUNDARY THEOREM [4]. *A convex hypersurface with $K \geq 0$ has at most two points at infinity in its asymptotic boundary, with the presence* *of two points being a rigidity condition that forces* M *to be an equidistant hypersurface (a tube of constant radius about a geodesic).*

A nonnegatively curved surface has finite total curvature by Cohn-Vossen's inequality, hence is "flat at infinity" in an integral sense. Since the only complete flat surfaces in hyperbolic spaces are horospheres and equidistant surfaces, one might wonder whether the behavior at infinity of these special surfaces is reflected in that of general superconvex surfaces. This geometric question turns out to be related to theorems of Hayman [20] and Arsove and Huber [6] on subharmonic functions h on the plane whose Riesz mass distributions $1/2\pi \, \Delta h \, dz$ have finite total mass. The following theorem states that a convex hypersurface with $K \geq 0$ has asymptotic rotational symmetry, and its behavior at infinity is "between" that of the horospheres and the equidistant surfaces.

THEOREM [4]. *Let* M *be a noncompact convex hypersurface with* $K \geq 0$. *Suppose* M *is not an equidistant hypersurface, so has a single point* p *in its asymptotic boundary. Then* (1) M *is either a horosphere, or asymptotic to an equidistant hypersurface, or weakly asymptotic to a rotation surface that supports every equidistant hypersurface at* p *and is supported by every horosphere at* p. *(Each of these can be shown by example to occur.) In particular,* (2) M *is contained in a horosphere.*

Asymptotic rotational symmetry means that M remains close to its inner rotation hypersurface M_0, that is, the height function of M in Busemann coordinates remains close to the function obtained by maximizing the height function on concentric spheres. A classification of nonnegatively curved, nonsmooth rotation hypersurfaces [3] implies that the height function of M is slowly growing. Thus analytic estimates on slowly growing subharmonic functions apply. Roughly, these estimates state that if the subharmonic function $h(z)$ on the plane grows as $\log|f(z)|$ where $f(z)$ is a polynomial, then the behavior of $h(z)$ reflects that of $\log|f(z)|$, with the mass distribution $1/2\pi \, \Delta h \, dz$ replacing the atomic distribution concentrated on the zeros of $f(z)$ [6, 20]. In particular, $h(z)$ remains close to its rotationally symmetric growth function outside a small set (where "close" and "small" can be interpreted variously). This analytic model lacks some of the available geometric information; further geometric estimates yield the theorem.

Asymptotic rotational symmetry has been shown to persist for nonnegatively curved imbedded ends of hypersurfaces. On the other hand, conclusion (2) fails in this case. That is, there are nonnegatively curved, incomplete rotation surfaces in H^3 that support every horosphere at some point at infinity (i.e. lie in the complement of every horoball centered at that point). These cannot be completed without introducing some points of negative curvature.

It is easy to construct examples of convex hypersurfaces satisfying Ric ≥ 0 but not $K \geq 0$. We conjecture that the asymptotic boundary theorem holds for Ric ≥ 0 also.

3. "Immersion ⇒ Imbedding" theorems in Euclidean space

In Euclidean space, the Sacksteder-van Heijenoort theorem guarantees that nonflat, nonnegatively curved immersed hypersurfaces are imbedded as convex hypersurfaces. What about immersed Euclidean hypersurfaces that are nonnegatively curved off a compact set?

An argument due to Okayasu [30] shows that the image of any end representative of such a hypersurface is unbounded. (By an *end representative* in a Riemannian manifold we mean an unbounded component of the complement of a compact set.) Moreover, we have the following theorem:

THEOREM 5. *Suppose M is an immersed hypersurface of R^{n+1}, $n \geq 3$, having a compact subset X such that all sectional curvatures are nonnegative on $M \backslash X$. (a) If the nullity of the second fundamental form is at most 1 on $M \backslash X$, then each end of M has an imbedded representative. (b) If $M \backslash X$ is strictly positively curved, then each end of M has a representative that is imbedded as the boundary in an open halfspace of a convex body.*

This is best possible in the sense that M need not have imbedded ends if the nullity of the second fundamental form is at most 2 on $M \backslash X$ and at most 1 somewhere outside every compact set. Similarly, (b) fails if one merely assumes that the nullity is at most 1 on $M \backslash X$ and vanishes somewhere outside every compact set.

The situation in R^3 is quite different. It is easy to construct an immersed surface in R^3 that is strictly positively curved off a compact set and has no imbedded end representative. For example, take a finite covering of a deformed cylinder. There are also examples for which no component of the intersection of the surface with an open halfspace is an end representative; that is, no end can be "cut off" by a plane. On the other hand, if an *imbedded* surface in R^3 is strictly positively curved off a compact set, then each end has a convex representative that may be cut off by a plane. This intuitively appealing fact follows from a special case of our argument.

The idea of the proof is to construct imbedded convex slices through given points p on M whose images lie far from the image of X. However, the portion of M that can be surveyed from p via imbedded convex slices may be compact, and so give no information about the ends of M. Thus the first concern is to locate in each end a noncompact slice. This is achieved by first constructing through every point p of M whose image lies outside some Euclidean ball, a "convex cap" in M having both p and a point of X in its boundary. The construction applies van Heijenoort's method to hypersurfaces that are not necessarily complete; this adaptation is necessary since slicing planes must be considered that intersect the set X of uncontrolled curvature. Once points of M outside a ball have been connected to points of a compact subset by convex caps, a convergence argument shows that each end of M contains a noncompact convex cap.

Now one can move to infinity on such a cap, looking for a way to cut off the corresponding end. In case (a), an imbedded end representative is constructed that has the form of a union of two convex caps whose boundary is compact. In case (b) the following lemma is then applied to the union of the recession cones of these two convex caps: *If the union of two closed convex subsets of the n-sphere contains no pair of antipodal points, then there is an open hemisphere that contains both subsets.* It follows that if M is positively curved off a compact set, each end of M may be cut off by a hyperplane.

Finally, we mention some related questions. In the above theorem, the immersion is a priori locally convex off a compact set. The proof of the theorem may suggest a formulation that does not involve a priori local convexity. As was explained in the introduction, such a formulation could be expected to have a considerably more intricate proof. Again, consider an immersed Euclidean hypersurface of dimension $n \geq 3$, nonnegatively curved off a compact set. If some end representative contains no open subset whose image splits off a line, does the end have a convex representative? If some end representative contains no open subset whose image splits off a two dimensional halfplane, does the end have an imbedded representative? These may be regarded as analogous to the combined theorems of Sacksteder-van Heijenoort and Hartman-Nirenberg, whereby if M is everywhere nonnegatively curved and its image does not split off an $(n-1)$-plane, then M is imbedded as a convex hypersurface.

4. "Immersion $\Rightarrow$ Imbedding" theorems in hyperbolic and other spaces

The analogue for spherical space of the Sacksteder-van Heijenoort theorem was proved by do Carmo and Warner [11]. They raised the question whether an analogue exists for noncompact hypersurfaces of hyperbolic space. As was mentioned in the introduction, it is easy to construct immersed, locally convex hypersurfaces of H^{n+1} that are not imbedded. These examples can be modified to obtain nonimbedded hypersurfaces all of whose sectional curvatures are at least C, where C is any negative constant. Thus an appropriate conjecture would be: Except for covering maps of equidistant surfaces in H^3, every nonnegatively curved immersed hypersurface in H^{n+1} is a proper imbedding.

M. Gromov has pointed out to us a Brouwer fixed point argument, showing that the asymptotic boundary theorem of §2 implies the truth of this conjecture when $n \geq 3$. The argument proceeds by setting up a continuous map on a closed disk D in the sphere at infinity, where D is determined by a local support element. Namely, to each point in D can be associated a slicing procedure that leads, by the asymptotic boundary theorem, to a single point in D on the first unbounded slice of M; or if not, then it can be shown that M is already imbedded as an equidistant hypersurface. A fixed point of this map corresponds to an imbedded exhaustion of the hypersurface by slices, so the Brouwer fixed point theorem guarantees that M is imbedded.

The conjecture has not yet been proved for $n = 2$, nor is it clear whether it is true for nonnegative Ricci curvature.

Finally, we briefly mention some theorems of the immersion-implies-imbedding type in spaces of variable curvature. The first author showed that a compact hypersurface immersed in a Hadamard manifold, having a continuous normal with respect to which its second fundamental form is positive semidefinite, is imbedded and hence convex [2]. (This also appears in [32].) An immersed hypersurface in a Hadamard manifold of curvature no greater than $-c^2$, the eigenvalues of whose second fundamental form satisfy $|\lambda| \le c$, is imbedded and diffeomorphic to R^n [2]. Menninga has shown that if a compact hypersurface M with sectional curvature at least c^2 is immersed in a Riemannian manifold N of sectional curvature at most $c^2/4$ and injectivity radius at least π/c, then M lies in some N-ball of radius $\pi/2c$ and is imbedded as a convex hypersurface [28]. Little seems to be known about noncompact hypersurfaces of nonnegative curvature in spaces of variable curvature.

REFERENCES

1. A. D. Aleksandrov, *Polnye vypuklye poverkhnosti v prostranstve lobachevskogo*, Izv. Akad. Nauk SSSR **9** (1945), 113–120.

2. S. B. Alexander, *Locally convex hypersurfaces of negative curved spaces*, Proc. Amer. Math. Soc. **64** (1977), 321–325.

3. S. B. Alexander and R. J. Currier, *Nonnegatively curved hypersurfaces in hyperbolic space*, Proceedings of the Meeting at Luminy: Geometry and Topology of Submanifolds, World Scientific, Singapore, New Jersey, London, Hong Kong, 1989, pp. 1–9.

4. ____, *Nonnegatively curved hypersurfaces of hyperbolic space and subharmonic functions*, J. London Math. Soc. (2) **41** (1991), 347–360.

5. ____, *Nonnegatively curved ends of Euclidean hypersurfaces*, Geometriae Dedicata, **40** (1991), 29–43.

6. M. Arsove and A. Huber, *Local behavior of subharmonic functions*, Indiana Univ. Math. J. **22** (1973), 1191–1199.

7. M. do Carmo and H. B. Lawson, Jr., *Spherical images of convex surfaces*, Proc. Amer. Math. Soc. **31** (1972), 635–636.

8. ____, *On Alexandrov-Bernstein theorems in hyperbolic space*, Duke Math. J. **50** (1983), 995–1003.

9. M. do Carmo and E. Lima, *Immersions of manifolds with non-negative sectional curvatures*, Bol. Soc. Brasil. Mat. **2** (1971), 9–22.

10. M. do Carmo, J. de Miranda Gomes, and G. Thorbergsson, *The influence of the boundary behavior on hypersurfaces with constant mean curvature in H^{n+1}*, Comment Math. Helv. **61** (1986), 429–441.

11. M. do Carmo and F. Warner, *Rigidity and convexity of hypersurfaces in spheres*, J. Differential Geom. **4** (1970), 133–144.

12. J. Cheeger and D. Gromoll, *The splitting theorem for manifolds of nonnegative Ricci curvature*, J. Differential Geom. **6** (1971), 119–128.

13. S. Y. Cheng and S. T. Yau, *Differential equations on Riemannian manifolds*, Comm. Pure Appl. Math. **28** (1975), 333–354.

14. S. S. Chern and R. K. Lashof, *On the total curvature of immersed manifolds*, Michigan Math. J. **5** (1958), 5–12.

15. R. J. Currier, *Hypersurfaces of hyperbolic space infinitesimally supported by horospheres*,

Trans. Amer. Math. Soc. **313** (1989), 419–431.

16. N. V. Efimov, *Hyperbolic problems in the theory of surfaces*, Proc. Internat. Congr. Math. Moscow (1966), Amer. Math. Soc. Transl. **70** (1968), 26–38.

17. C. Epstein, *The asymptotic boundary of a surface imbedded in H^3 with nonnegative curvature*, Michigan Math. J. **34** (1987), 227–239.

18. J. Hadamard, *Sur certaines propriétés des trajectoires en dynamique*, J. de Mathématiques **3** (1897), 331–387.

19. P. Hartman and L. Nirenberg, *On spherical image maps whose Jacobians do not change sign*, Amer. J. Math. **81** (1959), 901–920.

20. W. K. Hayman, *Slowly growing integral and subharmonic functions*, Comment. Math. Helv. **34** (1960), 75–84.

21. W. K. Hayman and P. B. Kennedy, *Subharmonic functions*, vol. 1, Academic Press, London, New York, San Francisco, 1976.

22. J. van Heijenoort, *On locally convex manifolds*, Comm. Pure Appl. Math. **5** (1952), 223–242.

23. H. Hopf, *Differential geometry in the large*, Lecture Notes in Math., vol. 1000, Springer-Verlag, Berlin, Heidelberg, New York, Tokyo, 1983.

24. W.-Y. Hsiang, *On generalization of theorems of A. D. Alexandrov and C. Delaunay on hypersurfaces of constant mean curvature*, Duke Math. J. **49** (1982), 485–496.

25. N. Korevaar, R. Kusner, and B. Solomon, *The structure of complete embedded surfaces with constant mean curvature*, J. Differential Geom. **30** (1989), 465–503.

26. N. Korevaar, R. Kusner, W. Meeks, III, and B. Solomon, *Constant mean curvature surfaces in hyperbolic space*, Amer. J. Math. **114** (1992), 1–43.

27. G. Levitt and H. Rosenberg, *Symmetry of constant mean curvature hypersurfaces in hyperbolic space*, Duke Math. J. **52** (1985), 53–59.

28. N. Menninga, *Immersions of positively curved manifolds into manifolds with curvature bounded above*, Trans. Amer. Math. Soc. **318** (1990), 809–821.

29. J. de Miranda Gomes, *On isometric immersions with semi-definite second quadratic forms*, An. Acad. Brasil. Ciênc. **55** (1983), 145–146.

30. T. Okayasu, *Some results in geometry of hypersurfaces*, Kodai Math. J. **9** (1986), 77–83.

31. R. Sacksteder, *On hypersurfaces with no negative sectional curvatures*, Amer. J. Math. **82** (1960), 609–630.

32. V. Schroeder and M. Strake, *Local rigidity of symmetric spaces of nonpositive curvature*, Proc. Amer. Math. Soc. **106** (1989), 481–487.

33. B. Smyth and F. Xavier, *Efimov's theorem in dimension greater than two*, Invent. Math. **90** (1987), 443–450.

34. M. Spivak, *A comprehensive introduction to differential geometry*, vol. 4, Publish or Perish, Boston, 1975.

35. J. J. Stoker, *Über die Gestalt der positiv gekrümmten offenen Flächen im dreidimensionalen Raume*, Compositio Math. **3** (1936), 55–89.

36. Yu. A. Volkov and S. M. Vladimirova, *Isometric immersions of the Euclidean plane in Lobachevskiĭ space*, Mat. Zametki **10** (1971), 327–332; Math. Notes **10** (1971), 619–622.

37. H. Wu, *The spherical images of convex hypersurfaces*, J. Differential Geom. **9** (1974), 279–290.

UNIVERSITY OF ILLINOIS AT URBANA-CHAMPAIGN

SMITH COLLEGE

Proceedings of Symposia in Pure Mathematics
Volume **54** (1993), Part 3

Shortening Space Curves

STEVEN J. ALTSCHULER

ABSTRACT. Singularities for space curves evolving by the curve shortening flow are studied. Space curves are used to give a flow through singularities of plane curves evolving by the curve shortening flow.

In recent years, much interesting work has been done in the area of mean curvature flow, and, in particular, the case of curve shortening for curves on surfaces (see the bibliography for an incomplete list). The evolution equation referred to as the curve shortening flow on the plane is

$$\frac{\partial \gamma}{\partial t} = kN$$

where $\gamma: S^1 \times [0, \omega) \rightarrow R^2$, $\gamma(\cdot, 0)$ is a smooth curve, and kN is the curvature times the normal to the curve. N is not defined at inflection points, but kN is always defined as being the second derivative of the positive vector with respect to arc-length. Of course, a plane curve is a special case of a space curve.

We are interested in considering the behavior of spaces curves evolving under the same evolution equation as above. We are in part motivated by a suggestion of E. Calabi's and a conjecture of Matt Grayson's which will be discussed later.

Below, we would like to compare and contrast the evolution of space curves with the evolution of planar curves. To set notation, the Frenet formulas for the derivates of the tangent T, normal N, and binormal B vectors are

$$\frac{\partial}{\partial s} \begin{pmatrix} T \\ N \\ B \end{pmatrix} = \begin{pmatrix} 0 & k & 0 \\ -k & 0 & \tau \\ 0 & -\tau & 0 \end{pmatrix} \cdot \begin{pmatrix} T \\ N \\ B \end{pmatrix}.$$

where τ is the torsion and the curvature $k \neq 0$.

1991 *Mathematics Subject Classification.* Primary 53A04, 35K22.
Research supported by an Alfred P. Sloan Doctoral Dissertation Fellowship.
The final version of this paper has been submitted for publication elsewhere.

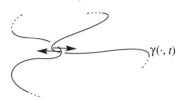

FIGURE 1. Clasp about to break

Short time existence. As in the case of plane curve evolution, short time existence of solutions to the space curve evolution is guaranteed for nice enough initial data [**G–H**].

Embedded curves. Although for curves on the plane, it has been shown that an embedded curve remains embedded [**G–H**], it is not hard to see that this is false in general for space curves. See Figure 1 for a picture of a clasp about to break.

Convex curves. The evolution equations for curvature and torsion are

$$\frac{\partial k}{\partial t} = \frac{\partial^2 k}{\partial s^2} + k^3 - \tau^2 k,$$

$$\frac{\partial \tau}{\partial t} = \frac{\partial^2 \tau}{\partial s^2} + 2\frac{1}{k}\frac{\partial k}{\partial s}\frac{\partial \tau}{\partial s} + 2\frac{\tau}{k}\left(\frac{\partial^2 k}{\partial s^2} - \frac{1}{k}\frac{\partial k^2}{\partial s} + k^3\right),$$

where τ is the torsion of the curve. In the case of plane curves, define $\tau \equiv 0$ and the maximum principle implies that a curve which is initially convex, $k > 0$, will remain convex. For space curves, however, the condition that $k \neq 0$ would only be preserved by the heat flow if τ remained bounded. As we will see below, this is not always true.

By considering the tangent indicatrix of a torus curve $\Gamma(\cdot, t) = T(\cdot, t) \in S^2$, and its induced evolution, it is easy to see that loops of Γ can pinch off and become singular. The geodesic curvature of the indicatrix on the sphere is τ/k. At such times, either the curvature goes to zero, the torsion goes to infinity, or both occur. In fact, both occur and the Frenet formula becomes undefined at such an inflection point $k = 0$.

Long time existence. Although one has short time existence of solutions on a small open interval in time [**G–H**], solutions do not exist for infinite time. In fact, as we have noted above, quantities such as the torsion may below up. We prove the result

THEOREM ([**Alt–Gr**]; Long time existence). *Solutions to the space curve shortening flow exist until the curvature becomes unbounded.*

This is the same result as for plane curves [**G–H**] but it is surprising that no other quantities (such as torsion) determine the long time existence of solutions.

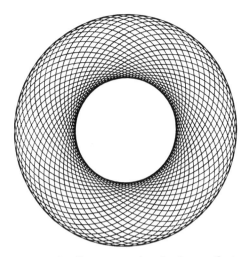

FIGURE 2. Curve moving by homothety

FIGURE 3. Yin-Yang curve moving by rotation

Solitons. We consider a soliton to be a solution of the curve shortening flow which moves by self-similarity. Soliton solutions turn out to be very important in understanding the limiting behavior of solutions near singularities.

Homothety. Abresch and Langer [A–L] and Epstein and Weinstein [E–W] found solutions to the curve shortening flow which move only by homothety (for example, the circle moves by dilation). See Figure 2 for a picture of a more complicated solution.

Rotation. The so-called Yin-Yang curve is a noncompact curve with infinite winding number which rotates for all eternity. See Figure 3.

Translation. The so-called Grim Reaper is a noncompact curve with $k \geq 0$ and total change in angle π which moves by translation for all eternity. As a graph, it looks like $y = -\log \cos x$. See Figure 4 on next page.

FIGURE 4. Grim Reaper curve moving by translation

In [AH-Su] we classify and explore nonplanar soliton curves. For example, Figure 5 is a curve which moves by rotation and translation.

Singularities. As we have noted, solutions to the curve shortening flow exist until the curvature becomes infinite. It is interesting to note what happens to the curve in a region where a singularity occurs. We prove [Alt] the rather surprising conjecture, due to Matt Grayson, that singularity formation is a planar phenomenon. We then give asymptotic descriptions of the solution.

Assume that a solution to $\partial\gamma/\partial t = kN$ exists on the maximal time interval $[0, \omega)$. Our main results, briefly stated, are

THEOREM ([Alt]; Type-I singularities). *If* $\lim_{t\to\omega} \|k^2(\cdot, t)\|_\infty (\omega - t)$ *is bounded, then* γ *is asymptotic to a solution which is moving by homothety.*

THEOREM ([Alt]; Type-II singularities). *If* $\lim_{t\to\omega} \|k^2(\cdot, t)\|_\infty (\omega - t)$ *is unbounded, then there exists a sequence of points and times* $\{p_n, t_n\}$ *on which the curvature blows up such that:*

(1) *a rescaling of the solution along this sequence converges in* C^∞ *to a planar, convex limiting solution* γ_∞;

(2) γ_∞ *is a solution which moves by translation.*

By rescaling the solution along a sequence $\{(p_n, t_n)\}$ we mean that we obtain new solutions γ_n to the curve shortening flow from γ by translating $t_n \mapsto 0$, $\gamma(p_n, t_n) \mapsto 0 \in R^3$, and dilating the solution in space and time so that $k^2(p_n, t_n) \mapsto 1$.

The result on type-I singularities is a generalization of the arguments given by Huisken [Hn] for codimension-1 mean curvature flow. Our argument makes use of the monotonicity formula developed by Huisken. Angenent [An1] has also obtained these results for plane curves.

A standard example for a type-II singularity is given by a loop pinching off to a cusp (Figure 6). Angenent [A1], in the case of convex-planar curves, showed that these singularities are asymptotic to Grim Reaper curves.

Our results for type-II singularities follow from first proving that a limit of solutions exists, then by showing that it is planar (Grayson's conjecture) and convex. We must then show that this solution exists for all time in the past and for all time in the future. It then follows, with some work, that the

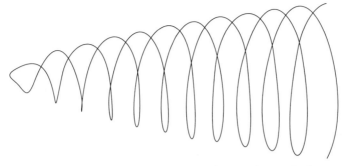

FIGURE 5. Curve moving by rotation and translation

FIGURE 6. Type II singularities developing

methods presented in [H] prove that this curve is the Grim Reaper.

Flow through plane curve singularities (work with Matt Grayson [Alt-Gr]). When a simple closed curve in the plane evolves by the curvature flow, it shrinks to a point in finite time, becoming round in the limit [G-H, Gr1]. When the curve is not simple, however, singularities form in finite time as loops pinch off to form cusps. The classical machinery for short time existence of solutions to the curvature flow breaks down when the curvature becomes unbounded. This is not to say that it cannot be continued. Angenent [An2] shows that the singular curves are nice enough that, with some possible trimming, they may be used as initial data for the curve shortening flow. Solutions after the singularity have fewer self-intersections than before.

Recently, E. Calabi suggested a method for flowing through planar singularities using space curves. The idea is to take a family Γ of embedded space curves limiting on the immersed plane curve, and then defining a flow through the singularity as the limit of the flows in Γ.

Several points must be checked.

(1) The space curves must be nonsingular for longer than the planar curve.
(2) The space curves must converge to a planar curve at later times.
(3) The limit planar curve should be independent of Γ.

DEFINITION. A *ramp* is a space curve which steadily gains height; that is, its tangent vector has positive vertical component at all points.

We use periodic ramps to approximate the planar curve.

DEFINITIONS. Let $\Gamma_0(t)$ be an evolving immersed closed curve in the plane, nonsingular for $t \in [0, \omega_0]$. Let $\Gamma_l(0)$, $l \in (0, 1]$, be any family of periodic ramps which project vertically onto $\Gamma_0(0)$ and have vertical period l. $\Gamma_l(t)$ then evolves by the curvature flow for space curves.

THEOREM ([**Alt-Gr**]; Flow through singularities). *Given Γ as above, we have*

(1) *For all $l > 0$, Γ_l exists and is smooth for all time.*

(2) *The limit as $l \to 0$ of $\Gamma_l(t)$ is a smooth curve at all but a finite number of times $\omega_i \in [\omega_0, \omega_n]$. For $t \geq \omega_n$, the limit is a point.*

(3) *The limit agrees with the planar evolution away from the singularities.*

(4) *The limit planar curve is independent of the choice of $\Gamma_l(0)$.*

Conclusion

It is of general interest to people in the subject of geometric evolution equations to obtain a better understanding of singularity formation. Type-II singularities, in particular, are quite elusive and one would like better asymptotic descriptions of the blow-up regions.

We are also interested in a generalization of the result that singularity formation for space curves is a planar phenomenon. To our knowledge, no other work has been done on singularity formation for mean curvature evolution in codimensions higher than one.

REFERENCES

[Alt] S. J. Altschuler, *Singularities of the curve shortening flow for space curves*, J. Diff. Geom. **34** (1991), 491–514.

[Alt-Gr] S. J. Altschuler and M. A. Grayson, *Shortening space curves and flow through singularities*, J. Diff. Geom. **35** (1992), 283–298.

[AH-Su] S. J. Altschuler and J. Sullivan, *Space Curve Solitons of the Mean Curvature Flow*, (1992), preprint.

[A-L] U. Abresch and J. Langer, *The normalized curve shrinking flow and homothetic solutions*, J. Differential Geom. **23** (1986), 175–196.

[An1] S. Angenent, *On the formation of singularities in the curve shrinking flow*, J. Diff. Geom. **33** (1991), 601–633.

[An2] ____, *Parabolic equations for curves on surfaces. II. Intersections, flow up and generalized solutions*, Ann. of Math. **131** (1991), 171–215.

[E-W] C. Epstein and M. Weinstein, *A stable manifold theorem for the curve shortening equation*, Comm. Pure Appl. Math. **40** (1987), 119–139.

[F-M] A. Friedman and B. McLeod, *Blow-up of solutions of nonlinear degenerate parabolic equations*, Arch. Rational Mech. Anal. **96** (1986), 55–80.

[G-H] M. Gage and R. S. Hamilton, *The heat equation shrinking convex plane curves*, J. Differential Geom. **23** (1986), 69–96.

[Gr1] M. A. Grayson, *The heat equation shrinks embedded plane curves to round points*, J. Differential Geom. **26** (1987), 285–314.

[Gr2] ____, *Shortening embedded curves*, Ann. of Math. (2) **129** (1989), 71–111.

[H] R. S. Hamilton, 1989 CBMS Conference, Hawaii, unpublished lecture notes.

[Hn] G. Huisken, *Asymptotic behaviour for singularities of the mean curvature flow*, J. Differential Geom. **31** (1990), 285–299.

[Shi] Wan-Xiong Shi, *Deforming the metric on complete Riemannian manifolds*, J. Differential Geom. **30** (1989), 223–301.

UNIVERSITY OF MINNESOTA

Proceedings of Symposia in Pure Mathematics
Volume **54** (1993), Part 3

Degeneration of Metrics with Bounded Curvature and Applications to Critical Metrics of Riemannian Functionals

MICHAEL T. ANDERSON

0. Introduction

During the past several years, there has been much progress in understanding the Ricci curvature of a Riemannian manifold M and its relations with the topology and other global aspects of the geometry of M. Many of these developments will be reported on by other authors in these volumes. For some previous surveys in this area, we refer to the comprehensive text [**13**], and also to [**53, 14**].

Here, our main focus is centered on a particular question, namely what light some of these developments shed on the existence of critical metrics for natural Riemannian functionals on the space of metrics on M. From a range of possibilities, we will concentrate on the following two functionals, namely

$$(0.1) \qquad \mathscr{S}(g) = \int_M s(g)\, dV_g,$$

where $s(g)$ is the scalar curvature of (M, g), and

$$(0.2) \qquad \mathscr{R}^p(g) = \int_M |R(g)|^p\, dV_g,$$

where R is the Riemann curvature of (M, g). These are considered as functionals on the space $\mathscr{M}_1$ of (equivalence classes) of Riemannian metrics on M with total volume 1, divided by the action of the diffeomorphism group; (the topology on $\mathscr{M}_1$ will be discussed later). Thus, we seek to understand the behavior of minimizing or minimax sequences of metrics with respect to these functionals.

1991 *Mathematics Subject Classification.* Primary 58E11, 58G03, 58G30.
Partially supported by NSF Grant DMS 89-01303 and an AMS Research Fellowship.
This paper is in final form and no version of it will be submitted for publication elsewhere.

It is well known that critical points of $\mathscr{S}$ are exactly the Einstein metrics on M. Although $\mathscr{S}$ does not admit any minimizing sequences, since $\inf_g \mathscr{S}(g) = -\infty$ if $\dim M \geq 3$, it does admit natural minimax sequences in many cases. Namely, one may first minimize $\mathscr{S}$ in a fixed conformal class, and then maximize $\mathscr{S}$ among all conformal classes; cf. [40] for further details. It is well known however that minimax sequences of $\mathscr{S}$ do not converge in general. For instance, there is a familiar obstruction to the existence of Einstein metrics in dimension 4 [13] given by the inequality

$$(0.3) \qquad\qquad |\sigma(M)| \leq \frac{2}{3}\chi(M),$$

where $\chi(M)$ is the Euler characteristic and $\sigma(M)$ is the signature of M. Similarly, there are of course also obstructions in dimension 3, where Einstein metrics are of constant curvature. However, so far it is not well understood how these obstructions relate to the behavior or degeneration of minimax sequences of $\mathscr{S}$ on M.

For the functional $\mathscr{R}^p$, if $p < n/2$ $(n = \dim M)$, then it was pointed out by Gromov, and is quite easy to see, that $\inf_g \mathscr{R}^p(g) = 0$, realized by rather degenerate metrics on M. (One concentrates all of the topology and curvature of M in a set of small volume, giving the complement an almost flat metric.) Thus, the only cases of general interest are when $p \geq n/2$, while certain arguments tend to suggest that the scale invariant case $p = n/2$ might be the most interesting.

Here there are no known obstructions to the existence of smooth minima, even for $\int |R|^2$ on 4-manifolds. Of course, the Gauss-Bonnet theorem in dimension 2 implies that constant curvature metrics are exactly the metrics on M^2 which minimize $\mathscr{R}^1$; one obtains the same statement for $p \geq 1$, by applying the Hölder inequality. In fact, it is not difficult to verify that the only critical metrics of $\mathscr{R}^2$ on a compact 2-dimensional surface M^2 are constant curvature metrics. However, in higher dimensions, the geometric properties of critical or minimal metrics for $\mathscr{R}^p$ are little understood, even in dimensions 3 or 4. Einstein metrics are known to be critical for $\mathscr{R}^2$, but there are many others as well [13]. In general, Einstein metrics are not critical metrics of $\mathscr{R}^p$, $p \neq 2$.

Now suppose $\{g_i\}$ is a minimizing sequence for $\mathscr{R}^p$ or a minimax sequence for $\mathscr{S}$ on $\mathscr{M}_1(M)$. If $\{g_i\}$ does not converge to a (smooth) critical metric of the appropriate functional, then $\{g_i\}$ must degenerate in some way. Our main purpose is to understand the possible degenerations of $\{g_i\}$ and, if possible, to see if one may suitably modify the sequence $\{g_i\}$ to obtain new sequences (also approaching a critical value) with better convergence properties.

Of course, the metric degeneration of $\{g_i\}$ must include the case of possible degenerations within the space of critical metrics itself, i.e. within the moduli space $\mathscr{C}$ of critical metrics of $\mathscr{F}$. The classical example of such degenerations is described by Teichmuller theory when $\dim M = 2$. Namely, one has exactly three possibilities for the behavior of sequences $\{g_i\} \in \mathscr{C}$, where $\mathscr{C}$ is the space of constant curvature metrics.

First, $\mathscr{C}$ may be compact, so that there are no degenerations. This happens exactly when $M = S^2$ and in fact $\mathscr{C} = \{pt\}$ in this case. Second, $\mathscr{C}$ may be noncompact and divergent sequences collapse, in the sense that the injectivity radius $\mathrm{inj}(x)$ satisfies

$$(0.4) \qquad\qquad \mathrm{inj}_{g_i}(x) \to 0 \quad \text{as } i \to \infty,$$

for every $x \in M^2$. This is the case for $M = T^2$. Third, $\mathscr{C}$ may be noncompact and divergent sequences $\{g_i\}$ converge to a finite number of cusps, i.e. complete noncompact manifolds N_i embedded in M, with $\sum \mathrm{vol}\, N_i = 1$, $\mathrm{diam}\, N_i = \infty$. The complement $M - \bigcup N_i$ is a finite collection of disjoint simple closed curves in M, which metrically are collapsed under $\{g_i\}$ to points as $i \to \infty$.

As we will see, many features of this trichotomy carry through to higher dimensions, for minimizing sequences or sequences approaching a critical value of $\mathscr{R}^p$, for $p > n/2$. However, it remains a challenge to carry out an analogous program for the functional $\mathscr{S}$.

In §1, we give an outline of the results of Cheeger-Gromov on the degeneration of metrics with bounded sectional curvature. In §2, some extensions of these results to metrics with bounded Ricci curvature are discussed, while in §3, similar questions are treated assuming integral bounds on the curvature. Finally, in §4 and §5, we apply these results to the study of the behavior of metrics approaching critical values of $\mathscr{R}^p$ and $\mathscr{S}$.

While this paper is partly intended as a brief survey on degenerations of metrics with bounded curvature, it also contains a number of new, previously unpublished results, in particular §3, 4, 5. These results were obtained prior to November 1990. The many developments that have taken place since then will be reported on elsewhere.

Discussions with Jeff Cheeger, Daryl Geller, Santiago Simanca and Deane Yang were helpful in the preparation of this work, and I would like to thank them.

1. Cheeger-Gromov theory

The theory developed by Cheeger and Gromov [17, 29, 18, 19] may be viewed as a vast generalization, to all dimensions and variable curvature, of the coarse aspects of Teichmuller theory described above. We describe some aspects of this theory relevant to our main concern above; in particular, the results are not described here in their full generality.

Let M be a fixed compact n-dimensional manifold. We will study the space $\mathscr{M}_1^\Lambda$ of Riemannian metrics on M such that

$$(1.0) \qquad\qquad |R| \leq \Lambda, \qquad \mathrm{vol}\, M = 1,$$

as well as all possible limits or degenerations of sequences in $\mathscr{M}_1^\Lambda$.

THEOREM 1.1 (Cheeger-Gromov compactness [17, 29]; cf. also [28, 31, 37]). *The space $\mathscr{M}_1^{\Lambda, D}$ of Riemannian metrics on M satisfying the bounds*

$$(1.1) \qquad |R| \le \Lambda, \quad \mathrm{vol}\, M = 1, \quad \mathrm{diam}\, M \le D,$$

for arbitrary constants Λ, D, is precompact in the $C^{1,\alpha}$ topology on $\mathscr{M}_1$, for any $\alpha < 1$.

This means that given any sequence $\{g_i\}$ of metrics on M, satisfying (1.1), there are diffeomorphisms f_i of M such that a subsequence of $\{f_i^* g_i\}$ converges, in the $C^{1,\alpha'}$ topology on M, to a $C^{1,\alpha}$ Riemannian metric g on M, for any $\alpha' < \alpha < 1$. Note that without further hypotheses on $\{g_i\}$, one cannot improve the convergence or the smoothness of the limit to say C^2, since bounded sequences in $C^2(M)$ are not compact (they are precompact in $C^{1,\alpha}$, $\alpha < 1$, by the Arzela-Ascoli theorem).

In particular, metrics satisfying these bounds do not degenerate in a significant way at all; most all Riemannian invariants, local and global, except curvature, vary within fixed controllable bounds under the bounds (1.1); for instance Sobolev and isoperimetric constants, eigenvalues of Laplace operators, etc.

A precursor of Theorem 1.1, and essentially an important special case, is Cheeger's finiteness theorem [17] that there are only finitely many diffeomorphism types of n-manifolds satisfying the bounds (1.1).

We now turn to the case where $\mathrm{diam}_{g_i} M \to \infty$. Then it is easily seen that the conclusion of Theorem 1.1 can never be valid. To understand the degeneration in this case, one decomposes the manifold (M, g) via a thick/thin decomposition. For reasons that will be apparent in later sections, we use a thick/thin decomposition relative to the volume of geodesic balls in (M, g).

DEFINITION. For a complete Riemannian manifold M and positive constant ε, write $M = M^\varepsilon \cup M_\varepsilon$, where

$$(1.2) \quad M^\varepsilon = \{x \in M : \mathrm{vol}\, B_x(1) \ge \varepsilon\}, \qquad M_\varepsilon = \{x \in M : \mathrm{vol}\, B_x(1) < \varepsilon\}.$$

The domain M^ε (resp. M_ε) is called the ε-thick (resp. thin) part of the manifold (M, g). If M satisfies (1.1), then $M = M^\varepsilon$, for all $\varepsilon \le C = C(\Lambda, D)$; this follows easily from the volume comparison theorem.

It is more common to define the thick and thin parts of M with respect to the size of the injectivity radius at $x \in M$ in place of $\mathrm{vol}\, B_x(1)$. Under the bounds $|R| \le \Lambda$, these definitions are basically equivalent, since by the results of [12] and [20] one has

$$c_1 \cdot \mathrm{inj}^n(x) \le \mathrm{vol}\, B_x(1) \le c_2 \cdot \mathrm{inj}^n(x),$$

for constants $c_1 = c_1(n, \Lambda)$, $c_2 = c_2(n, \Lambda)$.

We then have the following understanding of degenerations of (M, g_i), where $(M^\varepsilon, g_i) = \varnothing$, i.e. where M is thin everywhere.

THEOREM 1.2 (Cheeger-Gromov collapse [18, 19]). *Let M be a compact n-manifold M satisfying $|R| \leq \Lambda$. Then there is an $\varepsilon_0 = \varepsilon_0(n, \Lambda)$ such that if $M_{\varepsilon_0} = M$, i.e. $\operatorname{vol} B_x(1) < \varepsilon_0$, for all $x \in M$, then M admits an F-structure $\mathscr{F}$. If $\{g_i\}$ is a sequence of metrics in $\mathscr{M}_1$ with $\operatorname{vol}_{g_i} B_x(1) \to 0$, $\forall x \in M$, as $i \to \infty$, then there is a sequence of F-structures $\mathscr{F}_i$ such that $\{(M, g_i)\}$ collapses along $\mathscr{F}_i$ to a complete length space of lower dimension and infinite diameter.*

To paraphrase briefly, M admits an open covering $\{U_k\}$ such that a finite cover of each U_k admits a locally free torus action, with dimension possibly depending on k, but with a natural consistency condition on overlaps. The metrics $\{g_i\}$ collapse the orbits of these local toral actions (which may change with i) to points (with bounded curvature). An F-structure will always be taken to be an F-structure of positive rank.

Note that the existence of an F-structure is a topological condition on M, so that there are definite topological obstructions for M to admit a family of collapsing metrics. In particular, if (M, g_i) collapses, then $\chi(M) = 0$ (since M is compact). Conversely, if M admits an F-structure, then M admits a sequence of metrics of bounded curvature which collapse M along this F-structure.

REMARK. This structure theory has recently been significantly generalized by Cheeger-Fukaya-Gromov [21] to collapse along an N-structure (for nilpotent), which takes account of all the directions in which the manifold appears collapsed.

To account for the last possibility of the trichotomy, suppose $\operatorname{diam}_{g_i} M \to \infty$ and suppose there are points $x_i \in (M, g_i)$ such that

$$(1.3) \qquad \operatorname{vol}_{g_i} B_{x_i}(1) \geq \varepsilon,$$

for all i, for some $\varepsilon > 0$. For each i, choose a maximal disjoint family $B_{x_k}(1)$ satisfying (1.3). Since $\operatorname{vol}_{g_i} M = 1$, there is clearly a uniform bound on the number of such balls. By using the diffeomorphism group, we may assume (for convenience) that the center points $x_k \in M$ are fixed, independent of i. The volume comparison theorem implies that

$$(1.4) \qquad \operatorname{vol}_{g_i} B_y(1) \geq C \left(\min_k (\operatorname{dist}_{g_i}(y, x_k)) \right).$$

If we consider the pointed manifolds (M, g_i, x_k), it follows from a local version of Theorem 1.1 that a subsequence of $\{(M, g_i, x_k)\}$ converges in the $C^{1,\alpha}$ topology to a complete, noncompact manifold N_k, with $C^{1,\alpha}$ metric of finite volume. Of course, not all of the manifolds N_k are distinct, so we let $\{N_j\}$, $j = 1, \ldots, m$, be the resulting collection of distinct manifolds. By construction, each open manifold N_j is topologically embedded in M. With a slight abuse of terminology, we call the manifolds N_j cusps. Summarizing, we then have

THEOREM 1.3 (Formation of cusps). *Let $\{g_i\}$ be a sequence of metrics on M satisfying*

$$\text{(1.5)} \qquad \text{vol}_{g_i} M = 1, \quad |R| \leq \Lambda, \quad \text{diam}_{g_i} M \to \infty,$$

and

$$\text{(1.6)} \qquad \text{vol}_{g_i} B_{x_i}(1) \geq \varepsilon,$$

for some constant $\varepsilon > 0$ and sequence $x_i \in M$. Then a subsequence of (M, g_i) converges in the Gromov-Hausdorff topology to a finite number of cusps N_j, $1 \leq j \leq m$, $m \leq m(n, \Lambda, \varepsilon)$, together with a collapsed length space L of dimension $\leq n - 1$ and possibly of infinite diameter. The thin part M_{ε_0} of (M, g_i) (where ε_0 is as in Theorem 1.2) has an F-structure $\mathscr{F}_i$ which partially collapses under $\{g_i\}$, that is, part of (M_{ε_0}, g_i) converges to neighborhoods of infinity of the collection of cusps, while the remaining part collapses to the length space L.

In other words, there are F-structures on domains $\Omega_i \subset M$ and saturated closed subsets $T_i \Subset \Omega_i$, such that $(\Omega_i \backslash T_i) \cap N_j$ is a neighborhood of infinity of N_j, on which the sequence $\{g_i\}$ converges in the $C^{1,\alpha}$ topology. The subsets T_i are collapsed under $\{g_i\}$ to a complete lower-dimensional length space. Clearly, one has $\sum \text{vol} N_j \leq 1$. The simplest example is where M is a 2-dimensional surface and Ω is a collection of disjoint essential annuli, with T the set of core curves, or perhaps T essential subannuli of Ω.

We note that in contrast to the topological constraints imposed by the existence of a global F-structure on M as in Theorem 1.2, there are no topological constraints on the degenerations imposed by Theorem 1.3. To illustrate, let M be an arbitrary compact 3-manifold and let T be a solid torus $(D^2 \times S^1)$ embedded trivially in M. Then one may easily construct a sequence of metrics of bounded curvature on M, which converge to a cusp on $M \backslash T$ and collapse along the standard F-structure of T.

This concludes our description of the space $\mathscr{M}_1^\Lambda$ and its degenerations.

2. Degenerations with bounded Ricci curvature

In this section, we consider to what extent the description of possible degenerations given in §1 remains valid under the weaker hypotheses

$$\text{(2.1)} \qquad |\text{Ric}| \leq \lambda, \qquad \text{vol}\, M = 1.$$

Besides its intrinsic interest as a natural generalization of the results in §1, this is also of importance in studying the existence of critical metrics for natural Riemannian functionals on $\mathscr{M}_1$. We note that it is also of interest to study similar questions under the one-sided bound $\text{Ric} \geq -\lambda$; however, from the point of view of applications to the existence of critical metrics, the assumption of two-sided bounds on the Ricci curvature arises more naturally.

To begin then, consider the space $\mathcal{M}_1^{\lambda,D}$ of Riemannian metrics on M satisfying

(2.2) $|\mathrm{Ric}| \le \lambda, \quad \mathrm{vol}\, M = 1, \quad \mathrm{diam}\, M \le D.$

Unfortunately, there is no, even C^0, compactness theorem in this case, as in Theorem 1.1 (except when $\dim M = 2, 3$, which are covered by Theorem 1.1). As shown in [3], there are metrics satisfying (2.2), on $S^2 \times S^2$ for example, which converge in the Gromov-Hausdorff topology to the (double) suspension of $\mathbb{RP}^3$: the diagonal and antidiagonal S^2's in $S^2 \times S^2$ are collapsed to points. There are numerous further examples of this type of degeneration, namely degeneration to orbifolds, especially in dimension 4 and on complex surfaces. One may take for instance twisted products of examples of this type to obtain more complicated degenerations in dimensions > 4.

Thus, it is reasonable to first seek conditions under which one can control the possible degenerations under the bounds (2.2). One instance of this is the following result, proved in [4].

THEOREM 2.1. *If $\{g_i\}$ is a sequence of metrics on a closed n-manifold M satisfying*

(2.3) $|\mathrm{Ric}| \le \lambda, \quad \mathrm{vol}\, M = 1, \quad \mathrm{diam}\, M \le D,$

and

(2.4) $$\int_M |R|^{n/2} \le \Lambda,$$

then a subsequence converges, in the Gromov-Hausdorff topology, to a smooth orbifold (V, g) with a finite number of singular points, each homeomorphic to a cone $C(S^{n-1}/\Gamma)$, for a finite subgroup $\Gamma \subset O(n)$. The metric g is $C^{1,\alpha}$ on the regular set $V_0 = V - V_{\mathrm{sing}}$, and extends in a local uniformization of a singular point, to a C^0 metric Riemannian metric on V. Further, there are embeddings $F_i: V_0 \to M$ such that $F_i^*(g_i)$ converges in the $C^{1,\alpha}$ topology to the metric g.

Briefly, away from a finite number of points, near which the metrics are degenerating to orbifold singularities, one has good convergence as in Theorem 1.1. In fact, one has a (C^0) compactness theorem in the category of orbifolds satisfying (2.3), (2.4); cf. [4, 6]. We note that the number of singular points, as well as the orders of their local fundamental groups, are bounded in terms of (2.3), (2.4).

It is of importance however to understand how the metrics are degenerating, near the orbifold singular points, as bilinear forms on M, and not merely in the Hausdorff topology. From the analysis in [10, 6], cf. also [8], one finds that the degeneration set $D \subset M$ is a finite collection of closed submanifolds of M, of dimensions $\le n - 2$. Further, the homology of D injects in that of M. Thus, V is obtained topologically from M by collapsing the closed set D to a finite number of points.

REMARKS 2.1. (I) There are some special circumstances where in fact no degeneration occurs under the bounds (2.3), (2.4), and one has full $C^{1,\alpha}$ convergence as in Theorem 1.1. This holds for instance in any of the following cases; cf. [2, 4].

(i) $\dim M$ is odd. This arises from the fact that there are no nontrivial subgroups $\Gamma \subset O(n)$ with S^{n-1}/Γ orientable, if n is odd.

(ii) M is an integral homology sphere. This is due to the fact that a neighborhood of D cannot be acyclic.

(iii) $\int_M |R|^{n/2} \leq \varepsilon$, for $\varepsilon = \varepsilon(\lambda, D)$ sufficiently small. In fact, in this case, M is close in the $C^{1,\alpha}$ topology to a flat manifold, so that in particular, a finite cover of M is diffeomorphic to a torus. Similarly, if $\|R - R_c\|_{L^{n/2}} \leq \varepsilon' = \varepsilon'(\lambda, D)$, where R_c is the curvature tensor of a space form of constant curvature c, then M is close in the $C^{1,\alpha}$ topology to a space form of this constant curvature; cf. [6, 25, 26].

(iv) Tian [47] has recently shown that one does not have any degenerations under the bounds (2.3), (2.4), in the case of Kähler-Einstein manifolds M with definite c_1, provided $\dim_C M \geq 3$.

REMARK 2.2. In analogy to Cheeger's finiteness theorem in §1, it is proved in [8] that the class of manifolds admitting metrics satisfying the bounds (2.3), (2.4) has only finitely many diffeomorphism types.

We now turn to the case where $\operatorname{diam}_{g_i} M \to \infty$ under the bounds (2.1), (2.4). Basically, one has similar results to Theorem 1.2 and 1.3, with the exception that a finite number of singular points may develop in the limit.

First, the analogue of Theorem 1.2.

THEOREM 2.2. *There is an* $\varepsilon_0 = \varepsilon_0(n, \lambda, \Lambda) > 0$, *and a finite number of points* $\{q_k\}_1^N \in M$, *with* $N = N(n, \lambda, \Lambda)$, *such that if* M *is a compact n-manifold satisfying*

$$(2.5) \qquad \operatorname{vol} M = 1, \quad |\operatorname{Ric}_M| \leq \lambda, \quad \int_M |R|^{n/2} \leq \Lambda,$$

and

$$(2.6) \qquad v_M \equiv \sup_{x \in M}(\operatorname{vol} B_x(1)) \leq \varepsilon_0,$$

then M *admits an* (ε_0 *collapsed*) *F-structure on the domain* $\Omega \equiv M - \bigcup_{k=1}^N B_{q_k}(r_0)$, *where* $r_0 = r_0(n, \lambda, \Lambda)$ *with* $r_0 \to 0$ *as* $v_M \to 0$. *In particular,* $\operatorname{vol} \Omega \to 1$ *as* $v_M \to 0$.

Thus, if (M, g_i) is a sequence satisfying (2.5), with $v_M(g_i) \to 0$ as $i \to \infty$, then M collapses along a sequence of F-structures $\mathcal{F}_i$, (with $C^{1,\alpha}$ control), metrically on the complement of finitely many singular points. The structure of the degeneration of (M, g_i) near the singular points is not well understood. The sequence may develop orbifold singularities, in which case it can be handled as in Theorem 2.1. However, other degenerations may occur, which lack any coherent treatment to date; cf. [6] for some further discussion.

Despite the similarities in the statements of Theorems 2.2 and 1.2, there is a basic difference. Namely, the degeneration described by Theorem 2.2 is likely to occur on a much larger class of manifolds, since one is not able to conclude the existence of a global F-structure on M, $\dim M \geq 4$. A nontrivial example is the existence of degenerations as in Theorem 2.2 on Ricci-flat $K3$ surfaces X, with $\chi(X) = 24$; cf. [6].

One might conjecture for instance that any compact n-manifold with an effective S^1 action (not necessarily free) admits a sequence of metrics satisfying (2.5), with $v_M \to 0$.

Next, we turn to the analogue of Theorem 1.3.

THEOREM 2.3. *Let* (M, g_i) *be a sequence metrics satisfying* (2.5) *with*

$$(2.7) \qquad v_M(g_i) \geq \varepsilon, \qquad \operatorname{diam}_M(g_i) \to \infty,$$

for some $\varepsilon > 0$. *Then a subsequence converges, in the Gromov-Hausdorff topology, to a finite number of cusps* N_k, *together with a collapsed length space* L, *of dimension* $\leq n - 1$. *Each* N_k *is a smooth open* n-manifold, *with either a complete* $C^{1,\alpha}$ *metric or complete orbifold singular metric* g_k, *satisfying* $\sum \operatorname{vol} N_k \leq 1$ *and* $\operatorname{diam}_{g_k} N_k = \infty$. *The number of cusps, as well as the number of orbifold singularities on each cusp is uniformly bounded by* $\lambda, \Lambda, \varepsilon$. *The thin part* (M_{ε_0}, g_i) *admits the same structure as in Theorem 2.2.*

Although Theorems 2.2 and 2.3 do not appear in the literature, they are essentially implicit in, and not difficult to deduce from, existing results. Theorem 2.1 was first proved in the case of Einstein metrics, independently in [2] and [9]; cf. also [34] and [8]. Theorems 2.2 and 2.3 were proved for Einstein metrics in [6], where stronger results were obtained with respect to the L^2 topology on $\mathscr{M}_1(M)$. (Although stated on 4-manifolds in [6], as remarked in the introduction, the proofs remain valid in all dimensions under the bound (2.4).)

To generalize these results to metrics of bounded Ricci curvature (in place of constant Ricci curvature), one uses the concept of $C^{1,\alpha}$ harmonic radius of (M, g) introduced and developed in [4, 7]. The $C^{1,\alpha}$ harmonic radius at $x \in M$ is the radius $r_h(x)$ of the largest geodesic ball B about x, on which there are harmonic coordinates $\{u_j\}_1^n$ in which the metric coefficients $g_{ij} = g(\partial/\partial u_i, \partial/\partial u_j)$ are bounded in the $C^{1,\alpha}$ topology, i.e.

$$(2.8) \qquad e^{-C}\delta_{ij} \leq g_{ij} \leq e^{C}\delta_{ij} \quad \text{(as bilinear forms)},$$

$$(2.9) \qquad r_h(x)^{1+\alpha}\|g_{ij} - \delta_{ij}\|_{C^{1,\alpha}} \leq C,$$

for some fixed constant $C > 0$. The $C^{1,\alpha}$ harmonic radius $r_h(M)$ of M is then given by $r_h(M) = \inf r_h(x)$. From [4], one has the estimate that

$$(2.10) \qquad r_h(x) \geq C(|\mathrm{Ric}|_{B_x(1)}, \mathrm{vol}\, B_x(1)),$$

provided $\int_{B_x(1)} |R|^{n/2} \leq \varepsilon = \varepsilon(n, |\mathrm{Ric}|_{B_x(1)}, \mathrm{vol}\, B_x(1))$.

Under the hypotheses (2.5), one may then conclude a corresponding (local) $C^{1,\alpha}$ compactness, away from a finite number of singularities. Further, under (arbitrary) bounds on $|\mathrm{Ric}|_{B_x(1)}$, $\mathrm{vol}\, B_x(1)$ and $\int_{B_x(1)} |R|^{n/2}$, one shows as in [2, 4] that the singularities in $B_x(1/2)$ are all of orbifold type.

We point out that certain special cases of these results have also been proved by L. Z. Gao by quite different methods; compare with the proofs in [25, 26].

It is of basic interest to know to what extent the results above remain valid when one drops the hypothesis $\int |R|^{n/2} \leq \Lambda$, or replaces it with a weaker curvature integral.

We first note that in dimension 4, by the Chern-Gauss-Bonnet formula one has

$$(2.11) \qquad \chi(M) = \frac{1}{8\pi^2} \int_M (|R|^2 - 4|\mathrm{Ric}|^2 + s^2)\, dV,$$

so that

$$(2.12) \qquad \int_M |R|^2 \leq C(\chi(M), \|\mathrm{Ric}\|_{L^2}).$$

Thus, on a fixed 4-manifold, the results of Theorems 2.1–2.3 hold under the (natural) bounds (2.1).

Curiously however, although one thus has at least the beginnings of a sound understanding of the possible degenerations of metrics on a fixed 4-manifold under the bounds (2.1), there is no finiteness result known under the bounds (2.2), as in the case of sectional curvature. This leads to the following

CONJECTURE 2.1. If M is a compact 4-manifold, then

$$(2.13) \qquad \chi(M) \leq C(|\mathrm{Ric}|, \mathrm{vol}, \mathrm{diam}).$$

In fact, we conjecture that there are only finitely many homotopy types or homeomorphism types of 4-manifolds satisfying the bounds (2.2). Note that by [29], $b_1(M) \leq C(|\mathrm{Ric}|, \mathrm{diam})$, while by [3], the number of isomorphism classes of $\pi_1(M)$ is bounded by $C(|\mathrm{Ric}|, \mathrm{vol}, \mathrm{diam})$. In particular, an upper bound on $\chi(M)$ gives a bound on all Betti numbers of M.

It is even reasonable to believe that $\chi(M)$ is bounded by inf Ric, volume and diameter, i.e. that one only needs a lower bound on the Ricci curvature in (2.2). Note in particular that the examples of metrics of positive Ricci curvature on 4-manifolds with unbounded b_2 in [3, 43] do not give counterexamples here, since the volume goes to zero as $b_2 \to \infty$ (given diam $\leq D$).

In higher dimensions, there is much less evidence on which to make solid conjectures regarding the finiteness of topological type. Nevertheless, we venture

CONJECTURE 2.2. If M is a compact n-manifold, then there is a bound on the Betti numbers

$$(2.14) \qquad b_p(M) \leq C(|\mathrm{Ric}|, \mathrm{vol}, \mathrm{diam}).$$

If one considers sequences of metrics on a fixed n-manifold satisfying (2.1) or (2.2), then there are examples on which $\{g_i\}$ degenerates on a closed subset of Hausdorff codimension 4 in M. For instance, as indicated above, one may take the orbifold degenerations on a fixed 4-manifold and take locally twisted products.

More nontrivially, there are now a wealth of examples in higher dimensions, coming from algebraic and complex geometry. First, there are many recent constructions of degenerations of Calabi-Yau manifolds, i.e. Kähler 3-folds X^3 with $c_1 = 0$, which thus admit Ricci-flat Kähler metrics by [52]. Specific examples include cone degenerations (isolated singularities of the form $C(L(\Sigma^2))$, where L is circle bundle over a complex Kähler surface, e.g. $C(S^3 \times S^2))$, and degenerations along complex curves in X^3; cf. [16, 27]. Secondly, there are many examples constructed by Bando-Kobayashi [11] and Tian-Yau [47], of noncompact complete Ricci-flat Kähler metrics, which at infinity are smoothly asymptotic to $C(L(\Sigma^{n-1}))$, for a number of Kähler manifolds Σ^{n-1}. Performing for instance the scaling and doubling construction as in [3], one obtains examples of manifolds satisfying (2.2), and degenerating in the Gromov-Hausdorff topology to spaces with isolated (nonorbifold) singularities. In all these examples, one necessarily has $\int |R|^{n/2} \to \infty$, $n = \dim_{\mathbb{R}} M$.

In light of the above examples, it is not unreasonable to make the following

CONJECTURE 2.3. If $\{g_i\}$ is a sequence of metrics on M^n with

$$(2.15) \qquad |\mathrm{Ric}| \leq \lambda, \quad \mathrm{vol} \geq v, \quad \mathrm{diam} \leq D,$$

then there is an open set $\Omega \subset M$, with the Hausdorff dimension of the closed complement $S = M - \Omega \leq n - 4$, such that a subsequence of $\{(\Omega, g_i)\}$ converges in the $C^{1,\alpha}$ topology to a $C^{1,\alpha}$ metric on Ω.

Little is known regarding these conjectures even under the assumption of bounds on curvature integrals weaker than the $L^{n/2}$ norm of the curvature tensor. We point out that the results of Theorems 2.1–2.3 remain valid under the (scale-invariant) bound,

$$(2.16) \qquad \int_M \frac{|R|^2}{r^{n-4}} \leq \Lambda,$$

(in place of a bound on $\int |R|^{n/2}$), where r is the distance function from an arbitrary point in M. We will not go into details of this here, beyond pointing out that if the curvature integral (2.16) is sufficiently small, then it can be shown that the $L^{2,p}$ harmonic radius is bounded below, for any $p < \infty$, as in (2.10). We note that integrals of the type (2.16) are infinite on all nonflat cones, i.e. cones $C(\Sigma)$ on smooth manifolds Σ, with $\Sigma \neq S^{n-1}/\Gamma$, $\Gamma \subset O(n)$, exactly as in the case of the $L^{n/2}$ norm of R.

It would be significantly more interesting if one could obtain an understanding of degenerations (satisfying (2.15)), in the case of a bound on

$$(2.17) \qquad \frac{1}{r^{n-4}} \int_{B(r)} |R|^2 \leq \Lambda,$$

which is considerably weaker than (2.16). In particular, (2.17) is finite on all smooth cones, i.e. cones $C(\Sigma)$ which have an arbitrary smooth cross section Σ. Further, the condition (2.17) is closed under products. The main problem here is the lack of monotonicity of the integral (2.17), as a function of r. In fact, in the examples of Calabi [15] of complete Ricci-flat Kahler metrics on the canonical bundle over $\mathbb{CP}^n$, one may check that the integral (2.17) is not monotone in r.

Analogous problems have been treated much more successfully in the case of harmonic maps [39], and Yang-Mills fields [35], basically because the quantity corresponding to (2.17) is monotone in r in these cases (R is replaced by the energy in the case of harmonic maps).

We point out that Cheeger and the author have proved part of Conjecture 2.3 under the bounds (2.15), (2.17), namely that one has $C^{1,\alpha}$ control of the metric off *some* set S (not known to be necessarily closed) of Hausdorff codimension 4.

This lack of monotonicity also poses an obstacle to studying the space of metrics of M satisfying the bounds

$$|\mathrm{Ric}| \leq \lambda, \quad \mathrm{vol} = 1, \quad \mathrm{diam} \leq D,$$
$$(2.18) \qquad \int_M |R|^2 \leq \Lambda,$$

in dimensions ≥ 5. It is possible that some more restrictive hypotheses on the space of metrics considered would lead to more positive results. For instance, it would be very interesting to obtain results analogous to the above, on the space of Kahler metrics on a compact manifold satisfying the bounds (2.18).

3. Degenerations with integral bounds on curvature

In this section, we proceed with a generalization of the results of §2 to metrics of volume 1 on the compact manifold M satisfying either the bounds

$$(3.1) \qquad \int_M |R|^p \leq \Lambda_1,$$

or

$$(3.2) \qquad \int_M |R|^{n/2} \leq \Lambda_2, \qquad \int_M |\mathrm{Ric}|^p \leq \lambda,$$

for some $p > n/2$.

The main difficulty in obtaining the results of §2 here is the lack of a volume comparison theorem under the bounds (3.1) or (3.2), which one has of course if $\text{Ric} \geq -\lambda$. Consider for instance the following simple example, pointed out by D. Yang [50]; cf. also [24]. Let $M = (-1, 1) \times S^1$ and let g_ε be given by

$$(3.3) \qquad g_\varepsilon = dr^2 + (\varepsilon + r)^{2k} d\theta^2.$$

One easily calculates that the curvature of g_ε is uniformly bounded in L^p, for any p with $1 \leq p < \infty$, for appropriate $k = k(p) > 1$. As $\varepsilon \to 0$, these metrics degenerate along the center circle $\{r = 0\}$, so that one has a partial or local collapse of (M, g_ε) along an F-structrure (given by the S^1 action). One may do the same construction in higher dimensions for instance with T^{n-1} in place of S^1. In particular, one has here a local collapse of the manifold, while the diameter remains uniformly bounded. Of course, this local collapse cannot occur under L^∞ bounds on curvature, by the volume comparison theorem.

As we will see, this is essentially the only new phenomenon that arises under the bounds (3.1) or (3.2). Since there is no volume comparison result, we set

$$(3.4) \qquad v^\varepsilon(x) = \sup\left\{ r : \frac{\text{vol}\, B_y(s)}{s^n} \geq \varepsilon, \text{ for all } B_y(s) \subset B_x(r) \right\},$$

for a given $\varepsilon > 0$; compare with [49]. As in §2, define the $L^{2,p}$ harmonic radius $r_h(x)$ at x to be the radius of the largest geodesic ball about x, on which one has harmonic coordinates with $L^{2,p}$ control of the metric in these coordinates, i.e. (2.8), (2.9) hold, with the $L^{2,p}$ norm in place of $C^{1,\alpha}$. In this section, we will exclusively use the $L^{2,p}$ harmonic radius, so there will be no danger of confusion with the notation. Note that by the Sobolev embedding theorem, $L^{2,p} \subset C^\alpha$, for $\alpha = 2 - \frac{n}{p}$, so that the $L^{2,p}$ harmonic radius controls the correspondingly defined C^α harmonic radius. It is trivial to see that

$$(3.5) \qquad r_h(x) \leq v^\varepsilon(x),$$

provided ε is sufficiently small (depending on the C of (2.8)). Conversely, we have

PROPOSITION 3.1. *Let M be a compact n-manifold. Then there is a positive constant $c_0 = c_0(\Lambda_1, n, p, \varepsilon)$ such that the $L^{2,p}$ harmonic radius, $p > n/2$, satisfies*

$$(3.6) \qquad r_h(x) \geq c_0 \cdot v^\varepsilon(x),$$

provided

$$r_n(x) \leq c_0^{-1}.$$

PROOF. We will only sketch the proof; cf. [4, 5, 7] for further details. If (3.6) were false, there would exist a sequence (M_i, g_i) and points $x_i \in M_i$ such that

$$(3.7) \qquad \frac{r_h(x_i)}{v^\varepsilon(x_i)} \to 0, \quad \text{as } i \to \infty,$$

under uniform bounds on $\Lambda_1, n, p, \varepsilon$. We may assume that the points x_i realize the minimum of the ratio in (3.7) on M_i. Consider the rescaled metrics $\tilde{g}_i = r_h(x_i)^{-2} \cdot g_i$, so that $\tilde{r}_h(x_i) = 1$. The scale invariance of the ratio in (3.7) implies that $\tilde{v}^\varepsilon(x_i) \to \infty$. It follows from (3.4) that for all points $y \in M_i$, within $\tilde{g}_i$-uniformly bounded distance to x_i, one has

$$(3.8) \qquad \frac{\text{vol}_{\tilde{g}_i} B_y(s)}{s^n} \geq \varepsilon,$$

for all bounded s, for i sufficiently large. From this, it is easy to deduce from (3.7) that $\tilde{r}_h(y) \geq c(\text{dist}_{\tilde{g}_i}(y, x_i))$. It follows that the pointed sequence $\{(M_i, x_i, \tilde{g}_i)\}$ has a subsequence converging in the $C^{\alpha'}$ topology, $\alpha' < \alpha$, and weakly in the $L^{2,p}$ topology, to a smooth limit manifold N, with $C^\alpha \cap L^{2,p}$ metric $\tilde{g}$ and $x = \lim x_i$. In fact, one may use the equation for the Ricci curvature, as in [4], to show that the convergence to $(N, \tilde{g})$ is in the strong $L^{2,p}$ topology. Since $p > n/2$, the scaling property of the integral (3.1), and its lower semicontinuity in the $L^{2,p}$ topology imply that $R = 0$ in L^p. The regularity theory of elliptic operators then implies that $\tilde{g}$ is C^∞ and flat. Further, the condition (3.8), which passes to the limit $\tilde{g}$, implies that $(N, \tilde{g})$ is isometric to $\mathbb{R}^n$. Now the $L^{2,p}$ harmonic radius is continuous in the strong $L^{2,p}$ topology (cf. [4, 7]) so that $r_h(x) = 1$ on $(N, \tilde{g})$. Since $r_h(\mathbb{R}^n) = \infty$, this gives the required contradiction and proves (3.6). □

REMARKS 3.2. (i) Basically the same proof shows that a local version of Proposition 3.1 is also valid. Thus, if M is a manifold with boundary ∂M, then

$$(3.9) \qquad r_h(x) \geq c_0 \cdot v^\varepsilon(x) \cdot \frac{\text{dist}(x, \partial M)}{\text{diam } M}.$$

(ii) A simple modification of the argument above shows that (3.6) and (3.9) remain valid under the bounds $\int |\text{Ric}|^p \leq \lambda$, $\int |R|^{n/2} \leq \eta$, for some $\eta = \eta(\lambda, n, p, \varepsilon)$ sufficiently small (with $p > n/2$). One may then apply this to a covering of a compact manifold M by small balls to prove that if M satisfies the bounds (3.2), then $r_h(x)$ satisfies the estimate (3.6) away from a finite number K of points in M, where $K = K(\Lambda_2, \lambda, n, p, \varepsilon)$. Near these points, the metric is close to an orbifold singularity; cf. again [2, 4, 7, 8] for further details here.

Proposition 3.1 implies that one has C^α control of the geometry, and $C^{1,\alpha}$ control of the topology, on the domain where v^ε is bounded below.

As in §1.2, define a thick/thin decomposition of (M, g) as

$$(3.10) \qquad M^\delta = \{x \in M : v^\varepsilon(x) \geq \delta\}, \qquad M_\delta = \{x \in M : v^\varepsilon(x) < \delta\}.$$

Thus, the domain M^δ is precompact in the $C^\alpha \cap L^{2,p}$ topology, for any fixed $\delta > 0$. To understand the metric on the thin part M_δ of M, fix an arbitrary $x \in M_\delta$ and consider the rescaled pointed manifold $(M, \tilde{g}, x)$, where $\tilde{g} = (v^\varepsilon(x)^{-2}) \cdot g$. Note that then $\tilde{v}^\varepsilon(x) = 1$. Further, the scaling property of curvature implies

$$(3.11) \qquad \int_M |\widetilde{\mathrm{Ric}}|^p \, d\tilde{V} \leq \lambda \cdot (v^\varepsilon(x))^{2p-n} \leq \lambda \cdot \delta^{2p-n}.$$

If δ is sufficiently small, depending on n, p, ε, then the isoperimetric inequality of Gallot-Yang, cf. [24, 50], implies that

$$(3.12) \qquad \tilde{v}^\varepsilon(y) \geq c(\mathrm{dist}_{\tilde{g}}(x, y)) \cdot \tilde{v}^\varepsilon(x) = c(\mathrm{dist}_{\tilde{g}}(x, y)),$$

provided $\mathrm{dist}_{\tilde{g}}(x, y) \leq R = R(\delta)$, where $R \to \infty$ as $\delta \to 0$. Thus, the blown up manifold $(M, \tilde{g}, x)$ remains thick in bounded distances to x. By Proposition 3.1, we see then that the balls $B_x(R) \subset (M, \tilde{g})$ are precompact in the $C^\alpha \cap L^{2,p}$ topology. In particular, if $\delta > 0$ is sufficiently small, then the domains $B_x(R \cdot v^\varepsilon(x)) \subset (M, g)$ are close, in the scaled $C^\alpha \cap L^{2,p}$ topology (cf. (2.8), (2.9)) to balls in a complete noncontractible flat manifold $\mathbb{R}^n/\Gamma$. Thus, to any such $x \in M_\delta$ is associated an elementary F-structure [19] on the ball $B_x(R \cdot v^\varepsilon(x))$.

It would appear that these elementary F-structures can be pieced together to give an F-structure on M_δ (for δ sufficiently small). (One would need to extend the theory of [19] to the $C^\alpha \cap L^{2,p}$ topology, in place of the $C^{1,\alpha}$ topology used there.) By using a different technique, namely a smoothing of the metric obtained from a local Ricci flow, D. Yang [47] has shown that the metric on M_δ may be smoothed (locally) to satisfy the hypothesis of Cheeger-Gromov [19] (as in Theorem 1.2).

The discussion above then leads to the following result, noted independently in [51].

THEOREM 3.2. *Let M be a compact Riemannian manifold satisfying the bounds*

$$(3.13) \qquad \int_M |R|^p \leq \Lambda, \qquad \mathrm{vol}_M = 1,$$

for some $p > n/2$. Then M admits a thick/thin decomposition $M = M^\delta \cup M_\delta$, where

(i) *M^δ is precompact in the $C^\alpha \cap L^{2,p}$ topology, $\alpha = 2 - \frac{n}{p}$;*

(ii) *M_δ admits an F-structure.*

If $\{g_i\}$ is a sequence of metrics on M satisfying (3.13), then for any $\delta > 0$, the Riemannian manifolds (M^δ, g_i) have a subsequence converging in the

$C^\alpha \cap L^{2,p}$ topology to a $C^\alpha \cap L^{2,p}$ Riemannian manifold (N^δ, g), where N^δ is smoothly embedded in M. Let $N = \bigcup_{\delta>0} N^\delta \subset M$. Then there is a sequence $\delta_i \to 0$, such that $M \backslash N \subset M_{\delta_i}$ and M_{δ_i} admits an F-structure $\mathcal{F}_i$.

It appears very likely that the sequence $\{g_i\}$ collapses M_{δ_i} along the F-structures $\mathcal{F}_i$ to a lower dimensional length space. (This remains to be proved however.)

More generally, there is a similar version of Theorem 3.2 under the bounds (3.2) in place of (3.13), where one must then also take account of orbifold degenerations in (i), as well as the possibility of other isolated, point degenerations in (ii), as in §2.

4. Approach to critical metrics of $\mathcal{R}^p$

In this section, we apply the results of the previous sections in an attempt to understand the behavior of sequences of metrics which approach critical or minimal values of

$$(4.1) \qquad \mathcal{R}^p = \int_M |R|^p \, dV,$$

or related functionals on $\mathcal{M}_1(M)$, for M a compact n-manifold. Further, for these critical metrics to be of any significant interest, one needs to establish the regularity of weak solutions, i.e. the smoothness of metrics which satisfy the Euler-Lagrange equations of (4.1) weakly. It then remains further to understand the geometry of these critical metrics and their relation with the topology of M.

First, we must topologize $\mathcal{M}_1$ in a way compatible with the functional to be considered. For $\mathcal{R}^p$, it is natural to put the $L^{2,p}$ topology on $\mathcal{M}_1$, i.e. the normed topology given by

$$(4.2) \qquad \|h\|_{L^{2,p}} = \int_M (|D^2h|^p + |Dh|^p + |h|^p) \, dV_g,$$

for $h \in T_g\mathcal{M}_1 = \{h \in S^2(M): \int \mathrm{tr}_g(h) \, dV_g = 0\}$. Here D denotes the covariant derivative with respect to $g \in \mathcal{M}_1$. Although one can define this topology for all $p \in (1, \infty)$, it is well-behaved only when $p > n/2$, $n = \dim M$, since only in this range does one have the Sobolev embedding $C^\alpha \supset L^{2,p}$, $\alpha = 2 - \frac{n}{p}$. It is not difficult to verify that, if $p > n/2$, this induces the same topology as the $L^{2,p}$ topology defined by local coordinates, i.e.

$$\mathrm{dist}_{L^{2,p}}(g, g') < \varepsilon \Leftrightarrow \|g_{ij} - g'_{ij}\|_{L^{2,p}} < \varepsilon',$$

in the $L^{2,p}$ topology on functions on M, with respect to an arbitrary fixed coordinate atlas of M. Further $\mathcal{M}_1(M)$ is a smooth Banach manifold, for $p \in (n/2, \infty)$, on which $\mathcal{R}^p$ is a smooth functional; cf. [36].

Thus, from now on we will assume $p > n/2$. This rules out consideration of what might appear to be the most interesting case, namely the scale-invariant functional $\mathcal{R}^{n/2}$. This is especially true in dimension 4, where $\mathcal{R}^2$

carries topological information. One might try to study $\mathscr{R}^2$ in dimension 4 by perturbation, i.e. by considering the limiting behavior of critical metrics of $\mathscr{R}^p$ as $p \to 2$, in analogy to the case of harmonic maps, as so beautifully carried out by Sacks-Uhlenbeck [38]; cf. however the discussion below.

Since $\mathscr{R}^p$, for $p \neq n/2$, is not scale invariant, there are no critical metrics on $\mathscr{M}$, the space of all Riemannian metrics on M, modulo diffeomorphisms, and thus the restriction to $\mathscr{M}_1$ is necessary. By a well-known result of Moser, cf. [13], any two metrics g_1, g_2 with the same total volume have volume forms related by $dV_2 = f^* dV_1$, for some diffeomorphism f of M. Thus, we might as well consider $\mathscr{R}^p$ defined on the related space

$$\mathscr{M}_0 = \{g \in S_+^2 : dV_g = dV_{g_0}\}/\mathrm{Diff}_0(M),$$

where S_+^2 is the space of positive symmetric 2-tensors on M, g_0 is a fixed background metric and $\mathrm{Diff}_0(M)$ is the group diffeomorphisms of M preserving dV_{g_0}.

We consider then the following three quadratic functionals

(4.3)
$$\mathscr{R}^2(g) = \int_M |R(g)|^2 \, dV_g, \quad \mathscr{R}ic^2(g) = \int_M |\mathrm{Ric}(g)|^2 \, dV_g,$$
$$\mathscr{S}^2(g) = \int_M |s(g)|^2 \, dV_g.$$

Their L^2 gradients are computed in [13] to be

(4.4)
$$\nabla \mathscr{R}^2(g) = 4D^*D\,\mathrm{Ric} + 2D\,ds + 4\,\mathrm{Ric} \circ \mathrm{Ric} - 4\overset{\circ}{R}(\mathrm{Ric}) - 2\check{R} + \tfrac{1}{2}|R|^2 \cdot g,$$

(4.5) $\quad \nabla \mathscr{R}ic^2(g) = D^*D\,\mathrm{Ric} + D\,ds + \tfrac{1}{2}(\Delta s)g + \tfrac{1}{2}|\mathrm{Ric}|^2 \cdot g - 2\overset{\circ}{R}(\mathrm{Ric}),$

(4.6) $\qquad \nabla \mathscr{S}^2(g) = 2D\,ds + 2(\Delta s)g + \tfrac{1}{2}s^2 \cdot g - 2s \cdot \mathrm{Ric}.$

Here D is the covariant derivative of g, D^* its adjoint, $\overset{\circ}{R}$ is the action of R on $S^2(M)$, $\check{R}(x,y) = \sum R(x, e_i, e_j, e_k) \cdot R(y, e_i, e_j, e_k)$, and Δ is the Laplacian (with positive spectrum). We also note the following standard identities, cf. [13],

(4.7) $\quad DD^*\mathrm{Ric} = -\tfrac{1}{2}D\,ds, \quad DD^*(s \cdot g) = -D\,ds, \quad D^*D(s \cdot g) = (\Delta s) \cdot g.$

To obtain the gradients $\nabla \mathscr{F}$ restricted to $\mathscr{M}_0$, one subtracts $\frac{1}{n}\mathrm{tr}\,\nabla \mathscr{F} \cdot g$ from each gradient, and obtains

(4.8)
$$L_{\mathscr{R}^2} = \nabla|_{\mathscr{M}_0} \mathscr{R}^2(g) = 4\left(D^*D - \frac{2n}{2n-1}DD^*\right)\left(\mathrm{Ric} - \frac{s}{2n} \cdot g\right) + R_1,$$

(4.9) $\qquad L_{\mathscr{R}ic^2} = \nabla|_{\mathscr{M}_0} \mathscr{R}ic^2(g) = (D^*D - 2DD^*)(\mathrm{Ric}) + R_2,$

(4.10) $\qquad L_{\mathscr{S}^2} = \nabla|_{\mathscr{M}_0} \mathscr{S}^2(g) = \left(\frac{2}{n}D^*D - 2DD^*\right)(s \cdot g) + R_3,$

where R_i are the corresponding trace-free terms involving only quadratic terms of the curvature.

Finally, we consider the corresponding pth power functionals, but do this only for $\mathscr{R}^p$. A straightforward computation from the derivation in [13, Proposition 4.70] gives

$$(4.11) \quad \nabla|_{\mathscr{M}_0}\mathscr{R}^p(g) = -\delta^D D^*(|R|^{p-2}R) + \mathrm{tr}\left(\frac{\delta^D D^*(|R|^{p-2}R)}{n}\right) \cdot g + R_4,$$

where R_4 depends only on the curvature tensor (to the power p).

We now turn to the question of the regularity of metrics which are critical for these functionals. The operators L may all be viewed as 4th order systems of P.D.E., in the metric $g \in \mathscr{M}_0$. We have written them in the form (4.8)–(4.10), so that, to leading order, they may be treated as 2nd order systems in the unknowns $(\mathrm{Ric} - \frac{1}{2n}s \cdot g)$, etc.

We then have the following

PROPOSITION 4.1. *The linearized operators* $L'_{\mathscr{F}}: S^2(M) \to S^2(M)$, *given by* $L'_{\mathscr{F}}(h) = \frac{d}{dt}L_{\mathscr{F}}(u + th)$, *for* $\mathscr{F} = \mathscr{R}^2$, $\mathscr{R}ic^2$ *or* $\mathscr{S}^2$ *are elliptic 2nd order systems in the unknown* h, *for any* $u = \mathrm{Ric} - \frac{1}{2n}s \cdot g$, Ric *or* $s \cdot g$, *respectively.*

PROOF. We need to compute the symbol σ of $D^*D - \alpha DD^*$, for appropriate α, acting on $S^2(M) \subset TM \otimes TM$. Standard computations (cf. [13]) show that, for $\xi \in TM$,

$$\sigma_\xi(D^*D)(\eta \otimes v) = |\xi|^2\eta \otimes v, \qquad \sigma_\xi(DD^*)(\eta \otimes v) = \langle \xi, \eta\rangle\xi \otimes v.$$

The additivity of the symbol then gives

$$\sigma_\xi(D^*D - \alpha DD^*)(\eta \otimes v) = (|\xi|^2\eta - \alpha\langle\xi, \eta\rangle\xi) \otimes v.$$

This shows that σ_ξ is an isomorphism $S^2_x(M) \to S^2_x(M)$, $\forall \xi \neq 0$, provided $\alpha \neq 1$. Since this is the case for all three operators, we obtain the result. $\square$

Suppose the metric g is smooth enough so that M admits a covering by (sufficiently smooth) harmonic coordinates. Then it is well known [22] that the equation for the Ricci curvature of g is a 2nd order elliptic system (in g), given in these coordinates by

$$(4.12) \qquad g^{ij}\frac{\partial^2 g_{rs}}{\partial x_i \partial x_j} + Q_{rs}(g, \partial g) = -\Delta g_{rs} + Q_{rs}(g, \partial g) = 2\mathrm{Ric}_{rs}.$$

Here Q is a term, quadratic in g and its first derivatives, of the general form

$$Q = g^{-2}(\partial g)^2.$$

Similarly, $\mathrm{Ric} - \frac{1}{2n}s \cdot g$ is a 2nd order elliptic system in the unknown g, in harmonic coordinates, while $s \cdot g$ is an overdetermined 2nd order elliptic system in g. Since the composition of elliptic operators is elliptic, it follows that the operators $L_{\mathscr{F}}$ are 4th order elliptic systems (overdetermined for $\mathscr{S}^2$) in the metric g, in harmonic coordinates.

For the functional $\mathscr{R}^p$, for $p \neq 2$, the situation is not so favorable. The operator $\nabla|_{\mathscr{M}_0}\mathscr{R}^p$ cannot be written as a composition of 2nd order operators, and thus must be treated directly as a 4th order operator in g. There are many further difficulties. For instance, the operator is clearly not uniformly elliptic, due to the $|R|^{p-2}$ term. It is probably better to consider the related functional $\int(1+|R|^2)^{p/2}$, in analogy to [38]. In fact, it has not been verified that the operator $\nabla|_{\mathscr{M}_0}\mathscr{R}^p$ is an elliptic system in g. Even if this were true (for instance if $(p-2)$ is sufficiently small) the coefficients of the leading order terms of the linearized operator are at best in some L^q space, $q < \infty$. Unfortunately, there is no satisfactory regularity theory for elliptic systems with such weak control of the coefficients.

In sum, the Euler-Lagrange equation for $\mathscr{R}^2$ appears to be much more natural, analytically, than the other cases of $\mathscr{R}^p$. Thus it is reasonable to expect good behavior of the critical metrics of $\mathscr{R}^p$ only in low dimensions, namely in dimensions 2, 3, with $p = 2$, or possibly in dimension 4 with p sufficiently close to 2.

The following result confirms this expectation in dimension 3 (and thus also in dimension 2).

THEOREM 4.2 (Regularity of weak solutions). *Let* Ω *be a bounded domain in a 3-manifold* M *and suppose* g *is a* $L^{2,2}$ *weak solution of the Euler-Lagrange equation*

$$(4.14) \qquad L_{\mathscr{F}}(g) = 0,$$

for $\mathscr{F} = \mathscr{R}^2$ *or* $\mathscr{Ric}^2$ *on* Ω, *i.e.* g *is an* $L^{2,2}$ *critical metric for* $L_{\mathscr{F}}$. *Then* g *is* C^{∞}.

PROOF. We will only outline the main steps in the proof; full details will appear elsewhere. Both operators are treated in the same way, so only the (typical) case of $\mathscr{Ric}^2$ will be discussed.

The result is local, so that by using a smooth partition of unity, one may assume that Ω is a bounded domain in $\mathbb{R}^3$. First, note that by the Sobolev embedding theorem,

$$L^{2,2} \subset L^{1,6} \subset C^{1/2}.$$

Since the metric g is in $L^{2,2}$, it follows from [4, 5] that there are harmonic coordinates $\{x_k\}$ on any ball $B_x(r(x)) \subset \Omega$, of radius $r(x) \geq c(\text{dist}(x, \partial\Omega))$, in which the metric coefficients g_{ij} are bounded in $L^{2,2}$. Thus, we may assume $\Omega = B$, for some such ball.

Now the Ricci curvature Ric (and thus the full curvature tensor in dimension 3) is in L^2 and is a weak solution to the 2nd order elliptic system

$$(4.15) \qquad L_1(\text{Ric}) = (D^*D - 2DD^*)(\text{Ric}) - \underline{R} = 0,$$

where $\underline{R}$ is quadratic in the curvature of (M, g). In other words, when paired with any $\phi = \{\phi_{ij}\} \in L^{2,2}$, of compact support in B, $\int\langle\phi, L_1(\text{Ric})\rangle$ $\equiv \int\langle L_1(\phi), \text{Ric}\rangle = 0$.

Note that the curvature term $\underline{R}$ however is only in $L^1(B)$. If we put $\underline{R}$ on the right-hand side of (4.15), we wish to study the regularity of $L(u) = f$, with $f \in L^1$. Unfortunately, there is no regularity theory for elliptic equations with right-hand side $f \in L^1$.

Instead, we turn to the fractional Sobolev spaces $H^s(\Omega)$ of functions whose "sth derivatives" $(s \in \mathbb{R})$ are in L^2; cf. [23]. The Sobolev embedding theorem gives $L^1 \subset H^{s-2}$, for any $s < 1/2$. Thus, we consider regularity of solutions of the system

$$(4.16) \qquad\qquad L_2(u) = f,$$

with $L_2 = D^*D - 2DD^*$, mapping

$$(4.17) \qquad\qquad L_2: H^s \to H^{s-2},$$

for $s \in [0, 1/2)$. In the harmonic coordinates $\{x_k\}$, the operator L_2 takes the form

$$(4.18) \qquad (L_2 u)_{rs} = g^{ij}\partial_i\partial_j(u_{rs}) - 2g^{rs}\partial_r\partial_s(g^{ab}u_{ab}) + Q_{rs}(g, \partial g),$$

where Q is a term quadratic in g and ∂g. A computation shows that also $Q \in H^{s-2}$, for all $s < 1/2$, so that Q may also be put on the right-hand side of (4.16). Thus, we are reduced to studying the regularity of the system

$$(4.19) \qquad\qquad L: H^s \to H^{s-2},$$
$$(Lu)_{rs} = g^{ij}\partial_i\partial_j(u_{rs}) - 2g^{rs}\partial_r\partial_s(g^{ab}u_{ab}) = F_{rs} \in H^{s-2}.$$

We will outline the solution of the regularity problem in several steps. First, as usual, we view (4.19) as a linear system by freezing or uncoupling the coefficients g from the unknown u (the Ricci curvature).

First, we claim in order to prove smooth regularity, that it suffices to prove that any solution $u \in H^0$ of (4.19) is actually in H^s, $\forall s < 1/2$. Namely, $u \in H^s$, $\forall s < 1/2$ implies that $g \in H^{s+2}$ and thus by the Sobolev embedding $\underline{R} \in L^{(3/2)-\varepsilon}$, for any $\varepsilon > 0$. Similarly, one computes that $Q(g, \partial g) \in L^{(3/2)-\varepsilon}$, so that now we have the operator L mapping

$$(4.20) \qquad\qquad L: W^{2,p} \to W^{0,p},$$

for any $p < 3/2$. (Here $W^{k,p}$ is the Sobolev space of functions with k (weak) derivatives in L^p.) The coefficients of L are now in C^α, $\forall\alpha < 1$. Standard L^p regularity of elliptic systems, cf. [33], then implies that any solution u of $L(u) = f$, for $f \in W^{0,p}$ lies in $W^{2,p}$. Thus, the Ricci curvature of g lies in $W^{2,p}$, and thus $g \in W^{4,p}$, for $p < 3/2$. This gives by Sobolev embedding again that $g \in C^{1,\alpha}$, $\forall\alpha < 1$. Continuing in this fashion, we obtain $g \in C^\infty$.

Second, to prove the assertion above that $u \in H^s$, for $s < 1/2$, it suffices to prove the following estimate,

$$(4.21) \qquad \|v\|_{H^s} \leq C(\|g\|_{H^2})\{\|Lv\|_{H^{s-2}} + \|v\|_{H^0}\}, \qquad s \in [0, \tfrac{1}{2}),$$

for operators L of the form (4.19) with C^∞ coefficients, where C is a constant depending only on the H^2 norm of the coefficients. To see why this is so, note that the local solvability of (4.19) with C^∞ coefficients is standard, i.e. $\forall f \in H^{s-2}$, $\exists v \in H^s$ such that $L(v) = f$ (distribution solution); cf. [**32, 33**] for example. Then choose a sequence of operators $L_m: H^s \to H^{s-2}$, of the form (4.19) with C^∞ coefficients $\{g_m\}$, such that $g_m \to g$ in the $L^{2,2}$ topology on B. For a fixed $F \in H^{s-2}$ $(F = -Q - \underline{R})$, let v_m be a uniformly H^0-bounded solution to $L_m(v_m) = F$. The estimate (4.21) then implies that one has a uniform bound on $\|v_m\|_{H^s}$, so that $\{v_m\}$ has a subsequence converging (weakly) to a solution $\tilde{v} \in H^s$ of $L(\tilde{v}) = F$ (with the original coefficients g). Thus, to obtain $u \in H^s$, we need to prove regularity of the homogeneous equation

$$(4.22) \qquad\qquad L(w) = 0, \qquad w = \tilde{v} - v \in H^0.$$

In fact, (all) solutions to (4.22) are in H^1, following for instance the argument of [**1**, Lemma 6.1]. (Although the proof in [**1**] is for elliptic equations with Lipschitz coefficients, it is not difficult to verify the same technique yields the desired conclusion for $L^{2,2}$ coefficients, for elliptic systems of the form (4.19).) Thus, (4.21) implies $u \in H^s$, as required.

Finally, the estimate (4.21) is also standard for elliptic systems with C^∞ coefficients g_{ij}, with C depending on the C^∞ norm of $\{g_{ij}\}$; cf. [**23, 32**]. This is done by a local perturbation argument from the case of constant coefficient operators, where the estimate is easily established by Fourier transform methods. To see that only the $L^{2,2}$ norm of $\{g_{ij}\}$ is needed to carry out the perturbation argument (in dimension 3), one may follow for instance the argument in [**23**, Proposition 6.25] (with systems in place of equations). This requires one to establish the estimate (arising from perturbation)

$$(4.23) \qquad\qquad \|\phi z\|_{H^{s-2}} \leq C(\|\phi\|_{H^2})\{\|z\|_{H^{s-2}} + \|z\|_{H^{-2}}\},$$

for $\phi \in C_0^\infty(B)$. This is now easily proved by Fourier transform methods and interpolation in H^s spaces, following e.g. [**23**] or [**32**]. This completes the outline of the proof. $\square$

The results above lead easily to the following main result.

THEOREM 4.3. *Let M be a closed 3-manifold and let $\{g_i\}$ be a sequence of metrics, of volume 1, on M for which $\mathscr{R}^2(g_i)$ approaches a critical value of $\mathscr{R}^2$ (for example a minimizing sequence of the functional). Then for all $\varepsilon > 0$, the ε-thick part (M^ε, g_i) has a subsequence converging in the strong $L^{2,2}$ topology (and C^α topology, $\alpha \leq 1/2$), to a C^∞ Riemannian manifold*

$(\Omega^{\varepsilon}, g)$, $\Omega^{\varepsilon} \subset M$. *The metric g is a smooth solution of the Euler-Lagrange equation*

(4.24) $$\nabla|_{\mathscr{M}_0} \mathscr{R}^2 = 0.$$

Further $\mathscr{R}^2(\Omega^{\varepsilon}, g) \leq \lim \mathscr{R}^2(g_i)$.

REMARK. The same result holds if $\{g_i\}$ is a sequence for which $\mathscr{Ric}^2(g_i)$ approaches a critical value of $\mathscr{Ric}^2$.

PROOF. This is a simple consequence of the previous results. Namely, we may apply Theorem 3.2 to the sequence $\{g_i\}$ to obtain the convergence of a subsequence of $\{g_i\}$ in the weak $L^{2,2}$ topology on Ω^{ε} to a limit $L^{2,2}$ metric on Ω^{ε}. The fact that one has strong $L^{2,2}$ convergence follows from the fact that $\nabla\mathscr{R}^2(g_i) \to 0$, in the $L^{-2,2}$ topology. The metric g is a (weak) $L^{2,2}$ solution of the Euler-Lagrange equation (4.24) on Ω^{ε} and thus by Theorem 4.2, g is C^{∞}. The last statement also follows easily.   □

The manifolds Ω^{ε} are embedded as smooth domains in M, with $\Omega^{\varepsilon} \subset \Omega^{\varepsilon'}$ for $\varepsilon < \varepsilon'$. Thus, we may form $\Omega = \bigcup_{\varepsilon>0} \Omega^{\varepsilon} \subset M$, so that Ω is the largest domain in M on which the metrics $\{g_i\}$ do not degenerate. Theorem 3.2 describes the behavior of the metrics $\{g_i\}$ on $M\backslash\Omega$; namely for each i, $(M\backslash\Omega, g_i)$ admits an F-structure $\mathscr{F}_i$, along which the manifold $M\backslash\Omega$, presumably, is being collapsed to a lower-dimensional length space under $\{g_i\}$.

In particular, if M does not admit a global F-structure, then $\Omega \neq \varnothing$. In three dimensions, the closed manifolds which admit a global F-structure are fully classified. They are exactly the (locally) Seifert fibred spaces, i.e. 3-manifolds which admit a foliation by circles; cf. [18]. From a different point of view, M^3 does not admit a global F-structure if and only if the simplicial volume or Gromov invariant $\|M\|$ of M is positive.

It is particularly interesting to view these results in the context of Thurston's results and conjectures on the structure of 3-manifolds [45, 46]. Thurston conjectures that any (oriented), irreducible, closed 3-manifold M may be decomposed into domains, with boundaries incompressible tori in M, such that each domain admits a "geometric structure," that is a complete locally homogeneous metric. The possible locally homogeneous structures have been classified [46, 42]. There are eight possibilities, seven of which represent the possible geometries of (local) Seifert fibre spaces, while the remaining, most prevalent case, is hyperbolic geometry. Of course, Thurston [46] has proved this conjecture in a number of important cases.

There are numerous questions one can ask to relate this conjectured decomposition of Thurston with the decomposition obtained from minimizing sequences, and their limits, for $\mathscr{R}^2$ or $\mathscr{Ric}^2$ on M. For instance,

(i) If $\Omega \neq M$, are the boundary components of Ω essential tori in M?
(ii) Is any smooth minimizing metric on Ω complete hyperbolic?

(iii) If M has no essential S^2's or T^2's, is $\Omega = M$?

(iv) If M is not prime, what happens geometrically to the essential 2-spheres in M, under a minimizing sequence $\{g_i\}$? For instance, if M is a connected sum of hyperbolic manifolds, describe the resulting minimizing metric.

We will not address these interesting questions any further here. In particular, one needs to understand in detail the intrinsic geometry of the critical metrics.

5. Approach to critical metrics of $\mathscr{S}$

Here we address the same issues as in §4, but with respect to the total scalar curvature functional $\mathscr{S}$ in place of the curvature integrals $\mathscr{R}^p$.

Recall from §0 that although $\mathscr{S}$ does not admit any minimizing or maximizing sequences, it does admit in many cases natural minimax sequences. For instance, this is the case if M admits no metric of positive scalar curvature [40]; this is a topological condition that is now quite well understood (cf. [30, 41]).

In contrast to §4, there is no difficulty in identifying the critical metrics of $\mathscr{S}$ (even locally). They are exactly the Einstein metrics on M, and thus of constant curvature in dimensions ≤ 3. Similarly, the regularity theory is much simpler, since the Euler-Lagrange equation is now 2nd order in the metric g (and not 4th order as in §4). On the other hand, the functional $\mathscr{S}$ is rather weak, in the sense that bounds on $|\mathscr{S}(g)|$ or $|\nabla\mathscr{S}(g)|$ give little control over the behavior of g.

As before, one must choose a topology on $\mathscr{M}_1(M)$ (or $\mathscr{M}_0(M)$), for which $\mathscr{M}_1$ is a Banach manifold on which $\mathscr{S}$ is a C^1 functional, but which otherwise is as weak as possible, so that the condition $\|\nabla\mathscr{S}(g_i)\| \to 0$ (a condition in the dual topology) gives as strong information as possible.

A first natural choice would be the $L^{1,p}$ topology, defined as in §4 (with 1 in place of 2), for a fixed $p > n = \dim M$. By Sobolev embedding $L^{1,p} \subset C^\alpha$, $\alpha = 1 - \frac{n}{p}$, and $\mathscr{M}_1(M)$ or $\mathscr{M}_0(M)$ are smooth Banach manifolds [36]. Although we shall not verify it here, it is not difficult to prove that $\mathscr{S}$ is a C^1 functional on $\mathscr{M}_0$ in this topology. (One uses a partition of unity to localize, then integrates by parts with respect to harmonic coordinates.)

Now suppose $\{g_i\}$ is a sequence of metrics in $\mathscr{M}_0$ for which $\mathscr{S}(g_i)$ approaches a critical value of $\mathscr{S}$. With respect to the $L^{1,p}$ topology, it follows that

$$(5.1) \qquad \|\nabla\mathscr{S}_{g_i}\|_{L^{1,p}} := \sup_{|h|_{L^{1,p}}=1} \left| \int \left\langle \mathrm{Ric}_{g_i} - \frac{s(g_i)}{n} \cdot g_i, h \right\rangle_{g_i} dV \right| \to 0,$$

as $i \to \infty$. This means, by definition, that

$$(5.2) \qquad \left\| \mathrm{Ric}_{g_i} - \frac{s(g_i)}{n} \cdot g_i \right\|_{L^{-1,p'}} \to 0,$$

where $p' = \frac{p}{p-1}$. This is a rather weak condition. Suppose, for illustrative purposes, that $n = 3$. Then for $p > n = 3$, one may define an $L^{1,p}$ harmonic radius as in §3 and in a similar way prove that

$$(5.3) \qquad r_h^{L^{1,p}}(x) \geq c\left(n, \varepsilon, \left\|\mathrm{Ric} - \frac{s}{n} \cdot g\right\|_{L^{-1,p}}\right) \cdot \frac{\mathrm{dist}(x, \partial M)}{\mathrm{diam}\, M} \cdot v^\varepsilon(x),$$

at least for p relatively large; cf. [4, 7, 44] for example. Thus, the $L^{-1,p}$ norm of $\mathrm{Ric} - \frac{1}{n} \cdot sg$ controls the $L^{1,p}$ harmonic radius of M (given control of the local volumes of M), and thus the local $C^\alpha \cap L^{1,p}$ geometry of M.

Unfortunately, we have only control over the $L^{-1,p'}$ norm of $\mathrm{Ric} - \frac{1}{n} \cdot s \cdot g$ from (5.2), with $p' < \frac{n}{n-1} = \frac{3}{2}$. This appears too weak to be useful, for example to prove local existence and regularity of limits of minimax sequences.

From the results of §§3, 4, we see that these difficulties would not arise if one had control over $\int |R|^p$ for the sequence $\{g_i\}$, for some $p > n/2$. We will simply settle here by assuming this bound in order to close with the following positive result.

THEOREM 5.1. *Let M be a compact n-manifold and let $\{g_i\}$ be a sequence of metrics on M, of volume 1, which approach a critical value of $\mathcal{S}$. Suppose further that M does not admit an F-structure and that one has the uniform bound*

$$(5.4) \qquad \int_M |R|^p \leq \Lambda,$$

for some $\Lambda < \infty$ and $p > n/2$. Then one of the following occurs:

(i) A subsequence of $\{g_i\}$ converges, weakly in the $L^{2,p}$ topology, to a smooth Einstein metric on M, or

(ii) A subsequence of $\{g_i\}$ converges in the Gromov-Hausdorff topology to a finite number of (connected) cusps (N_k, g_k). Each (N_k, g_k) is a complete, noncompact Riemannian manifold, of volume ≤ 1, with smooth Einstein metric g_k. The convergence $g_i \to g_k$ is in the weak $L^{2,p}$ topology on N_k. The manifolds N_k are smoothly embedded in M, with complement $M - \bigcup N_k$ admitting an F-structure $\mathcal{F}_i$.

REMARKS. (i) We recall again that the first two assumptions can often be satisfied topologically. For instance, if $\chi(M) \neq 0$, then M admits no F-structure, while if M admits no metric of positive scalar curvature, e.g. M is spin and $\widehat{A}(M) \neq 0$), then $\mathcal{S}$ admits minimax sequences on M.

(ii) In dimension 3, any cusp that is formed is necessarily a complete, finite volume hyperbolic manifold, with toral boundary components.

PROOF. This is a simple consequence of the work in §§3, 4. Since we are assuming the bound (5.4), one may apply Theorem 3.2 to conclude that a subsequence of $\{g_i\}$ converges in the weak $L^{2,p}$ topology, on the thick part Ω of (M, g_i), to a limit Riemannian manifold (N, g), with $L^{2,p}$ metric

g. Since $\mathscr{S}$ is C^1 in the $L^{2,p}$ topology, it follows that locally, any limit g of $\{g_i\}$ is an $L^{2,p}$ weak solution of the Euler-Lagrange equation
(5.5)
$$L = \left(\mathrm{Ric} - \frac{s}{n} \cdot g\right)_{rs} = g^{ij}\frac{\partial^2 g_{rs}}{\partial x_i \partial x_j} - \frac{1}{n}g^{ab}g^{ij}\frac{\partial^2 g_{ab}}{\partial x_i \partial x_j} \cdot g_{rs} + Q_{rs}(g, \partial g) = 0.$$

The coefficients of g are in $C^\alpha \cap L^{2,p}$ and L is an elliptic operator. Thus, the elliptic regularity theory [33] implies that g is C^∞, so that g is in fact a smooth Einstein metric.

In case (i), it follows that g is a globally defined smooth Einstein metric on $\Omega = M$. In case (ii), note that (N, g) is Einstein, and so in particular is of bounded Ricci curvature. Thus, one has the usual volume comparison theorem on (N, g), which implies that

(5.6)
$$C(\mathrm{dist}(x, y))^{-1} \leq \frac{\mathrm{vol}\, B_x(1)}{\mathrm{vol}\, B_y(1)} \leq C(\mathrm{dist}(x, y)),$$

for all balls in (N, g). Thus, N can become thin only at infinity. It follows in particular that (N, g) is a complete Einstein n-manifold, possibly disconnected. Each component (N_k, g_k) of (N, g) is a complete cusp, in the sense of §1, of volume ≤ 1. This, together with Theorem 3.2, completes the proof. □

References

1. S. Agmon, *Lectures on elliptic boundary value problems*, Van Nostrand, Princeton, NJ, 1965.
2. M. Anderson, *Ricci curvature bounds and Einstein metrics on compact manifolds*, J. Amer. Math. Soc. **2** (1989), 455–490.
3. ____, *Short geodesics and gravitational instantons*, J. Differential Geom. **31** (1990), 265–275.
4. ____, *Convergence and rigidity of manifolds under Ricci curvature bounds*, Invent. Math. **102** (1990), 429–445.
5. ____, *Remarks on the compactness of isospectral sets in low dimensions*, Duke Math. Jour., **63:3** (1991), 699–711.
6. ____, *The L^2 structure of moduli spaces of Einstein metrics on 4-manifolds*, Geom. and Funct. Analysis **2:1** (1992), 29–89.
7. M. Anderson and J. Cheeger, *C^α compactness for manifolds with Ricci curvature and injectivity radius bounded below*, J. Differential Geom. **35** (1992), 265–281.
8. ____, *Diffeomorphism finiteness for manifolds with Ricci curvature and $L^{n/2}$ norm bounded*, Geom. and Funct. Analysis **1:3** (1991), 231–251.
9. S. Bando, A. Kasue, and H. Nakajima, *On a construction of coordinates at infinity on manifolds with fast curvature decay and maximal volume growth*, Invent. Math. **97** (1989), 313–349.
10. S. Bando, *Bubbling out of Einstein manifolds*, Tôhoku Math. J. **42** (1990), 205–216.
11. S. Bando and R. Kobayashi, *Ricci-flat Kähler metrics on affine algebraic manifolds*. II, Math. Ann. **287** (1990), 175–180.
12. M. Berger, *Une borne inferieure pour le volume d'une variete Riemannienne en fonction du rayon d'injectivite*, Ann. Inst. Fourier (Grenoble) **30** (1980), 259–265.
13. A. Besse, *Einstein manifolds*, Ergeb. Math., vol. 3, Springer-Verlag, New York, 1987.
14. J. P. Bourguignon, *A review of Einstein manifolds*, DD2 Proceedings, Shanghai-Hefei 1981, Science Press, Beijing, 1984.

15. E. Calabi, *Metriques kähleriennes et fibres holomorphes*, Ann. Sci. École Norm. Sup. **12** (1979), 269–294.

16. P. Candelas, P. Green, and T. Hubsch, *Finite distance between distinct Calabi-Yau vacua*, Phys. Rev. Lett. **62** (1989), 1956.

17. J. Cheeger, *Finiteness theorems for Riemannian manifolds*, Amer. J. Math. **92** (1970), 61–74.

18. J. Cheeger and M. Gromov, *Collapsing Riemannian manifolds while keeping their curvature bounded*. I, J. Differential Geom. **23** (1986), 309–346.

19. ___, *Collapsing Riemannian manifolds while keeping their curvature bounded*. II, J. Differential Geom. **32** (1990), 269–298.

20. J. Cheeger, M. Gromov, and M. Taylor, *Finite propagation speed, kernel estimates for functions of the Laplace operator and the geometry of complete Riemannian manifolds*, J. Differential Geom. **17** (1982), 15–53.

21. J. Cheeger, K. Fukaya, and M. Gromov, *Nilpotent structures and invariant metrics on collapsed manifolds*, J. Amer. Math. Soc. **5** (1992), 327–372.

22. D. DeTurck and J. Kazdan, *Some regularity theorems in Riemannian geometry*, Ann. Sci. École Norm. Sup. **14** (1980), 249–260.

23. G. Folland, *An introduction to partial differential equations*, Math. Notes, Princeton Univ. Press, Princeton, NJ, 1976.

24. S. Gallot, *Isoperimetric inequalities based on integral norms of Ricci curvature*, Asterisque **157–158** (1988), 191–216.

25. L. Gao, $L^{n/2}$ *curvature pinching*, J. Differential Geom. **32** (1990), 713–774.

26. ___, *Convergence of Riemannian manifolds: Ricci and $L^{n/2}$ curvature pinching*, J. Differential Geom. **32** (1990), 349–381.

27. P. Green and T. Hubsch, *Connecting moduli spaces of Calabi-Yau threefolds*, Comm. Math. Phys. **119** (1988), 931.

28. R. Greene and H. Wu, *Lipschitz convergence of Riemannian manifolds*, Pac. J. Math. **131** (1988), 119–141.

29. M. Gromov, *Structures metriques pour les varietes Riemanniennes*, Cedic-Fernand/Nathan, Paris, 1981.

30. M. Gromov and B. Lawson, *Positive scalar curvature and the Dirac operator on complete Riemannian manifolds*, Inst. Hautes Études Sci. Publ. Math. **58** (1983), 83–196.

31. A. Kasue, *A convergence theorem for Riemannian manifolds and some applications*, Nagoya Math. J. **114** (1989), 21–51.

32. J. L. Lions and E. Magenes, *Non-homogeneous boundary value problems and applications*. I, Grundlehren Series, vol. 181, Springer-Verlag, New York, 1972.

33. C. Morrey, Jr., *Multiple integrals in the calculus of variations*, Grundlehren Series, vol. 130, Springer-Verlag, New York, 1966.

34. H. Nakajima, *Hausdorff convergence of Einstein 4-manifolds*, J. Fac. Sci. Tokyo Univ. **35** (1988), 411–424.

35. ___, *Compactness of the moduli spaces of Yang-Mills connections in higher dimensions*, J. Math. Soc. Japan **40** (1988), 383–392.

36. R. Palais, *Foundations of global non-linear analysis*, Benjamin, New York, 1968.

37. S. Peters, *Convergence of Riemannian manifolds*, Compositio Math. **62** (1987), 3–16.

38. J. Sachs and K. Uhlenbeck, *The existence of minimal immersions of 2-spheres*, Ann. of Math. **113** (1981), 1–24.

39. R. Schoen, *Analytic aspects of the harmonic map problem*, MSRI Series, vol. 2, Springer-Verlag, New York, 1985, pp. 321–358.

40. ___, *Variational theory for the total scalar curvature functional for Riemannian metrics and related topics*, Lecture Notes in Math., vol. 1365, Springer-Verlag (1989), 120–154.

41. R. Schoen and S. T. Yau, *On the structure of manifolds with positive scalar curvature*, Manuscripta Math. **29** (1979), 159–183.

42. P. Scott, *The geometries of 3-manifolds*, Bull. London Math. Soc. **15** (1983), 401–487.

43. J. Sha and D. Yang, *Positive Ricci curvature on the connected sums of $S^n \times S^m$*, J. Differential Geom. (to appear).

44. M. Taylor, *Pseudodifferential operators and non-linear PDE*, Birkhäuser Verlag, Boston (1991).
45. W. Thurston, *The geometry and topology of 3-manifolds*, Preprint, Princeton.
46. _____, *Three-dimensional manifolds, Kleinian groups and hyperbolic geometry*, Bull. Amer. Math. Soc. **6** (1982), 357–381.
47. G. Tian, *Compactness theorem for Kahler-Einstein manifolds of dimension 3 and up*, J. Differential Geom. **35** (1992), 535–558.
48. G. Tian and S. T. Yau, *Complete Kahler manifolds with zero Ricci curvature. II*, Invent. Math. **106** (1991), 27–60.
49. D. Yang, *Riemannian manifolds with small integral norm of curvature*, Duke Math. J. **65** (1992), 501–510.
50. _____, *Convergence of Riemannian manifolds with integral bounds on curvature. I, II*, Preprints.
51. _____, *Existence and regularity of energy-minimizing Riemannian metrics*, Ann. Sci. Ecole Norm. Sup. (1992), (to appear).
52. S. T. Yau, *On the Ricci curvature of a compact Kahler manifold and the complex Monge-Ampère equation*, Comm. Pure Appl. Math. **31** (1978), 339–441.
53. S. T. Yau (Editor), *Seminar on differential geometry*, Ann. of Math. Stud., vol. 102, Princeton Univ. Press, Princeton, NJ, 1982.

SUNY AT STONY BROOK

Proceedings of Symposia in Pure Mathematics
Volume 54 (1993), Part 3

The Geometry of Totally Geodesic Hypersurfaces in Hyperbolic Manifolds

ARA BASMAJIAN

0. Introduction

In this paper, we consider a new type of spectrum associated to a hyperbolic manifold containing an embedded totally geodesic hypersurface. This spectrum is called the *orthogonal spectrum of the manifold relative to the hypersurface*. The purpose of this note is to describe some theorems about the orthogonal spectrum that appear in [**Ba1**] and to discuss some additional examples and consequences of these theorems.

1. The orthogonal spectrum

Let M^n be an hyperbolic manifold of dimension $n \geq 2$, and let S be an embedded oriented hypersurface which is either totally geodesic or flat. Throughout this paper by a flat hypersurface we mean the quotient of a horosphere in hyperbolic space by a subgroup of the ambient manifold fundamental group.

Once a normal orientation is chosen, since the fibers of the normal bundle are one-dimensional, there is a natural identification from S to its unit normal bundle given by $p \mapsto n_p$, where n_p is the unit normal based at p in the appropriate direction of the orientation. Lastly, we note that n_p in turn determines a normal ray emanating from p.

Suppose α is a (not necessarily simple) path from S to a totally geodesic hypersurface S_1 which is either disjoint from S or equal to S. We say that two paths from S to S_1 are *freely homotopic relative to S and S_1*, if they are homotopic through a homotopy which keeps the initial point in S and the terminal point in S_1. The equivalence class of α is called the *relative*

1991 *Mathematics Subject Classification*. Primary 53C20; Secondary 30F40.

Key words and phrases. Hyperbolic manifold, hypersurface, kleinian group, orthogonal spectrum, geodesic flow, ergodic.

This paper is in final form and no version of it will be submitted for publication elsewhere.

free homotopy class of α. The path α is said to have trivial relative free homotopy if $S = S_1$ and α can be homotoped (keeping its endpoints in S) to a single point in S. With the exception of a few cases, each nontrivial relative free homotopy class contains a shortest path which is the unique common orthogonal in that class. The exceptional cases occur when the non-trivial class α contains paths of arbitrarily short length. In these cases, we define the length of the class α to be zero. Otherwise, the length of a non-trivial class is defined to be the length of the common orthogonal in that class.

Let $\mathscr{C}$ be a (possibly infinite) disjoint set of embedded totally geodesic hypersurfaces on M. We do not allow the hypersurface S to intersect $\mathscr{C}$, unless S is actually one of the surfaces in $\mathscr{C}$. We define, for each nonnegative integer k, the kth-*orthogonal spectrum of M relative to S and $\mathscr{C}$*, denoted $\mathscr{O}_k(M; S, \mathscr{C})$, to be the ordered nondecreasing sequence of lengths of non-trivial relative free homotopy classes of paths which

(1) start in S and go in the direction of the normal to S,
(2) cross $\mathscr{C}$ along the way k times, and
(3) end in a hypersurface contained in $\mathscr{C}$.

We allow the possibility that $\mathscr{C}$ is empty in which case the spectrum is empty. The *full orthogonal spectrum* $\mathscr{O} = \mathscr{O}(M; S, \mathscr{C})$ is the ordered sequence of lengths of common orthogonals that start in S and end in $\mathscr{C}$.

We let m_h denote the hyperbolic measure (the measure inherited from the volume element) on $\mathbb{H}^n$ restricted to a lift of S. Since m_h is invariant under isometries of $\mathbb{H}^n$ (in particular under the covering group), we can think of m_h as being defined on S; that is, if X is a measurable subset of S and $\widetilde{S}$ is some lift of S, then the hyperbolic measure of X is $m_h(\widetilde{X})$ where $\widetilde{X}$ is a lift of X into some fundamental polygon inside $\widetilde{S}$. All of this, of course, is independent of the choice of lift for X or S. Hence, if S is a totally geodesic hypersurface then m_h is just the $(n-1)$-dimensional hyperbolic measure on S. If S is a flat manifold, then m_h is simply the $(n-1)$-dimensional lebesgue measure.

We define the function $r_S = r: \mathbb{R}_{\geq 0} \to \mathbb{R}^+ \cup \{\infty\}$ to be

$$r(x) = \begin{cases} \log \coth(\frac{x}{2}), & \text{if } S \text{ is an embedded} \\ & \text{totally geodesic hypersurface,} \\ e^{-x}, & \text{if } S \text{ is an embedded flat hypersurface.} \end{cases}$$

Finally, denote the (hyperbolic or euclidean) volume of the n-dimensional ball of radius r by $V_n(r)$. The choice of which of the above two geometries to use is dictated by whether S is (respectively) totally geodesic or flat.

We call the following theorem the orthogonal spectrum theorem.

THEOREM 1.1 [**Ba1**]. *Let $\mathscr{C}$ be a disjoint set of totally geodesic embedded hypersurfaces on the hyperbolic manifold M^n, and let S be an embedded oriented hypersurface which is either disjoint from $\mathscr{C}$ or one of the hypersurfaces* in $\mathscr{C}$. *Suppose no nontrivial relative free homotopy class from S to $\mathscr{C}$ has length zero. Then the kth-orthogonal spectrum,*

$$\mathscr{O}_k(M\,;S\,,\mathscr{C}) = \{d_i\}\,,$$

satisfies:

$$(*) \qquad\qquad \mathrm{Vol}_{n-1}(S) = m_h(F_k) + \sum_{i=1}^{\infty} V_{n-1}(r(d_i))\,,$$

Where F_k is the subset of S consisting of all points whose corresponding oriented normal ray to S crosses $\mathscr{C}$ at most k times. In particular, if the orthogonal spectrum $O_k(M\,;S\,,\mathscr{C})$ is nonempty and S has finite $(n-1)$-dimensional volume then the spectrum is discrete and goes to infinity. In this case, the number of $d_i \leq x$ is asymptotic to

$$\frac{\mathrm{Vol}_{n-1}(S) - m_h(F_k)}{V_{n-1}(r(x))}\,, \qquad \text{as } x \text{ goes to infinity.}$$

The sets F_k form a nested nondecreasing sequence. Suppose $\mathscr{C}$ is made up of totally geodesic boundary hypersurfaces. Since an orthogonal that crosses a geodesic boundary hypersurface never returns, we can conclude that the kth-orthogonal spectrum $\mathscr{O}_k(M\,;S\,,\mathscr{C})$ in this case is empty for $k \geq 1$. In particular,

$$S = F_1 = F_2 = \cdots = F_n = \cdots\,.$$

Formula $(*)$ holds even if S has infinite volume. In this case, either the series on the right-hand side of formula $(*)$ diverges or $m_h(F_k)$ is infinite. Of course the orthogonal spectrum may not be discrete any more (see the example in §3).

In the special case that M is a closed hyperbolic surface and S and $\mathscr{C}$ are closed geodesics, Wolpert [W] showed that the Weil-Petersson pairing on Teichmüller space of the twist fields corresponding to the closed geodesics is completely determined by, in our language, the full orthogonal spectrum $\mathscr{O}(M\,;S\,;\mathscr{C})$.

Bob Meyerhoff [Me] has studied a similar spectrum for hyperbolic 3-manifolds. His main focus is on the question of what extra information one can include to the length spectrum of the manifold to determine the manifold.

We remark that the first element of the orthogonal spectrum has implications to the volume of hyperbolic manifolds containing totally geodesic hypersurfaces. This will be investigated in a forthcoming paper.

In many situations, one can relate the set F_0 with the limit set of the discrete group representing the manifold M. We will content ourselves with a statement that applies only for dimension two, that is, hyperbolic surfaces with geodesic boundary (fuchsian groups of the second kind). In this case, a totally geodesic embedded hypersurface is a simple geodesic and a flat hypersurface is the boundary of a cusp. Denote either one of these by β.

In order to state the next result, we first need to define some terms. A lift $\tilde{\beta}$ of β divides the boundary circle at infinity into two disjoint intervals, if β is a geodesic. In this case, let $\mathscr{R}_+$ be the open interval for which the normal to $\tilde{\beta}$ points. If $\tilde{\beta}$ is a horocycle, then let $\mathscr{R}_+$ be the set of all points on the boundary $\partial \mathbb{H}^2$ except the basepoint at infinity of the horocycle. Consider the map

$$p_{\tilde{\beta}} : \mathscr{R}_+ \mapsto \beta$$

given by orthogonal projection to $\tilde{\beta}$ followed by the covering map into the quotient manifold M. In this context, by the *orthogonal projection* of a point x contained in $\mathscr{R}_+$ we mean the unique point on $\tilde{\beta}$ which is the endpoint of the orthogonal from the point x to $\tilde{\beta}$. Observe that $p_{\tilde{\beta}}$ is injective on a $\mathrm{Stab}_G(\tilde{\beta})$-fundamental polygon in $\mathscr{R}_+$, the group G being the covering group of isometries, that is, $\mathbb{H}^2/G = M$.

COROLLARY 1.2 [Ba1]. *Suppose $\mathbb{H}^2/G$ is a hyperbolic surface and let β be either a simple geodesic or the boundary of a cusp. Let $\mathscr{C}$ be the set of all boundary geodesics in $\mathbb{H}^2/G$, and let $\tilde{\beta}$ be a lift of β. Then with $\mathscr{R}_+$ as above, we have that*

$$p_{\tilde{\beta}}(\Lambda(G) \cap \mathscr{R}_+) = F_0.$$

Furthermore, the lebesgue one-dimensional measure of the set $\Lambda(G) \cap \mathscr{R}_+$ is zero if and only if $m_h(F_0) = 0$. In particular, if G is a finitely generated fuchsian group of the second kind, then $m_h(F_0) = 0$.

2. Ergodicity and totally geodesic hypersurfaces

Results related to the ergodicity of the geodesic flow date back to the works of Hedlund [He1, He3] and Hopf [Ho1, Ho2]. For the many equivalent conditions of ergodicity the reader is referred to the book of Peter Nicholls [N].

In the study of discrete groups the sequence of lengths $\{(a, g(a))\}_{g \in G}$ for some $a \in \mathbb{H}^n$ is of fundamental importance. One can equivalently think of this sequence as the lengths of (nonsmooth) geodesic loops based at some point on the manifold $\mathbb{H}^n/G$. The works of Patterson, Sullivan, and others [P1–P5, S1–S4] have shown that the Poincaré series $\sum_{g \in G} e^{-s(a, g(a))}$ is closely connected with the geometry of the hyperbolic manifold $\mathbb{H}^n/G$. One result in this direction [S1] is that the series $\sum_{g \in G} e^{-(n-1)(a, g(a))}$ diverges if and only if the geodesic flow acts ergodically on the unit tangent bundle of $\mathbb{H}^n/G$. Furthermore, if we define $\delta(G)$ to be the *critical exponent* of the above Poincaré series, that is

$$\delta(G) = \inf \left\{ s \in \mathbb{R}_{\geq 0} \Big| \sum e^{-s(a, g(a))} \text{ converges} \right\},$$

then it can be shown that $\delta(G)$ is related to the hausdorff dimension of the limit set of G. For example, if G is geometrically finite then the hausdorff dimension of the limit set is precisely $\delta(G)$.

Examples of hyperbolic manifolds containing totally geodesic hypersurfaces can be obtained through the works of Long, Millson, Maclachlan, and Reid [L, Mi, M-R]. In dimension three, one can geometrically construct examples by using the Klein-Maskit combination theorems [Ma].

The next theorem relates the Poincaré series mentioned above with one involving the full orthogonal spectrum of a totally geodesic hypersurface.

THEOREM 2.1 [Ba1]. *Let S be a closed embedded totally geodesic hypersurface in the hyperbolic manifold $\mathbb{H}^n/G$, and let $a \in \mathbb{H}^n$. Then for $s > (n-2)$ the series*

$$\sum_{g \in G} e^{-s(a,\,g(a))} \ \text{diverges if and only if} \ \sum_{d \in \mathscr{O}} e^{-sd} \ \text{diverges},$$

where $\mathscr{O}$ is the full orthogonal spectrum with respect to S and $\mathscr{C} = S$. If $s \le (n-2)$, then the Poincaré series on the left-hand side diverges.

Using Sullivan's ergodicity criterion (for $s = n-1$) with Theorem 1.1, and the fact that there exists a constant C so that the function $V_{n-1}(r(x))$ is asymptotic to $Ce^{-(n-1)x}$ as x goes to infinity, we have the following immediate corollary.

COROLLARY 2.2. *If M^n contains a closed embedded totally geodesic hypersurface S with a fixed normal orientation, then the geodesic flow acts ergodically on the unit tangent bundle of M if and only if the series*

$$\sum_{k=0}^{\infty} (2\,\mathrm{Vol}_{n-1}(S) - m_h(F_k^+) - m_h(F_k^-)),$$

diverges, where F_k^+ denotes the set of basepoints of oriented normal rays that cross S at most k times and F_k^- denotes the set of basepoints of negatively oriented normal rays that cross S at most k times.

3. Examples

The following example will show that there are (isometrically) different manifolds having the same othogonal spectrum.

Consider β a simple closed geodesic on the closed hyperbolic surface M_0^2. Twisting along β gives a deformation of M_0, through different hyperbolic structures (see [A]). Name these new hyperbolic surfaces M_t for $t \in [0, \ell(\beta)]$. Pick a normal direction to β, and observe that the orthogonal spectrum

$$\mathscr{O}_0(M_0; \beta, \beta) = \mathscr{O}_0(M_t; \beta, \beta),$$

for all t. Thus this spectrum is unaffected by twisting along β.

Our next example will show that the orthogonal spectrum is not necessarily discrete if the hypersurface has infinite area.

Once again, we choose to work in dimension two where totally geodesic embedded hypersurfaces are simple geodesics. Let P_i, for $i = 1, 2, 3, ...$, be an infinite number of copies of a fixed pair of pants having two of its boundary geodesics of the same length. Let γ_i and γ_{i+1} denote these geodesics for each copy P_i. Construct the hyperbolic surface M^2 by gluing, without any twist, P_i to P_{i+1} along the geodesic γ_{i+1}, for each i. The surface M^2 is complete and is known as a flute space; the reader is referred to the paper [Ba2] for the details of such a construction. Now, since we glued without twisting, there is a simple geodesic β of infinite length that runs orthogonal to all of the γ_i. Pick a normal direction to β and consider the Oth-orthogonal spectrum $\mathcal{O}_0(M; \beta, \beta)$. This is clearly not discrete, since the γ_i are all of equal length and orthogonal to β.

References

[A] W. Abikoff, *The real analytic theory of Teichmüller space*, Lecture Notes in Math., vol. 820, Springer-Verlag, 1980.

[Ba1] A. Basmajian, *The orthogonal spectrum of a hyperbolic manifold*, Amer. J. Math. (to appear).

[Ba2] ——, *Hyperbolic structures for surfaces of infinite type*, Trans. Amer. Math. Soc. (to appear).

[Be] A. F. Beardon, *The geometry of discrete groups*, Springer-Verlag, 1983.

[He1] G. A. Hedlund, *A metrically transitive group defined by the modular group*, Amer. J. Math. 57 (1935), 668–678.

[He2] ——, *Fuchsian groups and transitive horocycles*, Duke Math. J. 2 (1936), 530–542.

[He3] ——, *Fuchsian groups and mixtures*, Ann. of Math. (2) 40 (1939), 370–383.

[Ho1] E. Hopf, *Fuchsian groups and ergodic theory*, Trans. Amer. Math. Soc. 39 (1936), 299–314.

[Ho2] ——, *Statistik der geodatischen linien in mannigfaltigkeiten negativer krummung*, Ber. Verh. Sachs. Akad. Wiss. Leipzig 91 (1939), 261–304.

[L]   D. D. Long, *Immersions and embeddings of totally geodesic surfaces*, Bull. London Math. Soc. 19 (1987), 481–484.

[M-R] C. Maclachlan and A.W. Reid, *Parametrizing Fuchsian subgroups of the Bianchi groups*, Preprint.

[Ma] B. Maskit, *Kleinian groups*, Preprint, Springer-Verlag, 1987.

[Me] B. Meyerhoff, *The ortho-length spectrum for hyperbolic 3-manifolds*, Preprint.

[Mi] J. Millson, *On the first Betti number of a constant negatively curved manifold*, Ann. of Math. (2) 104 (1976), 235–247.

[N]   P. Nicholls, *The ergodic theory of discrete groups*, London Math. Soc. Lecture Note Ser., vol. 143, Cambridge Univ. Press, 1989.

[P1] S. J. Patterson, *A lattice point problem in hyperbolic space*, Mathematika 22 (1975), 81–88.

[P2] ——, *The limit set of a Fuchsian group*, Acta Math. 136 (1976), 241–273.

[P3] ——, *The exponent of convergence of Poincaré series*, Monatsh. Math. 82 (1976), 297–315.

[P4] ——, *Spectral theory and Fuchsian groups*, Proc. Cambridge Philos. Soc. 81 (1977), 59–75.

[P5] ——, *Lectures on measures on limit sets of Kleinian groups*, Analytical and Geometric Aspects of Hyperbolic Space, London Math. Soc. Lecture Note Ser., vol. 111, Cambridge Univ. Press, 1987, pp. 281–323.

[S1] D. Sullivan, *The density at infinity of a discrete group of hyperbolic motions*, Inst. Hautes Études Sci. Publ. Math. 50 (1979), 171–202.

[S2] ——, *On the ergodic theory at infinity af an arbitrary discrete group of hyperbolic motions*, Ann. of Math. Stud., vol. 97, Princeton Univ. Press, Princeton, NJ, 1981, pp. 465–496.

[S3]   _____, *Discrete conformal groups and measurable dynamics*, Bull. Amer. Math. Soc. (N.S.) **6** (1982), 57–73.

[S4]   _____, *Entropy, Hausdorff measures old and new, and limit sets of geometrically finite Kleinian groups*, Acta. Math. **153** (1984), 259–277.

[W]   S. A. Wolpert, *Thurston's riemannian metric for Teichmüller space*, J. Differential Geometry **23** (1986), 143–174.

UNIVERSITY OF OKLAHOMA
E-mail address: abasmajian@nsfuvax.math.uoknor.edu

Proceedings of Symposia in Pure Mathematics
Volume 54 (1993), Part 3

Finiteness of Diffeomorphism Types
of Isospectral Manifolds

ROBERT BROOKS, PETER PERRY, AND PETER PETERSEN V

1. Statement of results

A question which has received a great deal of attention over the years is the following: to what extent does the spectrum of the Laplacian of a manifold M determine the geometry of M?

In recent years, there have been a great many examples of pairs of manifolds M_1 and M_2 which are isospectral but not isometric. Such manifolds can even be taken to have different topological type, or to lie in continuous families of mutually isospectral manifolds, or to preserve various types of auxiliary structures. See [**Ber**] for a survey of such results, and [**S, GW, BG, BPY**] for some examples.

It was shown by Osgood, Phillips, and Sarnak [**OPS**] that 2-dimensional isospectral manifolds form compact sets in the C^∞ topology. A similar compactness phenomenon was proved for isospectral sets of metrics in a given conformal class of metrics on a 3-manifold of negative scalar curvature [**BPY**] or on a general 3-manifold ([**CY**]).

In this note, we announce some results concerning finiteness of diffeomorphism type for families of isospectral manifolds, under some auxiliary curvature assumptions:

THEOREM 1. *Let* $\{M_i\}$ *be a family of isospectral manifolds of negative sectional curvature of dimension* n.

Then the $\{M_i\}$'s contain only finitely many homeomorphism types. If $n \neq 4$, then the $\{M_i\}$'s contain only finitely many diffeomorphism types.

Received by the editors November, 1989; Revised April, 1990.
1991 *Mathematics Subject Classification.* Primary 58G25.
Partially supported by the National Science Foundation.
Work of the first and second authors supported in part by National Science Foundation grant RII-8610671 and the Commonwealth of Kentucky through the Kentucky EPSCoR program.
The detailed version of this paper has been submitted for publication elsewhere.

THEOREM 2. *Let* $\{M_i\}$ *be a family of isospectral manifolds of dimension* n, *whose sectional curvatures are uniformly bounded from below.*

Then the $\{M_i\}$'s contain only finitely many homeomorphism types. If $n \neq 4$, then the $\{M_i\}$'s contain only finitely many diffeomorphism types.

In dimension 3, we can say more:

THEOREM 3. *Let* $\{M_i\}$ *be a family of 3-manifolds which are isospectral, and which satisfy one of the following three sets of conditions:*
Either

(a) *The* $\{M_i\}$ *'s all have negative sectional curvature*

or

(b) *The* $\{M_i\}$ *'s have Ricci curvature uniformly bounded from below.*

Then the family $\{M_i\}$ contains only finitely many diffeomorphism types, and is precompact in the C^∞ topology.

It is a pleasure to thank Marc Burger and Peter Sarnak for extremely helpful conversations about this material.

The first and second authors would like to thank the National Science Foundation and the Commonwealth of Kentucky for their support under the Kentucky EPSCoR program. The first author would also like to express his gratitude to the UCLA Department of Mathematics for its hospitality and support.

See [BPP] for further details, and [An] for related work.

2. Spectral invariants

For M a Riemannian manifold, one may raise the question of how the spectrum of the Laplacian contains geometric information about M. There are essentially three families of such spectral invariants, given respectively by the heat asymptotics, wave asymptotics, and geometric bounds. We will use all three of these approaches below.

The first of these is the method of heat asymptotics. If we denote by $H_t^M(x, y)$ the fundamental solution to the heat equation

$$(\partial/\partial t + \Delta)(f_t) = 0,$$

then

$$\mathrm{tr}(H_t^M) = \int_M H_t^M(x, x)\, dx = \sum_i e^{-\lambda_i t}$$

has the asymptotic expansion

$$\mathrm{tr}(H_t^M) \sim \frac{1}{(4\pi t)^{n/2}}\{a_0 + a_1 t + a_2 t^2 + \cdots\}$$

as $t \to 0^+$, where $a_0, a_1, \ldots$ are integrals over M of local invariants of the geometry of M. For low values of i, one can compute the a_i's as follows:

$a_0 = \int_M 1 = \mathrm{vol}(M)$,

$a_1 = \frac{1}{6}\int_M S$,

$a_2 = \frac{1}{360} \int_M 5S^2 - 2|\rho|^2 + 2|R|^2$,

where S, ρ, and R are the scalar, Ricci, and Riemannian curvatures respectively.

As i gets large, the size of the expression for a_i grows exponentially, but the following theorem due to Gilkey allows us to handle such expressions.

THEOREM [**Gi**]. *For $i \geq 3$,*

$$a_i = \frac{(-1)^i}{2^{i+1} \cdot 1 \cdot 3 \cdots (2i+1)} \int_M (i^2 - i - 1)|\nabla^{i-2}S|^2 + 2|\nabla^{i-2}\rho|^2 + \cdots$$

where "$\cdots$" denotes terms which involve fewer derivatives of the metric.

The second source is given by the asymptotics of the wave equation. One has the following:

THEOREM [**DG, Do**]. *The singular support of the trace of the wave operator is contained in the closure of the set of lengths of closed geodesics of M.*

If the metric on M is "generic" in the sense of [**DG**], then the singular support of the trace of the wave operator is precisely the set of lengths of closed geodesics.

It is not difficult to see that if M has negative sectional curvature, then M is "generic" in this sense.

The third source of invariants are given by bounds on geometric data in terms of spectral data. We mention here one example, due to Cheng:

THEOREM [**Chn**]. *If M is a compact manifold whose Ricci curvatures are all bounded below by $-(n-1)k$, $k > 0$, then*

$$\lambda_l(M) \leq (n-1)^2 \frac{k}{4} + (\text{const}) \frac{l^2}{d^2}.$$

where (const) *depends only on the dimension, and d is the diameter of M.*

Rewriting this as

$$d^2 \leq \frac{(\text{const})l^2}{\lambda_l - (n-1)^2 k/4},$$

and choosing l large enough so that the denominator is positive, we obtain an upper bound for the diameter in terms of a lower bound for the Ricci curvature and spectral data.

3. Finiteness Theorems

A popular question in Riemannian geometry is to determine conditions on the geometry of a manifold which will determine the manifold up to finitely many topological types. The first example of such a theorem is the Cheeger Finiteness Theorem [**Chr**].

The following results are due to [**GPW**]; see also [**GP**] and [**Y**].

THEOREM [GPW]. *For any real number* k *and positive numbers* D *and* v, *the class of closed Riemannian* n-*manifolds* M *with sectional curvature* $\geq k$, *diameter* $\leq D$, *and volume* $\geq v$ *contains only finitely many homotopy types, only finitely many homeomorphism types when* $n \geq 4$, *and only finitely many diffeomorphism types when* $n \geq 5$.

THEOREM [GPW]. *For any positive numbers* c *and* v, *the class of closed Riemannian* n-*manifolds with injectivity radius* $\geq c$ *and volume* $\leq V$ *contains only finitely many homotopy types, only finitely many homeomorphism types when* $n \geq 4$, *and only finitely many diffeomorphism types when* $n \geq 5$.

We may now establish Theorems 1 and 2 in the case $n > 3$ in the following way: if M has negative curvature, then its injectivity radius is $\geq 1/2$ the length of the shortest closed geodesic. This in turn is spectrally determined by the asymptotics of the wave operator. Furthermore, the volume is determined by the asymptotics of the heat kernel.

If the curvature is bounded from below, then the diameter is bounded from above in terms of spectral data by Cheng's theorem, and again the volume is determined by the asymptotics of the heat kernel.

In either case, Theorems 1 and 2 for dimensions greater than 3 now follow directly.

We remark that when the manifolds carry metrics of negative curvature and when $n \geq 5$, we may bound the number of diffeomorphism types explicitly. This follows from explicit bounds on the number of homotopy types in [Y], together with the fact, due to Farrell and Jones [FJ], that for manifolds of negative curvature ($n \geq 5$), the homeomorphism type is determined by the homotopy type. Bounds on the diffeomorphism type from the homeomorphism type in these dimensions are standard.

4. Three-dimensional isospectral manifolds

The following theorem was proved by Deane Yang.

THEOREM [Ya]. *For positive constants* V, D, χ, *and* K, *and for* $p > n/2$, *let* M_i *be a sequence of smooth* n-*dimensional compact Riemannian manifolds such that*

$$\mathrm{vol}(M_i) > V, \qquad \mathrm{diam}(M_i) < D,$$

$$C_S(M_i) > \chi, \quad \text{and} \quad \|R\|_p < K\mathrm{vol}(M_i)^{1/p}.$$

Then

1. *There are a finite number of diffeomorphism classes represented in the sequence.*

2. *There exists a subsequence that converges in Hausdorff distance to a smooth compact manifold with a* C^0 *Riemannian metric.*

Here, $C_S(M)$ is the isoperimetric constant of M,

$$C_S(M) = \inf_N \frac{\text{area}(N)}{[\min(\text{vol}(A), \text{vol}(B)]^{1-1/n}},$$

where N runs over hypersurfaces which divide M into two pieces, A and B. It is well known that a lower bound on C_S is equivalent to an upper bound on the constants occurring in the Sobolev inequality.

The idea behind the proof of the theorem is to apply parabolic Moser iteration [Mo] to Hamilton's Ricci flow [Ham]. Roughly speaking, Moser iteration tells one how to turn L^p estimates, $p > n/2$, into L^∞ estimates.

For our purposes, dimension 3 enters as a number n for which $2 > n/2$. Then the a_0 and a_2 terms of the asymptotic expansion of the heat kernel bound $\text{vol}(M_i)$ and $\|R\|_2$. Any of the assumptions of Theorem 3 above now bound $\text{diam}(M_i)$, and a lower bound for $C_S(M_i)$ can be extracted from the work of Croke [Cr] and Gallot [Ga] (this was also observed in [Ya]).

With some care, one can extend Yang's theorem, using Gilkey's theorem, to show that the higher-order heat invariants insure that the C^0 convergence obtained by Yang can be improved to obtain C^∞ convergence. The main point is that, after obtaining the C^0 estimate, the lower-order terms in Gilkey's theorem can be a priori bounded—see [BPY] for a similar argument.

This then concludes the proof of Theorem 3.

REFERENCES

[An] M. Anderson, *Remarks on the compactness of isospectral sets in low dimensions*, Duke Math. J. **63** (1991), 699–711.

[Ber] P. Bérard, *Variétés Riemanniennes isospectrales non-isométriques*, Seminaire Bourbaki no. 705 (March 1989).

[BG] R. Brooks and C. Gordon, *Isospectral families of conformally equivalent Riemannian metrics*, Bull. Amer. Math. Soc. (N. S.) **23** (1990), 433–436.

[BPP] R. Brooks, P. Perry, and P. Petersen, *Compactness and finiteness theorems for isospectral manifolds*, J. Reine Angew. Math. **426** (1992), pp. 67–89.

[BPY] R. Brooks, P. Perry, and P. Yang, *Isospectral sets of conformally equivalent metrics*, Duke Math J. **58** (1989), 131–150.

[CY] S. Y. A. Chang and P. Yang, *Isospectral conformal metrics on 3-manifolds*, J. Amer. Math. Soc. **3** (1990), 131–150.

[Chr] J. Cheeger, *Finiteness theorems for Riemannian manifolds*, Amer. J. Math. **92** (1970), 61–74.

[Chn] S. Y. Cheng, *Eigenvalue comparison theorems and its geometric applications*, Math. Z. **143** (1975) 289–297.

[Cr] C.B. Croke, *Some isoperimetric inequalities and eigenvalue estimates*, Ann. Sci. École. Norm. Sup. (4) **13** (1980), 419–435.

[Do] H. Donnelly, *On the wave asymptotics of a compact negatively curved surface*, Invent. Math. **45** (1978), 115–137.

[DG] J. J. Duistermaat and V. Guillemin, *The Spectrum of positive elliptic operators and periodic bicharacteristics*, Invent. Math. **29** (1975), 39–79.

[FJ] T. Farrell and L. Jones, *Compact negatively curved manifolds (of dimension $\neq 3, 4$) are Topologically Rigid*, Proc. Nat. Acad. Sci. U.S.A. (to appear).

[Ga] S. Gallot, *Isoperimetric inequalities based on integral norms of Ricci curvature*, Astérisque no. 1 157–158, Soc. Math. France, Paris, 1988, pp. 191–216.

[Gi] P. B. Gilkey, *Functoriality and heat equation asymptotics*, to appear Coll. Math. Soc. János Bolyai, vol. 56 (to appear).

[GW] C. Gordon and E. Wilson, *Isospectral deformations of compact solvmanifolds*, J. Differential Geom. **24** (1986), 79–96.

[GP] K. Grove and P. Petersen V, *Bounding homotopy types by geometry*, Ann. of Math. (2) **128** (1988), 195–206.

[GPW] Karsten Grove, Peter Petersen V, and Jyh-Yang Wu, *Geometric Finiteness Theorems via Controlled Topology*, Invent. Math. **99** (1990), 205–213.

[Ham] R. S. Hamilton, *Three-manifolds with positive Ricci curvature*, J. Differential Geom. **17** (1982), 255–306.

[Mo] J. Moser, *A Harnack inequality for parabolic differential equations*, Comm. Pure Appl. Math. **17** (1964), 101–134.

[OPS] B. Osgood, R. Phillips, and P. Sarnak, *Compact Isospectral Sets of Surfaces*, J. Funct. Anal. **80** (1988), 212–234.

[S] T. Sunada, *Riemannian coverings and isospectral manifolds*, Ann. of Math. (2) **121** (1985), 169–186.

[Y] T. Yamaguchi, *Homotopy Type Finiteness Theorems for Certain Precompact Families of Riemannian Manifolds*, Proc. Amer. Math. Soc. **102** (1988), 660–666.

[Ya] D. Yang, L^p *pinching and compactness theorems for compact Riemannian manifolds*, Preprint.

UNIVERSITY OF SOUTHERN CALIFORNIA

UNIVERSITY OF KENTUCKY

UNIVERSITY OF CALIFORNIA AT LOS ANGELES

Proceedings of Symposia in Pure Mathematics
Volume **54** (1993), Part 3

Small Eigenvalues of the Laplacian on Negatively Curved Manifolds

P. BUSER, B. COLBOIS, AND J. DODZIUK

In this paper a *small eigenvalue* means one close to the bottom of the spectrum. The precise meaning will depend on the context. It is known (cf. [4]) that any finite sequence $0 < \lambda_1 \leq \lambda_2 \leq \cdots \leq \lambda_N$ can be realized as the beginning of the spectrum of the Laplace operator acting on functions on a sufficiently complicated compact Riemannian manifold M. It is equally well known that geometric assumptions on M impose severe restrictions on the spectrum. We discuss such restrictions in case of manifolds of negative curvature. Our results can be considered an update and a generalization of (a part of) [3]. The methods ultimately come down to elementary min-max techniques combined with a careful analysis of *thick and thin* decomposition [1, 6] of manifolds in the class under consideration. To carry out our estimates, we derive some new information about boundaries of cusps and tubes for finite volume manifolds of pinched negative curvature. This is of independent interest and is new even for compact manifolds of constant curvature. We give here the statements and brief indications of our methods. Full proofs will appear elsewhere.

We now state our results.

THEOREM 1. *Suppose M^n is a compact Riemannian manifold of dimension $n \geq 4$ and sectional curvatures in $[-1, 0)$. The smallest positive eigenvalue $\lambda_1(M)$ of the Laplace-Beltrami operator Δ on M satisfies*

$$\lambda_1(M) \geq \frac{c(n)}{\mathrm{vol}(M)^{b(n)}}.$$

The exponent $b(n) = 2$ for $n \geq 8$, $b(7) = b(6) = 3$, and $b(5) = b(4) = 6$.

1991 *Mathematics Subject Classification.* Primary 53C20, 58G25.
Research supported in part by the NSF Grant DMS-8902288.
The final version of this paper will be submitted for publication elsewhere.

The proof of this result depends on Gromov's volume-diameter estimate [6] and the technique of [5]. Since there are examples of sequences of three-dimensional manifolds (and of surfaces) of unbounded diameter, bounded volume and satisfying these curvature conditions, it is not surprising that Theorem 1 is false in two and three dimensions.

The next theorem requires different curvature assumptions.

THEOREM 2. *Suppose M^n is a complete Riemannian manifold of finite volume V with pinched sectional curvatures $-a^2 \leq K \leq -1$, $a > 1$. The number of eigenvalues of the Laplace-Beltrami operator in the interval $(0, (n-1)^2/4)$ does not exceed $c(n, a)V$.*

This was known previously for $n = 2$ [3, (3.1)], and for compact three-dimensional manifolds of constant curvature [3, (3.12)]. We remark that available examples based on [3, (3.10)] show that the constant $c(n, a)$ does depend on the pinching a.

The proof of Theorem 2 follows the argument used for proving [3, (3.12)]. However, there are essential technical difficulties arising if $n \geq 4$, even when M is compact and has constant sectional curvature. We will point out the source of these difficulties and state the technical result used to deal with them. As remarked already, these difficulties are present when M is compact and of constant curvature. Therefore, we shall assume, for simplicity of exposition only, that this is the case.

It is a consequence of Margulis's Lemma [1] that there exists a constant $\mu = \mu(n)$ with the property that the set of points $x \in M$ with injectivity radius $\iota(x) \geq \mu$ is nonempty. This set is connected and its complement consists of disjoint tubes surrounding simple closed geodesics of lengths less than 2μ. It is easy to see that it suffices to consider only short geodesics γ of length $l(\gamma) \leq \frac{\mu}{10}$. For such tubes we describe the geometry in more detail. For every $x \in \gamma$ and every unit vector $v \in T_x M$ perpendicular to γ, consider a geodesic ray $\delta : [0, \infty) \longrightarrow M$, $\delta(0) = x$, $\dot{\delta}(0) = v$. If t_0 is such that $\iota(\delta(t)) \leq \mu$ in the interval $[0, t_0]$, then $\iota(\delta(t))$ is strictly increasing in this interval. Let $R = R(v)$ be the smallest value of t such that $\iota(\delta(t)) = \mu$. R is finite and the arc $\delta([0, R])$ will be called a *maximal radial arc* or simply a *radial arc*. The radial arcs emanating from γ are pairwise disjoint, except possibly for their initial points. The tube T_γ is the set swept out by the maximal radial arcs of γ. We also introduce the unit vector field $\mathscr{R}$ on $T_\gamma \setminus \gamma$ tangent to radial arcs. T_γ is homeomorphic to $\gamma \times \mathbb{B}^{n-1}$, where $\mathbb{B}^{n-1}$ is the closed unit ball in $\mathbb{R}^{n-1}$, since M^n is assumed orientable for convenience. We stress that ∂T_γ is *not smooth* in general and that the lengths of maximal radial arcs may vary considerably. However, when $n = 3$ and $K \equiv -1$ every tube is *round*, i.e., $R(v) = \text{const}$. The reason for this can be seen as follows. Decompose the isometry of the universal covering $\mathbb{H}^n$ of M as $\tau \cdot \rho$, where τ is a pure translation whose axis projects onto γ and ρ is the rotation of the hyperplane perpendicular to the axis. In three dimensions ρ acts a rotation of S^1 and is determined by its action on *one* vector. In contrast, when $n \geq 4$, ρ is an element of $SO(n-1)$ and, if nontrivial, may treat different vectors very differently.

Even though ∂T_γ may not be smooth we have the following regularity result. We state it for tubes in manifolds satisfying curvature restrictions of Theorem 2.

PROPOSITION 3. *Let* $x, y \in \partial T_\gamma$ *and* $\text{dist}(x, y) \leq c$, *where c is a sufficiently small constant depending only on* n. *Let* ϑ *be the angle between the radial vector* $\mathscr{R}_x$ *and the geodesic arc from x to y. Then*

$$\frac{\mu}{10} \leq \vartheta \leq \pi - \frac{\mu}{10}.$$

The meaning of this is that the boundary ∂T_γ is not too steep if we think of $\mathscr{R}$ as the vertical direction.

The proposition above is the initial step in the proof of the following result, which, loosely speaking, says that a small perturbation of ∂T_γ is not too far from being round.

THEOREM 4. *There exists a smooth hypersurface H of M, contained in* $T_\gamma \setminus \gamma$, *with the following properties.*

(4.1) *The angle between the normal to H and* $\mathscr{R}$ *is less than* $\pi/2 - \alpha$ *for some* $\alpha = \alpha(n, a) \in (0, \pi/2)$.

(4.2) *The sectional curvatures of H with respect to the induced metric are bounded in absolute value by a constant depending only on a and n.*

(4.3) *Because of* (4.1), *H is homeomorphic to* ∂T_γ *by pushing along radial arcs. The distance between* $x \in H$ *and its image* $y \in \partial T_\gamma$ *satisfies*

$$\text{dist}(x, y) \leq \frac{\mu}{10}.$$

Actually, a somewhat stronger result than Theorem 4 is true. The hypersurface H can be made to be arbitrarily close to ∂T_γ. However, the curvature of H blows up as the approximation gets better.

A similar result can be proved for cusps. Once this is done, the method of proof of [3, (3.12)] can be applied to prove Theorem 2. The main idea is to cover the manifold by nonoverlapping domains whose number is controlled by the volume and whose first positive eigenvalue for Neumann boundary conditions is greater than or equal to $(n-1)^2/4$. Once we have such a covering, Theorem 2 follows from standard Neumann comparison.

It is easy to construct such a tiling for the thick part whose geometry is controlled. We describe here how to tile the thin part. First, using an old argument of Klingenberg [7] and Theorem 4, one can show that the injectivity radius, with respect to the induced metric, of the smoothed boundary H of every component of the thin part is bounded below by a constant depending only on the pinching constant a and the dimension n. We shall call such dependence *controlled*. Let $\epsilon > 0$ be a small parameter whose value will be fixed later in such a way that it depends only on a and n. Choose a maximal set of points $\{p_i\}$ whose pairwise distances in H are greater than or equal to ϵ. If ϵ is sufficiently small in relation to the injectivity radius of H then the Dirichlet domains $D_i = \{ x \in H \mid d(x, p_i) \leq d(x, p_j) \text{ for } i \neq j \}$, where d denotes the distance with respect to the induced metric on H, are quasi-isometric, in a controlled way, to Euclidean ϵ-balls. For every i, consider a hypersurface through p_i perpendicular to radial arcs. Project D_i onto this hypersurface along the radial arcs. Denote the image by E_i. By (4.1) this projection is a controlled quasi-isometry and, therefore E_i is quasi-isometric to a small Euclidean ball. Define the tile C_i, otherwise known as a piece of cheese, to be the union of all radial arcs emanating from E_i into the thin part of M. It is easy to see that the total number of all such pieces of cheese does not exceed $c(n, a) \operatorname{vol}(M)$. Next we have to estimate $\mu_1(C_i)$, the smallest positive Neumann eigenvalue of C_i. This is the crux of the matter. Fortunately, the estimate of Cheeger's constant given in [3, pp. 61–64] for the case of three-dimensional M of constant curvature -1 carries over almost verbatim. The only change required in the proof is to use the relative version of isoperimetric inequality [2] instead of the usual one. In any case, the argument of [3] proves that, if ε is sufficiently small, and this depends on a and n only,

$$\mu_1(C_i) \geq \frac{(n-1)^2}{4}.$$

REFERENCES

1. W. Ballmann, M. Gromov, and V. Schroeder, *Manifolds of nonpositive curvature*, Progr. Math. vol. 61, Birkhäuser, 1985.

2. Yu. D. Burago and V. A. Zalgaller, *Geometric inequalities*, Springer-Verlag, 1988.

3. P. Buser, *On Cheeger's inequality $\lambda_1 \geq h^2/4$*, Geometry of the Laplace Operator, (R. Osserman and A. Weinstein, eds.), Proc. Sympos. Pure Math., vol. 36, Amer. Math. Soc., Providence, RI, 1980, pp. 29–77.

4. Y. Colin de Verdière, *Constructions de laplaciens dont une partie finie du spectre est donnée.*, Ann. Sci. École Norm. Sup. **20** (1987), 599–615.

5. J. Dodziuk and B. Randol, *Lower bounds for λ_1 on a finite-volume hyperbolic manifold*, J. Differential Geom. **24** (1986), 133–139.

6. M. Gromov. *Manifolds of negative curvature*, J. Differential Geom. **13** (1978), 223–230.

7. W. Klingenberg, *Über Riemannsche Mannigfaltigkeiten mit positiver Krümmung.* Comment. Math. Helv. **35** (1961), 47–54.

ECOLE POLYTECHNIQUE FÉDÉRALE DE LAUSANNE, SWITZERLAND

MATHEMATISCHES INSTITUT DER UNIVERSITÄT BONN, GERMANY

CITY UNIVERSITY OF NEW YORK

Proceedings of Symposia in Pure Mathematics
Volume **54** (1993), Part 3

Geometrically Tame Hyperbolic 3-Manifolds

RICHARD D. CANARY

1. Introduction

In this paper we will survey some of the theory of infinite volume hyperbolic 3-manifolds. We will begin by recalling briefly some of the theory of geometrically finite hyperbolic 3-manifolds which was developed in the 1960s and 1970s by Ahlfors, Bers, Kra, Marden, Maskit and others. We will then describe Bonahon and Thurston's seminal work on ends of hyperbolic 3-manifolds which are not geometrically finite. They proved that large classes of hyperbolic 3-manifolds are geometrically tame. A hyperbolic 3-manifold N is said to be geometrically tame if all its ends are either geometrically finite (in which case the geometry is exponentially expanding) or simply degenerate (in which case the end is homeomorphic to $S \times [0, \infty)$ and is "filled up" by surfaces homotopic to S with bounded area.) We will also describe some more recent work in this direction. We will then focus on the analytic consequences of the above work. In particular, we will discuss the spectral theory of the Laplacian and the measure-theoretic properties of the limit set. The work of Patterson and Sullivan establishes a beautiful connection between these two subjects.

There are two related ways of thinking about a hyperbolic 3-manifold. As a differential geometer one defines a hyperbolic 3-manifold to be a complete Riemannian 3-manifold with constant sectional curvature -1. This naturally encourages one to do differential geometry on N, particularly spectral theory.

If one works in Kleinian groups, or more generally discrete groups, one defines a (orientable) hyperbolic 3-manifold to be the quotient of $\mathbf{H}^3$ by a discrete, torsion-free subgroup Γ of $\mathrm{Isom}_+(\mathbf{H}^3)$, the group of orientation-preserving isometries of $\mathbf{H}^3$. The upper half-space model for $\mathbf{H}^3$ identifies $\mathrm{Isom}_+(\mathbf{H}^3)$ with $\mathrm{PSL}(2, \mathbf{C})$ realized as the group of Möbius transformations

---

1991 *Mathematics Subject Classification.* Primary 30F40, 57M50, 58G25; Secondary 32G99.
Partially supported by National Science Foundation grant DMS 88-09085.
This paper is in final form and no version of it will be submitted for publication elsewhere.

of $\mathbf{C} \cup \{\infty\}$ and the sphere at infinity for $\mathbf{H}^3$ is identified with $\mathbf{C} \cup \{\infty\}$. This point of view encourages one to think of Γ as a conformal dynamical system. A basic object from this point of view is the *limit set* of Γ's action on $\mathbf{C} \cup \{\infty\}$, i.e., the smallest closed Γ-invariant subset of $\mathbf{C} \cup \{\infty\}$. (See Maskit [32] for the basic definitions in the theory of Kleinian groups.)

In this paper we will always assume that N has finitely generated fundamental group. This assumption is essential for almost all the work described. Also, in the interest of exposition, we will assume that N has *no cusps*, i.e., that every homotopically nontrivial curve in N is freely homotopic to a closed geodesic. In particular, when defining simply degenerate and geometrically finite ends and geometric tameness we will give the definitions in the restricted situation where there are no cusps. However, all the theorems will be stated in a manner which is true even if there are cusps. Those who are interested in the more general definitions are urged to consult the original papers referenced.

I would like to thank Peter Hislop, Darren Long, Rafe Mazzeo, and Peter Perry for their helpful comments on an early version of this article.

2. The convex core and geometric finiteness

The most basic geometric object associated to a hyperbolic 3-manifold N is its convex core $C(N)$. $C(N)$ is defined to be the smallest convex submanifold of N such that the inclusion map is a homotopy equivalence. Explicitly, $C(N)$ is the quotient of $CH(L_\Gamma)$ by Γ, where $CH(L_\Gamma)$ is the convex hull in $\mathbf{H}^3$ of L_Γ. $CH(L_\Gamma)$ is the union of all ideal tetrahedra in $\mathbf{H}^3$ with endpoints in L_Γ, so $\partial CH(L_\Gamma)$ is a union of geodesics and totally geodesic ideal triangles. In fact, the boundary of the convex core $\partial C(N)$ is a hyperbolic surface in the natural metric induced by lengths of arcs within $\partial C(N)$ (see Thurston [47] or Epstein-Marden [22]). (This is kind of amazing as, in general, $\partial C(N)$ is only a C^0-submanifold.) The following deep theorem in Kleinian groups is known as Ahlfors' finiteness theorem (see [3]).

THEOREM 2.1 (Ahlfors). *If N is a hyperbolic 3-manifold with finitely generated fundamental group, then $\partial C(N)$ is a complete, finite area hyperbolic surface. If N has no cusps, then $\partial C(N)$ is compact.*

There is a retraction $R : N \to C(N)$ called the nearest point retraction where $R(x)$ is defined to be the (unique) point of $C(N)$ which is nearest to x (see [12, 22]). R gives $N - C(N)$ a natural product structure as $\partial C_1(N) \times [0, \infty)$ where $C_1(N)$ is the neighborhood of radius 1 of $C(N)$ and the real coordinate is the distance to $\partial C(N)$. (Notice that $\partial C_1(N)$ is homeomorphic to $\partial C(N)$.) In these coordinates the metric on $N - C_1(N)$ is quasi-isometric to the metric $\cosh^2 t \, ds^2_{\partial C_1(N)} + dt^2$ where $ds^2_{\partial C_1(N)}$ is a hyperbolic metric on $\partial C_1(N)$. (See Perry [39] for a different, and more explicit, point of view on this fact.) Thus each component of $N - C(N)$ has exponentially expanding geometry. In particular, if S is a component of $\partial C_1(N)$ the diameter and area of $S \times \{t\}$ are growing exponentially in t, as is the volume of $S \times [0, t]$.

A hyperbolic 3-manifold N is said to be *geometrically finite* if $C(N)$ is compact (if we allow N to have cusps then N is geometrically finite if $C(N)$ has finite volume). In this case the geometry of each end of N is exponentially expanding. We now bring into play a topological property of geometrically finite hyperbolic 3-manifolds which will play a key role in the remainder of the paper. A 3-manifold is said to be *topologically tame* if it is homeomorphic to the interior of a compact 3-manifold. The retraction R allows one to construct a homeomorphism of N with the interior of its convex core and thus geometrically finite hyperbolic 3-manifolds are topologically tame.

Bers and Maskit [5, 31] (see also Greenberg [24]) proved that there exist hyperbolic 3-manifolds with finitely generated fundamental group which are not geometrically finite. The quickest examples to explain, and the first constructive examples, are due to Jorgenson [25] and Thurston [48]. Let M be a compact 3-manifold obtained from $S \times [0, 1]$ by gluing $S \times \{0\}$ to $S \times \{1\}$ by an attaching map ϕ. If ϕ is pseudo-Anosov (see [48] or Morgan [36]), M admits a hyperbolic metric. Let $\widehat{N}$ be the cover of M associated to $\pi_1(S)$ with the hyperbolic metric induced by the covering map. Since $\pi : \widehat{N} \to M$ is a regular covering (and $\widehat{N} \neq \mathbf{H}^3$) $C(\widehat{N}) = \pi^{-1}(C(M))$. Thus $C(\widehat{N}) = \widehat{N}$ and $\widehat{N}$ is clearly not geometrically finite. $\widehat{N}$ may be thought of as being constructed from infinitely many copies of M cut open along $S \times \{0\}$ stacked one on top of the other. Notice that $\widehat{N}$ has linear growth and admits a geometrically periodic action of $\mathbf{Z}$; this is at first glance extremely surprising.

REMARK. See Bowditch [8] for a discussion of the many equivalent definitions of geometric finiteness.

3. Geometric tameness

If N is not geometrically finite we have a topological replacement for the convex core called the compact core, due to Jaco, Scott and Shalen. A *compact core* is a compact submanifold M such that inclusion is a homotopy equivalence (see Scott [42]). The ends of N are in one-to-one correspondence with the components of $N - M$ (see Bonahon [6]). (In fact, readers unfamiliar with the theory of ends may simply take this as a definition.) An end E of N is said to be *geometrically finite* if there exists a neighborhood U of E such that $U \cap C(N) = \varnothing$. Thus, N is geometrically finite if and only if all its ends are geometrically finite.

In the late 1970s Bill Thurston recognized a property of ends which captures many useful topological and geometrical properties of the examples discussed at the end of §2, but is sufficiently broad, conjecturally, that all ends of N which are not geometrically finite might have this property. An end E of N is said to be *simply degenerate* if it has a neighborhood U which is homeomorphic to $S \times [0, \infty)$ (where S is a compact surface) and there exists a sequence of simplicial hyperbolic surfaces $\{f_n : S \to U\}$ such that $\{f_n(S)\}$ leaves every compact set and $f_n(S)$ is homotopic to $S \times \{0\}$ within U. (A map $f : S \to N$ is said to be a *simplicial hyperbolic surface* if S admits a triangulation T such that every face of T is mapped to a non-degenerate, totally geodesic triangle in N and the total angles of the triangles about any vertex of T is at least 2π. Notice that f need not be an embedding.) The key property of simplicial hyperbolic surfaces is that the geometry induced on S by f has curvature ≤ -1; in particular, it has constant curvature -1 away from the vertices and concentrated negative curvature at the vertices. One may make use of other classes of surfaces in place of simplicial hyperbolic surfaces as long as they share this property. Thurston originally made use of pleated surfaces in his definition (see [47]) and Minsky [35] has recently pioneered the use of harmonic maps.

N is said to be *geometrically tame* if all its ends are either geometrically finite or simply degenerate. Notice that N is topologically tame if N is geometrically tame. The key analytic property of geometrically tame hyperbolic 3-manifolds is:

$(*)$ $C(N)$ may be exhausted by compact submanifolds C_i such that $C_i \subset C_j$ if $i < j$, $\bigcup C_i^0 = C(N)$ (where C_i^0 is the interior of C_i as a subset of $C(N)$), and there exist constants K and L such that ∂C_i has area $\leq K$ and the neighborhood of radius one of ∂C_i has volume $\leq L$.

Each submanifold C_i may be taken to be the region "between" $\partial C(N)$ and a collection of simplicial hyperbolic surfaces (one for each simply degenerate end). We will call a hyperbolic 3-manifold *analytically tame* if it has property $(*)$. Another useful geometric property is that given a geometrically tame hyperbolic 3-manifold N there exists K such that every point in $C(N)$ has injectivity radius $\leq K$ (see [15]). However, even in the absence of cusps, there need not be any lower bound on injectivity radius (see Bonahon-Otal [7] or Thurston [48]).

A hyperbolic 3-manifold $N = \mathbf{H}^3/\Gamma$ is said to be the algebraic limit of geometrically finite hyperbolic 3-manifolds if there exists a sequence of discrete, faithful representation $\{\rho_i : \pi_1(N) \to \mathrm{Isom}_+(\mathbf{H}^3)\}$ such that $N_i = \mathbf{H}^3/\rho_i(\pi_1(N))$ is geometrically finite and $\{\rho_i\}$ converges as a sequence of representations to $\rho_\infty : \pi_1(N) \to \mathrm{Isom}_+(\mathbf{H}^3)$, a representation with image

Γ. Thurston [47] proved that if Γ is freely indecomposable (i.e., if the compact core of N has incompressible boundary) and N is the algebraic limit of geometrically finite hyperbolic 3-manifolds then N is geometrically tame. He used this in the proof of his geometrization theorem (see Morgan [36]).

In 1984, Francis Bonahon made a major breakthrough and proved:

THEOREM 3.1 [6]. *If* $N = \mathbf{H}^3/\Gamma$ *is a hyperbolic 3-manifold and* Γ *is finitely generated and freely indecomposable, then* N *is geometrically tame.*

One can use Bonahon's result to prove

THEOREM 3.2 [13]. *A hyperbolic 3-manifold* N *is topologically tame if and only if it is geometrically tame.*

The proof of Theorem 3.2 may be described very quickly. Given a topologically tame hyperbolic 3-manifold N we may choose a collection of geodesics γ in N which are "complicated enough" that the two-fold cyclic branched cover $\widehat{N}$ of N branched over γ has freely indecomposable fundamental group. $\widehat{N}$ inherits a singular Riemannian metric of constant sectional curvature -1, with concentrated negative curvature on the branching locus. A result of Gromov and Thurston [26] allows one to perturb the metric in a neighborhood of the branching locus to obtain an honest Riemannian metric of pinched negative curvature. Notice that any end E of N is isometric to some end $\widehat{E}$ of $\widehat{N}$. We then notice that the proof of Theorem 3.1 works for manifolds of pinched negative curvature, so if E is not geometrically finite, then $\widehat{E}$ is "simply degenerate" in $\widehat{N}$ and thus E itself is simply degenerate.

Theorem 3.2 reduces many questions in the theory of Kleinian groups, for example, Ahlfors' measure conjecture (see §5), to the purely topological question of whether or not every hyperbolic 3-manifold with finitely generated fundamental group is topologically tame. In this spirit, we make the following ambitious conjecture which was first posed as a question by Marden [30]:

CONJECTURE. *Every hyperbolic 3-manifold with finitely generated fundamental group is topologically, hence geometrically, tame.*

REMARKS. (1) Otal proved that if $N = \mathbf{H}^3/\Gamma$ is the algebraic limit of geometrically finite groups and Γ is isomorphic to $\pi_1(S_1) * \pi_1(S_2)$ or $\pi_1(S_1) * \mathbf{Z}$ (where S_1 and S_2 are closed surfaces) then N is geometrically tame.

(2) Culler and Shalen [17] proved, using work of McMullen, that there is a dense set of Kleinian groups in the boundary of the Schottky space of genus 2, whose associated hyperbolic 3-manifolds are analytically tame, but not geometrically finite. These hyperbolic 3-manifolds are not known to be topologically tame.

4. Spectral theory

In recent years a lot of work has been done on the spectrum of the Laplacian on hyperbolic manifolds. Much of this work has centered on either closed hyperbolic manifolds (see, e.g., Buser [11]) or hyperbolic surfaces (see e.g. Dodziuk-Pignataro-Randol-Sullivan [18]). We will briefly survey some of what is known about the spectrum of the Laplacian on infinite volume hyperbolic 3-manifolds. We will start at the bottom.

Let Δ denote the Laplacian acting on a complete Riemannian n-manifold M (we choose the convention $\Delta f = \text{div}(\text{grad } f)$). Let

$$\lambda_0(M) = \inf\{\text{spec}(-\Delta)\},$$

equivalently (see Cheng-Yau [16] or Sullivan [46])

$$\lambda_0(M) = -\inf\{\lambda | \exists f \in C^\infty(M) \text{ s.t. } \Delta f = \lambda f \text{ and } f > 0\}$$
$$= \inf_{f \in C_0^\infty(M)} \left(\frac{\int_M |\nabla f|^2}{\int_M f^2} \right).$$

We also recall that the Cheeger constant $h(M)$ of M is defined to be the infimum, over all compact n-submanifolds A of M, of $\text{vol}_{n-1}(\partial A)/\text{vol}(A)$. Buser [10] proved, in general, that if M is a noncompact Riemannian n-manifold whose Ricci curvature is bounded from below by $-(n-1)\kappa^2$, then $\lambda_0(M) \leq \kappa R h(M)$ where R depends only on n. Notice that if N is geometrically finite by considering the convex core we see that

$$h(N) \leq \frac{2\pi|\chi(\partial C(N))|}{\text{vol}(C(N))}$$

where $\text{vol}(C(N))$ denotes the volume of the convex core, while if N is geometrically tame and has a simply degenerate end, it follows from property $(*)$ that $h(N) = 0$. Therefore we obtain

THEOREM 4.1 [14]. *Let N be a geometrically tame hyperbolic 3-manifold which is not geometrically finite; then $\lambda_0(N) = 0$. Moreover, there exists R such that if N is any geometrically finite hyperbolic 3-manifold, then*

$$\lambda_0(N) \leq \frac{R|\chi(\partial C(N))|}{\text{vol}(C(N))}.$$

One may generalize work of Schoen [41] and Dodziuk-Randol [19], to obtain a lower bound for λ_0 which depends only on $\text{vol}(C_1(N))$, the volume of the neighborhood of radius one of the convex core $C_1(N)$.

THEOREM 4.2 [9]. *There exists $T > 0$ such that, if N is any infinite volume geometrically finite hyperbolic 3-manifold,*

$$\lambda_0(N) \geq \frac{T}{\text{vol}(C_1(N))^2}.$$

Theorems 4.1 and 4.2 tell us that in some sense λ_0 is controlled by the volume of the convex core. Moreover, they tell us that the spectrum knows whether or not a topologically tame hyperbolic 3-manifold is geometrically finite.

COROLLARY 4.3 [14]. *Let N be an infinite volume, topologically tame, hyperbolic 3-manifold. N is geometrically finite if and only $\lambda_0(N) \neq 0$.*

Lax and Phillips [27, 28] have extensively studied the entire spectrum of the Laplacian for geometrically finite hyperbolic n-manifolds. We summarize some of their more easily stated results

THEOREM 4.4 (Lax and Phillips). *Let N be an infinite volume geometrically finite hyperbolic 3-manifold. The intersection of $\text{spec}(-\Delta)$ with the interval $[0, 1)$ consists entirely of a finite number of point eigenvalues (of finite multiplicity) all lying in $(0, 1)$ and there are no point eigenvalues in $[1, \infty)$. Moreover, the spectrum is absolutely continuous and of infinite uniform multiplicity in $[1, \infty)$.*

REMARKS. (1) It is an easy consequence of Theorems 4.1 and 4.4 that $\lambda_0^{\text{ess}}(N)$, the bottom of the essential spectrum (see Donnelly [20]), is 1 if N is geometrically finite, and 0 if N is geometrically tame and not geometrically finite.

(2) C.L. Epstein [21] has done a detailed study of the spectrum for hyperbolic 3-manifolds which arise as the covers associated to the fiber subgroups of the fundamental groups of finite volume hyperbolic 3-manifolds which fiber over the circle (i.e., the examples mentioned at the end of §2).

(3) Agmon [2], Mandouvalos [29], Mazzeo-Melrose [33], and Perry [39] further studied the spectral resolution of the Laplace operator on geometrically finite hyperbolic manifolds with infinite volume and no cusps; these authors considered the analytic continuation of the Eisenstein series and the scattering operator. More recently, Froese, Hislop, and Perry [23] have studied similar questions for infinite volume hyperbolic 3-manifolds with cusps. (Mazzeo and Melrose have considered similar generalizations.)

(4) Patterson [38] obtained an analogue of Selberg's trace formula for certain geometrically finite hyperbolic manifolds without cusps. It relates spectral data (in terms of the scattering operator) to a zeta function defined in terms of "lengths" of geodesics. Patterson's formula was refined and extended to all infinite volume, geometrically finite hyperbolic manifolds without cusps by Perry [40].

(5) Mazzeo and Phillips [34] have analyzed the spectrum of the Laplacian acting on differential forms on a geometrically finite hyperbolic manifold.

5. The limit set

In [4] Ahlfors conjectured that if $N = \mathbf{H}^3/\Gamma$ has finitely generated fundamental group, then its limit set is either all of the sphere at infinity or has measure zero, and in that paper he proved his conjecture for geometrically finite hyperbolic 3-manifolds. One of the first consequences of geometric tameness is

THEOREM 5.1 [13, 47]. *If $N = \mathbf{H}^3/\Gamma$ is geometrically tame, then either L_Γ has measure zero or L_Γ is the entire sphere at infinity.*

The proof of Theorem 5.1 is based on the following minimum principle for positive superharmonic functions on analytically tame hyperbolic 3-manifolds. In particular, Ahlfors' measure conjecture holds for analytically tame hyperbolic 3-manifolds.

THEOREM 5.2 [13, 47]. *If N is an analytically tame hyperbolic 3-manifold, then for every nonconstant positive superharmonic (i.e., $\Delta h \leq 0$) function h on N,*

$$\inf_{C(N)} h = \inf_{\partial C(N)} h$$

where $C(N)$ denotes the convex core of N. In particular, if $C(N) = N$ (i.e. $L_\Gamma = S^2$) then there are no positive nonconstant superharmonic functions on N.

Another consequence of this same minimum principle and work of Sullivan [43] is

THEOREM 5.3 [13, 47]. *Let N be a geometrically tame hyperbolic 3-manifold. Then the geodesic flow of N is ergodic if and only if L_Γ is the entire sphere at infinity.*

The work of Patterson [37] and Sullivan [45] establishes a beautiful connection between the spectral theory of geometrically finite hyperbolic 3-manifolds and the measure-theoretic properties of their limit sets. This work culminates in the following theorem of Sullivan:

THEOREM 5.4 [45]. *If $N = \mathbf{H}^3/\Gamma$ is a geometrically finite hyperbolic 3-manifold and D denotes the Hausdorff dimension of L_Γ, then $\lambda_0(N) = 1$ if $D \leq 1$ and $\lambda_0(N) = D(2 - D)$ otherwise.*

Theorems 4.1, 4.2 and 5.4 make somewhat explicit the intuitive relationship between size (volume) of the convex core and fuzziness (Hausdorff dimension) of the limit set. They also suggest that if N is geometrically tame, but not geometrically finite, that its limit set has Hausdorff dimension 2. In fact,

THEOREM 5.5 [14]. *If* $N = \mathbf{H}^3/\Gamma$ *is topologically tame but not geometrically finite and there is a lower bound for the injectivity radius of* N *, then* L_Γ *has Hausdorff dimension* 2 .

The key element in the proof is establishing that the harmonic function on N associated to the Patterson-Sullivan measure on L_Γ has subexponential growth.

REMARKS. (1) Sullivan [44] proved Theorem 5.5 for "hyperbolic half-cylinders".

(2) Theorem 5.5 also holds for analytically tame hyperbolic 3-manifolds where there is a uniform upper bound on the diameter of each component of $\partial N_{\text{thin}(\varepsilon)} \cap C(N)$, where $\partial N_{\text{thin}(\varepsilon)}$ denotes the boundary of the ε-thin portion of N (see [14]).

(3) See Abikoff [1] for a discussion of more topological questions about the limit set.

REFERENCES

1. W. Abikoff, *Kleinian groups—geometrically finite and geometrically perverse*, Geometry of Group Representations, Contemp. Math., no. 74, Amer. Math. Soc., Providence, RI, 1988, pp. 1–50.

2. S. Agmon, *On the spectral theory of the Laplacian on non-compact hyperbolic manifolds*, Journées "Équations aux derivées partielles (Saint Jean de Monts, 1987), Exposée No. XVII, École Polytechnique, Palaiseau, 1987.

3. L. V. Ahlfors, *Finitely generated Kleinian groups*, Amer. J. of Math. **86** (1964), 413–429.

4. _____, *Fundamental polyhedrons and limit sets of Kleinian groups*, Proc. Nat. Acad. Sci. U.S.A. **55** (1966), 251–254.

5. L. Bers, *On boundaries of Teichmüller spaces and on Kleinian groups*. I, Ann. of Math. (2) **91** (1970), pp. 570–600.

6. F. Bonahon, *Bouts des variétés hyperboliques de dimension 3*, Ann. of Math. (2) **124** (1986), 71–158.

7. F. Bonahon and J. P. Otal, *Variétés hyperboliques à géodésiques arbitrairement courtes*, Bull. London Math. Soc. **20** (1988), 255–261.

8. B. Bowditch, *Geometrical finiteness for hyperbolic groups*, Preprint.

9. M. Burger and R. D. Canary, *A lower bound for* λ_0 *on hyperbolic n-manifolds*, preprint.

10. P. Buser, *A note on the isoperimetric constant*, Ann. Sci. École Norm. Sup. (4) **15** (1982), 213–230.

11. P. Buser, *On Cheeger's inequality* $\lambda_1 \geq h^2/4$, Geometry of the Laplace Operator, Amer. Math. Soc., Providence, RI, 1980, pp. 29–77.

12. R. D. Canary, D. B. A. Epstein, and P. Green, *Notes on notes of Thurston*, Analytical and Geometrical Aspects of Hyperbolic Spaces, Cambridge Univ. Press, 1987, pp. 3–92.

13. R. D. Canary, *Ends of hyperbolic 3-manifolds*, J. Amer. Math. Soc. (to appear).

14. _____, *The Laplacian and the geometry of hyperbolic 3-manifolds*, J. Differential Geom. (to appear).

15. _____, *A covering theorem for hyperbolic 3-manifolds and its applications*, preprint.

16. S. Y. Cheng and S. T. Yau, *Differential equations on Riemannian manifolds and their geometric applications*, Comm. Pure Appl. Math. **28** (1975), 333–354.

17. M. Culler and P. Shalen, *Paradoxical decompositions, 2-generator Kleinian groups, and volumes of hyperbolic 3-manifolds*, J. Amer. Math. Soc. **5** (1992), 231–288.

18. J. Dodziuk, T. Pignataro, B. Randol, and D. Sullivan, *Estimating small eigenvalues of Riemann surfaces*, The Legacy of Sonya Kovalevskaya, Contemp. Math., no. 64, Amer. Math. Soc., Providence, RI, 1987, pp. 93–121.

19. J. Dodziuk and B. Randol, *Lower bounds for* λ_1 *on a finite-volume hyperbolic manifold*, J. Differential Geom. **24** (1986), 133–139.

20. H. Donnelly, *On the essential spectrum of a complete Riemannian manifold*, Topology **20** (1981), 1–14.

21. C. L. Epstein, *The spectral theory of geometrically periodic hyperbolic 3-manifolds*, Memoirs Amer. Math. Soc. No. **335** (1985).

22. D. B. A. Epstein and A. Marden, *Convex hulls in hyperbolic spaces, a theorem of Sullivan, and measured pleated surfaces*, Analytical and Geometrical Aspects of Hyperbolic Spaces, Cambridge Univ. Press, 1987, pp. 113–253.

23. R. Froese, P. Hislop, and P. Perry, *The Laplace operator on hyperbolic three manifolds with cusps of non-maximal rank*, Invent.Math. **106** (1991), 295–333.

24. L. Greenberg, *Fundamental polyhedra for Kleinian groups*, Ann. of Math. (2) **84** (1966), 433–441.

25. T. Jorgenson, *Compact 3-manifolds of constant negative curvature fibering over the circle*, Ann. of Math. **98** (1977), 61–72.

26. M. Gromov and W. Thurston, *Pinching constants for hyperbolic manifolds*, Invent. Math. **89** (1987), 1–12.

27. P. Lax and R. S. Phillips, *The asymptotic disribution of lattice points in Euclidean and non-Euclidean spaces*, J. Funct. Anal. **46** (1982), 280–350.

28. _____, *Translation representations for automorphic solutions of the wave equation in non-Euclidean spaces*. I, II, III, Comm. Pure Appl. Math. **37** (1984), 303–324, 779–813; **38** (1985), 197–208.

29. N. Mandouvalos, *Spectral theory and Eisenstein series for Kleinian groups*, Proc. London Math. Soc. **57** (1988), 209–238.

30. A. Marden, *The geometry of finitely generated Kleinian groups*, Ann. of Math. (2) **99** (1974), 383–462.

31. B. Maskit, *On boundaries of Teichmüller spaces and on Kleinian groups*. II, Ann. of Math. (2) **91** (1970), 607–639.

32. _____, *Kleinian groups*, Springer-Verlag, 1988.

33. R. Mazzeo and R. Melrose, *Meromorphic extension of the resolvent on complete spaces with asymptotically constant negative curvature*, J. Funct. Anal. **75** (1987), 260–310.

34. R. Mazzeo and R.S. Phillips, *Hodge theory on hyperbolic manifolds*, Duke Math. J. **60** (1990), 509–559.

35. Y. Minsky, *Harmonic maps into hyperbolic 3-manifolds*, Trans. Amer. Math. Soc. **332** (1991), 607–632.

36. J. W. Morgan, *On Thurston's uniformization theorem for three-dimensional manifolds*, The Smith Conjecture, (J. Morgan and H. Bass, eds.), Academic Press, 1984, pp. 37–125.

37. S. J. Patterson, *Lectures on limit sets of Kleinian groups*, Analytical and Geometrical Aspects of Hyperbolic Spaces, Cambridge Unive. Press, 1987, pp. 281–323.

38. S. J. Patterson, *The Selberg zeta-function of a Kleinian group*, Number Theory, Trace Formulas, and Discrete Groups: Symposium in honor of Atle Selberg, Oslo, Norway, July 14–21, 1987, Academic Press, 1989, pp. 409–442.

39. P. Perry, *The Laplace operator on a hyperbolic manifold*. II. *Eisenstein series and the scattering matrix*, J. Reine Angew. Math. **398** (1989), 67–91.

40. _____, *The Selberg zeta function and a local trace formula for Kleinian groups*, J. Reine Angew. Math. **410** (1990), 116–152.

41. R. Schoen, *A lower bound for the first eigenvalue of a negatively curved manifold*, J. Differential Geom. **17** (1982), 233–238.

42. P. Scott, *Compact submanifolds of 3-manifolds*, J. London Math. Soc. **7** (1973), 246–250.

43. D. Sullivan, *The density at infinity of a discrete group of hyperbolic motions*, Inst. Hautes Études Sci. Publ. Math. **50** (1979), 419–450.

44. D. Sullivan, *Growth of positive harmonic functions and Kleinian group limit sets of zero planar measure and Hausdorff dimension* 2 , Geometry Symposium Utrecht 1980, Lecture Notes in Math., vol. 894, Springer-Verlag, 1981, pp. 127–144.

45. _____, *Entropy, Hausdorff measures old and new, and limit sets of geometrically finite Kleinian groups*, Acta Math. **153** (1984), 259–277.

46. _____, *Aspects of positivity in Riemannian geometry*, J. Differential Geom. **25** (1987), 327–351.

47. W. Thurston, *The geometry and topology of 3-manifolds*, lecture notes.

48. _____, *Hyperbolic Structures on 3-manifolds*. II. *Surface Groups and 3-manifolds which fiber over the circle*, Preprint.

STANFORD UNIVERSITY
Current address: University of Michigan

Proceedings of Symposia in Pure Mathematics
Volume 54 (1993), Part 3

Isoperimetric Constants and Large Time Heat Diffusion in Riemannian Manifolds

ISAAC CHAVEL AND EDGAR A. FELDMAN

We summarize, here, recent work to be published in [3] and [4], the background of which is as follows:

Let M be a noncompact Riemannian manifold of dimension $n \geq 2$, with associated Laplace-Beltrami operator Δ acting on functions on M, and with attendant minimal positive heat kernel $p(x, y, t)$, where $x, y \in M$, and $t > 0$. Thus if $M = \mathbb{R}^n$, then p is given by the classical Gauss kernel

$$p(x, y, t) = \mathbf{e}_n(x, y, t) =: (4\pi t)^{-n/2} e^{-|x-y|^2/4t}.$$

Our interest, for general M, is in those aspects of the geometry of M related to the inequalities of the type

$$(1) \qquad p(x, y, t) \leq \text{const}_\nu \, t^{-\nu/2}, \qquad \nu > 0,$$

for large $t > 0$.

One of the powerful geometric tools used to obtain such estimates is the apparatus of isoperimetric constants.

DEFINITION. To each $\nu > 1$ and open submanifold Ω with compact closure and smooth boundary, associate the *ν-isoperimetric quotient of Ω*, $\mathfrak{I}_\nu(\Omega)$, defined by

$$\mathfrak{I}_\nu(\Omega) = \frac{A(\partial\Omega)}{V(\Omega)^{(\nu-1)/\nu}},$$

where A denotes $(n-1)$-dimensional Riemannian measure, and V denotes n-dimensional Riemannian measure.

The *ν-isoperimetric constant of M*, $I_\nu(M)$, is defined to be the infimum of $\mathfrak{I}_\nu(\Omega)$ over all Ω described above.

The basic result which is the backdrop of our work is

1991 *Mathematics Subject Classification*. Primary 53C20, 58G11, 58G32; Secondary 31C15.
Supported in part by National Science Foundation Grant DMS 8704325 and PSC–CUNY FRAP awards.
The final form of this paper will be submitted for publication elsewhere.

THEOREM 1. *If* $I_\nu(M) > 0$, $\nu \geq 2$, *then* (1) *is valid of all of* $M \times M \times (0, +\infty)$.

Moreover, given any $\delta > 0$, *there exists* $\text{const}_{\nu, \delta} > 0$ *for which*

$$(2) \qquad p(x, y, t) \leq \text{const}_{\nu, \delta} \, t^{-\nu/2} e^{-d^2(x, y)/4(1+\delta)t}$$

for $(x, y, t) \in M \times M \times (0, +\infty)$. *In such a case we have, for* $\nu > 2$, *a minimal positive Green's function* $G(x, y)$ *satisfying*

$$(3) \qquad G(x, y) \leq \text{const}_\nu \, d^{2-\nu}(x, y)$$

for all $x \neq y$.

We note that from Theorem 1 and the work of [19] one may immediately derive

COROLLARY 1. *The 2-dimensional infinite "jungle gym" in* $\mathbb{R}^3$ *satisfies* (2) *for* $\nu = 3$ *for all* $t > 0$.

In what follows we first give a semilocalization of the heat diffusion in ends of the Riemannian manifold, wherein we seek results when only the geometry of the end is known, and the geometry of the rest of the manifold is essentially unknown. (See [3] for details.) Then we note that Theorem 1 cannot be exhaustive since I_ν must be 0 whenever $\nu < n$, and results of decay with $\nu < n$ should certainly exist for very natural situations. We announce a solution to this problem in §2, wherein we also solve a problem raised in [30]. (See [4] for details.)

1. Heat diffusion in ends

Fix Ω, a relatively compact domain in M with smooth boundary, and E a connected component of $M \backslash \overline{\Omega}$, $\overline{E}$ noncompact, and let

$$\Gamma = \partial \Omega \cap \overline{E}, \qquad E_R = \{y \in E : d(y, \Gamma) > R\}.$$

We assume that the exponential map of the normal bundle of Γ is defined on all elements in the normal bundle pointing into E. We refer to the minimal positive harmonic function Ψ on E, satisfying

$$\Psi | \Gamma = 1,$$

as the *elliptic capacitory potential of* E, and to the minimal positive solution ψ to the heat equation on E, satisfying

$$\psi | \Gamma \times (0, +\infty) = 1, \qquad \psi | E \times \{0\} = 0,$$

as the *parabolic capacitory potential of* E. We then have

THEOREM 2. *If* $I_\nu(E) > 0$ *for some given* $\nu > 2$, *then the elliptic capacitory potential* Ψ *of* E *satisfies*

$$(4) \qquad \Psi(y) \leq \text{const} \, d^{2-\nu}(y, \Gamma)$$

for all y in E_R sufficiently bounded away from Γ.

Therefore, for any $x \in M$ there exists a $\text{const}_x > 0$ for which

$$(5) \qquad G(x, y) \leq \text{const}_x d^{2-\nu}(y, \Gamma)$$

for all y in E sufficiently bounded away from Γ.

REMARK 1. The maximum principle and (4) imply that if $I_\nu(E) > 0$, $\nu > 2$, then Ψ is the unique solution to the Dirichlet problem on E with boundary data: $\Psi|\Gamma = 1$, $\Psi(\infty_E) = 0$, where ∞_E is the point at infinity of E.

THEOREM 3. *Suppose that $I_\nu(E) > 0$ for some given $\nu > 2$. Then, given any $R > 0$, there exist positive constants such that the parabolic capacitory potential $\psi(y, t)$ satisfies*

$$(6) \qquad \psi(y, t) \leq \text{const}_R d^{2-\nu}(y, \Gamma) e^{-d^2(y, \Gamma)/\text{const } t}$$

for all $y \in E_R$ sufficiently far from Γ, and all $t > 0$.

Also, we have the existence of positive constants such that

$$(7) \quad p(x, y, t) \leq \text{const}_R \{o(t^{-1}) d^{2-\nu}(y, \Gamma) + t^{-\nu/2} \ln t\} e^{-d^2(y, \Gamma)/\text{const } t}$$

for all $x \in M \backslash \overline{E_R}$, $y \in E$ sufficiently far from Γ, and for all $t > 2$.

REMARK 2. Of course,

$$t^{-\nu/2} \ln t = o(t^{-1}).$$

See [5].

We now consider the probabilistic aspects of Theorems 2 and 3. Consider Brownian motion associated to the heat kernel $p(x, y, t)$, and denote the random Brownian path by X_t, $t \geq 0$. For each $x \in M$, we denote the associated probability measure on the path space, concentrated on those paths starting at x, by $\mathscr{P}_x$. So for any Borel set B in M, we have

$$P_x(X_t \in B) = \int_B p(x, y, t) \, dV(y).$$

If B is a countable union of compacta in M, then we define $T_B(X)$, the *first hitting time of B by the Brownian path X*, by

$$T_B(X) = \inf\{t > 0 \colon X_t \in B\}.$$

Let Ω, E, and Γ be as in Theorems 2 and 3. Then (see [18]) the parabolic and elliptic capacitory potentials of E have the probabilistic interpretation:

$$\psi(y, t) = \mathscr{P}_y(T_\Gamma \leq t), \qquad \Psi(y) = \mathscr{P}_y(T_\Gamma < +\infty),$$

for $y \in E$.

If ζ denotes the explosion time of X_t, and (as above) ∞_E denotes the point at infinity of E, then to E we associate the *escape function of E*, esc, defined on M by

$$\text{esc } x = \mathscr{P}_x(X_\zeta = \infty_E).$$

Then esc is nonnegative bounded harmonic—and we say that E is a *transient component of* $M \backslash \overline{\Omega}$ if esc does not vanish identically on all of M. If esc does not vanish identically on all of M, then it is always positive, and cannot be a constant function unless that constant is equal to 1. If $M \backslash \overline{\Omega}$ has transient components $\{E_1, \dots\}$, then the associated probability escape functions $\{\mathrm{esc}_{E_1}, \dots\}$ are linearly independent.

Of course,

$$\mathrm{esc} \geq 1 - \Psi$$

on all of E. We immediately conclude

THEOREM 4. *If* $I_\nu(E) > 0$, *for some given* $\nu > 2$, *then*

$$(8) \qquad\qquad 1 - \mathrm{const}\, d^{2-\nu}(y, \Gamma) \leq \mathrm{esc}\, y \leq 1$$

for $y \in E$ *sufficiently far from* Γ.

For $\nu = \infty$, we may consider $I_\infty(M)$ to be the Cheeger constant of M. Then Theorems 2–4 above have corresponding formulations. See [3].

2. Heat diffusion and modified isoperimetric constants

DEFINITION. We say that the Riemannian manifold M has *bounded geometry* if the Ricci curvature of M is bounded uniformly from below, and if the injectivity radius of M is bounded uniformly away from 0 on all of M.

We note that the application of Theorem 1 is restricted by the fact that the inequality $I_\nu > 0$ itself is only possible for $\nu \geq n$. Let $\nu < n$, and consider small metric disks $B(x; \varepsilon)$, with center $x \in M$ and radius $\varepsilon > 0$. Then for the isoperimetric quotient of $B(x; \varepsilon)$ we have

$$\mathfrak{I}_\nu(B(x; \varepsilon)) \sim \mathrm{const}\, \varepsilon^{n-1-n(1-1/\nu)} = \mathrm{const}\, \varepsilon^{n/\nu - 1}$$

as $\varepsilon \downarrow 0$; so $I_\nu(M) = 0$ whenever $\nu < n$. Nevertheless, one expects results of the above type to be valid. First, for large t one does not expect local considerations to determine ν in (1). Second, one has a simple example to illustrate the phenomena we have in mind. Consider the Riemannian product $M = M_0 \times \mathbb{R}^k$, where M_0 is an $(n - k)$-dimensional compact Riemannian manifold. Then $I_k(M) = 0$. On the other hand, let p_0 denote the heat kernel of M_0; then

$$p((x_1, y_1), (x_2, y_2), t) = p_0(x_1, x_2, t)\mathbf{e}_k(y_1, y_2, t).$$

Since $p_0 \to 1/V(M_0)$ uniformly on M_0 as $t \uparrow +\infty$, we conclude that (1) is valid for $\nu = k$ and large t, and not valid for $\nu > k$ and large t. Naturally, (1) is not valid for $\nu = k$ and small t.

In order to avoid the difficulty posed by strictly local phenomena, one introduces a modified isoperimetric constant $I_{\nu, \rho}(M)$ in place of the usual isoperimetric constant $I_\nu(M)$.

DEFINITION. For each $\nu > 1$, $\rho > 0$, the *modified ν-isoperimetric constant* of M, $I_{\nu,\rho}(M)$, is defined to be the infimum of $\mathfrak{I}_{\nu}(\Omega)$ where Ω is not also required to contain a closed geodesic disk of radius ρ.

Note that in our example above, $M = M_0 \times \mathbb{R}^k$, where M_0 is compact, we do have (by Propositon 2 below) $I_{\nu,\rho}(M) > 0$ if and only if $\nu \in [1, k]$.

We note the following facts about modified isoperimetric constants:

A. It follows fom the work of [19] that, when M has bounded geometry, the nonvanishing of $I_{\nu,\rho}(M)$ is independent of the choice of $\rho > 0$. (Of course, with no hypotheses whatsoever, if $I_{\nu,\rho}(M) > 0$, then $I_{\nu,\rho_0}(M) > 0$ for all $\rho_0 > \rho$.)

B. Let $V(x; r)$ denote the volume of the metric disk $B(x; r)$ in M centered at x, with radius r. If M is Riemannian complete and $I_{\nu,\rho}(M) > 0$, for some $\nu > 1$, $\rho > 0$, then

$$(9) \qquad \liminf_{r \to +\infty} r^{-\nu} V(x; r) > 0$$

for all $x \in M$.

C. If $I_{\nu,\rho}(M) > 0$ for some $\nu > 1$, $\rho > 0$, then for any $\varepsilon > 0$, $V(x; \rho + \varepsilon)$ is uniformly bounded from below for all x with $d(x, \partial M) > \rho + \varepsilon$. Therefore $I_{\mu, \rho + \varepsilon}(M) > 0$ for all μ in $(1, \nu)$.

D. Let Ω be a relatively compact domain in M with smooth boundary Γ, Ω' an n-dimensional Riemannian manifold with compact closure and smooth boundary Γ, such that M', given by

$$M' = \{M \backslash \Omega\} \cup \Omega',$$

is smooth Riemannian. Then, if $I_{\nu,\rho}(M) > 0$, one has the existence of $\rho' \geq \rho$ for which $I_{\nu,\rho'}(M') > 0$.

THEOREM 5. *Assume that M is Riemannian complete and that, for some given $\nu > 2$, there exists $\rho > 0$ such that $I_{\nu,\rho}(M) > 0$; then M possesses a Green's function $G(x, y)$ (equivalently, it has transient Brownian motion). Furthermore,*

$$(10) \qquad V(\{y: G(x, y) > \sigma\}) \leq \mathrm{const}_x \, \sigma^{-\nu/(\nu-2)}$$

for each $x \in M$.

Furthermore, let Ω, E, and Γ be as in Theorems 2 and 3. Assume that for some given $\nu > 2$, there exists $\rho > 0$ such that $I_{\nu,\rho}(E) > 0$; then M possesses a Green's function $G(x, y)$, and

$$(11) \qquad V(\{y \in E: G(x, y) > \sigma\}) \leq \mathrm{const}_x \, \sigma^{-\nu/(\nu-2)}.$$

for each $x \in M$.

THEOREM 6. *Let M be complete Riemannian of dimension $n \geq 2$, with bounded geometry. If for any $\nu > 1$ and $\rho > 0$ we have $I_{\nu,\rho}(M) > 0$, then*

(2) *is valid for sufficiently large* $t > 0$ *on all of* $M \times M$, *and* (3) *is valid for large* $d(x, y) > 0$.

We note that the result is valid for all $\nu > 1$—as it should be—in contrast to the restriction to $\nu \geq 2$ in Theorem 1. Of course, it might very well be the case that given bounded geometry and nonintegral $1 < \nu < 2$, examples for which the isoperimetric constant does not vanish, and for which the heat kernel satisfies (1) for large time, are very hard to come by. However, if one expands the theory to weighted Riemannian spaces, then examples of this type are easily manufactured.

It is important to note that one may define the modified isoperimetric constant for $\nu = 1$; namely, given $\rho > 0$ we define

$$I_{1,\rho}(M) = \inf_{\Omega} A(\partial \Omega),$$

where Ω ranges over relatively compact domains in M with smooth boundary and inradius greater than ρ. We then have

LEMMA 1. *If* M *is Riemannian complete with bounded geometry, then* $I_{1,\rho}(M) > 0$ *for every* $\rho > 0$.

But the arguments for Theorem 6 will apply to the hypothesis $I_{1,\rho} > 0$, that is, to $\nu = 1$. We will automatically have

THEOREM 7. *Let* M *be a complete noncompact Riemannian manifold with bounded geometry. Then*

$$(12) \qquad\qquad p(x, y, t) \leq \mathrm{const}_T \, t^{-1/2}$$

on all of $M \times M \times [T, +\infty)$.

REMARK 3. This gives the best rate of decay possible, for such generality. The first such result was proved by Varopoulos, in [30], with exponent $-1/2 + \varepsilon$ for any $\varepsilon > 0$. Our result confirms Varopoulos's conjecture, there, that one can sharpen the result there to $\varepsilon = 0$.

The intuition with which to approach Theorems 1 and 6 is greatly enhanced by considering the corresponding result for infinite connected graphs, namely, networks in which any pair of vertices is connected by at most one edge.

We are given a countable set $\mathbf{G}_0$, such that to each $\xi \in \mathbf{G}_0$ we have a finite nonempty subset $\mathsf{N}(\xi) \subseteq \mathbf{G}_0 \backslash \{\xi\}$, of cardinality $m(\xi)$, each element of which is referred to as a *neighbor of* ξ. Furthermore, we require that $\eta \in \mathsf{N}(\xi)$ if and only if $\xi \in \mathsf{N}(\eta)$. Then one determines a graph structure $\mathbf{G}$ by postulating the existence of precisely one oriented edge from any ξ to each of its neighbors, that is, the elements of $\mathsf{N}(\xi)$. We refer to $m(\xi)$ as the *valence of* $\mathbf{G}$ *at* ξ. A sequence of points $(\xi_0, \ldots, \xi_k)$ is a *combinatorial path of length* k if $\xi_j \in \mathsf{N}(\xi_{j-1})$ for all $j = 1, \ldots, k$. For any two vertices ξ and η in the graph $\mathbf{G}$, one defines their distance $\mathrm{d}(\xi, \eta)$ to be the minimum of the length of all combinatorial paths connecting ξ to η.

Any finite subset $\mathbf{K}_0$ in $\mathbf{G}_0$ determines a finite subgraph $\mathbf{K}$ of $\mathbf{G}$, for which one can describe a variety of suitable definitions for its boundary. We prefer a slight variant to the Dodziuk-Kendall regime [12], namely, the *boundary of* $\mathbf{K}$, $\partial\mathbf{K}$, will be the subset of $\mathbf{G}_1$ consisting of those oriented edges which connect points of $\mathbf{K}$ to the complement of $\mathbf{K}$ in $\mathbf{G}_0$.

On the collection of vertices $\mathbf{G}_0$ we let $d\iota$ denote the counting measure, and define the *volume measure on* $\mathbf{G}_0$, $d\mathsf{V}$, by

$$d\mathsf{V}(\xi) = m(\xi)\, d\iota(\xi).$$

We define the *area measure* $d\mathsf{A}$ *on* $\mathbf{G}_1$ to be equal to the counting measure for the oriented edges. Thus, for any finite subset of vertices, the area of its boundary will be equal to the number of unoriented edges in the boundary. Furthermore, all L^p-spaces of functions on $\mathbf{G}_0$ and $\mathbf{G}_1$, respectively, will be given relative to the measures $d\mathsf{V}$ and $d\mathsf{A}$, respectively.

One now easily defines, for each $\nu \geq 1$, the isoperimetric constant $\mathsf{I}_\nu(\mathbf{G})$, of $\mathbf{G}$. Then one can define the Sobolev constant $\mathsf{S}_\nu(\mathbf{G})$, for which the Federer-Fleming theorem (see [15]; also p. 91 of [2]) remains valid (in the discrete setting) in the sense that $\mathsf{I}_\nu(\mathbf{G}) > 0$ if and only if $\mathsf{S}_\nu(\mathbf{G}) > 0$. Thus, if $\mathsf{I}_\nu(\mathbf{G}) > 0$ then one has the Nash-Sobolev inequality (see (0.3) of [1]) for all functions of finite support on $\mathbf{G}_0$. (When $\nu = 1$ one always has $\mathsf{I}_1(\mathbf{G}) > 0$ and $\mathsf{S}_1(\mathbf{G}) > 0$.) Similarly, one has the combinatorial Laplacian $\Delta_\mathbf{G}$ acting on functions on $\mathbf{G}_0$, defined by

$$(\Delta_\mathbf{G}f)(\xi) = \frac{2}{m(\xi)} \sum_{\eta \in N(\xi)} \{f(\eta) - f(\xi)\}.$$

One easily sees that Δ_G, acting on $L^2(\mathbf{G}_0)$, is bounded nonpositive selfadjoint with minimal positive heat semigroup P_t and symmetric heat kernel

$$\mathsf{p}(\xi, \eta, t) = (\mathsf{P}_t\delta_\eta)(\xi)$$

(where δ denotes the delta function on $\mathbf{G}_0$) for the heat equation on $\mathbf{G}_0$

$$\frac{1}{2}\Delta_\mathbf{G} = \frac{\partial}{\partial t}$$

associated to the Laplacian $\Delta_\mathbf{G}$—with *continuous* time.

THEOREM 8. *If, for some given* $\nu \geq 2$, *we have* $\mathsf{I}_\nu(\mathbf{G}) > 0$, *then we have*

(13) $$\mathsf{p}(\xi, \eta, t) \leq \mathrm{const}\, t^{-\nu/2}$$

for all vertices ξ, η *and* $t > 0$.

For the off-diagonal large time estimate we have: Given any $\delta > 0$, *there exists a positive constant* $k = k(\delta)$ *(depending only on* δ*), for which* $k(\delta) \to 0$ *as* $\delta \downarrow 0$, *such that*

(14) $$\frac{\mathsf{d}(\xi, \eta)}{t} \leq k(\delta)$$

implies

$$(15) \qquad p(\xi, \eta, t) \leq \text{const}_{\nu, \delta} \, t^{-\nu/2} \exp - \frac{d^2(\xi, \eta)}{2(1+\delta)t},$$

where $\text{const}_{\nu, \delta} \to +\infty$ *as* $\delta \downarrow 0$. *(See* [1].*)*

For the off-diagonal short time estimate we have: Given any $\delta > 0$, then

$$(16) \qquad \frac{d(\xi, \eta)}{t} \geq 2e(1+\delta)$$

implies

$$(17) \qquad p(\xi, \eta, t) \leq \text{const}_{\nu, \delta} \, t^{-\nu/2} \exp - \frac{d(\xi, \eta)}{2} \left\{ \ln \frac{d(\xi, \eta)}{(1+\delta)t} - 2 \right\},$$

where $\text{const}_{\nu, \delta} \to +\infty$ *as* $\delta \downarrow 0$.

In particular, when $I_\nu(\mathbf{G}) > 0$, $\nu > 2$, we also have the existence of the minimal positive Green's function $G(\xi, \eta)$, satisfying

$$(18) \qquad G(\xi, \eta) \leq \text{const}_\nu \, d^{2-\nu}(\xi, \eta)$$

for large $d(\xi, \eta)$.

REMARK 4. As mentioned in [1], one can only expect a Gaussian correction to $t^{-\nu/2}$ when time is large with respect to distance, for only then will the central-limit effects come into play. (Just think of when time is discrete.) The short time estimate (17) is of the right order (except for the 1/2) as can be verified for heat diffusion on $\mathbb{Z}$.

The relation of the continuous to the discrete has been developed in [19] as follows: Given a complete Riemannian manifold M, a subset G in M, and $\varepsilon > 0$, we say that G is *ε-separated* if for any x, y in G we have $d(x, y) > \varepsilon$, where (as mentioned earlier) d denotes the distance metric on M induced by the given Riemannian metric. For any $x \in M$ and $r > 0$ we denote (also, as mentioned earlier) the metric disk in M, centered at x with radius r, by $B(x; r)$.

A *discretization of M* is a graph $\mathbf{G}$ with $\mathbf{G}_0 \subseteq M$, for which there exist $0 < \varepsilon \leq R$ such that

(i) $\mathbf{G}_0$ is ε-separated,

(ii) $\bigcup_{\xi \in \mathbf{G}_0} B(\xi; R) = M$,

(iii) for each $x \in M$, the number of elements of $\mathbf{G}_0$ in $B(x; 2R)$ is finite, and

(iv) to each $\xi \in \mathbf{G}_0$, the neighbors of ξ in $\mathbf{G}_0$ are given by

$$\mathbf{N}(\xi) = \{B(\xi; 2R) \cap \mathbf{G}_0\} \setminus \{\xi\}.$$

PROPOSITION 1. *If the Ricci curvature of M is bounded from below, then $m(\xi)$ is uniformly bounded from above for any discretization $\mathbf{G}$ of M.*

DEFINITION. We say that the graph $\mathbf{G}$ has bounded geometry if $m(\xi)$ is bounded uniformly from above on all of $\mathbf{G}_0$.

PROPOSITION 2. *If M is Riemannian complete with bounded geometry, then, for $\nu \geq n$, one has $I_\nu(M) > 0$ only if $I_\nu(\mathbf{G}) > 0$, for any discretization $\mathbf{G}$ of M.*

For the discretization of Riemannian manifolds, as above, one can easily adjust Kanai's arguments (for Proposition 2, [19]) to obtain

PROPOSITION 3. *If M is Riemannian complete with bounded geometry, then for $\nu > 1$ one has $I_{\nu,\rho}(M) > 0$ if and only if $I_\nu(\mathbf{G}) > 0$, for any discretization $\mathbf{G}$ of M.*

Note that in the discretization of M the local phenomena disappear, and no local difficulties occur. Consider the example we discussed earlier, $M_0 \times \mathbb{R}^k$, where M_0 is a compact Riemannian manifold. Pick one point $x_0 \in M_0$ and consider the integer lattice $\mathbb{Z}^k$ in $\mathbb{R}^k$. Then

$$\mathbf{G}_0 =: \{x_0\} \times \mathbb{Z}^k$$

is a discretization of M, and $I_k(\mathbf{G}) > 0$. So the respective asymptotics of the continuous and discrete heat kernels coincide.

Somehow, one expects here that the discrete phenomena point to the heart of the continuous phenomena. After all, since we are interested exclusively in large time asymptotics, the heat flow on the Riemannian manifold M should depend only on the coarse macroscopic geometry of M—once we assume some version of local uniformity of M.

3. Remarks

We note that the Li-Yau Harnack inequalities [24] provide efficient estimates on heat diffusion for long time only when the Ricci curvature is nonnegative; see [24, 22, 23].

See [27, 6, 8, 14] for details of Theorem 1. Also, see [32] and the recent [20, 1]. Finally, see the most recent [10].

For the equivalence of (1) for all $t > 0$ with the validity of a log-Sobolev inequality for all C^∞ functions of compact support, see [9]. For the equivalence of (1) for all $t > 0$ with the validity of a Sobolev inequality valid for all C^∞ functions of compact support—thereby reducing the question to the study of geometric conditions for which the appropriate Sobolev inequalities are valid—see [32, 1].

To our knowledge, escape functions were first considered in [21]. They have since been discussed (without being identified as such) in [13, 23] for Riemannian manifolds with nonnegative curvature, except for at most a compact subset. Early remarks in a more abstract setting can be found in [29].

Our Theorem 5 is contained in the arguments of [17], albeit without the explicit statement that one works with a modified isoperimetric constant, and without the refinement for ends. We also refer to the recent [25], in which the author also considers a modified isoperimetric constant (which he refers to as a *truncated isoperimetric constant*), for which he obtains the existence of the Green's function when the ν-truncated isoperimetric constant does not vanish for $\nu > 2$.

The result (13) was first proved in [31] when $\nu > 2$, using a discrete version of Moser iteration.

The existence of the Green's function for the discrete case was first proven in [11, 12] for the discrete Cheeger constant $I_\infty(G)$. The estimate corresponding to (13) for nonvanishing Cheeger constant can be found in [16]. The estimate (18), for nonvanishing Cheeger constant, can be found in [28].

N. Th. Varopoulos recently sent us a reprint of his [33], in which he shows that if M covers a compact manifold, and G is the discretization of M given by an orbit of the deck transformation group of the covering, then for any $\nu > 0$, we have (1) for large t if and only if (13) if valid for large t (actually, in [33] the time for the diffusion on the lattice is discrete). It is not yet clear whether such a strong theorem is true in our more general case—although one certainly expects it to be true. T. Coulhon has suggested (private communication) the possibility that his (more explicitly analytic semigroup theoretic) methods [7] might be employed successfully in our more general case.

We add here that our approach is very close to the ideas in [33].

Finally, we refer to [26] where the discretization procedure and isoperimetric constants are used to derive the hyperbolicity of Scherk's minimal surface.

REFERENCES

1. E. A. Carlen, S. Kusuoka, and D. W. Stroock, *Upper bounds for symmetric Markov transition densities*, Ann. Inst. H. Poincaré Probab. Statist. **23** (1987), 245–287.

2. I. Chavel, *Eigenvalues in Riemannian geometry*, Academic Press, 1984.

3. I. Chavel and E. A. Feldman, *Isoperimetric constants, the geometry of ends, and large time heat diffusion in Riemannian manifolds*, Proc. London Math. Soc. (3) **62** (1991), 427–448.

4. ____, *Modified isoperimetric constants, and large time heat diffusion in Riemannian manifolds*, Duke Math. J. **64** (1991), 473–499.

5. I. Chavel, E. A. Feldman, and L. Karp, *Large time decay of the heat kernel of λ-transient Riemannian manifolds*, Preprint.

6. S. Y. Cheng and P. Li, *Heat kernel estimates and lower bounds of eigenvalues*, Comment. Math. Helv. **56** (1981), 327–338.

7. T. Coulhon, *Dimension à l'infini d'un semigroup analytique*, Bull. Soc. Math. France **114** (1990).

8. E. B. Davies, *Explicit constants for Gaussian upper bounds on heat kernels*, Amer. J. Math. **109** (1987), 319–334.

9. E. B. Davies and B. Simon, *Ultracontractivity and the heat kernel for Schrödinger operators and Dirichlet Laplacians*, J. Funct. Anal. **59** (1984), 335–395.

10. E. B. Davies, *Heat kernels and spectral theory*, Cambridge Univ. Press, 1989.

11. J. Dodziuk, *Difference equations, isoperimetric inequality, and transience of certain random walks*, Trans. Amer. Math. Soc. **284** (1984), 787–794.

12. J. Dodziuk and W. S. Kendall, *Combinatorial Laplacians and isoperimetric inequality*, From Local Times to Global Geometry, Control, and Physics (K. D. Elworthy, ed.), Res. Notes in Math., vol. 150, Pitman, Boston, 1986, pp. 68–74.

13. H. Donnelly, *Bounded harmonic functions and positive Ricci curvature*, Math. Z. **191** (1986), 559–565.

14. E. B. Fabes and D. W. Stroock, *A new proof of Moser's parabolic inequality via the old ideas of Nash*, Arch. Rational Mech. Anal. **96** (1986), 327–338.

15. H. Federer and W. H. Fleming, *Normal integral currents*, Ann. of Math. (2) **72** (1960), 458–520.

16. P. Gerl, *Random walks on graphs with a strong isoperimetric property*, J. Theoret. Probab. **1** (1988), 171–187.

17. A. A. Grigor'yan, *On the existence of positive fundamental solutions of the Laplace equation on Riemannian manifolds*, Mat. Sb. **128** (1985); English transl., Math. USSR-Sb. **56** (1987), 349–358.

18. G. A. Hunt, *Some theorems concerning Brownian motion*, Trans. Amer. Math. Soc. **81** (1956), 294–319.

19. M. Kanai, *Rough isometries, and combinatorial approximations of geometries of noncompact Riemannian manifolds*, J. Math. Soc. Japan **37** (1985), 391–413.

20. S. Kusuoka and D. W. Stroock, *Long time estimates for the heat kernel associated with a uniformly subelliptic symmetric second order operator*, Ann. of Math. (2) **127** (1988), 165–189.

21. Yu. T. Kuz'menko and S. A. Molchanov, *Counterexamples to Liouville-type theorems*, Vestnik Moskov. Univ. Ser. I Mat. Mekh. **34** (1979); English transl., Moscow Univ. Math. Bull. **34** (1979), 35–39.

22. P. Li, *Large time behavior of the heat equation on complete manifolds with nonnegative Ricci curvature*, Ann. of Math. (2) **124** (1986), 1–21.

23. P. Li and L.-F. Tam, *Positive harmonic functions on complete manifolds with nonnegative curvature outside a compact set*, Ann. of Math. (2) **125** (1987), 171–207.

24. P. Li and S. T. Yau, *On the parabolic kernel of the Schrödinger operator*, Acta Math. **156** (1986), 153–201.

25. S. Markvorsen, *Truncated isoperimetric constants and transience of Riemannian manifolds*, Preprint.

26. S. Markvorsen, S. McGuinness, and C. Thomassen, *Transient random walks on graphs and metric spaces with applications to hyperbolic surfaces*, Proc. London Math. Soc. (3) **64** (1992), 1–20.

27. J. Nash, *Continuity of solutions of parabolic and elliptic equations*, Amer. J. Math. **80** (1958), 931–954.

28. S. Northshield, *Geodesics and bounded harmonic functions on infinite planar graphs*, Proc. Amer. Math. Soc., **113** (1991), 229–233.

29. J. C. Taylor, *The Martin boundaries of equivalent sheaves*, Ann. Inst. Fourier (Grenoble) **20** (1970), 433–456.

30. N. Th. Varopoulos, *Brownian motion and random walks on manifolds*, Ann. Inst. Fourier (Grenoble) **34** (1984), 243–269.

31. ____, *Isoperimetric inequalities and Markov chains*, J. Funct. Anal. **63** (1985), 240–260.

32. ____, *Hardy-Littlewood theory for semigroups*, J. Funct. Anal. **66** (1986), 406–431.

33. ____, *Random walks and Brownian motion on manifolds*, Symposia Mat. **29** (1986), 97–109.

CITY UNIVERSITY OF NEW YORK

Proceedings of Symposia in Pure Mathematics
Volume **54** (1993), Part 3

Large Time Behavior of Solutions
of the Heat Equation

ISAAC CHAVEL AND LEON KARP

In this note we abstract some of the results of [6] and [7], to which papers we refer for proofs as well as additional results and remarks.

1. Movement of hot spots

Let (M^n, ds^2) be an n-dimensional complete Riemannian manifold, dV the Riemannian measure, and $p(x, y, t)$ the minimal positive heat kernel of (M^n, ds^2). Then for any $\varphi \geq 0$ in $L_c^\infty(M)$,

$$(P_t\varphi)(x) =: \int_M p(x, y, t)\varphi(y) \, dV(y)$$

is the minimal positive solution of the heat equation on (M^n, ds^2) satisfying

$$\lim_{t \downarrow 0} P_t\varphi = \varphi$$

at all points of continuity of φ. Thus, for $M^n = \mathbb{R}^n$ with canonical Riemannian metric, we have

$$(P_t\varphi)(x) = \int_{\mathbb{R}^n} (4\pi t)^{-n/2} e^{-|x-y|^2/4t} \varphi(y) \, dy$$

where dy is the usual Lebesgue measure. (See [2, Chapters VI–VIII].)

Our primary interest is in Riemannian manifolds where, for every non-negative φ in $L_c^\infty(M)$, the locus

$$H(t) = H(t; \varphi) =: \left\{ x \colon (P_t\varphi)(x) = \max_y (P_t\varphi)(y) \right\}$$

1991 *Mathematics Subject Classification*. Primary 53C20, 58G11, 58G25.

The final version of this paper will be submitted for publication elsewhere.

The first author was supported in part by National Science Foundation grant DMS 8704325 and PSC-CUNY FRAP awards, and the second, and the second in part by National Science Foundation grant DMS 8506636 and PSC-CUNY FRAP awards.

of "hot spots" is compact for all $t > 0$. The collection of such Riemannian manifolds is reasonably large. Besides the compact ones, the class contains those noncompact Riemannian manifolds for which for all $t > 0$, $\varphi \in L_c^\infty(M)$, we have $(P_t\varphi)(x) \to 0$ as $x \to \infty$ (see [15, 8, 9] for results on the geometric hypotheses guaranteeing this condition). Here we initiate a study of the location of $H(t)$ relative to the support of φ, S_φ, and, more precisely, the behavior of $H(t)$ as $t \uparrow +\infty$.

We emphasize that this is not the study of the maximum *values* of $P_t\varphi$. Rather, our interest is actually in the function

$$\Phi_t = P_t\varphi / \max P_t\varphi,$$

which we refer to as the *profile of* $P_t\varphi$. Thus our study of $H(t)$ as $t \uparrow +\infty$ reduces to the study of $\{\Phi_t = 1\}$ as $t \uparrow +\infty$. If $\Phi_t \to \Phi$ everywhere on M, as $t \uparrow +\infty$, for some function Φ, then we refer to Φ as the *limiting profile of* $P_t\varphi$. Of course, Φ might very well exist, and not even have maximum value equal to 1. However, in the those cases where we know in advance that there exists a compact K containing $H(t)$ for all $t > 0$, and $\Phi_t \to \Phi$ as $t \uparrow +\infty$, then the limit points of $H(t)$, as $t \uparrow +\infty$, are located in $\{\Phi = 1\}$.

Our main results here are as follows:

1. If M is Euclidean space $\mathbb{R}^n$, hyperbolic space $\mathbb{H}^n$, or the sphere $\mathbb{S}^n$, of dimension $n \geq 1$ with standard Riemannian metric, then $H(t)$ is always contained in the closed convex hull, C_φ, of the support of φ, S_φ. (Of course, for the sphere $\mathbb{S}^n$, if S_φ cannot be contained in any closed hemisphere then $C_\varphi = \mathbb{S}^n$.)

2. For $M^n = \mathbb{R}^n$, let m_φ denote the Euclidean center of mass of φ. Then

$$H(t) \to \{m_\varphi\}$$

as $t \uparrow +\infty$.

3. For hyperbolic space, there exists a single point h_∞ for which

$$H(t) \to \{h_\infty\}$$

if S_φ is sufficiently small. Otherwise, one gives an easy counterexample to $H(t)$ converging to a single point.

4. If $M = \mathbb{S}^n$ then we have a special case of the study of $H(\infty)$ (the limit set of $H(t)$ as $t \uparrow +\infty$) in general compact Riemannian manifolds, the analysis of which is obtained from the Sturm-Liouville decomposition of L^2. For $M = \mathbb{S}^n$ the results is: if φ is not $L^2(\mathbb{S}^n)$-orthogonal to the restriction of the coordinate functions of $\mathbb{R}^{n+1}$ to $\mathbb{S}^n$, then $H(\infty)$ consists of one point. Of course, since $\mathbb{S}^n$ is compact, $P_t\varphi$ has a locus of "cold spots." In the case under consideration, the limiting position of the "cold spots" as $t \uparrow +\infty$ is $-H(\infty)$.

5. One is tempted to believe that the general result would be: Given noncompact (M^n, ds^2) for which $P_t\varphi \to 0$ as $x \to \infty$ for all $t > 0$ and compactly supported φ, then there exists a compact set K in M such that

$H(t) \subseteq K$ for all $t > 0$. We give a rather elementary counterexample to this proposed result. So to extend the theory beyond the classical spaces requires geometric hypotheses on the spaces in question. Even in those cases where $H(t)$ is not uniformly bounded, one might ask for an upper bound on the rate at which "hot spots" escape to ∞. We note that in our counterexample, which has constant curvature equal to -1, the rate at which $H(t)$ escapes to ∞ is essentially equal to t; however, when M is Riemannian complete with nonnegative Ricci curvature, then for any $x \in H(t)$ we have

$$(1) \qquad d(x, S_\varphi) \leq \text{const} \sqrt{t \ln t}$$

for sufficiently large t. The $\sqrt{t \ln t}$ may be improved to $\sqrt{t}$ if we add the assumption of "maximal volume growth," i.e., that

$$V(x_0; r) \geq \text{const}\, r^n$$

as $r \uparrow +\infty$, where $V(x_0; r)$ denotes the volume of the metric disk of radius r centered at some fixed x_0 in M. However, it seems reasonable to conjecture that, for manifolds of nonnegative curvature, $H(t)$ is contained in a fixed compact set for all positive t.

We note that one might wish to view our results as "inverse results" relating (especially in $\mathbb{R}^n$) the asymptotics of $P_t\varphi$ to the geometry of φ. In this spirit one might ask an even stronger question, viz., to what extent does the asymptotic behavior of $P_t\varphi$ determine the initial data φ? Some results in this direction are discussed in [6].

2. Large time behavior of the heat kernel

Let $\lambda =: \lambda(M)$ the bottom of spec. In this section we discuss the following result:

THEOREM. *For all x, y in M we have the existence of the limit*

$$(2) \qquad \lim_{t \uparrow +\infty} e^{\lambda t} p(x, y, t) =: \mathcal{F}(x, y),$$

for which we have the following alternative:

Either $\mathcal{F}$ vanishes identically on all of $M \times M$, in which case λ possesses no L^2 eigenfunctions; or $\mathcal{F}$ is strictly positive on all of $M \times M$ in which case λ possesses a positive normalized L^2 eigenfunction ϕ (normalized in the sense that its L^2 norm is equal to 1) for which

$$(3) \qquad \lim_{t \uparrow +\infty} e^{\lambda t} p(x, y, t) = \phi(x)\phi(y)$$

locally uniformly on all of $M \times M$. Furthermore, if M is noncompact Riemannian complete with Ricci curvature bounded from below, and with positive injectivity radius, then

$$(4) \qquad \lim_{x \to \infty} \phi(x) = 0.$$

The simplest example of the case $\mathscr{F} = 0$ is $\mathscr{R}^n$, $n \geq 1$, in which case we have

$$\lambda = 0, \qquad p(x, y, t) = (4\pi t)^{-n/2} e^{-|x-y|^2/4t}.$$

So

$$e^{\lambda t} p(x, y, t) = p(x, y, t) \to 0$$

as $t \uparrow +\infty$.

When M has compact closure one always has $\mathscr{F}$ strictly positive, and (3) follows from the Sturm-Liouville expansion of the heat kernel.

We also have the following easy consequences of our theorem.

COROLLARY 1 (P. Li [10]). *We always have*

$$(5) \qquad \lim_{t \uparrow +\infty} \frac{\ln p(x, y, t)}{t} = -\lambda$$

locally uniformly on $M \times M$; *and when* M *has finite volume* V *we have*

$$(6) \qquad \lim_{t \uparrow +\infty} p(x, y, t) = 1/V$$

locally uniformly on $M \times M$.

COROLLARY 2. *For any* M *we have*

$$(7) \qquad \lim_{t \uparrow +\infty} p(x, y, t) = 0$$

if and only if M *has infinite volume.*

Indeed, if M has finite volume then (6) implies that $\lim p$, as $t \uparrow +\infty$, is nonzero. If, on the other hand, M has infinite volume, then (i) for $\lambda > 0$ simply use (2); and for (ii) $\lambda = 0$, then if $\mathscr{F}$ were positive, we would have the existence of an L^2 harmonic function on M, which is impossible by a theorem of Yau [14]. □

COROLLARY 3. *Suppose* M *noncompact is a covering of a compact Riemannian manifold. Then* $\mathscr{F}$ *is identically equal to zero. Consequently, if the covering is nonamenable—by* [1], $\lambda > 0$—*then* p *tends to 0 faster than* $e^{-\lambda t}$.

Indeed, if $\lambda = 0$, one uses the above corollary, since M has infinite volume. If $\lambda > 0$ and $\mathscr{F} > 0$ then the L^2 eigenspace of λ, which is nontrivial, is 1-dimensional by Theorem 2.8 of [12]. But this is impossible, by the invariance of the eigenspace under the action of the deck transformation group. □

REMARK 1. We note that there are very few results valid, with any sharpness, for large time diffusion. The only cases we know, of sharp estimates from above and below, are those of [11] when the Ricci curvature of M is nonnegative. The only case of a precise limit (the result of sharp estimates from above and below), of which we are aware, is that of nonnegative Ricci curvature with maximal volume growth [10].

REMARK 2. Our theorem gives no discussion of the rate of convergence to $\mathscr{F}$ as $t \uparrow +\infty$. For some general results, see [5, 13]. In [5] it is shown that when M is λ-transient, then

$$e^{\lambda t} p(x, y, t) = o(t^{-1})$$

as $t \uparrow +\infty$; and in [13] is it shown that if M is complete noncompact with bounded geometry, then

$$p(x, y, t) = O(t^{-1/2+\varepsilon})$$

for every $\varepsilon > 0$, as $t \uparrow +\infty$. This last result was improved in [4] to

$$p(x, y, t) = O(t^{-1/2})$$

under the explicit bounded geometry hypothesis of Ricci curvature bounded from below, and positive injectivity radius. For a sampling of results under more explicit geometric hypotheses, we refer the reader to [10, 3, 4] and the references therein.

REMARK 3. When λ possesses an L^2 eigenfunction, then (3) has potential application to the "movement of hot spots" [6]. Simply put, for any $\psi \geq 0$ in $L_c^\infty(M)$,

$$(P_t\psi)(x) =: \int_M p(x, y, t)\psi(y)\, dV(y)$$

is the minimal positive solution of the heat equation on M satisfying

$$\lim_{t \downarrow 0} P_t\psi = \psi$$

at all points of continuity of ψ. If M possesses a compact set K such that $H(t)$ is contained in K for all $t > 0$, and if $\mathscr{F} > 0$, then (3) implies that the limiting locus of $H(t)$, as $t \uparrow +\infty$, is contained within the locus of maxima of ϕ—independent of the location of the support of the initial data ψ. (Of course, which maxima of ϕ are realized might very well depend on ψ.) This situation is in stark contrast with the examples of Euclidean and hyperbolic spaces discussed above.

REFERENCES

1. R. Brooks, *The fundamental group and the spectrum of the Laplacian*, Comment. Mat. Helv. **56** (1981), 581–598.
2. I. Chavel, *Eigenvalues in Riemannian geometry*, Academic Press, New York, 1984.
3. I. Chavel and E. A. Feldman, *Isoperimetric constants, the geometry of ends, and large time heat diffusion in Riemannian manifolds*, Proc. London Math. Soc. (3) **62** (1991), 427–448.
4. ____, *Modified isoperimetric constants, and large time heat diffusion in Riemannian manifolds*, Duke Math. J. **64** (1991), 473–499.
5. I. Chavel, E. A. Feldman, and L. Karp, *Large time decay of the heat kernel of λ-transient Riemannian manifolds*, Preprint.
6. ____, *Movement of hot spots in Riemannian manifolds*, J. Analyse Math. **55** (1990), 271–286.
7. I. Chavel and L. Karp, *Large time behavior of the heat kernel: the parabolic λ-potential alternative*, Comment. Math. Helv. **66** (1991), 541–556.

8. J. Dodziuk, *Maximum principle for parabolic inequalities and heat on open manifolds*, Indian Univ. Math. J. **32** (1983), 703–716.

9. P. Hsu, *Heat semigroup on a complete Riemannian manifold*, Ann. Prob., **17** (1989), 1248–1254.

10. P. Li, *Large time behavior of the heat equation on complete manifolds with nonnegative Ricci curvature*, Ann. of Math. (2) **124** (1986), 1–21.

11. P. Li and S. T. Yau, *On the parabolic kernel of the Schrödinger operator*, Acta Math. **156** (1986), 153–201.

12. D. Sullivan, *Related aspects of positivity in Riemannian geometry*, J. Differential Geom. **25** (1987), 327–351.

13. N. Th. Varopoulos, *Brownian motion and random walks on manifolds*, Ann. Inst. Fourier (Grenoble) **34** (1984), 243–269.

14. S. T. Yau, *some function-theoretic properties of complete riemannian manifolds, and their applications to geometry*, Indiana Univ. Math. J. **25** (1976), 659–670; also cf. ibid. **31** (1982), 307.

15. ____, *On the heat kernel of a complete Riemannian manifold*, J. Math. Pures Appl. (9) **57** (1978), 191–201.

CITY COLLEGE OF THE CITY UNIVERSITY OF NEW YORK

LEHMAN COLLEGE OF THE CITY UNIVERSITY OF NEW YORK

Proceedings of Symposia in Pure Mathematics
Volume **54** (1993), Part 3

Manifolds with 2-Nonnegative Curvature Operator

HAIWEN CHEN

This paper is part of an effort to understand how geometry restrains topology. It is closely related to the sphere theorems, on which D. Gromoll has given a survey at this conference [G].

Suppose M^n is an n-dimensional manifold with a riemannian metric $\langle \ , \ \rangle$ and Levi-Civita connection ∇. We can identify the tangent bundle TM with the cotangent bundle T^*M via the given metric. We also define the curvature tensor R by

$$\langle R_{v,w} y, z \rangle = -\langle \nabla_v \nabla_w y, z \rangle + \langle \nabla_w \nabla_v y, z \rangle + \langle \nabla_{[v,w]} y, z \rangle.$$

Since the curvature tensor R is anti-symmetric on its first entry with its second one, and is symmetric on its first two entries with the last two entries, R can be regarded as a linear symmetric map

$$R_M : \bigwedge\nolimits^2 TM \to \bigwedge\nolimits^2 TM.$$

DEFINITION 1. R_M is said to be k-nonnegative (resp. k-positive) if the sum of the first k eigenvalues of R_M is ≥ 0 (resp. > 0).

MAIN THEOREM. *If M^n ($n > 2$) is a compact manifold with R_M 2-nonnegative, then the universal covering space is*

$$(\widetilde{M}, \tilde{g}) = (R^d, g_{\text{flat}}) \times (M_1, g_1) \times \cdots \times (M_k, g_k),$$

where (M_i, g_i) is one of the following:

(1) *a compact symmetric space,*
(2) *a manifold homeomorphic to a sphere,*
(3) *a Kähler manifold biholomorphic to a complex projective space.*

REMARK. This theorem is a generalization of B. Chow and D. Yang's work [C-Y], where they required that R_M be nonnegative.

1991 *Mathematics Subject Classification.* Primary 53C21; Secondary 58G11.
This paper is in final form and no version of it will be submitted for publication elsewhere.

The idea of the proof is to investigate the properties of R_M under the Ricci flow. By working on the universal covering space, we may assume that M is a simply-connected irreducible manifold. Then we reduce the problem to several cases.

Under the unnormalized evolution equation

$$\frac{\partial}{\partial t} g = -2Rc,$$

Hamilton [H2] showed that the curvature operator R_M satisfies the heat equation

(1) $$\frac{\partial}{\partial t} R_M = \Delta R_M + R_M^2 + R_M^{\#},$$

where $R_M^{\#}$ is an abbreviation of $R_M \# R_M$ and generally $A\#B$ is defined by

$$(A\#B)_{\alpha\beta} = C_{\alpha\gamma\eta} C_{\beta\delta\theta}(A)_{\gamma\delta}(B)_{\eta\theta},$$

where the $C_{\alpha\beta\gamma}$'s are the Lie structure constants related to an orthonormal basis of $\bigwedge^2 TM$, and are fully anti-symmetric, i.e., $C_{\alpha\beta\gamma} = -C_{\beta\alpha\gamma} = C_{\beta\gamma\alpha}$, etc.

PROPOSITION 1. *If R_M is 2-nonnegative at $t = 0$, then either R_M becomes 2-positive everywhere for $t > 0$, or R_M is nonnegative, and the holonomy group of M is restricted.*

Roughly speaking, the sum of the first two eigenvalues of R_M, $r_1 + r_2$, "satisfies" the heat equation

$$\frac{\partial}{\partial t}(r_1 + r_2) = \Delta(r_1 + r_2) + r_1^2 + r_2^2 + (r_1 + r_2)h + k$$

where h and k are some nonnegative functions. By the maximum principle, either $r_1 + r_2$ becomes positive everywhere when $t > 0$, or $r_1^2 + r_2^2 \equiv 0$ which means that R_M is nonnegative. Then the Ambrose-Singer theorem shows that the holonomy group of M is restricted. For details, see [C1].

THEOREM 1 [M-M]. *If R_M is 2-positive, then the universal covering space of M is a topological sphere.*

On the other hand, Berger has classified all possible holonomy groups for manifolds which are not locally symmetric. They are listed below as (e.g., see [B])

$$O_n, \quad U_m, \quad Sp_k \cdot Sp_1, \quad SU_m, \quad Sp_k, \quad G_2, \quad Spin_7.$$

O_n is the holonomy group for a generic metric. Manifolds with U_m are called Kähler manifolds. Those with $Sp_k \cdot Sp_1$ are called quaternionic-Kähler manifolds. Manifolds whose holonomy groups are any of the last four groups are Ricci flat. If in addition R_M is nonnegative, then they are flat. Hence, we only need to consider Kähler manifolds and quaternionic-Kähler manifolds.

THEOREM 2 [C-C]. *If M is a compact Kähler manifold with R_M nonnegative, then either M is a Hermitian symmetric space, or M is biholomorphic to $\mathbf{CP}^m$.*

REMARK 1. There is a natural global decomposition of

$$\bigwedge^2 TM = u(m) \oplus N,$$

where N is contained in the kernel of R_M. The theorem still holds if we only require that R_M is 2-nonnegative on $u(m)$.

REMARK 2. Cao and Chow's theorem relies heavily on Mori, Siu and Yau's results ([Mr], [S-Y], which answer Frankel's conjecture affirmatively. Later, Mok classified all compact Kähler manifolds under the much weaker, but more natural, assumption that M has only nonnegative holomorphic bisectional curvature [Mo].

THEOREM 3. *If M is a quaternionic-Kähler manifold with $R_M \geq 0$, then M is locally symmetric.*

REMARK 1. Tachibana [T] proved that an Einstein manifold with $R_M \geq 0$ is locally symmetric, and we know that quaternionic-Kähler manifolds are Einstein manifolds.

REMARK 2. Chow and Yang claimed Theorem 3 in [C-Y]. However, it is questionable in their proof that M is symmetric if and only if the twist space SM is symmetric.

Here we will give the sketch of an alternative proof. First, we can show that

$$\bigwedge^2 TM = A \oplus B \oplus T$$

where A is an sp_1-bundle, B is an sp_k-bundle, and

$$[A, B] = 0, \qquad [B, T] \subset T, \qquad [B, B] \subset B.$$

Furthermore, we can show that

$$R_M |_A = C_k r I \quad \text{and} \quad R_M |_T = 0,$$

where $r = \text{tr}(R_M)$ is the scalar curvature of M, C_k is a constant depending only on the dimension of the manifold ($n = 4k$). The computation is lengthy but simple [C2], or one may derive it from Alekseevskiǐ's work [A]. Then it is easy to see that these subspaces are perpendicular under the action of R_M.

We denote $R_M |_B$ by R_B. Then with some computation we have

$$(2) \qquad \frac{\partial}{\partial t} R_B = \Delta R_B + R_B^2 + R_B^{\#}.$$

We can repeat the argument mentioned before to show that either R_B is positive when $t > 0$, or the holonomy of M is restricted further. Thus we only need to consider the case that $R_B > 0$.

On the other hand, by the uniqueness of the solution of equation (1), we see that if the initial metric g_0 is Einstein, then $g_t = h(t)g_0$, i.e., the metric in the Ricci flow has only been rescaled. Hence,

$$(3) \qquad 0 = \frac{\partial}{\partial t}\frac{R_B}{r} = \frac{r\frac{\partial}{\partial t}R_B - \frac{\partial r}{\partial t}R_B}{r^2}.$$

Since

$$(4) \qquad \frac{\partial r}{\partial t} = \frac{2}{n}r^2 = \frac{1}{2k}r^2,$$

after substitution and simplification, we obtain

$$(5) \qquad \Delta R_B + R_B^2 + R_B^\# = \frac{r}{2k}R_B.$$

Let $\{r_i\}$ be the set of eigenvalues of R_B. Multiplying by R_B^{-1} on both sides of (5), and taking the traces, we have

(6)

$$\Delta \log \det(R_B) + \|R_B^{-1}\nabla R_B\|^2 + \operatorname{tr}(R_B) + \sum_{r_i, r_j, r_k \in R_B} C_{ijk}^2 \frac{r_j r_k}{r_i} = \frac{2k+1}{2}r.$$

It is a easy computation to show that:

$$(7) \quad \operatorname{tr}(R_B) + \sum_{r_i, r_j, r_k \in R_B} C_{ijk}^2 \frac{r_j r_k}{r_i} = \frac{2k+1}{2}r + \frac{1}{2}\sum_{r_i, r_j, r_k \in R_B} C_{ijk}^2 \frac{(r_i - r_j)^2}{r_i r_j}r_k.$$

Comparing (6) and (7), we have

$$\nabla R_B = 0.$$

(With a little more computation, we can show that $R_B = \frac{C_k}{k}rI$. Hence M is isometric to $\mathbf{HP}^k$.)

REMARK 1. Similarly, we can give an alternative proof for Tachibana's theorem when the holonomy group is either O_n or U_m.

REMARK 2. For quaternionic-Kähler manifolds, the condition can be weakened as R_B 2-nonnegative.

QUESTION 1. Hamilton and the author show that M^n with 2-positive curvature operator is diffeomorphic to a spherical space form when $n = 3$ [H1] or $n = 4$ [C1]. Is it true for all dimensions? Right now, we do not even have an answer for M^n with positive curvature operator, when $n > 4$.

QUESTION 2. Find weaker curvature conditions on a quaternionic-Kähler manifold M^{4k} which imply that M^{4k} is symmetric. At present, all known examples of quaternionic-Kähler manifolds with positive scalar curvature are symmetric. Actually, there are no other examples when the dimension is 4

[Hi] or 8 [P-S]. Are there any nontrivial examples in higher dimensions? Can one classify such manifolds in higher dimensions?

REFERENCES

[A]   D. Alekseevskiĭ, *Riemannian spaces with exceptional holonomy groups*, Funct. Anal. Appl. **2** (1968), 97–105.

[A-S]   W. Ambrose and I. Singer, *A theorem on holonomy*, Trans. Amer. Math. Soc. **75** (1953), 428–443.

[B]   R. Bryant, *A survey of riemannian metrics with special holonomy groups*, Proc. Internat. Congr. Math. **1** (1986), 505–514.

[C1]   H. Chen, *Pointwise $\frac{1}{4}$-pinched 4-manifolds*, Ann. Global Anal. Geom., **2** (1991), 161–176.

[C2]   H. Chen, *Some properties of quaternionic-Kähler manifolds*, Preprint.

[C-C]   H.-D. Cao and B. Chow, *Compact Kähler manifolds with nonnegative curvature operator*, Invent. Math. **83** (1986), 553–556.

[C-Y]   B. Chow and D. Yang, *Rigidity of nonnegatively curved compact quaternionic-Kähler manifolds*, J. Differential Geom. **29** (1989), 361–372.

[G]   D. Gromoll, *Spaces on nonnegative curves*, Amer. Math. Soc. Providence (1993), PSPUM **54** (3) 337–356.

[H1]   R. Hamilton, *Three-manifolds with positive Ricci curvature*, J. Differential Geom. **17** (1982), 255–306.

[H2]   _____ , *Four-manifolds with positive curvature operator*, J. Differential Geom. **24** (1986), 153–179.

[Hi]   N. Hitchin, *Kählerian twistor spaces*, Proc. London Math. Soc. (3) **43** (1981), 133–150.

[M-M]   M. Micallef and J. Moore, *Minimal two-spheres and the topology of manifolds with positive curvature on totally isotropic two-planes*, Ann. of Math. **127** (1988), 199–227.

[Mo]   N. Mok, *The uniformisation theorem for compact Kähler manifolds of nonnegative holomorphic bisectional curvature*, J. Differential Geom. **27** (1988), 179–214.

[Mr]   S. Mori, *Projective manifolds with ample tangent bundles*, Ann. of Math. **110** (1979), 593–606.

[P-S]   Y. S. Poon and S. Salamon, *Quaternionic-Kähler δ-dimensional manifolds with positive scalar curvature* **33** (1991), 363–378.

[S-Y]   Y. T. Siu and S. T. Yau, *Compact Kähler manifolds of positive bisectional curvature*, Invent. Math. **59** (1980), 189–204.

[T]   S. Tachibana, *A theorem on riemannian manifolds of positive curvature operator*, Proc. Japan Acad. **50** (1974), 301–302.

MICHIGAN STATE UNIVERSITY

Proceedings of Symposia in Pure Mathematics
Volume **54** (1993), Part 3

The Geometry of Isospectral Deformations

DENNIS DeTURCK, HERMAN GLUCK,
CAROLYN GORDON, AND DAVID WEBB

Introduction

We continue to apply the method of calibrated geometries to the study of isospectral deformations of closed Riemannian manifolds. Recall that an *isospectral deformation* of a Riemannian manifold is a one-parameter family of metrics on the manifold such that the eigenvalues and multiplicities of the Laplace-Beltrami operators of the metrics (acting on scalar functions) are the same for all the metrics in the family. The main thrust of our work is to find specific geometric quantities which change during isospectral deformations; such quantities can then be used to detect the nontriviality of the deformations. In concert with the "Can you hear the shape of ... " fashion of describing spectral invariants, we call our geometric quantities *inaudible*.

We review the construction of isospectral deformations of metrics on nilmanifolds in §1, after which we survey the way in which we shall use the method of calibrated geometries to track the volumes of minimizing currents in various homology classes of the manifold. In particular, we show that it is sufficient to consider only "left-invariant" forms and currents in order to apply these methods (Proposition 4.1). Much of our previous work **[DGGW1, DGGW2]** had indicated that the minimum volume of a current in some homology class of a nilmanifold would be an inaudible geometric property which suffices to distinguish isospectral metrics. However, in §6, we give an example of a five-step nilmanifold (i.e., a compact quotient of a simply-connected five-step nilpotent Lie group), and a nontrivial isospectral deformation on it whose nontriviality *cannot* be detected simply by measuring the minimum volume in each homology class (see also **[DGGW3]**). We do

1991 *Mathematics Subject Classification.* Primary 53C20, 49F22, 58G25.

This paper is in final form and no version of it will be submitted for publication elsewhere.

The authors gratefully acknowledge the support of the National Science Foundation. The first two authors would like to thank the Institute for Advanced Study, and the second two authors would like to thank Cornell University (particularly Keith Dennis), for their hospitality.

show that a refinement of these measurements, namely one which takes into account the geometry of the volume-minimizing cycles in certain homology classes of the manifold, does detect the nontriviality of the deformation.

In the final two sections, we demonstrate that this finer set of calibration-invariants is *always* sufficient to detect the nontriviality of an isospectral deformation of a *two-step* nilmanifold (Theorem 8.1). This result depends on an algebraic lemma about codimension-one ideals of two-step nilpotent Lie algebras (Lemma 8.2) which is of some interest in its own right.

1. Isospectral deformations of nilmanifolds

We review the construction of isospectral deformations given in [GW] and generalized in [G] and [DG].

Let G be a simply-connected nilpotent Lie group with Lie algebra $\mathfrak{g}$. The exponential map from $\mathfrak{g}$ to G is a diffeomorphism, and we call its inverse "log." If G admits a uniform discrete subgroup Γ, then $\log \Gamma$, while not necessarily a lattice itself, generates a lattice of maximal rank in $\mathfrak{g}$. The rational span $\mathfrak{g}_Q$ of $\log \Gamma$ is closed under the bracket operation of $\mathfrak{g}$ and thus has the structure of a rational Lie algebra (see [R]). Moreover, $\mathfrak{g} \simeq \mathfrak{g}_Q \otimes_Q \mathbb{R}$. Elements X of $\mathfrak{g}_Q$ (i.e., elements of $\mathfrak{g}$ of the form $X \otimes 1$) will be called *rational elements* of $\mathfrak{g}$; likewise, a subalgebra of $\mathfrak{g}$ spanned by rational elements will be called a *rational subalgebra*. We remark that if H is a connected subgroup of G, then $(H \cap \Gamma) \backslash H$ is compact if and only if the Lie algebra $\mathfrak{h}$ of H is a rational subalgebra of $\mathfrak{g}$. The rational elements of $\mathfrak{g}$, relative to the rational structure defined by Γ, are precisely the rational linear combinations of elements of $\log \Gamma$.

In [GW], an automorphism Φ of G was defined to be *almost-inner* if Φ maps every element of G to a conjugate. In [G], the notion of almost-inner automorphism was extended: Given a cocompact discrete subgroup Γ of G, an automorphism Φ of G is *almost-inner relative to* Γ if Φ maps every element of Γ to a conjugate; i.e., if for each $\gamma \in \Gamma$, there exists $a \in G$ such that $\Phi(\gamma) = a\gamma a^{-1}$. When there is no possibility of confusion, we will refer to an almost-inner automorphism Φ relative to Γ as a *rationally almost-inner automorphism*. An example was given in [G] of a family of rationally almost-inner automorphisms which are not almost-inner (note that an almost-inner automorphism is necessarily rationally almost-inner).

1.1. PROPOSITION [G]. *Given a cocompact discrete subgroup* Γ *of* G, *the collection of rationally almost-inner automorphisms is a Lie subgroup of* $\mathrm{Aut}(G)$. *Its Lie algebra is the set of all "rationally almost-inner derivations of* $\mathfrak{g}$," *i.e., all derivations* D *of* $\mathfrak{g}$ *such that* $D(X) \in [X, \mathfrak{g}]$ *for all* $X \in \log \Gamma$ (*or, equivalently, for all* $X \in \mathfrak{g}_Q$).

1.2. REMARK. Let D be a rationally almost-inner derivation. Then D generates a one-parameter family $\{\Phi_t\}$ of rationally almost-inner automorphisms of G. Indeed, we have $\Phi_{t*} = e^{tD} = \mathrm{Id} + D + \frac{1}{2}D^2 + \cdots$. If $\gamma \in \Gamma$, we have $D(\log \gamma) = [A, \log \gamma]$ for some $A \in \mathfrak{g}$. Set $a(t) = \exp(tA) \in G$. Since $\Phi_{t*}(\log \gamma) = e^{t \operatorname{ad} A}(\log \gamma) = \operatorname{Ad}(a(t))(\log \gamma)$ (where $\operatorname{Ad}(a_t)$ is the differential of conjugation by $a(t)$), we obtain $\Phi_t(\gamma) = a(t)\gamma a(t)^{-1}$.

1.3. DEFINITION. We will say that a rationally almost-inner derivation D of $\mathfrak{g}$ has *inner index* k if (i) for each k-dimensional subspace V of $\mathfrak{g}$, there exists $A \in \mathfrak{g}$ such that $D|_V = \operatorname{ad} A|_V$, and (ii) k is the largest integer such that (i) holds.

The following family of examples, constructed in [**DGGW2**], shows that a rationally almost-inner derivation can have arbitrary inner index.

1.4. EXAMPLE. We exhibit, for each integer $k > 1$, a two-step nilpotent Lie algebra $\mathfrak{g}_k$ admitting a derivation D_k which has inner index $k-1$. The algebra $\mathfrak{g}_k$ has a basis

$$\{X_2, X_3, \dots, X_k\} \cup \{X_{ij} : 1 \leq i < j \leq k\} \cup \{Y_1, \dots, Y_k\} \cup \{Z_1, \dots, Z_k\}.$$

Brackets of basis elements are zero except for the following:

$$\begin{aligned} [X_i, Y_1] &= Z_1, & 2 \leq i \leq k, \\ [X_i, Y_i] &= Z_i, & 2 \leq i \leq k, \\ [X_{ij}, Y_i] &= Z_j, & 1 \leq i < j \leq k. \end{aligned}$$

The derivation D_k vanishes on all basis vectors except Y_1, and sends $Y_1 \mapsto Z_1$. One verifies that D_k has inner index $\leq k-1$ by showing that the system of linear equations obtained by requiring that D_k agree with some inner derivation on the span of $\{Y_1, \dots, Y_k\}$ is inconsistent. Now we show that D_k *does* agree on every $(k-1)$-dimensional subspace with some inner derivation. It clearly suffices to show that D_k agrees on every $(k-1)$-dimensional subspace of the span of $\{Y_1, \dots, Y_k\}$ with an inner derivation. Let S be such a subspace, with basis $\{E_1, \dots, E_{k-1}\}$. The matrix expressing $E_1, \dots, E_{k-1}$ as row vectors of coefficients with respect to the basis vectors $Y_1, \dots, Y_k$ is row equivalent to a row echelon matrix, so S is spanned by a set of vectors of the form

$$\{Y_1 + \alpha_1 Y_m, Y_2 + \alpha_2 Y_m, \dots, Y_{m-1} + \alpha_{m-1} Y_m, Y_{m+1}, Y_{m+2}, \dots, Y_k\}$$

for some $m \leq k$. One easily checks that D_k agrees with

$$\operatorname{ad}\left(X_m - \sum_{i=1}^{m-1} \alpha_i X_{im}\right)$$

on the above basis of S, so that D_k has inner index $k-1$.

To obtain isospectral deformations, let M be the compact nilmanifold $M = \Gamma \backslash G$. A left-invariant metric g on G descends to a metric (also denoted by g) on M. For simplicity, we will refer to g as a *left-invariant metric* on M, even though G does not act on M on the left. We will call a nilmanifold with a left-invariant metric a *Riemannian nilmanifold*. If $\Phi \in \operatorname{Aut}(G)$, then $\Phi^* g$ is again a left-invariant metric on G and M. In

[GW] it was shown that almost-inner automorphisms give rise to isospectral metrics. The following slightly stronger theorem was given in **[G]**; see **[DG]** for another proof:

1.5. THEOREM. *Let G be a simply-connected nilpotent Lie group, Γ a uniform discrete subgroup, Φ a rationally almost-inner automorphism of G, and g any left-invariant metric. Then the metrics g and $\Phi^* g$ on $M = \Gamma \backslash G$ are isospectral. If $\{\Phi_t\}$ is a one-parameter subgroup of rationally almost-inner automorphisms which are not all inner, then the isospectral deformation $\Phi_t^* g$ of g is nontrivial (i.e., the metrics $g_t = \Phi_t^* g$ on M are not all isometric).*

The original proofs of nontriviality of the deformations were abstract and shed no light on the changing geometry. The purpose of this paper, as well as **[DGGW1–DGGW3]**, is to describe the deformations geometrically.

2. Real homology and cohomology via invariant currents and forms

Let $X_1, \dots, X_n$ be a basis for the left-invariant vector fields on the n-dimensional nilpotent Lie group G. These left-invariant vector fields on G descend to well-defined vector fields of the same name on the right coset space $M = \Gamma \backslash$. By abuse of language, we refer to these as *left-invariant vector fields* on M, even though G does not have a left-action on M.

Let $\alpha_1, \dots, \alpha_n$ denote the dual basis of left-invariant 1-forms on G. They likewise descend to "left-invariant" 1-forms on M.

On either G or M, the exterior derivatives of these 1-forms can be read off from the Lie brackets of the vector fields via the formula

$$d\phi(X, Y) = -\phi([X, Y]),$$

in which ϕ is any left-invariant 1-form and X and Y are any left-invariant vector fields. The left-invariant 1-forms may be combined via exterior multiplication to yield the left-invariant k-forms. The exterior derivative on k-forms is determined, via the Leibniz rule, by its values on the 1-forms.

We will use "k-current" in the sense of deRham to denote a continuous linear functional on the space of smooth k-forms in the C^∞ topology. Exterior products of vector fields define currents by evaluation:

$$X_1 \wedge \cdots \wedge X_k(\phi) = \int_M \phi(X_1 \wedge \cdots \wedge X_k) \, d\,\text{vol}$$

where ϕ is a k-form. We will call a current "left-invariant" if it is a linear combination of exterior products of left-invariant vector fields. The boundary map ∂ on the space of k-currents is the adjoint of the exterior derivative on $(k-1)$-forms. The boundary of a left-invariant k-current is a left-invariant $(k-1)$-current, specifically,

$$\partial(X_1 \wedge \cdots \wedge X_k) = \sum_{1 \leq i < j \leq k} (-1)^{i+j} [X_i, X_j] \wedge X_1 \wedge \cdots \wedge \hat{X}_i \wedge \cdots \wedge \hat{X}_j \wedge \cdots \wedge X_k.$$

2.1. PROPOSITION (Nomizu [**No**]). *The cohomology of left-invariant forms on any nilpotent Lie group G is isomorphic in the obvious way to the real cohomology of the coset space $M = \Gamma \backslash G$. By duality, the homology of left-invariant currents on G is isomorphic to the real homology of M.*

3. Mass, comass, and calibrations

We define the "comass" of a form and the "mass" of a current, following Federer [**Fe**] (see also [**M**]), and begin in a linear algebra setting.

Let V be a finite-dimensional real vector space with an inner product. The inner product extends in a natural way to the space $\bigwedge^k V$ of k-vectors, and to the space $\bigwedge^k V^*$ of k-forms. In particular, it provides norms on these spaces.

Given a k-form ϕ, its *comass* is

$$\text{Comass}(\phi) = \sup\{\phi(U) : U \text{ is a } simple \text{ } k\text{-vector of norm } 1\},$$

"simple" meaning "decomposable as an exterior product of vectors." For example, let $V = \mathbb{R}^4$, with orthonormal basis $e_1, \ldots, e_4$, and dual orthonormal basis $e_1^*, \ldots, e_4^*$ for V^*. Then the 2-form $e_1^* e_2^* + e_3^* e_4^*$ has comass 1, and takes this maximum value on the 2-vector $e_1 e_2$, as well as on any other 2-vector corresponding to a complex line in $\mathbb{C}^2$.

Given a k-vector U, its *mass* is

$$\text{Mass}(U) = \sup\{\phi(U) : \phi \text{ is a } k\text{-form of comass } 1\}.$$

For example, the mass of the 2-vector $e_1 e_2 + e_3 e_4$ is 2, and this maximum is achieved when the 2-vector is evaluated against the 2-form $e_1^* e_2^* + e_3^* e_4^*$ of comass 1.

These ideas carry over from the linear algebra setting to that of forms and currents on a compact Riemannian manifold M.

Given a smooth k-form ϕ on M, its *comass* is

$$\text{Comass}(\phi) = \sup\{\text{Comass}(\phi_x) : x \in M\}.$$

Given a k-current U on M, its *mass* is

$$\text{Mass}(U) = \sup\{U(\phi) : \phi \text{ is a smooth } k\text{-form of comass } 1\}.$$

If S is a smooth k-chain, and U_S is the current defined by $U_S(\phi) = \int_S \phi$ for any k-form ϕ, then $\text{Mass}(U_S) = \text{vol}(S)$; thus mass is a generalization to currents of the notion of volume of a classical k-chain.

If we restrict ourselves to currents on M of finite mass, whose boundaries also have finite mass (the so-called *normal currents*), then their homology coincides with the real homology $H_*(M; \mathbb{R})$, by a theorem of Federer and Fleming [**Fe-Fl**].

By the *mass* of a real homology class, we mean the minimum mass of any closed current in that class. Note that "mass" is a norm on homology: it is subadditive and it is linear on rays.

Frequently, the mass of a homology class and the corresponding minimizing currents therein can be found with the aid of a "calibrating" form.

A closed k-form ϕ of comass 1 on a Riemannian manifold M is called a *calibration*. A closed k-current U on M for which $U(\phi)$ coincides with the mass of U is said to be *calibrated* by ϕ. The simplest example of such a U is a smooth oriented k-dimensional submanifold of M on which ϕ restricts to the volume form.

The principal observation is:

A closed k-current U which is calibrated by some form ϕ must be mass-minimizing in its homology class.

For if U' is another closed k-current in the same class, then

$$\text{Mass}(U) = U(\phi) = U'(\phi) \leq \text{Mass}(U').$$

The first equality is because ϕ calibrates U. The second is because ϕ is closed, and hence Stokes' theorem may be applied. The final inequality is because ϕ has comass one. Note that equality holds if and only if U' is also calibrated by ϕ.

Standard examples of calibrations are provided by the normalized powers of the Kähler form on a Kähler manifold. The classical cycles so calibrated are just the complex subvarieties, which are thereby seen to be mass-minimizing in their homology classes. Many more examples are given in [HL].

4. Computing the mass of a homology class via invariant currents

Let $M = \Gamma \backslash G$ be a compact nilmanifold. Let $\mathscr{E}^k(M)$ denote the space of smooth k-forms on M, in the C^∞ topology. Let $\bigwedge^k(\mathfrak{g}^*)$ denote the space of left-invariant k-forms on M. The inclusion map $\bigwedge^k(\mathfrak{g}^*) \to \mathscr{E}^k(M)$ induces a map $H^k(\mathfrak{g}; \mathbb{R}) \to H^k(M; \mathbb{R})$, which by Proposition 2.1 is an isomorphism.

Let $\mathscr{E}_k(M)$ denote the space of k-currents on M. Each k-current $U \in \mathscr{E}_k(M)$ is a continuous linear functional $U: \mathscr{E}^k(M) \to \mathbb{R}$. Restricting U to the subspace $\bigwedge^k(\mathfrak{g}^*)$ of $\mathscr{E}^k(M)$, we obtain an element in the dual space $(\bigwedge^k(\mathfrak{g}^*))^*$ of $\bigwedge^k(\mathfrak{g}^*)$. Naturally, we identify this double dual space with $\bigwedge^k(\mathfrak{g})$, the space of k-vectors on $\mathfrak{g}$, or equivalently, the space of left-invariant k-currents on M. Thus we have defined a surjection $\pi: \mathscr{E}_k(M) \to \bigwedge^k(\mathfrak{g})$. This in turn induces a map $H_k(M; \mathbb{R}) \to H_k(\mathfrak{g}; \mathbb{R})$, which by Proposition 2.1 is an isomorphism.

Note that from this point of view, elements of $\bigwedge^k(\mathfrak{g})$ are treated as equivalence classes of currents. That is, $W \in \bigwedge^k(\mathfrak{g})$ is identified with the set (W) of all currents whose restriction to $\bigwedge^k(\mathfrak{g}^*)$ coincides with W. At the same time, elements of $\bigwedge^k(\mathfrak{g})$ can be viewed as left-invariant currents themselves:

$W \in \bigwedge^k(\mathfrak{g})$ acts on $\omega \in \mathscr{E}^k(M)$ via integration:

$$W(\omega) = \int_M \omega(W)\, d\,\text{vol}.$$

Assuming the metric is normalized so that the volume of M is 1, the left-invariant current W lies in the equivalence class (W). Thus we can view $\bigwedge^k(\mathfrak{g})$ as a subspace of $\mathscr{E}_k(M)$, and the map π as a projection.

Let $U \in \mathscr{E}_k(M)$ be a k-current on M. We saw in the preceding section that the "mass" of U was defined by taking the supremum of the numbers $U(\omega)$, where ω runs over all smooth k-forms of comass 1. If we restrict ω to run only over the left-invariant k-forms $\omega \in \bigwedge^k(\mathfrak{g}^*)$, then the resulting supremum of the numbers $U(\omega)$ will be called the *restricted mass* of U, denoted $\text{Rmass}(U)$. Naturally, $\text{Rmass}(U) \le \text{Mass}(U)$ for every current $U \in \mathscr{E}_k(M)$.

4.1. PROPOSITION.

(i) *For any left-invariant current* $U \in \bigwedge^k(\mathfrak{g})$, *we have*

$$\text{Rmass}(U) = \text{Mass}(U).$$

In other words, the mass of a left-invariant current is the supremum of the values it takes on all left-invariant *forms of comass 1.*

(ii) *The mass of a real homology class is the infimum of the masses of all closed* left-invariant *currents in that class. Thus, only left-invariant currents and left-invariant forms are needed to compute the mass of a homology class in* M.

PROOF. (i) Let $U \in \bigwedge^k(\mathfrak{g})$ be a left-invariant k-current on M, and suppose $\omega \in \mathscr{E}^k(M)$ with $\text{Comass}(\omega) = 1$. Then

$$U(\omega) = \int_M \omega(U)\, d\,\text{vol} = \omega_x(U_x),$$

for some point $x \in M$ by the mean-value theorem, since we are assuming that the volume of M is 1. Let $\omega' \in \bigwedge^k(\mathfrak{g}^*)$ be the left-invariant k-form which satisfies $\omega'_x = \omega_x$. Then $\text{Comass}(\omega') \le 1$, and $U(\omega') = U(\omega)$, since U is also left-invariant. Hence $\text{Rmass}(U) \ge \text{Mass}(U)$, and since the opposite inequality is obvious, the two are equal.

(ii) The projection map $\pi \colon \mathscr{E}_k(M) \to \bigwedge^k(\mathfrak{g})$ takes a smooth k-current to a left-invariant one which equals it on left-invariant k-forms. Observe that π is mass nonincreasing. For if we start with a smooth k-current U, then

$$\text{Mass}(U) \ge \text{Rmass}(U) = \text{Rmass}(\pi(U)) = \text{Mass}(\pi(U)),$$

with the last equality by (i). Now in determining the mass of a real homology class in $H_k(M; \mathbb{R})$, we are attempting to minimize mass among *all* cycles in the class. Since $\text{Mass}(U) \ge \text{Mass}(\pi(U))$, we can restrict attention to the left-invariant k-cycles, such as $\pi(U)$. Finally, since restricted mass and mass are the same on left-invariant currents, we need only measure our left-invariant currents against left-invariant forms.

5. Effect of automorphisms on mass

Let $M = \Gamma\backslash G$ be a compact nilmanifold endowed with a "left-invariant" metric g_2 with respect to which M has unit volume. Let Φ be a volume-form-preserving automorphism of G, and set $g_1 = \Phi^* g_2$.

Let $\phi: \mathfrak{g} \to \mathfrak{g}$ denote the differential of Φ. Thus ϕ is an automorphism of the Lie algebra $\mathfrak{g}$ of G. It induces an isomorphism $\phi_*: \bigwedge^k(\mathfrak{g}) \to \bigwedge^k(\mathfrak{g})$, which in turn induces an isomorphism $\phi_*: H_k(\mathfrak{g}; \mathbb{R}) \to H_k(\mathfrak{g}, \mathbb{R})$. Using the identification of $H_k(\mathfrak{g}; \mathbb{R})$ with $H_k(M; \mathbb{R})$ afforded by Proposition 2.1, we obtain an induced isomorphism

$$\phi_*: H_k(M; \mathbb{R}) \to H_k(M; \mathbb{R}).$$

Denote by Mass_1 and Mass_2 the mass functions on $H_*(M; \mathbb{R})$ defined by the left-invariant metrics g_1 and g_2.

5.1. PROPOSITION. *The isomorphism*

$$\phi_*: (H_*(M; \mathbb{R}), \text{Mass}_1) \to (H_*(M; \mathbb{R}), \text{Mass}_2)$$

is mass-preserving.

PROOF. The left-invariant metrics g_1 and g_2 on G correspond to inner products and their associated norms $\| \ \|_1$ and $\| \ \|_2$ on the Lie algebra $\mathfrak{g}$. The isomorphism $\phi: (\mathfrak{g}, \| \ \|_1) \to (\mathfrak{g}, \| \ \|_2)$ is by definition norm-preserving, as is its transpose $\phi^t: (\mathfrak{g}^*, \| \ \|_2) \to (\mathfrak{g}^*, \| \ \|_1)$, using the dual norms. The induced isomorphisms

$$\phi_*: (\bigwedge^k \mathfrak{g}, \| \ \|_1) \to (\bigwedge^k \mathfrak{g}, \| \ \|_2)$$

and

$$\phi_*^t: (\bigwedge^k \mathfrak{g}^*, \| \ \|_2) \to (\bigwedge^k \mathfrak{g}^*, \| \ \|_1)$$

are then also norm-preserving.

Now suppose we are given a left-invariant k-form $\omega \in \bigwedge^k \mathfrak{g}^*$. Let $\text{Comass}_i(\omega)$ denote its comass in the norm $\| \ \|_i$, that is, the largest value ω takes on a decomposable k-vector which is of unit length in the corresponding norm $\| \ \|_i$. Since ϕ_* takes decomposable k-vectors of unit length in the norm $\| \ \|_1$ to the same in the norm $\| \ \|_2$, it follows that

$$\text{Comass}_1(\phi_*^t(\omega)) = \text{Comass}_2(\omega).$$

That is,

$$\phi_*^t: (\bigwedge^k \mathfrak{g}^*, \text{Comass}_2) \to (\bigwedge^k \mathfrak{g}^*, \text{Comass}_1)$$

is a comass-preserving isomorphism.

Suppose we have a left-invariant k-current $U \in \bigwedge^k \mathfrak{g}$. Then its restricted mass, $\text{Rmass}_i(U)$, will be the largest value that U takes when applied to a left-invariant k-form ω with $\text{Comass}_i(\omega) = 1$. It follows immediately that

$$\text{Rmass}_1(U) = \text{Rmass}_2(\phi_*(U)).$$

That is

$$\phi_*: (\textstyle\bigwedge^k \mathfrak{g}, \text{Rmass}_1) \to (\textstyle\bigwedge^k \mathfrak{g}, \text{Rmass}_2)$$

is a mass-preserving isomorphism.

But we saw in the previous section that for left-invariant currents, Rmass and Mass coincide. Hence

$$\phi_*: (\textstyle\bigwedge^k \mathfrak{g}, \text{Mass}_1) \to (\textstyle\bigwedge^k \mathfrak{g}, \text{Mass}_2)$$

is a mass-preserving isomorphism.

Passing to homology via Proposition 2.1, and using the observation of the preceding section that the mass of a homology class can be computed by just minimizing the masses of its left-invariant cycles, we get a mass-preserving isomorphism

$$\phi_*: (H_k(M; \mathbb{R}), \text{Mass}_1) \to (H_k(M; \mathbb{R}), \text{Mass}_2),$$

as asserted.

5.2. REMARKS. (1) Define the k-area spectrum of a compact Riemannian manifold as the collection (counting multiplicities) of masses of the k-dimensional *integral* homology classes. Although

$$\phi_*: (H_k(M; \mathbb{R}), \text{Mass}_1) \to (H_k(M; \mathbb{R}), \text{Mass}_2)$$

is mass-preserving, it does *not* follow that (M, g_1) and (M, g_2) are k-area isospectral; indeed, ϕ_* need not preserve the image of $H_k(M; \mathbb{Z})$ in $H_k(M; \mathbb{R})$, since ϕ_* is not induced by a diffeomorphism of M, but rather of its cover G. Indeed, the crux of [DGGW1] was that a certain metric deformation of a 6-dimensional manifold, although Laplace isospectral, is not 2-area isospectral, so the changing 2-area spectrum detects the deformation.

(2) If ϕ_* happens to be the identity on $H_k(M; \mathbb{R})$, then (since ϕ_* is mass-preserving) it follows that no k-dimensional homology class changes mass during the deformation. To find an isospectral deformation in which *no* homology class changes mass, then, we seek an almost-inner derivation D of a nilpotent Lie algebra $\mathfrak{g}$ which is not inner but which nonetheless induces the zero map on $H_*(\mathfrak{g}; \mathbb{R})$. The rest of this paper is devoted to finding such examples, and to exhibiting a finer geometric invariant with which to detect the nontriviality of the resulting isospectral deformations.

We conclude this section with an algebraic lemma which will simplify our calculations.

5.3. LEMMA. *Let $\mathfrak{g}$ be a Lie algebra of dimension n. Suppose that $\mathfrak{g}$ has a basis $\{X_1, \ldots, X_n\}$ relative to which the structure constants c_{ij}^k (defined* by $[X_i, X_j] = \sum_k c_{ij}^k X_k)$ satisfy $c_{ij}^k = 0$ whenever $k = i$ or $k = j$. Let $\{\alpha_1, \dots, \alpha_n\}$ be the basis of $\mathfrak{g}^*$ dual to $\{X_1, \dots, X_n\}$. Then the map $P: \bigwedge^k(\mathfrak{g}) \to \bigwedge^{n-k}(\mathfrak{g}^*)$ given by $P(U) = \iota_U(\alpha_1 \wedge \cdots \wedge \alpha_n)$ induces a "Poincaré duality" isomorphism $P: H_k(\mathfrak{g}; \mathbb{R}) \to H^{n-k}(\mathfrak{g}; \mathbb{R})$.

PROOF. It suffices to show that P is a chain map, for certainly its inverse is given (up to sign) by the interior product with $X_1 \wedge \cdots \wedge X_n$. This is a straightforward, albeit tedious, calculation; the point is that the monomials in $P\partial(U) - dP(U)$ all have coefficients of the form c_{ij}^i or c_{ji}^i, and by hypothesis, these coefficients are zero.

5.4. COROLLARY. Let D be a derivation of an n-dimensional nilpotent Lie algebra $\mathfrak{g}$. If the induced mapping on homology $D_*: H_k(\mathfrak{g}; \mathbb{R}) \to H_k(\mathfrak{g}; \mathbb{R})$ is zero for all $k \le n/2$, then it is zero for all $0 \le k \le n$.

PROOF. Being nilpotent, $\mathfrak{g}$ certainly satisfies the hypothesis of the lemma. It is also easy to see that, since $D_* = 0$ on $H_1(\mathfrak{g}; \mathbb{R})$, D is not a surjective linear endomorphism of $\mathfrak{g}$. Thus D^* maps $\alpha_1 \wedge \cdots \wedge \alpha_n$ to zero, and so it commutes with the duality map P, therefore $D^*: H^{n-k}(\mathfrak{g}; \mathbb{R}) \to H^{n-k}(\mathfrak{g}; \mathbb{R})$ is the zero map for $0 \le k \le n/2$. But $H^p(\mathfrak{g}; \mathbb{R})$ is canonically isomorphic to $H_p(\mathfrak{g}; \mathbb{R})^*$, and D^* is the transpose of D_*, so the conclusion is immediate.

We emphasize the technical point that this result does not follow from Poincaré duality for the manifold $M = \Gamma\backslash G$ in our setting, since the map $D_*: H_k(M; \mathbb{R}) \to H_k(M; \mathbb{R})$ is not induced by a diffeomorphism of M.

6. An isospectral deformation in which no homology class changes mass

By Remark 5.2, to find an isospectral deformation in which no homology class changes mass, it is sufficient to find a (noninner) almost-inner derivation of a nilpotent Lie algebra which acts as zero on homology. We now give such an example. The philosophy behind the example is to begin with a Lie algebra having as little homology as possible, so that its almost-inner derivations have a better chance to kill it all.

Dixmier [Di] proved that for any n-dimensional nilpotent Lie algebra $\mathfrak{g}$, we have the lower bound $\dim H_k(\mathfrak{g}; \mathbb{R}) \ge 2$ for $1 \le k \le n-1$ (and of course $\dim H_k(\mathfrak{g}; \mathbb{R}) = 1$ for $k = 0, n$). He then constructed a six-dimensional five-step nilpotent algebra $\mathfrak{g}$ all of whose real homology groups realize this minimum dimension. We will show that $\mathfrak{g}$ admits a noninner almost-inner derivation which induces the zero map on homology.

Dixmier's Lie algebra has basis $\{X_1, \dots, X_6\}$ and bracket relations

$$[X_1, X_2] = X_3, \qquad [X_2, X_3] = X_5,$$
$$[X_1, X_3] = X_4, \qquad [X_5, X_2] = X_6,$$
$$[X_1, X_4] = X_5, \qquad [X_3, X_4] = X_6.$$

Routine computations show that the space of all derivations of $\mathfrak{g}$ is eight-dimensional, and is generated by the inner derivations and three outer derivations. The space of almost-inner derivations is six-dimensional, generated by the inner derivations together with the derivation D which maps X_2 to X_5 and the other basis elements to zero.

By Corollary 5.4, the derivation D will be homologically trivial if it is trivial on the homology up through dimension 3. A basis for $H_1(\mathfrak{g};\mathbb{R})$ is given by the homology classes of X_1 and X_2. Since D maps X_2 to the exact cycle X_5, D is zero on H_1. The two-dimensional real homology is spanned by the homology classes of $X_1 X_5$ and $X_1 X_4 - X_2 X_3$. Again, since $X_3 X_5$ is exact, D kills $H_2(\mathfrak{g};\mathbb{R})$. The three-dimensional homology of $\mathfrak{g}$ is spanned by the homology classes of the cycles $X_1 X_4 X_5$ and $X_1 X_4 X_6 - X_2 X_3 X_6$. The latter cycle is mapped to the exact cycle $X_5 X_3 X_6$ by D, so all of $H_3(\mathfrak{g};\mathbb{R})$ is killed by D. We conclude that D induces the zero map on all of the homology of $\mathfrak{g}$.

Now let G be the simply-connected nilpotent Lie group whose Lie algebra is $\mathfrak{g}$, let Γ be the cocompact discrete subgroup of G generated by the exponentials of the Lie algebra elements $X_1, \ldots, X_6$, let M be the nilmanifold $\Gamma\backslash G$ and endow M with the left-invariant metric g_0 for which $\{X_1, \ldots, X_6\}$ is an orthonormal basis of $\mathfrak{g}$. The derivation D generates a one-parameter group of almost-inner automorphisms Φ_t of G, and we define a one-parameter family of metrics on G and hence on M by $g_t = \Phi_t^* g_0$. Theorem 1.5 shows that this is an isospectral deformation. Since each Φ_t induces the identity on the homology of M and hence by Proposition 5.1 is mass-preserving, *no* homology class of M changes mass as the metric is deformed.

However, we can still detect the change of shape of the Dixmier manifold M be examining the changing geometry of the mass-minimizing cycles in a certain codimension-one homology class. To do this, note that $\mathfrak{g}$ contains a five-dimensional ideal $\mathfrak{h}$ spanned by X_2, X_3, X_4, X_5 and X_6. Let H be the simply-connected subgroup of G whose Lie algebra is $\mathfrak{h}$; then $\Gamma' = \Gamma \cap H$ is a cocompact discrete subgroup of H, and the nilmanifold $M' = \Gamma'\backslash H$ is a codimension-one submanifold of M. The projection $G \to G/H \cong \mathbb{R}$ is a fibration whose fibers are the cosets $H \cdot \exp(sX_1)$ for $s \in \mathbb{R}$. This fibration induces a fibration of $M = \Gamma\backslash G$ over the circle $S^1 = \mathbb{Z}\backslash\mathbb{R}$ whose fiber over $s \in \mathbb{Z}\backslash\mathbb{R}$ is $M_s' = \Gamma' \cdot \exp(sX_1)\backslash H \cdot \exp(sX_1)$. The right translation R_s of G by $\exp(sX_1)$ descends to a diffeomorphism of M which maps M_0' diffeomorphically to M_s'. We will use R_s to identify fibers near M_0' with the model fiber M_0'. The time-t metric $g_t = \Phi_t^* g_0$ on M restricts to a metric $g_{s,t}$ on M_s'. We set $h_{s,t} = R_s^* g_{s,t}$ to obtain a two-parameter family of metrics on the model fiber M_0'. Note that for fixed t, the family $\{h_{s,t}\}_s$ keeps track of the way the time-t metric g_t varies as one moves from fiber to fiber, whereas for fixed s, the family $\{h_{s,t}\}_t$ reflects the way in which the restriction of the deforming metric g_t to the fiber M'_s is changing with time.

We now use the fact that the homology of the fibers M'_s is richer than that of M in order to show that the deformation $\{g_t\}_t$ is nontrivial, i.e., there is no continuous family of isometries $\sigma_t : (M, g_0) \to (M, g_t)$ with $\sigma_0 = \mathrm{Id}$. To do this, we will show that

(1) The fibers M'_s are all homologous, and for each t, they are the only g_t-mass-minimizing cycles in their common homology class. This implies that each σ_t must carry fibers to fibers; in particular, $\sigma_t(M'_0) = M'_{s(t)}$ for some continuous function s of t with $s(0) = 0$, and the composition of $R^{-1}_{s(t)}$ and σ_t will be an isometry from $(M'_0, h_{0,0})$ to $(M'_0, h_{s(t),t})$.

(2) There is an integer homology class $U \in H_2(M'_0; \mathbb{Z})$ whose mass in the metric $h_{s,t}$ is $\mathrm{Mass}_{s,t}(U) = \sqrt{1+s^2}$, independent of t. But since σ_t is homotopic to the identity, it preserves the homology class U, and since σ_t is an isometry, it is mass-preserving. Therefore, $\mathrm{Mass}_{0,0}(U) = \mathrm{Mass}_{s(t),t}(U)$, and it follows that $s(t)$ is identically zero, i.e., each σ_t carries M'_0 to itself.

(3) The mass of a certain integer class $V \in H_2(M'_0; \mathbb{Z})$ changes continuously as t changes.

By (1) and (2), σ_t restricts to an isometry from $(M'_0, h_{0,0})$ to $(M'_0, h_{0,t})$, so we must have $\mathrm{Mass}_{0,t}(V) = \mathrm{Mass}_{0,t}(\sigma_t(V)) = \mathrm{Mass}_{0,0}(V)$, contradicting (3). Therefore, the family $\{\sigma_t\}_t$ of isometries cannot exist.

We begin by proving (1). Since R_s carries M'_0 diffeomorphically to M'_s and is homotopic to the identity, we conclude that M'_0 and M'_s are homologous cycles in M. To show that these fibers are the only g_t-mass-minimizing cycles in their common homology class, recall that D is the almost-inner derivation of the Lie algebra $\mathfrak{g}$ which takes X_2 to X_5 and the remaining basis elements to zero, and that g_0 is the left-invariant metric on the six-dimensional manifold M which makes the basis $X_1, X_2, X_3, X_4, X_5, X_6$ of left-invariant vector fields orthonormal. Then the left-invariant metric g_t associated with the almost-inner derivation tD has the orthonormal basis

$$X_1, \quad X_2(t) = X_2 - tX_5, \quad X_3, \quad X_4, \quad X_5, \quad X_6.$$

The dual 1-forms will then also be orthonormal in the metric g_t:

$$\alpha_1, \quad \alpha_2, \quad \alpha_3, \quad \alpha_4, \quad \alpha_5(t) = \alpha_5 + t\alpha_2, \quad \alpha_6.$$

It follows that the decomposable 5-form $\alpha_2\alpha_3\alpha_4\alpha_5\alpha_6$ has g_t-norm equal to 1, and therefore also g_t-comass equal to 1. Likewise, the closed left-invariant 5-current $X_2X_3X_4X_5X_6$ has g_t-norm equal to 1, and g_t-mass equal to 1. Hence the 5-form calibrates the 5-current, showing it to be mass-minimizing in its homology class. Since this current is tangent to the fibers M'_s, we conclude that combinations of fibers are the only g_t-mass-minimizing cycles in this homology class.

We now turn to the proof of (2). Let $\overline{X}_2, \ldots, \overline{X}_6$ denote the restrictions of $X_2, \ldots, X_6$ to H (or to $\Gamma'\backslash H$). Note that $\overline{X}_3$ and $\overline{X}_5$ commute, and hence they span an involutive distribution whose integral submanifolds are tori. Therefore $\overline{X}_3\overline{X}_5$ represents an *integer* homology class U of M_0'. The t-parameter deformation $\{h_{s,t}\}_t$ arises from the restriction to $\mathfrak{h}$ of the derivation D which sends $\overline{X}_2 \mapsto \overline{X}_5$ and which annihilates the other basis vectors, while a computation (which will be done in general in the proof of Proposition 7.4) shows that the s-parameter deformation $\{h_{s,t}\}_s$ arises from the restriction of $-\operatorname{ad} X_1$ to $\mathfrak{h}$, which is a noninner derivation of $\mathfrak{h}$. The effects of D and $-\operatorname{ad} X_1$ on U are quite different. Indeed, D annihilates U, so that by Remark 5.2, the $h_{s,t}$-mass of U is independent of t. On the other hand, the map induced on the homology of M_0' by $-\operatorname{ad} X_1$ carries the class of $\overline{X}_3\overline{X}_5$ to that of $\overline{X}_4\overline{X}_5$. Since these classes are nonzero and distinct in $H_2(M_0';\mathbb{R})$, the mass of U might change with s. In fact, a calculation using calibrations (see [DGGW3]) shows that its $h_{s,t}$-mass is $\sqrt{1+s^2}$.

Finally, to prove (3), consider the class $V \in H_2(M_0';\mathbb{R})$ represented by $\overline{X}_2\overline{X}_4$. One can show (using calibrations) that the left-invariant two-current $\overline{X}_2(t)\overline{X}_4 - (t/2)\overline{X}_4\overline{X}_5 - (t/2)\overline{X}_2(t)\overline{X}_6 - (t^2/4)\overline{X}_5\overline{X}_6$ has minimum mass $1+(t^2/4)$ in this homology class, so the mass of V changes with t. This completes the proof of the nontriviality of the deformation $\{g_t\}_t$.

This example and similar ones suggested to us that perhaps all nontrivial isospectral deformations on Riemannian nilmanifolds can be detected by the changing geometry of volume-minimizing cycles in some codimension-one homology class. In §8, we shall prove this conjecture in the case of two-step nilmanifolds.

7. Codimension-one homology classes

In this section we prepare the groundwork for showing that, for two-step nilmanifolds, the changing geometry of the fibers in a codimension-one fibration always reveals the nontriviality of isospectral deformations induced by almost-inner automorphisms. However, the results of the present section are valid for nilmanifolds of arbitrary step-size.

Let G be an n-dimensional nilpotent Lie group with Lie algebra $\mathfrak{g}$, and let Γ be a cocompact discrete subgroup. As usual, set $M = \Gamma\backslash G$. Every $(n-1)$-dimensional left-invariant current $U \in \bigwedge^{n-1}(\mathfrak{g})$ is decomposable, say $U = X_1 \wedge \cdots \wedge X_{n-1}$. Write $\mathfrak{h} = \operatorname{span}(U)$ for the subspace of $\mathfrak{g}$ spanned by the X_i's.

7.1. LEMMA. *The map $\partial: \bigwedge^n(\mathfrak{g}) \to \bigwedge^{n-1}(\mathfrak{g})$ is identically zero. In other words, no nonzero left-invariant $(n-1)$-current is exact. For $U \in \bigwedge^{n-1}(\mathfrak{g})$ and $\mathfrak{h} = \operatorname{span}(U)$, the following are equivalent*:

(i) $[\mathfrak{g},\mathfrak{g}] \subset \mathfrak{h}$.

(ii) $\mathfrak{h}$ *is an ideal of* $\mathfrak{g}$.

(iii) U *is a closed current.*

The proof is elementary and uses only the fact that the central series of $\mathfrak{g}$ satisfies $[\mathfrak{g}^{(i)}, \mathfrak{g}^{(j)}] \subset \mathfrak{g}^{(i+j+1)}$.

By Lemma 7.1 and Proposition 2.1, we find that $H_{n-1}(M; \mathbb{R})$ is isomorphic to the subspace of $\bigwedge^{n-1}(\mathfrak{g})$ consisting of all currents U such that $\operatorname{span}(U)$ is an ideal of $\mathfrak{g}$. Each such ideal $\mathfrak{h}$ defines an integrable distribution on M, i.e., a foliation of M.

7.2. PROPOSITION. *If $\mathfrak{h}$ is a rational $(n-1)$-dimensional ideal of $\mathfrak{g}$, then the leaves of the associated foliation of M are compact; in fact, they are all right translates of the compact nilmanifold $(H \cap \Gamma)\backslash H$, where H is the connected subgroup of G with Lie algebra $\mathfrak{h}$. Moreover, the leaves are the fibers of a submersion $p: M \to S^1$.*

PROOF. Since $\mathfrak{h}$ is a left-invariant distribution, the leaves of the associated foliation of G are the cosets $aH = Ha$, $a \in G$, of the normal subgroup H. Recalling that the right translation R_a descends to a diffeomorphism R_a of M, we see that the leaves of the foliation of M are the right translates of $(H \cap \Gamma)\backslash H$. An element X of $\log \Gamma$ is *prime* if it is not a nontrivial integer multiple of an element of $\log \Gamma$. Choosing a prime element X of $\log \Gamma$ which is not in $\mathfrak{h}$, we obtain a commutative diagram:

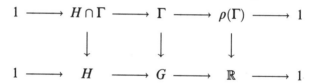

where $\rho(\exp(sX + U)) = s$ for all $s \in \mathbb{R}$, $U \in \mathfrak{h}$. Since X is prime, $\exp(sX) \in \Gamma$ precisely when $s \in \mathbb{Z}$, so $\rho(\Gamma) = \mathbb{Z}$. Thus ρ descends to a submersion $\rho: M \to \mathbb{Z}\backslash\mathbb{R} = S^1$, whose fibers are the leaves of the foliation determined by $\mathfrak{h}$.

Let $F = F_0 = (H \cap \Gamma)\backslash H$ denote the model fiber of the fibration $F \to M \xrightarrow{\pi} S^1$ constructed in Proposition 7.2. For $s \in S^1$, let $F_s = \pi^{-1}(s)$ denote the fiber above s. We now turn our attention to the masses of the fibers.

7.3. PROPOSITION. *The fibers F_s all represent the same homology class $[F] \in H_{n-1}(M; \mathbb{R})$. For any left-invariant metric g on M, each fiber F_s is g-mass-minimizing in the class $[F]$, and these fibers are the only mass-minimizing cycles in $[F]$.*

PROOF. Choose a g-orthonormal basis $\{V_1, \dots, V_{n-1}\}$ of $\mathfrak{h}$, so that $\|V_1 \wedge \cdots \wedge V_{n-1}\| = 1$. Extend to an orthonormal basis $\{V_1, \dots, V_{n-1}, W\}$ of $\mathfrak{g}$, and let $\{\alpha^1, \dots, \alpha^{n-1}, \beta\}$ denote the dual basis of $\mathfrak{g}^*$. Let U be the $(n-1)$-current $\operatorname{vol}(F) \cdot V_1 \wedge \cdots \wedge V_{n-1}$. All the F_s are compact, and are integral manifolds of the closed current U, hence $\int_{F_s} \omega = \int_{F_0} \omega$ for any left-invariant $(n-1)$-form ω. Since every cohomology class is represented by a left-invariant form (Nomizu's theorem), it follows that $[F_s] = [F_0] = [F]$. To show that F_0 (and hence each F_s) is volume-minimizing in $[F]$, note that $\alpha^1 \wedge \cdots \wedge \alpha^{n-1}$ has comass 1, is closed and calibrates the sumbanifolds F_s for all s. Since the fibers F_s are the only cycles calibrated by $\alpha^1 \wedge \cdots \wedge \alpha^{n-1}$, we conclude that the fibers F_s are mass-minimizing, and are the only mass-minimizing cycles in $[F]$.

We now study how the Riemannian geometry induced on the fibers F_s by the metric g varies with s.

7.4. PROPOSITION. *Let X be an element of $\log \Gamma$ which is not in $\mathfrak{h}$, as in the proof of Proposition 7.2, and let $\{\Psi_s\}_s$ be the one-parameter subgroup of automorphisms of F given by $\Psi_s = \mathrm{Ad}(\exp(-sX))$ (i.e., conjugation by $\exp(-sX)$). Then the restriction of the metric g to the fiber F_s is isometric to the restriction of the pullback metric $\Psi_s^* g$ to the model fiber F_0.*

PROOF. Using the notation of Proposition 7.2, recall that the one-parameter subgroup $\{\exp(sX)\}$ of G is transverse to the cosets of H, and that the fibers F_s (resp. H_s) of the submersion $M \to S^1$ (resp. of $G \to \mathbb{R}$) are labelled by the value of s for which $\exp(sX) \in F_s$ (resp. H_s, note that H_s is just the right coset $H(\exp(sX))$. Let L_s, R_s and I_s be the diffeomorphisms of G given respectively by left translation, right translation and conjugation by $\exp(sX) \in G$. We first show that R_s is an isometry from $(H_0, (\Psi_s^* g)|_{H_0})$ to $(H_s, g|_{H_s})$. For this purpose, it suffices to show that $R_s^* = \Psi_s^*$ on G. Since g is left-invariant,

$$R_s^* g = R_s^* \circ (L_s^{-1})^* g = (I_s^{-1})^* g,$$

where $I_s = L_s \circ R_s^{-1}$. But $(I_s)_* = \exp(s(\mathrm{ad}\, X))$, so $(I_s^{-1})_* = \exp(-s(\mathrm{ad}\, X)) = \Psi_s *$ and thus $R_s^* g = \Psi_s^* g$. Now the metric on G descends to the metric on M upon factoring out the lattice Γ, and in particular the metric on H_s descends to that of F_s. The right translations descend to diffeomorphisms of M, so we also have that R_s is an isometry from $(F_0, (\Psi_s^* g)|_{F_0})$ to $(F_s, g|_{F_s})$, as claimed.

7.5. PROPOSITION. *Let Φ_t be a one-parameter group of almost-inner automorphisms of G and let $g_t = \Phi_t^* g$ be an associated isospectral family of metrics on M. Then the g_t-mass of the codimension-one homology class $[F]$ described above is independent of t, i.e., the mass of $[F]$ does not change during the deformation.*

PROOF. Note that since any almost-inner derivation maps $\mathfrak{g}$ into $[\mathfrak{g}, \mathfrak{g}]$, each Φ_t restricts to a nilpotent automorphism of H. Moreover, denoting by $\bar{g}_t$ the restriction of g to a left-invariant metric on $F_0 = (H \cap \Gamma)\backslash H$, we have $\bar{g}_t = \Phi_t^* \bar{g}$. Since Φ_t is unipotent, $\bar{g}_t$ and $\bar{g}$ have the same volume; i.e., F_0 does not change mass during the deformation. But all the fibers have the same g_t-mass as F_0.

Our goal now is to show that the nontriviality of the deformation g_t can be revealed by the changing geometry of the minimal cycles in certain $(n-1)$-dimensional homology classes, even though Proposition 7.5 says that their volumes do not change. In other words, during the deformation, the geometry of each fiber changes in a way different from the way the geometry changes as we move from fiber to fiber for a fixed metric. We now make this notion precise.

7.6. DEFINITION. Let $\mathfrak{h}$ be a rational ideal in $\mathfrak{g}$, let F_s, $s \in S^1$, be the fibers of the associated fibration, and let $[F]$ be their common homology class. We will say that a continuous family of metrics $\{g_t\}_{t \geq 0}$ induces a *trivial metric deformation of* $[F]$ if for each t_0 there exists a continuous map $\sigma: [t_0, t_0 + \epsilon) \to \mathbb{R}$ for some $0 < \epsilon < 1$ such that $\sigma(t_0) = 0$ and there exists a continuous family of isometries $\{\tau_t\}$ from (F_0, g_{t_0}) to $(F_{\sigma(t)}, g_{t, \sigma(t)})$ for $t_0 \leq t < t_0 + \epsilon$, where $g_{t,s}$ denotes the restriction of g_t to F_s. Otherwise, we say that $\{g_t\}$ induces a *nontrivial metric deformation of* $[F]$.

7.7. PROPOSITION. *Let* $M = \Gamma \backslash G$ *be a nilmanifold, and* $[F]$ *a codimension-one integer homology class as in Definition 7.6. If* $\{g_t\}$ *is a trivial left-invariant deformation of* $g = g_0$ *on* M *(i.e., if there exists a continuous family of diffeomorphisms* τ_t *of* M *such that* $g_t = \tau_t^* g$ *and* $\tau_0 = \mathrm{Id}$*), then* $\{g_t\}$ *induces a trivial metric deformation of* $[F]$.

PROOF. The diffeomorphisms τ_t, being homotopic to the identity, preserve the homology class $[F]$. Thus τ_t maps g-mass-minimizing cycles in $[F]$ to g_t-mass-minimizing cycles in $[F]$. But by Proposition 7.3, the g_t-mass-minimizing cycles for any t are precisely the fibers. Hence τ_t maps F_0 to some fiber $F_{\sigma(t)}$ with $\sigma(0) = 0$ and σ continuous, and τ_t restricts to an isometry from $g_{0,0}$ to $g_{t,\sigma(t)}$. Thus the triviality condition of Definition 7.6 holds for $t_0 = 0$. The argument is easily adjusted for arbitrary t_0. Thus $\{g_t\}$ induces a trivial metric deformation of $[F]$, as claimed.

We conclude from Proposition 7.7 that in order to reveal the nontriviality of a deformation $\{g_t\}$, it suffices to find a codimension-one homology class on which $\{g_t\}$ induces a nontrivial metric deformation.

8. Detecting the changing geometry of isospectral two-step nilmanifolds

We now show that for every nontrivial isospectral deformation of a two-step nilmanifold, the change in geometry can be detected by a change in the geometry of the volume-minimizing submanifolds in some codimension-one homology class. More precisely, in the language of Definition 7.6, we prove:

8.1. THEOREM. *Let* G *be a simply-connected, two-step nilpotent Lie group, and* $M = \Gamma \backslash G$ *a compact nilmanifold. Let* g *be a left-invariant metric on* G *and* Φ_t *a one-parameter group of almost-inner automorphisms of* G. *Set* $g_t = \Phi_t^* g$. *Suppose that* $\{\Phi_t\}_{t \in \mathbb{R}}$ *is not contained in* $\mathrm{Inn}(G)$. *Then the* *deformation $\{g_t\}$ of g on M induces a nontrivial metric deformation of some codimension-one homology class $[F]$ associated with some rational ideal $\mathfrak{h}$.*

The ideal $\mathfrak{h}$ is obtained via the following:

8.2. LEMMA. *Let $\mathfrak{g}$ be a two-step nilpotent Lie algebra of dimension ≥ 2, and $D: \mathfrak{g} \to \mathfrak{g}$ a derivation. Suppose that on each codimension-one ideal of $\mathfrak{g}$, the derivation D agrees with some inner derivation of $\mathfrak{g}$. Then D is inner.*

In particular, Lemma 8.2 says that if D is a noninner derivation of $\mathfrak{g}$, then there exists a codimension-one rational ideal $\mathfrak{h}$ of $\mathfrak{g}$ such that $D|_{\mathfrak{h}} \neq \operatorname{ad} X|_{\mathfrak{h}}$ for any $X \in \mathfrak{g}$. Thus, not only is D not inner on $\mathfrak{h}$, but it is not even the restriction of an inner derivation of $\mathfrak{g}$.

PROOF OF THEOREM 8.1. We assume Lemma 8.2 for now and prove Theorem 8.1. We will prove our theorem inductively, and in order that the inductive hypothesis be readily applicable, we reformulate the theorem slightly more generally. Fix a subspace $\mathfrak{z}$ of the center of $\mathfrak{g}$ containing $[\mathfrak{g}, \mathfrak{g}]$. Define

$$\mathscr{D}_{\mathfrak{z}}(\mathfrak{g}) = \{D \in \operatorname{Der} \mathfrak{g}) : D(\mathfrak{g}) \subseteq \mathfrak{z}, \ D(\mathfrak{z}) = 0\}.$$

Thus $\mathscr{D}_{\mathfrak{z}}(\mathfrak{g})$ consists of nilpotent derivations whose squares are zero. It is easily verified that $\mathscr{D}_{\mathfrak{z}}(\mathfrak{g})$ is an abelian subalgebra of $\operatorname{Der}(\mathfrak{g})$, thus the matrix exponential $\exp: \operatorname{Der}(\mathfrak{g}) \to \operatorname{Aut}(\mathfrak{g})$ carries $\mathscr{D}_{\mathfrak{z}}(\mathfrak{g})$ diffeomorphically onto the abelian subgroup $\mathscr{A}_{\mathfrak{z}}(\mathfrak{g}) = \exp(\mathscr{D}_{\mathfrak{z}}(\mathfrak{g})) = \{\operatorname{Id} + D : D \in \mathscr{D}_{\mathfrak{z}}(\mathfrak{g})\}$. Since G is simply-connected, each element of $\mathscr{A}_{\mathfrak{z}}(\mathfrak{g})$ is the differential of an automorphism of G. We denote by $\mathscr{A}_{\mathfrak{z}}(G)$ the abelian subgroup of all automorphisms of G whose differentials lie in $\mathscr{A}_{\mathfrak{z}}(\mathfrak{g})$. Since every almost-inner derivation lies in $\mathscr{D}_{\mathfrak{z}}(\mathfrak{g})$, we see that $\mathscr{A}_{\mathfrak{z}}(G)$ contains the group of almost-inner automorphisms. We now prove Theorem 8.1 under the weaker hypothesis that $\{\Phi_t\}_t$ is a curve in $\mathscr{A}_{\mathfrak{z}}(G)$ with $\Phi_0 = \operatorname{Id}$ (note that $\{\Phi_t\}$ need not be a one-parameter group).

We prove the theorem by induction on $\dim G$. Thus, assume that the theorem holds for groups of dimension less than n. Notice that the induction hypothesis, along with Proposition 7.3, implies that any deformation $g_t = \Phi_t^* g$ of metrics on a lower-dimensional nilmanifold $\Gamma' \backslash G'$, with $\{\Phi_t\}_t$ a curve not completely contained in $\operatorname{Inn}(G')$, is nontrivial.

Now assume that $\dim G = n$. Set $\phi_t = \Phi_{t*} \in \mathscr{A}_{\mathfrak{z}}(\mathfrak{g})$. Suppose that the curve ϕ_t strays away from $\operatorname{Int}(\mathfrak{g})$ (the inner *automorphisms* of $\mathfrak{g}$), say $\phi_{t_0} \notin \operatorname{Int}(\mathfrak{g})$; then $\{\phi_t\}$ determines a curve $\{D_t = \log \phi_t\}$ (so $\phi_t = \operatorname{Id} + D_t$) with $D_{t_0} \notin \operatorname{ad}(\mathfrak{g})$. Choose a codimension-one ideal $\mathfrak{h}$ as in Lemma 8.2 so that $D_{t_0}|_{\mathfrak{h}} \notin (\operatorname{ad} \mathfrak{g})|_{\mathfrak{h}}$. Since $(\operatorname{ad} \mathfrak{g})|_{\mathfrak{h}}$ is a closed subset of $\operatorname{Der} \mathfrak{h}$, we see that $D_t|_{\mathfrak{h}} \notin (\operatorname{ad} \mathfrak{g})|_{\mathfrak{h}}$ for all t in some interval $(t_0 - \epsilon, t_0 + \epsilon)$.

Let $g_{t,s}$ be the restriction of g_t to F_s as before. By Proposition 7.4,

$$(8.3) \qquad (F_s, g_{t,s}) \simeq (F_0, \Psi_s^* g_t|_{F_0}) = (F_0, (\Phi_t \circ \Psi_s)^* g_t|_{F_0})$$

with $\Psi_s = \mathrm{Ad}(\exp(-sX))$ for some $X \in \mathfrak{g}$. Now observe that the two-parameter family of automorphisms $\{\Phi_t \circ \Psi_s : s \in S^1, \ t \in (t_0 - \epsilon, t_0 + \epsilon)\}$ of the nilpotent Lie group H (whose Lie algebra is $\mathfrak{h}$) lies completely outside of $\mathrm{Inn}(H)$. Indeed, $\Phi_t|_H$ and $\Psi_s|_H$ lie in the abelian group $\mathscr{A}_3(H)$ and $\log((\Phi_t \circ \Psi_s)_*) = D_t|_{\mathfrak{h}} - s\,\mathrm{ad}\,X|_{\mathfrak{h}} \notin (\mathrm{ad}\,\mathfrak{g})|_{\mathfrak{h}}$ for any $s \in S^1$, $t \in (t_0 - \epsilon, t_0 + \epsilon)$, since $D_t \notin (\mathrm{ad}\,\mathfrak{g})|_{\mathfrak{h}}$. In particular, by the inductive hypothesis, if we take any curve $\sigma : (t_0 - \epsilon, t_0 + \epsilon) \to \mathbb{R}$ with $\sigma_{t_0} = 0$, then $(\Phi_t \circ \Psi_{\sigma(t)})^* g_t|_{F_0}$ is a nontrivial deformation of metrics on F_0. Equivalently, by (8.3), there does *not* exist such a curve σ together with isometries $\eta_t : (F_0, g_{t_0}) \to (F_{\sigma(t)}, g_{t, \sigma(t)})$. Thus $\{g_t\}_t$ induces a nontrivial metric deformation of $[\bar{F}]$, and the theorem is proved.

PROOF OF LEMMA 8.2. Assume that D is a rationally almost-inner derivation of $\mathfrak{g}$ and that for any codimension-one rational ideal $\mathfrak{h}$, there is some $U_{\mathfrak{h}} \in \mathfrak{g}$ such that $D|_{\mathfrak{h}} = (\mathrm{ad}\,U_{\mathfrak{h}})|_{\mathfrak{h}}$. We must show that D is inner. Let $\mathfrak{a}$ be a rational subspace complementary to $[\mathfrak{g}, \mathfrak{g}]$, and let $r = \dim \mathfrak{a}$. Note that $D([\mathfrak{g}, \mathfrak{g}]) = 0$, since $[\mathfrak{g}, \mathfrak{g}]$ is contained in the center of $\mathfrak{g}$, and D, being rationally almost-inner, vanishes on the center of $\mathfrak{g}$. The codimension-one rational ideals of $\mathfrak{g}$ are precisely the codimension-one rational subspaces of $\mathfrak{g}$ containing $[\mathfrak{g}, \mathfrak{g}]$, and hence correspond bijectively to the codimension-one rational subspaces of $\mathfrak{a}$.

We now establish the following claim for repeated use later:

(*)   If D is not inner, then given any rational elements $X_1, \ldots, X_l, Y \in \mathfrak{a}$ with $l < r$, and $Y \notin \mathrm{span}\{X_1, \ldots, X_l\}$, there is some $U \in \mathfrak{a}$ such that D agrees with $\mathrm{ad}\,U$ on $\mathrm{span}\{X_1, \ldots, X_l\}$ but $D(Y) \neq [U, Y]$.

Indeed, let $\mathfrak{h}$ be a codimension-one rational ideal containing $X_1, \ldots, X_l$ but not Y. By hypothesis, D does agree with $\mathrm{ad}\,U_{\mathfrak{h}}$ on $\mathfrak{h}$; it cannot agree with $\mathrm{ad}\,U_{\mathfrak{h}}$ on Y as well without being inner ($= \mathrm{ad}\,U_{\mathfrak{h}}$).

Next, we use (*) to show that rational elements of $\mathfrak{g}$ have large centralizers. Let $C_{\mathfrak{a}}(X)$ denote the set of elements of $\mathfrak{a}$ which commute with X. We claim that for every rational vector $X \in \mathfrak{a}$, $\dim C_{\mathfrak{a}}(X) \geq r - 1$, i.e., $\mathrm{rk}(\mathrm{ad}\,X) \leq 1$.

To prove this, let $X_1 = X$, and extend X_1 to a rational basis $\{X_1, \ldots, X_r\}$ of $\mathfrak{a}$. Use (*) to produce an element $U_1 \in \mathfrak{a}$ such that $D(X_1) = [U_1, X_1]$, but $D(X_2) \neq [U_1, X_2]$. Now use (*) again to produce $U_2 \in \mathfrak{a}$ such that D agrees with $\mathrm{ad}\,U_2$ on $\mathrm{span}\{X_1, X_2\}$ but $D(X_3) \neq [U_2, X_3]$. Continue using (*) in this manner to produce U_i such that D agrees with $\mathrm{ad}\,U_i$ on $\mathrm{span}\{X_1, \ldots, X_i\}$ but $D(X_{i+1}) \neq [U_i, X_{i+1}]$ for $i = 1, \ldots, r-1$. Then note that $Y_i = U_{i+1} - U_i$ commutes with X_1 for $i = 1, \ldots, r-2$.

Note that we can now assume, without loss of generality, that D is zero on $\mathrm{span}\{X_1, \ldots, X_{r-2}\}$, since the derivation $D - \mathrm{ad}(U_{r-2})$ differs from D by an inner derivation, which affects neither the hypotheses nor the conclusion of the lemma. We will assume that D has been so replaced in what follows.

We need to show that $X = X_1$ commutes with one more independent element of $\mathfrak{a}$. For each ordered pair $(a, b) \in \mathbb{Q}^2$, the codimension-one subspace $\text{span}\{X_1, \dots, X_{r-2}, aX_{r-1} + bX_r\}$ of $\mathfrak{a}$ determines a codimension-one rational ideal $\mathfrak{h}$, so there is some $U_{a,b} \in \mathfrak{a}$ such that D agrees with $\text{ad}(U_{a,b})$ on this subspace. Since D vanishes on $\text{span}\{X_1, \dots, X_{r-2}\}$, $U_{a,b}$ commutes with $X_1, \dots, X_{r-2}$.

If the $U_{a,b}$ are all the same, then D is inner. If the $U_{a,b}$ are all multiples of a fixed vector U (we will write $U_{a,b} = c(a, b)U$, with $c(a, b) \in \mathbb{R}$), then we consider two cases:

CASE 1. $D(X_{r-1})$ and $D(X_r)$ are linearly independent. Then

$$0 \neq D(X_{r-1}) = [U_{1,0}, X_{r-1}] = c(1, 0)[U, X_{r-1}],$$

so $c(1, 0) \neq 0$. Similarly, $c(0, 1) \neq 0$. Now for $a, b \in \mathbb{R}$,

$$\begin{aligned}
aD(X_{r-1}) + bD(X_r) &= [U_{a,b}, aX_{r-1} + bX_r] \\
&= c(a, b)[U, aX_{r-1} + bX_r] \\
&= c(a, b)\left(\frac{a}{c(1, 0)}D(X_{r-1}) + \frac{b}{c(0, 1)}D(X_r)\right),
\end{aligned}$$

i.e., $a\left(1 - \frac{c(a,b)}{c(1,0)}\right)D(X_{r-1}) + b\left(1 - \frac{c(a,b)}{c(0,1)}\right)D(X_r) = 0$. By linear independence of $D(X_{r-1})$ and $D(X_r)$ we conclude that $c(a, b) = c(0, 1) = c(1, 0)$ for all a, b, so D is inner in this case.

CASE 2. $D(X_{r-1})$ and $D(X_r)$ are linearly dependent. First assume that $[U, X_{r-1}]$ and $[U, X_r]$ are linearly independent; then the linear dependence of $D(X_{r-1})$ and $D(X_r)$ implies that at least one of $c(1, 0)$ and $c(0, 1)$ is zero. Unless D is inner $(= 0)$, the other must be nonzero, so we assume that $c(1, 0) = 0$ and $c(0, 1) \neq 0$, i.e., $D(X_{r-1}) = 0$ and $D(X_r) \neq 0$. But then $c(0, 1)[U, X_r] = D(X_r) = D(X_{r-1} + X_r) = c(1, 1)[U, X_{r-1} + X_r]$, which contradicts the linear independence of $[U, X_{r-1}]$ and $[U, X_r]$.

On the other hand, if $[U, X_{r-1}]$ and $[U, X_r]$ are linearly dependent, there is some nonzero $W_1 \in \text{span}\{X_{r-1}, X_r\}$ such that $[U, W_1] = 0$. Extend to a basis $\{W_1, W_2\}$ of $\text{span}\{X_{r-1}, X_r\}$. Thus $D(W_1) = 0$ and $D(W_2) = c[U, W_2]$ for some $c \in \mathbb{R}$. But now D agrees with $\text{ad } cU$, and hence is inner.

We now know that if $\dim \text{span}\{U_{a,b} : (a, b) \in \mathbb{Q}^2\} \leq 1$, then D is inner. Now suppose Y_{r-2} and Y_{r-1} are linearly independent elements of $\text{span}\{U_{a,b} : (a, b) \in \mathbb{Q}^2\}$, and recall the elements $Y_1, \dots, Y_{r-3}$ already constructed. For $k < r - 2$, Y_k commutes with $X_1, \dots, X_k$ but not with X_{k+1}. Since for $i \leq r - 2$, each X_i commutes with $Y_i, \dots, Y_{r-1}$ but not with Y_{i-1}, we see easily that $Y_1, \dots, Y_{r-1}$ are linearly independent elements of $C_{\mathfrak{a}}(X_1)$, from which the conclusion $\dim C_{\mathfrak{a}}(X) \geq r - 1$ follows. We now know that $\text{rk}(\text{ad } X) \leq 1$ for all rational $X \in \mathfrak{a}$.

Finally, we claim that $\dim[\mathfrak{g}, \mathfrak{g}] \leq 1$. Otherwise, there are rational vectors $X_1, Y_1, X_2, Y_2 \in \mathfrak{a}$ with $[X_1, Y_1] = Z_1$ and $[X_2, Y_2] = Z_2$, with Z_1 and Z_2 linearly independent. But then $[X_1, Y_2] \in \mathrm{im}(\mathrm{ad}\, X_1) \cap \mathrm{im}(\mathrm{ad}\, Y_2) = \mathbb{R} \cdot Z_1 \cap \mathbb{R} \cdot Z_2 = \{0\}$, and likewise $[X_2, Y_1] = 0$. It follows that $[X_1 + X_2, Y_1] = Z_1$ and $[X_1 + X_2, Y_2] = Z_2$, which contradicts the fact that $\mathrm{rk}\,\mathrm{ad}(X_1 + X_2) \leq 1$. Thus $\dim[\mathfrak{g}, \mathfrak{g}] \leq 1$.

Hence $\mathfrak{g}$ is a direct product of a Heisenberg algebra and an abelian Lie algebra. But every rationally almost-inner derivation of such an algebra is inner. Thus D is inner, and the proof of Lemma 8.2 is complete.

REFERENCES

[DGGW1] D. DeTurck, H. Gluck, C. Gordon, and D. Webb, *You can not hear the mass of a homology class*, Comment. Math. Helv. **64** (1989), 589–617.

[DGGW2] _____, *How can a drum change shape, while sounding the same?*, Differential Geometry: A Symposium in Honor of Manfredo do Carmo (B. Lawson and K. Tenenblat, eds.), Pitman Surveys in Pure and Appl. Math., vol. 52, Longman Scientific and Technical, 1991, pp. 111–122.

[DGGW3] _____, *How can a drum change shape, while sounding the same?*, Part 2, Mechanics, Analysis and Geometry, 200 Years after Lagrange (M. Francaviglia, ed.), Elsevier Science Publishers, 1991, pp. 335–358.

[DG] D. DeTurck and C. Gordon, *Isospectral deformations* II: *Trace formulas, metrics and potentials*, Comm. Pure Appl. Math. **40** (1989), 1067–1095.

[Di] J. Dixmier, *Cohomologie des algèbres de Lie nilpotentes*, Acta Sci. Math. **16** (1955), 246–250.

[Fe] H. Federer, *Geometric measure theory*, Springer-Verlag, New York, 1969.

[Fe-Fl] H. Federer and W. Fleming, *Normal and integral currents*, Ann. of Math. **72** (1960), 458–520.

[G] C. Gordon, *The Laplace spectra versus the length spectra of Riemannian manifolds*, Contemp. Math. **51** (1986), 63–79.

[GW] C. Gordon and E. Wilson, *Isospectral deformations of compact solvmanifolds*, J. Differential Geom. **19** (1984), 241–256.

[HL] R. Harvey and B. Lawson, *Calibrated geometries*, Acta Math. **147** (1981), 47–157.

[M] F. Morgan, *Geometric measure theory, a beginner's guide*, Academic Press, New York, 1988.

[No] K. Nomizu, *On the cohomology of homogeneous spaces of nilpotent Lie groups*, Ann. of Math. **59** (1954), 531–538.

[R] M. S. Raghunathan, *Discrete subgroups of Lie groups*, Springer-Verlag, New York, 1972.

[S] T. Sunada, *Riemannian coverings and isospectral manifolds*, Ann. of Math. **121** (1985), 169–186.

UNIVERSITY OF PENNSYLVANIA

DARTMOUTH COLLEGE

Proceedings of Symposia in Pure Mathematics
Volume **54** (1993), Part 3

Quadric Representation of a Submanifold
and Spectral Geometry

IVKO DIMITRIĆ

ABSTRACT. This paper is of a survey character. In the first section we give
some results on finite type submanifolds, i.e., isometric immersions of Riemannian manifolds into Euclidean spaces which are decomposable into sums
of finitely many vector eigenfunctions of the Laplacian. For an immersion
$x\colon M^n \to E^m$ we let $\tilde{x} = xx^t$ and call this map into the set of symmetric
matrices the quadric representation of M. In the second section we discuss
some recent results on submanifolds with finite type quadric representation.
In the last section it is indicated how study of finite type submanifolds helps
determine few eigenvalues of the Laplacian and produces sharp estimates on
low eigenvalues for certain kinds of submanifolds.

1. Finite type submanifolds

It is generally acknowledged that the study of minimal submanifolds (i.e.,
immersions that are critical points for the volume functional) occupies an
important place in differential geometry. Minimal submanifolds are also
characterized by the vanishing of the mean curvature vector, $H = 0$. Now if
$x\colon M^n \to E^m$ is an isometric immersion of a smooth Riemannian manifold
into a Euclidean space then since $\Delta x = -nH$ we see that for a minimal
immersion, x is a vector eigenfunction of the Laplacian for the eigenvalue
0. Similarly if $x\colon M^n \to S^{m-1} \subset E^m$ is a minimal immersion into a unit
hypersphere centered at the origin then $H = -x$ and $\Delta x = nx$, so that x
is again an eigenfunction. Actually, the converse is also true; i.e., $\Delta x = \lambda x$
if and only if either M is minimal in E^m (if $\lambda = 0$) or M is minimal
in the hypersphere S^{m-1} (if $\lambda \neq 0$, actually $\lambda > 0$) [T1]. We see that in
the above two cases the immersion x is built using only one eigenspace of
the Laplacian, $x = x_t$, $\Delta x_t = \lambda_t x_t$. More generally, if the immersion x is

1991 *Mathematics Subject Classification.* Primary 53C40; Secondary 53-02, 53A07, 53A10,
53C42, 58G25.
This paper is in final form and no version of it will be submitted for publication elsewhere.

decomposable into a sum of finitely many vector functions on M,

$$x = x_0 + x_{t_1} + \cdots + x_{t_k} \tag{1}$$

where x_0 is a constant mapping and $\Delta x_{t_i} = \lambda_{t_i} x_{t_i}$ for every $i = 1, \ldots, k$, then x is said to be a finite type immersion. More precisely a submanifold M is of k-type (via x) if there are exactly k nonconstant terms in (1) from different eigenspaces. For a compact manifold M, we call the collection of indices $\{t_1, \ldots, t_k\}$ the order of M. If x is not decomposable into a finite sum (1) then it is an infinite type immersion. For example a cylinder $x(\theta, t) = (\cos\theta, \sin\theta, t)$ is of 2-type since $x = (0, 0, t) + (\cos\theta, \sin\theta, 0) = x_p + x_q$ where $\Delta x_p = 0$, $\Delta x_q = x_q$. It follows easily from (1) that if M is a compact submanifold then x_0 is the center of mass; i.e., $x_0 = (1/\operatorname{vol}M)\int_M x\, dV$. If M is immersed into a hypersphere $S_c^{m-1}(r)$, the immersion is said to be mass-symmetric if the center of mass of M coincides with the center c of the hypersphere. From (1) by taking successive Laplacians and eliminating x_{t_i}'s one obtains

$$P(\Delta)(x - x_0) = 0, \quad \text{where } P(T) = \prod_{i=1}^{k}(T - \lambda_{t_i}). \tag{2}$$

For a compact manifold M the condition (2) is equivalent to M being of finite type (a fact following from the existence of natural L^2 structure on M; cf. [C3]) but in the noncompact case things are not clear; for some classes of submanifolds (e.g., curves) (2) implies finite type, but seemingly that statement is not true in general [CP]. We note that the decomposition (1) is global; i.e., all the maps are defined on entire manifold. It is moreover possible that different portions of M are of different type. E.g., take a piece of cylinder (2-type) and glue smoothly to a piece of sphere (1-type); the smoothing part, and entire submanifold, is of infinite type.

One of the fundamental questions here is: How large is the set of finite type submanifolds? Namely, it is clear that in a certain way finite type immersions are a generalization of minimal immersions into a Euclidean space or a sphere so one wants to know as many examples as possible of finite type immersions that are not minimal. Even though there is an abundance of such examples it appears that in the low codimension case they do not occur as often. For example, B. Lawson [L] constructs minimal (1-type) surfaces in S^3 of arbitrary genus, but it was shown in [BCG] that the only compact 2-type surface in S^3 is $S^1(a)\times S^1(b)$, $a^2+b^2 = 1$, $a \neq b$ (see also [HV]). However, for any k (and any preassigned order) there are submanifolds of k type (and the given order). They are easily constructed using the diagonal immersion of standard immersions of a compact, irreducible symmetric space [C4]. In the same vein, every equivariant isometric immersion $x: G/H \to E^m$ of a compact homogeneous space into E^m is of finite type and the type is $\leq m$ [C3]. Actually, Takahashi [T2] shows that if G/H is irreducible then such an immersion is of 1-type. This is also proved by Deprez in [De] where he investigated the relationship between the conditions "to be of finite type" and "to have finite type geodesics" (the two conditions are equivalent for irreducible symmetric spaces of compact type).

Going back to the spherical case, if a compact hypersurface $M^n \subset S^{n+1}$ is mass-symmetric and of 2-type in E^{n+2} then its mean curvature in the sphere is constant $\neq 0$ and its scalar curvature is constant. The converse of this is also true with the possible inclusion of a small hypersphere (which is of 1-type) [C6]. As a consequence every isoparametric hypersurface of S^{n+1} which is neither minimal nor a small hypersphere is mass-symmetric and of 2-type and we know that there are infinitely many of those (with four principal curvatures [FKM]). There can be no mass-symmetric 2-type compact surface M^2 lying fully in S^4 but there are infinitely many compact surfaces (flat tori) immersed fully in S^5 that are mass-symmetric and 2-type [BC1, 3]. Moreover, M. Kotani [Ko2] shows that any mass-symmetric and proper 2-type immersion of a topological 2-sphere into S^n is the direct sum of two minimal immersions into spheres (as a corollary she derives that in this case n is odd and greater than 5). R. Bryant [Br] classifies connected minimal (1-type) surfaces $M^2(K)$ of constant Gaussian curvature K in S^n and Y. Miyata [M], building on Bryant's ideas, classifies mass-symmetric 2-type surfaces $M^2(K)$ of constant curvature in S^n . In particular, he shows that there is no mass-symmetric 2-type immersion of $M^2(K)$ in S^n if $K < 0$, and if $K > 0$ every 2-type immersion of $M^2(K)$ in S^n is a diagonal immersion of two 1-type immersions. He also shows that if $M^2(K)$ is mass-symmetric and a 2-type surface fully immersed in S^n then n is necessarily odd. For a compact surface with constant curvature K in S^3 the following result [CD] is relevant: A compact surface $M^2(K) \subset S^3$ is of finite type if and only if M is totally umbilical thus of 1-type) or $M^2 = S^1(a) \times S^1(b)$ (thus of 2-type or 1-type).

Complete classification of low type (in particular 2-type) hypersurfaces in E^{n+1} or S^{n+1} is not known. Also no example of 3-type hypersurface in E^{n+1} or S^{n+1} is known. However, B. Y. Chen and S. J. Li [CL] prove that a 3-type hypersurface in S^{n+1} if it exists has nonconstant mean curvature. Of course, ideally one wants to classify all finite type submanifolds in E^m or S^m but the problem is impossible to handle in this generality. Understandably most results are obtained for low type submanifolds of low dimension or codimension. For example a closed curve in E^m is of finite type if and only if Fourier series expansions *of its coordinate* functions have finitely many nonzero terms [C3]. This fact enables us to classify finite type curves in E^2 as pieces of either straight lines or circles (thus they are actually of 1-type). It is interesting to note that every finite type closed curve is a rational curve [CDVV]. For study of finite type curves in general see [CDDVV]. The next simplest case would be a study of surfaces in E^n but even for $n = 3$ we do not know how large is the set of finite type immersions. For example, other than the standard sphere (1-type) no other examples of compact finite type surfaces in E^3 are known and B. Y. Chen makes the following conjecture:

CONJECTURE. Ordinary spheres are the only compact finite type surfaces in E^3.

If it is true it would give a nice characterization of a standard sphere in terms of its spectral behavior. This conjecture is confirmed for some special classes of surfaces, e.g., tubes, quadrics [C7], [CDS]. Other classification results on finite type surfaces (not necessarily compact) can be found in [DPV, G1, G2, Di, CDVV].

As a historical note, the notion of finite type submanifolds was first introduced by B. Y. Chen in the late seventies [C1] through his attempts to find the best possible estimate for the total mean curvature of the compact manifold immersed into a Euclidean space. Namely he obtained the following

$$\frac{\lambda_p}{n} \operatorname{vol}(M) \le \int_M |H|^2 \, dV \le \frac{\lambda_q}{n} \operatorname{vol}(M)$$

where $|H|^2$ is the square of the length of the mean curvature vector, λ_p and λ_q are eigenvalues of the Laplacian corresponding to the first and last nonzero eigenvectors in the decomposition (1) (if M is of infinite type then $\lambda_q = \infty$). Finite type submanifolds can naturally be defined in pseudo Euclidean spaces and they can be related to some notions of mechanics (total tension, energy) and conformal geometry [C5, C8, C10]. For more results and up to date references see [C11].

2. Quadric representation of a submanifold

Given an isometric immersion $x : M^n \to E^m$ one of the first questions one might ask is: "What are the natural maps related to x?" It seems there are only few (notably the Gauss map $\nu : M^n \to \bigwedge^n E^m$). If we consider $x = (x_1, \ldots, x_m)^t$ as a column matrix then the map $\tilde{x} = xx^t$ from M into the set $SM(m)$ of $m \times m$ real symmetric matrices is also one natural map. This map is given by

$$(3) \qquad \tilde{x} = \begin{pmatrix} x_1^2 & x_1 x_2 & \cdots & x_1 x_m \\ x_2 x_1 & x_2^2 & \cdots & x_2 x_m \\ \cdot & \cdot & \cdot & \cdot \\ \cdot & \cdot & \cdot & \cdot \\ \cdot & \cdot & \cdot & \cdot \\ x_m x_1 & x_m x_2 & \cdots & x_m^2 \end{pmatrix}.$$

Note that the matrix in (3) has rank 1 at each point (actually all square symmetric matrices of rank 1 can be represented this way). If the manifold in question is the unit hypersphere centered at the origin then $\tilde{x}$ is the second standard immersion of the sphere. It is well known that $\tilde{x}$ immerses S^{m-1} as a minimal submanifold of a hypersphere of $SM(m)$ centered at I/m, where I is the $m \times m$ identity matrix, with radius $r = \sqrt{(m-1)/2m}$

(Veronese submanifold). (See e.g. **[C3, R1, S]**.) Obviously, $\tilde{x}$ gives also an isometric immersion of any submanifold $M^n \subset S^{m-1}$ but in general $\tilde{x}$ is just a smooth $SM(m)$-valued map on M^n (actually if $n \geq 2$ it is an isometric immersion only for submanifolds of S^{m-1}; see **[D2]**). We call $\tilde{x}$ *the quadric representation* of a submanifold M since the coordinates of $\tilde{x}$ depend on coordinates of x in a quadratic manner.

The set $SM(m)$ becomes the standard Euclidean space of dimension $N = m(m-1)/2$ when equipped with the metric $\langle P, Q \rangle = \frac{1}{2}\operatorname{tr}(PQ)$, so it makes sense to study submanifolds of E^m which are of finite type in $SM(m)$ via $\tilde{x}$. Studying submanifolds with finite type quadric representation amounts to studying spectral behavior of products of coordinate functions $x_i x_j$.

For an isometric immersion $x: M^n \to E^m$ we have

$$\langle \tilde{x} - \lambda I, \tilde{x} - \lambda I \rangle = \tfrac{1}{2}[(\langle x, x \rangle - \lambda)^2 + (m-1)\lambda^2].$$

Therefore, if $\langle x, x \rangle = \text{const}$, i.e., M^n belongs to a sphere centered at the origin, then $\tilde{x}(M^n)$ belongs to a hypersphere of $E^N = SM(m)$ centered at λI for any $\lambda \in \mathbf{R}$. Conversely if $\tilde{x}(M^n)$ belongs to a hypersphere of E^N centered at λI for some $\lambda \in \mathbf{R}$ then $\langle x, x \rangle = \text{const}$; i.e., $x(M^n)$ is spherical. However there are nonspherical examples of submanifolds M^n in E^m whose image $\tilde{x}(M^n)$ is spherical. That can be seen from the following reasoning. Let K be any symmetric matrix not proportional to the identity; e.g., K is diagonalized as $K = \operatorname{diag}(\lambda_1, \ldots, \lambda_m)$. Then $\tilde{x}(M^n)$ belongs to a hypersphere of E^N centered at K with radius r if and only if $\langle \tilde{x} - K, \tilde{x} - K \rangle_{SM(m)} = r^2$. This condition is equivalent to

$$(4) \qquad f(x_1, \ldots, x_m) = \langle x, x \rangle^2 - 2 \sum_t \lambda_t x_t^2 + \sum_t \lambda_t^2 - 2r^2 = 0.$$

The gradient is found to be

$$\nabla f = \sum_t \frac{\partial f}{\partial x_t} \frac{\partial}{\partial x_t} = 4 \sum_t [\langle x, x \rangle - \lambda_t x_t] \frac{\partial}{\partial x_t}.$$

If we choose K so that $2r^2 > \sum_t \lambda_t^2$ then $\nabla f \neq 0$ and $f^{-1}(0)$ defines a smooth nonspherical hypersurface of E^m whose image via $\tilde{x}$ belongs to the sphere centered at K with radius r.

For a compact submanifold $M^n \subset E^m$ its quadric representation $\tilde{x}$ has the spectral decomposition $\tilde{x} = \tilde{x}_0 + \sum_t \tilde{x}_t$, where the subscript t denotes the orthogonal projection onto the eigenspace V_t of the Laplacian. Because $\operatorname{tr} \tilde{x}_t = (\operatorname{tr} \tilde{x})_t$ and $\operatorname{tr} \tilde{x} = 1$ for a submanifold of the unit sphere S^{m-1} we have $\operatorname{tr} \tilde{x}_0 = 1$ and $\operatorname{tr} \tilde{x}_t = 0$, $t \geq 1$. Actually in the class of compact submanifolds these conditons on traces characterize spherical submanifolds; i.e., for an isometric immersion $x: M^n \to E^m$ we have $x(M^n) \subset S^{m-1}$ if and only if $\operatorname{tr} \tilde{x}_0 = 1$ and $\operatorname{tr} \tilde{x}_t = 0$, $t \geq 1$. (The converse follows from $\langle x, x \rangle = \operatorname{tr} \tilde{x} = \operatorname{tr} \tilde{x}_0 = 1$.)

Study of immersions $x : M^n \to E^m$ whose quadric representation $\tilde{x}$ is a finite type map is difficult in general. Some results on low type quadric representation are as follows: $\tilde{x}$ is a 1-type map if and only if M^n is a totally geodesic submanifold of the hypersphere $S^{m-1}(r)$ of radius r centered at the origin. The only constant mean curvature hypersurface $x : M^n \to E^{n+1}$ with 2-type quadric representation $\tilde{x}$ is (a piece of) a hypersphere which is not centered at the origin. Minimal submanifolds of E^m have quadric representation of infinite type. At first it appears that examples with low type $\tilde{x}$ that turn up are all spherical. However it is easy to construct a nonspherical example with finite type $\tilde{x}$: Given two nonspherical finite type curves C_1, C_2 (they exist by [C3, p. 289]) consider their product $C_1 \times C_2$. Such a product does not belong to any sphere and its quadric representation is a finite type map. These results can be found in [D1]. We conjecture that ordinary hyperspheres are the only hypersurfaces whose quadric representation is of finite type.

If a submanifold of E^m is assumed to belong to a unit hypersphere S^{m-1} centered at the origin (from now on such submanifolds will be called spherical) then the quadric representation is simply the restriction of the second standard immersion of the sphere to a submanifold and things seem to be just a bit easier to handle. Clearly spherical submanifolds with 1-type quadric representation are totally geodesic in that sphere. As for the spherical submanifolds with 2-type quadric representation the first result is that of A. Ros [R2] who shows that a compact submanifold M immersed fully and minimally in S^{m-1} is of 2-type via $\tilde{x}$ if and only if M is Einstein and $\operatorname{tr}(A_\xi A_\eta) = k\langle \xi, \eta \rangle$ for every ξ, η in the normal bundle of M ($k = $ const). In the course of the proof he shows that such submanifolds are mass-symmetric via $\tilde{x}$. As we know, for a spherical submanifold $M^n \subset S^{m-1}$ its image $\tilde{x}(M^n)$ belongs to the hypersphere of $SM(m)$ centered at I/m and to say that a compact M^n is mass-symmetric in $SM(m)$ via $\tilde{x}$ means that the center of mass is $\tilde{x}_0 = I/m$. Ros's result was generalized by M. Barros and B. Y. Chen [BC2] who obtained a characterization of compact spherical submanifolds that are mass-symmetric and of 2-type via $\tilde{x}$ in terms of certain conditions on the mean curvature vector, Weingarten maps and the Ricci tensor. They show that for such spherical hypersurfaces $M^n \subset S^m$ there is some restriction on dimensions, namely $m \leq n(n+3)/2$, and they provide an example of a submanifold of a hypersphere which is not mass-symmetric but which is of 2-type in the ambient Euclidean space. Most importantly they obtain the classification of compact hypersurfaces in S^{n+1} whose quadric representation is mass-symmetric and of 2-type. This result was subsequently extended by the author [D3] who classified spherical hypersurfaces which are of 2-type via $\tilde{x}$ (without assuming mass-symmetry a priori). Those hypersurfaces are products of spheres $S^p(r_1) \times S^{n-p}(r_2)$ with only three different possiblities for the radii (r_1, r_2). Surfaces that are mass-symmetric and of 2-type via $\tilde{x}$ are also classified by Barros and Chen, namely

THEOREM 1 [BC2]. *Let* $x: M^2 \to S^m$ *be a full isometric immersion of a compact surface M into S^m. Then $\tilde{x}$ is mass-symmetric and of 2-type if and only if one of the following cases occurs:*

(1) $m = 3$ *and M is immersed as a small hypersphere* $S^2(\sqrt{3}/2)$;

(2) $m = 3$ *and M is immersed as the Clifford* (*minimal*) *torus* $S^1(1/\sqrt{2}) \times S^1(1/\sqrt{2})$ *in* S^3;

(3) $m = 4$ *and M is immersed as the Veronese* (*minimal*) *surface in* S^4;

(4) $m = 5$ *and M is immersed as a Veronese* (*minimal*) *surface in a small hypersphere* $S^4(\sqrt{5/6})$ *of* S^5.

A study of spherical submanifolds with 3-type quadric representation is more difficult in general and understandably there are some classification results only for surfaces or hypersurfaces (and then assuming extra conditions such as minimality). For example Barros and Urbano have the following result for surfaces:

THEOREM 2 [BU]. *Let* $x: M \to S^n(1)$ *be a minimal isometric immersion of a compact surface in the sphere, which is assumed to be full. Then the immersion $\tilde{x}$ is of 3-type if and only if either*

(1) *M has constant Gaussian curvature $K = 1/6$ and x is the Veronese surface in* $S^6(1)$ *or*

(2) *M is flat and x is an equilateral torus in* $S^5(1)$.

In [D3] the author undertook a study of constant mean curvature spherical hypersurfaces with 3-type quadric representation. For example the following characterization is obtained:

THEOREM 3. *Let* $x: M^n \to S^{n+1}$ *be an isometric immersion of a compact manifold M^n as a minimal hypersurface of S^{n+1}. If $\tilde{x}$ is mass-symmetric and of 3-type then*

(1) $\operatorname{tr} A = \operatorname{tr} A^3 = 0$,

(2) $\operatorname{tr} A^2$ *and* $\operatorname{tr} A^4$ *are constant,*

(3) $\operatorname{tr}(\nabla_X A)^2 = \langle A^2 X, A^2 X \rangle + p\langle AX, AX \rangle + q\langle X, X \rangle$, *for every tangent vector $X \in TM$, where p and q are constants. Conversely, if (1)–(3) hold then M is mass-symmetric and of 1, 2 or 3-type via $\tilde{x}$.*

Condition (1)–(3) above also imply $\operatorname{tr} A^5 = 0$ and that is used in the following characterization of the Cartan hypersurface (for other characterizations see [KN, TV] and references there).

THEOREM 4 [D3]. *Let* $x: M^n \to S^{n+1}$ *be a compact minimal hypersurface of S^{n+1} of dimension $n \le 5$. Then $\tilde{x}$ is mass-symmetric and of 3-type if and only if $n = 3$ and $M^3 = SO(3)/Z_2 \times Z_2$ is the Cartan hypersurface.*

Actually, all minimal spherical isoparametric hypersurfaces with 3 curvatures are mass-symmetric and of 3-type via $\tilde{x}$ (they are homogeneous spaces

$SO(3)/Z_2 \times Z_2$, $SU(3)/T^2$, $Sp(3)/Sp(1)^3$, $F_4/Spin(8))$, and no other such example is known. It would be of interest to resolve if any minimal spherical hypersurface which is of 3-type and mass-symmetric via $\tilde{x}$ is necessarily isoparametric and if the only such hypersurfaces are those four examples with three principal curvatures.

For a spherical submanifold $x: M^n \to S^{m-1} \subset E^m$ the quadric representation is defined to be the composition of x with the second standard immersion ι of the sphere, $\tilde{x} = \iota x$. A similar situation arises for submanifolds of other compact rank-1 symmetric spaces. In particular, if $x: M^n \to CP^m$ is an isometric immersion of a compact manifold into the complex projective space, then the quadric representation is defined as $\tilde{x} = \phi x$, where ϕ is the first standard embedding of CP^m. ϕ is most easily realized by the following construction [Tai]. Let $\pi: S^{2m+1} \to CP^m$ be the Hopf fibration and let $p \in CP^m$. Pick any $z \in \pi^{-1}(p) \subset S^{2m+1} \subset \mathbf{C}^{m+1}$ and let $\phi(p) = z\bar{z}^t$ (z is regarded as a column vector in $\mathbf{C}^{m+1}$). Then ϕ is a well-defined map that embeds CP^m into the set $H(m+1)$ of Hermitian matrices of degree $m+1$, the latter being a Euclidean space of dimension $N = (m+1)^2$ with the metric $\langle P, Q \rangle = \frac{1}{2} \operatorname{tr}(PQ)$. Thus the quadric representation $\tilde{x} = \phi x$ immerses M^n into a Euclidean space $H(m+1)$ and therefore it makes sense to study submanifolds of CP^m which are of finite type via $\tilde{x}$. However things are more difficult here; for example we do not have complete classification of even 1-type submanifolds of CP^m. The best result in that direction so far is the following theorem of the author:

THEOREM 5 [D4]. *Let* $x: M^n \to CP^m$ *be an isometric immersion of a compact Riemannian manifold with parallel mean curvature vector (or let M be a CR submanifold). Then* $\tilde{x} = \phi x$ *is of 1-type if and only if one of the following cases occurs*:

(1) *n is even, and M^n is congruent to the complex projective space $CP^{n/2}$ immersed as a totally geodesic complex submanifold of CP^m.*

(2) *M^n is a totally real minimal submanifold of a complex totally geodesic CP^n in CP^m.*

(3) *n is odd, and M^n is congruent to the geodesic hypersphere*

$$M_{0,(n-1)/2}^{\mathbf{C}} = \pi(S^1(\sqrt{1/(n+3)}) \times S^n(\sqrt{(n+2)/(n+3)}))$$

of the complex projective space $CP^{(n+1)/2}$ immersed as a totally geodesic complex submanifold of CP^m.

Actually, if the dimension or codimension is small ($= 1, 2$ or 3) the above classification holds even without assumption on the mean curvature vector being parallel (or M being CR). This generalizes results of [R1] and [MR]. Some examples of 2-type submanifolds of the complex projective space are as follows: The complex quadric $Q^n \subset CP^{n+1}$ is of 2-type in $H(n+2)$ of order $\{1, 2\}$ [R3]. If $M_{n,n}^{\mathbf{C}} \subset CP^{2n+1}$ denotes the image under the Hopf projection of the generalized Clifford torus

$$M_{2n+1, 2n+1} = S^{2n+1}(\sqrt{1/2}) \times S^{2n+1}(\sqrt{1/2})$$

then $M^C_{n,n}$ is of 2-type in $H(2n+2)$ [C3]. Furthermore, S. Udagawa classifies compact real hypersurfaces with constant mean curvature in CP^m with 2-type quadric representation; they are the Hopf projections of certain products of spheres. I believe however that his assumption on constant mean curvature is superfluous. A. Ros gives the following characterization of 2-type compact complex submanifolds of CP^m.

THEOREM 6 [R3]. *Let* $x : M^n \to CP^m$ *be a full isometric immersion of a compact Kähler manifold. Then* M *is of 2-type in* $H(m+1)$ *via* $\tilde{x}$ *if and only if* (1) M *is an Einstein submanifold and* (2) $\operatorname{tr}(A_\xi A_\eta) = k\langle \xi, \eta \rangle$ (k *is a constant and* ξ, η *are in the normal bundle of* x).

Then Udagawa [U3] notices that the condition (2) above is equivalent to the existence of Kähler-Einstein metric in the normal bundle of M and subsequently characterizes compact Kähler 2-type submanifolds as Einstein parallel submanifolds (non totally geodesic ones). They are necessarily of order $\{1, 2\}$ and they are classified as $CP^n(1/2)$, Q^n, $CP^n \times CP^n$, $U(s+2)/U(s) \times U(2)$, $s \geq 3$, $SO(10)/U(5)$ and $E_6/\operatorname{Spin}(10) \times T$ [U2]. Udagawa also gives examples of 3-type submanifolds of CP^m; namely, compact irreducible Hermitian symmetric submanifolds of degree 3 are of order $\{1, 2, 3\}$ [U1].

Some of the related problems to consider are as follows:

1. If x is an isometric immersion of a manifold into S^m or CP^m and $\tilde{x}$ its quadric representation, is it true that $\tilde{x}$ of finite type (k) implies that x is of finite type $(\leq k)$? The converse of this is seemingly not true. A result of this kind is a theorem in [B2] proving that if $\tilde{x}$ is mass-symmetric and of 2-type then x is of 1- or 2-type (S^m case).

2. There are also natural embeddings of Grassmann manifolds, classical groups and some symmetric spaces of higher rank into Euclidean spaces constructed by Kobayashi [K]. Therefore for a submanifold of any such space, the composition map can be studied in terms of finite type property (see [BN]).

3. For an oriented surface $x : M^2 \to E^n$ we have the Gauss map $\nu : M^2 \to G_{2,n}$ into the Grassmannian of oriented 2-planes. $G_{2,n}$ can naturally be identified with the complex quadric Q^{n-2} in CP^{n-1} which in turn can be embedded into the Euclidean space $H(n)$. The composition of these maps gives a map from M^2 into $H(n)$ given by $\nu = \frac{1}{2}(e_1 + ie_2)(e_1 + ie_2)^*$, where $\{e_1, e_2\}$ is a tangent frame for M^2 and $*$ denotes the conjugate transpose. It seems to be interesting to study surfaces which have finite type "quadric Gauss map" ν. The imaginary part of ν is the usual Gauss map $e_1 \wedge e_2$ which was studied from the point of view of finite type maps in [CPi].

3. Eigenvalues and spectral inequalities

Study of finite type immersions often provides some information on the spectrum of the Laplace operator of a submanifold. For example for a compact minimal spherical submanifold of dimension n we have $\Delta x = nx$ from which we see that the first eigenvalue of the Laplacian satisfies $\lambda_1 \leq n$. On the other hand $\lambda_1 \geq n/2$ for a compact embedded spherical hypersurface M^n [CW]. It is still an open question as to which minimal spherical submanifolds (in particular hypersurfaces) have $\lambda_1 = n$. This equality is shown to be true for some isoparametric hypersurfaces (see [Mu, Ko1]). Regarding spectra of cubic isoparametric hypersurfaces the author has been able to find three eigenvalues of the Laplacian as a byproduct of studying minimal spherical submanifolds with 3-type quadric representation [D1, 3]. Namely for those we have a formula like (2),

$$\Delta^3 \tilde{x} + a\Delta^2 \tilde{x} + b\Delta \tilde{x} + c(\tilde{x} - I/(3m+2)) = 0$$

with $a = -(10 + 24m)$, $b = 4[(3m + 1)(15m + 6) - 2]$, $c = -48m(3m + 1) \cdot (3m+2)$ from which we find three eigenvalues to be $\lambda_p = 6m$, $\lambda_q = 2(3m + 1)$ and $\lambda_r = 4(3m + 2)$ (m denotes the common multiplicity of the principal curvatures). However, the spectrum of the Cartan hypersurface was computed in [MOU] and spectra of all cubic isoparametric minimal hypersurfaces in a recent paper by B. Solomon [So]. For a compact 2-type hypersurface of a unit hypersphere of E^{n+2} B. Y. Chen also proved $\lambda_1 < n$ (see [C9, C6]). In principle, every time we are able to show that a submanifold is of 2, 3, etc. type we can extract 2, 3, etc. eigenvalues of the Laplacian from the equations involved. For example, the submanifold $M_{n,n}^C \subset CP^{2n+1}(4)$ mentioned earlier satisfies $P(\Delta)(\tilde{x} - \tilde{x}_0) = 0$ with $P(t) = (t - 4(2n + 1))(t - 4(2n + 2))$. Consequently, $4(2n + 1)$, $4(2n + 2) \in \mathrm{Spec}(M_{n,n}^C)$. The quadric representation helps give sharp upper bounds on the first eigenvalue of compact submanifolds of projective spaces. In [R3], Ros gives the following sharp estimate for compact n-dimensional Kähler submanifolds of $CP^m(4)$: $\lambda_1 \leq 2(n + 2)$. This generalizes the estimate of Yang and Yau for holomorphic curves and the same estimate is actually valid for any minimal submanifold [C2]. In general for a compact n-dimensional submanifold of a projective space we have

$$\lambda_1 \leq 2(n + d) + \frac{n}{\mathrm{vol}(M)} \int_M \alpha^2 \, dV$$

where α is the mean curvature of the immersion and d is $1, 2$ or 4 according to the case RP^m, $CP^m(4)$ or $QP^m(4)$. This follows from

$$(\Delta \tilde{x}, \Delta \tilde{x}) - \lambda_1(\Delta \tilde{x}, \tilde{x}) = \sum_{t \geq 1} \lambda_t(\lambda_t - \lambda_1)(\tilde{x}_t, \tilde{x}_t) \geq 0$$

after computation of terms on the left-hand side. Here $(\ ,\ )$ denotes the natural L^2 inner product on submanifold and $\tilde{x} = \sum_t \tilde{x}_t$ is the spectral decomposition of $\tilde{x}$ (cf. [D4]). For a compact, n-dimensional, minimal

CR submanifold of CP^m, Ejiri and Ros independently show that $\lambda_1 \leq 2(n^2 + n + 2k)/n$, where k is the complex dimension of the holomorphic distribution (see [R1]). In [C8] B. Y. Chen obtains inequalities involving the first two eigenvalues λ_1, λ_2 of the Laplacian. Namely for a compact n-dimensional minimal submanifold of $S^m(1)$ we have $m\lambda_1\lambda_2 \geq 2n(m+1)[\lambda_1 + \lambda_2 - 2n - 2]$, and for a compact n-dimensional minimal submanifold of a projective space FP^m we have $(m/2(m+1))\lambda_1\lambda_2 \geq n(\lambda_1 + \lambda_2 - 2n - 2d)$, where $d = 1, 2$ or 4 corresponds respectively to $F = R, C$ or Q. In addition, there are many other interesting spectral inequalities arising from the study of finite type submanifolds; see [BU, C3, C4, C6, MR, R1, R2, R3, U1, U2, U4].

BIBLIOGRAPHY

[BC1] M. Barros and B. Y. Chen, *Finite type spherical submanifolds*, Proc. II Internat. Sympos. Differential Geom., Lecture Notes in Math. vol. 1209, Springer-Verlag, 1986, pp. 73–93.

[BC2] ____, *Spherical submanifolds which are of 2-type via the second standard immersion of the sphere*, Nagoya Math. J. **108** (1987), 77–91.

[BC3] ____, *Stationary 2-type surfaces in a hypersphere*, J. Math. Soc. Japan **39** (1987), 627–648.

[BU] M. Barros and F. Urbano, *Spectral geometry of minimal surfaces in the sphere*, Tôhoku Math. J. **39** (1987), 575–588.

[BCG] M. Barros, B.Y. Chen and O. Garray, *Spherical finite type hypersurfaces*, Algebras, Groups, Geom. **4** (1987), 58–72.

[BN] D. Brada and L. Niglio, *Connected compact minimal Chen-type-1 submanifolds of Grassmannian manifold*, C.R. Acad. Sc. Paris, **311** (1990), 899–902.

[Br] R. L. Bryant, *Minimal surfaces of constant curvature in S^n*, Trans. Amer. Math. Soc. **290** (1985), 259–271.

[C1] B. Y. Chen, *On total curvature of immersed manifolds. IV*, Bull. Inst. Math. Acad. Sinica **7** (1979), 301–311.

[C2] ____, *On the first eigenvalue of Laplacian of compact minimal submanifolds of rank one symmetric spaces*, Chinese J. Math. **11** (1983), 259–273.

[C3] ____, *Total mean curvature and submanifolds of finite type*, World Scientific, Singapore, 1984.

[C4] ____, *Finite type submanifolds and generalizations*, Instituto Matematico "Guido Castelnuovo", Rome, 1985.

[C5] ____, *Finite type submanifolds in pseudo-Euclidean spaces and applications*, Kodai Math. J. **8** (1985), 359–374.

[C6] ____, *2-type submanifolds and their applications*, Chinese J. Math. **14** (1986), 1–14.

[C7] ____, *Surfaces of finite type in Euclidean 3-space*, Bull. Soc. Math. Belg. Ser. B **39** (1987), 243–254.

[C8] ____, *Some estimates of total tension and their applications*, Kodai Math. J. **10** (1987), 93–101.

[C9] ____, *Mean curvature of 2-type spherical submanifolds*, Chinese J. Math. **16** (1988), 1–9.

[C10] ____, *Submanifolds of finite type in hyperbolic spaces*, Chinese J. Math. **20** (1992), 5–21.

[C11] ____, *Submanifolds of finite type and applications*, Proc. of Intern. Geom. Top. Workshop 1992, Taegu, Korea.

[CD] B. Y. Chen and F. Dillen, *Surfaces of finite type and constant curvature in the 3-space*, C. R. Math. Rep. Acad. Sci. Canada **12** (1990), 47–49.

[CDDVV] B. Y. Chen, J. Deprez, F. Dillen, L. Verstraelen, and L. Vrancken, *Curves of finite type*, Geometry and Topology of Submanifolds II, 1990, pp. 76–110.

[CDVV] B. Y. Chen, F. Dillen, L. Verstraelen, and L. Vrancken, *Ruled surfaces of finite type*, Bull. Austral. Math. Soc. **42** (1990), 447–453.

[CDS] B. Y. Chen, F. Dillen, and H. Z. Song, *Quadric hypersurfaces of finite type*, Colloq. Math **64** (1992).

[CL] B. Y. Chen and S. J. Li, *3-type hypersurfaces in a hypersphere*, Bull. Soc. Math. Belg. **43** (1991), 135–141.

[CP] B. Y. Chen and M. Petrovic, *On spectral decomposition of immersions of finite type*, Bull. Austral. Math. Soc. **44** (1991), 117–129.

[CPi] B. Y. Chen and P. Piccinni, *Submanifolds with finite type Gauss map*, Bull. Austral. Math. Soc. **35** (1987), 161–186.

[CW] H. I. Choi and A. N. Wang, *A first eigenvalue estimate for minimal hypersurfaces*, J. Differential Geom. **18** (1983), 559–562.

[DPV] F. Dillen, J. Pas and L. Verstraelen, *On surfaces of finite type in Euclidean 3-space*, Kodai Math. J. **13** (1990), 10–21.

[De] J. Deprez, *Immersions of finite type of compact homogeneous Riemannian manifolds*, Thesis, K. U. Leuven, 1988.

[Di] F. Dillen, *Ruled submanifolds of finite type*, Proc. Amer. Math. Soc. **114** (1992), 795–798.

[D1] I. Dimitrić, *Quadric representation and submanifolds of finite type*, Thesis, Michigan State Univ. 1989.

[D2] ——, *Quadric representation of a submanifold*, Proc. Amer. Math. Soc. **114** (1992), 201–210.

[D3] ——, *Spherical hypersurfaces with low type quadric representation*, Tokyo J. Math. **13** (1990), 469–492.

[D4] ——, *1-type submanifolds of the complex projective space*, Kodai Math. J. **14** (1991), 281–295.

[FKM] D. Ferus, H. Karcher, and H. F. Münzner, *Cliffordalgebren und neue isoparametrische Hyperflachen*, Math. Z. **177** (1981), 479–502.

[G1] O. J. Garay, *Finite type cones shaped on spherical submanifolds*, Proc. Amer. Math. Soc. **104** (1988), 868–870.

[G2] ——, *On a certain class of finite type surfaces of revolution*, Kodai Math. J. **11** (1988), 25–31.

[HV] T. Hasanis and T. Vlachos, *A local classification of 2-type surfaces in S^3* , Proc. Amer. Math. Soc. **122** (1991), 533–538.

[KN] U. H. Ki and H. Nakagawa, *A characterization of the Cartan hypersurface in a sphere*, Tôhoku Math. J. **39** (1987), 27–40.

[K] S. Kobayashi, *Isometric imbeddings of compact symmetric spaces*, Tôhoku Math. J. **20** (1968), 21–25.

[Ko1] M. Kotani, *The first eigenvalue of homogeneous minimal hypersurfaces in a unit sphere $S^{n+1}(1)$* , Tôhoku Math. J. **37** (1985), 523–532.

[Ko2] ——, *A decomposition theorem of 2-type immersions*, Nagoya Math. J. **118** (1990), 55–64.

[L] H. B. Lawson, Jr., *Complete minimal surfaces in S^3* , Ann. of Math. **92** (1970), 335–374.

[MR] A. Martinez and A. Ros, *On real hypersurfaces of finite type of CP^m* , Kodai Math. J. **7** (1984), 304–316.

[M] Y. Miyata, *2-type surfaces of constant curvature in S^n* , Tokyo J. Math. **11** (1988), 157–203.

[Mu] H. Muto, *The first eigenvalue of the Laplacian of an isoparametric minimal hypersurface in a unit sphere*, Math. Z. **197** (1988), 531–549.

[MOU] H. Muto, Y. Ohnita, and H. Urakawa, *Homogeneous minimal hypersurfaces in the unit spheres and the first eigenvalue of their Laplacian*, Tôhoku Math. J. **36** (1984), 253–267.

[R1] A. Ros, *Spectral geometry of CR-minimal submanifolds in the complex projective space*, Kodai Math. J. **6** (1983), 88–99.

[R2] ——, *Eigenvalue inequalities for minimal submanifolds and P-manifolds*, Math. Z. **187** (1984), 393–404.

[R3] ——, *On spectral geometry of Kaehler submanifolds*, J. Math. Soc. Japan **36** (1984), 433–447.

[S] K. Sakamoto, *Planar geodesic immersions*, Tôhoku Math. J. **291** (1977), 25–56.

[So] B. Solomon, *The harmonic analysis of cubic isoparametric minimal hypersurfaces. I and II*, Amer. J. Math. **112** (1990), 157–241.

[Tai] S. S. Tai, *Minimum imbeddinngs of compact symmetric spaces of rank one*, J. Differential Geom. **2** (1968), 55–66.

[T1] T. Takahashi, *Minimal immersions of Riemannian manifolds*, J. Math. Soc. Japan **18** (1966), 380–385.

[T2] ___, *Isometric immersion of Riemannian homogeneous manifold*, Tsukuba J. Math. **12** (1988), 79–107.

[TV] F. Tricerri and L. Vanhecke, *Cartan Hypersurfaces and reflections*, Nihonkai Math. J. **1** (1990), 203–208.

[U1] S. Udagawa, *Spectral geometry of compact Hermitian symmetric submanifolds*, Math. Z. **192** (1986), 57–72.

[U2] ___, *Spectral geometry of Kaehler submanifolds of a complex projective space*, J. Math. Soc. Japan **38** (1986), 453–471.

[U3] ___, *Einstein parallel Kaehler submanifolds in a complex projective space*, Tokyo J. Math. **9** (1986), 335–340.

[U4] ___, *Bi-order real hypersurfaces in a complex projective space*, Kodai Math. J. **10** (1987), 182–196.

PENNSYLVANIA STATE UNIVERSITY - FAYETTE

Proceedings of Symposia in Pure Mathematics
Volume **54** (1993), Part 3

Embedded Eigenvalues
for Asymptotically Flat Surfaces

HAROLD DONNELLY

1. Introduction

Consider a complete Riemannian manifold M of dimension n. One has a second-order elliptic Laplace operator Δ acting on smooth functions of compact support. Moreover, the Laplacian Δ extends to a unique unbounded selfadjoint operator on $L^2 M$. Since Δ is nonnegative, its spectrum is contained in $[0, \infty)$. However, more refined statements appear to require special hypotheses upon M.

Suppose that M is simply connected and nonpositively curved. If the curvature of M approaches a constant $-c$, with $c \geq 0$, at infinity, then the Laplacian has essential spectrum $[(n-1)^2 c/4, \infty)$. This was proved in [**4**]. The present paper contains a simplified argument for the case $c = 0$. If there exists $f \in L^2 M$ with $\Delta f = \lambda f$ and $\lambda > (n-1)^2 c/4$, then we say that Δ has an embedded eigenvalue. One would like to prove the absence of embedded eigenvalues. If $c > 0$, decay conditions on the curvature and its first two covariant derivatives [**3**] or stronger decay conditions on the curvature alone [**5**] prevent the appearance of embedded eigenvalues.

Assume one has a surface M with Gaussian curvature K approaching zero at infinity. In [**10**], L. Karp gave decay conditions on K and its first derivative which preclude the possibility of positive eigenvalues for Δ. The present paper formulates more stringent decay hypotheses for K alone and reaches the same conclusion: the Laplacian has no embedded eigenvalues. It follows easily that the operator Δ possesses a purely continuous spectrum. One naturally inquires about analogous results for higher-dimensional manifolds with curvature decaying to zero. Unfortunately, the results for surfaces

1991 *Mathematics Subject Classification.* Primary 58G25; Secondary 58G03.
This paper is in final form and no version of it will be submitted for publication elsewhere.

require faster than quadratic decay for K. Similar hypotheses, in dimensions greater than two, would force M to be isometric with R^n [8].

2. Essential spectrum

Let M^n denote a complete Riemannian manifold of dimension n. The Laplacian Δ of M extends uniquely from $C_0^\infty(M)$ to an unbounded self-adjoint operator on $L^2 M$. One defines the essential spectrum of Δ to be those real numbers which are either (i) cluster points of the spectrum or (ii) eigenvalues of infinite multiplicity. Since the Laplace operator is non-negative, one has $\operatorname{Ess\,Spec}\Delta \subset \operatorname{Spec}\Delta \subset [0, \infty)$.

Suppose now that M is simply connected and has nonpositive sectional curvatures everywhere. Let $K(x, \pi)$ denote the sectional curvature of a plane π in the tangent space at $x \in M$. Fix a basepoint $p \in M$, and define $d(x, p)$ to be the geodesic distance from p. As $d(x, p) \to \infty$, we suppose that $K(x, \pi)$ decays to a constant $-c$, $c \geq 0$. More precisely, let $\Phi(t) = \sup |K(x, \pi) + c|$, for $\pi \subset T_x M$ and $d(x, p) \geq t$. Our hypothesis is that $\lim_{t \to \infty} \Phi(t) = 0$. In [4], it was shown consequently that $\operatorname{Ess\,Spec}\Delta = [(n-1)^2 c/4, \infty)$. Our purpose here is to simplify the proof in the special case $c = 0$. In particular, it is possible to avoid reliance upon properties of Bessel functions.

Assume $\lambda \in (0, \infty)$. To show that λ lies in the essential spectrum, it suffices to produce infinitely many approximate eigenfunctions with eigenvalue λ. That is, for any given $\varepsilon > 0$, there should exist an infinite-dimensional subspace S_ε, of the domain of Δ, so that for any $f \in S_\varepsilon$, we have $\|(\Delta - \lambda)f\| \leq \varepsilon\|f\|$. Here $\| \ \|$ denotes the global L^2 norm on M. This criterion for locating the essential spectrum is a simple consequence of the spectral theorem [6, pp. 13–15].

The remainder of this section will be devoted to the proof of

THEOREM 2.1. *Let M be a complete simply connected Riemannian manifold having nonpositive sectional curvatures K. Assume that these curvatures K approach zero at infinity. Then $\operatorname{Ess\,Spec}\Delta = [0, \infty)$.*

Since the essential spectrum is a closed set, it suffices to show $\operatorname{Ess\,Spec}\Delta \supset (0, \infty)$. Fix $\lambda \in (0, \infty)$. Let x be suitably far away from our base-point p, and use the letter r to denote the geodesic distance from x. We transplant the function $\cos(\sqrt{\lambda}\,r)$ into a large geodesic ball B centered at x. Assume that the radius of B is at least $2\pi(2R + 1)/\sqrt{\lambda}$, where R is a large number. Choose a cut-off function of $\phi(r)$ which belongs to $C_0^\infty(2\pi(R - 1)/\sqrt{\lambda}, 2\pi(2R + 1)/\sqrt{\lambda})$, with $\phi(r) = 1$ for $2\pi R/\sqrt{\lambda} < r < 4\pi R/\sqrt{\lambda}$. We may suppose that both $|\phi'(r)|$ and $|\phi''(r)|$ are bounded by a constant c_1, which may depend upon λ but not upon R. One calculates

$$(\Delta - \lambda)(\phi(r)\cos\sqrt{\lambda}\,r) = \Delta\phi(r)\cos(\sqrt{\lambda}\,r) + 2\sqrt{\lambda}\phi'(r)\sin(\sqrt{\lambda}\,r)$$
$$+ \phi(r)(\Delta - \lambda)\cos(\sqrt{\lambda}\,r).$$

If $\theta(r, w)$ denotes the volume element in geodesic spherical coordinates (r, w), and $\theta' = \partial\theta/\partial r$, then

$$
(2.2) \quad \begin{aligned}
(\Delta - \lambda)(\phi(r)\cos\sqrt{\lambda}\,r) &= -\phi''(r)\cos(\sqrt{\lambda}\,r) - \frac{\theta'}{\theta}\cos(\sqrt{\lambda}\,r)\phi'(r) \\
&\quad + 2\sqrt{\lambda}\sin(\sqrt{\lambda}r)\phi'(r) + \frac{\theta'}{\theta}\sqrt{\lambda}\phi(r)\sin(\sqrt{\lambda}r).
\end{aligned}
$$

Suppose that the absolute values of the sectional curvatures in B are less than a constant $b_1 > 0$. Then a standard comparison theorem [2, p. 254] gives $|\theta'/\theta| \leq \sqrt{b_1}\coth(\sqrt{b_1}r)(n-1)$. Suppose that E denotes the subset of B where $2\pi(R-1)/\sqrt{\lambda} \leq r \leq 2\pi(2R+1)/\sqrt{\lambda}$. Thus E contains the support of ϕ. The comparison theorem implies

$$
\sup_E |\theta'(r, w)/\theta(r, w)| \leq \frac{c_2}{R} + \gamma_1(x, R).
$$

The functions $\gamma_i(x, R)$ approach zero as x approaches infinity, for fixed R. All constants labeled c_i may depend upon λ, but not upon x or R. Using the related comparison $|\theta| \leq (\sinh\sqrt{b_1}r/\sqrt{b_1})^{n-1}$, we obtain

$$
\sup_E |\theta(r, w)| \leq c_3 R^{n-1} + \gamma_2(x, R).
$$

It is now elementary to estimate the L^2 norm of each of the four terms in (2.2):

   (i) $\|\phi''(r)\cos(\sqrt{\lambda}r)\| \leq c_3 R^{(n-1)/2}[1 + \gamma_3(x, R)]$,
   (ii) $\|\frac{\theta'}{\theta}\cos(\sqrt{\lambda}r)\phi'(r)\| \leq c_4[\gamma_1(x, R) + c_2/R]R^{(n-1)/2}[1 + \gamma_3(x, R)]$,
   (iii) $\|2\sqrt{\lambda}\sin(\sqrt{\lambda}r)\phi'(r)\| \leq c_5 R^{(n-1)/2}[1 + \gamma_3(x, R)]$,
   (iv) $\|\frac{\theta'}{\theta}\sqrt{\lambda}\phi(r)\sin(\sqrt{\lambda}r)\| \leq c_6[\gamma_1(x, R) + c_2/R]R^{n/2}[1 + \gamma_3(x, R)]$.

Adding these four estimates together yields

$$
\|(\Delta - \lambda)(\phi(r)\cos\sqrt{\lambda}r)\| \leq c_7\left[\gamma_4(x, R) + \frac{c_8}{R^{1/2}}\right]R^{n/2}.
$$

We also need a lower bound for the L^2 norm of our approximate eigenfunction. Since $\phi(r)$ is identically one on an interval of length $2\pi R/\sqrt{\lambda}$, it follows that

$$
\|\phi(r)\cos\sqrt{\lambda}r\| \geq c_9 R^{n/2}.
$$

Combining the last two estimates yields

$$
\begin{aligned}
\|(\Delta - \lambda)&(\phi(r)\cos\sqrt{\lambda}r)\|/\|\phi(r)\cos\sqrt{\lambda}r\| \\
&\leq c_{10}(\gamma_4(x, R) + c_8 R^{-1/2}).
\end{aligned}
$$

Now choose R sufficiently large so that $c_8 c_{10} R^{-1/2} < \varepsilon/2$. For this fixed R, we suppose x is sufficiently far from our basepoint to guarantee $c_{10}\gamma_1(x, R) < \varepsilon/2$.

Our construction has produced $\phi(r)\cos(\sqrt{\lambda}r) \in C_0^\infty(B)$ with

$$\|(\Delta - \lambda)(\phi(r)\cos\sqrt{\lambda}r)\| \leq \varepsilon\|\phi(r)\cos\sqrt{\lambda}r\|.$$

Clearly M contains an infinite number of such balls B, which are pairwise disjoint. This provides an infinite dimensional space of ε-approximate eigenfunctions. Theorem 2.1 follows since ε is arbitrary.

3. Geodesic polar coordinates

Suppose that M is a complete simply connected surface with nonpositive Gauss curvature. According to the theorem of Hadamard-Cartan, the exponential map $\exp : T_p M \to M$ is a diffeomorphism, for any basepoint $p \in M$. Thus we may endow M with a global system of geodesic polar coordinates, where the metric is given by $ds^2 = dr^2 + g^2(r, \theta)d\theta^2$. Along geodesics emanating from p, the function g satisfies the ordinary differential equation $\partial^2 g/\partial r^2 + Kg = 0$, with initial conditions $g(0) = 0$ and $\partial g/\partial r(0) = 1$.

We assume the curvature $K(r, \theta)$ decays rapidly to zero. More precisely, suppose that for large r, $|K(r, \theta)| \leq \psi(r)$, with

$$(3.1) \qquad \int_0^\infty t^{\beta_1}\psi(t)\,dt < \infty, \qquad \lim_{t\to\infty} t^{\beta_2}\psi(t) = 0.$$

Here $\beta_1 \geq 2$ and $\beta_2 \geq 3$ are constants. Then one has

PROPOSITION 3.2. *For a constant* $c_\theta \geq 1$,
 (i) $g(t, \theta) = c_\theta t + O(1)$,
 (ii) $g_t(t, \theta) = c_\theta + o(t^{1-\beta_1})$,
 (iii) $g_{tt}(t, \theta) = O(t^{1-\beta_2})$.

PROOF. We apply the method of asymptotic integrations [9, pp. 304–306] to the Jacobi equation $g_{tt} + Kg = 0$. The given decay conditions, for $K \to 0$, guarantee that all solutions are closely approximated by solutions of the unperturbed equation, $g_{tt} = 0$. That is, for each fixed θ, $g(t, \theta)$ approaches a linear function of the form $c_\theta t + d_\theta$, as $t \to \infty$. Moreover, $\partial g/\partial t(t, \theta)$ is near to the corresponding derivative c_θ. Since M has nonpositive curvature, elementary comparison theorems [2] guarantee $c_\theta \geq 1$. This proves (i) and (ii). For (iii), one simply notes that $g_{tt} = -Kg$, and then invokes (i) along with the decay conditions on K.

Using the standard formula, valid for any coordinate system, the Laplacian in geodesic polar coordinates becomes

$$\Delta f = -g^{-1}\frac{\partial}{\partial r}\left(g\frac{\partial f}{\partial r}\right) - g^{-1}\frac{\partial}{\partial\theta}\left(g^{-1}\frac{\partial f}{\partial\theta}\right).$$

We want to renormalize the measure in our Hilbert space $L^2 M$. Consequently, we conjugate Δ by a radial function $e^{p(r)}$, defined for large r. Set $H = e^p \Delta e^{-p}$. One calculates

$$Hw = -w'' + Sw' + Vw - g^{-1}\frac{\partial}{\partial\theta}\left(g^{-1}\frac{\partial w}{\partial\theta}\right).$$

Here w' denotes the radial derivative $\partial w/\partial r$. Moreover,

$$S = -g^{-1}g' + 2p', \qquad V = g^{-1}p'g' + p'' - (p')^2.$$

Assume that $p(r) = \frac{1}{2}\log r + c_1/2r$, for a suitable constant c_1. Elementary computation then yields, for large r, the following consequence of our Proposition 3.2:

PROPOSITION 3.3. *Under the decay conditions* (3.1),

 (i) $S = O(r^{-2})$,
 (ii) $V = O(r^{-2})$,
 (iii) *if* c_1 *is sufficiently large, then* $(rS)'$ *is positive.*

4. Eigenvalues embedded in the continuum

Let M^2 be a complete simply connected surface having nonpositive Gaussian curvature K. Assume that $K \to 0$ at infinity. As shown in §2, the essential spectrum of Δ then consists of the entire half line $[0, \infty)$. If there exists a nonzero $f \in L^2 M$ with $\Delta f = \lambda f$, $\lambda > 0$, then one has an eigenvalue embedded in the continuum. Our main result is the following criterion:

THEOREM 4.1. *If the decay conditions* (3.1) *hold, then* Δ *has no positive eigenvalues.*

This section is dedicated to the proof of Theorem 4.1. Assume that $\Delta f = \lambda f$, $\lambda > 0$, and $f \in L^2 M$. We prove that f is identically zero. Let $p(r) = \frac{1}{2}\log r + c_1/2r$, where c_1 is sufficiently large. Set $w = e^p f$ and $H = e^p \Delta e^{-p}$. Then $Hw = \lambda w$. In geodesic polar coordinates (r, θ), centered at a given basepoint, we let $\langle w, w \rangle = \int w^2(r, \theta)\, d\theta$ denote the standard L^2 norm on S^1. Moreover, our metric may be written $ds^2 = dr^2 + g^2(r, \theta)\, d\theta^2$. Proposition 3.2(i), and our choice of $p(r)$, guarantee that $\int_1^\infty \langle w, w \rangle\, dr$ is finite. Define γ as the logarithmic derivative of gr^{-1}, $\gamma = \partial/\partial r \log(gr^{-1}) = g'/g - 1/r$. Since M has nonpositive curvatures, a basic comparison theorem [2] yields $\gamma \geq 0$.

Now suppose that f is not identically zero and reason by contradiction. The unique continuation theorem of Aronszajn [1] implies that f, and thus also w, cannot vanish identically on any open set. We exploit this fact along with suitable differential inequalities. Define $w' = \partial w/\partial r$ and $w_\theta = \partial w/\partial \theta$. One may set

$$\mathscr{G}(r) = \langle w', w'g \rangle r^{-1} - r^{-1}\langle w_\theta, w_\theta g^{-1} \rangle + \lambda\langle w, wg \rangle r^{-1}.$$

We begin by establishing

LEMMA 4.2. *For* $r > R_0$, $\frac{d}{dr}(r\mathscr{G}(r)) \geq 0$ *and thus* $\mathscr{G}(r) \leq 0$. *Here* R_0 *just denotes a suitably large number.*

PROOF. It is straightforward to calculate

$$\frac{d}{dr}(r\mathcal{G}) = \langle w', w'g\rangle r^{-1} + \lambda\langle w, wg\rangle r^{-1} + \langle w', w'\gamma g\rangle$$
$$+ \lambda\langle w, w\gamma g\rangle + 2\langle w'', w'g\rangle + 2\lambda\langle w, w'g\rangle$$
$$+ \langle w_\theta, w_\theta g'g^{-2}\rangle - 2\langle w_\theta', w_\theta g^{-1}\rangle.$$

As noted above $\gamma \geq 0$. Furthermore $g'/g = \gamma + 1/r \geq 0$. Thus

$$\frac{d}{dr}(r\mathcal{G}) \geq \langle w', w'g\rangle r^{-1} + \lambda\langle w, wg\rangle r^{-1}$$
$$+ 2\left\langle w'' + g^{-1}\frac{\partial}{\partial\theta}(g^{-1}w_\theta) + \lambda w, gw'\right\rangle.$$

This rewriting includes partial integration in θ.

Since $Hw = \lambda w$, this eigenfunction property yields

$$\frac{d}{dr}(r\mathcal{G}) \geq \langle w', w'g\rangle r^{-1} + \lambda\langle w, wg\rangle r^{-1}$$
$$+ 2\langle Sw', gw'\rangle + 2\langle Vw, gw'\rangle.$$

Proposition 3.3 gives both S and V to be of order r^{-2}. By the Schwarz inequality, $2\langle w, w'g\rangle \leq \langle w, wg\rangle + \langle w', w'g\rangle$.

So, for $r > R_0$, appropriately large,

$$\frac{d}{dr}(r\mathcal{G}) \geq \frac{1}{2}\langle w', w'g\rangle r^{-1} + \frac{1}{2}\lambda\langle w, wg\rangle r^{-1} \geq 0.$$

Recall that f is in the domain of the Laplacian Δ. So $df = f' dr + f_\theta d\theta$ is a square integrable 1-form on M. This means that $\langle f', f'g\rangle$ and $\langle f_\theta, f_\theta g^{-1}\rangle$ are integrable functions of r, as is $\langle f, fg\rangle$ for $f \in L^2 M$. Elementary computation verifies that consequently $\mathcal{G}(r)$ is integrable, for large r. Moreover, the basic inequality $\frac{d}{dr}(r\mathcal{G}(r)) \geq 0$ thereby forces $\mathcal{G}(r) \leq 0$, for $r > R_0$.

To proceed further, we define $w_m = r^m w$, for nonnegative integers m. Since $Hw = \lambda w$, one sees that

$$(4.3) \qquad w_m'' + g^{-1}\frac{\partial}{\partial\theta}\left(g^{-1}\frac{\partial w_m}{\partial\theta}\right) - 2mr^{-1}w_m' - Sw_m'$$
$$+ (m(m+1)r^{-2} + mSr^{-1} + \lambda - V)w_m = 0.$$

Our next differential inequality will involve

$$\mathcal{L}(m, r) = \langle w_m', w_m'g\rangle r^{-1} + (\lambda - \lambda R_0 r^{-1} + m(m+1)r^{-2})\langle w_m, w_m gr^{-1}\rangle$$
$$- r^{-1}\langle g^{-1}w_{m,\theta}, w_{m,\theta}\rangle + mr^{-1}\langle Sw_m, w_m gr^{-1}\rangle.$$

Specifically, one has

LEMMA 4.4. *For $m > m_0$ and $r > R_1 > R_0$, $\frac{d}{dr}(r^2 \mathcal{L}) \geq 0$. There is some $m_1 > m_0$ so that for $r \geq R_2 > R_1$, $\mathcal{L}(m_1, r) > 0$. As usual, the symbols m_0, m_1 and R_1, R_2 denote large constants.*

PROOF. It is straightforward to compute

$$\frac{d}{dr}(r^2 \mathcal{L}) = 2\langle w'_m, w'_m g \rangle + (2\lambda - \lambda R_0 r^{-1})\langle w_m, w_m g \rangle$$

$$+ r\langle w'_m, w'_m \gamma g \rangle + (\lambda r - \lambda R_0 + m(m+1)r^{-1})\langle w_m, w_m \gamma g \rangle$$

$$+ 2\langle w''_m, w'_m g r \rangle + 2(\lambda r - \lambda R_0 + m(m+1)r^{-1})\langle w_m, w'_m g \rangle$$

$$+ \langle g^{-1} \gamma w_{m,\theta}, w_{m,\theta} \rangle r - 2r\langle g^{-1} w_{m,\theta}, w'_{m,\theta} \rangle$$

$$+ m\langle (rS)' w_m, w_m g r^{-1} \rangle + m\langle Sw_m, w_m \gamma g \rangle + 2m\langle Sw_m, w'_m g \rangle.$$

We now invoke the positivity conditions $\gamma \geq 0$ and $(rS)' \geq 0$, from Proposition 3.3. Combining these with an integration by parts in θ, one finds, for large r,

$$\frac{d}{dr}(r^2 \mathcal{L}) \geq 2\langle w'_m, w'_m g \rangle + \lambda\langle w_m, w_m g \rangle - 2\lambda R_0 \langle w_m, w'_m g \rangle$$

$$+ 2\left\langle w''_m + g^{-1} \frac{\partial}{\partial \theta}(g^{-1} w_{m,\theta}) \right.$$

$$\left. + (\lambda + m(m+1)r^{-2} + mSr^{-1})w_m, w'_m g \right\rangle r.$$

Using the partial differential equation (4.3) for w_m, we deduce

$$\frac{d}{dr}(r^2 \mathcal{L}) \geq (4m + 2)\langle w'_m, w'_m g \rangle + \lambda\langle w_m, w_m g \rangle$$

$$- 2\lambda R_0 \langle w_m, w'_m g \rangle + 2\langle Sw'_m, w'_m g \rangle r + 2\langle V w_m, w'_m g \rangle r.$$

One now applies Proposition 3.3 and the Schwarz inequality yielding

$$\frac{d}{dr}(r^2 \mathcal{L}) \geq (2m + 1)\langle w'_m, w'_m g \rangle + \frac{1}{2}\lambda\langle w_m, w_m g \rangle - 2\lambda R_0 \langle w_m, w'_m g \rangle.$$

Clearly $|w_m||w'_m| \leq \varepsilon^{-1}|w'_m|^2 + \varepsilon|w_m|^2$, for any $\varepsilon > 0$. Choose ε sufficiently small and $m > m_0$ sufficiently large to force $\frac{d}{dr}(r^2 \mathcal{L}(m, r)) \geq 0$, when $r > R_1 > R_0$.

The formula defining $\mathcal{L}(m, r)$ shows that $\mathcal{L}(m_1, R_2) > 0$ for some $m_1 > m_0$ and $R_2 > R_1$. In fact,

$$r^{-2m} \mathcal{L}(m, r) = \left\langle \left(w' + \frac{m}{r}w\right)^2, gr^{-1} \right\rangle + mr^{-1}\langle Sw, wgr^{-1} \rangle$$

$$+ (\lambda - \lambda R_0 r^{-1} + m(m+1)r^{-2})\langle w, wgr^{-1} \rangle$$

$$- r^{-1}\langle g^{-1} w_\theta, w_\theta \rangle.$$

So one need only require $\langle w, wgr^{-1} \rangle(R_2) > 0$, as implied by unique continuation [1] for some $R_2 > R_1$, and then choose m_1 sufficiently large. Since the derivative $\frac{d}{dr}(r^2\mathscr{L}(m, r)) \geq 0$ for $r > R_1$, $m = m_1 > m_0$, we have $\mathscr{L}(m_1, r) > 0$, when $r \geq R_2$. This proves the Lemma 4.4.

Since $\int_1^\infty \langle w, wgr^{-1}\rangle\, dr$ is finite, there exists an arbitrarily large $R_3 > R_2$ with $2\langle w, w'gr^{-1}\rangle + \langle w, w\gamma gr^{-1}\rangle$ negative at $r = R_3$. Since $\gamma > 0$, one has $\langle w, w'gr^{-1}\rangle(R_3) < 0$. Fixing m_1, we may assume $-\lambda R_0 r^{-1} + m_1 r^{-1} S + m_1(2m_1 + 1)r^{-2} < 0$, for $r \geq R_3$.

One may write

$$
\begin{aligned}
R_3^{-2m_1}\mathscr{L}(m_1, R_3) &= \langle w', w'gR_3^{-1}\rangle + 2m_1 R_3^{-2}\langle w', wg\rangle + \lambda\langle w, wgR_3^{-1}\rangle \\
&\quad + (-\lambda R_0 R_3^{-1} + m_1(2m_1 + 1)R_3^{-2})\langle w, wgR_3^{-1}\rangle \\
&\quad + m_1 R_3^{-1}\langle Sw, wgR_3^{-1}\rangle - R_3^{-1}\langle g^{-1}w_\theta, w_\theta\rangle.
\end{aligned}
$$

Thus

$$
R_3^{-2m_1}\mathscr{L}(m_1, R_3) \leq \langle w', w'gR_3^{-1}\rangle + \lambda\langle w, wgR_3^{-1}\rangle - R_3^{-1}\langle g^{-1}w_\theta, w_\theta\rangle.
$$

Equivalently $R_3^{-2m_1}\mathscr{L}(m_1, R_3) \leq \mathscr{G}(R_3)$. However, Lemma 4.2 requires $\mathscr{G}(R_3) \leq 0$ and Lemma 4.4 constrains $\mathscr{L}(m_1, R_3) > 0$. This contradiction shows that w, and therefore f, vanish for large r. By unique continuation [1], f is identically zero. The proof of Theorem 4.1 is complete.

It is easy to deduce

COROLLARY 4.5. *Under the hypotheses of Theorem* 4.1, *the Laplacian* Δ *has purely continuous spectrum consisting of the half line* $[0, \infty)$.

PROOF. We have already observed that $\operatorname{Spec}\Delta = \operatorname{Ess}\operatorname{Spec}\Delta = [0, \infty)$. It suffices to show that the point spectrum is empty. The Laplacian is nonnegative by definition. Theorem 4.1 guarantees the absence of positive eigenvalues. However, zero cannot occur as an eigenvalue either. In fact, it is well known [11] that complete Riemannian manifolds of infinite volume never admit L^2 harmonic functions.

BIBLIOGRAPHY

1. N. Aronszajn, *A unique continuation theorem for solutions of elliptic partial differential equations or inequalities of second order*, J. Math. Pures Appl. **36** (1957), 235–250.
2. R. Bishop and R. Crittenden, *Geometry of manifolds*, Academic Press, 1964.
3. H. Donnelly, *Eigenvalues embedded in the continuum for negatively curved manifolds*, Michigan Math. J. **2** (1981), 53–62.
4. ____, *On the essential spectrum of a complete Riemannian manifolds*, Topology **20** (1981), 1–14.
5. ____, *Negative curvature and embedded eigenvalues*, Math. Z. **203** (1990), 301–308.
6. I. M. Glazman, *Direct methods of qualitative spectral analysis of singular differential operators*, Daniel Davey, New York, 1965.
7. R. Greene and H. Wu, *Function theory on manifolds which possess a pole*, Lecture Notes in Math., vol. 699, Springer-Verlag, (1979).
8. ____, *Gap theorems for noncompact Riemannian manifolds*, Duke Math. J. **49** (1981), 731–756.
9. P. Hartman, *Ordinary differential equations*, Wiley, 1964.

10. L. Karp, *Noncompact Riemannian manifolds with purely continuous spectrum*, Michigan Math. J. **31** (1984), 339–347.
11. S. T. Yau, *Some function theoretic properties of complete Riemannian manifolds and their applications to geometry*, Indiana Univ. Math. J. **25** (1976), 659–670.

PURDUE UNIVERSITY

Proceedings of Symposia in Pure Mathematics
Volume **54** (1993), Part 3

Manifolds of Nonpositive Curvature

PATRICK EBERLEIN, URSULA HAMENSTÄDT,
AND VIKTOR SCHROEDER

Introduction

This article is intended to be a sketch of some of the major developments over the past thirty years of research on manifolds with nonpositive sectional curvature. In the early 1960s this area was revived after some years of dormancy by parallel developments in semisimple and algebraic groups and hyperbolic dynamical systems. At the same time geometric and analytic methods appeared in the work of Mostow and Karpelevic [**Kar**] that were to be important for future results in nonpositive curvature. Advances in semisimple and algebraic groups not only provided many examples of locally symmetric manifolds of nonpositive curvature and finite volume but also suggested problems, particularly those related to the Mostow Rigidity Theorem, whose investigation provided to be fruitful [**Ba3, Ham3, BuS, BGS**]. Progress in hyperbolic dynamical systems had a similar effect, and this framework proved to be well suited to problems involving the geodesic flow (cf. §§7 and 8), especially for compact manifolds with negative or nearly negative sectional curvature [**Ano, Pe1, Pe2, Pe3, Pe4**]. Geometric and analytic methods arose from several sources including

(1) synthetic and nondifferentiable geometric methods due originally to Alexandrov [**Al**] and Busemann [**Bus1**] and pushed much farther by Gromov [**BGS, GLP**];

(2) differential geometric methods involving convexity, first used systematically in the paper of Bishop and O'Neill [**BO**];

(3) harmonic mappings begun by Eells and Sampson in [**ES**] and recently assuming an important role in many questions [**Co, GS, Her, YZ**].

One cannot really separate the influence of the different contributing areas

1991 *Mathematics Subject Classification*. Primary 53C20; Secondary 53C35.
This paper is in final form and no version of it will be submitted for publication elsewhere.

mentioned above. For example, prior to its general development in [EO] the boundary sphere of a simply connected, nonpositively curved manifold was studied in detail for symmetric spaces by Karpelevic [Kar], who used Lie algebraic methods combined with geometric insight to study the Laplace operator. The boundary sphere also played an important role in Mostow's proof of the Rigidity Theorem, but its treatment varied from analytic to algebraic depending on whether one was considering the rank 1 or higher rank case. This dichotomy of methods persists in the differential geometric generalizations of the Rigidity Theorem. On the other hand the boundary sphere lends itself naturally to nondifferentiable geometric methods of study, as may be seen in §4, and is closely related to the stable and unstable Anosov foliations of the geodesic flow in compact manifolds of strictly negative sectional curvature.

The article is divided roughly into two parts. In the first part we present and discuss some basic objects, and in the second part we outline some of the major results. It is impossible to do justice adequately to the many developments in the past thirty years, and in particular we have little or no treatment of problems related to partial differential equations or complex differential geometry. Nevertheless, we hope that our discussion and bibliography will be a helpful guide.

For the benefit of the reader we now give a more detailed description of the two parts of this article, which consist of five and six sections respectively. In §1 we discuss basic properties of negatively curved manifolds including convexity properties, the boundary sphere $\widetilde{M}(\infty)$, geodesic symmetries and holonomy, Busemann functions, the duality condition, and the visibility axiom. In §2 we describe relevant geometry of the tangent and unit tangent bundles including the Sasaki metric, contact 1-form and symplectic 2-form, stable and unstable foliations, and the relationship between the geodesic flow and Jacobi vector fields. In §3 we discuss Anosov foliations and the Kanai connection in the unit tangent bundle of a compact negatively curved manifold. In §4 we consider pseudometrics in the boundary sphere $\widetilde{M}(\infty)$ introduced by Gromov and Hamenstädt. In §5 we describe some important geodesic flow invariant measures on the unit tangent bundle of a compact negatively curved manifold M and on the boundary sphere $\widetilde{M}(\infty)$ of the universal cover $\widetilde{M}$. These include the Lebesgue-Liouville, harmonic, and Bowen-Margulis measures. Section 5 also contains a discussion of quasiconformal structures on $\widetilde{M}(\infty)$ that are important for the study of manifolds of strictly negative curvature.

In §6 we begin our discussion of results by considering the relationship of the geometry of a compact nonpositively curved manifold and the algebraic structure of its fundamental group. We describe the classical results of Bieberbach, Cartan, and Preissmann as well as more recent work of Gromoll-Wolf, Bangert-Schroeder, and others. In §7 we look at the problem of entropy rigidity, characterizing locally symmetric spaces by properties of the metric and topological entropy functionals on suitably normalized spaces of homotopically equivalent, compact negatively curved manifolds. Topics include results of Hamenstädt and Pansu, and conjectures of Gromov and Katok. In §8 we discuss generalizations of the Mostow Rigidity Theorem due to Gromov, Ballmann, Burns-Spatzier, Hamenstädt, and others. We also describe the characterization of locally symmetric spaces up to geodesic conjugacy by the assumption that the Anosov foliations be C^∞. This work was begun by Kanai and continued by Feres-Katok, Benoist-Foulon-Labourie, and others. In §9 we discuss the problems of geodesic conjugacy and marked length spectrum, and present the results of Croke and Otal. In §10 we describe some applications of the Margulis Lemma including the structure of ends of noncompact manifolds of finite volume and negative curvature. In §11 we discuss the question of arithmeticity of lattices in semisimple Lie groups with no compact factors, including the work of Margulis for the higher rank case and many others for the rank 1 case.

1. Preliminaries [BGS, BO, EO]

(1.1) Notation. In this article $\widetilde{M}$ will denote a complete, simply connected Riemannian manifold of nonpositive sectional curvature $(K \leq 0)$. We use M to denote a complete manifold with $K \leq 0$ and finite volume; usually M will be compact. The unit tangent bundles of $\widetilde{M}$, M will be denoted by $S\widetilde{M}$, SM and the unit vectors at a point p in $\widetilde{M}$, M will be denoted by $S_p\widetilde{M}$, S_pM. The geodesic flow in either $S\widetilde{M}$ or SM will be denoted by $\{g^t\}$. All geodesics of $\widetilde{M}$, M will have unit speed. The Riemannian distance function will be denoted by d. We let γ_v denote the geodesic with initial velocity $v \in S\widetilde{M}$, SM.

(1.2) Basic facts and convexity [BO]. It is known that any two points of a simply connected manifold $\widetilde{M}$ with $K \leq 0$ are joined by a unique geodesic, and it follows that $\widetilde{M}$ is diffeomorphic to a Euclidean space of the same dimension. In particular, the homotopy groups $\pi_k(M^*)$ of any quotient manifold M^* of $\widetilde{M}$ vanish for $k \geq 2$.

Like the Euclidean spaces, the simply connected spaces $\widetilde{M}$ of nonpositive curvature admit many convex functions $f \colon \widetilde{M} \to \mathbb{R}$, and this richness in convexity makes it possible to study nonpositively curved manifolds in a systematic way. The importance of convexity was emphasized and exploited in [BO] and is also used in [Kar] and [Mo3].

A continuous function $g \colon \mathbb{R} \to \mathbb{R}$ is convex if $g(tx + (1 - t)y) \leq tg(x) + (1-t)g(y)$ for $t \in [0, 1]$ and $x, y \in \mathbb{R}$. If g is C^2, then g is convex if and only if $g''(x) \geq 0$ for all x. A continuous function $f \colon \widetilde{M} \to \mathbb{R}$ is convex if its restriction to any complete geodesic $\gamma \colon \mathbb{R} \to \widetilde{M}$ is a convex function on $\mathbb{R}$.

We present some important examples of convex functions that arise in the study of nonpositively curved manifolds. Other examples may be found in [**BO**].

(1) If A is a closed convex subset of $\widetilde{M}$ (e.g., a complete, totally geodesic submanifold of $\widetilde{M}$), then $f(p) = d(p, A)$ is a continuous convex function from $\widetilde{M}$ to $\mathbb{R}$. The most important cases occur when A is a single point or a complete geodesic of $\widetilde{M}$.

(2) If $\varphi: \widetilde{M} \to \widetilde{M}$ is an isometry, then the displacement function $d_\varphi: \widetilde{M} \to \mathbb{R}$ given by $d_\varphi(p) = d(p, \varphi p)$ is a continuous convex function.

(3) Busemann functions $f: \widetilde{M} \to \mathbb{R}$ (see (1.5)).

(4) If $Y(t)$ is a Jacobi vector field on a complete geodesic γ of an arbitrary complete manifold M^* with $K \leq 0$, then $f(t) = |Y(t)|^2$ is a C^∞ convex function on $\mathbb{R}$.

(1.3) Boundary sphere $\widetilde{M}(\infty)$ [Kar, EO]. Two unit speed geodesics γ, σ of $\widetilde{M}$ are said to be *asymptotic* or *asymptotes* if there exists a constant $c > 0$ such that $d(\gamma t, \sigma t) \leq c$ for all $t \geq 0$ [**BO**]. This is equivalent to saying that the convex functions $f(t) = d(\gamma t, \sigma)$ or $g(t) = d(\gamma t, \sigma t)$ are nonincreasing in t. The *boundary sphere* $\widetilde{M}(\infty)$ is defined to be the set of equivalence classes of asymptotic unit speed geodesics of $\widetilde{M}$. The space $\widetilde{M}(\infty)$ was first constructed for symmetric spaces $\widetilde{M}$ by F. Karpelevic [**Kar**] and later for arbitrary spaces $\widetilde{M}$ in [**EO**].

Given a point p of $\widetilde{M}$ and a point x of $\widetilde{M}(\infty)$ there is a unique unit speed geodesic γ_{px} of $\widetilde{M}$ such that $\gamma_{px}(0) = p$ and γ_{px} belongs to the equivalence class x. The map $f_p: \widetilde{M}(\infty) \to S_p \widetilde{M}$ given by $f_p(x) = \gamma'_{px}(0)$ is a bijection, and the topology on $\widetilde{M}(\infty)$ that makes f_p a homeomorphism is independent of the point p. This topology will be called the *sphere topology*. More generally, there is a natural (*cone*) topology on $\widetilde{M} \cup \widetilde{M}(\infty)$ that agrees with the sphere topology when restricted to $\widetilde{M}(\infty)$ and such that $\widetilde{M}$ with its own topology is dense and open in $\widetilde{M} \cup \widetilde{M}(\infty)$. Natural geometric operations are continuous with respect to these topologies. For example, the angular measurement function $(p, x, y) \to \measuredangle_p(x, y) =$ angle subtended at $p \in \widetilde{M}$ by points x, y in $\widetilde{M}(\infty) = \measuredangle(\gamma'_{px}(0), \gamma'_{py}(0))$ depends continuously on p, x and y.

If φ is an isometry of $\widetilde{M}$, then φ extends to a homeomorphism of $\widetilde{M}(\infty)$ by defining $\varphi(\gamma(\infty)) = (\varphi \circ \gamma)(\infty)$, where $\gamma(\infty)$ denotes the asymptote equivalence class in $\widetilde{M}(\infty)$ of a unit speed geodesic γ of $\widetilde{M}$. If Γ is a discrete group of isometries of $\widetilde{M}$ such that the quotient space $M = \widetilde{M}/\Gamma$ is a smooth manifold with finite volume, then the topological properties of the action of Γ on $\widetilde{M}(\infty)$ reflect strongly the geometric properties of M. See the discussion below in (1.6).

(1.4) Geodesic symmetries, holonomy and symmetric spaces. Fix a point p in a simply connected space $\widetilde{M}$ with $K \leq 0$. There is a unique diffeomorphism S_p of $\widetilde{M}$ of order two such that S_p fixes p and $S_p(\gamma(t)) = \gamma(-t)$

for every $t \in \mathbb{R}$ and every geodesic γ of $\widetilde{M}$ with $\gamma(0) = p$. The diffeomorphism S_p is called the *geodesic symmetry at* p. We let G^*, G_e^* denote the groups of diffeomorphisms of $\widetilde{M}$ generated respectively by $\{S_p : p \in \widetilde{M}\}$ and $\{S_p \circ S_q : p, q \in \widetilde{M}\}$. The groups G^*, G_e^* are called respectively the *symmetry diffeomorphism group of* $\widetilde{M}$ and the *even symmetry diffeomorphism group of* $\widetilde{M}$. Each geodesic symmetry S_p extends in a natural way to a homeomorphism of order two of the boundary $\widetilde{M}(\infty)$.

If every geodesic symmetry S_p, $p \in \widetilde{M}$, is actually an isometry of $\widetilde{M}$, then $\widetilde{M}$ is said to be a *symmetric space*. Symmetric spaces are the model spaces for the simply connected spaces $\widetilde{M}$ with $K \leq 0$, and examples include the Euclidean spaces with $K \equiv 0$ and the real hyperbolic spaces with $K \equiv -1$. Elie Cartan completely classified the Riemannian symmetric spaces, which include spaces with $K \geq 0$ as well as those with $K \leq 0$ [**Ca1, Ca2, Ca3, Ca4**]. If $\widetilde{M}$ is symmetric with $K \leq 0$ and if $G = I_0(\widetilde{M})$, the identity component of the isometry group of $\widetilde{M}$, then G acts transitively on $\widetilde{M}$. Hence $\widetilde{M}$ may be regarded as a coset space G/K, where $K = G_p = \{g \in G : gp = p\}$. If $\widetilde{M}$ has no Euclidean de Rham factor, then G is a semisimple Lie group; that is, the Killing form B is nondegenerate on the Lie algebra $\mathscr{G}$.

A *k-flat* in a simply connected space $\widetilde{M}$ with $K \leq 0$ is a complete, totally geodesic, k-dimensional submanifold of $\widetilde{M}$ that with its induced metric is isometric to the k-dimensional Euclidean space with its standard flat metric. The *rank* of a symmetric space $\widetilde{M}$ is the largest integer $k \geq 1$ such that every geodesic of $\widetilde{M}$ lies in at least one k-flat.

There is a close correspondence between the action of the holonomy group Φ_p on $S_p\widetilde{M}$, the unit vectors at p, and the action of the even symmetry diffeomorphism group G_e^* on $\widetilde{M}(\infty)$. If $f_p : \widetilde{M}(\infty) \to S_p\widetilde{M}$ is the homeomorphism defined above and if A is a closed subset of $\widetilde{M}(\infty)$ invariant under G_e^*, then $f_p(A)$ is a closed subset of $S_p\widetilde{M}$ invariant under Φ_p (Theorem A of [**E10**]). In particular if $\widetilde{M}$ is irreducible and if G_e^* leaves invariant a proper, closed subset A of $\widetilde{M}(\infty)$, then a result of M. Berger on holonomy implies that $\widetilde{M}$ must be symmetric with rank at least two [**Ber, E10**].

(1.5) Busemann functions and horospheres [Bus1, EO, HI]. Let $\widetilde{M}$ be simply connected with $K \leq 0$, and let $v \in S\widetilde{M}$ be any unit vector. We define the *Busemann function* $f_v : \widetilde{M} \to \mathbb{R}$ by

$$f_v(p) = \lim_{t \to \infty} d(p, \gamma_v t) - t,$$

where γ_v denotes the geodesic with initial velocity v. Busemann functions have the following properties:

(1) f_v is C^2 and convex for every $v \in S\widetilde{M}$.

(2) If v, w are asymptotic unit vectors in $S\widetilde{M}$ (i.e., γ_v, γ_w are asymptotic geodesics in $\widetilde{M}$), then $f_v - f_w$ is constant in $\widetilde{M}$. Moreover, $\mathrm{grad}\, f_v = \mathrm{grad}\, f_w$ is a unit vector field in M.

Now let x be any point of $\widetilde{M}(\infty)$, and let $v \in S\widetilde{M}$ be any unit vector such that γ_v belongs to the equivalence class x.

The *horosphere at x through p*, denoted $H(p, x)$, is defined to be the level set of f_v that contains p. The unit vector field $V_x = \mathrm{grad}(f_v)$ is an outward unit normal vector field to the horospheres at x.

Note that the definitions of $H(p, x)$ and V_x do not depend on the choice of v by property (2) above.

Given $v \in S\widetilde{M}$ we also define $H(v)$ to be the level set of f_v that contains the footpoint of v. We call $H(v)$ the *horosphere determined by v*.

(1.6) Duality condition [Ba1, CE, E5]. Let $\widetilde{M}$ be any simply connected space with $K \leq 0$, and let Γ be any subgroup (not necessarily discrete) of $I(\widetilde{M})$, the isometry group of $\widetilde{M}$. The subgroup Γ is said to satisfy the *duality condition* if for every point p of $\widetilde{M}$ and every geodesic γ of $\widetilde{M}$ there exists a sequence $\{\varphi_n\} \subseteq \Gamma$ such that $\varphi_n(p) \to \gamma(\infty)$ and $\varphi_n^{-1}(p) \to \gamma(-\infty)$ as $n \to \infty$, where convergence takes place in the cone topology of $\widetilde{M} \cup \widetilde{M}(\infty)$ and $\gamma(\infty)$, $\gamma(-\infty)$ denote the asymptote equivalence classes of the geodesics γ and $\gamma^{-1} \colon t \to \gamma(-t)$.

There is an equivalent definition of the duality condition that seems more natural and appealing from a dynamical point of view. A unit vector $v \in S\widetilde{M}$ is said to be *nonwandering modulo the geodesic flow $\{g^t\}$ and a group $\Gamma \subseteq$ $I(\widetilde{M})$* if for any neighborhood O of v in $S\widetilde{M}$ there exist sequences $\{t_n\} \subseteq \mathbb{R}$ and $\{\varphi_n\} \subseteq \Gamma$ such that $t_n \to +\infty$ and $(d\varphi_n \circ g^{t_n})(O) \cap O$ is nonempty for every n. One can then show that Γ satisfies the duality condition if and only if every unit vector $v \in S\widetilde{M}$ is nonwandering modulo $\{g^t\}$ and Γ. From this formulation it is clear that if $\Gamma \subseteq I(\widetilde{M})$ is a *lattice*, i.e., $\widetilde{M}/\Gamma$ is smooth with finite Riemannian volume, then Γ satisfies the duality condition. The lattice subgroups Γ of $I(\widetilde{M})$ are the most important examples of groups Γ that satisfy the duality condition.

Groups Γ of $I(\widetilde{M})$ that satisfy the duality condition have nice properties, and it is often more convenient to consider problems in this context even when one is really interested in the behavior of lattices (see property (3) for example). We list some basic properties.

(1) If $\Gamma \subseteq I(\widetilde{M})$ satisfies the duality condition and if Γ^* is any subgroup of $I(\widetilde{M})$ that contains Γ, then Γ^* satisfies the duality condition.

(2) If $\Gamma \subseteq I(\widetilde{M})$ satisfies the duality condition and if Γ^* is a subgroup of Γ with finite index in Γ, then Γ^* satisfies the duality condition.

(3) Let $\widetilde{M} = \widetilde{M}_1 \times \widetilde{M}_2$ be a Riemannian product, and let $\Gamma \subseteq I(\widetilde{M})$ be a group that preserves the splitting (i.e., leaves invariant the foliations of $T\widetilde{M}$ induced by $\widetilde{M}_1$, $\widetilde{M}_2$). If Γ satisfies the duality condition in $\widetilde{M}$,

then $\Gamma_i = p_i(\Gamma)$ satisfies the duality condition in $\widetilde{M}_i$ for $i = 1, 2$, where $p_i: \Gamma \to I(\widetilde{M}_i)$ denotes the natural projection homomorphism (note that Γ_i may not be discrete for $i = 1, 2$ even if Γ is discrete).

Let $\Gamma \subseteq I(\widetilde{M})$ be a discrete group such that the quotient space $\widetilde{M}/\Gamma$ is smooth. If Γ satisfies the duality condition, then the topological action of Γ on $\widetilde{M}(\infty)$ reflects strongly the geometry of $\widetilde{M}/\Gamma$. We give two examples. The first is part of Theorem 4.14 of [**E3**].

THEOREM. *Let $\Gamma \subseteq I(\widetilde{M})$ be a discrete subgroup such that $M^* = \widetilde{M}/\Gamma$ is smooth and Γ satisfies the duality condition. Then the following are equivalent*:

(1) *The geodesic flow has a dense orbit in SM^*.*

(2) *Γ has a dense orbit in $\widetilde{M}(\infty)$.*

(3) *Every Γ-orbit in $\widetilde{M}(\infty)$ is dense in $\widetilde{M}(\infty)$.*

A complement to this result is Theorem 4.1 of [**E10**].

THEOREM. *Let $\widetilde{M}$ be simply connected and irreducible with $K \leq 0$, and let $\Gamma \subseteq I(\widetilde{M})$ be any subgroup that satisfies the duality condition. If Γ leaves invariant a proper closed subset of $\widetilde{M}(\infty)$, then $\widetilde{M}$ is symmetric of rank $k \geq 2$.*

(1.7) Visibility axiom. A manifold $\widetilde{M}$ (or any quotient manifold of $\widetilde{M}$) is said to satisfy the *visibility axiom* if for every point p in $\widetilde{M}$ and every positive number ε there exists a positive number $R = R(p, \varepsilon)$ such that for any geodesic σ of $\widetilde{M}$ with $d(p, \sigma) \geq R$, the angle subtended at p by $\sigma = \sup\{\measuredangle_p(\sigma t, \sigma s): s, t \in \mathbb{R}\}$ is at most ε. Briefly stated, $\widetilde{M}$ satisfies the visibility axiom if distant geodesics look small. If the sectional curvature of $\widetilde{M}$ is bounded above by a negative constant, then $\widetilde{M}$ satisfies the visibility axiom [**EO**]. There are a number of equivalent formulations of the visibility axiom (cf. [**BGS**, pp. 54–55]. We list some of these. The following properties are equivalent in $\widetilde{M}$:

(1) $\widetilde{M}$ satisfies the visibility axiom.

(2) Given distinct points x, y in $\widetilde{M}(\infty)$ there exists a geodesic γ of $\widetilde{M}$ that *joins* x to y; that is, $\gamma(\infty) = x$ and $\gamma(-\infty) = y$ or $\gamma(\infty) = y$ and $\gamma(-\infty) = x$.

(3) The Tits metric Td is degenerate in $\widetilde{M}(\infty)$ (cf. §4); that is, $Td(x, y) = +\infty$ whenever x, y are distinct points of $\widetilde{M}(\infty)$.

If $\widetilde{M}$ admits a compact quotient manifold $M = \widetilde{M}/\Gamma$, then the properties above are equivalent to the following properties:

(4) As a metric space with respect to the Riemannian metric the space $(\widetilde{M}, d)$ is *hyperbolic* in the sense of Gromov [**Gro4**, **Gh-Ha**]; that is, there exists a constant $\delta > 0$ such that for any geodesic triangle Δ in $\widetilde{M}$ the distance of a point p on one edge of Δ to the union of the other two sides of Δ is at most δ.

(5) $\widetilde{M}$ admits no 2-flat [**E2**].

Note that in (4) the definition of hyperbolic makes sense in any metric space (X, d) for which minimal geodesics exist between any two points of X.

2. Geometry of the tangent and unit tangent bundles

We sketch some basic facts about the tangent and unit tangent bundles of an arbitrary Riemannian manifold and then specialize to the case of nonpositive curvature.

(2.1) Connection map [GKM, E11]. Let X denote a complete Riemannian manifold, and let π denote the base point projection from either the tangent bundle TX or the unit tangent bundle SX onto X. The kernel of $d\pi$ defines the *vertical* distribution on TX or SX. Next we define a connection map [**GKM**] $K^*: T(TX) \to TX$ such that for each $v \in TX$ the restriction $K^*: T_v(TX) \to T_{\pi(v)}X$ is linear and surjective. The kernel of K^* then defines the *n*-dimensional *horizontal* distribution in TX. If $V(v)$, $H(v)$ denote the vertical and horizontal subspaces of $T_v(TX)$, then $T_v(TX) = V(v) \oplus H(v)$, direct sum, for every $v \in TX$.

Given a vector $\xi \in T_v(TX)$ let $Z: (-\varepsilon, \varepsilon) \to TX$ be a smooth curve with initial velocity ξ. Let $\alpha: (-\varepsilon, \varepsilon) \to X$ denote the projection $\pi \circ Z$. We define

$$K^*(\xi) = Z'(0)$$

where $Z'(0)$ denotes the covariant derivative of Z along α evaluated at $t = 0$. The definition of $K^*(\xi)$ does not depend on the choice of curve Z. For a local coordinate definition and further properties of K^* see [**GKM**].

(2.2) Sasaki metric. The inner product $\langle\,,\,\rangle$ in X induces in a natural way the Sasaki inner product in TX or SX. Given vectors ξ, η in $T_v(TX)$ or $T_v(SX)$ we define

$$\langle \xi, \eta \rangle_v = \langle d\pi(\xi), d\pi(\eta) \rangle_{\pi v} + \langle K^*(\xi), K^*(\eta) \rangle_{\pi v}.$$

Note that the vertical and horizontal subspaces $V(v)$ and $H(v)$ are orthogonal and the restrictions of the maps $d\pi$, K^* in $T_v(TX)$ to $H(v)$, $V(v)$ respectively are linear isometries onto $T_{\pi(v)}X$. The projection $\pi: TX \to X$ is a Riemannian submersion. The vertical fibers of $\pi: TX \to X$ and $\pi: SX \to X$ are complete, totally geodesic submanifolds.

(2.3) Geodesic flow and Jacobi vector fields. The geodesic flow transformations $\{g^t: t \in \mathbb{R}\}$ are 1-parameter groups of diffeomorphisms of TX or SX defined by $g^t v = \gamma_v'(t)$, where v is a vector in TX or SX and $\gamma_v'(t)$ is the velocity at time t of the geodesic γ_v with initial velocity v. The *geodesic vector field* or *geodesic spray* X^0 is the vector field in TX or SX whose flow transformations are the diffeomorphisms $\{g^t\}$.

A Jacobi vector field $Y(t)$ on a geodesic $\gamma(t)$ is uniquely determined by the values $Y(0)$, $Y'(0)$, where Y' denotes the covariant derivative of Y along γ. Hence the $2n$-dimensional vector space of Jacobi vector fields on γ, denoted by $J(\gamma)$, may be identified with $(T_pX) \times (T_pX)$, where $p = \gamma(0)$.

One may also define a direct isomorphism between $T_v(TX)$ and $J(\gamma_v)$ for any vector v in TX. Given $\xi \in T_v(TX)$ let Y_ξ be the unique Jacobi vector field on γ_v such that $Y_\xi(0) = d\pi(\xi)$ and $Y'_\xi(0) = K^*(\xi)$. It is not difficult to show that

$$Y_\xi(t) = d\pi(dg^t\xi) \quad \text{and} \quad Y'_\xi(t) = K^*(dg^t\xi) \quad \text{for all } t \in \mathbb{R}.$$

The map $\xi \to Y_\xi$ is a linear isomorphism of $T_v(TX)$ onto $J(\gamma_v)$ for all $v \in TX$. If $v \in SX$ is a unit vector, then the restriction of this map sends $T_v(SX)$ onto $J^*(\gamma_v) = \{Y \in J(\gamma_v): \langle Y'(t), \gamma'_v(t)\rangle \equiv 0\}$.

(2.4) Contact 1-form and symplectic 2-form in TX. Let ω denote the 1-form in TX that is dual to the geodesic vector field X^0 relative to the Sasaki metric; that is,

$$\theta(\xi) = \langle \xi, X^0(v)\rangle = \langle d\pi(\xi), v\rangle \quad \text{for } \xi \in T_v(TX) \text{ and } v \in TX.$$

The 1-form ω has the following properties:

(1) ω is invariant under the geodesic flow transformations $\{g^t: t \in \mathbb{R}\}$.

(2) $(d\omega)^n$ is a nonzero $2n$-form in TX and equals the Riemannian volume element in TX relative to the Sasaki metric.

(3) $\omega \wedge (d\omega)^{n-1}$ is a nonzero $(2n-1)$ form when restricted to SX (contact property) and equals the Riemannian volume element in SX relative to the induced Sasaki metric.

(4) If ξ, η are arbitrary elements of $T_v(TX)$, then

$$\begin{aligned}
d\omega(\xi, \eta) &= \langle K^*(dg^t\xi), d\pi(dg^t\eta)\rangle - \langle K^*(dg^t\eta), d\pi(dg^t\xi)\rangle \\
&= \langle Y'_\xi(t), Y_\eta(t)\rangle - \langle Y'_\eta(t), Y_\xi(t)\rangle \quad \text{for all } t \in \mathbb{R}.
\end{aligned}$$

where Y_ξ, Y_η are the Jacobi vector fields on γ_v as defined in (2.3).

The 2-form $\Omega = d\omega$ is a symplectic form in TX. The forms Ω^n and $\omega \wedge \Omega^{n-1}$ in TX and SX are the *Liouville measures* in those spaces. Clearly, ω, Ω and the Liouville measures are invariant under the geodesic flow transformations $\{g^t: t \in \mathbb{R}\}$. The Liouville measures in TX, SX can also be expressed as $\text{vol}_{\text{vert}} \wedge \pi^*(\text{vol}_X)$, where vol_{vert} denotes the Riemannian volume element of the fibers of π and vol_X denotes the Riemannian volume element of X.

(2.5) Stable and unstable foliations in $\widetilde{SM}$. We now specialize to the case that $X = \widetilde{M}$, a simply connected manifold of nonpositive curvature, or a quotient manifold of $\widetilde{M}$. We define the stable and unstable foliations in $\widetilde{M}$ only; those in a quotient manifold are obtained by projection under the covering map.

For each unit vector v in $S\widetilde{M}$ we define

$$W^u(v) = \{w \in S\widetilde{M} : -v \text{ and } -w \text{ are asymptotic}\},$$

$$W^s(v) = \{w \in S\widetilde{M} : v \text{ and } w \text{ are asymptotic}\}.$$

The sets $W^u(v)$, $W^s(v)$ are C^2 submanifolds of $S\widetilde{M}$ that are mapped diffeomorphically onto $\widetilde{M}$ by the projection $\pi : S\widetilde{M} \to \widetilde{M}$. They are called the *unstable* and *stable* manifolds through v. We let W^u, W^s denote the unstable, stable foliations in $S\widetilde{M}$ whose leaves through v are $W^u(v)$, $W^s(v)$.

The foliations W^u, W^s are continuous with C^2 leaves, but in general they are not C^1 or even Hölder continuous. They are invariant under the geodesic flow $\{g^t\}$. More precisely, $g^t W^u(v) = W^u(v)$ and $g^t W^s(v) = W^s(v)$ for all $v \in S\widetilde{M}$, all $t \in \mathbb{R}$.

Stable and unstable Jacobi vector fields. It is useful to describe the stable and unstable foliations in terms of Jacobi vector fields by using the isomorphism $\xi \in T_v(T\widetilde{M}) \to Y_\xi \in J(\gamma_v)$ discussed in (2.3). Let $\xi \in T_v(S\widetilde{M})$ be arbitrary. Then

(1) $\qquad \xi \in T_v(W^s(v)) \Leftrightarrow |Y_\xi(t)|$ is bounded above for $t \geq 0$

$\qquad\qquad\qquad \Leftrightarrow |Y_\xi(t)|$ is nonincreasing in t;

(2) $\qquad \xi \in T_v(W^u(v)) \Leftrightarrow |Y_\xi(t)|$ is bounded above for $t \leq 0$

$\qquad\qquad\qquad \Leftrightarrow |Y_\xi(t)|$ is nondecreasing in t;

(3) $\qquad \xi \in T_v(W^u(v)) \cap T_v(W^s(v)) \Leftrightarrow |Y_\xi$ is a parallel Jacobi vector

$\qquad\qquad\qquad$ field on γ_v

The Jacobi vector fields described in (1) and (2) are called respectively *stable* and *unstable* Jacobi vector fields

Strong stable and strong unstable foliations. Now let the sectional curvatures of $\widetilde{M}$ be bounded above and below by negative constants. Using the terminology of (1.5) we define

$$W^{su}(v) = \{w \in W^u(v) : -v \text{ and } -w \text{ determine the same horosphere}$$

$$H(-v) = H(-w) \text{ in } \widetilde{M}\}.$$

$$W^{ss}(v) = \{w \in W^s(v) : v \text{ and } w \text{ determine the same horosphere}$$

$$H(v) = H(w) \text{ in } \widetilde{M}\}.$$

(These definitions make sense for arbitrary nonpositively curved manifolds $\widetilde{M}$ but are not useful in this wider context.) The sets $W^{su}(v)$, $W^{ss}(v)$ are submanifolds of codimension 1 in $W^u(v)$, $W^s(v)$ and are called the *strong unstable* and *strong stable* manifolds through v. We let W^{su}, W^{ss} denote the strong unstable, strong stable foliations in $S\widetilde{M}$ whose leaves through v are $W^{su}(v)$, $W^{ss}(v)$. Geometrically, the manifolds $W^{su}(v)$, $W^{ss}(v)$ can be regarded respectively as the outward unit normals to the horosphere $H(-v)$, inward unit normals to the horosphere $H(v)$.

The foliations W^{su}, W^{ss} are also called the *Anosov* foliations. If $\widetilde{M}$ has a compact quotient, then they are Hölder continuous with C^{∞} leaves that depend continuously in the C^{∞} topology on the points [**Shu**]. They are also invariant under the geodesic flow. More precisely $g^t W^{su}(v) = W^{su}(g^t v)$ and $g^t W^{ss}(v) = W^{ss}(g^t v)$ for all $v \in S\widetilde{M}$, all $t \in \mathbb{R}$.

The foliations W^{su}, W^{ss} can also be described in the following manner, where d^* denotes the Sasaki metric:

$$W^{su}(v) = \{w \in S\widetilde{M} : d^*(g^t v, g^t w) \to 0 \text{ as } t \to -\infty\},$$
$$W^{ss}(v) = \{w \in S\widetilde{M} : d^*(g^t v, g^t w) \to 0 \text{ as } t \to +\infty\}.$$

The rate at which $d^*(g^t v, g^t w)$ converges to zero in each of the two statements above is exponential, where the exponent depends on the upper bound for sectional curvature in $\widetilde{M}$.

Strong stable and strong unstable Jacobi vector fields. In terms of the isomorphism $\xi \in T_v(T\widetilde{M}) \to Y_{\xi} \in J(\gamma_v)$ discussed in (2.3) we can describe W^{su}, W^{ss} as follows:

(1) $\xi \in T_v(W^{ss}(v)) \Leftrightarrow |Y_{\xi}(t)|$ is nonincreasing in t (bounded above

  for $t \geq 0$) and Y_{ξ} is orthogonal to γ_v';

(2) $\xi \in T_v(W^{su}(v)) \Leftrightarrow |Y_{\xi}(t)|$ is nondecreasing in t (bounded above

  for $t \leq 0$) and Y_{ξ} is orthogonal to γ_v'.

The *strong stable* and *strong unstable* Jacobi vector fields defined in (1) and (2) can be defined more generally for an arbitrary manifold $\widetilde{M}$ of nonpositive sectional curvature [**BBE**].

3. Anosov foliations [Ano]

(3.1) Definitions and a key example. Let $\{\Phi^t\}$ be a flow on a compact Riemannian manifold X, and let E^0 denote the 1-dimensional foliation of TX spanned by the vector field determined by $\{\Phi^t\}$. The flow $\{\Phi^t\}$ is said to be *Anosov* if there exist $\{\Phi^t\}$ invariant bundles E^s, E^u in TX such that

(1) $TX = E^s \oplus E^u \oplus E^0$ (direct sum)

(2) There exist positive constants a, b such that

$$|d\Phi^t(\xi)| \geq be^{at}|\xi| \quad \text{for all } \xi \in E^u(x),$$
$$\text{all } x \in X, \text{ all } t \geq 0;$$
$$|d\Phi^t(\xi)| \leq be^{-at}|\xi| \quad \text{for all } \xi \in E^s(x),$$
$$x \in X, \text{ all } t \geq 0.$$

If $\{\Phi^t\}$ preserves a smooth volume form on X, then the flow $\{\Phi^t\}$ is ergodic in X. The definition of Anosov flow $\{\Phi^t\}$ on X depends only on X and

$\{\Phi^t\}$, not on the choice of Riemannian metric on X, although the choices of a, b in (2) depend on the metric.

One of the most important examples of an Anosov flow, and the example of interest to us, occurs when $X = SM$, the unit tangent bundle of a compact Riemannian manifold with strictly negative sectional curvature, and $\{\Phi^t\} = \{g^t\}$, the geodesic flow in SM. In this case the bundles E^s, E^u are the strong stable and strong unstable foliations W^{ss}, W^{su}. The geodesic flow preserves the Liouville measure in SM.

If $\widetilde{M}$ is the universal cover of M (compact, $K < 0$) we also refer to the geodesic flow in $S\widetilde{M}$ as being Anosov even though $S\widetilde{M}$ is noncompact. The growth estimates in (2) above on W^{ss}, W^{su} are the same in $S\widetilde{M}$ as in SM since the Sasaki metrics of these two spaces are locally isometric.

(3.2) Growth estimates from curvature bounds [HI]. The exponential growth rates in (2) above can be estimated from upper and lower bounds on the sectional curvature. Let a, b be positive constants such that $-b^2 \leq K \leq -a^2 < 0$ on M and $\widetilde{M}$. If $\xi \in W^{su}(v)$ or $W^{ss}(v)$, then $|K^*(\xi)| \leq b|d\pi(\xi)|$ (cf. Proposition 2.11 of [E11]), where K^* denotes the connection map of (2.1) and $\pi: SM \to M$ is the projection. Let $|\ |$ denote the Sasaki norm in SM as well as the Riemannian norm in M. It follows that

$$|Y_\xi(t)| = |d\pi(dg^t\xi)| \leq |dg^t(\xi)|$$
$$\leq (1+b^2)^{1/2}|d\pi(dg^t\xi)| = (1+b^2)^{1/2}|Y_\xi(t)|$$

for all $t \in \mathbb{R}$. Hence $|Y_\xi(t)|$ and $|dg^t(\xi)|$ have essentially the same growth rate for $\xi \in W^{su}(v)$ or $W^{ss}(v)$.

Next, by using comparison theorems we obtain the following estimates [HI]:

If $\xi \in W^{su}(v)$, then $e^{at}|Y_\xi(0)| \leq |Y_\xi(t)| \leq e^{bt}|Y_\xi(0)|$ for all $t \in \mathbb{R}$,

If $\xi \in W^{ss}(v)$, then $e^{-bt}|Y_\xi(0)| \leq Y_\xi(t)| \leq e^{-at}|Y_\xi(0)|$ for all $t \in \mathbb{R}$.

(3.3) Regularity of the Anosov foliations. Let M be compact with $K < 0$. In general the Anosov splitting $T(SM) = W^{su} \oplus W^{ss} \oplus E^0$ is Hölder continuous but not differentiable except in special cases. If M has dimension 2 or if the sectional curvature is strictly $\frac{1}{4}$-pinched, then the foliations W^{su}, W^{ss} are C^1 [Gre, Ho, HP]. These foliations are "almost" C^1 if the metric on M is weakly $\frac{1}{4}$-pinched [Ham1]. The Anosov foliations are C^∞ if M is a rank 1 locally symmetric space. Conversely, if the Anosov foliations are C^∞, then the geodesic flow in SM is smoothly conjugate to the geodesic flow in SM^*, where M^* is a rank 1 locally symmetric space [BFL]. This result and the problem of geodesic conjugacy will be discussed later in (8.3) in more detail. For further results on regularity of Anosov foliations see [Has].

(3.4) Kanai connection. Let M be a compact Riemannian manifold with strictly negative sectional curvature. If ω denotes the canonical contact 1-form, then $\omega \equiv 0$ on $W^{su} \oplus W^{ss}$. In fact, ω may be characterized as the unique 1-form on SM such that

$$\omega(X^0) \equiv 1, \quad \omega \equiv 0 \quad \text{on } W^{su} \oplus W^{ss} = W.$$

Here X^0 denotes the geodesic vector field on $S\widetilde{M}$. The symplectic 2-form $\Omega = d\omega$ is nondegenerate on W but vanishes when restricted to W^{su} or W^{ss}.

If the Anosov splitting of $T(SM)$ is C^1 then one may define an important $\{g^t\}$ invariant connection ∇ on SM first introduced by Kanai [**Kan1**]. If the Anosov splitting is C^k, then ∇ is of class C^{k-1}. The Kanai connection plays a key role in the study of manifolds M with C^∞ Anosov foliations [**Kan1, FK, Fer, BFL**].

We first introduce a pseudo-Riemannian structure g on $S\widetilde{M}$. Let W denote the direct sum of the bundles W^{su} and W^{ss}, and let c be the transformation on W that equals the identity on W^{su} and minus the identity on W^{ss}. Let $\Omega = d\omega$ be the symplectic 2-form on $S\widetilde{M}$. Define a symmetric, nondegenerate, bilinear form g on $S\widetilde{M}$ by

$$g(x, y) = \Omega(x, cy) \quad \text{if } x, y \text{ are tangent to } W,$$
$$g(X^0, X^0) = 1,$$
$$g(X^0, x) = 0 \quad \text{if } x \text{ is tangent to } W,$$

Let ∇' denote the Levi-Civita connection on $S\widetilde{M}$ determined by g and let $p_W: T(S\widetilde{M}) \to W$ denote the projection onto W.

We now define the Kanai connection ∇ on $S\widetilde{M}$ as follows:

$$\nabla_X Y = p_w(\nabla'_X Y) \quad \text{if } X, Y \text{ are sections of } W,$$
$$\nabla X^0 \equiv 0,$$
$$\nabla_{X^0}\xi = [X^0, \xi] \quad \text{for any section } \xi \text{ of } T(S\widetilde{M}).$$

(The definition of $\nabla_{X^0}\xi$ is slightly different in [**BFL**].) The Kanai connection ∇ satisfies the following properties and is uniquely characterized by the first 3 of these:

(1) $\nabla\omega = 0$, $\nabla\Omega = 0$.

(2) W^{ss}, W^{su} are invariant under ∇; that is, if ξ^s, ξ^u are arbitrary sections of W^{ss}, W^{su} and if η is any section of $T(S\widetilde{M})$, then $\nabla_\eta\xi^s$, $\nabla_\eta\xi^u$ are sections of W^{ss}, W^{su}.

(3) The torsion tensor $T(\xi_1, \xi_2) = \nabla_{\xi_1}\xi_2 - \nabla_{\xi_2}\xi_1 - [\xi_1, \xi_2]$ satisfies $T(\xi_1, \xi_2) = -p_0([\xi_1, \xi_2])$ for any sections ξ_1, ξ_2 of $T(S\widetilde{M})$, where $p_0: T(S\widetilde{M}) \to E^0$ denotes projection onto E^0, the 1-dimensional foliation of $T(S\widetilde{M})$ spanned by X^0.

(4) ∇ is invariant relative to the geodesic flow $\{g^t\}$.

(5) The leaves of W^{su}, W^{ss} are flat relative to ∇; that is, the tangent bundles of the leaves admit local trivializations by parallel vector fields.

4. Metrics on $\widetilde{M}(\infty)$

For any complete, simply connected manifold $\widetilde{M}$ with $K \leq 0$ the boundary sphere $\widetilde{M}(\infty)$ admits a natural sphere topology, and the maps $f_p : \widetilde{M}(\infty) \to S_p\widetilde{M}$ given by $f_p(x) = \gamma'_{px}(0)$ are homeomorphisms for every point p of $\widetilde{M}$. In general the space $\widetilde{M}(\infty)$ admits no natural differentiable structure; for arbitrary points p, q of $\widetilde{M}$ the homeomorphisms $\pi_{pq} = f_q \circ f_p^{-1} : S_p\widetilde{M} \to S_q\widetilde{M}$ need not be differentiable. If the sectional curvature of $\widetilde{M}$ is strictly $\frac{1}{4}$-pinched, then $\widetilde{M}(\infty)$ admits a C^1 structure by the discussion in (3.2), but it is possible that if $\widetilde{M}$ covers a compact manifold M and admits a $C^{1,\alpha}$ structure on $\widetilde{M}(\infty)$ for all α with $0 < \alpha < 1$, then $\widetilde{M}$ must be symmetric of rank 1.

Despite the lack of a natural differentiable structure $\widetilde{M}(\infty)$ admits a variety of other structures that are useful in studying the geometry of $\widetilde{M}$ and its quotient manifolds M of finite volume. In this section we discuss two pseudometric space structures on $\widetilde{M}(\infty)$ that correspond to the cases where the maximum sectional curvature of $\widetilde{M}$ is zero or a negative number. In each case the finiteness of the pseudometric on distinct points x, y of $\widetilde{M}(\infty)$ is a kind of asymptotic measurement of the amount of the maximum sectional curvature (normalized to be zero or -1) on triangular sectors in $\widetilde{M}$ with vertices at x, y and a point of $\widetilde{M}$.

(4.1) The pseudometric Td. In [BGS] Gromov introduced a pseudometric Td on $\widetilde{M}(\infty)$ that he called the *Tits metric*, which is defined for every simply connected space $\widetilde{M}$ with $K \leq 0$. In the special case that $\widetilde{M}$ is a symmetric space with no Euclidean de Rham factor and rank at least two the pseudometric Td is actually a finite metric and carries geometric information that is equivalent to that provided by the Tits building structure in the semisimple Lie group $G = I_0(\widetilde{M})$. We shall give two equivalent definitions of the pseudometric Td. For further discussion see [BGS].

Inner metrics. Before giving the first definition of Td we recall some metric space constructions due originally to A. D. Alexandrov [Al]. See also [GLP, Ri] and the bibliography of [Ri] for further references. Let (X, d) be a metric space, and let $\gamma : [a, b] \to X$ be a continuous curve. For each partition $a = t_0 < t_1 < \cdots < t_n = b$ of $[a, b]$ we consider the polygonal approximation length $\sum_{i=1}^{n} d(\gamma t_i, \gamma t_{i-1})$, and we define the length $L(\gamma)$ of γ to be the supremum of the polygonal approximation lengths over all partitions of $[a, b]$. It may be the case that $L(\gamma) = +\infty$. For points p, q in X we then define $d_i(p, q)$ to be the infimum of the lengths of all continuous curves in X from p to q. The pseudometric d_i is called the *inner metric* determined by d, and it may be the case that $d_i(p, q) = +\infty$ for some or all pairs of distinct points p, q in X.

In general $d_i \geq d$ and $(d_i)_i = d_i$. If $d_i = d$ then X is called a *length space*. Riemannian manifolds are length spaces, and in general many geometric constructions that one associates with Riemannian manifolds may be carried out for length spaces. For example, if (X, d) is a complete, locally compact length space, then between any two points p, q of X there exists a minimizing unit speed *geodesic* γ; that is, γ is defined on $[0, c]$, where $c = d(p, q)$, $\gamma(0) = p$, $\gamma(c) = q$ and $d(\gamma s, \gamma t) = |t - s|$ for all $s, t \in [0, c]$. In particular the length of γ is $d(p, q)$ [**CV, GLP**]. The existence of minimizing geodesics does not require local compactness of the metric space (X, d) as the example of $(\widetilde{M}(\infty), Td)$ shows in the case that $\widetilde{M}$ is symmetric of rank at least two [**BGS**].

Inner metric definition of Td [**BGS**]. We now give the inner metric definition of the Tits metric Td in $\widetilde{M}(\infty)$, where $\widetilde{M}$ is any simply connected manifold with $K \leq 0$. Given arbitrary points x, y in $\widetilde{M}(\infty)$ we define

$$\angle(x, y) = \sup\{\angle_p(x, y): p \in \widetilde{M}\},$$

where $\angle_p(x, y) = \angle(\gamma'_{px}(0), \gamma'_{py}(0))$ is the angle at p subtended by x and y. The *angle metric* $\angle$ is a complete metric on $\widetilde{M}(\infty)$.

We define the Tits metric Td on $\widetilde{M}(\infty)$ to be the inner metric on $\widetilde{M}(\infty)$ determined by $\angle$. The pseudometric space $(\widetilde{M}(\infty), Td)$ is not locally compact, but nevertheless there is a minimizing Td-geodesic in $\widetilde{M}(\infty)$ joining any two points x, y of $\widetilde{M}(\infty)$ for which $Td(x, y)$ is finite. This geodesic is unique if $Td(x, y) < \pi$.

Td as a limit of Riemannian metrics. We now present the second definition of the Tits metric Td as an asymptotic limit of Riemannian metrics. Given a point p in $\widetilde{M}$ we set $S_r(p) = \{q \in \widetilde{M}: d(p, q) = r\}$, the sphere of radius r centered at p. For each number $r > 0$ we define a Riemannian metric d_r on $\widetilde{M}(\infty)$ as follows: given points x, y in $\widetilde{M}(\infty)$ we let

$$d_r(x, y) = \frac{1}{r} d_{S_r}(\gamma_{px}(r), \gamma_{py}(r))$$

where d_{S_r} denotes the Riemannian distance along $S_r(p)$ and γ_{px}, γ_{py} denote the unit speed geodesics starting at p that belong to x, y. The fact that sectional curvature in $\widetilde{M}$ is nonpositive means that the metrics d_r are nondecreasing in r (constant if $\widetilde{M}$ is flat Euclidean space). One can prove (cf. [**BGS**, p. 43]) that independent of the point p

$$Td(x, y) = \lim_{r \to \infty} d_r(x, y).$$

This definition reveals an important feature of the Tits metric; the finiteness of $Td(x, y)$ says that $\widetilde{M}$ is asymptotically flat with respect to x and y in some sense. More precisely, if $\sigma: [a, b] \to \widetilde{M}(\infty)$ is a minimizing Tits geodesic in $\widetilde{M}(\infty)$ joining x to y and if p is a point of $\widetilde{M}$, then p and σ determine an infinite triangular sector $\Delta(p, x, y) = \bigcup_{t \in [a, b]} \gamma_{p\sigma(t)}[0, \infty)$ that is asymptotically flat and totally geodesic; that is, as a point q diverges in $\Delta(p, x, y)$ to infinity, the Gaussian curvature at q and the norm of the second fundamental form of Δ at q converge to zero.

This definition of the Tits metric as a limit of Riemannian metrics d_r on $\widetilde{M}(\infty)$ as $r \to \infty$ also implies that the geodesic behavior in $\widetilde{M}(\infty)$ of the Tits metric is a limit of the geodesic behaviors of the Riemannian metrics d_r as $r \to \infty$. Let $x, y \in \widetilde{M}(\infty)$ be points such that $Td(x, y)$ is finite. Fix an interval $I = [a, b]$ and let $\{\gamma_r : I \to \widetilde{M}(\infty)\}$ be minimizing geodesics in the metrics d_r that join $x = \gamma_r(a)$ to $y = \gamma_r(b)$ for every $r > 0$. The family $\{\gamma_r\}$ is equicontinuous with respect to any one of the metrics, say d_1, and any sequence $\{\gamma_{r_n}\}$ where $r_n \to +\infty$ has a subsequence converging to a minimizing Tits geodesic $\gamma : I \to \widetilde{M}(\infty)$ that joins $x = \gamma(a)$ to $y = \gamma(b)$. Conversely, if $\gamma : I \to \widetilde{M}(\infty)$ is any unit speed Tits geodesic (possibly not minimizing) joining distinct points x, y in $\widetilde{M}(\infty)$, let $a = a_0 < a_1 < \cdots < a_N = b$ be a partition of I such that $\gamma|[a_{i-1}, a_i]$ is minimizing and of length $< \pi$ for $1 \le i \le N$. Let $\gamma_r : I \to \widetilde{M}(\infty)$ be any (possibly) broken geodesic in the metric d_r such that $\gamma_r(a_i) = \gamma(a_i)$ and γ_r is minimizing on $[a_{i-1}, a_i]$ for all i. The uniqueness of minimizing Tits geodesics of length $< \pi$ implies that γ_r converges to γ as $r \to \infty$.

It is easy to see that if φ is any isometry of $\widetilde{M}$, then the extension of φ to $\widetilde{M}(\infty)$ is an isometry of the Tits metric Td.

(4.2) The pseudometrics Hd_v, $v \in S\widetilde{M}$. For manifolds $\widetilde{M}$ whose sectional curvatures are bounded above by a negative constant U. Hamenstädt in [**Ham3**] has defined pseudometrics $\{Hd_v : v \in S\widetilde{M}\}$ that in many respects are analogous to the Tits pseudometric Td. We again present two equivalent definitions, one as an inner metric and a second as a limit of Riemannian metrics.

Inner metric definition of Hd_v. Let the sectional curvatures of $\widetilde{M}$ satisfy $K \le -c^2 < 0$ for some positive constant c. For each positive number r and each unit vector $v \in S\widetilde{M}$ one defines a distance function η_v^r on $\widetilde{M}(\infty)$ as follows. Let $z = \gamma_v(-\infty)$ and let $\pi_v : W^{su}(v) \to \widetilde{M}(\infty) - \{z\}$ be the homeomorphism given by $\pi_v(w) = \gamma_w(\infty)$ for $w \in W^{su}(v)$. Let d^{su} denote Riemannian distance on the horospheres $\pi \circ W^{su}(w) = H(-w)$ in $\widetilde{M}$, where $w \in S\widetilde{M}$ is arbitrary and $\pi : S\widetilde{M} \to \widetilde{M}$ is the projection. For distinct points x, y in $\widetilde{M}(\infty) - \{z\}$ the function $d_{xy}(t) = d^{su}(g^t(\pi_v^{-1}x), g^t(\pi_v^{-1}y))$ is strictly monotone increasing on $\mathbb{R}$, and the range of $d_{xy}(t)$ is $(0, \infty)$. We define

$$\eta_v^r(x, y) = e^{-\tau}$$

where τ is the unique number with $d_{xy}(t) = r$. The triangle inequality for

η_v^r might be violated, but only up to a controlled factor determined by the curvature maximum [**Ham5**]. If we normalize the upper bound for sectional curvature to be -1, then η_v^r satisfies the triangle inequality for every $r > 0$ and becomes a metric on $\widetilde{M}(\infty) - \{z\}$. Even without this normalization the functions η_v^r without the triangle inequality are sufficient for many purposes.

If we assume that $\widetilde{M}$ has upper sectional curvature bound -1 and we define $\eta_v = \eta_v^1$ as above, then the inner metric Hd_v of η_v plays a role similar to that of the Tits metric Td. The metrics η_v and Hd_v depend on the choice of the unit vector v, but there is a good compatibility between any two of them. See [**Ham3**] for further details. For example, if $Hd_v(x, y)$ is finite for two points x, y in $\widetilde{M}(\infty) - \{\gamma_v(-\infty)\}$ and some vector $v \in S\widetilde{M}$, then $Hd_w(x, y)$ is finite for all $w \in S\widetilde{M}$ such that $\gamma_w(-\infty)$ is distinct from both x and y.

Hd_v as a limit of Riemannian metrics. We now present a definition of the pseudometric Hd_v as a limit of Riemannian metrics. Again let -1 be the upper bound for the sectional curvature of $\widetilde{M}$ and let $v \in S\widetilde{M}$ be given. Let x, y be points in $\widetilde{M}(\infty) - \{z\}$, where $z = \gamma_v(-\infty)$, and let $\pi_v : W^{su}(v) \to \widetilde{M}(\infty) - \{z\}$ and d^{su} have the meanings given above. For each number $r > 0$ we define a Riemannian metric d_v^r on $\widetilde{M}(\infty) - \{z\}$ by

$$d_v^r(x, y) = e^{-r} d^{su}(g^r(\pi_v^{-1}x), g^r(\pi_v^{-1}y))$$
$$= e^{-r} d_{xy}(r).$$

The metric d_v^r is essentially a rescaling of the Riemannian metric on the horosphere $H(-g^r v)$. The metrics $\{d_v^r\}$ are nondecreasing in r since the sectional curvature of $\widetilde{M}$ is at most -1, and one can show that

$$Hd_v(x, y) = \lim_{r \to \infty} d_v^r(x, y).$$

If $Hd_v(x, y)$ is finite and $r_n \to +\infty$ is any sequence, then any sequence $\{\gamma_n : I \to \widetilde{M}(\infty) - \{\gamma_v(-\infty)\}\}$ of minimizing geodesics from x to y in the metrics $d_v^{r_n}$ has a subsequence converging to a minimizing Hd_v geodesic $\gamma : I \to \widetilde{M}(\infty) - \{\gamma_v(-\infty)\}$ that joins x to y.

Unlike the case of the Tits metric Td the isometries of $\widetilde{M}$ acting on $\widetilde{M}(\infty)$ are not isometries of the metrics $\{Hd_v : v \in S\widetilde{M}\}$ or of the metrics $\{\eta_v : v \in S\widetilde{M}\}$. Indeed, there is no natural way to single out any one of the metrics η_v or Hd_v. However, the isometries of $\widetilde{M}$ acting on $\widetilde{M}(\infty)$ are generalized conformal maps in the sense of the next section. See (5.5).

5. Measures and quasiconformal structures

In this section we assume that $\widetilde{M}$ has sectional curvature bounded above and below by negative constants, and for simplicity we assume the upper bound to be -1. Moreover, we assume that $\widetilde{M}$ admits a compact quotient manifold M. There are three natural measures on $S\widetilde{M}$ and associated measure classes on $\widetilde{M}(\infty)$. The measures on $S\widetilde{M}$ are the Lebesgue-Liouville, harmonic and Bowen-Margulis measures, which coincide if $\widetilde{M}$ is a rank 1 symmetric space. Moreover, these measures in $S\widetilde{M}$ are invariant under the isometries of $\widetilde{M}$ and hence descend to the unit tangent bundle SM^* of any quotient manifold M^* of $\widetilde{M}$. If $M^* = M$ is compact, then all three measures in SM are ergodic with respect to the geodesic flow $\{g^t\}$. It is an important question to determine if M must be rank 1 locally symmetric whenever any two of these measures coincide. We discuss progress on this question in greater detail in (7.4). In this section we also describe a natural quasiconformal structure on $\widetilde{M}(\infty)$ introduced in [**Ham3**].

(5.1) Measures quasi-invariant under the stable foliation. Let M be compact with upper sectional curvature bound -1, and let ν be any $\{g^t\}$ invariant probability measure on SM. If $\tilde{\nu}$ denotes the lift of ν to $S\widetilde{M}$, then $\tilde{\nu}$ defines a family $\{\tilde{\nu}^{su}\}$ of conditional measures on the leaves of W^{su} in $S\widetilde{M}$. These conditional measures are uniquely determined by $\tilde{\nu}$ up to a family of leaves of $\tilde{\nu}$-measure zero. The measure ν is called *quasi-invariant under the stable foliation* if the following conditions are satisfied:

(a) For every vector $v \in S\widetilde{M}$ a subset E of $W^{su}(v)$ is $\tilde{\nu}^{su}$-neglectable if and only if $\tilde{\nu}(\bigcup\{W^s(w): w \in E\}) = 0$.

(b) There is a subset $E \subseteq S\widetilde{M}$ of full measure with the property that for all v, $w \in E$ the map $\pi_w^{-1} \circ \pi_v: W^{su}(v) - \pi_v^{-1}(\gamma_w(-\infty)) \to W^{su}(w)$ is absolutely continuous with respect to the conditional measures $\tilde{\nu}^{su}$.

Property (b) implies that $\tilde{\nu}$ defines a unique measure class ν_∞ on $\widetilde{M}(\infty)$ that is invariant under the action of $I(\widetilde{M})$. All *Gibbs equilibrium states* on SM are quasi-invariant under the stable foliation (see for example [**Led1**]). Among these the most interesting geometrically are the Lebesgue-Liouville, harmonic and Bowen-Margulis measures, which we now define.

(5.2) Lebesgue-Liouville measure λ. Recall from (2.4) that the Lebesgue-Liouville measure λ in $S\widetilde{M}$ can be expressed as a volume form $\lambda = \mathrm{vol}_{\mathrm{vert}} \wedge \pi^*(\mathrm{vol}_{\widetilde{M}})$, where $\mathrm{vol}_{\mathrm{vert}}$ is the $(n-1)$ form in $S\widetilde{M}$ that equals the Riemannian volume form on the spherical fibers of $\pi: S\widetilde{M} \to \widetilde{M}$ and $\mathrm{vol}_{\widetilde{M}}$ is the Riemannian volume element of $\widetilde{M}$.

In addition to the measure class of $\widetilde{M}(\infty)$ determined by λ and the leaves of the strong unstable foliation W^{su}, we can also define a family of measures $\{\lambda_p: p \in \widetilde{M}\}$ on $\widetilde{M}(\infty)$ that determine a single measure class on $\widetilde{M}(\infty)$. For each point $p \in \widetilde{M}$ we define λ_p to be the measure on $\widetilde{M}(\infty)$ that corresponds to the canonical Riemannian measure on $S_p\widetilde{M}$ under the natural homeomorphism $f_p: \widetilde{M}(\infty) \to S_p\widetilde{M}$.

(5.3) Harmonic measure ω. When $\widetilde{M}$ has sectional curvatures bounded above and below by negative constants, then the Dirichlet problem on $\widetilde{M}(\infty)$ has a solution; that is, if $f: \widetilde{M}(\infty) \to \mathbb{R}$ is any continuous function, then there exists a unique harmonic function $\tilde{f}: \widetilde{M} \to \mathbb{R}$ that extends continuously to the boundary $\widetilde{M}(\infty)$ and whose boundary values on $\widetilde{M}(\infty)$ are given by f. See [**Anc**, **And**, **ARS**, **Ki1**, **Sul**]. For manifolds $\widetilde{M}$ of rank 1 (as defined in 8.1) that cover a compact manifold M the Dirichlet problem on $\widetilde{M}(\infty)$ has been solved by W. Ballmann [**Ba2**].

For each point $p \in \widetilde{M}$ (with sectional curvatures bounded by negative constants) the linear functional $L_p(f) = \tilde{f}(p)$, where $f: \widetilde{M}(\infty) \to \mathbb{R}$ is continuous, is realized by a measure ω_p on $\widetilde{M}(\infty)$; that is,

$$\tilde{f}(p) = \int_{\widetilde{M}(\infty)} f(x)\, d\omega_p$$

for all continuous functions $f: \widetilde{M}(\infty) \to \mathbb{R}$. The measures $\{\omega_p : p \in \widetilde{M}\}$, which can also be defined in terms of Brownian motion, are called the *harmonic measures* and form a single measure class on $\widetilde{M}(\infty)$. Identifying $\widetilde{M}(\infty)$ with $S_p\widetilde{M}$ by means of the homomorphism $f_p: \widetilde{M}(\infty) \to S_p\widetilde{M}$ we may also regard each measure ω_p as a *spherical harmonic measure* on $S_p\widetilde{M}$. The measures ω_p are equivariant with respect to the action of isometries of $\widetilde{M}$, and hence the measures ω_p descend to a family of spherical harmonic measures on S_pM^* for any quotient manifold M^* of $\widetilde{M}$. If $M^* = M$ is compact, then F. Ledrappier in [**Led1**] has shown that there exists a unique $\{g^t\}$-invariant Borel probability measure ω on SM that admits the spherical harmonic measures $\{\omega_p : p \in M\}$ as a family of transverse measures. The measure ω is called the *harmonic measure* on SM or $S\widetilde{M}$. For other results on harmonic measure see [**Ki3**, **KL**].

(5.4) Bowen-Margulis measure μ. If M is compact with strictly negative sectional curvature, then there is a unique $\{g^t\}$-invariant probability measure μ on SM for which the metric entropy h_μ is a maximum; that is, $h_\mu = h$, where h is the topological entropy of $\{g^t\}$. (For a discussion of entropy see §7). This measure is called the *Bowen-Margulis measure*. See for example [**Wa**] for further discussion and literature citations.

We let $\tilde{\mu}$ denote the lift of the Bowen-Margulis measure μ to $S\widetilde{M}$, where $\widetilde{M}$ is the universal Riemannian cover of $\tilde{M}$. The measure $\tilde{\mu}$ induces conditional measures $\tilde{\mu}^{\mathrm{su}}$ on the leaves of W^{su} that satisfy

$$\tilde{\mu}^{\mathrm{su}}_{g^t v}(g^t A) = e^{ht} \tilde{\mu}^{\mathrm{su}}_v(A)$$

for any measurable subset $A \subseteq W^{\mathrm{su}}(v)$, $v \in S\widetilde{M}$, where again h denotes the topological entropy of $\{g^t\}$. Identifying the unstable horosphere $W^{\mathrm{su}}(v)$ with $\widetilde{M}(\infty) - \{\gamma_v(-\infty)\}$ under the homeomorphism $\pi_v: W^{\mathrm{su}}(v) \to \widetilde{M}(\infty) - \{\gamma_v(-\infty)\}$ we obtain a collection $\{\tilde{\mu}^{\mathrm{su}}_v : v \in S\widetilde{M}\}$ of measures on $\widetilde{M}(\infty)$ that define a single measure class.

For each $v \in S\widetilde{M}$ the conditional measure $\tilde{\mu}_v^{su}$ can also be described up to a constant factor as the h-dimensional spherical measure σ_v on $W^{su}(v)$ determined by the distance function η_v^r defined in (4.2), where $r > 0$ is arbitrary [**Ham5**]. We assume that -1 is the upper bound for the sectional curvatures of $\widetilde{M}$ so that the distance functions $\{\eta_v^r\}$ satisfy the triangle inequality. For convenience we recall the definition of the h-dimensional spherical measure σ_v. Fix a positive number r and a vector $v \in S\widetilde{M}$. For any subset A of $W^{su}(v)$ we define

$$\sigma_v(A) = \sup_{\varepsilon \to 0} \sigma_\varepsilon(A)$$

where

$$\sigma_\varepsilon(A) = \inf\left\{\sum_{j=1}^\infty \varepsilon_j^h : \varepsilon_j \le \varepsilon \text{ and } A \subseteq \bigcup_{j=1}^\infty B(x_j, \varepsilon_j)\right\}.$$

Here $\{x_j\}_{j=1}^\infty$ lies in A and $B(x, \varepsilon)$ denotes the ball of radius ε and center x with respect to the metric η_v^r. See for example [**Fa**] or [**Fed**] for a further discussion of such measures arising from a metric.

(5.5) Quasiconformal structures on $\widetilde{M}(\infty)$. Let $f: X \to Y$ be a homeomorphism of topological spaces, and let d, δ be distance functions (not necessarily satisfying the triangle inequality) on X, Y that induce the topologies of those spaces. For a point x in X and a positive number r we define

$$q_f(d, \delta, x, r) = \inf\{s_2/s_1 : 0 < s_1 \le s_2 < \infty\}$$

and

(*) $$B_\delta(fx, s_1) \subseteq f(B_d(x, r)) \subseteq B_\delta(fx, s_2)$$

We define $q_f(d, \delta, x, r) = +\infty$ if there do not exist finite positive numbers s_1, s_2 satisfying the property (*) above. Here $B_d(x, r)$, $B_\delta(y, s)$ denote the closed balls with centers $x \in X$, $y \in Y$ and radii r, s with respect to the distance functions d, δ respectively.

We say that f is q-*quasiconformal* for a finite number $q \ge 1$ if

$$\limsup_{r \to 0} q_f(d, \delta, x, r) \le q \quad \text{for all } x \in X,$$

$$\limsup_{r \to 0} q_{f^{-1}}(\delta, d, y, r) \le q \quad \text{for all } y \in Y.$$

We say that f is *conformal* if f is q-quasiconformal with $q = 1$. Two distance functions d, δ on the same space X are said to be *quasiconformally equivalent* if the identity function $i: (X, d) \to (X, \delta)$ is q-quasiconformal for some finite $q \ge 1$, and they are said to be *conformally equivalent* if $i: (X, d) \to (X, \delta)$ is conformal. A *quasiconformal structure* on X (respectively a *generalized conformal structure* on X) is an equivalence class of quasiconformally equivalent (respectively conformally equivalent) metrics on X.

We apply these ideas to the space $\widetilde{M}(\infty)$, where $\widetilde{M}$ is simply connected with sectional curvatures bounded below and at most equal to -1. For every positive number $r > 0$ and every $v \in S\widetilde{M}$ the distance functions η_v^r are actually metrics on $\widetilde{M}(\infty) - \{\gamma_v(-\infty)\}$ by the discussion in (4.2). Now fix $r > 0$ and let v, w be any vectors in $S\widetilde{M}$. The metrics η_v^r, η_w^r are conformally equivalent on $\widetilde{M}(\infty) - \{\gamma_v(-\infty), \gamma_w(-\infty)\}$ by [**Ham3**]. Hence one obtains a generalized conformal structure on $\widetilde{M}(\infty)$ that is invariant under the action of $I(\widetilde{M})$. However, this generalized conformal structure does depend on r. Different choices of r yield quasiconformally equivalent metrics, and hence $\widetilde{M}(\infty)$ carries a well-defined $I(\widetilde{M})$-invariant quasiconformal structure. See [**Ham3**] and [**Pa1, Pa2, Pa5**].

In general the quasiconformal structure defined above is not standard; that is, the projections $\pi_v \colon W^{\mathrm{su}}(v) \to \widetilde{M}(\infty) - \{\gamma_v(-\infty)\}$ are not quasiconformal, where $W^{\mathrm{su}}(v)$ is equipped with the quasiconformal structure induced by a Riemannian metric. However, this is the case if the sectional curvature of $\widetilde{M}$ is constant. It may be conjectured that the converse of this statement is true. For a partial answer see the discussion in (8.2) below.

6. Fundamental group and geometry

Let M be compact with $K \leq 0$. We observed earlier that the homotopy groups $\pi_k(M)$ vanish for $k \geq 2$ so that the homotopy information of M is entirely contained in $\pi_1(M)$, the fundamental group of M. In this section we discuss the relationship between the topology and geometry of M and the fundamental group $\pi_1(M)$, particularly the algebraic structure of $\pi_1(M)$.

(6.1) **Topology and fundamental group.** In 1989, T. Farrell and L. Jones obtained two powerful results relating the topology of a compact nonpositively curved manifold M and $\pi_1(M)$.

THEOREM [**FJ2**]. *Let M_1, M_2 be compact manifolds with $K \leq 0$ whose fundamental groups are isomorphic. Then M_1 and M_2 are homeomorphic.*

A priori one can only say that M_1 and M_2 are homotopy equivalent if they have isomorphic fundamental groups.

In a companion paper [**FJ1**] Farrell and Jones showed that one can find homeomorphic compact nonpositively curved manifolds that are not diffeomorphic. More precisely, for each integer $n \geq 5$ they construct two sequences $\{M_k\}$, $\{M_k^*\}$ of compact n-dimensional manifolds such that

(a) M_k is homeomorphic but not diffeomorphic to M_k^* for each positive integer k.

(b) M_k has sectional curvature $K \equiv -1$ for each k.

(c) M_k^* has sectional curvature satisfying $-1 - \varepsilon_k \leq K \leq -1$, where $\{\varepsilon_k\}$ is a sequence of positive numbers that decrease to zero as $k \to \infty$.

The diameters of $\{M_k\}$ and $\{M_k^*\}$ are unbounded as they must be by the Cheeger-Gromov compactness theorem. The manifolds M_k^* do not admit any metric with $K \equiv -1$ since if they did they would be diffeomorphic to the manifolds M_k by the Mostow Rigidity Theorem. Earlier Gromov and Thurston [GT] constructed examples of compact manifolds with curvature satisfying $-1 - \varepsilon \leq K \leq -1$ that do not admit metrics with $K \equiv -1$, where $\varepsilon > 0$ is given arbitrarily.

Geometry and fundamental group. We begin, more or less in chronological order, with some relationships between the geometry of a compact nonpositively curved manifold M and the algebraic structure of $\pi_1(M)$.

(6.2) THEOREM (Bieberbach). *Let M be a compact n-dimensional flat manifold $(K \equiv 0)$. Then $\pi_1(M)$ contains $\mathbb{Z}^n$ as a finite index subgroup.*

(6.3) THEOREM (E. Cartan). *Let M be a complete manifold with $K \leq 0$. Then every nonidentity element of $\pi_1(M)$ has infinite order.*

(6.4) THEOREM (Preissmann). *Let M be compact with $K < 0$. Then every nonidentity abelian subgroup is infinite cyclic.*

A generalization of Preissmann's theorem was obtained by J. Wolf and Lawson-Yau [LY].

(6.5) FLAT TORUS THEOREM. *Let M be compact with $K \leq 0$, and let A be a free abelian subgroup of $\pi_1(M)$ of rank $k \geq 2$. Then the universal cover $\widetilde{M}$ admits a k-flat F such that A leaves F invariant and F/A is compact. In particular M admits a totally geodesic, isometrically immersed flat k-torus $T^k \approx F/A$.*

Note that the flat torus theorem yields the converse of the Bieberbach theorem as a corollary: if M is a compact n-manifold with $K \leq 0$ such that $\pi_1(M)$ contains $\mathbb{Z}^n$, a free abelian group of rank n, then M is flat.

The first three results quoted use geometric assumptions to reach algebraic conclusions. Beginning with the flat torus theorem the emphasis has been in the opposite direction: use algebraic conditions on the fundamental group of a compact nonpositively curved manifold M to draw conclusions about the geometric structure of M. The next advances of this sort were due to Yau and Gromoll-Wolf.

(6.6) THEOREM [Ya1]. *Let M be a compact manifold with $K \leq 0$ and solvable fundamental group. Then M is flat.*

This result confirmed a conjecture of J. Wolf. In turn J. Wolf and D. Gromoll strengthened this result with the following:

(6.7) THEOREM [GW]. *Let Γ be a solvable group of isometries of a complete, simply connected manifold $\widetilde{M}$ with $K \leq 0$ such that every element φ of Γ is* semisimple; *that is, the displacement function $d_\varphi : \widetilde{M} \to \mathbb{R}$ given by $d_\varphi(p) = d(p, \varphi p)$ assumes a minimum value (which may be zero) in $\widetilde{M}$.*

Then $\widetilde{M}$ admits a k-flat F for some integer $k \geq 0$ such that Γ leaves F invariant and the quotient space F/Γ is compact.

By definition a k-flat F is a point if $k = 0$ and a complete geodesic if $k = 1$. A noteworthy feature of this result is that Γ is not assumed to be discrete and in particular is not assumed to be the fundamental group of some smooth quotient manifold of $\widetilde{M}$.

If M is a Riemannian product $M_1 \times M_2$, then clearly the fundamental group $\pi_1(M)$ is a direct product $\pi_1(M_1) \times \pi_1(M_2)$. What seems astonishing is that the converse is essentially true for compact nonpositively curved manifolds. The next result was proved by Gromoll-Wolf [GW] in the C^∞ case and independently by Lawson-Yau [LY] in the real analytic case.

(6.8) SPLITTING THEOREM. *Let M be a compact manifold with $K \leq 0$ such that $\pi_1(M)$ has trivial center and $\pi_1(M)$ is a direct product $A_1 \times A_2$. Then M is a Riemannian product $M_1 \times M_2$ where $\pi_1(M_i) = A_i$ for $i = 1, 2$.*

If M has no local Euclidean de Rham factor, then $\pi_1(M)$ has trivial center. If $\pi_1(M)$ has nontrivial center, then the splitting theorem is false; see [LY] for an example. Extensions of the splitting result have been made in [E9] and [Sc4].

(6.9) Euclidean de Rham factor. It is not difficult to show that if M is compact with $K \leq 0$, then $\pi_1(M)$ admits a unique maximal normal abelian subgroup A, the *Clifford subgroup* of $\pi_1(M)$. If one regards $\pi_1(M)$ as a discrete subgroup of isometries of the universal cover $\widetilde{M}$, then A is the subgroup of elements φ whose displacement function d_φ is constant in $\widetilde{M}$ [Wol, CE].

THEOREM [E1, E9]. *Let M be compact with $K \leq 0$. The rank of the Clifford subgroup of $\pi_1(M)$ equals the dimension of the local Euclidean de Rham factor of M.*

(6.10) Algebraic rank of $\pi_1(M)$. By modifying slightly a definition of G. Prasad and M. Raghunathan [PR] one may assign to each compact manifold M with $K \leq 0$ an integer, $\mathrm{rank}(\pi_1 M)$, that is defined entirely by algebraic properties of $\pi_1(M)$. This integer has the property that $1 \leq \mathrm{rank}(\pi_1 M) \leq \dim M$ and $\mathrm{rank}(\pi_1(M_1 \times M_2)) = \mathrm{rank}(\pi_1(M_1)) + \mathrm{rank}(\pi_1(M_2))$. Moreover, $\mathrm{rank}(\pi_1 M) = 1$ if $K < 0$ and $\mathrm{rank}(\pi_1 M) = \dim M$ if and only if M is flat.

The concept of $\mathrm{rank}(\Gamma)$ makes sense for any group Γ although it may have no value outside the context of fundamental groups of nonpositively curved manifolds. Given an abstract group Γ and an integer $k \geq 1$ we define $A_k(\Gamma)$ to be the set of elements φ in Γ such that the centralizer of φ in Γ contains $\mathbb{Z}^r$ as a finite index subgroup for some integer r with

$1 \leq r \leq k$. Note that $A_k(\Gamma) \subseteq A_{k+1}(\Gamma)$ for every integer k. Now define

$$r(\Gamma) = \inf\left\{k \geq 1 : \Gamma = \bigcup_{i=1}^{N}\{\varphi_i \cdot A_k(\Gamma)\}, \text{ where } \{\varphi_1, \ldots, \varphi_N\} \subseteq \Gamma \text{ is finite}\right\}.$$

Define

$$\text{rank}(\Gamma) = \sup\{r(\Gamma^*) : \Gamma^* \text{ is a subgroup of } \Gamma \text{ with finite index in } \Gamma\}.$$

In [**BE**] it is shown that if M is compact with $K \leq 0$, then $\text{rank}(\pi_1 M)$ equals the geometric rank of M as defined below in (8.1). As a consequence one obtains

THEOREM [**BE**]. *Let M be compact with $K \leq 0$, and let $\pi_1(M)$ satisfy the following properties:*

(1) *$\pi_1(M)$ admits no direct product subgroup $A \times B$ of finite index.*
(2) *$\text{rank}(\pi_1 M) = k \geq 2$.*

Then the universal cover $\widetilde{M}$ is a symmetric space of rank k with no Euclidean de Rham factor, and no finite cover of M splits as a Riemannian product.

The converse of this assertion is known to be true. This assertion is also true for complete manifolds M with $K \leq 0$ and finite volume if one adds a third condition that $\pi_1(M)$ be finitely generated.

QUESTION. Let Γ be the fundamental group of a compact, locally symmetric, irreducible manifold M with rank $k \geq 2$. Let $G = I_0(\widetilde{M})$, where $\widetilde{M}$ is the universal cover of M. Can one determine the group G purely from algebraic properties of the lattice Γ? An answer to this question seems quite difficult.

(6.11) Visibility axiom and the Preissmann property (P). A result from [**E4**] suggests that the visibility axiom may have an equivalent formulation as an algebraic condition or set of algebraic conditions on the fundamental group of a compact nonpositively curved manifold M.

(6.11A) THEOREM. *Let M_1, M_2 be compact manifolds with $K \leq 0$ and isomorphic fundamental groups. If the universal cover $\widetilde{M}_1$ of M satisfies the visibility axiom, then the universal cover $\widetilde{M}_2$ of M_2 also satisfies the visibility axiom.*

The first explicit formulation of the visibility axiom as an algebraic condition on the fundamental group was obtained by M. Anderson and V. Schroeder.

(6.11B) THEOREM [AVS]. *Let $M = \widetilde{M}/\Gamma$ be a compact Riemannian manifold with $K \leq 0$. Then $\widetilde{M}$ contains a k-flat for some integer $k \geq 2$ if and only if there exists a quasiisometry $\varphi : \mathbb{Z}^k \to \Gamma$; that is,*

$$(1/B)d_E(x, y) \leq d_\Gamma(\varphi(x), \varphi(y)) \leq Bd_E(x, y)$$

for some positive number B, *where* d_E *denotes the standard Euclidean distance on the integer lattice* $\mathbb{Z}^k \subseteq \mathbb{R}^k$ *and* d_Γ *is the word length metric on* Γ *with respect to some system of generators.*

Recall from the discussion in (1.7) that if $\widetilde{M}$ admits a compact quotient $M = \widetilde{M}/\Gamma$, $\Gamma \subseteq I(\widetilde{M})$, then $\widetilde{M}$ satisfies the visibility axiom if and only if $\widetilde{M}$ admits no 2-flat. Hence by the result above $\widetilde{M}$ satisfies the visibility axiom if and only if there exists no quasi-isometry $\varphi : \mathbb{Z}^2 \to \Gamma$, where $M = \widetilde{M}/\Gamma$ is a smooth compact quotient of $\widetilde{M}$.

The Preissmann property (P) gives a second and somewhat simpler approach to formulating the visibility axiom in $\widetilde{M}$ as an algebraic condition on the fundamental group of a compact quotient M. We say that an abstract group Γ satisfies *property* (P) if every abelian subgroup of Γ is infinite cyclic. If Γ is the fundamental group of a compact manifold with $K < 0$, then Preissmann's theorem (6.4) says that Γ satisfies property (P). More generally, if M is compact with $K \leq 0$ and if the universal cover $\widetilde{M}$ satisfies the visibility axiom, then $\pi_1(M)$ satisfies property (P).

Let $M = \widetilde{M}/\Gamma$ be compact with $K \leq 0$. A k-flat F in $\widetilde{M}$ for some integer $k \geq 2$ is said to be *closed* (relative to Γ) if the subgroup of Γ that leaves F invariant acts with compact quotient on F. If $\widetilde{M}$ admits a closed k-flat F, then $\pi_1(M)$ contains $\mathbb{Z}^k$ as a subgroup. Hence $\pi_1(M)$ satisfies property (P) if and only if $\widetilde{M}$ admits no *closed* 2-flats, and $\widetilde{M}$ satisfies the visibility axiom if and only if $\widetilde{M}$ admits no 2-flats.

If $M = \widetilde{M}/\Gamma$ is a compact real analytic manifold with $K \leq 0$, then the discussion above and the next result of V. Bangert and V. Schroeder show that property (P) for $\pi_1(M)$ and the visibility axiom for $\widetilde{M}$ are equivalent. It is not known if these properties are equivalent for C^∞ manifolds.

(6.11C) THEOREM [**BaS**]. *Let* $M = \widetilde{M}/\Gamma$ *be a compact real analytic manifold with* $K \leq 0$. *If* $\widetilde{M}$ *contains a* k-*flat for some integer* $k \geq 2$, *then* $\widetilde{M}$ *also contains a closed* k-*flat.*

If $M = \widetilde{M}/\Gamma$ is C^∞ and has bounded nonpositive sectional curvature and finite volume, then Schroeder has shown in [**Sc1**] that $\widetilde{M}$ contains a closed $(n-1)$-flat if it contains an $(n-1)$-flat. See also [**Buy**] when $n = 3$. In the real analytic category Schroeder has extended this result to $(n-2)$-flats [**Sc2**].

7. Entropy rigidity

To each homeomorphism f of a compact metric space X one can assign a number $h(f) \geq 0$ called the *topological entropy of* f. We omit a precise definition since we shall present equivalent definitions in the case that X is the unit tangent bundle of a compact manifold with $K < 0$ and f is the time-1 map g^1 of the geodesic flow $\{g^t\}$. For formal definitions and a discussion see for example [**Wa**].

If ν is a Borel probability measure on X that is left invariant by f, then one may assign in similar fashion a number $h_\nu(f) \geq 0$ called the *metric entropy of f with respect to ν*. In general $h_\nu(f) \leq h(f)$ for any f-invariant Borel probability measure ν, and $h(f) = \sup\{h_\nu(f): \nu$ is an f-invariant Borel probability measure on $X\}$. If $h_\nu(f) = h(f)$, then ν is called a *measure of maximal entropy*. See [**Wa**] for further discussion.

(7.1) Topological entropy for negatively curved manifolds. Let M be compact with $K < 0$ and define the topological entropy h of the geodesic flow $\{g^t\}$ to be the topological entropy of the time-1 map g^1 acting on SM. In the present context there are several formulas for h.

(1) Closed geodesics [**Mar1**]. Let $N(t)$ be the number of free homotopy classes of closed curves in M that are represented by closed geodesics of length $\leq t$. Then

$$h = \lim_{t \to \infty} \frac{1}{t} \log N(t).$$

(2) Growth of spheres [**Man2**]. Fix a point p in the universal cover $\widetilde{M}$ and let $S(t, p)$ denote the sphere with center p and radius t. Then

$$h = \lim_{t \to \infty} \frac{1}{t} \log(\operatorname{vol} S(t, p)).$$

(3) Growth of balls [**Man2**]. Fix a point p in the universal cover $\widetilde{M}$ and let $B(t, p)$ denote the closed ball with center p and radius t. Then

$$h = \lim_{t \to \infty} \frac{1}{t} \log(\operatorname{vol} B(t, p)).$$

(4) [**Ham5**]. Let the upper curvature bound for M be normalized to equal -1. For each vector v in $S\widetilde{M}$, $\widetilde{M}$ the universal cover of M, let $\eta_v = \eta_v^1$ be the metric on $\widetilde{M}(\infty)$ defined in (4.2). Then h is the Hausdorff dimension of $\widetilde{M}(\infty)$ for each metric η_v.

For a discussion of Hausdorff measure and dimension see for example [**Fa**].

(7.2) Metric entropy for negatively curved manifolds. Let M be compact with $K < 0$. If no $\{g^t\}$-invariant Borel probability measure ν is specified, then the *metric entropy* of the geodesic flow $\{g^t\}$ is defined to be the metric entropy of the time-1 map g^1 with respect to Lebesgue-Liouville measure. The metric entropy will be denoted by h_λ.

Now let -1 be the least upper bound for the sectional curvatures of M, and let ν be a Borel probability measure on SM that is quasi-invariant under the stable foliation. Then the metric entropy h_ν of the geodesic flow with respect to ν equals the infimum of the Hausdorff dimensions relative to the metrics η_v, $v \in S\widetilde{M}$, of Borel subsets of $\widetilde{M}(\infty)$ with full ν_∞ measure [**Ham6, Km**]. Recall that ν_∞ denotes the measure class on $\widetilde{M}(\infty)$ defined by ν (cf. (5.1)).

(7.3) Curvature and metric entropy. Let M be compact with $K \leq 0$. A formula of Pesin [**Pe2**] states that the metric entropy h_λ (of the Lebesgue-Liouville measure λ) is given by

$$h_\lambda = \int_{SM} h(v)\, d\lambda,$$

where $h(v)$ denotes the mean curvature at the base point $\pi(v)$ of the horosphere $H(v)$ in $\widetilde{M}$ determined by v. Considerable effort has been spent on obtaining sharp estimates for h_λ in terms of curvature. We describe one of the most comprehensive of these results.

For each vector $v \in SM$ let $R_v : T_{\pi v}M \to T_{\pi v}M$ denote the curvature transformation given by $R_v(w) = R(w, v)v$, where $R(\cdot, \cdot)$ denotes the curvature tensor of M. The transformation R_v is selfadjoint and negative semidefinite, and $-R_v$ therefore admits a well-defined selfadjoint square root that we denote by $\sqrt{-R_v}$.

THEOREM. *Let M be compact with $K \leq 0$. Then*

$$h_\lambda \geq \int_{SM} \operatorname{tr}\left(\sqrt{-R_v}\right) d\lambda$$

and equality holds if and only if M is locally symmetric.

This result was proved by W. Ballmann and M. Wojtkowski in [**BW**]. For strictly negative sectional curvature R. Osserman and P. Sarnak had obtained this result earlier [**OS**]. Related work prior to this result can be found in [**FM, Sa, Man1**].

(7.4) Katok entropy conjecture. In [**Kat1**] A. Katok proved that if M is a compact 2-dimensional manifold with $K < 0$ and if the metric entropy h_λ equals the topological entropy h, then the curvature of M is constant and negative. Katok also conjectured that if an n-dimensional compact manifold M with $K < 0$ has metric entropy h_λ equal to topological entropy h, then M must be rank 1 locally symmetric. This conjecture is not yet resolved, but it lies in the center of present research on negatively curved manifolds and it has led to investigations of many related problems. We describe some of these.

Recall that the topological entropy is the metric entropy h_μ of the Bowen-Margulis measure μ. Hence the Katok conjecture is closely related to the following:

CONJECTURE. Let M be compact with $K < 0$, and suppose that any two of the measures {Lebesgue-Liouville, harmonic and Bowen-Margulis} coincide. Then M must be a rank 1 locally symmetric space.

This conjecture is known to be true for surfaces. Katok proved it in [**Kat1**] if Lebesgue-Liouville measure λ equals Bowen-Margulis measure μ and in [**Kat2**] if $\lambda = \omega$ (harmonic measure). The case $\mu = \omega$ is due to Ledrappier in [**Led2**] and a different proof may be found in [**Ham6**].

Next we discuss some partial results related to the conjecture above. We say that a compact manifold M with $K < 0$ is *asymptotically harmonic* if there exists a negative number h^* such that the mean curvature $h(v)$ at the footpoint of v for the horosphere $H(v)$ in $\widetilde{M}$ equals h^* for all unit vectors $v \in S\widetilde{M}$. As usual $\widetilde{M}$ denotes the universal cover of M. If M is rank 1 and locally symmetric, then M is asymptotically harmonic. The converse assertion may also be true, which would be sufficient to prove the Katok conjecture in view of the next result.

(7.4A) THEOREM [**Ham4**]. *Let M be compact with $K < 0$. If the metric and topological entropies, h_λ and h_μ, coincide, then M is asymptotically harmonic.*

A number of results on asymptotically harmonic manifolds have been obtained recently; see [**Yu1, Yu2, Yu3, Yu4, Led3, Kn1, FL**].

One can also use asymptotic harmonicity to define secondary characteristic classes for the unstable foliations W^u, which in general require foliations of class C^2 to carry out such constructions. These classes can be computed explicitly, and they coincide with the corresponding ones for locally symmetric metrics [**Ham7**]. For secondary characteristic classes in this context see [**HK**]. Using these secondary classes one can prove the Katok conjecture in dimension 4.

(7.4B) THEOREM. *Let M be a compact 4-dimensional manifold with $K < 0$. If the metric and topological entropies of the geodesic flow coincide, then M is locally symmetric.*

The Katok conjecture in dimensions 2 and 3 follows immediately from (7.4A).

(7.5) Locally symmetric metrics as minima for topological entropy. For compact Riemannian manifolds with $K < 0$ that lie in a fixed homotopy equivalence class, normalized by volume or an upper bound for sectional curvature, there is evidence to suggest that locally symmetric metrics correspond to absolute minima for the topological entropy. It is unknown if locally symmetric metrics correspond to extreme values of topological entropy for metrics in a homotopy equivalence class of manifolds normalized by a lower bound for the sectional curvature.

The next conjecture is due to Gromov [**Gro2**].

(7.5A) CONJECTURE. Let S be a compact, locally symmetric space with $K < 0$ and $\dim S \geq 3$. Let M be compact with $K < 0$ such that $\mathrm{vol}(M) = \mathrm{vol}(S)$ and $\pi_1(M)$ is isomorphic to $\pi_1(S)$. Then the topological entropies h_M, h_S for M, S satisfy $h_M \geq h_S$ with equality if and only if M and S are isometric.

In [**Kat1**] Katok proved the 2-dimensional analogue of this conjecture. Here one has to consider conformally equivalent metrics due to the fact that the Mostow Rigidity Theorem does not apply in dimension 2. More precisely Katok proved

(7.5B) THEOREM. *Let g and $\tilde{g}$ be conformally equivalent metrics of negative curvature on a compact surface. If the volumes of g and $\tilde{g}$ coincide and if the curvature of $\tilde{g}$ is constant, then the topological entropies h_g, $h_{\tilde{g}}$ satisfy $h_g \geq h_{\tilde{g}}$ with equality if and only if g and $\tilde{g}$ are isometric.*

If one normalizes upper sectional curvature bounds in a fixed homotopy equivalence class, then one obtains the following.

(7.5C) THEOREM. *Let S be compact and locally symmetric with $K < 0$ and $\dim S \geq 3$. Let M be compact with $K < 0$ such that $\pi_1(M)$ is isomorphic to $\pi_1(S)$ and M, S have the same upper bounds for sectional curvature. Then the topological entropies h_M, h_S satisfy $h_M \geq h_S$ with equality if and only if M and S are isometric.*

The inequality $h_M \geq h_S$ was first derived by Pansu in [**Pa5**]. The theorem is proved in [**Ham2**].

8. Other rigid properties of compact locally symmetric spaces

In this section we discuss a variety of geometric conditions on a complete manifold M with $K \leq 0$ that force M to be locally symmetric.

(8.1) Mostow rigidity and its generalizations. Let M_1, M_2 be compact manifolds with $K \leq 0$ such that the universal covers $\widetilde{M}_1$, $\widetilde{M}_2$ are symmetric spaces. We assume furthermore that

(a) M_i is irreducible for $i = 1, 2$; that is, no finite cover of M_i splits as a Riemannian product.

(b) $\widetilde{M}_i$ does not admit a Euclidean space as one of its de Rham factors for $i = 1, 2$.

(8.1A) THEOREM (Mostow [**Mo3**, Theorems 18.1 and 24.1]). *Let M_1, M_2 satisfy the conditions above and assume that M_1, M_2 have isomorphic fundamental groups. Then M_1 and M_2 are isometric up to a rescaling of the metric by constant multiples on each local de Rham factor of M_1 or M_2.*

Mostow's theorem was extended by G. Prasad in [**Pra2**] to the case that M_1, M_2 are both noncompact but with finite volume.

Mostow's theorem also has an algebraic formulation in terms of lattices in semisimple Lie groups. In this form it has been generalized by G. Margulis, R. Zimmer and others. See [**Mar4, Mar5, Z**] for further details.

Mostow's proof of the rigidity theorem falls into two distinct cases: (1) the case where M_1 and M_2 have rank at least two; (2) the case where M_1 and M_2 both have rank 1. In the first case one uses the theory of Tits buildings and in the second one uses results from quasiconformal analysis. Gromov realized that Mostow's ideas could be extended to arbitrary compact manifolds with $K \leq 0$ and carried out this extension in [**BGS**] for compact manifolds that admit a significant amount of zero sectional curvature, those for which the Tits metric in $\widetilde{M}(\infty)$ is nontrivial. An extension of Mostow's ideas to the case of strictly negative sectional curvature has been carried out primarily by P. Pansu and U. Hamenstädt. In the work of each, a quasiconformal structure on $\widetilde{M}(\infty)$ plays an important role, as in Mostow's result. Hamenstädt's quasiconformal structure is described in (5.5) while that of Pansu is described in [**Pa2**].

(8.1B) Gromov rigidity. In [**BGS**] Gromov extended Mostow's theorem in the higher rank case to the situation where only one of the manifolds M_1, M_2 is locally symmetric.

THEOREM (Gromov). *Let M_1, M_2 be compact manifolds with $K \leq 0$ that are irreducible and have isomorphic fundamental groups. Let the universal cover $\widetilde{M}_1$ be a symmetric space of rank at least two that does not admit a Euclidean space as one of its de Rham factors. Then M_1 and M_2 are isometric up to a rescaling of the metric by constant multiples on each local de Rham factor of M_1 (or M_2).*

An independent proof of this result in the special case that $\widetilde{M}_1$ or $\widetilde{M}_2$ is reducible can be found in [**E8**].

COROLLARY. *Let (M, g_0) be compact and irreducible with $K \leq 0$ such that the universal cover $\widetilde{M}$ is symmetric with rank at least two and does not admit a Euclidean space as one of its de Rham factors. Let g be any other metric on M with $K \leq 0$. Then (M, g) is isometric to a manifold obtained by multiplying the original metric g_0 by positive constants on each local de Rham factor of (M, g_0).*

(8.1C) Higher rank rigidity [Ba3, BuS]. Let M be complete with $K \leq 0$, not necessarily simply connected. For each unit vector $v \in SM$ define $r(v)$ to be the dimension of the space of parallel Jacobi vector fields on the geodesic $\gamma_v \colon \mathbb{R} \to M$ determined by v. Note that $r(v) \geq 1$ for every v since the tangent vector field $\gamma'_v(t)$ is a parallel Jacobi vector field on γ_v. If $E(t)$ is a parallel vector field on γ_v that is orthogonal to $\gamma'_v(t)$, then $E(t)$ is a Jacobi vector field if and only if the sectional curvatures $K(E(t), \gamma'_v(t))$ are zero for all t.

Define $\operatorname{rank}(M) = \min\{r(v) \colon v \in SM\}$. This is the *geometric rank* of M and gives some kind of global measurement of flatness in view of the remarks above. It is not hard to show that $\operatorname{rank}(M) = 1$ if all sectional curvatures at one point of M are negative; $\operatorname{rank}(M) = \dim M$ if and only if M is flat; $\operatorname{rank}(M_1 \times M_2) = \operatorname{rank}(M_1) + \operatorname{rank}(M_2)$ if M is a Riemannian product $M_1 \times M_2$ and $\operatorname{rank}(\widetilde{M})$ agrees with the usual rank of $\widetilde{M}$ if $\widetilde{M}$ is a symmetric space. The next result was obtained independently by W. Ballmann [**Ba3**] and K. Burns and R. Spatzier [**BuS**].

THEOREM. *Let M be complete with finite volume and assume furthermore that $-a^2 \leq K \leq 0$ for some positive number a. Let the universal cover $\widetilde{M}$ be irreducible with $\mathrm{rank}(\widetilde{M}) = k \geq 2$. Then $\widetilde{M}$ is a symmetric space of rank k.*

This result yields another proof of the Gromov Rigidity Theorem. In [EH] a generalization of the theorem above was obtained by using the Tits metric on $\widetilde{M}(\infty)$.

THEOREM [EH]. *Let $\widetilde{M}$ be irreducible with $\mathrm{rank}(\widetilde{M}) = k \geq 2$ and assume that $I(\widetilde{M})$ satisfies the duality condition. Then $\widetilde{M}$ is symmetric of rank k.*

Recently J. Heber in [Heb] has obtained an analogue of this result for homogeneous spaces. Although the proof uses ideas from [Ba3] and [EH] it is essentially different from the proofs of these earlier results; if $I(\widetilde{M})$ satisfies the duality condition for a simply connected homogeneous space $\widetilde{M}$ with $K \leq 0$, then $\widetilde{M}$ must be symmetric by Theorem 5.4 of [CE] or Proposition 4.1 of [E5].

THEOREM [Heb]. *Let $\widetilde{M}$ be irreducible and homogeneous with $\mathrm{rank}(\widetilde{M}) = k \geq 2$. Then $\widetilde{M}$ is symmetric.*

(8.1D) Tits metric rigidity. In [Ba1] W. Ballmann shows that if $\widetilde{M}$ has rank 1 and $I(\widetilde{M})$ satisfies the duality condition, then there exists a nonempty subset A of $\widetilde{M}(\infty)$ with the property that if $x \in A$ then x can be joined to any point y in $\widetilde{M}(\infty)$ distinct from x; that is, there exists a geodesic γ of $\widetilde{M}$ such that $\gamma(\infty) = x$ and $\gamma(-\infty) = y$. (In fact one can show that A is dense in $\widetilde{M}(\infty)$ in the sphere topology.) In particular, if $x \in A$ and $y \in \widetilde{M}(\infty)$, $y \neq x$, then $Td(x, y) \geq \angle(x, y) = \pi$ and by the discussion in (4.1) it follows that $Td(x, y) = +\infty$; if $Td(x, y)$ were finite then there would exist a minimizing Tits geodesic σ in $\widetilde{M}(\infty)$ from x to y and σ would contain points at arbitrarily small Tits distance from x. This discussion together with results stated earlier yields the following.

THEOREM. *Let $\widetilde{M}$ be irreducible with $K \leq 0$. Assume that $I(\widetilde{M})$ satisfies the duality condition and $Td(x, y)$ is finite for all points x, y in $\widetilde{M}(\infty)$. Then $\widetilde{M}$ is symmetric with $\mathrm{rank}\, k \geq 2$.*

The discussion above shows that the rank of $\widetilde{M}$ cannot be 1, and one now applies the result of [EH] from (8.1C).

REMARKS. (1) It is natural to try to extend the result above to the case of sectional curvature bounded above by -1, replacing the finiteness of the Tits metric Td by the finiteness of the metrics $\{Hd_v, v \in S\widetilde{M}\}$. This remains an open problem.

(2) The attempt to extend the Tits finiteness theorem above to homogeneous spaces fails. J. Heber in [Heb] has constructed examples of simply

connected homogeneous spaces $\widetilde{M}$ that are not symmetric but have Tits diameter in $\widetilde{M}(\infty)$ arbitrarily close to π; that is, given a number $\alpha > \pi$ there exists a homogeneous, nonsymmetric manifold $\widetilde{M}$ such that $Td(x, y) \leq \alpha$ for all x, y in $\widetilde{M}(\infty)$. In general, if $\widetilde{M}$ is any simply connected manifold with $K \leq 0$ and if $x \in \widetilde{M}(\infty)$ is any point, then there is a point $y \in \widetilde{M}(\infty)$ with $Td(x, y) \geq \pi$. If $\widetilde{M}$ is symmetric with rank at least two, then the Tits diameter of $\widetilde{M}(\infty)$ is exactly π. See [**BGS**, p. 46].

In another direction, one can recognize a symmetric space $\widetilde{M}$ of rank at least two by the geometry of its Tits metric [**BGS**, Appendix 4].

THEOREM. *Let $\widetilde{M}_1$ be a symmetric space with $K \leq 0$ and rank $k \geq 2$, and let Td_1 denote its Tits metric in $\widetilde{M}_1(\infty)$. Let $\widetilde{M}_2$ be any simply connected space with $K \leq 0$ and finite Tits metric Td_2 on $\widetilde{M}_2(\infty)$. Suppose there exists a map $F_\infty: \widetilde{M}_1(\infty) \to \widetilde{M}_2(\infty)$ that is a homeomorphism relative to the sphere topologies and an isometry relative to the Tits metrics. Then, after multiplying the metric of $\widetilde{M}_1$ by positive constants on de Rham factors, there is an isometry F of $\widetilde{M}_1$ onto $\widetilde{M}_2$ that induces the boundary isometry F_∞.*

(8.1E) Rank 1 rigidity [Ham3]. Hamenstädt in [**Ham3**] has obtained a rank 1 version of the rigidity theorem of Ballmann and Burns-Spatzier discussed above. The statements are similar but the methods are entirely different, corresponding to the difference in Mostow's Rigidity Theorem between the rank 1 and the higher rank cases. One begins with a simply connected manifold $\widetilde{M}$ whose (normalized) sectional curvatures have -1 as a least upper bound. For each vector $v \in S\widetilde{M}$ one defines an integer $hr(v)$, replacing parallel Jacobi vector fields in the definition of $r(v)$ by Jacobi vector fields of the form $e^t E(t)$, where $E(t)$ is a parallel vector field orthogonal to $\gamma_v'(t)$ such that $K(E, \gamma_v')(t) = -1$ for all $t \in \mathbb{R}$. Finally one defines the *hyperbolic rank* of $\widetilde{M}$, $h\text{-rank}(\widetilde{M}) = \inf\{hr(v): v \in S\widetilde{M}\}$. If M^* is any quotient manifold of $\widetilde{M}$, then we define $h\text{-rank}(M^*)$ in analogous fashion. Clearly $h\text{-rank}(M^*) = h\text{-rank}(\widetilde{M})$.

THEOREM [**Ham3**]. *Let M be compact with $K \leq -1$ and $h\text{-rank}(M) \geq 1$. Then M is locally symmetric of rank 1.*

This result raises the following questions:

(1) Can one relax the assumption that M be compact. For example, is it sufficient to assume as in [**EH**] that $I(\widetilde{M})$ satisfies the duality condition, where $\widetilde{M}$ is the universal cover of M?

(2) Does the result above remain true if in the definition of hyperbolic rank one replaces the curvature maximum by the curvature minimum?

(3) If $h\text{-rank}(M) \geq 1$, then M has the property

(*)   Along every geodesic γ of M there exists a perpendicular Jacobi vector field $J(t)$ such that $E(t) = J(t)/|J(t)|$ is parallel along γ.

If M is a compact manifold with $K < 0$ that satisfies (∗) must M be locally symmetric?

(8.2) Conformal rigidity. Let M be compact with $K < 0$ and universal cover $\widetilde{M}$. Although $\widetilde{M}$ admits generalized conformal and quasiconformal structures as described in (5.5) these are rarely conformal in the classical sense. Indeed Kanai in [**Kan2**] has proved the following.

THEOREM. *Let M be compact with $K < 0$, and assume that one of the following conditions holds*:

(1) *TW^{su} admits a continuous $\{g^t\}$ invariant conformal structure in the classical sense.*

(2) *The sectional curvature of M is strictly $\frac{1}{4}$-pinched and TW^{su} admits a measurable $\{g^t\}$ invariant conformal structure.*

Then there is a compact manifold M^* of constant negative curvature and a diffeomorphism $\Lambda: SM \to SM^*$ that commutes with the geodesic flows $\{g^t\}$, $\{g^{*t}\}$ of SM, SM^*.

We call the diffeomorphism Λ above a *time-preserving conjugacy* of the geodesic flows.

One easy consequence of assumption (1) in the result above is that the quasiconformal structure on $\widetilde{M}(\infty)$ constructed in (5.5) is standard. The conclusion of the theorem then implies that the curvature of M is constant [**Ham4**].

Locally symmetric metrics of nonconstant negative curvature induce rather special quasiconformal structures on $\widetilde{M}(\infty)$. Indeed they are defined by a Carnot-Carathéodory metric on the leaves of W^{su} that induces an $I(\widetilde{M})$-invariant distribution on the smooth manifold $\widetilde{M}(\infty)$ (see [**Ham3**] and [**Pa3**]). It is possible that such a quasiconformal structure uniquely characterizes locally symmetric metrics.

(8.3) Regularity of Anosov foliations. We saw in (3.3) that for a complete manifold M with bounded strictly negative sectional curvatures the Anosov foliations W^{su}, W^{ss} are in general only C^0 and are C^1 if the sectional curvature of M is $\frac{1}{4}$-pinched. E. Ghys in [**Gh**] proved that if M is a compact 2-manifold with $K < 0$ and C^∞ Anosov foliations in SM, then M has constant negative curvature. More generally, Ghys considered contact Anosov flows on 3-manifolds with C^∞ Anosov foliations, and he showed that after a certain very specific reparametrization the Anosov flow must arise as an algebraic flow on a homogeneous space. In [**Kan1**] Kanai extended the first part of Ghys's result to higher dimensions. One of the major components of his proof was the *Kanai connection* described in (3.4), and this connection remained a major tool in all subsequent improvements of his result.

(8.3A) THEOREM [**Kan1**]. *Let* M *be compact with* $-\frac{9}{4} < K < -1$, *and let the Anosov foliations in* SM *be* C^∞. *Then the geodesic flow in* SM *is time-preserving conjugate to the geodesic flow in* SM^*, *where* M^* *is compact with constant negative sectional curvature.*

With more subtle dynamical arguments Feres and Katok in [**FK**] later proved the same conclusion under the optimal pinching condition $-4 < K < -1$. Feres in [**Fer**] further extended the methods of [**FK**] to obtain results very similar to those of [**BFL**] valid for compact negatively curved manifolds M of dimension $4k + 2$, where $k \geq 1$ is any integer. The best result in this direction is due to Benoist-Foulon-Labourie.

(8.3B) THEOREM [**BFL**]. *Let* M *be compact with* $K < 0$, *and let the Anosov foliations in* SM *be* C^∞. *Then the geodesic flow in* SM *is time-preserving conjugate to the geodesic flow in* SM^*, *where* M^* *is a compact, rank* 1, *locally symmetric space.*

If M_1, M_2 are compact manifolds with $K \leq 0$ whose geodesic flows are the same up to a time-preserving conjugacy, then the geodesic flows in SM_1, SM_2 have the same metric (topological) entropies. Hence from (7.4B) we obtain the following.

(8.3C) COROLLARY. *Let* M *be compact with* $K < 0$ *and* $\dim M \leq 4$. *If the Anosov foliations in* SM *are* C^∞, *then* M *is a rank* 1, *locally symmetric space.*

More generally, if the Katok entropy conjecture stated in (7.4) were known to be true, then the corollary above would be true without any restriction on the dimension of M.

This result of [**BFL**] is actually only a special case of their more general result. They show that if $\{\varphi^t\}$ is a contact Anosov flow on an odd-dimensional manifold M such that the Anosov foliations are C^∞, then after a very specific reparametrization (already described in [**Gh**] by Ghys) the flow $\{\varphi^t\}$ is time-preserving conjugate to an algebraic flow on a homogeneous space of the type $\Gamma\backslash G/H$, where G is a semisimple connected Lie group, H is a closed subgroup of G and Γ is a discrete cocompact subgroup of G.

All of the contributors to these results on C^∞ Anosov foliations have noted that the C^∞ condition is merely a convenience and that far less smoothness of the foliations would be sufficient to obtain the same conclusions. The work of Hurder and Katok in [**HK**] on surfaces suggests that a $C^{1+\mathrm{Lip}}$ Anosov splitting should be optimal: for any number $\alpha \in (0, 1)$ there are compact surfaces of nonconstant negative curvature with Anosov foliations of class $C^{1,\alpha}$, but an Anosov foliation of class $C^{1+\mathrm{Lip}}$ is only possible if the curvature is constant.

(8.4) **Quarter pinching.** Let M be compact with $K < 0$. M is said to be *pointwise quarter pinched* if for each point p in M there exists a positive number $a = a(p)$ such that $-a^2 \leq K(\pi) \leq -\frac{1}{4}a^2$ for all 2-planes π in T_pM. M is said to be *quarter pinched* if the positive number a can be chosen independent of p. For a long time differential geometers dreamed of a negative curvature version of the differentiable sphere theorem, which says that if M is compact with sectional curvature satisfying $0 < \lambda < K \leq 1$, where $\lambda \approx .80$, then M is diffeomorphic to a standard sphere. The best possible value of λ for the sphere is still unknown and possibly equals $\frac{1}{4}$.

In the negative curvature version one would replace "standard sphere" by "compact manifold with $K \equiv -1$". This dream in full generality was exploded by the examples of Gromov and Thurston [GT] and later by those of Farrell and Jones [FJ1]. However, a special case is true.

THEOREM. *Let (M, g) and (M^*, g^*) be compact manifolds with $K < 0$ and isomorphic fundamental groups. Let (M, g_0) be rank 1, locally symmetric with nonconstant sectional curvature, and let the sectional curvature of (M^*, g^*) be $\frac{1}{4}$-pinched. Then (M, g) is isometric to (M^*, g^*) after multiplying the metric g or g^* by a suitable positive constant.*

This result was proved by M. Ville in [Vil] in dimension 4. For compact manifolds M covered by the complex hyperbolic spaces the result was proved independently by L. Hernández [Her] and S. Yau and F. Zheng [YZ]. In fact, for these manifolds the result remains true under the weaker pointwise $\frac{1}{4}$-pinching condition. For compact manifolds M covered by the quaternionic hyperbolic spaces or the Cayley hyperbolic plane the theorem follows from the work of Hernández [Her], K. Corlette [Co] and Gromov [Gro3].

(8.5) Kähler manifolds. The compact quotients of quaternionic hyperbolic space are examples of compact locally quaternionic Kähler manifolds of negative curvature, i.e., of compact negatively curved manifolds with holonomy contained in $Sp(n) \cdot Sp(1)$. Recently C. Lebrun [Leb] has constructed examples of simply connected quaternionic Kähler manifolds $\widetilde{M}$ of negative curvature that are not symmetric; in fact the moduli space of these metrics is infinite dimensional. However, none of these spaces admits compact quotients. This situation is in contrast to the well-known fact, due to M. Berger (see [Bes]), that the only simply connected quaternionic Kähler manifold of *positive* sectional curvature is the quaternionic projective space. In fact it has been conjectured that the only compact locally quaternionic Kähler manifolds of negative sectional curvature are the compact quotients of quaternionic hyperbolic space.

Every compact quotient of quaternionic hyperbolic space is automatically Einstein. Related to the discussion above is the following.

QUESTION. If M is a compact Kähler-Einstein manifold of negative sectional curvature, then must M be locally symmetric?

(8.6) The duality condition and rigidity. For a simply connected space $\widetilde{M}$ with $K \leq 0$ whose isometry group $I(\widetilde{M})$ satisfies the duality condition there are a number of geometric characterizations of symmetric spaces. Recall from (1.6) that if $\widetilde{M}$ covers a complete manifold $M = \widetilde{M}/\Gamma$ with finite volume, then the fundamental group $\Gamma \subseteq I(\widetilde{M})$ satisfies the duality condition and hence so does $I(\widetilde{M})$.

The first characterizations of symmetric spaces are corollaries of the next result, which is Proposition 4.1 of [E5].

THEOREM. *Let $\widetilde{M}$ be simply connected with $K \leq 0$ and let $I(\widetilde{M})$ satisfy the duality condition. Then there exist nonpositively curved spaces $\widetilde{M}_0$, $\widetilde{M}_1$, $\widetilde{M}_2$, two of which may have dimension zero, such that*

(1) *$\widetilde{M}$ is isometric to the Riemannian product $\widetilde{M}_0 \times \widetilde{M}_1 \times \widetilde{M}_2$.*

(2) *$\widetilde{M}_0$ is a flat Euclidean space.*

(3) *$\widetilde{M}_1$ is a symmetric space with no Euclidean de Rham factor.*

(4) *$I(\widetilde{M}_2)$ is discrete but satisfies the duality condition.*

(8.6A) COROLLARY 1. *Let $\widetilde{M}$ be simply connected and homogeneous with $K \leq 0$. If $I(\widetilde{M})$ satisfies the duality condition, then $\widetilde{M}$ is symmetric.*

(8.6B) COROLLARY 2. *Let $\widetilde{M}$ be simply connected and irreducible. If $I(\widetilde{M})$ is not discrete and satisfies the duality condition, then $\widetilde{M}$ is a symmetric space.*

The first corollary is also proved in Theorem 5.4 of [CE]. It extends a result of Heintze in [Hei3] that if M is compact and locally homogeneous with $K < 0$, then M is locally symmetric.

Lattices in $I(\widetilde{M})$. A second set of characterizations of symmetric spaces involves the existence of lattices of various types in $I(\widetilde{M})$, and these results also follow from the theorem stated above. Before stating the results we need some terminology and basic facts about lattices.

Let $\widetilde{M}$ be simply connected with $K \leq 0$. A subgroup Γ of $I(\widetilde{M})$ will be called a *lattice* if Γ is discrete and if the quotient space $M = \widetilde{M}/\Gamma$ is a smooth manifold with finite volume. This definition of lattice is slightly more restrictive than the usual definition of lattice Γ in a Lie group G, which permits Γ to have elements of finite order. However, if G is semisimple with trivial center, then any lattice Γ of G has a subgroup Γ^* of finite index in Γ that is a lattice in our sense.

A lattice Γ in $I(\widetilde{M})$ is said to be *uniform* or *cocompact* if $\widetilde{M}/\Gamma$ is compact, and otherwise a lattice Γ is said to be *nonuniform* or *noncocompact*. A lattice Γ in $I(\widetilde{M})$ is *irreducible* if the quotient manifold $\widetilde{M}/\Gamma$ has no finite cover that splits as a Riemannian product.

For certain reducible symmetric spaces $\widetilde{M}$ (in particular, Riemannian products of n hyperbolic planes) it has been known for a long time that $G = I_0(\widetilde{M})$ admits irreducible lattices Γ, both cocompact and noncocompact. A sketch of one classical arithmetic procedure for constructing irreducible lattices Γ in reducible Lie groups G may be found in [Shi, p. 64] for the case $G = SL(2, \mathbb{R}) \times \cdots \times SL(2, \mathbb{R})$. See also [J].

However, not all reducible symmetric spaces $\widetilde{M} = \widetilde{M}_1 \times \cdots \times \widetilde{M}_k$ admit irreducible lattices. A certain compatibility of the factors $\widetilde{M}_i$ is necessary. F. Johnson in [J] provides necessary and sufficient conditions for the existence of irreducible lattices Γ in $G = I_0(\widetilde{M})$, where $\widetilde{M}$ is a reducible symmetric space with no Euclidean factor.

(8.6C) THEOREM [J]. *Let $G = G_1 \times \cdots \times G_n$ be a product of noncompact simple Lie groups G_i, and let $\mathscr{G}$ be the real Lie algebra of G. Then the following conditions on G are equivalent:*

(1) *G contains an irreducible lattice Γ.*

(2) *G is isomorphic as a Lie group to $H_{\mathbb{R}}^0$, the identity component of the group of real points of a $\mathbb{Q}$-simple algebraic group H.*

(3) *The complexification $\mathscr{G} \otimes \mathbb{C}$ is isomorphic to $\mathscr{A} \oplus \cdots \oplus \mathscr{A}$ for some simple Lie algebra $\mathscr{A}$ over $\mathbb{C}$ (G is called* isotypic *of type $\mathscr{A}$).*

Moreover, any such isotypic group G contains both cocompact and noncocompact irreducible lattices.

If G is a semisimple noncompact Lie group, then G admits both cocompact and noncocompact lattices Γ [**Bo**; **Ra**, Theorem 14.1].

The next two characterizations of symmetric spaces may now be regarded as complements to the facts stated above.

(8.6D) THEOREM [E5, Proposition 4.7]. *Let $\widetilde{M}$ be simply connected with $K \leq 0$, and let $\widetilde{M}$ be a Riemannian product $\widetilde{M}_1 \times \cdots \times \widetilde{M}_k$, where $k \geq 2$ and none of the manifolds $\widetilde{M}_i$ has a Euclidean de Rham factor. If $I(\widetilde{M})$ contains an irreducible lattice subgroup Γ, then $\widetilde{M}$ is a symmetric space.*

The next result follows easily from Proposition 4.4 of [E5].

(8.6E) THEOREM. *Let $\widetilde{M}$ be simply connected and irreducible with $K \leq 0$. If $I(\widetilde{M})$ admits both cocompact and noncocompact lattice subgroups Γ, then $\widetilde{M}$ is a symmetric space.*

Action of $I(\widetilde{M})$ on $\widetilde{M}(\infty)$. We conclude this section by recalling a result from (1.6).

(8.6F) THEOREM [E10, Theorem 4.1]. *Let $\widetilde{M}$ be simply connected and irreducible with $K \leq 0$. Let $\Gamma \subseteq I(\widetilde{M})$ be a subgroup that satisfies the duality condition. If Γ leaves invariant a proper closed subset A in $\widetilde{M}(\infty)$, then $\widetilde{M}$ is symmetric with $\operatorname{rank} k \geq 2$.*

9. Geodesic conjugacy and marked length spectrum

Let M_1, M_2 be compact manifolds with $K \leq 0$. The unit tangent bundles SM_1, SM_2 are said to be *geodesically conjugate* if there exists a diffeomorphism of SM_1 onto SM_2 that commutes with the geodesic flow transformations $\{g^t\}$ in SM_1, SM_2. It is not difficult to see that if SM_1, SM_2 are geodesically conjugate, then the fundamental groups of M_1, M_2 are isomorphic. Hence M_1, M_2 are homeomorphic by the work of Farrell and Jones [FJ2]. We have already seen in (8.2) and (8.3) that geodesic conjugacy to a model space arises in a natural way as a consequence of various conditions on the unit tangent bundle of a compact negatively curved manifold. One would like to give an affirmative answer to the following:

(9.1) QUESTION. If M_1, M_2 are compact manifolds with $K \leq 0$ whose unit tangent bundles SM_1, SM_2 are geodesically conjugate, then must M_1, M_2 be isometric?

(9.2) **Marked length spectrum.** Before discussing partial answers to the question above we relate geodesic conjugacy to marked length spectrum and then reformulate the question in (9.1).

The *length spectrum* of a compact Riemannian manifold X is the collection of all lengths (counted with multiplicities) of the closed geodesics in X. The *marked length spectrum* of X is a collection of pairs $(\mathscr{C}, \lambda)$ for which the first element of the pair is a free homotopy class $\mathscr{C}$ of closed curves in X and the second element λ is the length of a closed geodesic in X that belongs to $\mathscr{C}$. If X has nonpositive sectional curvature, then a free homotopy class $\mathscr{C}$ of closed curves in X may contain infinitely many distinct closed geodesics, but each of these closed geodesics has the same length $\lambda = \lambda(\mathscr{C})$. If X has negative sectional curvature, then each free homotopy class $\mathscr{C}$ of closed curves in X contains a unique closed geodesic.

Free homotopy classes of closed curves in a compact Riemannian manifold X correspond in a natural way to conjugacy classes of elements in the fundamental group of X. Hence if X_1, X_2 are compact manifolds with isomorphic fundamental groups then any isomorphism $\varphi: \pi_1(X_1) \to \pi_1(X_2)$ induces a bijection φ_* between the free homotopy classes of closed curves in X_1 and those in X_2.

Now let M be compact with $K \leq 0$, and for each free homotopy class $\mathscr{C}$ of closed curves in M let $\lambda(\mathscr{C})$ denote the length of any closed geodesic in M that belongs to $\mathscr{C}$. We say that compact manifolds M_1, M_2 with $K \leq 0$ have the *same marked length spectrum* if there exists an isomorphism $\varphi: \pi_1(M_1) \to \pi_2(M_2)$ such that $\lambda(\mathscr{C}) = \lambda(\varphi_*(\mathscr{C}))$ for any free homotopy class $\mathscr{C}$ of closed curves in M_1.

It is known that if M_1, M_2 are compact manifolds with $K \leq 0$, then the unit tangent bundles SM_1, SM_2 are geodesically conjugate if and only if M_1, M_2 have the same marked length spectrum. Hence we may rephrase the question in (9.1) as follows:

QUESTION. If M_1, M_2 are compact manifolds with $K \leq 0$ that have the same marked length spectrum, then must M_1, M_2 be isometric?

We note that for any integer $g \geq 4$ there exist compact surfaces M_1, M_2 with genus g and $K \equiv -1$ that have the same (unmarked) length spectrum but are not isometric. For compact surfaces with $K \equiv -1$ the unmarked length spectrum determines the Laplace spectrum and vice versa by means of the Selberg Trace Formula. For every integer $g \geq 4$ it is known that there exist pairs of nonisometric compact surfaces M_1, M_2 with genus g, $K \equiv -1$ and the same Laplace spectrum. See for example [**Buser, Sun, Vig**].

(9.3) **Partial results.** The geodesic conjugacy problem as stated above in (9.1) or (9.2) has only beeen solved completely for compact 2-manifolds with $K \leq 0$ by J. P. Otal in [**O1**] and C. Croke in [**Cr**].

THEOREM. *Let M_1 be a compact surface with $K \leq 0$, and let M_2 be another compact surface whose unit tangent bundle is geodesically conjugate to SM_1. Then M_1 and M_2 are isometric.*

In the result above no assumption is made a priori on the curvature of M_2. In Otal's version of the result both surfaces M_1, M_2 are assumed to have strictly negative curvature. The theorem above is a combination of Theorems B and C of Croke in [**Cr**].

In higher dimensions the optimal result occurs when one of the manifolds M_1, M_2 has constant negative curvature.

THEOREM [**Ham4**]. *Let M_1, M_2 be compact, n-dimensional manifolds with $K < 0$ whose unit tangent bundles SM_1, SM_2 are geodesically conjugate. If one of the manifolds M_1, M_2 has constant negative sectional curvature, then M_1 and M_2 are isometric.*

Combining this result with the result of [**BFL**] from (8.3) we obtain

COROLLARY. *Let M be a compact n-dimensional manifold with $K < 0$ whose Anosov foliations in SM are C^∞. Then M is a rank 1, locally symmetric space under any of the following conditions:*
   (1) *the dimension of M is odd.*
   (2) *the sectional curvature of M is strictly $\frac{1}{4}$-pinched.*
   (3) *M has dimension ≤ 4.*

The third case has already been discussed in (8.3). In the first two cases the rank 1 locally symmetric space to which M is geodesically conjugate by [**BFL**] (cf. (8.3B)) must have constant negative sectional curvature (cf. (8.4)).

Recently, P. Foulon and F. Labourie [**FL**] have obtained the following description of asymptotically harmonic manifolds (cf. (7.4)).

THEOREM. *Let M be an asymptotically harmonic, compact, C^∞ manifold of negative sectional curvature. Then the unit tangent bundle SM is geodesically conjugate to the unit tangent bundle SM^* of a compact, locally symmetric manifold M^* of rank 1.*

10. Margulis Lemma and applications

In this section we discuss some results for manifolds of nonpositive curvature coming from commutator estimates for small geodesic loops. All of these results are based on the *Margulis Lemma*. The origin of this lemma goes back to the Bieberbach theorems on flat manifolds and the results of Zassenhaus, Kazhdan-Margulis for symmetric spaces (cf. [**Ra**]). In the case of pinched sectional curvature $-b^2 \leq K \leq -a^2 < 0$ a version of this result was proved by E. Heintze [**Hei2**]. We state the lemma in a form due to Gromov, where we specialize to the case $K \leq 0$ (see [**Gro1**] and [**BK**]).

(10.1) MARGULIS LEMMA. *Let an integer $n \geq 2$ and a positive number b be given. Then there exists a positive number $\mu = \mu(n, b)$ with the following property: Let $\widetilde{M}$ be a complete, simply connected manifold with $-b^2 \leq K \leq 0$, Γ a discrete group of isometries of $\widetilde{M}$, and p a point of $\widetilde{M}$. Let $\Gamma_\mu(p)$ denote the subgroup of Γ generated by those elements φ in Γ with $d_\varphi(p) := d(p, \varphi p) < \mu$. Then $\Gamma_\mu(p)$ is almost nilpotent; that is, $\Gamma_\mu(p)$ contains a nilpotent subgroup of finite index.*

We remark that related results are fundamental for the theory of collapsing Riemannian manifolds (cf. the lectures of Cheeger and Fukaya in this conference). The main consequence of the Margulis Lemma is a nice decomposition of a complete manifold $M = \widetilde{M}/\Gamma$ into a thick and a thin part. If we denote by Inj the injectivity radius function on M, then $\mathrm{Inj}(p) = \min\{\frac{1}{2}d_\varphi(\tilde{p}): \varphi \in \Gamma, \varphi \neq 1 \text{ and } \tilde{p} \in \widetilde{M} \text{ lies above } p\}$. We decompose M into a thin part $M_{\mathrm{thin}} = \{p \in M: \mathrm{Inj}(p) < \mu/2\}$ and a thick part $M_{\mathrm{thick}} = \{p \in M: \mathrm{Inj}(p) \geq \mu/2\}$. The thin part is the set of points $p \in M$ such that for a corresponding $\tilde{p} \in \widetilde{M}$ the group $\Gamma_\mu(\tilde{p})$ is not trivial. By the Margulis Lemma we have an algebraic description of the group $\Gamma_\mu(\tilde{p})$ in this case, and consequently we obtain a geometric description of the metric balls of radius $\mu/2$ whose centers lie in the thin part of M. Locally the geometry is described by $\widetilde{M}/\Delta$, where $\Delta = \Gamma_\mu(\tilde{p})$ is an almost nilpotent group. A detailed study of the action of almost nilpotent groups on $\widetilde{M}$ then yields the following finiteness results:

(10.2) THEOREM. *Let M be a complete n-dimensional manifold with finite volume such that one of the following conditions holds:*
   (1) [**Hei2, Gro6**] $-b^2 \leq K < 0$.
   (2) [**BGS**] $-b^2 \leq K \leq 0$ *and the metric is real analytic.*
Then
   (a) *There exists a positive constant $c_1(n, b)$ depending only on n and b such that $\mathrm{vol}(M) \geq c_1(n, b)$.*
   (b) *M is diffeomorphic to the interior of a compact manifold with boundary.*
   (c) *The topology of M is bounded by the volume; that is; $\sum_i b_i(M) \leq c_2(n, b) \cdot \mathrm{vol}(M)$ where $c_2(n, b)$ is a positive constant depending only on n and b and $\{b_i(M)\}$ denote the Betti numbers with respect to an arbitrary coefficient field.*

REMARK. If $-b^2 \leq K \leq 0$ and M has finite volume but is not real analytic, then there are examples of infinite topological type [**Gro6**].

Structure of ends of manifolds of finite volume. Another application of the Margulis Lemma is a description of the ends of a complete manifold M with finite volume and pinched sectional curvature $-b^2 \leq K \leq -a^2 < 0$.

(10.3) THEOREM [**E6**]. *Let M be complete with finite volume such that $-b^2 \leq K \leq 0$ and the universal cover $\widetilde{M}$ satisfies the visibility axiom. Then M has finitely many ends and any end of M has a neighborhood diffeomorphic to $N \times (0, \infty)$ such that the curves $t \to (p, t)$ are unit speed minimizing geodesic rays intersecting each $N \times \{t\}$ orthogonally. The universal cover $\widetilde{N}$ of N is diffeomorphic to $\mathbb{R}^{n-1}$ and $\pi_1(N)$ is an almost nilpotent group of rank $n - 1$.*

In the case of pinched negative sectional curvature Heintze in [**Hei2**] had previously shown that M has only finitely many ends.

(10.4) **Remarks.** (1) This description of the ends of manifolds of finite volume makes it possible to characterize the finiteness of the volume of M in terms of algebraic properties of $\pi_1(M)$. More precisely, for an integer $n \geq 3$ there is an algebraic condition A_n such that the following holds: let M be a complete, n-manifold such that $-b^2 \leq K \leq -a^2 < 0$. Then M has finite volume if and only if $\pi_1(M)$ satisfies A_n [**Sc3**]. In the case of locally symmetric spaces this property is due to G. Prasad [**Pra1**].

(2) In the case $-b^2 \leq K \leq 0$ a real analytic n-manifold M with finite volume has a much more complicated end structure; see for example the ends of products $M_1 \times M_2$ or locally symmetric spaces with rank at least two. Even in the case of bounded negative curvature $-b^2 \leq K < 0$ there is a great variety of possible end structures for a manifold M with finite volume [**Fu, AbS**].

11. Arithmeticity of lattices in semisimple Lie groups

Let $\widetilde{M}$ be a symmetric space with $K \leq 0$ and no Euclidean de Rham factor, and let $M = \widetilde{M}/\Gamma$ be a quotient manifold of finite volume. Then $G = I_0(\widetilde{M})$ is a semisimple Lie group with trivial center and no compact factors, and $\Gamma \approx \pi_1(M)$ is a lattice in G; that is, Γ is a discrete subgroup of G and the quotient space G/Γ has finite Haar measure. Since the symmetric spaces are the model spaces for the simply connected spaces $\widetilde{M}$ with $K \leq 0$ it is of great interest to describe the various ways in which lattices Γ in G can be constructed and, if possible, to give a complete description. A. Borel in [**Bo**] has shown that every noncompact semisimple Lie group G contains a cocompact lattice, and a slight modification of his proof shows that G also contains a non-cocompact lattice [**Ra**, Theorem 14.1].

(11.1) **Arithmeticity in higher rank groups.** We recall some known facts and examples. See, for example, [**Z**] for further discussion. First, one may essentially reduce the study of semisimple groups to those that arise as algebraic groups over $\mathbb{Q}$.

(11.1A) PROPOSITION [Z, p. 35]. *Let H be a connected semisimple Lie group with trivial center. Then there is a connected semisimple algebraic group $G \subseteq GL(n, \mathbb{C})$ defined over $\mathbb{Q}$ such that as Lie groups H and $G_{\mathbb{R}}^0$ are isomorphic.*

We recall that $G_{\mathbb{R}}$ denotes those elements of $G \subseteq GL(n, \mathbb{C})$ whose entries are real numbers, and $G_{\mathbb{R}}^0$ denotes the connected component of $G_{\mathbb{R}}$ in the Hausdorff topology. $G_{\mathbb{R}}^0$ has finite index in $G_{\mathbb{R}}$.

The following result of Borel-Harish Chandra gives a very useful way to construct lattices.

(11.1B) THEOREM [BHC]. *Let $G \subseteq GL(n, \mathbb{C})$ be a semisimple algebraic group defined over $\mathbb{Q}$. Then $\Gamma = G_{\mathbb{Z}}$ is a lattice in $G_{\mathbb{R}}$, where $G_{\mathbb{Z}}$ denotes the subgroup of $G_{\mathbb{R}}$ whose entries are all integers.*

The result above generalizes the classical example where $G = SL(n, \mathbb{R})$ and $\Gamma = G_{\mathbb{Z}} = SL(n, \mathbb{Z})$. Here are two other ways to construct lattices [Z, Chapter 6].

(1) Let Γ be a lattice in a Lie group G, and let Γ' be a subgroup of G that is *commensurable* with Γ; that is, $\Gamma \cap \Gamma'$ has finite index in both Γ and Γ'. Then Γ' is also a lattice in G.

(2) Let H, G be Lie groups, and let Γ be a lattice in H. If $\varphi: H \to G$ is a surjective homomorphism with compact kernel, then $\varphi(\Gamma)$ is a lattice in G.

DEFINITION. Let G be a connected semisimple Lie group with trivial center and no compact factors. Let $\Gamma \subseteq G$ be a lattice. Then Γ is called *arithmetic* if Γ can be constructed from the three procedures described above. More precisely, Γ is arithmetic if there exist a semisimple $\mathbb{Q}$-group H and a surjective homomorphism $\varphi: H_{\mathbb{R}}^0 \to G$ with compact kernel such that $\varphi(H_{\mathbb{Z}} \cap H_{\mathbb{R}}^0)$ and Γ are commensurable.

In [**Mar4**] G. Margulis proved the following result, a proof of which also may be found in [**Z**].

(11.1C) THEOREM. *Let G be a connected semisimple Lie group with trivial center and no compact factors. Let $\Gamma \subseteq G$ be an irreducible lattice, and assume that G has $\mathbb{R}$-rank at least two. Then Γ is arithmetic.*

If $\widetilde{M}$ is symmetric with $K \leq 0$ and no Euclidean factor, then the rank of $\widetilde{M}$ as a symmetric space equals the $\mathbb{R}$-rank of $G = I_0(\widetilde{M})$.

(11.2) Arithmeticity in $\mathbb{R}$-rank 1 groups. The result of Margulis leaves unanswered the question of the arithmeticity of a lattice $\Gamma \subseteq G = I_0(\widetilde{M})$ when the symmetric space $\widetilde{M}$ has rank 1 (i.e., $K \leq -a^2 < 0$). By the classification of E. Cartan the consideration of rank 1 symmetric spaces $\widetilde{M}$ falls into 4 distinct cases:

(1) $\widetilde{M}$ is the real hyperbolic space with $K \equiv -1$.

(2) $\widetilde{M}$ is the complex hyperbolic space.

(3) $\widetilde{M}$ is the quaternionic hyperbolic space.

(4) $\widetilde{M}$ is the Cayley hyperbolic space.

In case (1) geometric constructions of lattices Γ in $G = SO(n, 1)$ have existed for a long time [**Lö, SW**]. See also [**Best, Mak, Vin1–Vin6**] for other examples. In particular Vinberg in [**Vin1**] studies Coxeter groups in $SO(n, 1)$, and for certain classes of these groups, which he shows to be lattices, he obtains necessary and sufficient criteria for these lattices to be arithmetic. Gromov and Piatetski-Shapiro have shown in [**GPS**] that if $\widetilde{M}$ is the real hyperbolic space of any dimension $n \geq 2$, then $I_0(\widetilde{M}) = SO(n, 1)$ admits nonarithmetic lattices.

In case (2) Mostow in [**Mo1, Mo2**] and Deligne-Mostow in [**DM**] have constructed examples of nonarithmetic lattices in low dimensions. The existence of nonarithmetic lattices in all dimensions remains an open problem. Recently K. Corlette [**Co**] and Gromov-Schoen [**GS**] have used harmonic maps to prove that all lattices in cases (3) and (4) are arithmetic.

REFERENCES

[Al] A. D. Alexandrov, *Die innere Geometrie der konvexe Flächen*, Akademie-Verlag, Berlin, 1955 (German transl. of Russian original, Moscow-Leningrad 1948).

[Anc] A. Ancona, *Negatively curved manifolds, elliptic operators and the Martin boundary*, Ann. of Math. (2) **125** (1987), 495–536.

[And] M. Anderson, *The Dirichlet problem at infinity for manifolds of negative curvature*, J. Differential Geom. **18** (1983), 701–721.

[Ano] D. V. Anosov, *Geodesic flows on closed Riemann manifolds with negative curvature*, Proc. Steklov Inst. Math. **90** (1967).

[AbS] U. Abresch and V. Schroeder, *Graph manifolds, ends of negatively curved spaces and the hyperbolic 120-cell space*, J. Differential Geom. **35** (1992), 299–336.

[AB] S. Alexander and R. Bishop, *The Hadamard-Cartan theorem in locally convex metric spaces*, Enseign. Math. (2) **36** (1990), 309–320.

[ARS] M. Anderson and R. Schoen, *Positive harmonic functions on complete manifolds of negative curvature*, Ann. of Math. (2) **121** (1985), 429–461.

[AVS] M. Anderson and V. Schroeder, *Existence of flats in manifolds of nonpositive curvature*, Invent. Math. **85** (1986), 303–315.

[Ba1] W. Ballmann, *Axial isometries of manifolds of nonpositive curvature*, Math. Ann. **259** (1982), 131–144.

[Ba2] ____, *On the Dirichlet problem at infinity for manifolds of nonpositive curvature*, Forum Math. **1** (1989), 201–213.

[Ba3] ____, *Nonpositively curved manifolds of higher rank*, Ann. of Math. (2) **122** (1985), 597–609.

[Ba4] ____, *Structure of manifolds of nonpositive sectional curvature*, Curvature and Topology of Riemannian Manifolds, Lecture Notes in Math., vol. 1201, Springer-Verlag, 1986, pp. 1–13.

[Ber] M. Berger, *Sur les groupes d'holonomie des variétés riemanniennes*, Bull. Soc. Math. France **83** (1953), 279–330.

[Bes] A. Besse, *Einstein manifolds*, Springer-Verlag, 1987.

[Best] L. A. Best, *On torsion free discrete subgroups of* PSL(2, ℂ) *with compact orbit space*, Canad. J. Math. **23** (1971), 451–460.

[Bo] A. Borel, *Compact Clifford-Klein forms of symmetric spaces*, Topology **2** (1963), 111–122.

[Bus1] H. Busemann, *The geometry of geodesics*, Academic Press, 1955.

[Bus2] ____, *Spaces with nonpositive curvature*, Acta. Math. **80** (1948), 259–310.

[Buser] P. Buser, *Isospectral Riemann surfaces*, Ann. Inst. Fourier (Grenoble) **36** (1986), 167–192.

[Buy] S. V. Buyalo, *Euclidean planes in 3-dimensional manifolds of nonpositive curvature*, Mat. Zametki **43** (1988), 103–114; English transl. in Math. Notes **43** (1988), 60–66.

[BaS] V. Bangert and V. Schroeder, *Existence of flat tori in analytic manifolds of nonpositive curvature*, Ann. Sci. École Norm. Sup. **24** (1991), 605–634.

[BBE] W. Ballmann, M. Brin, and P. Eberlein, *Structure of manifolds of nonpositive curvature.* I, Ann. of Math. (2) **122** (1985), 171–203.

[BBS] W. Ballmann, M. Brin, and R. Spatzier, *Structure of manifolds of nonpositive curvature.* II, Ann. of Math. (2) **122** (1985), 205–235.

[BE] W. Ballmann and P. Eberlein, *Fundamental groups of manifolds of nonpositive curvature*, J. Differential Geom. **25** (1987), 1–22.

[BFL] Y. Benoist, P. Foulon, and F. Labourie, *Flots d'Anosov à distributions stable et instable différentiables*, J. Amer. Math. Soc. **5** (1992), 33–74..

[BGS] W. Ballmann, M. Gromov, and V. Schroeder, *Manifolds of nonpositive curvature*, Progr. Mathematics, vol. 61, Birkhäuser, 1985.

[BHC] A. Borel and Harish-Chandra, *Arithmetic subgroups of algebraic groups*, Ann. of Math. **75** (1962), 485–535.

[BK] P. Buser and H. Karcher, *Gromov's almost flat manifolds*, Astérisque, vol. 81, Soc. Math. France, Paris, 1981.

[BO] R. Bishop and B. O'Neill, *Manifolds of negative curvature*, Trans. Amer. Math. Soc. **145** (1969), 1–49.

[BuS1] K. Burns and R. Spatzier, *On topological Tits buildings and their classification*, Inst. Hautes Études Sci. Publ. Math. **65** (1987), 5–34.

[BuS2] ____, *Manifolds of nonpositive curvature and their buildings*, Inst. Hautes Études Sci. Publ. Math. **65** (1987), 35–59.

[BW] W. Ballmann and M. Wojtkowski, *An estimate for the measure theoretic entropy of geodesic flows*, Ergodic Theory Dynamical Systems **9** (1989), 271–279.

[Ca1] E. Cartan, *Sur une classe remarquable d'espaces de Riemann*, Bull. Soc. Math. France **54** (1926), 214–264; **55** (1927), 114–134.

[Ca2] ____, *La géométrie des groupes de transformations*, J. Math. Pures Appl. **6** (1927), 1–119.

[Ca3] ____, *La géométrie des groupes simples*, Ann. Math. Pura Appl. **4** (1927), 209–256; **5** (1928), 253–260.

[Ca4] ____, *Sur certaines formes riemanniennes remarquables des géométries a groupe fondamental simple*, Ann. Sci. École Norm. Sup. **44** (1927), 345–367.

[CV] S. Cohn-Vossen, *Existenz kurzester Wege*, Dokl. Akad. Nauk SSSR **8** (1935), 339–342.

[Co] K. Corlette, *Archimedean superrigidity and hyperbolic geometry*, Ann. of Math. (2) **135** (1992), 165–182.

[Cr] C. Croke, *Rigidity for surfaces of nonpositive curvature*, Comment. Math. Helv. **65** (1990), 150–169.

[CE] S. Chen and P. Eberlein, *Isometry groups of simply connected manifolds of nonpositive curvature*, Illinois J. Math. **24** (1980), 73–103.

[DM] P. Deligne and G. D. Mostow, *Monodromy of hypergeometric functions and non-lattice integral monodromy*, Inst. Hautes Études Sci. Publ. Math. **63** (1986), 5–90.

[E1] P. Eberlein, *Euclidean de Rham factor of a lattice of nonpositive curvature*, J. Differential Geom. **18** (1983), 209–220.

[E2] ____, *Geodesic flow in certain manifolds without conjugate points*, Trans. Amer. Math. Soc. **167** (1972), 151–170.

[E3] ____, *Geodesic flows on negatively curved manifolds.* II, Trans. Amer. Math. Soc. **178** (1973), 57–82.

[E4] ____, *Geodesic rigidity in compact nonpositively curved manifolds*, Trans. Amer. Math. Soc. **268** (1981), 411–443.

[E5] ____, *Isometry groups of simply connected manifolds of nonpositive curvature*. II, Acta Math. **149** (1982), 41–69.

[E6] ____, *Lattices in manifolds of nonpositive curvature*, Ann. of Math. (2) **111** (1980), 435–476.

[E7] ____, *Manifolds of nonpositive curvature*, Global Differential Geometry, (S. S. Chern, ed.), MAA Stud. Math., vol. 27, pp. 223–258.

[E8] ____, *Rigidity of lattices of nonpositive curvature*, Ergodic Theory Dynamical Systems **3** (1983), 47–85.

[E9] ____, *Symmetry diffeomorphism group of a manifold of nonpositive curvature*. I, Trans. Amer. Math. Soc. **309** (1988), 355–373.

[E10] ____, *Symmetry diffeomorphism group of a manifold of nonpositive curvature*. II, Indiana Math. J. **37** (1988), 735–752.

[E11] ____, *When is a geodesic flow of Anosov type?* I, J. Differential Geom. **8** (1973), 437–463.

[EH] P. Eberlein and J. Heber, *A differential geometric characterization of symmetric spaces of higher rank*, Inst. Hautes Études Sci. Publ. Math. **71** (1990), 33–44.

[EO] P. Eberlein and B. O'Neill, *Visibility manifolds*, Pacific J. Math. **46** (1973), 45–109.

[ES] J. Eells and J. Sampson, *Harmonic mappings of Riemannian manifolds*, Amer. J. Math. **86** (1964), 109–160.

[Fa] K. Falconer, *The geometry of fractal sets*, Cambridge Tracts in Math., vol. 85, Cambridge Univ. Press, 1985.

[Fed] H. Federer, *Geometric measure theory*, Springer-Verlag, 1969.

[Fer] R. Feres, *Geodesic flows on manifolds of negative curvature with smooth horospheric foliations*, Ergodic Th. Dynam. Syst. **11** (1991), 653–686.

[Fu] K. Fugiwara, *A construction of negatively curved manifolds*, Proc. Japan Acad. Ser. A Math. Sci. **64** (1988), 352–355.

[FJ1] T. Farrell and L. Jones, *Negatively curved manifolds with exotic smooth structures*, J. Amer. Math. Soc. **2** (1989), 899–908.

[FJ2] ____, *A topological analogue of Mostow's rigidity theorem*, J. Amer. Math. Soc. **2** (1989), 257–370.

[FK] R. Feres and A. Katok, *Invariant tensor fields of dynamical systems with pinched Lyapunov exponents and rigidity of geodesic flows*, Ergodic Theory Dynamical Systems **8** (1989), 427–432.

[FL] P. Foulon and F. Labourie, *Sur les variétés asymptotiquement harmoniques*, Invent. Math. **109** (1992), 97–111.

[FM] A. Freire and R. Mañé, *On the entropy of the geodesic flow in manifolds without conjugate points*, Invent. Math. **69** (1982), 375–392.

[Gh] E. Ghys, *Flots d'Anosov dont les feuilletages stables sont différentiables*, Ann. Sci. École Norm. Sup. **20** (1987), 251–270.

[Gre] L. Green, *Generalized geodesic flow*, Duke J. Math. **41** (1974), 115–126; correction, **42** (1975), 381.

[Gro1] M. Gromov, *Almost flat manifolds*, J. Differential Geom. **13** (1978), 231–241.

[Gro2] ____, *Filling Riemannian manifolds*, J. Differential Geom. **18** (1983), 1–147.

[Gro3] ____, *Foliated plateau problem*, Geom. Funct. Anal. (to appear).

[Gro4] ____, *Hyperbolic groups*, Essays in Group Theory (S. M. Gersten, ed.), Math. Sci. Res. Inst. Publ., vol. 8, Springer-Verlag, 1987, pp. 75–263.

[Gro5] ____, *Hyperbolic manifolds, groups and actions*, Riemann Surfaces and Related Topics, (I. Kra and B. Maskit, ed.), Annals of Math. Stud. vol. 97, Princeton, Univ. Press, Princeton, NJ, 1981, pp. 183–213.

[Gro6] ____, *Manifolds of negative curvature*, J. Differential Geom. **13** (1978), 223–230.

[Gro7] ____, *Rigid transformation groups*, Géométrie Différentielle (D. Bernard and Y. Choquet-Bruhat, eds.), Travaux en cours 33, Hermann, Paris, 1988, pp. 65–139.

[Gh-Ha] E. Ghys and P. de la Harpe, *Sur les groupes hyperboliques d'apres Mikhail Gromov*, Progr. Math. vol. 83, Birkhäuser, 1990.

[GKM] D. Gromoll, W. Klingenberg, and W. Meyer, *Riemannsche Geometrie in Grossen*, Lecture Notes in Math., vol. 55, Springer-Verlag, 1968.

[GLP] M. Gromov, J. Lafontaine, and P. Pansu, *Structures métriques pour les variétés riemanniennes*, Cedic/Fernand Nathan, Paris, 1981.

[GPS] M. Gromov and I. Piatetski-Shapiro, *Nonarithmetic groups in Lobachevsky spaces*, Inst. Hautes Études Sci. Publ. Math. **66** (1988), 93–103.

[GS] M. Gromov and R. Schoen, *Harmonic maps into singular spaces and p-adic superrigidity for lattices in groups of rank* 1, preprint.

[GT] M. Gromov and W. Thurston, *Pinching constants for hyperbolic manifolds*, Invent. Math. **89** (1987), 1–12.

[GW] D. Gromoll and J. Wolf, *Some relations between the metric structure and the algebraic structure of the fundamental group in manifolds of nonpositive curvature*, Bull. Amer. Math. Soc. **77** (1971), 545–552.

[Ham1] U. Hamenstädt, *Compact manifolds with $\frac{1}{4}$-pinched negative curvature*, in Global Diff. Geom. and Global Anal., Proceedings Berlin 1990, Lecture Notes in Math. 1481, Springer-Verlag (1991) 74–79.

[Ham2] ____, *Entropy rigidity of compact locally symmetric spaces*, Ann. of Math. (2) **131** (1990), 35–51.

[Ham3] ____, *A geometric characterization of compact locally symmetric spaces*, J. Differential Geom. **34** (1991), 193–221.

[Ham4] ____, *Harmonic measures for compact negatively curved manifolds and rigidity*, preprint

[Ham5] ____, *A new description of the Bowen-Margulis measure*, Ergodic Theory Dynamical Systems **9** (1989), 455–464.

[Ham6] ____, *Time preserving conjugacies of geodesic flows*, Ergodic Th. Dyn-Syst. **12** (1992), 67–74.

[Ham7] ____, in preparation.

[Has] B. Hasselblatt, *Regularity of the Anosov splitting and a new description of the Margulis measure*, Ph.D. dissertation, California Inst. of Technology, 1989.

[Heb] J. Heber, *Tits Metrik und geometrischer Rang homogener Räume nicht-positiver Krümmung*, dissertation, University of Augsburg, 1991.

[Hei1] E. Heintze, *On homogeneous manifolds of negative curvature*, Math. Ann. **211** (1974), 23–34.

[Hei2] ____, *Mannigfaltigkeiten negativer Krümmung*, Habilitation, University of Bonn, 1976.

[Hei3] ____, *Compact quotients of homogeneous negatively curved Riemannian manifolds*, Math. Z. **140** (1974), 79–80.

[Her] L. Hernández, *Kähler manifolds and $\frac{1}{4}$-pinching*, Duke J. Math. **62** (1991), 601–611.

[Ho] E. Hopf, *Statistik der geodätischen Linien in Mannigfaltigkeiten negativer Krümmung*, Ber. Verh. Sächs. Akad. Wiss. Leipzig **91** (1939), 261–304.

[HI] E. Heintze and H. C. ImHof, *On the geometry of horospheres*, J. Differential Geom. **12** (1977), 481–491.

[HK] S. Hurder and A. Katok, *Differentiability, rigidity and Godbillon-Vey classes for Anosov flows*, Preprint, revised, 1990.

[HP] M. Hirsch and C. Pugh, *Smoothness of horocycle foliations*, J. Differential Geom. **10** (1975), 225–238.

[HPS] M. Hirsch, C. Pugh, and M. Shub, *Invariant manifolds*, Lecture Notes in Math., vol. 583, Springer-Verlag, 1977.

[J] F. E. A. Johnson, *On the existence of irreducible discrete subgoups in isotypic Lie groups of classical type*, Proc. London Math. Soc. **56** (1988), 51–77.

[Kan1] M. Kanai, *Geodesic flows on negatively curved manifolds with smooth stable and unstable foliations*, Ergodic Theory Dynamical Systems **8** (1988), 215–239.

[Kan2] ____, *Differential geometric studies on dynamics of geodesic and frame flows*, Preprint.

[Kar] F. Karpelevic, *The geometric of geodesics and the eigenfunctions of the Beltrami-Laplace operator on symmetric spaces*, Trans. Moscow Math. Soc. **14** (1965), 51–199.

[Kat1] A. Katok, *Entropy and closed geodesics*, Ergodic Theory Dynamical Systems **2** (1982), 339–367.

[Kat2] ____, *Four applications of conformal equivalence to geometry and dynamics*, Ergodic Theory Dynamical Systems **8** (1988), 139–152.

[Ki1] Y. Kifer, *Brownian motion and harmonic functions on manifolds of negative curvature*, Theory Probab. Appl. **21** (1976), 81–95.

[Ki2] ____, *Brownian motion and positive harmonic functions on complete manifolds of nonpositive curvature*, Res. Notes in Math., vol. 150, Longman, Harlow, 1986, pp. 187–232.

[Ki3] ____, *A lower bound for Hausdorff dimensions of harmonic measures on negatively curved manifolds*, Israel J. Math. **71** (1990), 339–348.

[Km] V. Kaimanovich, *Invariant measures of the geodesic flow and measures at infinity of negatively curved manifolds*, Annales de l'Instit. Henri Poincaré Physique Théorique **53** (1990), 361–393.

[Kn1] G. Knieper, *Spherical means on compact Riemannian manifolds of negative curvature*, Habilitationsschrift, Augsburg, 1992.

[Kn2] ____, *A second derivative formula of the measure theoretic entropy at spaces of constant curvature*, Preprint, 1989.

[KKPW] A. Katok, G. Knieper, M. Pollicott, and H. Weiss, *Differentiability and analyticity of topological entropy for Anosov and geodesic flows*, Invent. Math. **98** (1989), 581–597.

[KL] Y. Kifer and F. Ledrappier, *Hausdorff dimension of harmonic measures on negatively curved manifolds*, Trans. Amer. Math. Soc. **318** (1990), 685–704.

[KW] G. Knieper and H. Weiss, *Regularity of measure theoretic entropy for geodesic flows of negative curvature. I*, Invent. Math. **95** (1989), 579–589.

[Leb] C. LeBrun, *On complete quaternionic-Kähler manifolds*, Duke Math. J. **63** (1991), 723–743.

[Led1] F. Ledrappier, *Ergodic properties of Brownian motion on covers of compact negatively curved manifolds*, Bol. Soc. Brasil Mat. **19** (1988), 115–140.

[Led2] ____, *Harmonic measures and Bowen-Margulis measures*, Israel J. Math. **71** (1990), 275–287.

[Led3] ____, lecture, Penn State Univ., March 1991.

[Lö] F. Löbell, *Beispiele geschlossener dreidimensionaler Clifford-Kleinsche Räume negativer Krümmung*, Bericht Verh. Sächs. Akad. Wiss. Leipzig **83** (1931), 167–174.

[LY] H. B. Lawson and S.-T. Yau, *Compact manifolds of nonpositive curvature*, J. Differential Geom. **7** (1972), 211–228.

[Mag] W. Magnus, *Noneuclidean tesselations and their groups*, Academic Press, 1964, pp. 151–156.

[Mak] V. Makarov, *On a certain class of discrete Lobachevsky space groups with infinite fundamental domain of finite measure*, Soviet Math. Dokl. **7** (1966), 328–331.

[Man1] A. Manning, *Curvature bounds for the entropy of the geodesic flow on a surface*, J. London Math. Soc. **24** (1981), 351–357.

[Man2] ____, *Topological entropy for geodesic flows*, Ann. of Math. (2) **110** (1979), 567–573.

[Mar1] G. Margulis, *Applications of ergodic theory to the investigation of manifolds of negative curvature*, Functional Anal. Appl. **3** (1969), 335–336.

[Mar2] ____, *Arithmeticity of the irreducible lattices in the semisimple groups of rank greater than* 1, Invent. Math. **76** (1984), 93–120.

[Mar3] ____, *Certain measures associated with U-flows on compact manifolds*, Functional Anal. Appl. **4** (1970), 55–67.

[Mar4] ____, *Discrete groups of motions of manifolds of nonpositive curvature*, Amer. Math. Soc. Transl. **109** (1977), 33–45.

[Mar5] ____, *Nonuniform lattices in semisimple algebraic groups*, Lie Groups and Their Representations (I. M. Gelfand, ed.), Wiley, 1975.

[Mo1] G. D. Mostow, *Discrete subgroups of Lie groups*, Elie Cartan et les Mathématiques d'aujourdhui, Astérisque, numéro hors serie Soc. Math. France, Paris, 1985, pp. 289–309.

[Mo2] ____, *On a remarkable class of polyhedra in complex hyperbolic space*, Pacific J. Math. **86** (1980), 171–276.

[Mo3] ____, *Strong rigidity of locally symmetric spaces*, Ann. of Math. Stud., vol. 78, Princeton Univ. Press, Princeton, NJ, 1973.

[O1] J. P. Otal, *Le spectre marqué des surfaces à courbure negative*, Ann. of Math. (2) **131** (1990), 151–162.

[O2] ____, *Sur les longuers des géodésiques d'une métrique a courbure négative dans le disque*, Comment. Math. Helv. **65** (1990), 334–347.

[OS] R. Osserman and P. Sarnak, *A new curvature invariant and entropy of geodesic flows*, Invent. Math. **77** (1984), 455–462.

[Pa1] P. Pansu, *Dimension conforme et sphère à l'infini des variétés à courbure négative*, Ann. Acad. Sci. Fenn. Ser. A I Math. **14** (1989), 177–212.

[Pa2] ____, *Géométrie conforme grossière*, Preprint, 1987.

[Pa3] ____, *Métriques de Carnot-Carathéodory et quasiisométries des espaces symétriques de rang un*, Ann. of Math. (2) **129** (1989), 1–60.

[Pa4] ____, *Quasiconformal mappings and manifolds of negative curvature*, Curvature and Topology of Riemannian Manifolds, Lecture Notes in Math., vol. 1201, Springer-Verlag, 1986, pp. 212–229.

[Pa5] ____, *Quasiisométries des variétés à courbure négative*, Thesis, University of Paris 7, 1987.

[Pe1] Ya. Pesin, *Characteristic Lyapunov indicators and smooth ergodic theory*, Russian Math. Surveys **32** (1977), no. 4, 55–114.

[Pe2] ____, *Formulas for the entropy of a geodesic flow on a compact Riemannian manifold without conjugate points*, Math. Notes **24** (1978), 796–805.

[Pe3] ____, *Geodesic flows on closed Riemannian manifolds without focal points*, Math. USSR-Izv. **11** (1977), 1195–1228.

[Pe4] ____, *Geodesic flows with hyperbolic behavior of the trajectories and objects connected with them*, Russian Math Surveys **36** (1981), no. 4, 1–59.

[Pra1] G. Prasad, *Discrete subgroups isomorphic to lattices in Lie groups*, Amer. J. Math. **98** (1976), 853–863.

[Pra2] ____, *Strong rigidity of Q-rank 1 lattices*, Invent. Math. **21** (1973), 255–286.

[Pre] A. Preissman, *Quelques propriétés des espaces de Riemann*, Comment. Math. Helv. **15** (1942–43), 175–216.

[PR] G. Prasad and M. Raghunathan, *Cartan subgroups and lattices in semisimple Lie groups*, Ann. of Math. (2) **96** (1972), 296–317.

[Ra] M. Raghunathan, *Discrete subgroups of Lie groups*, Springer-Verlag, 1972.

[Ri] W. Rinow, *Die innere Geometrie der metrischen Räume*, Springer-Verlag, 1961.

[Sa] P. Sarnak, *Entropy estimates for geodesic flows*, Ergodic Theory Dynamical Systems, **2** (1982), 513–524.

[Sc1] V. Schroeder, *Codimension one flats in manifolds of nonpositive curvature*, Geom. Dedicata **33** (1990), 251–263.

[Sc2] ____, *Existence of immersed tori in manifolds of nonpositive curvature*, J. Reine Angew. Math. **390** (1988), 32–46.

[Sc3] ____, *Finite volume and fundamental group in manifolds of negative curvature*, J. Differential Geom. **20** (1984), 175–183.

[Sc4] ____, *A splitting theorem for spaces of nonpositive curvature*, Invent. Math. **79** (1985), 323–327.

[Shi] H. Shimizu, *On discontinuous groups operating on the product of upper half planes*, Ann. of Math. (2) **77** (1963), 33–71.

[Shu] M. Shub, *Global stability of dynamical systems*, Springer-Verlag, 1987.

[Si] Y. Sinai, *The asymptotic behavior of the number of closed geodesics on a compact manifold of negative curvature*, Izv. Akad. Nauk SSSR Ser. Mat. **30** (1966), 1275–1295; English transl., Amer. Math. Soc. Transl. **73** (1968), 229–250.

[Sul] D. Sullivan, *The Dirichlet problem at infinity for a negatively curved manifold*, J. Differential Geom. **18** (1983), 723–732.

[Sun] T. Sunada, *Riemannian coverings and isospectral manifolds*, Ann. of Math. (2) **121** (1985), 169–186.

[ST] H. Seifert and W. Threlfall, *Lehrbuch der Topologie*, Teubner, Leipzig-Berlin, 1934, p. 218.

[SW] H. Seifert and C. Weber, *Die beiden Dodekaederräume*, Math. Z. **37** (1933), 237–253.

[Vig] M. Vignéras, *Variétés riemanniennes isospectrales et non isométriques*, Ann. of Math. (2) **112** (1980), 21–32.

[Vil] M. Ville, *On $\frac{1}{4}$-pinched 4-dimensional Riemannnian manifolds of negative curvature*, Ann. Global Anal. Geom. **3** (1985), 329–336.

[Vin1] E. Vinberg, *Discrete groups generated by reflections in Lobachevsky spaces*, Math. USSR-Sb. **1** (1967), 429–444.

[Vin2] ____, *Some examples of crystallographic groups in Lobachevskiĭ spaces*, Math. USSR-Sb. **7** (1969), 617–622.

[Vin3] ____, *Some arithmetical discrete groups in Lobachevskiĭ spaces*, Discrete Subgroups of Lie Groups, Oxford Univ. Press, 1975, pp. 323–348.

[Vin4] ____, *Absence of crystallographic groups of reflections in Lobachevskiĭ spaces of large dimension*, Functional Anal. Appl. **15** (1981), 128–130.

[Vin5] ____, *The absence of crystallographic groups of reflections in Lobachevskiĭ spaces of large dimension*, Trans. Moscow Math. Soc. **47** (1985), 75–112.

[Vin6] ____, *Hyperbolic reflection groups*, Uspekhi Mat. Nauk. **40** (1985), 29–66.

[Wa] P. Walters, *An introduction to ergodic theory*, Graduate Texts in Math., Springer-Verlag, 1982.

[Wo1] J. Wolf, *Homogeneity and bounded isometries in manifolds of negative curvature*, Illinois J. Math. **8** (1964), 14–18.

[Wo2] ____, *Spaces of constant curvature*, 3rd ed., Publish or Perish, Boston, 1974.

[Ya1] S.-T. Yau, *On the fundamental group of compact manifolds of nonpositive curvature*, Ann. of Math. (2) **93** (1971), 579–585.

[Ya2] ____, *Seminar on differential geometry*, Ann. of Math. Stud., vol. 102, Princeton Univ. Press, Princeton, NJ, University of Tokyo Press, Tokyo, 1982, problem section.

[Yu1] C. Yue, *Brownian motion on Anosov foliations, integral formula and rigidity*, Preprint, 1991.

[Yu2] ____, *Contribution to Sullivan's conjecture*, Preprint, 1990.

[Yu3] ____, *Integral formulas for the Laplacian along the unstable foliation and applications to rigidity problems for manifolds of negative curvature*, Erg. Th. and Dyn. Syst. **11** (1991), 803–813.

[Yu4] ____, lecture, Penn State Univ., March 1991.

[YZ] S.-T. Yau and F. Zheng, *Negatively $\frac{1}{4}$-pinched riemannian metric on a compact Kähler manifold*, Invent. Math. **103** (1991), 527–535.

[Z] R. Zimmer, *Ergodic theory and semisimple groups*, Birkhäuser, 1984.

UNIVERSITY OF NORTH CAROLINA AT CHAPEL HILL

UNIVERSITÄT BONN, GERMANY

UNIVERSITÄT FREIBURG, GERMANY

Proceedings of Symposia in Pure Mathematics
Volume **54** (1993), Part 3

Topological Rigidity
for Compact Nonpositively Curved Manifolds

F. T. FARRELL AND L. E. JONES

0. Introduction

In this paper we use the term *closed manifold* to denote a compact connected manifold with empty boundary. A closed manifold M is *aspherical* if $\pi_n M = 0$ for all $n \neq 1$; this is equivalent to the universal cover of M being contractible. A *nonpositively curved manifold* will mean a Riemannian manifold whose sectional curvatures are all less than or equal to zero. It is a consequence of the Cartan-Hadamard theorem that closed nonpositively curved manifolds are aspherical. Locally symmetric spaces of noncompact type and flat Riemannian manifolds are important classes of nonpositively curved manifolds. Recall that a complete Riemannian manifold is locally symmetric if its sectional curvatures are constant under parallel translation. And it has noncompact type if its universal cover does not have a flat or compact metric factor. A complete Riemannian manifold is flat if its sectional curvatures are identically zero. (Real) hyperbolic manifolds are special cases of locally symmetric spaces of noncompact type; namely, they are the ones whose sectional curvatures are all equal to -1. The main aim of this paper is to prove the following result.

TOPOLOGICAL RIGIDITY THEOREM 0.1 (cf. Theorem 3.2). *Let M and N be closed aspherical manifolds such that M is nonpositively curved with $\dim M \neq 3, 4$. (But N is only assumed to be a topological manifold.) If $\pi_1 M$ is isomorphic to $\pi_1 N$, then this isomorphism is induced by a homeomorphism between M and N.*

This result is an analogue of Mostow's rigidity theorem [**39**] which states that if the above assumptions are strengthened by requiring that both M and

1991 *Mathematics Subject Classification*. Primary 53C20, 57R55; Secondary 18F25.
This paper is in final form and no version of it will be submitted for publication elsewhere.
Both authors were supported in part by the National Science Foundation.

N are locally symmetric spaces of noncompact type (and the universal cover of N has no two-dimensional metric factor projecting to a closed subset of N), then there is a stronger conclusion. Namely, the isomorphism between fundamental groups is induced by an isometry of Riemannian manifolds. (Mostow does not require that $\dim M \neq 3, 4$.)

Theorem 0.1 extends earlier results from [15] and [18]. Namely, the conclusion of 0.1 was there shown to hold in the two special cases when M is Riemannian flat [15] and when M is (real) hyperbolic [18]. A version of 0.1 was proven in [18] even for noncompact hyperbolic manifolds of $\dim \neq 3, 4$ as long as they are complete.

REMARK 0.1.1. The wording of results listed in the introduction is slightly different from that used in the main body of the paper. Therefore, we have enumerated them differently and put in parentheses a reference to the result from some later section which directly implies it.

Theorem 0.1 yields a topological characterization of closed locally symmetric spaces of noncompact type.

THEOREM 0.2 (cf. Theorem 3.3). *Let M be a closed (topological) manifold with $m = \dim M \neq 3, 4$. Then M supports the structure of a locally symmetric space of noncompact type if and only if M is aspherical (i.e., $\pi_n M = 0$ for $n \neq 1$) and $\pi_1 M$ is isomorphic to a discrete cocompact subgroup of a (virtually connected) linear semisimple Lie group.*

REMARK 0.2.1. A similar characterization of closed flat Riemannian manifolds (of $\dim \neq 3$) was given a [15].

Let $\operatorname{Top} M$ denote the topological group of all homeomorphisms of M. A slightly extended form of 0.1 (namely, Addendum 2.4) is combined with the main result of [25] to yield the following result.

THEOREM 0.3 (cf. Theorem 3.7). *Let M be a closed nonpositively curved manifold with $m = \dim M > 10$. Then,*

$$\pi_n(\operatorname{Top} M) \otimes \mathbb{Q} = 0$$

provided $2 \leq n \leq (m - 7)/3$. Furthermore, $\pi_0(\operatorname{Top} M)$ is a group extension of $\operatorname{Out}(\pi_1 M)$ by the normal subgroup $\mathbb{Z}_2^\infty$ where $\mathbb{Z}_2^\infty$ denotes the countably infinite direct sum of cyclic groups of order 2 and $\operatorname{Out}(\pi_1 M)$ is the group of all outer automorphisms of $\pi_1 M$. In addition,

$$\pi_1(\operatorname{Top} M) \otimes \mathbb{Q} \quad and \quad \operatorname{Center}(\pi_1 M) \otimes \mathbb{Q}$$

are isomorphic.

REMARK 0.3.1. In the case that M is a hyperbolic manifold, this result was proven in [18]. When M is closed flat Riemannian, the above calculation for $n > 0$ is implicit in [13] except that the necessary stability result (see Lemma 3.11) had not been proven at that time.

REMARK 0.3.2. This paper together with [25] contain the proof of the results announced in [21].

We complete the introduction by outlining the proof of Theorem 0.1. Our intention is to make as transparent as possible the idea of the proof. Hence this outline is a bit heuristic. The precise argument is contained in §§1 and 2. In fact, Theorem 1.7 of §1 is the key result of the paper. We assume that M is orientable. (The nonorientable case is deduced in §2 from a strengthened version of the orientable case via splitting theory.)

OUTLINE OF THE PROOF OF THEOREM 0.1. Let $f\colon N \to M$ be a homotopy equivalence realizing the isomorphism $\pi_1 N \simeq \pi_1 M$. It must be shown that f is homotopic to a homeomorphism. It was shown in [25] that $\mathrm{Wh}(\pi_1 M \oplus A) = 0$ for any finitely generated free abelian group A. So by results from surgery theory [47, 35, 12] it follows that $f\colon N \to M$ is homotopic to a homeomorphism if $f \times \mathrm{id}\colon N \times S^1 \to M \times S^1$ is homotopic to a homeomorphism, where $\mathrm{id}\colon S^1 \to S^1$ is the identity map of the circle. Note that $M \times S^1$ is also a closed nonpositively curved manifold. So to complete the proof of Theorem 0.1 it will suffice to show that $f \times \mathrm{id}\colon N \times S^1 \to M \times S^1$ is homotopic to a homeomorphism when $\dim M$ is odd.

There is the following reduction of this last problem which plays a motivating role in this proof. Let $\pi\colon X \to M \times S^1$ denote a fiber bundle over $M \times S^1$ having for fiber a closed connected oriented $4k$-dimensional manifold with signature equal one, such that $X \to M \times S^1$ is the trivial bundle over the one-skeleton of a triangulation K for $M \times S^1$. Let $Y \to N \times S^1$ denote the pull-back of the bundle $X \to M \times S^1$ along the map $f \times \mathrm{id}$, and let $\hat{f}\colon Y \to X$ denote the induced bundle map. Results from surgery theory [47, 35] imply that $f \times \mathrm{id}\colon N \times S^1 \to M \times S^1$ is homotopic to a homeomorphism if $\hat{f}\colon Y \to X$ is homotopic to a map $\hat{f}_1\colon Y \to X$ which is split over the triangulation K for $M \times S^1$; i.e. for each $\Delta \in K$, $\hat{f}_1$ is in transverse position to $\pi^{-1}(\Delta)$ and $\hat{f}_1\colon \hat{f}_1^{-1}(\pi^{-1}(\Delta)) \to \pi^{-1}(\Delta)$ is a homotopy equivalence.

(Before proceeding with the proof of Theorem 0.1, it seems appropriate to indicate the history of this reduction result. It has its origins in the result which states that for any closed connected oriented $4k$-dimensional manifold L the signature of L and of $L \times \mathbb{CP}^2$ are equal, where $\mathbb{CP}^2$ is 2-dimensional complex projective space. In the context of surgery theory this signature result becomes Wall's periodicity theorem [47] which states that for any surgery normal map $h\colon L' \to L$ the surgery obstructions to completing surgery on $h\colon L' \to L$ and on $h \times \mathrm{id}\colon L' \times \mathbb{CP}^2 \to L \times \mathbb{CP}^2$ are equal. There is also a version of Wall's theorem for surgery classifying spaces [40]. Finally using ideas of Sullivan (cf. [35]) and of Kirby and Siebenmann [35], the space version of Wall's theorem implies the reduction result formulated above for $X \to M \times S^1$ equals the standard projection $M \times S^1 \times \mathbb{CP}^2 \to M \times S^1$. The derivation of this reduction result for more general fiber bundles $X \to M \times S^1$ uses this same set of ideas.)

This reduction result will only be useful in this proof if some aspects of the geometry of M appear in the fiber bundle $X \to M \times S^1$ in a manner which makes splitting $\hat{f}: Y \to X$ over the triangulation K a more accessible problem then showing (directly) that $f \times \text{id}: N \times S^1 \to M \times S^1$ is homotopic to a homeomorphism. A fiber bundle $X \to M \times S^1$ having this desired geometric structure is described in detail below, where it is denoted by $\mathscr{F}M \to M \times S^1$ and where $\hat{f}: Y \to X$ is denoted by $\hat{f}: \mathscr{F}_f \to \mathscr{F}M$. Unfortunately there is the added complication that $\mathscr{F}M$ is not a manifold, but rather is a compact stratified space having the three strata $\mathbb{A}$, $\mathbb{B}$, $\mathbb{T}$ described below. This stratification induces one on $\mathscr{F}_f$ and $\hat{f}: \mathscr{F}_f \to \mathscr{F}M$ preserves the strata. The main technical result of this paper (cf. 0.4) states that there is a strata-preserving homotopy of $\hat{f}$ to a map $\hat{f}_1$ such that $\hat{f}_1$, restricted to each stratum, is split over the triangulation K of $M \times S^1$.

What is needed to complete the proof of Theorem 0.1 is the following modified form of the reduction result previously cited: $f \times \text{id}: N \times S^1 \to M \times S^1$ is homotopic to a homeomorphism if there is a strata-preserving homotopy of $\hat{f}: \mathscr{F}_f \to \mathscr{F}M$ to a map $\hat{f}_1$ such that $\hat{f}_1$, restricted to each stratum, is split over the triangulation K of $M \times S^1$. Such a modified reduction result is implied by the surgery result of [47; 35; 18; Theorem 4.3]. Note that to apply [18; Theorem 4.3] in the present situation it is required that $\mathscr{F}M \to M \times S^1$ be a trivial bundle over the one-skeleton of K (which is a consequence of M being orientable), and that the fiber of $\mathscr{F}M \to M \times S^1$ have dimension congruent to $0 \mod 4$ (which is a consequence of $\dim M$ being odd arranged at the outset of this proof). In fact, the fiber of $\mathscr{F}M \to M \times S^1$ is a compact stratified space having three strata A, B, T (which are respectively the fibers of $\mathbb{A} \to M \times S^1$, $\mathbb{B} \to M \times S^1$, $\mathbb{T} \to M \times S^1$) which satisfy the following properties (cf. [18; §4]): $B \cup T$ is a closed connected oriented $4k$-dimensional $\mathbb{Z}(\frac{1}{2})$-homology manifold having signature equal to one; B is a k-dimensional sphere and $A \cup B$ is a $(k+1)$-dimensional ball which bounds B. Were $B \cup T$ actually a manifold then a splitting of $\hat{f}_1|(\mathbb{B} \cup \mathbb{T})$ over K would be sufficient information to conclude (from the first reduction result) that $f \times \text{id}: N \times S^1 \to M \times S^1$ is homotopic to a homeomorphism. The splitting of $\hat{f}_1|(\mathbb{A} \cup \mathbb{B})$ over K is used to effectively ignore the problems caused by the existence of the singular set B in the $\mathbb{Z}(\frac{1}{2})$-homology manifold $B \cup T$.

This completes the discussion of how the proof of Theorem 0.1 reduces to proving an alternate statement about $\hat{f}: \mathscr{F}_f \to \mathscr{F}M$. We now proceed to formulate this alternate statement Proposition 0.4.

There is a bundle $\mathscr{F}M \to M \times S^1$ whose fiber over a point $(x, \theta) \in M \times S^1$ consists of *unordered pairs* of unit length vectors $\langle u, v \rangle$ tangent to $M \times S^1$ at (x, θ) and satisfying the following two contraints:

    1. if $u \neq v$, then both u and v are tangent to the level surface $M \times \theta$;

    2. if $u = v$, then the projection $\bar{u}$ of u onto $\tau_\theta S^1$ points in a counterclockwise direction (or is 0).

(The manifold $M \times S^1$ has the Cartesian product Riemannian metric and the symbol $\langle a, b \rangle$ denotes an unordered pair of elements. The tangent space of a manifold N at a point $x \in N$ is denoted $\tau_x N$.) The total space $\mathscr{F}M$ is stratified with three strata:

$$\mathbb{B} = \{\langle u, u \rangle | \bar{u} = 0\},$$
$$\mathbb{A} = \{\langle u, u \rangle | \bar{u} \neq 0\},$$
$$\mathbb{T} = \{\langle u, v \rangle | u \neq v\}.$$

Note that $\mathbb{B}$ is the bottom stratum and $\mathscr{F}M - \mathbb{B}$ is the union of two open sets $\mathbb{A}$ and $\mathbb{T}$. Let $f: N \to M$ be a homotopy equivalence inducing the given isomorphism $\pi_1 N \to \pi_1 M$. Let $\mathscr{F}_f \to N \times S^1$ denote the pullback of $\mathscr{F}M \to \mathscr{M} \times S^1$ via $f \times \mathrm{id}: N \times S^1 \to M \times S^1$ and let $\hat{f}: \mathscr{F}_f \to \mathscr{F}M$ be the induced bundle map. Note that the stratification of $\mathscr{F}M$ induces one on $\mathscr{F}_f$ and that $\hat{f}$ preserves strata.

KEY PROPOSITION 0.4. (cf. Theorem 1.7). *There exists a homotopy* $\hat{f}_t$, $t \in [0, 1]$, *with* $\hat{f}_0 = \hat{f}$ *and satisfying the following.*

1. *Each* $\hat{f}_t: \mathscr{F}_f \to \mathscr{F}M$ *preserves strata.*

2. *Over some closed "tublar neighborhood"* $\mathscr{N}_0$ *of* $\mathbb{B}$ *in* $\mathbb{B} \cup \mathbb{T}$, $\hat{f}_t$ *is a homotopy of bundle maps (in particular, each* $\hat{f}_t$ *maps fibers homeomorphically).*

3. *There exists a larger closed "tubular neighborhood"* $\mathscr{N}_1$ *of* $\mathbb{B}$ *in* $\mathbb{B} \cup \mathbb{T}$ *such that* $\hat{f}_1$ *is a homeomorphism over* $\mathbb{B} \cup \mathbb{T} - \mathrm{Int}\,\mathscr{N}_1$ *as well over* $\mathbb{B} \cup \mathbb{A}$.

4. *There exists a triangulation of* $M \times S^1$ *such that* $\hat{f}_1$ *is a split homotopy equivalence over* $\mathscr{N}_1$ *relative to this triangulation; i.e., for each of its simplexes* σ, $\hat{f}_1$ *is transverse to* $\rho^{-1}(\sigma)$ *and* $\hat{f}_1: \hat{f}_1^{-1}(\rho^{-1}(\sigma)) \to \rho^{-1}(\sigma)$ *is a homotopy equivalence. (Here* $\rho: \mathscr{N}_1 \to M \times S^1$ *is the composite of the two bundle projections* $\mathscr{N}_1 \to \mathbb{B}$ *and* $\mathbb{B} \to M \times S^1$.)

OUTLINE OF PROOF OF PROPOSITION 0.4. The proof of 0.4 is a consequence of several applications of both ordinary and foliated topological control theory as developed in [7, 42, 16, 17, 24]. Let $g: M \to N$ be a homotopy inverse of f and let h_t and k_t be homotopies of the composite $f \circ g$ to id_M and $g \circ f$ to id_N, respectively. Because of [14], we may assume that N is a smooth manifold, both h_t and k_t are smooth homotopies, and the maps f and g are also smooth. The crucial point is to construct "good" transfers of the map g and the homotopies h_t, k_t to a map $\hat{g}: \mathscr{F}M \to \mathscr{F}_f$ and homotopies $\hat{h}_t$, $\hat{k}_t$ from $\hat{f} \circ \hat{g}$ to $\mathrm{id}_{\mathscr{F}M}$, and $\hat{g} \circ \hat{f}$ to $\mathrm{id}_{\mathscr{F}_f}$, respectively, so that control theory can be applied to prove 0.4. We proceed to describe what a good transfer is and then indicate how to construct one. The first requirement is that $\hat{g}$, $\hat{h}_t$ and $\hat{k}_t$ be bundle maps covering $g \times \mathrm{id}$, $h_t \times \mathrm{id}$, $k_t \times \mathrm{id}$, respectively. (Here id is the identity map on S^1.) Second, each map

$\hat{g}$, $\hat{h}_t$ and $\hat{k}_t$ should preserve strata. Finally, it is necessary that a certain family $\mathscr{T}$ of paths determined by the lift is sufficiently "shrinkable." A path $\alpha\colon [0,1] \to \mathscr{F}M$ is in $\mathscr{T}$ if either

$$\alpha(t) = \hat{h}_t(\omega) \quad \text{for some } \omega \in \mathscr{F}M, \text{ or}$$

$$\alpha(t) = \hat{f}(\hat{k}_t(\omega)) \quad \text{for some } \omega \in \mathscr{F}_f.$$

(The family $\mathscr{T}$ is called the tracks of the transfer.) Note that each track is contained in a single stratum of $\mathscr{F}M$.

There are control maps $p_i\colon X_i \to Y_i$ $(i = 1, 2, 3, 4)$ where each X_i is a region in $\mathscr{F}M$ and each Y_i $(i = 2, 3, 4)$ is equipped with a foliation $\mathscr{F}_i$ with one-dimensional leaves. The maps p_3, p_4 are fiber bundle projections while p_1 and p_2 are homeomorphisms. The control theorems are applied relative to these control maps so it is necessary that if a track α is contained in X_i, then the composite $p_i \circ \alpha$ is small (when $i > 1$) and skinny with respect to the foliation $\mathscr{F}_i$ (when $i > 1$). This is what is meant by the tracks being shrinkable. We proceed to describe X_i and Y_i. The spaces X_2 and Y_2 are both $\mathbb{A} \cup \mathbb{B}$. While $X_3 = \mathscr{N}_1 - \mathrm{Int}(\mathscr{N}_0)$ and Y_3 is $\mathbb{B}$. (Here $\mathscr{N}_0$ and $\mathscr{N}_1$ are from 0.4 and $\mathrm{Int}(\mathscr{N}_0)$ denotes the interior of $\mathscr{N}_0$.) To describe the remaining spaces, we introduce the *core* $\mathbb{P}$ of the stratum $\mathbb{T}$; it consists of all (unordered) pairs $\langle u, -u \rangle \in \mathbb{T}$. Then $Y_4 = \mathbb{P}$ while X_4 is a closed tubular neighborhood $\mathscr{N}_2$ of $\mathbb{P}$ in $\mathbb{T}$. The spaces $\mathscr{N}_1$ and $\mathscr{N}_2$ are disjoint and X_1 is the closure of $\mathbb{T} - (\mathscr{N}_1 \cup \mathscr{N}_2)$. There is a second pair of (disjoint) closed tubular neighborhoods $\mathscr{N}_1'$ and $\mathscr{N}_2'$ with $\mathscr{N}_1 \subset \mathscr{N}_1'$ and $\mathscr{N}_2' \subset \mathscr{N}_2$ such that Y_1 is the closure of $\mathbb{T} - (\mathscr{N}_1' \cup \mathscr{N}_2')$.

We now describe the foliations $\mathscr{F}_i$ $(i = 2, 3, 4)$. The flow lines of the geodesic flow give the leaves of foliations $\mathscr{F}_2$ and $\mathscr{F}_3$. Note that $\mathbb{A}$ and $\mathbb{B}$ are subspaces of unit sphere tangent bundle $S(M \times S^1)$ of $M \times S^1$ which are left invariant under the geodesic flow g^t. In fact, $\mathbb{B}$ can be canonically identified with $SM \times S^1$ and $\mathbb{P}$ with $PM \times S^1$ where PM is the real projective line bundle associated to the tangent bundle of M. Hence there is a canonical two-sheeted covering map $\mathbb{B} = SM \times S^1 \to PM \times S^1 = \mathbb{P}$ and the leaves of $\mathscr{F}_4$ are the images of the leaves of $\mathscr{F}_3$ under this map.

The control maps themselves are defined via dynamical properties. For example, $p_2 = g^{t_0}$ for a sufficiently large fixed positive real number t_0. Here g^{t_0} denotes the time t_0 diffeomorphism of the total space of the tangent bundle of $M \times S^1$ induced by the geodesic flow. More precisely stated p_2 is g^{t_0} restricted to the invariant subspace $\mathbb{A} \cup \mathbb{B}$. The other three control maps are determined by a variant of what we referred to in [20] and [22] as the *incomplete radial flow* on $\mathbb{T}$ towards its core $\mathbb{P}$. Consider the following construction.

CONSTRUCTION 0.5. Here X denotes the universal cover of M and $\omega = \langle u, v \rangle$ is an arbitrary unordered pair of unit length vectors tangent to X at a common "foot" $x \in X$ subject to the constraint that $u \neq v$. Fix a large positive real number r and let y and z be the foot of $g^r(u)$ and $g^r(v)$, respectively. (Here g^t is geodesic flow on the tangent bundle of X.) Let ℓ be the unique geodesic line passing through both y and z, and let x' denote the unique closest point on ℓ to x. Let $\langle U, -U \rangle$ be the unordered pair of unit length vectors tangent to ℓ at x'. If $u \neq -v$, then this construction determines another unit length vector W with foot x'; namely, W is tangent to the geodesic line connecting x and x', and W points away from x. Note W is defined even when $u = v$.

The control maps p_4 and p_3 are essentially the assignment $\langle u, v \rangle \mapsto \langle U, -U \rangle$ and $\langle u, v \rangle \mapsto \langle W, W \rangle$, respectively. More precisely, if $\langle u_1, v_1 \rangle \in \mathbb{T}$ or $\mathscr{N}_1$, then u_1, v_1 are tangent to the level surface $M \times \theta$ at a common foot (x_1, θ). Let $\langle u, v \rangle$ be a lift of $\langle u_1, v_1 \rangle$ to X at a common foot x. Then $p_4 \langle u_1, v_1 \rangle = \langle U_1, -U_1 \rangle$ or $p_3 \langle u_1, v_1 \rangle = \langle W_1, W_1 \rangle$ where U_1 or W_1 is the image of U or W, respectively, under the differential of the covering projection $X \mapsto M \times \theta$.

To describe the final control map p_1 we must say more about the radial "flow." With respect to the above notation, we can define an incomplete flow $\bar{r}^t(\omega) = \langle u_t, v_t \rangle$, where the common foot x_t of the $\langle u_t, v_t \rangle$ lies on the geodesic segment connecting x to x' at a distance t from x. (We assume that $0 \leq t \leq d(x, x')$ where $d(,)$ is the distance function on X.) The vectors u_t, v_t are the inward pointing tangents to the two geodesic segments of length r whose other endpoints lie on ℓ. This induces an incomplete flow r^t on $\mathbb{T}$ together with a function $\rho : \mathbb{T} \cup \mathbb{B} \to [0, r]$ as follows. Let $\omega_1 = \langle u_1, v_1 \rangle \in \mathbb{T}$ be tangent to the level surface $M \times \theta$ at (x, θ) and $\langle u, v \rangle$ be a lift of $\langle u_1, v_1 \rangle$ to X at a common foot x. Then $\rho(\omega_1) = d(x, x')$ and $r^t(\omega_1) = \langle u'_t, v'_t \rangle$, provided $t \in [0, \rho(\omega_1)]$, where u'_t and v'_t are the images of u_t and v_t, respectively, under the differential of the covering projection $X \to M \times \theta$. Note that $\rho^{-1}(0) = \mathbb{P}$, $\rho^{-1}(r) = \mathbb{B}$ and $r^t(\omega_1) \in \mathbb{P}$ if and only if $t = \rho(\omega_1)$. Furthermore, each "flow" line is contained in a fibre of $p_4 : \mathbb{T} \to \mathbb{P}$ and $p_4(\omega_1) = r^{\rho(\omega_1)}(\omega_1)$.

The tubular neighborhoods $\mathscr{N}_1$, $\mathscr{N}'_1$, $\mathscr{N}_2$, $\mathscr{N}'_2$ are determined from a choice of numbers r_i, $i = 1, 2, 3, 4$, satisfying the inequalities

$$0 < r_1 < r_2 < r_3 < r_4 < r,$$

via the formulas

$$\mathscr{N}'_2 = \rho^{-1}([0, r_1]), \qquad \mathscr{N}_2 = \rho^{-1}([0, r_2]),$$
$$\mathscr{N}'_1 = \rho^{-1}([r_3, r]), \qquad \mathscr{N}_1 = \rho^{-1}([r_4, r]).$$

(Our outline now becomes less accurate. But, it is still a good first approximation to the actual argument in §1.) The control map p_1 is constructed from a choice of a smooth diffeomorphism $\phi : [r_2, r_4] \to [r_1, r_3]$, with $\phi(x) \leq x$ for all $x \in [r_2, r_4]$, via the formula

$$p_1(\omega) = r^t(\omega), \quad \text{with } t = \rho(\omega) - \phi(\rho(\omega)).$$

We are now ready to discuss the construction of the *good* transfers $\hat{g}$, $\hat{h}_t$, $\hat{k}_t$. These are actually determined by constructing their tracks $\mathcal{T}$. Let $\mathcal{T}_1$ be the tracks determined by f, g, h_t, k_t; i.e., a curve $\alpha\colon [0, 1] \to M$ is in $\mathcal{T}_1$ if for all $t \in [0, 1]$ either

$$\alpha(t) = h_t(x) \quad \text{for some } x \in M\,; \text{ or}$$
$$\alpha(t) = f(k_t(y)), \quad \text{for some } y \in N\,.$$

Given $\alpha \in \mathcal{T}_1$ and $\omega = \langle u, v \rangle \in \mathcal{F}M$ with foot $(\alpha(0), \theta) \in M \times S^1$, we associate a lift $\omega\alpha$ of α to a path in $\mathcal{F}M$ covering $t \mapsto (\alpha(t), \theta)$ under the bundle projection $\mathcal{F}M \to M \times S^1$. The set of all these lifts $\omega\alpha$ constitute $\mathcal{T}$. The definition of $\omega\alpha$ depends on where ω is located in $\mathcal{F}M$. We here examine three locations:

1. $\mathbb{A} \cup \mathbb{B}$,
2. $r_2 \le \rho(\omega) \le r_4$,
3. $\rho(\omega) \le r_1$.

The construction in each of the two remaining locations is done by tapering between the constructions in the two adjacent locations. The construction (in locations 1–3) is via what we call a *focal lift* or variant of it. In our earlier papers [18] and [20], we used instead an *asymptotic lift*. The asymptotic lift is fine when M is negatively curved but accomplishes *nothing* when M is flat.

To construct a focal lift, first consider the following situation. Let v be a unit length vector tangent to the Riemannian product $X \times \mathbb{R}$ at $\gamma(0)$ where $\gamma\colon [0, 1] \to X \times \mathbb{R}$ is a curve with $\operatorname{diam}(\gamma) < r$, where $\operatorname{diam} \gamma$ is the diameter of γ. Let x' be the foot of $g^r(v)$. (The number t_0, used to define p_2, is equated with r.) A lift $v\gamma$ to $S(X \times \mathbb{R})$ is given by $v\gamma(t) = v_t$ where v_t is the inward pointing vector tangent to the geodesic segment connecting $\gamma(t)$ to x'. Let $\omega_1 = (v_1, v_1) \in \mathbb{A} \cup \mathbb{B}$, γ be a lift of $t \mapsto (\alpha(t), \theta)$ to the covering space $p\colon X \times \mathbb{R} \to M \times S^1$, and let v be a lift of v_1 with foot $\gamma(0)$. Then

$$\omega_1\alpha(t) = \langle v'_t, v'_t \rangle \quad \text{where } v'_t = dp(v_t)\,.$$

When either $\rho(\omega_1) \le r_1$ or $r_2 \le \rho(\omega_1)$ we refer back to Construction 0.5 to define $\omega_1\alpha$. Let $\gamma\colon [0, 1] \to X$ be a lift of α and let $\omega = \langle u, v \rangle$ be a lift of $\omega_1 = \langle u_1, v_1 \rangle$ with common foot $x = \gamma(0)$. When $\rho(\omega_1) \le r_1$, let $u(t)$, $v(t)$ be the inward pointing tangent vectors to the two geodesic segments of length r connecting $\gamma(t)$ to ℓ. Then $\omega_1\alpha(t) = \langle u'(t), v'(t) \rangle$ where $u'(t)$ and $v'(t)$ are the images of $u(t)$ and $v(t)$, respectively, under the differential of the covering projection $X \to M \times \theta$. When $r_2 \le \rho(\omega_1) \le r_4$, let ℓ_t be the unique geodesic line containing x' satisfying

1. ℓ_t is perpendicular to the geodesic line ℓ'_t which contains both $\gamma(t)$ and x';

2. the tangent space to ℓ_t at x' is in the linear span of the tangent spaces to ℓ'_t and ℓ at x'.

Let $U(t)$, $V(t)$ be the inward pointing unit length vectors tangent to the two geodesic segments of length r connecting $\gamma(t)$ to ℓ_t. Then $\omega_1 \alpha(t) = \langle U'(t), V'(t) \rangle$ where $U'(t)$ and $V'(t)$ are the images of $U(t)$ and $V(t)$, respectively, under the differential of the covering projection $X \to M \times \theta$.

It is not hard to see how the set $\mathscr{T}$ determines *good* transfers $\hat{g}$, $\hat{h}_t$, $\hat{k}_t$ having $\mathscr{T}$ for tracks and that the tracks $\mathscr{T}$ are shrinkable in the sense described above. This is done in §1 (under the heading *Focal Transfers*). We conclude this introduction by briefly indicating now the control theorems are applied to prove 0.4. First independently apply control theorems relative to the maps $p_1 \colon X_1 \to Y_1$ and $p_2 \colon X_2 \to Y_2$. Since there is an upper bound c for the arc lengths of the paths in $\mathscr{T}_1$, we can choose the numbers r_i and the diffeomorphism ϕ so that the numbers $\operatorname{diam}(p_1 \circ \alpha)$ are uniformly small for each $\alpha \in \mathscr{T}$ with $\alpha(0) \in X_1$. Also by picking r large enough and t_0 appropriately relative to r, we have that each path $p_2 \circ \alpha$, where $\alpha \in \mathscr{T}$ and $\alpha(0) \in X_2$, is uniformly as skinny as we like; i.e., is $(\bar{c}, \varepsilon)$ controlled relative to the foliation $\mathscr{F}_2$, where $\bar{c}$ is fixed and ε can be arbitrarily small. (Recall $p_2 = g^{t_0}$ and that a path γ is $(\bar{c}, \varepsilon)$-controlled if there exists a path $\bar{\gamma}$ contained in a leaf L of $\mathscr{F}_2$ such that $d(\gamma(t), \bar{\gamma}(t)) \le \varepsilon$, for all $t \in [0, 1]$, and the diameter of $\bar{\gamma}$, measured in L, is no greater than $\bar{c}$.) By applying control theorems relative to these two control maps, we can construct a homotopy $\hat{f}_t \colon \mathscr{F}_f \to \mathscr{F}M$ satisfying properties 1 and 2 of 0.4 and so that $\hat{f}_1$ is a homeomorphism over $X_1 \cup X_2$. Finally we apply the foliated (and fibered) control theorem independently relative to $p_3 \colon X_3 \to Y_3$ and $p_4 \colon X_4 \to Y_4$. Since these control maps are fiber bundle projections rather than homeomorphisms, structure set properties of the fibers F_3 and F_4, respectively, are important for applying the control theorem. In particular, since $|\mathscr{S}(F_4 \times T^n \operatorname{rel} \partial)| = 1$ for all n, the control theorem applied to p_4 allows us to improve $\hat{f}_1$ so that property 3 of 0.4 holds. (The fiber F_4 is a closed ball.) On the other hand, $F_3 = \mathbb{RP}^{m-1} \times [0, 1]$ (where $\mathbb{RP}^{m-1}$ is real projective $(m-1)$-dimensional space) and there exist integers n with $|\mathscr{S}(\mathbb{RP}^{m-1} \times [0, 1] \times T^n \operatorname{rel} \partial)| > 1$. The control theorem in this situation does not yield that $\hat{f}_1$ can be improved to be a homeomorphism over Y_3; but it does imply that we can satisfy property 4 of 0.4.

REMARK 0.6. Note that the set of all closed leaves (circles) of length less than a fixed constant for each foliation $\mathscr{F}_i$ need not be discrete. (For example, consider the case where M is a flat torus.) Our earlier foliated control theorems in [16] and [17] needed such a fact to be immediately applicable. But the control theorem in [24] does not. Hence it can be applied in the present situation. On the other hand, a close look at the geometry of the space of closed geodesics in M, as was done in [25], allows one also to use the earlier papers [16, 17] to prove Theorem 1.7. (See [25] where the second approach is used in a similar situation.)

We have included an appendix to this paper which is a collection of errata to our earlier paper [18]. None of the errata are particularly deep; but the appendix does rectify some misprints and sloppy arguments in [18].

1. Focal transfer and Theorem 1.7

We start by listing some notations used throughout this section. The letter M denotes a closed nonpositively curved manifold and m is the dimension of M. The total space of the universal cover of M is denoted by X and Γ is the fundamental group of M which we also identify with the group of all deck transformations of $X \to M$; in particular, $X/\Gamma = M$. Let S^1 be the circle of radius one and give $M \times S^1$ the product Riemannian metric. The total space of the universal cover of $M \times S^1$ is $X \times \mathbb{R}$ and $\pi_1(M \times S^1) = \Gamma \times \mathfrak{C}$ where $\mathfrak{C}$ denotes the infinite cyclic group. (We also abbreviate $\Gamma \times \mathfrak{C}$ by $\bar{\Gamma}$.)

The symbol $\langle u, v \rangle$ denotes an unordered pair of elements u, v. With respect to the standard inner product $u \cdot v$ on vectors u, $v \in \mathbb{R}^{m+1}$ we associate several topological spaces. The letter S denotes the unit sphere in $\mathbb{R}^{m+1}$, S^+ its closed "northern hemisphere" and E its equator. Put more precisely

$$S = \{u \in \mathbb{R}^{m+1} \mid |u| = 1\},$$
$$S^+ = \{u = (x_1, \ldots, x_{m+1}) \in S \mid x_{m+1} \geq 0\},$$
$$E = \{(x_1, x_2, \ldots, x_m, 0) \in S\}.$$

We are especially interested in the space $\mathscr{F}$ defined by

$$\mathscr{F} = \{\langle u, v \rangle \mid u, v \in S^+, \text{ and either } u = v \text{ or both } u, v \in E\}.$$

The space $\mathscr{F}$ has a stratification with three strata B, A and T defined by

$$B = \{\langle u, u \rangle \mid u \in E\},$$
$$A = \{\langle u, u \rangle \mid u \in S^+ - E\},$$
$$T = \{\langle u, v \rangle \mid u \neq v\}.$$

We refer to B, T and A as the bottom, top and auxiliary stratum, respectively. Note that $B \cup T$ and $B \cup A$ are closed (stratified) subspaces of $\mathscr{F}$ which are denoted by F and D, respectively. The top stratum contains an interesting subspace P, called its *core*, defined by

$$P = \{\langle u, -u \rangle \mid u \in E\}.$$

Note that P is homeomorphic to real projective space $\mathbb{R}P^{m-1}$.

Identify the orthogonal group $O(m)$ with a subgroup of $O(m+1)$ via the matrix representation

$$A \mapsto \begin{pmatrix} A & O \\ O & 1 \end{pmatrix}.$$

Then $O(m)$ acts on $\mathscr{F}$ by

$$A\langle u, v \rangle = \langle Au, Av \rangle$$

where $A \in O(m)$ and $\langle u, v \rangle \in \mathscr{F}$. This action leaves invariant each stratum of $\mathscr{F}$ as well as the core P. Also note that the natural action of $O(m)$ on S (via $O(m) \subset O(m+1)$) leaves invariant both E and S^+ and hence induces actions on these spaces.

Given a real vector bundle η with base space $\mathscr{B}$, structure group $O(m)$ and fiber $\mathbb{R}^m$, we can form its associated bundle $\mathscr{F}\eta$ with fiber $\mathscr{F}$. Also there are associated bundles $S^+\eta$ and $E\eta$ with fibers S^+ and E, respectively. (We will denote a bundle and its total space with the same symbol when no confusion can result. Also η_x denotes the fiber of the bundle η over the base point x.) We can explicitly describe the fibers $E\eta_x$, $S^+\eta_x$ and $\mathscr{F}\eta_x$ by

$$E\eta_x = \{u \in \eta_x | \; |u| = 1\},$$
$$S^+\eta_x = \{u \in \eta_x \times \mathbb{R} | \; |u| = 1 \text{ and } u = (v, t) \text{ with } t \geq 0\},$$
$$\mathscr{F}\eta_x = \{\langle u, v \rangle | u, v \in S^+\eta_x, \text{ and either } u = v \text{ or both } u, v \in E\eta_x\}.$$

(In this description, we have identified η_x with $\eta_x \times 0$ in $\eta_x \times \mathbb{R}$.) Since each stratum B, A, T and the core P is left invariant by $O(m)$, we also have associated bundles $B\eta$, $A\eta$, $T\eta$ which give a stratification to $\mathscr{F}\eta$ and a *core* $P\eta$ to the top stratum $T\eta$. Note that we can explicitly describe the fibers $P\eta_x$ by

$$P\eta_x = \{\langle u, -u \rangle | u \in \eta_x \text{ and } |u| = 1\}.$$

There are also bundles $F\eta$ and $D\eta$ associated to the $O(m)$-invariant subspaces F and D of $\mathscr{F}$. Note that $B\eta$ and $D\eta$ can be canonically identified with $E\eta$ and $S^+\eta$, respectively. Consider the special case where η is the pull back of τM—the tangent bundle of M—via the canonical projection of $M \times S^1$ onto its first factor; i.e., $\eta = (\tau M) \times S^1$. In this case, we abbreviate our notation as follows:

$$\mathscr{F}M = \mathscr{F}\eta, \quad \mathbb{B} = B\eta, \quad \mathbb{A} = A\eta, \quad \mathbb{T} = T\eta,$$
$$\mathbb{F} = F\eta, \quad \mathbb{D} = D\eta, \quad \mathbb{P} = P\eta.$$

Let N be a second closed manifold with $\dim N = \dim M = m$ and let $f: N \to M$ be a homotopy equivalence. Consider $f \times \mathrm{id}: N \times S^1 \to M \times S^1$ and construct the pullback bundle $(f \times \mathrm{id})^* \eta$, where $\eta = (\tau M) \times S^1$. Denote this bundle by η' and the associated bundles, with base $N \times S^1$, as follows:

$$\mathscr{F}_f = \mathscr{F}\eta', \quad \mathbb{B}_f = B\eta', \quad \mathbb{A}_f = A\eta', \quad \mathbb{T}_f = T\eta',$$
$$\mathbb{F}_f = F\eta', \quad \mathbb{D}_f = D\eta', \quad \mathbb{P}_f = P\eta'.$$

Note that each of these bundles is canonically identified with the pullback via $f \times \mathrm{id}$ of the corresponding bundle $\mathscr{F}M$, $\mathbb{B}$, $\mathbb{A}$, $\mathbb{T}$, $\mathbb{F}$, $\mathbb{D}$, $\mathbb{P}$, respectively. In each case let $\hat{f}$ denote the induced bundle map between total spaces covering $f \times \mathrm{id}$. This map is, of course, strata and core preserving. We next

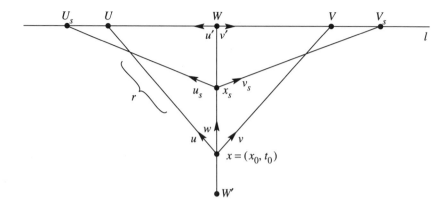

FIGURE 1

construct a pair of fiber bundle projections

(1.0.1) $p: \mathbb{T} \to \mathbb{P}$ and $q: (\mathbb{F} - \mathbb{P}) \to \mathbb{B}$.

The fiber of p is homeomorphic to $\mathbb{R}^{m-1}$ while the fiber of q is homeomorphic to $c(\mathbb{R}P^{m-2})$—the *open* cone on $\mathbb{R}P^{m-2}$. The constructions of p and q use the fact that M is nonpositively curved and depend on an a priori choice of a positive constant r called the *focusing radius*. (Usually r will be chosen very large depending on $f: N \to M$.)

Consider the diagram above where u and v are unit length vectors tangent to the level surface $X \times t_0$ of $X \times \mathbb{R}$ at the point $x = (x_0, t_0) \in X \times \mathbb{R}$. We assume that $u \neq v$ and $u \neq -v$. The points U and V in Figure 1 are constructed by traveling along the geodesic α_u and α_v, respectively, the focusing radius r. (We use the notation α_w for the unique geodesic in $X \times \mathbb{R}$ such that $\dot{\alpha}_w(0) = w$ where $w \in \tau(X \times \mathbb{R})$.) That is,

$$U = \alpha_u(r)   \text{and}   V = \alpha_v(r).$$

Note that $x = \alpha_u(0) = \alpha_v(0)$. The geodesic line ℓ is the unique one passing through both U and V. (Note $U \neq V$ since $u \neq v$.) The point W is the (unique) closest point on ℓ to x. Note that x itself is not on ℓ since $u \neq -v$. Hence there is a unique unit length vector w tangent to $X \times t_0$ at x such that the geodesic α_w passes through W in positive time. Notice that this entire construction is contained in the totally geodesic level surface $X \times t_0$ of $X \times \mathbb{R}$. Note that W lies strictly between U and V on ℓ. Let u' and v' be the unit length vectors tangent to ℓ at W and pointing towards U and V, respectively. (Clearly, $u' = -v'$.)

In the case where η is the pullback of τX via the canonical projection $X \times \mathbb{R} \to X$, we denote the corresponding bundles with base $X \times \mathbb{R}$ by $\mathscr{F}X$, $\overline{\mathbb{B}}$, $\overline{\mathbb{A}}$, $\overline{\mathbb{T}}$, $\overline{\mathbb{F}}$, $\overline{\mathbb{D}}$, and $\overline{\mathbb{P}}$, respectively. These bundles can also be viewed as the pullbacks via the canonical covering projection $X \times \mathbb{R} \to M \times S^1$ of the bundle $\mathscr{F}M$, etc., respectively. Now we define fiber bundle projections

$$\bar{p}\colon \overline{\mathbb{T}} \to \overline{\mathbb{P}} \quad \text{and} \quad \bar{q}\colon (\overline{\mathbb{F}} - \overline{\mathbb{P}}) \to \overline{\mathbb{B}}$$

by

(1.0.2)

$$\bar{p}(\langle u, v\rangle) = \begin{cases} \langle u', v'\rangle & \text{if } u \neq -v, \\ \langle u, v\rangle & \text{if } u = -v, \end{cases}$$

$$\bar{q}(\langle u, v\rangle) = \begin{cases} \langle w, w\rangle & \text{if } u \neq v, \\ \langle u, v\rangle & \text{if } u = v. \end{cases}$$

Since the constructions in Figure 1 are equivariant with respect to the group $\overline{\Gamma}$ of all deck transformations of $X \times \mathbb{R} \to M \times S^1$, we see that $\bar{p}$ and $\bar{q}$ are also $\overline{\Gamma}$-equivariant. Hence they induce the desired bundle projections p and q.

The bundle $p\colon \mathbb{T} \to \mathbb{P}$ is a smooth bundle with zero section $\mathbb{P} \subseteq \mathbb{T}$ and can be identified (as a smooth bundle with zero section) with an open ball sub-bundle $\mathscr{L}\mathscr{P}^0(r)$ in a vector bundle $\mathscr{L}\mathscr{P}$ over $\mathbb{P}$. The total space of $\mathscr{L}\mathscr{P}$ consists of all pairs (L, v) where L and v are a line (passing through 0) and a vector, respectively, which both lie in a common (but arbitrary) fiber of $\tau M \times S^1$ and such that v is perpendicular to L. The point $(L, v) \in \mathscr{L}\mathscr{P}^0(r)$ if and only if $|v| < r$. We construct this identification later in §1. (See the paragraph following formula 1.25 and the one containing Figure 9.)

The bundle $\bar{p}\colon \overline{\mathbb{T}} \to \overline{\mathbb{P}}$ has additionally a $\overline{\Gamma}$-equivariant open cone structure (of radius r) inducing a cone structure on $p\colon \mathbb{T} \to \mathbb{P}$. This is also seen by examining Figure 1. Consider the geodesic segment $[W', W]$ in $X \times \mathbb{R}$ of length r and containing x. It determines the cone line through $\langle u, v\rangle$ in $\overline{\mathbb{T}}$ of "length" r which terminates at $\bar{p}(\langle u, v\rangle)$ as follows. The point $\langle u_s, v_s\rangle$ on this line of "distance" s, $0 < s < r$, from $\overline{\mathbb{P}}$ is based at $x_s \in X \times \mathbb{R}$ where x_s is on $[W', W]$ and its distance, in $X \times \mathbb{R}$, from W is s. There are unique points U_s, V_s on ℓ of distance r from x_s in $X \times \mathbb{R}$ and such that U_s, U and V_s, V (respectively) lie in the same component of $\ell - W$. The unit length vector u_s is tangent to the geodesic segment $[x_s, U_s]$ at x_s and points towards U_s. Likewise v_s is tangent to $[x_s, V_s]$ at x_s and points towards V_s. Under the identification of $p\colon \mathbb{T} \to \mathbb{P}$ with the vector bundle $\mathscr{L}\mathscr{P}$ of the last paragraph, these cone lines map bijectively to the straight line rays emanating from zero in the vector space fibers.

ASIDE A. This cone structure on $p\colon \mathbb{T} \to \mathbb{P}$ induces the *incomplete radial flow* r^t on $\mathbb{T}$ which was referred to in §0 and used importantly in [20]. But in proving Theorem 1.7, the map $\hat{\phi}_\sigma$ plays a crucial role. This map $\hat{\phi}_\sigma$ can be constructed from r^t via formula (1.3.1). But its most convenient definition (1.2.10) does not explicitly use r^t. We first define $\bar{r}^t$ on $\overline{\mathbb{T}}$ by

(1.0.3) $$\bar{r}^t(\langle u, v\rangle) = \langle u_s, v_s\rangle \quad \text{where } t + s = d(x, W)$$

provided $t \in [0, d(x, W)]$. Since $\bar{r}^t$ is $\overline{\Gamma}$-equivariant, we induce r^t on

$\mathbb{T}$ from it. (Note that if $\langle u, v \rangle \notin \overline{\mathbb{P}}$, definition (1.0.3) is meaningful for $t \in (d(x, W) - r, d(x, W)]$.)

ASIDE B. We comment in passing that the bundle $q: (\mathbb{F} - \mathbb{P}) \to \mathbb{B}$ is equivalent to the fiberwise open cone on the bundle over $\mathbb{B}$ with fiber $\mathbb{R}P^{m-2}$ whose total space consists of all pairs (v, L) based at a common point $x = (x_0, t_0)$ in $M \times S^1$, where v is a *unit* length vector and L is a line (through 0) in the tangent space to $M \times t_0$ at (x_0, t_0). Furthermore, L must be perpendicular to v. But we will not need this particular formulation of the bundle q.

A (*level*) *path* α in $M \times S^1$ (or $X \times \mathbb{R}$) is a continuous function α with domain $[0, 1]$ whose image is contained in a level surface $M \times s_0$ on $M \times S^1$ (or $X \times \mathbb{R}$); i.e., the S^1 (or $\mathbb{R}$) coordinate of $\alpha(t)$ is constant as t varies. We will sometimes supress the word "level" when it is obvious from the context. We proceed to describe a few specific ways of lifting α to $\mathbb{F}$ (or $\overline{\mathbb{F}}$). Each lift depends on first fixing a focal radius $r > 0$ and an "initial condition" $\omega = \langle u, v \rangle$ in $\mathbb{F}$ (or $\overline{\mathbb{F}}$) based at $\alpha(0)$. We denote the lifted path by $(\omega\alpha)_i$ where i is a positive integer ($i = 1, 2, 3, 4, 5$) denoting the type of the lift. The lifts $(\omega\alpha)_i$, $i < 5$, are only defined when ω is in a certain region of $\mathbb{F}$ which depends on the index i. The lift $(\omega\alpha)_5$ is a splicing together of the lifts $(\omega\alpha)_i$, $i < 5$, and is defined for arbitrary $\omega \in \mathbb{F}$. We will then construct the *focal lift* $\omega\alpha$, for $\omega \in \mathscr{F}M$, by extending the definition of $(\omega\alpha)_5$ from $\mathbb{F}$ to the rest of $\mathscr{F}M$. This lift is used later to construct the *focal transfer*. The construction of each $(\omega\alpha)_i$, ($i = 1, 2, 3, 4, 5$) will be done $\overline{\Gamma}$-equivariantly for level curves in $X \times \mathbb{R}$ thus inducing the construction for level curves in $M \times S^1$.

Type 1 lifts. We must assume that $u \neq v$ and $u \neq -v$; i.e., $\omega \in \mathbb{T} - \mathbb{P}$. (Other constraints on ω will be listed presently.) The lift $(\omega\alpha)_1$ is determined by looking at Figure 2 in which $\alpha(0) = x = (x_0, t_0)$. The points U, V, W and the line ℓ are determined as in Figure 1. We will $\overline{\Gamma}$-equivariantly construct $(\omega\alpha)_1$ for the case α is a level curve in $X \times \mathbb{R}$, thus determining the construction in the case α is in $M \times S^1$. Provided $\alpha(t) \notin \ell$, there is a unique geodesic line ℓ_t containing W and satisfying

(1.0.4)
1. the geodesic segment $[\alpha(t), W]$ meets ℓ_t perpendicularly, and
2. the tangent space to ℓ_t at W is in the span of the tangent spaces of ℓ and $[W, \alpha(t)]$ at W.

Provided $d(\alpha(t), W) < r$, there is a unique (unordered) pair of (distinct) points $\langle U_t, V_t \rangle$ on ℓ_t such that

(1.0.5) $d(U_t, \alpha(t)) = d(V_t, \alpha(t)) = r$.

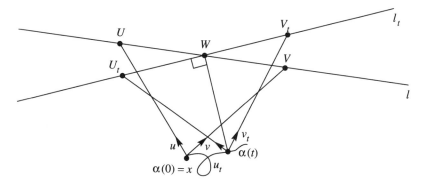

FIGURE 2

Let u_t, v_t be the unit length vectors tangent to the geodesic segments $[\alpha(t), U_t]$, $[\alpha(t), V_t]$, respectively, at $\alpha(t)$ and pointing towards U_t, V_t. We define the first lifting of α by

$$(1.0.6) \qquad (\omega\alpha)_1(t) = \langle u_t, v_t \rangle .$$

Note that $(\omega\alpha)_1$ is a continuous path in $\mathbb{T}$ and is certainly defined when

$$(1.0.7) \qquad \operatorname{diam}(\alpha) < d(\alpha(0), l) < r - \operatorname{diam}(\alpha)$$

where $\operatorname{diam}(\alpha)$ denotes the *real* diameter of α.

DEFINITION 1.1. Associated to a curve α in $M \times S^1$ are two real numbers $\operatorname{Diam}\alpha$ and $\operatorname{diam}\alpha$—called the *nominal diameter* and *real diameter* of α; respectively. The first is the usual notion of diameter; i.e.,

$$\operatorname{Diam}\alpha = \sup\{d(\alpha(s), \alpha(t)) | s, t \in [0, 1]\} .$$

(It is also defined for curves in $X \times \mathbb{R}$.) The second is defined by lifting α to a curve $\tilde{\alpha}$ in $X \times \mathbb{R}$; then

$$\operatorname{diam}\alpha = \operatorname{Diam}\tilde{\alpha} .$$

Note that $\operatorname{Diam}\tilde{\alpha}$ is independent of the lift $\tilde{\alpha}$ since $\overline{\Gamma}$ is a group of isometries of $X \times \mathbb{R}$; consequently, $\operatorname{diam}\alpha$ is well defined.

Type 2 lifts. We now assume that $\omega \in \mathbb{T}$ and $d(\alpha(0), \ell) < r - \operatorname{diam}(\alpha)$. (Note that ℓ is defined even when $\omega \in \mathbb{P}$.) Look at Figure 3 on next page. (Recall the construction will again be done $\overline{\Gamma}$-equivariantly for level curves in $X \times \mathbb{R}$ thus inducing the construction for level curves in $M \times S^1$.) The line ℓ and the points U, V are defined as in Figures 1 and 2. Since $d(\alpha(t), \ell) < r$, there is a unique (unordered) pair of distinct points $\langle U_t, V_t \rangle$ lying on ℓ such that $d(\alpha(t), U_t) = d(\alpha(t), V_t) = r$. As before, let u_t, v_t be the inward pointing unit length tangent vectors to the geodesic segments $[\alpha(t), U_t]$, $[\alpha(t), V_t]$ at $\alpha(t)$. The second lift of α is defined by the formula

$$(1.1.1) \qquad (\omega\alpha)_2(t) = \langle u_t, v_t \rangle .$$

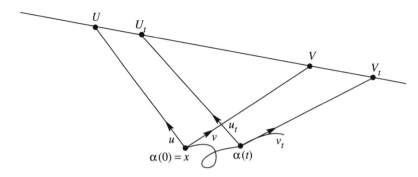

FIGURE 3

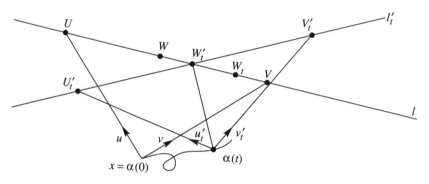

FIGURE 4

Type 3 lifts. These lifts depend on a choice of positive real numbers $r_2 > r_1$, and $(\omega\alpha)_3$ is constructible provided $r_1 > \text{diam}(\alpha)$ and $d(\alpha(0), \ell) < r - \text{diam}(\alpha)$. The construction involves splicing together the two lifts $(\omega\alpha)_1$ and $(\omega\alpha)_2$ depending on $d(\alpha(0), \ell)$. For instance,

$$(1.1.2) \qquad (\omega\alpha)_3 = \begin{cases} (\omega\alpha)_1 & \text{if } d(\alpha(0), \ell) \geq r_2, \\ (\omega\alpha)_2 & \text{if } d(\alpha(0), \ell) \leq r_1. \end{cases}$$

Assume now that $r_1 \leq d(\alpha(0), \ell) \leq r_2$; then the splicing uses Figure 4. In this figure U, V, W and ℓ are as in Figure 1 and W_t denotes the unique closest point on ℓ to $\alpha(t)$. The point W_t' on the geodesic segment $[W_t, W]$ divides it in the same proportions as $d(\alpha(0), \ell)$ divides $[r_1, r_2]$. The line ℓ_t' is the unique geodesic line containing W_t' and satisfying

$(1.1.3)$

    1. the geodesic segment $[\alpha(t), W_t']$ meets ℓ_t' perpendicularly, and

    2. the tangent space to ℓ_t' at W_t' is in the span of the tangent spaces of ℓ and $[W_t', \alpha(t)]$ at W_t'.

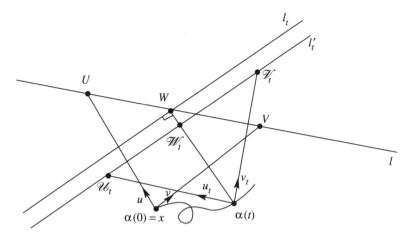

FIGURE 5

There is a unique (unordered) pair of distinct points $\langle U_t', V_t' \rangle$ on ℓ_t' such that

(1.1.4) $$d(U_t', \alpha(t)) = d(V_t', \alpha(t)) = r.$$

Let u_t', v_t' be the inward pointing (unit length) tangent vectors to $[\alpha(t), U_t']$, $[\alpha(t), V_t']$, respectively, at $\alpha(t)$. We complete the definition of the type 3 lift by the formula

(1.1.5) $$(\omega\alpha)_3(t) = \langle u_t', v_t' \rangle, \quad \text{provided } r_1 \le d(\alpha(0), \ell) \le r_2.$$

Type 4 lifts. The symbols U, V, W, ℓ and ℓ_t have the same significance as in Figure 2. We now assume that $d(\alpha(0), \ell) > \operatorname{diam}(\alpha)$. To define the lift $(\omega\alpha)_4$ consider Figure 5. The point $\mathscr{W}_t$ is on the geodesic ray emanating from $\alpha(t)$ which contains W and is uniquely defined by requiring that $d(\mathscr{W}_t, \alpha(t)) = d(W, \alpha(0))$. The geodesic line ℓ_t' is the parallel translate of ℓ_t along the geodesic segment $[W, \mathscr{W}_t]$ to $\mathscr{W}_t$. The unordered pair of points $\langle \mathscr{U}_t, \mathscr{V}_t \rangle$ on ℓ_t' is uniquely defined by requiring that $d(\mathscr{U}_t, \alpha(t)) = d(\mathscr{V}_t, \alpha(t)) = r$. The vectors u_t, v_t are the inward pointing unit length tangents at $\alpha(t)$ to the geodesic segments $[\alpha(t), \mathscr{U}_t], [\alpha(t), \mathscr{V}_t]$, respectively. Define the type 4 lift of α by the formula

(1.1.6) $$(\omega\alpha)_4(t) = \langle u_t, v_t \rangle.$$

REMARK 1.2. The type 4 lifts extend uniquely to initial conditions $\omega \in \mathbb{B}$ (provided $r > \operatorname{diam}(\alpha)$) as follows. Let $\omega = \langle u, u \rangle$ and consider Figure 6 on next page. The point W is uniquely determined by the two conditions
  1. $d(\alpha(0), W) = r$,
  2. u is the inward pointing tangent to the geodesic segment $[\alpha(0), W]$.

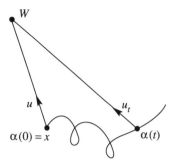

FIGURE 6

Let u_t be the inward pointing unit length tangent vector to the geodesic segment $[\alpha(t), W]$. Then $(\omega\alpha)_4$ is defined by

$$(\omega\alpha)_4(t) = \langle u_t, u_t \rangle, \quad \text{when } \omega = \langle u, u \rangle.$$

Type 5 lifts. We now splice together types 3 and 4 lifts to get lifts $(\omega\alpha)_5$ defined for all $\omega \in \mathbb{F}_{\alpha(0)}$ (or in $\overline{\mathbb{F}}_{\alpha(0)}$) provided $2\operatorname{diam}(\alpha) < r$. The construction depends on the choice of two more parameters r_3 and r_4 satisfying the inequalities

(1.2.1) $(r - \operatorname{diam}(\alpha)) > r_4 > r_3 > r_2 > r_1 > \operatorname{diam}(\alpha)$.

The tapering of $(\omega\alpha)_3$ to $(\omega\alpha)_4$ depends on the number $d(\alpha(0), \ell)$. In particular,

(1.2.2) $(\omega\alpha)_5(t) = \begin{cases} (\omega\alpha)_3(t), & \text{if } d(\alpha(0), \ell) \leq r_3, \\ (\omega\alpha)_4(t), & \text{if } d(\alpha(0), \ell) \geq r_4 \text{ (or } \omega \in \mathbb{B}). \end{cases}$

We now assume that $d(\alpha(0), l) \in [r_3, r_4]$. The tapering is determined by considering Figure 7 in which U, V, W, ℓ, ℓ_t and $\mathcal{W}_t$ have the same signification as in Figure 5. The point $\mathcal{W}_t^{\#}$ is on the geodesic segment $[W, \mathcal{W}_t]$ and divides it in the same proportions as the number $d(\alpha(0), \ell)$ divides the interval $[r_3, r_4]$. The parallel translate of ℓ_t along $[W, \mathcal{W}_t]$ to $\mathcal{W}_t^{\#}$ is the geodesic line denoted $\ell_t^{\#}$. The unordered pair $\langle \mathcal{U}_t^{\#}, \mathcal{V}_t^{\#} \rangle$ of distinct points of $\ell_t^{\#}$ is defined by

(1.2.3) $d(\mathcal{U}_t^{\#}, \alpha(t)) = d(\mathcal{V}_t^{\#}, \alpha(t)) = r$.

The vectors $u_t^{\#}$ and $v_t^{\#}$ are the inward pointing unit length vectors tangent at $\alpha(t)$ to the geodesic segments $[\alpha(t), \mathcal{U}_t^{\#}]$ and $[\alpha(t), \mathcal{V}_t^{\#}]$, respectively. The definition of $(\omega\alpha)_5$ is completed by the formula

(1.2.4) $(\omega\alpha)_5(t) = \langle u_t^{\#}, v_t^{\#} \rangle$, \quad if $d(\alpha(0), \ell) \in [r_3, r_4]$.

Note that $(\omega\alpha)_5(t) \in \mathbb{B}$ if and only if $\omega \in \mathbb{B}$ and that if $(\omega\alpha)_5(t) \in \mathbb{P}$, then $d(\alpha(0), \ell) \leq r_1$. There is also the following useful identity:

(1.2.5) $q \circ (\omega\alpha)_5 = (q(\omega)\alpha)_5$, \quad provided $d(\alpha(0), \ell) \geq r_2$.

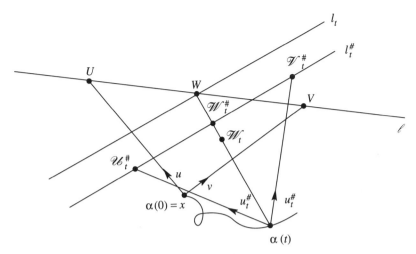

FIGURE 7

Before defining the focal lift $\omega\alpha$, let us first develop some metric properties of the lifts $(\omega\alpha)_i$. The total space of the bundle $\mathscr{LP}$ (described shortly after formula (1.0.2)) will be used as a control space; hence we need to examine it in more detail. Recall a point in $\mathscr{LP}$ is a triple (x, L, v) where $x = (x_0, t_0)$ is a point in $M \times S^1$ (respectively, $X \times \mathbb{R}$), L is a one-dimensional subspace and v is a vector with both tangent to the hypersurface $M \times t_0$ ($X \times t_0$) at x such that L and v are perpendicular. The Riemannian metric on $M \times S^1$ naturally induces a Riemannian metric (and hence a metric) on $\mathscr{LP}$. It is also interesting to note that the points of $\mathscr{LP}$, in the case of $X \times \mathbb{R}$, can be identified with the set of all pairs (ℓ, y) where $y = (y_0, t_0)$ is a point in $X \times \mathbb{R}$ and ℓ is a geodesic line in the hypersurface $X \times t_0$ (hence the notation $\mathscr{LP}$, $\mathscr{L}$ for line, $\mathscr{P}$ for point). To see this identification, consider Figure 8 (see next page) in which x is the closest point on ℓ to y, L is the tangent line to ℓ at x, and v is the inward pointing tangent vector to the geodesic segment $[x, y]$ at x whose length is $d(x, y)$. When $y \notin \ell$, w denotes the inward pointing unit length vector tangent to $[y, x]$ at y.

There are two bundle projections

(1.2.6) $p: \mathscr{LP} \to (PM) \times S^1 = \mathbb{P}$ and $q: \mathscr{LP} - \mathbb{P} \to (SM) \times S^1 = \mathbb{B}$;

p is given by the formula

(1.2.7) $$p(x, L, v) = L.$$

The map q is induced by the $\overline{\Gamma}$-equivariant map $\overline{q}$ defined by

(1.2.8) $$\overline{q}(\ell, y) = w.$$

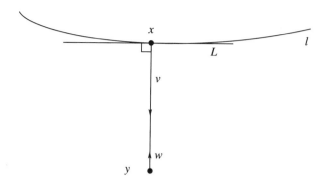

FIGURE 8

The fiber of p is $\mathbb{R}^{m-1}$ and the fiber of q is $\mathbb{R}P^{m-2} \times \mathbb{R}$. Let $\mathscr{LP}(r)$ denote the closed ball subbundle of radius $r > 0$ in $\mathscr{LP}$; i.e., it consists of all points (x, L, v) such that $|v| \leq r$. In the $X \times \mathbb{R}$ case, this can be identified with all pairs (ℓ, y) such that the distance from y to ℓ is less than or equal to r.

Let $\sigma > 1$ be a number such that $\sigma(2 + r_2) < r_3$. It determines a unique piecewise linear self-homeomorphism $\phi_\sigma \colon [0, +\infty) \to [0, +\infty)$ satisfying the following properties:

(1.2.9)
1. $\phi_\sigma(t) = t/\sigma$, if $t \in [\sigma(1 + r_2), \sigma(2 + r_2)]$;
2. $\phi_\sigma(t) = t$, if $t \in [0, r_2] \cup [r_3, +\infty)$;
3. ϕ_σ is linear on $[r_2, \sigma(1 + r_2)]$ and on $[\sigma(2 + r_2), r_3]$.

This determines a self-homeomorphism $\hat{\phi}_\sigma \colon \mathscr{LP} \to \mathscr{LP}$ by the formula

(1.2.10) $$\hat{\phi}_\sigma(x, L, v) = (x, L, \phi_\sigma(|v|)v/|v|), \quad \text{if } v \neq 0.$$

Note that $\hat{\phi}_\sigma$ is fiber preserving (relative to p) and induces $\mathrm{id} \colon \mathbb{P} \to \mathbb{P}$ (on the base space of p).

There is another bundle structure of interest whose total space is $\mathscr{LP}$. Composing $p \colon \mathscr{LP} \to \mathbb{P}$ with the natural bundle projection $\mathbb{P} \to M \times S^1$ gives a bundle projection $\rho \colon \mathscr{LP} \to M \times S^1$ defined explicitly by

(1.2.11) $$\rho(x, L, v) = x.$$

Recall this is the bundle relative to which the *natural metric* d' on $\mathscr{LP}$ is defined. There is an alternate metric (called the *fiber metric*) on each fiber $F_x = \rho^{-1}(x)$ of ρ, where $x = (x_0, t_0) \in M \times S^1$. We proceed to describe this metric. Recall a point in F_x is a pair (L, v) where L is a one-dimensional subspace and v is a perpendicular vector in $\tau_{x_0} M$. Let $\omega_i = (L_i, v_i)$, $i = 1, 2$, be two points in F_x. The distance between them in the fiber metric is the sum of $|v_1 - v_2|$ and the angle between L_1 and

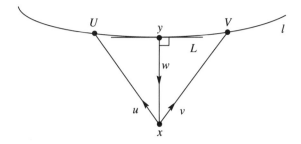

FIGURE 9

L_2. An elementary exercise in Euclidean geometry shows that this metric d is uniformly commensurable with the natural one d' on $\mathscr{LP}$, i.e., there exists a constant $c \geq 1$, independent of x, such that

(1.2.12) $$\tfrac{1}{c} d(\omega_1, \omega_2) \leq d'(\omega_1, \omega_2) \leq c\, d(\omega_1, \omega_2)$$

for all $\omega_1, \omega_2 \in F_x$. Consider the two *fiber* semimetrics d_1, and d_2 such that $d = d_1 + d_2$; i.e., $d_1(\omega_1, \omega_2) = |v_1 - v_2|$ and $d_2(w_1, \omega_2)$ is the angle between L_1 and L_2 where $\omega_i = (L_i, v_i)$. For any curve γ in a fiber of ρ, $\mathrm{diam}_i(\gamma)$ denotes its diameter relative to the semimetric d_i, $i = 1, 2$. The homeomorphism $\hat{\phi}_\sigma$ has the following important property.

PROPERTY 1.3. Let γ be a curve in a fiber of $\rho \colon \mathscr{LP} \to M \times S^1$; then

$$\mathrm{diam}_2(\hat{\phi}_\sigma \circ \gamma) = \mathrm{diam}_2(\gamma).$$

Also, provided $\mathrm{image}(\gamma) \subseteq \mathscr{LP}(\sigma(2 + r_2)) - \mathscr{LP}(\sigma(1 + r_2))$,

$$\mathrm{diam}_1(\hat{\phi}_\sigma \circ \gamma) = \tfrac{1}{\sigma} \mathrm{diam}_1(\gamma).$$

Let $\mathscr{LP}^0(r)$ denote the interior of $\mathscr{LP}(r)$; i.e., $\mathscr{LP}^0(r)$ is the open ball subbundle of $\rho \colon \mathscr{LP} \to \mathbb{P}$ consisting of all fiber vectors whose length is less than r. We identify $\mathbb{T}$ with $\mathscr{LP}^0(r)$ as illustrated in Figure 9. (Here things are as usual done $\overline{\Gamma}$-equivariantly in $X \times \mathbb{R}$.) The point $(x, \ell) \in \mathscr{LP}^0(r)$ consists of a line ℓ in $X \times t_0$ and a point $x = (x_0, t_0)$ in $X \times \mathbb{R}$ such that x is less than distance r from ℓ. Let $\langle U, V \rangle$ be the unordered pair of distinct points on ℓ such that $d(x, U) = d(x, V) = r$. Then u and v are the inward pointing unit length vectors tangent to the geodesic segment $[x, U]$ and $[x, V]$, respectively, at x. We identify (x, ℓ) and $\langle u, v \rangle$. Note that if α is a (level) curve in $M \times S^1$, then its lift $(\omega\alpha)_1$ is contained in a fiber of ρ. Recall that (x, ℓ) and (y, L, w) are also identified where y is the closest point on ℓ to x, w is the inward pointing vector tangent to the geodesic segment $[y, x]$ at y such that $|w| = d(x, y)$, and L is the one-dimensional subspace tangent to ℓ at y. These identifications provide an alternate method of constructing $\hat{\phi}_\sigma$ in terms of the incomplete radial flow r^t as follows:

(1.3.1) $$\hat{\phi}_\sigma(y, L, w) = r^t(\langle u, v \rangle)$$

where $t = |w| - \phi_\sigma(|w|)$ and (y, L, w) is identified with $\langle u, v \rangle$ via Figure 9.

The next property is easily verified.

PROPERTY 1.4. Given a pair of positive real numbers b and ε, there exists a number $c > 0$ such that the following inequalities are true for any level curve α in $M \times S^1$ whose $\mathrm{diam}(\alpha) \leq b$:

$$\mathrm{diam}_1(\omega\alpha)_1 \leq b,$$

and, provided $d(\alpha(0), \ell) > c$, where $\omega = \langle u, v \rangle$ and (x, ℓ) are identified as above,

$$\mathrm{diam}_2(\omega\alpha)_1 \leq \varepsilon.$$

The path lift $\omega\alpha$. We now combine the specific lifts $(\omega\alpha)_i$ to define the *focal* lift $\omega\alpha$ of a level path $\alpha \in M \times S^1$ relative to a fiber element $\omega \in \mathscr{F}_{\alpha(0)}(M)$ thus obtaining a curve $\omega\alpha$ in $\mathscr{F}(M)$ covering α. This will be the lift of primary interest to us. The focal lift depends on a fixed choice of real parameters $0 < r_1 < r_2 < r_3 < r_4 < r$ and is defined for all (level) paths α such that $\mathrm{diam}(\alpha) < \min\{r_1, r - r_4\}$. If $\omega \in \mathbb{F}$, we define $\omega\alpha = (\omega\alpha)_5$. If $\omega \in \mathbb{D}$, it is defined by considering Figure 10. In this case, the construction is done in a $\overline{\Gamma}$-equivariant fashion in $\overline{\mathbb{D}}$. Hence we change notation and assume that $\omega \in \overline{\mathbb{D}}_{\alpha(0)}$ and that α is a level curve in $X \times \mathbb{R}$. Now $\omega = \langle u, u \rangle$ where $u \in S^+_{\alpha(0)}(X \times \mathbb{R})$; i.e., u is a unit length vector tangent to $X \times \mathbb{R}$ at $\alpha(0)$ having nonnegative component in the $\mathbb{R}$-direction. The point U is determined by the conditions that $d(U, \alpha(0)) = r$ and u is the inward pointing unit length vector tangent to the geodesic segment $[\alpha(0), U]$ at $\alpha(0)$. The unit length vector u_t is tangent to the geodesic segment $[\alpha(t), U]$ at $\alpha(t)$ and is also inward pointing. We set $\omega\alpha(t) = \langle u_t, u_t \rangle$. This completes the definition of $\omega\alpha$.

Focal transfers. Let $f: N \to M$ be a homotopy equivalence where N is a closed manifold. A (*strong*) *homotopy inverse structure* for f consists of a map $g: M \to N$ together with homotopies $r^+: N \times [0, 1] \to N$ and $r^-: M \times [0, 1] \to M$ satisfying the following properties:

1. $r_0^- = \mathrm{id}_M$ and $r_1^- = fg$. (We denote $r^-(x, t)$ by $r_t^-(x)$. Likewise, $r_t^+(\alpha) = r^+(x, t)$.)
2. $r_0^+ = \mathrm{id}_N$ and $r_1^+ = gf$.
3. For each $x \in N$, the paths $t \mapsto f(r_t^+(x))$ and $t \mapsto r_t^-(f(x))$ are homotopic holding their endpoints fixed.

Note that if condition (3) is satisfied at a *single* point $x \in N$, then it is satisfied for *all* points $x \in N$. (This follows from a continuity argument.) Using this observation together with a mapping cylinder construction, it is easy to show that strong homotopy inverse structures exist for every $f: N \to$

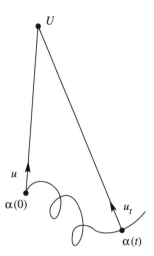

FIGURE 10

M. (If condition (3) is dropped, the resulting object is called, more briefly, a *homotopy inverse structure*.)

Recall that f induces a bundle map $\hat{f}: \mathscr{F}_f \to \mathscr{F}M$ covering $f \times \mathrm{id}: N \times S^1 \to M \times S^1$. Let $(g, h^{\pm})$ be a strong homotopy inverse structure for f. We proceed to use the focal path lifts $\omega\alpha$, described above, to construct a homotopy inverse structure $(\hat{g}, \hat{h}^{\pm})$ for $\hat{f}$, called a *focal transfer* of $(g, h^{\pm})$. It will depend on a choice of real parameters $0 < r_1 < r_2 < r_3 < r_4 < r$. Define the *real diameter* of $(g, h^{\pm})$, denoted by $\mathrm{diam}(g, h^{\pm})$, to be $\sup_\alpha \mathrm{diam}\,\alpha$ where α varies over all paths in M of either the form $t \mapsto h_t^-(y)$, where $y \in M$, or $t \mapsto f(h_t^+(x))$, where $x \in N$. (See Definition 1.1 for $\mathrm{diam}\,\alpha$ which is the *real* diameter of α. There it was defined for paths in $M \times S^1$; thus we must include $M \subset M \times S^1$ as an arbitrary level set.) The choice of parameters must satisfy the following constraint:

$$\mathrm{diam}(g, h^{\pm}) < \min(r_1, r - r_4).$$

(Note that a simple compactness argument shows that $\mathrm{diam}(g, h^{\pm}) < \infty$.) Let $\mathscr{P} = (p_1, p_2)$ denote the canonical "coordinates" of the bundle projection $\mathscr{P}: \mathscr{F}M \to M \times S^1$ with $p_1: \mathscr{F}M \to M$ and $p_2: \mathscr{F}M \to S^1$. For each $\omega \in \mathscr{F}M$ and $t \in [0, 1]$, let

$$\hat{h}_t^-(\omega) = \omega\alpha(t)$$

where $\alpha(t)$ is the level curve in $M \times S^1$ defined by

$$\alpha(t) = (h_t^-(p_1(\omega)), p_2(\omega)).$$

Recall that the bundle $\mathscr{Q}: \mathscr{F}_f \to N \times S^1$ is the pullback via $f \times \mathrm{id}: N \times S^1 \to M \times S^1$ of $\mathscr{P}: \mathscr{F}M \to M$. Hence a point $\eta \in \mathscr{F}_f$ is a triple (x, θ, ω) with

$(x, \theta) \in N \times S^1$ and $\omega \in \mathscr{F}M$ satisfying

$$(f(x), \theta) = \mathscr{P}(\omega).$$

The projection $\mathscr{Q}$ is described by $\mathscr{Q}(\eta) = (x, \theta)$ and $\hat{f}: \mathscr{F}_f \to \mathscr{F}M$ is given by $\hat{f}(\eta) = \omega$. We now define $\hat{h}^+$. Let $\eta = (x, \theta, \omega) \in \mathscr{F}_f$ and $t \in [0, 1]$; then

$$\hat{h}_t^+(\eta) = (h_t^+(x), \theta, \omega\alpha(t))$$

where $\alpha(t)$ is the level curve in $M \times S^1$ defined by

$$\alpha(t) = (f(h_t^+(x)), \theta).$$

We define $\hat{g}$ by

$$\hat{g}(\omega) = (g(p_1(\omega)), p_2(\omega), \hat{h}_1^-(\omega))$$

for each $\omega \in \mathscr{F}M$.

It is obvious that $\hat{h}^-$ is a homotopy from $\mathrm{id}_{\mathscr{F}M}$ to the composite $\hat{f}\hat{g}$. But, to show that $\hat{h}^+$ is a homotopy from $\mathrm{id}_{\mathscr{F}_f}$ to the composite $\hat{g}\hat{f}$ uses two facts. First, the curves

$$t \mapsto (f(h_t^+(x)), \theta) \quad \text{and} \quad t \mapsto (h_t^-(f(x)), \theta)$$

are homotopic holding their endpoints fixed. (Recall $(g, h^\pm)$ is a *strong homotopy inverse structure* for f.) Second, the focal lifts $\omega\alpha$ have the following property.

PROPERTY 1.5. If α is a null homotopic level closed loop in $M \times S^1$, then $\omega\alpha(0) = \omega\alpha(1)$.

Property 1.5 is an immediate consequence of the definition of $\omega\alpha$. The following is also clearly true.

REMARK 1.6. Each map $\hat{f}$, $\hat{g}$, $\hat{h}_t^+$ and $\hat{h}_t^-$ is a bundle map covering $f \times \mathrm{id}$, $g \times \mathrm{id}$, $h_t^+ \times \mathrm{id}$ and $h_t^- \times \mathrm{id}$ (respectively) and each preserves the stratifications of $\mathscr{F}M$ and $\mathscr{F}_f$.

In [18, p. 304], the term *admissible homotopy of* $\hat{f}$ was defined. (See Definitions 5.2 and 5.3.) We make the following slight modifications in these definitions for the purposes of this paper. In [18], τ denotes the bundle $\rho_\tau: \mathbb{F} - \mathbb{P} \to \mathbb{B}$ where ρ_τ is defined by equation 3.24 in [18, p. 289]. In this paper, τ will denote the bundle $q: \mathbb{F} - \mathbb{P} \to \mathbb{B}$, defined in this section via (1.0.2). (Note that q depends on the choice of the real parameter r.) After identifying $\mathbb{T}$ with $\mathscr{L}\mathscr{P}^0(r)$, via Figure 9, we redefine $_\delta\tau$ (for $\delta > 0$) by

$$_\delta\tau = \mathbb{B} \cup (\mathbb{T} - \mathscr{L}\mathscr{P}^0(r - \delta)).$$

We keep the old definition of $_0\tau$ as $\mathbb{B}$. Note that the spaces $C(\xi)$, $S(\xi)$ and $D(\xi)$ of [18] are, in the notation of the present paper, $\mathbb{F}$, $\mathbb{B}$ and $\mathbb{D}$ (respectively) except that instead of being bundles over $M' \times \mathbb{R}$ (as in [18]), they are here bundles over $M \times S^1$. There is also a slight change in Definition 5.4 of [18] where *split admissible map* is defined. The change is necessitated by the above fact that "$M = M' \times \mathbb{R}$" is replaced by $M \times S^1$. Hence, in the present paper, K denotes a triangulation for $M \times S^1$ (instead of for $M' \times \mathbb{R}$ as in [18]).

We now state the main (working) result of this paper.

THEOREM 1.7. *Let* $f: N \to M$ *be a homotopy equivalence where* N *is a closed manifold. Then* $\hat{f}: \mathscr{F}_f \to \mathscr{F}M$ *is admissibly homotopic to a split admissible map.*

REMARK 1.8. It is an old result that f is homotopic to a homeomorphism when $\dim M \leq 2$. In which case, $\hat{f}$ is admissibly homotopic to an (admissible) homeomorphism. And admissible homeomorphisms are split maps. Therefore, we may assume that $\dim M \geq 3$ when proving Theorem 1.7.

The proof will consist in choosing the real parameters $0 < r_1 < r_2 < r_3 < r_4 < r$ in such a way that the control theorems [7, 42, 24] can be applied to obtain the admissible homotopy of $\hat{f}$ to a split map. Let $(g, h^{\pm})$ be a strong homotopy inverse structure for $f: N \to M$. At some point in the construction we will use the fibered foliated control theorem [24] where the target space is $\mathscr{L}\mathscr{P}(\sigma(r_2 + 1.2))$ which fibers over $\mathbb{P}$ via the composite map $p \circ \hat{\phi}_\sigma$. Recall p is defined by (1.0.2) and has fiber a closed ball. Note that $\mathbb{B}$ is canonically identified with $SM \times S^1$ and the geodesic flow g^t on $S(M \times S^1)$ leaves $SM \times S^1$ invariant. Restricted to $SM \times S^1$ it is $g_M^t \times \text{id}$ where g_M^t denotes the geodesic flow on SM. Also $\mathbb{P}$ is canonically identified with $\mathbb{R}PM \times S^1$. The flow lines of the geodesic flow g^t determines a one-dimensional foliation of $\mathbb{B}$, and also of $\mathbb{P}$ via the canonical two-sheeted cover $\mathbb{B} \to \mathbb{P}$. These foliations are called the *geodesic line foliations* of $\mathbb{B}$ and $\mathbb{P}$, respectively. There is moreover a one-dimensional foliation of $\mathbb{B} \times [0, 1]$ with leaves $\ell \times t$ where ℓ is a leaf of $\mathbb{B}$ and $t \in [0, 1]$. It is called the geodesic line foliation of $\mathbb{B} \times [0, 1]$. The geodesic line foliations are the ones with respect to which the foliated control theorem of [24] will be applied. To use it with respect to $p \circ \hat{\phi}_\sigma: \mathscr{L}\mathscr{P}(\sigma(r_2 + 1.2)) \to \mathbb{P}$ and the geodesic line foliation on $\mathbb{P}$ requires that $r_1 \gg \text{diam}(g, h^{\pm})$, and we set $r_2 = r_1 + 1$. Property 1.4 is relevant to how large r_1 must be chosen.

We also set $r_4 = r_3 + 1$ and $r = r_4 + 3 \text{diam}(g, h^{\pm})$. Therefore the only variables are r_1, r_3, and σ.

There is a real-valued function

$$\psi: \mathbb{T} = \mathscr{L}\mathscr{P}^0(r) \to [0, r)$$

defined by $\psi(x, L, v) = |v|$; cf. (1.2.7). Our first step is to homotope $\hat{f}$ over

$$\psi^{-1}[\sigma(1 + r_2), \sigma(2 + r_2)] = (\phi_\sigma \circ \psi)^{-1}([1 + r_2, 2 + r_2])$$

by making use of the ordinary structure space control theorem of [7]. If σ is sufficiently large, then the homotopy inverse structure $(\hat{g}, \hat{h}^{\pm})$ has (by

Property 1.3) arbitrarily fine pointwise control relative to the control map

$$\hat{\phi}_\sigma : \mathscr{L}\mathscr{P}(\sigma(2+r_2)) - \mathscr{L}\mathscr{P}^0(\sigma(1+r_2)) \to \mathscr{L}\mathscr{P}(2+r_2) - \mathscr{L}\mathscr{P}^0(1+r_2).$$

Hence there is an admissible homotopy of $\hat{f}$ to $\hat{f}_1$, where $\hat{f}_1$ is a "homeomorphism" (embedding) over $(\phi_\sigma \circ \psi)^{-1}([r_2+1.1, r_2+1.9])$. Furthermore, the new homotopy inverse structure $({}_1\hat{g}, {}_1\hat{h}^\pm)$ is different from the old one only over $(\phi_\sigma \circ \psi)^{-1}([r_2+1, r_2+2])$, and here it is within a preassigned positive number ε of the old one, as measured via the control map $\hat{\phi}_\sigma$ (and $\hat{f}$). (In fact, ${}_1\hat{g} = \hat{g}$, and $\hat{f} = \hat{f}_1$ except over $(\phi_\sigma \circ \psi)^{-1})([1+r_2, 2+r_2])$.)

The number ε is chosen small enough (see also Property 1.4) so that the foliated (and fibered) structure space control theorem of [24] can be applied to $(\hat{g}, {}_1\hat{h}^\pm)$ over $(\phi_\sigma \circ \psi)^{-1}[(0, r_2+1.2])$ relative to the control map

$$p \circ \hat{\phi}_\sigma : \mathscr{L}\mathscr{P}(\sigma(r_2+1.2)) \to \mathbb{P}$$

and the geodesic line foliation of $\mathbb{P}$. (Recall $p \circ \hat{\phi}_\sigma$ is a fiber bundle projection and its fiber is a closed ball.) Hence after a second admissible homotopy of $\hat{f}_1$ to $\hat{f}_2$, which is constant except over $(\phi_\sigma \circ \psi)^{-1}[0, r_2+1.2]$, we have that $\hat{f}_2$ is a homeomorphism over $(\phi_\sigma \circ \psi)^{-1}[0, r_2+1.9]$. And we obtain a new homotopy inverse structure $(\hat{g}, {}_2\hat{h}^\pm)$ which agrees with $(\hat{g}, {}_1\hat{h}^\pm)$ except over $(\phi_\sigma \circ \psi)^{-1}([0, r_2+1.5])$. (Hence, it agrees with $(\hat{g}, \hat{h}^\pm)$ except over $(\phi_\sigma \circ \psi)^{-1}([0, r_2+2])$.)

The number ε must also be chosen small enough so that for any curve α in $\mathbb{B}$ with $\mathrm{Diam}(\alpha) \leq \varepsilon$, we have that $\mathrm{Diam}(g^t\alpha) < \varepsilon'/2$ for all $t \in [0, r_2+2]$, where ε' is a small positive number discussed below.

The geodesic flow determines a smooth map

$$g : \mathbb{B} \times J \to \mathbb{B}$$

by the equation $g(x, t) = g^t(x)$, where $J \subseteq \mathbb{R}$ is any interval (including $J = \mathbb{R}$). Let η denote the composite of the following maps:

(1.8.1) $$\mathbb{T} - \mathscr{L}\mathscr{P}^0(\sigma(r_2+1.8)) \xrightarrow{\hat{\phi}_\sigma} \mathbb{T} - \mathscr{L}\mathscr{P}^0(r_2+1.8)$$
$$\xrightarrow{q \times \psi} \mathbb{B} \times [r_2+1.8, r) \xrightarrow{g} \mathbb{B}.$$

It is easily seen that η can also be expressed as the composite

(1.8.1.1) $$\mathbb{T} - \mathscr{L}\mathscr{P}^0(\sigma(r_2+1.8)) \xrightarrow{q \times \psi} \mathbb{B} \times [\sigma(r_2+1.8), r) \xrightarrow{g} \mathbb{B}.$$

Note that η is a fiber bundle projection with fiber $\mathbb{R}P^{m-2} \times [0, 1)$. Given $\varepsilon' > 0$, we can obtain $(3\,\mathrm{diam}(\hat{g}, h^\pm), \varepsilon')$-control for $(\hat{g}, {}_2\hat{h}^\pm)$ over $\mathbb{B}$ relative to η and the geodesic line foliation of $\mathbb{B}$. This can be done by picking r_1 and σ large enough.

Before making use of the last statement, we will admissibly homotope $\hat{f}_2$ to $\hat{f}_3$ so that $\hat{f}_3$ is a homeomorphism over $\mathbb{D}$ and over a neighborhood of

$\mathbb{B}$ in $\mathbb{F}$, and so that $\hat{f}_3$ agrees with $\hat{f}_2$ over $\mathscr{L}\mathscr{P}(r_3)$. To accomplish this, restrict attention to $\hat{f}\colon \mathbb{D}_f \to \mathbb{D}$. (Recall $\hat{f}$ and $\hat{f}_2$ agree over $\mathbb{D}$.) Note that $\mathbb{D}$ is identified with $S^+(M \times S^1)$ which is left invariant by the geodesic flow g^t. Given $\delta > 0$, then $(\hat{g}, {}_2\hat{h}^\pm)$ has $(3\,\mathrm{diam}(\hat{g}, h^\pm), \delta)$-control over $\mathbb{D}$ relative to the control map $g^r\colon \mathbb{D} \to \mathbb{D}$, and the one-dimensional folia-tion of $\mathbb{D}$ consisting of the orbits of the geodesic flow g^t, *provided r_3 is sufficiently large*. Given $\varepsilon'' > 0$ and provided δ is sufficiently small, we can apply [24] to obtain a $(\beta\,\mathrm{diam}(\hat{g}, h^\pm), \varepsilon'')$-controlled homotopy of $\hat{f}$ to a homeomorphism f_3, over $\mathbb{D}$. (Here $\beta > 3$ is a number which depends only on $\dim \mathbb{D} = 2m + 1$.) Using the homotopy lifting theorem for the bundle

$$q\colon \mathbb{F} - \mathscr{L}\mathscr{P}^0(r_4) \to \mathbb{B},$$

we obtain an admissible homotopy of $\hat{f}_2$ to a map $\hat{f}_3$ (defined over all of $\mathscr{F}M$) with the following properties:

(1.8.2)

1. $\hat{f}_3$ is a homeomorphism (i.e., embedding) ex-cept over $\psi^{-1}(\sigma(r_2 + 1.9), r_4 + 1)$;

2. $\hat{f}_3$ agrees with $\hat{f}_2$ over $\psi^{-1}([0, r_4])$;

3. $(\hat{g}, {}_2\hat{h}^\pm)$ agrees with $(\hat{g}, {}_3\hat{h}^\pm)$ to $\psi^{-1}([0, r_4])$;

4. the restriction of $(\hat{g}, {}_2\hat{h}^\pm)$ over $\psi^{-1}([r_4, r))$ has $(3\beta\,\mathrm{diam}(\hat{g}, \hat{h}^\pm), C(\varepsilon'+\varepsilon''))$-control over $\mathbb{B}$ relative to η and the geodesic line foliation of $\mathbb{B}$. Here C denotes $\max \|dg^t(v)\|$ where v varies over all unit length vectors tangent to $\mathbb{B}$ and t varies over all real numbers such that $|t| \le 5\beta\,\mathrm{diam}(\hat{g}, \hat{h}^\pm)$.

Note that $C(\varepsilon' + \varepsilon'')$ can be made arbitrarilly small by picking σ and r_3 sufficiently large.

Letting $\mu\colon [\sigma(r_2 + 1.8), r_4 + 1] \to [0, 1]$ denote the (unique) increasing linear homeomorphism, then the following statement is a consequence of this fact, the control property listed after (1.8.1.1) for $(\hat{g}, {}_2\hat{h}^\pm)$, and properties 3, 4 of (1.8.2). Given $\varepsilon_0 > 0$, if σ and r_3 are sufficiently large, then $(\hat{g}, {}_3\hat{h}^\pm)$ restricted to $\psi^{-1}(\sigma(r_2 + 1.8), r_4 + 1)$ has $(3\beta\,\mathrm{diam}(\hat{g}, \hat{h}^\pm), \varepsilon_0)$-control relative to

$$\eta \times (\mu \circ \psi)\colon \mathscr{L}\mathscr{P}(r_4 + 1) - \mathscr{L}\mathscr{P}^0(\sigma(r_2 + 1.8)) \to \mathbb{B} \times [0, 1]$$

and the geodesic line foliation for $\mathbb{B}\times[0, 1]$. Note $\eta\times(\mu\circ\psi)$ is a fiber bundle whose fiber is $\mathbb{R}P^{m-2}$ and $\hat{f}_3$ is a homeomorphism over the boundary of $\mathscr{L}\mathscr{P}(r_4+1)-\mathscr{L}\mathscr{P}^0(\sigma(r_2+1.9))$. If ε_0 is sufficiently small, then the structure set foliated control theorem [24] yields that $\hat{f}_3$ restricted to $\mathscr{L}\mathscr{P}(r_4 + 1) - \mathscr{L}\mathscr{P}^0(\sigma(r_2+1.9))$ can be homotopied rel boundary to a map $\hat{f}_4$ which is "split" over $\mathbb{B} \times [0, 1]$ relative to a given triangulation $\mathscr{K}$ of $\mathbb{B} \times [0, 1]$. In doing this, we use the fact that

$$\mathrm{Wh}(\pi_1 \mathbb{R}P^{m-2} \times A) = 0$$

where A is any finitely generated free abelian group; cf. [22, p. 43]. We see by (1.8.1.1) that the control map $\eta \times (\mu \circ \psi)$ is the composite of $q \times \psi$ with a diffeomorphism ν. Pick the triangulation $\mathscr{K}$ to be subordinate to a triangulation K of $M \times S^1$, i.e., so that $\nu(\xi^{-1}(\sigma))$ is a subcomplex of $\mathscr{K}$ for each simplex σ of K. Here ξ denotes the composite of $\mathbb{B} \times J \to \mathbb{B}$, where $J = [\sigma(r_2 + 1.8), r_4 + 1]$, and the bundle projection $\mathbb{B} \to M \times S^1$. Define a map $\bar{f} : \mathscr{F}_f \to \mathscr{F}M$ by the formula

$$\bar{f}(x) = \begin{cases} \hat{f}_4(x), & \text{if } x \in \mathscr{L}\mathscr{P}(r_4 + 1) - \mathscr{L}\mathscr{P}^0(\sigma(r_2 + 1.9)), \\ \hat{f}_3(x), & \text{otherwise}. \end{cases}$$

Then $\bar{f}$ is admissibly split and admissibly homotopic to $\hat{f}$. Thus, Theorem 1.7 is proven.

2. The main result

Throughout this section, M denotes a closed, nonpositively curved Riemannian manifold, and m denotes the dimension of M.

Let W be a compact topological manifold with possibly nonempty boundary ∂W; but such that $\mathrm{Wh}\,\pi_1 W = 0$. We recall the definition of the (homotopy-topological) structure set rel ∂ of W. (see [47] and [35] for more details.) A *homotopy-topological structure rel ∂* on W is a homotopy equivalence $f_1 : W_1 \to W$, where W_1 is another manifold with boundary, such that f_1 maps ∂W_1 homeomorphically onto ∂W. Two structures (W_1, f_1) and (W_2, f_2) are equivalent if there exists a homeomorphism $g : W_2 \to W_1$ such that the composite $f_1 \circ g$ is homotopic to f_2 rel ∂. The set of equivalence classes of these structures is the (*homotopy-topological structure set rel ∂ of W* and is denoted $\mathscr{S}(W \,\mathrm{rel}\, \partial)$. (We abbreviate this notation to $\mathscr{S}(W)$ when $\partial W = \varnothing$.) This set is nonempty since id: $W \to W$ is a homotopy-topological structure rel ∂ on W. In fact $\mathscr{S}(W \,\mathrm{rel}\, \partial)$ has an abelian group structure when $\dim W > 4$, and id: $W \to W$ represents its zero element (cf. [35]). The cardinality of this set is denoted $|\mathscr{S}(W \,\mathrm{rel}\, \partial)|$. In general, $\mathscr{S}(W \,\mathrm{rel}\, \partial)$ contains more than one element. For example, Novikov showed that $|\mathscr{S}(\mathbb{C}P^n)| = \infty$ when $n > 2$. (Here $\mathbb{C}P^n$ denotes complex projective n-space.) On the other hand, we now prove the following result. (Let $\mathbb{D}^n = \{x \in \mathbb{R}^n \,||x| \leq 1\}$; i.e., $\mathbb{D}^n$ is a closed ball in $\mathbb{R}^n$. In particular, $\mathbb{D}^0$ is a point.)

THEOREM 2.1. *If M is orientable and $m + n \neq 3, 4$, then $|\mathscr{S}(M \times \mathbb{D}^n \,\mathrm{rel}\, \partial)| = 1$.*

REMARK 2.1.1. We will show later in this section (cf. Addendum 2.4) that the conclusion of Theorem 2.1 remains true even when M is nonorientable.

Also, when all the sectional curvatures of M are 0 (i.e., M is flat), the conclusion of Theorem 2.1 is true even when $m + n = 4$ (cf. [15] and [16]); this uses work of Freedman and Quinn [27]. In fact there are no known counterexamples to the conclusion of Theorem 2.1 if we totally drop the constraint $m + n \neq 3, 4$. On the other hand, note its truth would imply that the Poincaré Conjecture is also true (cf. [18, p. 258]).

We make more explicit the conclusion of Theorem 2.1 when $n = 0$.

COROLLARY 2.2. *Let* $f: N \to M$ *be a homotopy equivalence where* N *is a closed orientable (topological) manifold and* $\dim M \neq 3, 4$. *Then,* f *is homotopic to a homeomorphism.*

REMARK 2.2.1. The conclusion of Corollary 2.2 is also true even when N is *nonorientable* as will be shown later in this section; cf., Addendum 2.4.

We now start to prove Theorem 2.1. The result is well known when $m + n \leq 2$; hence, we may assume $m + n \geq 5$. Let T^s denote the flat s-dimensional torus $(s \geq 0)$; i.e., T^s is the (Riemannian) Cartesian product of s-copies of the circle S^1. (Our convention is that T^0 is a point.) Note that $M \times T^s$ is also closed, orientable and nonpositively curved. Hence $\mathrm{Wh}\,\pi_1(M \times T^s) = 0$ by [25]. Because of [12] and [35], $\mathscr{S}(M \times \mathbb{D}^n \,\mathrm{rel}\,\partial)$ is in one-to-one correspondence with a subset of $\mathscr{S}(M \times T^{n+s})$ for any integer $s \geq 1$. In particular, choose s so that $m + n + s$ is odd. Therefore while proving Theorem 2.1, we may assume that $n = 0$, $m \geq 5$ and m is odd.

Using [12] and [35] (cf. [5]), the set $\mathscr{S}(M)$ is mapped in a one-to-one fashion onto a subset of $\mathscr{S}(M \times S^1)$ by a function Σ defined as follows. Let the homotopy equivalence $f: N \to M$ represent an arbitrary element x in $\mathscr{S}(M)$; then $\Sigma(x)$ is represented by $f \times \mathrm{id}: N \times S^1 \to M \times S^1$. Note that $\Sigma(0) = 0$. To complete our proof, it suffices to show that $\Sigma(x) = 0$ for every $x \in \mathscr{S}(M)$. Because of the main result of [14], the surgery exact sequence of [47, 35] for $\mathscr{S}(M \times S^1)$ reduces to the following short exact sequence of abelian groups:

$$0 \to [M \times S^1 \times [0, 1] \,\mathrm{rel}\,\partial, \, G/\mathrm{Top}] \xrightarrow{\theta} L_{m+2}(\pi_1(M \times S^1)) \xrightarrow{d} \mathscr{S}(M \times S^1) \to 0.$$

In particular, there exists an element $y \in L_{m+2}(\pi_1(M \times S^1))$ with $d(y) = \Sigma(x)$. Now y can be represented by a degree one normal map.

$$h: (W, \partial_0 W, \partial_1 W) \to (M \times S^1 \times [0, 1], \, M \times S^1 \times 0, \, M \times S^1 \times 1)$$

where W is a compact manifold with two boundary components $\partial_0 W = M \times S^1$ and $\partial_1 W = N \times S^1$. Furthermore, $h|\partial_0 W = \mathrm{id}$ and $h|\partial_1 W = f \times \mathrm{id}$. Since Theorem 1.7 states that $\hat{f}: \mathscr{F}_f \to \mathscr{F}M$ is admissibly homotopic to a split map, we can apply a slightly modified version of the argument of [18, p. 337] to show that y is in the image of θ, thus completing the proof of Theorem 2.1.

REMARK 2.2.2. The appendix of this paper contains some corrections to [18] and one of these relates to p. 337. Also the modification mentioned above relates to the fact that the argument in [18] applies to $M \times \mathbb{R}$ instead of $M \times S^1$. But this requires only a minor change in the argument. In the present case $E = M \times S^1$, p is the identity map, $L^{-\infty}_{\ell+1+m}(\pi_1 E, w) = L_{m+2}(\pi_1(M \times S^1))$ and $\eta(\tau; L^{-\infty}(p))$ can be identified with $[M \times S^1 \times [0, 1] \operatorname{rel} \partial, G/\operatorname{Top}]$.

We now embark on extending Theorem 2.1 to *nonorientable* M. The reason the above argument fails to cover this situation is that it depends on [18, Theorem 4.3] to make the reasoning on [18, p. 337] work. (See the appendix below. In particular, showing the composite $r_2 \circ \sigma_2$ is an isomorphism uses [18, Theorem 4.3], and in this theorem ξ^{k+1} is an orientable vector bundle. But in the proof of Theorem 2.1, $\xi^{k+1} = \tau M \times S^1$ which is orientable only when M is orientable.)

To get around this difficulty, we need to generalize Theorem 1.7. Let $p: E \to M$ denote a fiber bundle whose fiber is a compact manifold F. (It is allowed that $\partial F \neq \varnothing$.) Note that E is a compact manifold and consider the pullback of the bundle $\mathscr{F}M \to M \times S^1$ via $p \times \operatorname{id}: E \times S^1 \to M \times S^1$. Denote this bundle (as well as its total space) by $(\mathscr{F}M)_p$. Let $f: N \to E$ be a homotopy-topological structure $\operatorname{rel} \partial$ on E. Let $(\mathscr{F}M)_{p,f}$ denote the pullback of $(FM)_p$ via $f \times \operatorname{id}$ and

$$\hat{f}: (\mathscr{F}M)_{p,f} \to (\mathscr{F}M)_p$$

denote the induced bundle map between total spaces which covers $f \times \operatorname{id}$. (When $p = \operatorname{id}_M$, $\hat{f}$ is the map occurring in Theorem 1.7.) Notice that both $(\mathscr{F}M)_p$ and $(\mathscr{F}M)_{p,f}$ are stratified spaces and $\hat{f}$ is strata preserving. The argument proving Theorem 1.7 is easily adapted to yield the following extension (cf. [18, Theorem 5.5]).

ADDENDUM 2.3. *Let* $\mathfrak{C}_2$ *denote the cyclic group of order* 2. *Assume that* $\operatorname{Wh}(p_{\#}^{-1}(S) \times \mathfrak{C}_2 \times A) = 0$ *for every cyclic subgroup* S *of* $\pi_1(M \times S^1)$ *and every finitely generated free abelian group* A. *Then* $\hat{f}$ *is a admissibly homotopic to a split map.*

We are now ready to prove the extension of Theorem 2.1 to nonorientable M.

ADDENDUM 2.4. *Even when* M *is nonorientable,* $|\mathscr{S}(M \times \mathbb{D}^n \operatorname{rel} \partial)| = 1$ *provided* $m + n \neq 3, 4$.

As remarked in proving Theorem 2.1, it is sufficient to consider the case where $n = 0$ and m is an *even* integer greater than 5. We may also assume that M is nonorientable. Let $\widetilde{M} \to M$ denote its two-sheeted orientable cover and identify its group of deck transformations with $\mathfrak{C}_2$. Identifying $\mathfrak{C}_2$ with $O(1)$, we have an associated line bundle ℓ with closed interval $[-1, 1]$ subbundle denoted $E \to M$. Note that $\partial E = \widetilde{M}$ and denote the double of E (along $\widetilde{M}$) by M_1. We thus see that M_1 is a closed orientable odd-dimensional manifold. We proceed to give an alternate description of M_1 which will show that M_1 has a Riemannian metric with nonpositive sectional curvatures. Let $\mathfrak{D}$ be the ∞-dihedral group generated by the two isometries α and γ of the Euclidean line $\mathbb{R}$ where

$$\alpha(x) = -x \quad \text{and} \quad \gamma(x) = 1 - x$$

for all $x \in \mathbb{R}$. There is a canonical epimorphism of $\mathfrak{D}$ onto $\mathfrak{C}_2$ obtained by sending both α and γ to the generator of $\mathfrak{C}_2$. This gives a free diagonal action via isometries of $\mathfrak{D}$ on the (Riemannian) Cartesian product $\widetilde{M} \times \mathbb{R}$ whose orbit space can be identified with M_1. Since $\widetilde{M} \times \mathbb{R}$ is nonpositively curved, this gives the same structure to M_1.

Let η be a vector bundle over M such that the Whitney sum $\eta \oplus \ell = \Theta^n$ where Θ^n denotes the trivial n-dimensional vector bundle. We may assume that n is divisble by 4; i.e., $n = 4k$ for some integer $k > 0$. Let $r: M_1 \to E$ be the retraction map given by the fact that M_1 is the double of E. And let $r_1: M_1 \to M$ be the composite of r and the bundle projection $E \to M$. Let $\mathscr{E}$ be the total space of the closed n-ball bundle associated to the pullback $r_1^* \eta$. Note that the two bundles $\mathscr{E} \to M_1$ and $\mathscr{E} \times [0, 1] \to M_1$ both satisfy the hypotheses of Addendum 2.3. (The bundle projection $\mathscr{E} \times [0, 1] \to M_1$ is the composite of $\mathscr{E} \times [0, 1] \to \mathscr{E}$ and $\mathscr{E} \to M_1$.)

For any compact manifold W (with $\mathrm{Wh}\,\pi_1 W = 0$) let $\mathscr{S}_0(W \operatorname{rel} \partial)$ denote the image of

$$d: L_{s+1}(\pi_1 W, w_1) \to \mathscr{S}(W \operatorname{rel} \partial)$$

where d is the homomorphism occurring in the surgery exact sequence for W, $s = \dim W$ and $w_1: \pi_1 W \to \mathbb{Z}_2$ is the homomorphism determined by the first Steifel-Whitney class of W.

Since M_1 is orientable, we can argue as in the proof of Theorem 2.1, using Addendum 2.3 in place of Theorem 1.7, to prove the following fact.

LEMMA 2.5. *Let $\mathscr{E}$ be the manifold defined above; then both $|\mathscr{S}_0(\mathscr{E} \operatorname{rel} \partial)| = 1$ and $|\mathscr{S}_0(\mathscr{E} \times [0, 1] \operatorname{rel} \partial)| = 1$.*

Let $p: \mathscr{E} \to M_1$ denote the bundle projection; define $\mathscr{M}$ and $\mathscr{E}'$ by $\mathscr{M} = p^{-1}(\widetilde{M})$ and $\mathscr{E}' = p^{-1}(E)$. Note that $\mathscr{E}'$ is a compact codimension-zero submanifold of $\mathscr{E}$, and $\mathscr{M}$ is a compact codimension-zero submanifold of $\partial \mathscr{E}'$. It is easy to see that $\mathscr{E}$ is the "double" of $\mathscr{E}'$ along $\mathscr{M}$ (cf. Figure 11 on next page).

We can now state a consequence of Lemma 2.5.

COROLLARY 2.6. *Let $\mathscr{E}'$ be the manifold defined above.   Then $|\mathscr{S}_0(\mathscr{E}' \operatorname{rel} \partial)| = 1$.*

Before proving Corollary 2.6, we use it to complete the proof of Addendum 2.4. Note that $\mathscr{E}'$ is homeomorphic to $M \times \mathbb{D}^{4k}$ since $\eta \oplus \ell = \Theta^{4k}$. (Recall $n = 4k$.) Because of [35, p. 283], $\mathscr{S}_0(M \times \mathbb{D}^{4k} \operatorname{rel} \partial)$ is isomorphic to $\mathscr{S}_0(M)$;

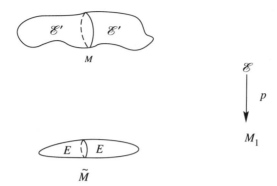

FIGURE 11

consequently, $|\mathscr{S}_0(M)| = 1$. But $\mathscr{S}_0(M) = \mathscr{S}(M)$ because of [14]. This proves Addendum 2.4.

PROOF OF COROLLARY 2.6. This is a consequence of Cappell's codimension-one splitting theorem as follows. Let $f_1: N_1 \to \mathscr{E}'$ represent an element $x_1 \in \mathscr{S}_0(\mathscr{E}' \operatorname{rel} \partial)$. Its "double" along $\mathscr{M}$, denoted $\overline{f}_1: \mathscr{N}_1 \to \mathscr{E}$, represents an element in $\mathscr{S}_0(\mathscr{E} \operatorname{rel} \partial)$. This defines a map $D: \mathscr{S}_0(\mathscr{E}' \operatorname{rel} \partial) \to \mathscr{S}_0(\mathscr{E} \operatorname{rel} \partial)$. We need only show that D is monic since $|\mathscr{S}_0(\mathscr{E} \operatorname{rel} \partial)| = 1$, because of Lemma 2.5. To do this, suppose $f_2: N_2 \to \mathscr{E}'$ represents a second element $x_2 \in \mathscr{S}_0(\mathscr{E}' \operatorname{rel} \partial)$ such that $D(x_1) = D(x_2)$. Then there exists a homeomorphism $g: \mathscr{N}_1 \to \mathscr{N}_2$ and a "homotopy $\operatorname{rel} \partial$"

$$h: \mathscr{N}_1 \times [0, 1] \to \mathscr{E} \times [0, 1]$$

such that $h|\mathscr{N}_1 \times 0$ is $\overline{f}_1 \times 0$, and $h|\mathscr{N}_1 \times 1$ is $(\overline{f}_2 \circ g) \times 1$. We wish to split $h \operatorname{rel} \partial$ along the codimension-one submanifold $\mathscr{M} \times [0, 1]$ of $\mathscr{E} \times [0, 1]$. Cappell [6] shows this is possible if a certain abelian group $\mathrm{UNil}_{m+n+2}(\phi)$ vanishes. Here, ϕ denotes the amalgamated free product decomposition of $\pi_1 \mathscr{E}$, determined by the fact that $\mathscr{E}$ is the double of $\mathscr{E}'$ along $\mathscr{M}$, together with $w_1: \pi_1 \mathscr{E} \to \mathbb{Z}_2$ determined by the first Stiefel-Whitney class of $\mathscr{E}$. On the other hand, he shows that $\mathrm{UNil}_{m+n+2}(\phi)$ is isomorphic to a subgroup of $\mathscr{S}_0(\mathscr{E} \times [0, 1] \operatorname{rel} \partial)$. Consequently, it vanishes in the current situation because of Lemma 2.5.

Therefore, we may assume that h is split along $\mathscr{M} \times [0, 1]$. The restriction of h over this set determines an element in $\mathscr{S}(\mathscr{M} \times [0, 1] \operatorname{rel} \partial)$. Since the pullback of ℓ to $\widetilde{M}$ via r_1 is the trivial line bundle, $\mathscr{M} \times [0, 1]$ is homeomorphic to $\widetilde{M} \times \mathbb{D}^n$. Consequently, $|\mathscr{S}(M \times [0, 1] \operatorname{rel} \partial)| = 1$ by Theorem 2.1. We may therefore assume that h is a homeomorphism over $\mathscr{M} \times [0, 1]$. By considering h over $\mathscr{E}' \times [0, 1]$, it is now easy to show that f_1 and f_2 are equivalent structures (i.e., $x_1 = x_2$). Hence D is monic which proves Corollary 2.6.

3. Applications

We start this section by recalling Mostow's Rigidity Theorem [39]. A *locally symmetric space* is a complete (connected) Riemannian manifold whose sectional curvatures stay constant under parallel translation. And it is of *noncompact type* if its universal cover does not have a (nontrivial) compact or flat metric factor.

EXAMPLES. Compact flat, spherical and hyperbolic manifolds are all locally symmetric spaces since their sectional curvatures are identically 0, 1 and -1, respectively. Of these examples, only the hyperbolic manifolds are of noncompact type. In fact, Cartan showed that the sectional curvatures of a locally symmetric space of noncompact type must be nonpositive. (A locally symmetric space of noncompact type can have zero as a sectional curvature value; but the values cannot be identically zero.)

THEOREM 3.1 (Mostow). *Let M and N be compact locally symmetric spaces of noncompact type such that the universal cover of N has no two-dimensional metric factor projecting to a closed subset of N. Then any isomorphism of $\pi_1 M$ to $\pi_1 N$ is induced by a unique isometry (after adjusting the normalizing constants for N).*

The following topological analogue of Mostow's Rigidity Theorem is an immediate consequence of Addendum 2.4.

THEOREM 3.2 (Topological Rigidity). *Let N be a closed (connected) Riemannian manifold with nonpositive sectional curvatures and such that $\dim N \neq 3, 4$. Let M be a closed (topological) manifold which is aspherical (i.e., $\pi_n M = 0$ for $n \neq 1$). Then any isomorphism from $\pi_1 M$ to $\pi_1 N$ is induced by a homeomorphism from M to N.*

PROOF. Let $\phi\colon \pi_1 M \to \pi_1 N$ be an isomorphism. Since both M and N are aspherical, it is induced by a homotopy equivalence $f\colon M \to N$. But Addendum 2.4 shows that f is homotopic to a homeomorphism. (See also Corollary 2.2 and Remark 2.2.1.)

REMARK 3.2.1. In the two cases where N is a real hyperbolic or a flat Riemannian manifold, this result was previously proven in [18] and [15], respectively. Also the same result was proven in [16] when N is an infrasolvmanifold, extending the case when N is an almost flat Riemannian manifold which was proven in [15].

REMARK 3.2.2. Theorem 3.2 is a partial affirmative answer to a question posed by Lawson and Yau (cf. [52, Problem 12]). They asked whether any isomorphism f between the fundamental groups of two closed negatively curved Riemannian manifolds M and N is induced by a homeomorphism, or perhaps even a diffeomorphism. Eells and Sampson [11], Al'ber [1] and Hartman [29] showed that f is induced by a unique harmonic map $\overline{f}$. Schoen and Yau [45] and Sampson [44] showed that $\overline{f}$ is a diffeomorphism when $\dim M = 2$, and it is a consequence of Mostow's Rigidity Theorem

3.1 that $\overline{f}$ is a diffeomorphism when M and N are locally symmetric spaces and $\dim M > 2$. Cheeger showed that the total spaces of the bundles of orthonormal two-frames of M and N are homeomorphic, and Gromov showed that the total spaces of the unit sphere bundles of M and N are homeomorphic via a homeomorphism preserving the flow lines of the geodesic flows. Mishchenko [37] showed that $\overline{f}$ pulls the rational Pontryagin classes of N back to those of M, and Farrell and Hsiang [14] showed that $\overline{f} \times \mathrm{id} \colon M \times \mathbb{R}^3 \to N \times \mathbb{R}^3$ is properly homotopic to a homeomorphism, where $\mathrm{id} \colon \mathbb{R}^3 \to \mathbb{R}^3$ denotes the identity map. Theorem 3.2 shows that $\overline{f}$ is homotopic to a homeomorphism, provided $\dim M \neq 3, 4$. On the other hand, there are examples constructed in [19] where f cannot be induced by a diffeomorphism since M and N are not diffeomorphic. Ye [53] has recently used these examples to construct a counterexample to Problem 13 in the list complied by Yau [52]. He shows there are two compact negatively curved Einstein manifolds M and N with $\pi_1 M \simeq \pi_1 N$ and $\dim M > 6$; but M and N are not isometric. In fact, M is not diffeomorphic to N.

REMARK 3.2.3. Metric rigidity and topological rigidity both fail for compact positively curved locally symmetric spaces as can be seen by considering Lens spaces. (Bieberbach [2] proved affine rigidity for compact flat locally symmetric spaces, and topological rigidity was proven in [15] except in dimension 3.) On the other hand, it is still interesting to calculate $\mathscr{S}(M)$ where M is an arbitrary compact locally symmetric space. Much is known on this problem when M has compact type (i.e., when $\pi_1 M$ is finite) except when $\dim M = 3$. (See Wall [47, 48, 49] for information on $\mathscr{S}(M)$ when M has compact type.) It now seems appropriate to investigate the general question. (See [10] and [28] for related results.)

REMARK 3.2.4. Mostow's and Bieberbach's rigidity theorems hold for compact "orbifolds." Bieberbach showed for example that any two properly discontinuous (effective) actions of a group Γ on Euclidean n-dimensional space $\mathbb{R}^n$ with compact quotient are conjugate via an affine diffeomorphism of $\mathbb{R}^n$. (This is true even when Γ contains nontrivial elements of finite order.) On the other hand, Mike Davis (unpublished result, 1977) has shown this is not true topologically; i.e., if the actions are via homeomorphisms, there is not always a topological conjugacy. In Davis's example, Γ is the Cartesian product of four-copies of the ∞-dihedral group $\mathscr{D}$ and one of the actions of Γ on $\mathbb{R}^4$ is the Cartesian product of four copies of the crystallographic action of $\mathscr{D}$ on $\mathbb{R}$ generated by the two reflections in 0 and 1, respectively. The orbit space of this action is $[0, 1] \times [0, 1] \times [0, 1] \times [0, 1]$. Davis constructs a second properly discontinuous (effective) smooth action of Γ on $\mathbb{R}^4$ whose orbit space is a compact contractible four-dimensional manifold $\mathscr{B}$ whose boundary is *not* simply connected. Hence the two actions cannot be topologically conjugate. Quinn [43], cognizant of this example, has formulated a more refined version of what perhaps implies topological rigidity in the orbifold case. Connolly and Koźniewski [8, 9] have significant results in the flat orbifold case. The question of topological rigidity for (compact) locally symmetric orbifolds of noncompact type should of course also be investigated.

A topological characterization of compact flat and almost flat Riemannian manifolds of dim $\neq 3$ was given in [15]. Compact locally symmetric spaces of noncompact type and dim $\neq 3, 4$ are similarly characterized in the following result.

THEOREM 3.3. *Let* M *be a closed* (*topological*) *manifold with* dim $M \neq$ $3, 4$. *Then* M *supports the structure of a locally symmetric space of noncompact type if and only if*

1. M *is aspherical* (*i.e.,* $\pi_n M = 0$ *for* $n \neq 1$), *and*

2. $\pi_1 M$ *is isomorphic to a cocompact discrete subgroup of a linear* (*virtually connected*) *semisimple Lie group.*

PROOF. E. Cartan (cf. Helgason [32]) showed that the compact (connected) locally symmetric spaces of noncompact type are precisely the double coset spaces $K\backslash G/\Gamma$ where G is a linear (virtually connected) semisimple Lie group; K is a maximal compact subgroup of G; and Γ is a discrete, torsion-free, cocompact subgroup of G. This clearly demonstrates the necessity of conditions 1 and 2 (even when dim $M = 3, 4$). To show the sufficiency of these conditions, let Γ be a cocompact subgroup of a linear (virtually connected) semisimple Lie group G such that $\pi_1 M$ is isomorphic to Γ. Note that Γ is torison-free since M is aspherical. Let K be a maximal compact subgroup of G. Recall that $K\backslash G$ is simply connected since it is diffeomorphic to a Euclidean space; consequently,

$$\pi_1(K\backslash G/\Gamma) \simeq \Gamma \simeq \pi_1 M.$$

Let $N = K\backslash G/\Gamma$ which is a compact (connected) locally symmetric space of noncompact type by E. Cartan's result. Hence Theorem 3.2 shows that M and N are homeomorphic. We pull the locally symmetric structure on N back to M via this homeomorphism completing the proof of Theorem 3.2.

Specific types of locally symmetric spaces of noncompact type are topologically characterized by specifying the semisimple Lie group. The next result illustrates this comment. It is proven in the same way as Theorem 3.3 but uses only the earlier topological rigidity result contained in [18]. Let $O(n, 1, \mathbb{R})$ denote the Lie group consisting of all linear transformations of an $(n + 1)$-dimensional $\mathbb{R}$-vector space which preserves a fixed symmetric nondegenerate bilinear form of signature $n - 1$.

ADDENDUM 3.4. *Let* M *be a closed* (*topological*) *manifold with* dim $M \neq$ $3, 4$. *Then* M *supports a hyperbolic structure if and only if*

1. M *is aspherical, and*

2. $\pi_1 M$ *is isomorphic to a cocompact discrete subgroup of* $O(n, 1, \mathbb{R})$.

REMARK 3.4.1. Addendum 3.4 was proven in [**18**, Corollary 10.4]; cf. [**22**, 5.3, Theorem]. Also Theorem 3.3 was announced in [**22**, 6.2 Corollary]. We wish here to correct a misprint in the statement of 6.2 Corollary; namely, the adjective " linear" should be inserted before the word "semisimple" in [**22**, p. 47, line 19].

REMARK 3.4.2. Recall that Mostow's Rigidity Theorem implies that at most one differentiable manifold in a homeomorphism class can support a hyperbolic structure. Also given any $\varepsilon > 0$, there exists a hyperbolic manifold M and a homeomorphic but *not* diffeomorphic Riemannian manifold N whose sectional curvatures are pinched with ε of -1. (This was shown in [**19**].) A characterization of differentiable manifolds supporting hyperbolic structures (or locally symmetric space structures of noncompact type) must therefore be more complicated.

REMARK 3.4.3. It would be interesting to know if Theorem 3.3 or Addendum 3.4 is true in dimension 3 or 4. The corresponding result for flat, almost flat and infrasolvmanifolds is true when $\dim M = 4$. This uses the work of Freedman and Quinn [**27**] showing the topological surgery theory is valid for four-dimensional manifolds with virtually poly-$\mathbb{Z}$ fundamental groups. But it is unknown whether topological surgery theory works for four-manifolds whose fundamental groups are isomorphic to discrete, torsion-free, cocompact subgroups of $O(4, 1, \mathbb{R})$. If it does, then Addendum 3.4 would still be true when $\dim M = 4$. A similar comment can be made about Theorem 3.3. It is easily shown that the truth of Addendum 3.4 when $\dim M = 3$ (which is equivalent Theorem 3.3 in this case) or of the corresponding topological characterizations of three-dimensional compact flat, infranil, or infrasolv-manifolds would imply the truth of the Poincaré Conjecture. On the other hand, we recall that Thurston has conjectured an even stronger characterization of compact hyperbolic three-manifolds. The following is one case of his Geometrization Conjecture (cf. [**46**]).

THURSTON'S CONJECTURE. Let M be a closed three-dimensional manifold. Then M has a hyperbolic structure if and only if

1. M is aspherical, and
2. every abelian subgroup of $\pi_1 M$ is cyclic.

Note in particular that Thurston's Conjecture would imply that any closed three-dimensional Riemannian manifold with negative sectional curvatures is homeomorphic to a hyperbolic manifold.

For the remainder of this section M denotes a closed (connected) Riemannian manifold of nonpositive sectional curvatures and $m = \dim M$. Let $\operatorname{Top} M$ and $\operatorname{Diff} M$ denote the topological groups of all self-homeomorphisms of M and all self-diffeomorphism of M, respectively. The last topic of this section is to obtain results about $\pi_n \operatorname{Top} M$ and $\pi_n \operatorname{Diff} M$ using Addendum 2.4 together with [**25**]. Let $\operatorname{Out}(\pi_1 M)$ denote the outer automorphism group of $\pi_1 M$. Recall that $\operatorname{Out}(\pi_1 M)$ is naturally identified (since M is aspherical) with the (discrete) group of homotopy classes of self-homotopy

equivalences of M. Hence, there are clearly canonical continuous homo-
morphisms

$$\alpha: \operatorname{Top} M \to \operatorname{Out}(\pi_1 M) \quad \text{and} \quad \gamma: \operatorname{Diff} M \to \operatorname{Out}(\pi_1 M).$$

Note that γ factors through α and the kernels of both α and γ are normal subgroups denoted by $\operatorname{Top}_0 M$ or $\operatorname{Diff}_0 M$, respectively. (Caveat: Although $\operatorname{Out}(\pi_1 M)$ is discrete, neither $\operatorname{Top}_0 M$ nor $\operatorname{Diff}_0 M$ is (in general)
the connected component of the identity element in $\operatorname{Top} M$ or $\operatorname{Diff} M$, respectively.)

Note that Mostow's Rigidity Theorem has the following corollary where
$\operatorname{Iso} M$ denotes the group of all isometries of M.

PROPOSITION 3.5 (Mostow). *Let M be a compact (connected) locally symmetric space of noncompact type such that its universal cover has no two-
dimensional metric factor which projects to a closed subset of M. Then the
topological group $\operatorname{Iso} M$ is isomorphic to the discrete group $\operatorname{Out}(\pi_1 M)$ via
the composite of γ and the inclusion of $\operatorname{Iso} M$ into $\operatorname{Diff} M$. Consequently,
both $\operatorname{Top} M$ and $\operatorname{Diff} M$ are (in this way) semidirect products*

$$\operatorname{Top} M = \operatorname{Top}_0 M \rtimes \operatorname{Out}(\pi_1 M)$$

and

$$\operatorname{Diff} M = \operatorname{Diff}_0 M \rtimes \operatorname{Out}(\pi_1 M).$$

Also, $\operatorname{Out}(\pi_1 M)$ is a finite group as was earlier shown by Borel [3].

This result topologically partially extends to general M of $\dim \neq 3, 4$;
but not differentiably.

COROLLARY 3.6. *Let M be a closed nonpositively curved Riemannian
manifold with $\dim M \neq 3, 4$. Then α is an epimorphism. But there are
examples where γ is not an epimorphism.*

PROOF. That α is an epimorphism is an immediate consequence of Theorem 3.2. On the other hand, given $\varepsilon > 0$, there are manifolds whose sectional
curvatures are pinched within ε of -1 such that γ is *not* an epimorphism.
These examples were constructed in [23]. They are a subset of the examples
in [19].

We next analyze $\pi_n \operatorname{Top}_0 M$ for $0 \leq n \leq (m - 7)/3$ where $m = \dim M >
10$. Let $\mathbb{Z}_2^\infty$ denote the direct sum of a countably infinite number of copies
of the cyclic group of order 2.

THEOREM 3.7. *Let M be a closed nonpositively curved Riemannian manifold with $m = \dim M > 10$. Let n be any integer satisfying $1 \leq n \leq
(m - 7)/3$ and $\hat{n} = [(n + 4)/2]!$. Then*

$$\pi_0 \operatorname{Top}(M) = \mathbb{Z}_2^\infty,$$

and

$$\pi_n \operatorname{Top}(M) \otimes \mathbb{Z}\left[\frac{1}{\hat{n}}\right] = \begin{cases} 0, & \text{if } n > 1, \\ \operatorname{Center}(\Gamma) \otimes \mathbb{Z}\left[\frac{1}{2}\right], & \text{if } n = 1. \end{cases}$$

COROLLARY 3.8. *Let M be a compact (connected) locally symmetric space of noncompact type such that its universal cover has no two-dimensional metric factor which projects to a closed subspace of M. Assume that $m = \dim M > 10$. Then,*

$$\pi_0 \operatorname{Top} M = \mathbb{Z}_2^\infty \rtimes \operatorname{Out}(\pi_1 M),$$

and

$$\pi_n \operatorname{Top} M \otimes \mathbb{Z}\left[\frac{1}{\hat{n}}\right] = 0 \quad \text{if } 1 \le n \le (n-7)/3$$

where $\hat{n} = [(n+4)/2]!$. Also, $\operatorname{Out}(\pi_1 M)$ is a finite group.

The proof of Theorem 3.7 will be given presently, but first we derive Corollary 3.8 from it.

PROOF OF COROLLARY 3.8. Lawson and Yau [36] showed that Center $\pi_1 M = 1$. Combining this fact with Theorem 3.7 and Proposition 3.5 yields Corollary 3.8.

REMARK 3.8.1. In the special case when M is a hyperbolic manifold, more detailed information on the weak homotopy type of $\operatorname{Top} M$ (through the stable range of dimensions $n \le (m-7)/3$) was given in [18]. It would be interesting to prove analogous results for general M satisfying the hypotheses of Theorem 3.7. Addendum 2.4 and [25] should be useful for this purpose.

PROOF OF THEOREM 3.7. For this we need to introduce the auxiliary spaces $G(M)$, $\overline{\operatorname{Top}}(M)$, $P(M)$ and $\mathscr{P}(M)$. We follow the general program developed by Hatcher [30] for analyzing $\pi_n \operatorname{Top}(M)$. Let $G(M)$ denote the H-space of all self-homotopy equivalences of M; note that $\operatorname{Top}(M)$ is a subspace of $G(M)$. The semisimplicial group $\overline{\operatorname{Top}}(M)$ of blocked homeomorphisms of M can be interpolated between $\operatorname{Top} M$ and $G(M)$. A typical k-simplex of $\overline{\operatorname{Top}}(M)$ consists of a homeomorphism $h: \Delta^k \times M \to \Delta^k \times M$ such that $h(\Delta \times M) = \Delta \times M$ for each face Δ of Δ^k, where Δ^k is the standard k-simplex.

Let W be a compact manifold (with possibly $\partial W \ne \varnothing$). Recall that a self-homeomorphism $f: W \times [0, 1] \to W \times [0, 1]$ is called a pseudoisotopy of W if its restriction to $W \times 0$ is the inclusion map. Let $P(W)$ denote the space of all topological pseudoisotopies of W. There is an obvious inclusion of $P(W \times I^k)$ into $P(W \times I^{k+1})$ (where I^k denotes the Cartesian product of k-copies of $I = [0, 1]$) by sending f to $f \times \operatorname{id}$. Let $\mathscr{P}(M)$ be the direct limit of the sequence

$$P(M) \to P(M \times I) \to \cdots \to P(M \times I^k) \to \cdots.$$

Let $G(M)/\operatorname{Top}(M)$, and $\overline{\operatorname{Top}} M/\operatorname{Top} M$ denote the homotopy fiber of the map $B\operatorname{Top}(M) \to BG(M)$, and $B\operatorname{Top}(M) \to B\overline{\operatorname{Top}}(M)$, respectively. Hatcher [30, Proposition 2.1] proved the following very useful result.

LEMMA 3.9 (Hatcher). *There is a spectral sequence with $E'_{pq} = \pi_q P(M \times I^p)$ converging to $\pi_{p+q+1}(\overline{\operatorname{Top}} M/\operatorname{Top} M)$.*

Because of Quinn's function space interpretation of the surgery exact sequence [40, 41], the relative homotopy groups of the map

$$\overline{\mathrm{Top}}(M) \to G(M)$$

can be identified with the groups

$$\mathscr{S}(M \times I^n \operatorname{rel} \partial)$$

which all vanish because of Addendum 2.4. A consequence of this is that $G(M)/\mathrm{Top}\,M$ and $\overline{\mathrm{Top}}\,M/\mathrm{Top}M$ have the same weak homotopy type and therefore Hatcher's spectral sequence has the following nice form in the situation under consideration.

COROLLARY 3.10. *Let* M *be a closed nonpositively curved Riemannian manifold with* $m = \dim M \geq 5$. *Then there is a spectral sequence with* $E^1_{pq} = \pi_q P(M \times I^p)$ *converging to* $\pi_{p+q+1}(G(M)/\mathrm{Top}\,M)$.

Two ingredients are used to analyze the groups $\pi_q P(M \times I^p)$. First, there is a stability result which depends on deep work of Igusa [34].

LEMMA 3.11 (Igusa [34], Burghelea and Lashof [4], Goodwillie cf. [34]). *The inclusion map induces isomorphisms*

$$\pi_q P(M \times I^p) \to \pi_q \mathscr{P}(M)$$

provided $m + p > 10$ *and* $q \leq (m + p - 7)/3$.

Second, we recall the following calculation proven in [25, Corollary 0.6].

LEMMA 3.12. *Let* M *be a closed nonpositively curved Riemannian manifold with* $m = \dim M \geq 5$. *Then*

$$\pi_q \mathscr{P}(M) \otimes \mathbb{Z}\left[\frac{1}{\hat{q}}\right] = 0$$

where $\hat{q} = [(q+4)/2]!$ *and furthermore*

$$\pi_0 \mathscr{P}(M) = \mathbb{Z}_2^\infty.$$

REMARK 3.12.1. That $\pi_0 \mathscr{P}(M) = \mathbb{Z}_2^\infty$ depends on the important fact that $\pi_0 \mathscr{P}(S^1) = \mathbb{Z}_2^\infty$. In [18] we attributed this fact to [31]; but the first complete proof of it is due to Igusa [33].

It is an immediate consequence of Corollary 3.10, Lemma 3.11 and Lemma 3.12 that

$$(3.12.2) \qquad \pi_{q+1}(G(M)/\mathrm{Top}\,M) \otimes \mathbb{Z}\left[\frac{1}{\hat{q}}\right] = 0$$

provided $q \leq (p-7)/3$ and $\dim M > 10$. The space $G(M)$ is (weakly) homotopically equivalent with the Cartesian product

$$\mathrm{Out}(\pi_1 M) \times K(\operatorname{center} \pi_1 M, 1).$$

(For any discrete group Γ, $K(\Gamma, 1)$ denotes an aspherical space with $\pi_1 K(\Gamma, 1) = \Gamma$, and Out($\Gamma$) is given the discrete topology.) This fact, together with formula (3.12.2), directly yields all the calculations posited in Theorem 3.7 except for $\pi_0 \operatorname{Top}_0(M) = \mathbb{Z}_2^\infty$.

To do this calculation, recall $\pi_0 \mathscr{P}(M) = \mathbb{Z}_2^\infty$ because of Lemma 3.12 and that Hatcher [**30**, Proposition 2.2] showed for his spectral sequence that

$$E_{00}^2 = H_0(\mathbb{Z}_2, \pi_0 \mathscr{P}(M)).$$

Since $E_{00}^2 = E_{00}^\infty = \pi_1(G(M)/\operatorname{Top} M)$, there is an exact sequence

(3.12.3) $\operatorname{Center} \pi_1 M \to H_0(\mathbb{Z}_2, \pi_0 \mathscr{P}(M)) \to \pi_0 \operatorname{Top} M \to \operatorname{Out}(\pi_1 M) \to 0$

induced from the fibration

$$G(M)/\operatorname{Top} M \to B \operatorname{Top} M \to BG(M).$$

Since the kernel of $\pi_0 \operatorname{Top} M \to \operatorname{Out}(\pi_1 M)$ is $\pi_0 \operatorname{Top}_0(M)$, this exact sequence can be rewritten as

(3.12.4) $\operatorname{Center}(\pi_1 M) \to H_0(\mathbb{Z}_2, \mathbb{Z}_2^\infty) \to \pi_0 \operatorname{Top}_0(M) \to 0.$

Define a homomorphism $d: \mathbb{Z}_2^\infty \to \mathbb{Z}_2^\infty$ by

$$d(x) = x + \overline{x}$$

where $x \mapsto \overline{x}$ denotes the action of the generator of $\mathbb{Z}_2$ on $\mathbb{Z}_2^\infty$. Then the formula

$$H_0(\mathbb{Z}_2, \mathbb{Z}_2^\infty) = \mathbb{Z}_2^\infty / \operatorname{image} d$$

is the definition of $H_0(\mathbb{Z}_2, \mathbb{Z}_2^\infty)$. We claim that $\mathbb{Z}_2^\infty / \operatorname{image} d$ cannot be a finite group. If it were, then $\mathbb{Z}_2^\infty / \ker d$ would also be finite since image $d \equiv \ker d$. (Note $d^2 = 0$ since $\mathbb{Z}_2^\infty$ has exponent 2.) The finiteness of both image d and $\mathbb{Z}_2^\infty / \operatorname{image} d$ would imply that $\mathbb{Z}_2^\infty$ is finite, which is a contradiction. Since $H_0(\mathbb{Z}_2, \mathbb{Z}_2^\infty)$ is thus a countable infinite group of exponent 2, it must be isomorphic to $\mathbb{Z}_2^\infty$. The sequence (3.12.4) can therefore be rewritten

(3.12.5) $\operatorname{Center}(\pi_1 M) \to \mathbb{Z}_2^\infty \to \pi_0 \operatorname{Top}(M) \to 0.$

Lawson and Yau [**36**] showed that $\operatorname{Center}(\pi_1 M)$ is finitely generated; hence, (3.12.5) implies that $\pi_0 \operatorname{Top}_0(M)$ is a countably infinite group of exponent 2, and therefore it is isomorphic to $\mathbb{Z}_2^\infty$. This completes the proof of Theorem 3.7.

Addendum 2.4 also yields a calculation of the surgery obstruction groups $L_n(\pi_1 M, w_1)$ defined by Wall in [**47**], where $w_1: \pi_1 M \to \mathbb{Z}_2$ is the homomorphism determined by the first Stiefel-Whitney class of M and n is an arbitrary integer. (Recall that these groups are periodic of period four in the index n.) There is a surgery homomorphism

$$\Theta: [M \times I^k \operatorname{rel} \partial, G/\operatorname{Top}] \to L_{m+k}(\pi_1 M, w_1)$$

defined in [**35**]. (When Top is replaced by PL, Θ was earlier defined in [**47**].)

THEOREM 3.13. *Let M be a closed Riemannian manifold with nonpositive sectional curvatures and $m = \dim M$. Then*

$$\Theta: [M \times I^k \operatorname{rel}\partial, G/\operatorname{Top}] \to L_{m+k}(\pi_1 M, w_1)$$

is an isomorphism provided both $m + k > 4$ and $k > 0$.

This result is an immediate consequence of Addendum 2.4 and the surgery exact sequence of [47, 35].

Our final application of Addendum 2.4 uses the general theory from [13, Theorem 4.5] together with [25, 0.7 Corollary] and Igusa's stability theorem for the space of smooth psuedoisotopies of M [34].

THEOREM 3.14. *Let M be a closed Riemannian manifold with nonpositive sectional curvatures and $\dim M = m > 10$. If n is any integer satisfying $1 < n \leq (m - 7)/3$, then*

$$\pi_n \operatorname{Diff}(M) \otimes \mathbb{Q} = \begin{cases} \bigoplus_{j=1}^{\infty} H_{(n+1)-4j}(M, \mathbb{Q}), & \text{if } m \text{ is odd}, \\ 0, & \text{if } m \text{ is even}. \end{cases}$$

Furthermore, $\pi_1 \operatorname{Diff}(M) \otimes \mathbb{Q} = \operatorname{Center}(\pi_1 M) \otimes \mathbb{Q}$.

REMARK 3.14.1. In the case that M is a hyperbolic manifold, Theorems 3.13 and 3.14 were previously proven in [18]. While if M is flat Riemannian, they were proven in [15] and [13], respectively. Also recall that Lawson and Yau [36] showed that $\operatorname{Center}(\pi_1 M)$ is a finitely generated free abelian group which vanishes if M is a (compact) locally symmetric space of non-compact type.

REMARK 3.14.2. Ferry and Weinberger [26] prove that $\pi_n \operatorname{Top}(M)$ is *always* a direct summand of $\pi_n \operatorname{Diff}(M)$ provided $n \geq 2$ and M is a closed Riemannian manifold of nonpositive sectional curvatures. Note that this important result *does not* require a stable range assumption. Getting unstable results about $\pi_n \operatorname{Top}(M)$ is clearly an important area for future research.

Appendix

This section contains corrections to our earlier paper [18] on which this paper is based. These corrections do not involve any major changes in [18] but rather correct some misprints and sloppy arguments in it. The page and line numbers as well as other numeration in this section refer to those used in [18]. We now make the corrections.

The statements of Theorems 6.2 and 5.5 should be modified as follows:

1. Replace the condition "where $\pi = \pi_1(F)$ or $\pi_1(\partial F)$" in the first sentence of Theorem 6.2 by the apparently stronger condition "where $\pi = \pi_1(F)$, $\pi_1(\partial F)$, $\pi_1(E)$ or $\pi_1(\partial_2 E)$."
This change necessitates the following change in Theorem 5.5:

2. Replace the second sentence of Theorem 5.5 with the following two new sentences. "Let C_2 denote the cyclic group of order 2, let $q: \partial E \to M$ be the restriction of p, and assume that $\dim F \geq 6$. Suppose that

Wh$(\pi \times C_2 \times A) = 0$ for any finitely generated free abelian group A and any subgroup π of either $p_\#^{-1}(S)$ or $q_\#^{-1}(S)$ where S is any finitely generated, virtually abelian subgroup of $\pi_1 M$."

These two changes do not affect the other results in [18]. This is because the other results depend only on the "stable versions" of Theorems 5.5 and 6.2, namely, on Remarks 5.6 and 6.3, and there are no changes in these remarks.

Since writing [18], we have observed that Remark 6.3 (when M and E are both orientable) can be deduced from Yamasaki's two papers [50] and [51]. The nonorientable situation can also be handled with a slight modification. The new version of Theorem 6.2 is a consequence of Remark 6.3.

We wish to thank Frank Connolly for pointing out an error on p. 337; namely, the maps σ_1 and σ_2 are not isomorphisms as asserted there. This necessitates a modification to lines 16–21 and -9 of p. 337 (the rest of the paper is unaffected). Replace the three sentences beginning, on line 16, with "It follows from" and ending, on line 21, with "also an isomorphism" by the following:

"Define r_1 and r_2 to be the composite maps

$$H_c^{q+4k}(T; L^{-\infty}(\bar{p})) \xrightarrow{s_1} H_c^{q+4k}(T; L^{-\infty}(p)) \xrightarrow{w_1} H_c^q(T; L^{-\infty}(p))$$

and

$$L_{q+4k+m}^{-\infty}(\pi_1(\bar{E}), \bar{w}) \xrightarrow{s_2} L_{q+4k+m}^{-\infty}(\pi_1(E), w) \xrightarrow{w_2} L_{q+m}^{-\infty}(\pi_1(E), w),$$

respectively, where w_1, w_2 are the Wall-Quinn periodicity isomorphisms, and s_1, s_2 are induced by the projection $\bar{E} \to E$. Note that $A^q \circ r_1 = r_2 \circ \bar{A}^{q+4k}$. Note also that Corollary 4.3 and Remark 4.3 imply that $r_2 \circ \sigma_2$ is an isomorphism. On the other hand, the transfer construction of 4.2 gives rise to a mapping between spectral sequences—associated to $H_c^q(T, L^{-\infty}(p))$ and $H_c^{q+4k}(T, L^{-\infty}(p))$ in 9.6—which (by 4.3) is an isomorphism on the $E_{*,*}^2$ terms and which abuts to $s_1 \circ \sigma_1$. Thus $r_1 \circ \sigma_1$ is also an isomorphism."

Also replace line -9 of p. 337 with the following: "(b) $r_2 \circ \sigma_2$ and $(I^{-1} \circ r_1 \circ \bar{I}) \circ (\bar{I}^{-1} \circ \sigma_1 \circ I)$ are isomorphisms; $(A^{\ell+1} \circ I) \circ (I^{-1} \circ r_1 \circ \bar{I}) = r_2 \circ (\bar{A}^{\ell+1+4k} \circ \bar{I})$." The argument given also does not handle the case when M is nonorientable. To do this requires inserting the following additional argument between lines three and four on p. 339:

"*Step* IV. Finally, we prove Theorem 9.5 when M is nonorientable. Let $\widetilde{M} \to M$ be the orientation cover of M and let $\widetilde{E} \to E$ be the cover of E induced by p. Letting the generator of $\mathbb{Z}_2$ correspond to the function $x \mapsto -x$ determines an action of $\mathbb{Z}_2$ on $\mathbb{R}$, S^2 and $\mathbb{R} \times S^2$. Let

$$\bar{M} = \widetilde{M} \times_{\mathbb{Z}_2} \mathbb{R} \quad \text{and} \quad \bar{E} = \widetilde{E} \times_{\mathbb{Z}_2} (\mathbb{R} \times S^2).$$

One easily sees the $\overline{M}$ supports the structure of a complete *orientable* (real) hyperbolic manifold. Consider the commutative diagram

$$
\begin{array}{ccc}
\overline{E} & \xrightarrow{\ \overline{q}\ } & E \\
{\scriptstyle\overline{p}}\big\downarrow & & \big\downarrow{\scriptstyle p} \\
\overline{M} & \xrightarrow[\ q\ ]{} & M
\end{array}
$$

where $\overline{p}$, q and $\overline{q}$ are the obvious maps. Note that $\overline{p} \colon \overline{E} \to \overline{M}$ is a fiber bundle which satisfies the hypotheses of Step III. Hence its amalgamation map

$$
\overline{A}^q \colon H^q_c(\overline{M}\,;\, L^{-\infty}(\overline{p})) \to L^{-\infty}_{q+m+1}(\pi_1\overline{E},\,\overline{w})
$$

is an isomorphism. Arguing as in Step III, there is a commutative diagram

$$
\begin{array}{ccc}
H^{q+1}_c(M\,;\, L^{-\infty}(p)) & \xrightarrow{\ A^{q+1}\ } & L^{-\infty}_{q+m+1}(\pi_1\overline{E},\,w) \\[4pt]
{\scriptstyle S_1}\big\uparrow & & \big\uparrow{\scriptstyle S_2} \\[4pt]
H^q_c(\overline{M}\,;\, L^{-\infty}(\overline{p})) & \xrightarrow{\ \overline{A}^q\ } & L^{-\infty}_{q+m+1}(\pi_1\overline{E},\,\overline{w})
\end{array}
$$

such that both S_1 and S_2 are isomorphisms. Thus A^{q+1} is also an isomorphism."

The following minor changes should also be made where indicated. (The changes on pp. 335, 337 and 338, line 15 relate to the added Step IV.)

p. 260, line −12 and p. 348, line 14.	Insert the phrase "equipped with the box topology" after "$\mathrm{Top}(S^1 \times \mathbb{R}^{m-1})$" and after "multiplication."
p. 260, line −2	Insert the word "orientable" after "compact."
p. 263, last line	Change "C'_{2k+m+1}" to "it."
p. 296, line −6	Change "continuous" to "proper continuous."
p. 297, line 4	Change "homotopy" to "proper homotopy."
p. 335, line 4	Change "three" to "four."
p. 337, line 2	Change "another" to "an orientable."
p. 338, line 15	Change "9.10" to "9.10 except M is still orientable."
p. 338, last line	Change "(X_e, X_e)" to "(X_e, Y_e)."
p. 348, line 12	Insert the phrase "but not a topological subgroup" after "$\mathrm{Top}_0 M$."
p. 350, line −15	Change "$\mathscr{P}_{\mathrm{Diff}}M$" to "$\pi_q\mathscr{P}_{\mathrm{Diff}}M$."

p. 352, line 10   Change this line to "where ψ is induced by the map $\phi\colon X \times \mathscr{D}(*) \to \mathscr{D}(X)$ with $\phi/p \times \mathscr{D}(*)$ equal to the composite map $p \times \mathscr{D}(*) \simeq \mathscr{D}(p) \xrightarrow{\text{inclusion}} \mathscr{D}(X)$ for each $p \in X$."

p. 362, line 4   Change this line to "Let $\mathscr{G}(\ )$ denote any contravariant functor from smooth manifolds to Ω-spectra which satisfies the continuity and inverse limit axioms of F. Quinn [65, 8.5]. Then"

Finally change "$(m-4)/3$" to "$(m-7)/3$" in the following places: p. 261, lines 4, 12 and 23; p. 346, line 15; p. 347, line 3; p. 348, line 2; p. 349, lines 14 and -16; p. 350, lines 2 and 15.

References

1. S. I. Al'ber, *Spaces of mappings into manifold of negative curvature*, Dokl. Akad. Nauk. USSR **178** (1968), 13–16.

2. L. Bieberbach, *Über die Bewgungsgruppen der Euklidischen Räume. II, Die Gruppen mit einem endlichen Fundamentalbereich*, Math. Ann. **72** (1912), 400–412.

3. A. Borel, *On the automorphisms of certain subgroups of semi-simple Lie groups*, Proc. Bombay Colloq. on Algebraic Geometry, 1968, pp. 43–73.

4. D. Burghelea and R. Lashof, *Stability of concordances and suspension homeomorphism*, Ann of Math. (2) **105** (1977), 449–472.

5. D. Burghelea, *The structures of block-automorphisms of $M \times S^1$*, Topology **16** (1977), 65–78.

6. S. Cappell, *Manifolds with fundamental group a generalized free product. I*, Bull. Amer. Math. Soc. **80** (1974), 1193–1198.

7. T. A. Chapman and S. Ferry, *Approximating homotopy equivalences by homeomorphisms*, Amer. J. Math. **101** (1979), 567–582.

8. F. X. Connolly and T. Koźniewski, *Rigidity and crystallographic groups. I*, Invent. Math. **99** (1990), 25–48.

9. ____, *Rigidity and crystallographic groups. II*, in prepration.

10. F. X. Connolly and Stratos Prassidis, *Groups which act freely on $\mathbb{R}^m \times S^{n-1}$*, Topology **28** (1989), 133–148.

11. J. Eells and J. H. Sampson, *Harmonic mappings of Riemannian manifolds*, Amer. J. Math. **86** (1964), 109–160.

12. F. T. Farrell, *The obstruction to fibering a manifold over a circle*, Indiana Univ. Math. J. **21** (1971), 315–346.

13. F. T. Farrell and W. C. Hsiang, *On the rational homotopy groups of the diffeomorphism groups of discs, spheres and aspherical manifolds*, Proc. Sympos. Pure Math., vol. 32, Amer. Math. Soc., Providence, R.I., 1978, pp. 325–337.

14. ____, *On Novikov's conjecture for nonpositively curved manifolds. I*, Ann. of Math. (2) **113** (1981), 199–209.

15. ____, *Topological characterization of flat and almost flat Riemannian manifolds M^n* $(n \neq 3, 4)$, Amer. J. Math. **105** (1983), 641–672.

16. F. T. Farrell and L. E. Jones, *The surgery L-groups of poly-(finite or cyclic) groups*, Invent. Math. **91** (1988), 559–586.

17. ____, *Foliated control theory. II*, K-theory **2** (1988), 401–430.

18. ____, *A topological analogue of Mostow's rigidity theory*, J. Amer. Math. Soc. **2** (1989), 257–370.

19. ____, *Negatively curved manifolds with exotic smooth structures*, J. Amer. Math. Soc. **2** (1989), 899–908.

20. ____, *Compact negatively curved manifolds (of* dim $\neq 3$, 4) *are topologically rigid*, Proc. Nat. Acad. Sci. U.S.A. **86** (1989), 3461–3463.

21. ____, *Rigidity and other topological aspects of compact non-positively curved manifolds*, Bull. Amer. Math. Soc. (N.S.) **22** (1990), 59–64.

22. ____, *Classical aspherical manifolds*, CBMS Regional Conf. Ser. in Math, vol. 75, Amer. Math. Soc., Providence, R.I., 1990.

23. ____, *Smooth non-representability of* Out $\pi_1 M$, Bull. London Math. Soc. **22** (1990), 485–488.

24. ____, *Foliated control without radius of injectivity restrictions*, Topology **30** (1991), 117–142.

25. ____, *Stable pseudoisotopy spaces of non-positively curved manifolds*, J. Differential Geom. **34** (1991), 769–834.

26. S. Ferry and S. Weinberger, *Curvature, tangentiality, and controlled topology*, Invetn. Math. **105** (1991), 401–414.

27. M. Freedman and F. Quinn, *Topology, of four-manifolds*, Princeton Univ. Press, Princeton, N.J., (1990).

28. I. Hambleton and E. K. Pedersen, *Bounded surgery and dihedral group actions on spheres*, J. Amer. Math. Soc. **4** (1991), 105–126.

29. P. Hartman, *On homotopic harmonic maps*, Canad. J. Math. **19** (1967), 673–687.

30. A. E. Hatcher, *Concordance spaces, higher simple homotopy theory, and applications*, Proc. Sympos. Pure Math., vol. 32, Amer. Math. Soc., Providence, R.I., 1978, pp. 3–21.

31. A. Hatcher and J. Wagoner, *Pseudo-isotopies of compact manifolds*, Astérisque, no. 6, Soc. Math. France, Paris, 1973.

32. S. Helgason, *Differential geometry and symmetric spaces*, Academic Press, New York, 1962.

33. K. Igusa, *What happens to Hatcher and Wagoner's formula for* $\pi_0 \mathscr{C}(M)$ *when the first Postnikov invariant of* M *is nontrivial?*, Lecture Notes in Math., vol. 1046, Springer-Verlag, New York, 1984, pp. 104–177.

34. ____, *The stability theorem for pseudoisotopies*, K-theory **2** (1988), 1–355.

35. R. C. Kirby and L. C. Siebenmann, *Foundational essays on topological manifolds, smoothings, and triangulations*, Ann. of Math. Stud., Princeton Univ. Press, Princeton, N.J., 1977.

36. B. Lawson and S.-T. Yau, *Compact manifolds of nonpositive curvature*, J. Differential Geom. **7** (1972), 211–228.

37. A. S. Mishchenko, *Infinite dimensional representations of discrete groups and higher signatures*, Izv. Akad. Nauk SSSR Ser. Mat. **38** (1974), 81–106.

38. G. D. Mostow, *Quasi-conformal mappings in n-space and the rigidity of hyperbolic space forms*, Inst. Hautes Études Sci. Publ. Math. **34** (1967), 53–104.

39. ____, *Strong rigidity of locally symmetric spaces*, Princeton Univ. Press, Princeton, N.J., 1973.

40. F. Quinn, *A geometric formulation of surgery*, Ph.D. Thesis, Princeton Univ., 1969.

41. ____, *A geometric formulation of surgery*, Topology of manifolds, Markham, Chicago, 1970, pp. 500–511.

42. ____, *Ends of maps. I*, Ann. of Math. (2) **110** (1979), 275–331.

43. ____, *Applications of topology with control*, Proc. Internat. Congr. Math . (Berkeley, 1986), Amer. Math. Soc., Providence, R.I., 1987, pp. 598–606.

44. J. H. Sampson, *Some properties and applications of harmonic mappings*, Ann. Sci. École Norm. Sup. **11** (1978), 201–218.

45. R. Schoen and S.-T. Yau, *On univalent harmonic maps between surfaces*, Invent. Math. **44** (1978), 265–278.

46. W. Thurston, *Three-dimensional manifolds, Kleinian groups and hyperbolic geometry*, Bull. Amer. Math. Soc. (N.S.) **6** (1982), 357–381.

47. C. T. C. Wall, *Surgery on compact manifolds*, Academic Press, London, 1970.
48. ____, *Classifications of hermitian forms. VI. Group rings*, Ann. of Math. (2) **103** (1976), 1–80.
49. ____, *Free actions of finite groups on spheres*, Proc. Sympos. Pure Math, vol. 32, Amer. Math. Soc., Providence, R.I., 1978, pp. 115–124.
50. M. Yamasaki, *L-groups of crystallographic groups*, Invent. Math. **88** (1987), 571–602.
51. ____, *L-groups of certain virtually solvable groups*, Topology and its Applications **33** (1989), 223–233.
52. S.-T. Yau, *Seminar on differential geometry*, Ann. of Math. Stud., no. 102, Princeton Univ. Press, Princeton, N.J., 1982, 669–706.
53. R. Ye, *Ricci flow and manifolds of negatively pinched curvature*, Preprint.

STATE UNIVERSITY OF NEW YORK AT BINGHAMTON

Proceedings of Symposia in Pure Mathematics
Volume **54** (1993), Part 3

Almost Nonnegatively Curved Manifolds

KENJI FUKAYA AND TAKAO YAMAGUCHI

In this note, we announce some topological properties of almost nonnegatively curved manifolds and related results.

Let M be a compact Riemannian manifold of dimension n. It is a common sense among geometers that if the curvature of M is big, then its fundamental group is small. For instance, when the curvature is nonnegative, the following results are well known. Let Ric_M denote the Ricci curvature of M. A classical theorem of Myers [**My**] states that if $\mathrm{Ric}_M > 0$, then the fundamental group $\pi_1(M)$ is finite. By Bochner [**BY**] if $\mathrm{Ric}_M \geq 0$, then the first Betti number $b_1(M; \mathbf{R})$ is less than or equal to n, where the equality occurs if and only if M is isometric to a flat torus. Milnor [**Mi**] obtained that if $\mathrm{Ric}_M \geq 0$, then $\pi_1(M)$ has polynomial growth. Cheeger and Gromoll [**CG2**] extended these pioneering works as follows: If $\mathrm{Ric}_M \geq 0$, then $\pi_1(M)$ is almost abelian. More precisely, $\pi_1(M)$ contains a finite index free abelian subgroup of rank r, where $b_1(M; \mathbf{R}) \leq r \leq n$.

These theorems are results in the rigid case of nonnegative curvature. In the case when curvature takes both positive and negative signs, Gromov [**GLP**] found that if $\mathrm{Ric}_M \, \mathrm{diam}(M)^2 > -D^2$, then $b_1(M; \mathbf{R}) \leq n - 1 + C_n^D$. (See also Gallot [**Ga**].) In particular, if $\mathrm{Ric}_M \, \mathrm{diam}(M)^2 > -\varepsilon_n$, then $b_1(M; \mathbf{R}) \leq n$.

We do not know any other topological information on manifolds whose Ricci curvatures are bounded from below by a negative constant. Here we consider manifolds M of almost nonnegative sectional curvature such as

$$K_M \, \mathrm{diam}(M)^2 > -\varepsilon.$$

Our purpose is to find more topological properties of such manifolds.

In [**Y2**], the second-named author obtained the following result related with fibering manifolds over tori.

1991 *Mathematics Subject Classification.* Primary 53C21.

This paper is in final form and no version will be submitted for publication elsewhere.

THEOREM b_1 [Y2]. *There exists a positive number* ε_n *depending only on* n *such that if* $K_M \operatorname{diam}(M)^2 > -\varepsilon_n$, *then*

(1) *a finite cover of* M *fibers over a* $b_1(M; \mathbf{R})$-*dimensional torus,*
(2) *if* $b_1(M; \mathbf{R}) = n$, *then* M *is diffeomorphic to an* n-*torus.*

By Theorem b_1, we have topological informations on an almost nonnegatively curved manifold from its first Betti number. The next question arises here quite naturally.

QUESTION. What can be said about $\pi_1(M)$?

In the more recent work [FY2], we have the following result for π_1.

NILPOTENCY THEOREM [FY2]. *There exists a positive number* ε_n *such that if the curvature and diameter satisfy* $K_M \operatorname{diam}(M)^2 > -\varepsilon_n$, *then* $\pi_1(M)$ *is almost nilpotent.*

Namely, $\pi_1(M)$ contains a finite index nilpotent subgroup Λ.

REMARK 1. We can also show that the nilpotent subgroup Λ can be generated by at most n elements, and that the nilpotency of Λ is not greater than n.

REMARK 2. Recall the almost flat manifold theorem due to Gromov [G1]: If $|K_M| \operatorname{diam}(M)^2 < \varepsilon_n$, then a finite cover of M is diffeomorphic to the quotient of a simply connected nilpotent Lie group by its lattice. Thus our Nilpotency Theorem generalizes the almost flat manifold theorem in the π_1-level.

As a local version, we have a generalization of Margulis's Lemma [G1].

GENERALIZED MARGULIS'S LEMMA [FY2]. *There exists* $\varepsilon_n > 0$ *such that if* M *is complete* n-*manifold with* $K_M \geq -1$, *then the image under the inclusion homomorphism*

$$\operatorname{Im}[\pi_1(B_p(\varepsilon_n, M)) \to \pi_1(M)]$$

is almost nilpotent for every $p \in M$.

In the Margulis lemma [G1], the curvature bound $|K_M| \leq 1$ was needed to get the same conclusion.

So far, we cannot prove the existence of the uniform bound for the index of the nilpotent subgroup unfortunately. However if we take a solvable subgroup in place of nilpotent subgroup, we can do it as follows.

SOLVABILITY THEOREM [FY2]. *There exists a positive number* ε_n *and* $w_n \in \mathbf{Z}_+$ *such that if* $K_M \operatorname{diam}(M)^2 > -\varepsilon_n$, *then* $\pi_1(M)$ *contains a solvable subgroup* H *satisfying the following.*

(1) $[\pi_1(M) : H] < w_n$.
(2) $H = C_1 \propto C_2 \propto \cdots \propto C_s$, $s \leq n$, *where each* C_i *is a cyclic group and* $\propto$ *denotes a group extension.*
(3) *If* $s = n$, *then* M *is a* $K(\pi, 1)$-*space and* H *is a poly-*$\mathbf{Z}$ *group.*

REMARK 3. No bounds w_n are known for the index of an abelian subgroup of the fundamental group of a nonnegatively curved manifold. We also remark that our Solvability Theorem is new even for nonnegatively curved manifolds.

As an immediate consequence of the Solvability Theorem, we have the following corollary.

COROLLARY 1. *There exists* $p_n \in \mathbf{Z}_+$ *such that if* $K_M \operatorname{diam}(M)^2 > -\varepsilon_n$, *then* $b_1(M; \mathbf{Z}_p) \leq n$ *for all prime numbers* $p \geq p_n$.
In the case of $b_1(M; \mathbf{Z}_p) = n$, *M is diffeomorphic to an n-torus.*

In the case when $\pi_1(M)$ is finite, we have the following.

COROLLARY 2. *Suppose* $\pi_1(M)$ *is finite. Then if* $K_M \operatorname{diam}(M)^2 > -\varepsilon_n$, *then we have*

$$\frac{\operatorname{diam}(\widetilde{M})}{\operatorname{diam}(M)} < C_n,$$

for some uniform bound C_n, *where* $\widetilde{M}$ *is the universal cover of* M.

This corollary looks interesting especially when $K_M > 0$. In that case, Myers's theorem ensures $\#\pi_1(M) < \infty$, and the corollary tells us something related to the fundamental group.

For lense spaces S^n/Γ for example, we have

$$\frac{\operatorname{diam}(S^n)}{\operatorname{diam}(S^n/\Gamma)} = 2.$$

It would be interesting to find a realistic bound C_n.

In the case when $\pi_1(M)$ is infinite, we have

COROLLARY 3. *Suppose* $\pi_1(M)$ *is infinite. Then if* $K_M \operatorname{diam}(M)^2 > -\varepsilon_n$, *then the Euler characteristic* $\chi(M)$ *vanishes.*

Corollaries 2 and 3 follow from the Solvability Theorem and a method in Theorem b_1.

By the following, we have many examples of almost nonnegatively curved manifolds which do not admit metrics of nonnegative curvature.

EXAMPLE. Let $F \to M \to N$ be a fibration with a compact Lie group G as structure group. Suppose that N is in infranilmanifold and that F has G-invariant metric of nonnegative sectional curvature. Then for every $\varepsilon > 0$, M admits a metric g_ε such that $K_{g_\varepsilon} \operatorname{diam}(g_\varepsilon)^2 > -\varepsilon$.

A significance of the study of almost nonnegatively curved manifolds is also in the following observation. Let us consider the family of compact n-manifolds M with geometric bounds

$$K_M \geq -1, \qquad \operatorname{diam}(M) \leq D,$$

for a fixed constant D. It is easily seen that a manifold can be collapsed to a point in this family if and only if it admits a metric of almost nonnegative sectional curvature. Thus, the study of almost nonnegatively curved manifolds is important to understand a typical example of the collapsing phenomenon in this family.

In fact, we have the following finiteness theorem (modulo almost nilpotent group) for the fundamental groups of manifolds in that family.

FINITENESS THEOREM [FY3]. *For given* n *and* D, *there exist finitely many groups* $G_1, \ldots, G_m$, *which are finitely presented, satisfying the following: If* M *satisfies* $K_M \operatorname{diam}(M)^2 > -D^2$, *then* $\pi_1(M)$ *admits an exact sequence,*

$$1 \to \Lambda \to \pi_1(M) \to G_i \to 1,$$

for some $1 \le i \le m$, *where* Λ *is an almost nilpotent group.*

REMARK 4. In the case of $|K_M| \operatorname{diam}(M)^2 < D^2$, the first-named author [F3] proved the following: For given n and D, there exist finitely many $C^{1,\alpha}$-Riemannian manifolds $Y_1, \ldots, Y_l$ such that if $|K_M| \operatorname{diam}(M)^2 < D^2$, then the orthonormal frame bundle FM of M admits a fibration,

$$N \to FM \to Y_i,$$

for some $1 \le i \le l$, where N is an infranilmanifold. Hence when $|K_M| \operatorname{diam}(M)^2 < D^2$, the Finiteness Theorem follows from this result.

REMARK 5. In [G1], Gromov proved that if $K_M \operatorname{diam}(M)^2 > -D^2$, then the minimal number of generators of $\pi_1(M)$ has a uniform upper bound $C_n(D)$. Gromov [G4] also proved that if $K_M \operatorname{diam}(M)^2 > -D^2$, then the Betti number sum has a uniform bound; $\sum_{i=1}^n b_i$ $(M; \text{field}) < C_n(D)$.

For the proofs of our results, we use the convergence and collapsing theory [GLP, F2, Y2]. In particular, in the course of the proof of Finiteness Theorem, we obtained the following result.

ISOMETRY THEOREM [FY3]. *Let* X *be a locally compact complete inner metric space. Assume that the Hausdorff dimension of* X *is finite and the Toponogov curvature of* X *is bounded from below as* $K_X \ge -c$. *Then the group of isometries of* X *is a Lie group.*

REMARK 6. (1) This is closely related with the Hilbert fifth problem. (See [MZ].) (2) The assumption of the Isometry Theorem is satisfied if X is a pointed Hausdorff limit of complete Riemannian n-manifolds M_i with $K_{M_i} \ge -1$.

Next we present two main methods employed in our work.

Recall that the splitting theorem due to Toponogov [T], Cheeger and Gromoll [CG2] played an essential role in the study of nonnegative curvature. The first main method is the splitting theorem for the limit of almost nonnegatively curved manifolds.

SPLITTING THEOREM [GP2, Y2]. *Let* M_i, $i = 1, 2, \ldots$, *be a sequence of complete manifolds with* $K_{M_i} \ge -\varepsilon_i \to 0$. *Suppose* (M_i, p_i) *converges to a pointed space* (X, x_0) *with respect to the pointed Hausdorff distance. Then if* X *contains a line, then* X *is isometric to a product* $X_1 \times \mathbf{R}$.

The other main method is the following result.

FIBRATION THEOREM [Y2]. *There exists a positive number* $\varepsilon_n(\mu_0)$ *satisfying the following. Let* M *be a compact* n*-manifold with* $K_M \geq -1$, *and* N *be such that* $|K_N| \leq 1$, $\mathrm{inj} \cdot \mathrm{rad}(N) \geq \mu_0 > 0$. *Suppose the Hausdorff distance between* M *and* N *is less than* $\varepsilon_n(\mu_0)$. *Then there exists a fibration* $f: M \to N$ *such that the fibre* $f^{-1}(\text{point})$ *has almost nonnegative curvature in a suitable generalised sense.*

In the case when $|K_M| \leq 1$, the corresponding fibration theorem was first proved in [F2], where the fibre $f^{-1}(point)$ has almost flat metrics.

By the Splitting Theorem, we can study the limit of almost nonnegatively curved manifolds, and by the Fibration Theorem, we can study topological relations between collapsed manifolds and the limit at least when the limit is a manifold.

Next we give a very rough idea of the proof of the Solvability Theorem. The proof is done by contradiction and induction on the dimension n. Suppose the theorem does not hold for dimension n. Then we have a sequence M_i of compact n-manifolds with $K_{M_i} > -\varepsilon_i \to 0$, $\mathrm{diam}(M_i) = 1$ such that $\pi_1(M_i)$ is not almost solvable. By Gromov's precompactness theorem [GLP], for a subsequence, M_i converges to a compact inner metric space X. We consider the simplest case.

Special case X *is a manifold.*

By the Fibration Theorem, we have a fibration

$$F_i \to M_i \to X$$

for large i, and the exact sequence

$$\pi_1(F_i) \to \pi_1(M_i) \to \pi_1(X) \to 1.$$

Here we can apply the induction hypothesis to the fibre F_i because F_i has almost nonnegative curvature in a generalized sense. Notice that X is nonnegatively curved, and hence $\pi_1(X)$ is almost abelian. Hence $\pi_1(M_i)$ is almost solvable, a contradiction.

In the general case when X is not necessarily a manifold, we have no fibration $M \to X$. However we can modify the basic idea in the special case by characterizing local geometric structure of X. To do this, we use the Splitting Theorem. Since the procedure is not so simple, we omit the details.

For the proof of the Nilpotency Theorem, we use the Solvability Theorem and an argument of covering spaces along fibres developed in [FY1]. The proof of the Finiteness Theorem needs the Isometry Theorem and the idea of the proof of the Nilpotency Theorem.

Finally we briefly describe the problem for manifolds of almost nonnegative Ricci curvature. We believe that our results still hold for such manifolds.

In fact, our argument does work if the Splitting Theorem holds for the limit of manifolds of almost nonnegative Ricci curvature, and if we have a weaker form of the Fibration Theorem for manifolds whose Ricci curvatures are bounded from below. (See [FY2] for details.)

REFERENCES

[BY] S. Bochner and K. Yano, *Curvature and Betti numbers*, Princeton Univ. Press, Princeton, N. J., 1953.

[BG] Yu. D. Burago, M. Gromov, and G. Perelman, *A. D. Alexandrov's spaces with curvatures bounded from below*. I, Preprint.

[BK] P. Buser and H. Karcher, *Gromov's almost flat manifolds*, Asterisque, no. 81, Soc. Math. France, Paris, 1981, pp. 1–148.

[CE] J. Cheeger and D. Ebin, *Comparison theorems in Riemannian geometry*, North-Holland, 1975.

[CG1] J. Cheeger and D. Gromoll, *On the structure of complete manifolds of nonnegative curvature*, Ann. of Math. (2) **96** (1972), 413–443.

[CG2] ____, *The splitting theorem for manifolds of nonnegative Ricci curvature*, J. Differential Geom. **6** (1971), 119–128.

[F1] K. Fukaya, *Theory of convergence for Riemannian orbifolds*, Japan J. Math. **12** (1986), 121–160.

[F2] ____, *Collapsing Riemannian manifolds to one of lower dimension*, J. Differential Geom. **25** (1987), 139–156.

[F3] ____, *A boundary of the set of the Riemannian manifolds with bounded curvatures and diameters*, J. Differential Geom. **28** (1988), 1–21.

[F4] ____, *Hausdorff convergence of Riemannian manifolds and its applications* in Recent topics in Differential and Analytic Geometry, (T. Ochiai ed.) Kinokuniya, Tokyo, 1990.

[FY1] K. Fukaya and T. Yamaguchi, *Almost nonpositively curved manifolds*, J. Differential Geom. J. Diff. Geometry **33** (1961) 69–90.

[FY2] K. Fukaya and T. Yamaguchi, *The fundamental groups of almost nonnegatively curved manifolds*, to appear in Ann. of Math.

[FY3] K. Fukaya and T. Yamaguchi, *Isometry group of singular spaces*, preprint.

[Ga] S. Gallot, *A Sobolev inequality and some geometric applications*, Spectra of Riemannian Manifolds, Kaigai, Tokyo, 1983, pp. 45–55.

[G1] M. Gromov, *Almost flat manifolds*, J. Differential Geom. **13** (1978), 231–241.

[G2] ____, *Synthetic geometry in Riemannian manifolds*, Proc. Internat. Cong. Math. (Helsinki 1978), pp. 415–419.

[G3] ____, *Groups of polynomial growth and expanding maps*, Inst. Hautes Études Sci. Publ. Math. **53** (1981), 53–73.

[G4] ____, *Curvature, diameter and Betti numbers*, Comment. Math. Helv. **56** (1981), 179–195.

[GLP] ____, *Structure métrique pour les variétés riemanniennes*, (rédige par J. Lafontaine et P. Pansu), Cedic/Fernand Nathan, Paris, 1981.

[GP1] Grove and Petersen, *Manifolds near the boundary of existence*, J. Differential Geom. (to appear).

[GP2] ____, *On the excess of the metric spaces and manifolds*, Preprint.

[Mi] J. Milnor, *A note on curvature and fundamental group*, J. Differential Geom. **2** (1968), 1–7.

[MZ] D. Montgomery and L. Zippen, *Topological transformation groups*, Interscience, 1955.

[My] S. Myers, *Riemannian manifolds in the large*, Duke Math. J. **1** (1935), 39–49.

[O] B. O'Neill, *Fundamental equations for a submersion*, Michigan J. Math. **13** (1966), 459–469.

[P] C. Plaut, *Extending local actions to global actions with applications to geometry*, Preprint.

[SW] Z. Shen and G. Wei, *On Riemannian manifolds of almost nonnegative curvature*, Preprint.

[T] V. Toponogov, *Spaces with straight lines*, Amer. Math. Soc. Transl., vol. 37, Amer. Math. Soc., Providence, RI (1964), pp. 287–290.

[We] G. Wei, *On the fundamental groups of manifolds with almost nonnegative Ricci curvature*, Proc. Amer. Math. Soc. (to appear).

[Wo] J. A. Wolf, *Spaces of constant curvature*, McGraw-Hill, New York, 1966.

[Y1] T. Yamaguchi, *Manifolds of almost nonnegative Ricci curvature*, J. Differential Geom. **28** (1988), 157–167.

[Y2] ——, *Collapsing and pinching under a lower curvature bound*, Ann. of Math. (to appear).

UNIVERSITY OF TOKYO, JAPAN

KYUSHU UNIVERSITY, JAPAN

Proceedings of Symposia in Pure Mathematics
Volume **54** (1993), Part 3

Local Structure of Framed Manifolds

PATRICK GHANAAT

This note describes recent results [**11, 12, 13**] on the local behaviour of a smooth manifold M carrying a coframe $\omega : TM \to \mathbb{R}^n$, i.e., an $\mathbb{R}^n$-valued one-form that maps each tangent space $T_p M$ isomorphically onto $\mathbb{R}^n$. Pulling back the standard metric on $\mathbb{R}^n$ to M via ω, one obtains a Riemannian metric g_ω on M. The term "local" means that we are considering small distance balls in M, whose size is bounded from above and below in terms of the dimension n of M and the exterior derivative $\|d\omega\|$, but does not depend on the injectivity radius of M.

The coframe ω determines a connection D^ω on the tangent bundle of M as follows: Let $(X_1, \ldots, X_n)$ denote the frame field dual to ω; then $D^\omega X_i = 0$ for all i. D^ω is flat and its torsion tensor T is given by $\omega(T(X, Y)) = d\omega(X, Y)$. ω is called complete if the exponential map exp of the connection D^ω is defined on all of TM.

We list three types of examples:

1. Let $M = \Gamma \backslash G$ be a quotient of a Lie group G with Lie algebra $\mathcal{G}$ by a discrete subgroup Γ. The left invariant Maurer-Cartan form on G descends to a form $\overline{\omega} : M \to \mathcal{G}$ which after choosing a basis for $\mathcal{G}$ becomes a coframe. The corresponding connection $D^{\overline{\omega}}$ will be denoted by $\overline{D}$. In this case exp coincides with the group exponential map, followed by the projection of G onto M. According to a theorem of Lie (see [**24**, p.220]), the existence of a complete coframe satisfying $D^\omega d\omega = 0$ actually characterizes such quotients. If M is compact and connected and if G nilpotent, then M is called a *nilmanifold*.

2. If α is a connection form on a subbundle P of the frame bundle of a manifold, then α and the canonical one-form θ together provide a coframe ω on P. The exterior derivative of ω and hence the torsion of D^ω are given by the Cartan structure equations $d\alpha = -\alpha \wedge \alpha + \Omega$ and $d\theta = -\alpha \wedge \theta + \Theta$.

1991 *Mathematics Subject Classification*. Primary 53C20.
This paper is in final form and no version will be submitted for publication elsewhere.

This indicates that even for applications in Riemannian geometry it is useful to work with the torsion $d\omega$ instead of the curvature of the associated metric g_ω. The exponential map exp, when restricted to horizontal vectors in TP, projects to the exponential map of the connection in the base.

3.   According to a result of Stiefel [25], any orientable 3-manifold M admits a coframe. Hardorp [17] showed that ω can be chosen such that the three distributions $\ker(\omega^i)$ defined by the component one-forms of ω are completely integrable. Hirsch [18, 16] proved that, in the open case, a stronger integrability condition can be imposed without assumptions on the dimension of M: If M is open, n-dimensional and has trivial tangent bundle, then M can be immersed into $\mathbb{R}^n$. The differential of such an immersion $x : M \to \mathbb{R}^n$ is an exact coframe $\omega = dx$, and in fact any homotopy class of coframes contains such a dx. This means that the condition $d\omega = 0$ by itself does not impose restrictions on M. However, by Lie's theorem such ω cannot be complete unless M is diffeomorphic to a cylinder.

In Riemannian geometry, the fact that nilmanifolds provide the only examples of nontrivial local objects was first proved by Gromov [14, 1]. Gromov showed that a closed Riemannian manifold M which is almost flat in the sense that its sectional curvature K and diameter d satisfy $|K|d^2 < \varepsilon(n)$ has a finite Galois covering diffeomorphic to a nilmanifold $\Gamma\backslash G$. Here $\varepsilon(n)$ is a positive constant depending only on the dimension n of M. Ruh [21, 22], using part of Gromov's proof, showed that the covering group acts by affine automorphisms of the Maurer-Cartan connection $\overline{D}$ on $\Gamma\backslash G$. Such smooth quotients of nilmanifolds by a finite group of affine automorphisms of $\overline{D}$ are called *infranilmanifolds*. Conversely, starting from a left invariant metric on the universal covering G, it is not hard to prove [14, 5] that any infranilmanifold admits Riemannian metrics satisfying the condition $|K|d^2 < \varepsilon$ for arbitrary small positive ε.

For a complete Riemannian manifold one can try to distinguish between directions with small and large injectivity radius in order to obtain a local decomposition of M. Part of this structure was described by Cheeger and Gromov in their study of collapsing [2, 3]. Fukaya [8], using Gromov compactness and an equivariant version of his fiber bundle theorem [6, 7], which in turn is based on the results of Gromov and Ruh, proved that every point p in a complete Riemannian manifold with sectional curvature bound $|K| \le 1$ has a neighborhood that contains the distance ball of radius $\varepsilon(n)$ around p and is diffeomorphic to a vector bundle over an infranilmanifold. Cheeger, Fukaya and Gromov have since combined and sharpened part of their results to obtain, among other things, a detailed local picture [9, 4] of a general Riemannian manifold up to quasi-isometry.

We restrict our attention to framed manifolds. It turns out that a decomposition similar to Fukaya's theorem can be obtained rather directly, with torsion playing the role of curvature, by investigating the exponential map exp of the connection D^ω. When applied to the frame bundle of a Riemannian manifold, this leads to a new proof of the theorem of Gromov and Ruh on almost flat manifolds.

For a connected framed manifold (M, ω) let d denote the diameter of (M, g_ω) and define $\kappa := \|d\omega\|_\infty = \|T\|_\infty$, where $\|\cdot\|_\infty$ is the maximum norm. The following result deals with the case of small diameter. A version of it was announced by Ruh in [22], where a proof under additional assumptions was outlined.

THEOREM A. *There is a constant* $\varepsilon(n) > 0$, *depending only on* n *such that the following is true. If* M *is compact and the coframe* $\omega : TM \to \mathbb{R}^n$ *satisfies* $\kappa d < \varepsilon(n)$, *then there exist a nilmanifold* $\overline{M} = \Gamma \backslash G$, *a Maurer-Cartan form* $\overline\omega$ *on* $\overline{M}$ *and a diffeomorphism* $\phi : M \to \overline{M}$ *such that*

$$d^{-1}\|\omega - \phi^*\overline\omega\|_\infty + \|d\omega - \phi^* d\overline\omega\|_\infty \le c(n)\|\mathrm{pr}_\perp d\omega\|_\infty.$$

Here $\mathrm{pr}_\perp$ denotes the orthogonal projection onto the second factor in the decomposition $L^2(\Lambda^* T^* M \otimes \mathbb{R}^n) = \ker(D^\omega) \oplus \ker(D^\omega)^\perp$ of the space of square integrable $\mathbb{R}^n$-valued forms, and $c(n)$ depends only on n. A proof of Theorem A is given in [11] (the estimate follows from §§7 and 8 of that article). To prove the result, one constructs $\omega' = \phi^* \omega_0$ as a solution of the Maurer-Cartan equation $D^{\omega'} d\omega' = 0$.

The coframe ω' is obtained as the limit of a sequence (ω_k) with $\omega_0 = \omega$, each of whose members ω_k is the unique solution of an elliptic equation invariant under affine isometries[1] of ω_{k-1}. As a consequence, the deformation of ω into ω' preserves affine isometries.

Basic for the iteration is a lower bound for the Laplacian on $\ker(D^\omega)^\perp$. We illustrate this bound by sketching a Bochner-method proof for the following result of Nomizu [20]:

THEOREM (Nomizu). *The real cohomology of a nilmanifold* $\Gamma \backslash G$ *satisfies* $H^*(\Gamma \backslash G; \mathbb{R}) = H^*(\mathscr{G}; \mathbb{R})$, *where* $\mathscr{G}$ *is the Lie algebra of* G.

PROOF. For a left invariant Riemannian metric g (pushed down to $\Gamma \backslash G$) consider the L^2-orthogonal splitting of the space of differential forms $\Omega^*(\Gamma \backslash G) = \ker(\overline{D}) \oplus \ker(\overline{D})^\perp$. By definition, $\ker(\overline{D}) = \Lambda^*(\mathscr{G})$. The exterior derivative d as well as $\delta = \pm * d*$ and Δ preserve the splitting, so it suffices to show that Δ is positive on $\ker(\overline{D})^\perp$ at least for a suitable choice of g.

Define a componentwise Laplacian Δ^0 on forms by

$$\Delta^0(\alpha_{i_1 \cdots i_k} \overline\omega^{i_1} \wedge \cdots \wedge \overline\omega^{i_k}) = (\Delta \alpha_{i_1 \cdots i_k})\overline\omega^{i_1} \wedge \cdots \wedge \overline\omega^{i_k},$$

where $\overline\omega^{i_s}$ are component 1-forms of $\overline\omega$. A Bochner formula [12] comparing

---
[1] An affine isometry of a coframe ω is an isometry of g_ω preserving D^ω.

Δ and Δ^0 shows that for α in $\ker(\overline{D})^\perp$ one has

$$(\Delta\alpha, \alpha) \geq (\lambda_1 - c(n)\kappa\lambda_1^{1/2} - c(n)\kappa^2)(\alpha, \alpha)$$

where $(\cdot, \cdot)$ is the L^2-inner product, $\kappa = \|T\|_\infty$, T the torsion of $\overline{D}$ and λ_1 the first positive eigenvalue of the Laplacian on functions. Choose [14] a sequence of metrics g_k such that $\kappa d \to 0$ as $k \to \infty$. By [11, Theorem 3.1], $\lambda_1 d^2$ stays bounded from below by $c_1(n) > 0$, and the result follows.

A local decomposition for framed manifolds is achieved in the following theorem [13] which describes small regions in M up to quasi-affine-isometry. It shows that if exp has small injectivity radius at $p \in M$, then one can find a compact connected local soul manifold N passing through p such that exp restricted to the normal bundle of N has large injectivity radius. As a consequence of Theorem A, N turns out to be a nilmanifold, and after rotating ω by a constant orthogonal matrix, a Maurer-Cartan coframe for the nilpotent structure on N can be obtained as a small perturbation of the pullback of some components of ω to N.

THEOREM B. *There is a constant $\varepsilon(n) > 0$ such that the following is true. Suppose $0 \leq \kappa\varepsilon_1 \leq \varepsilon(n)$. If $\omega : TM \to \mathbb{R}^n$ is a coframe and $p \in M$ and if exp is defined on the ball $B(0, 100\varepsilon_1) \subseteq T_pM$, then there exists a radius R such that $(100n)^{-n}\varepsilon_1 \leq R \leq \varepsilon_1$ and such that*

(1) the distance ball $B(p, R) \subseteq M$ contains a nilmanifold N, imbedded with trivial normal bundle ν,

(2) exp maps the subset $\nu_{10R} := \{X \in \nu : \|X\| < 10R\}$ diffeomorphically onto a neighborhood of $B(p, R)$, and

(3) there is a nilpotent Maurer-Cartan coframe η_0 on N and an element $a \in SO(n)$ such that the product $\omega_0 = \pi_1^*\eta_0 \oplus \pi_2^*dx$ on $\nu_{10R} \cong N \times D_{10R}$ satisfies $\|\exp_* \omega_0 - a\omega\| \leq \delta(\kappa\varepsilon_1)$, where $\lim_{t\to 0} \delta(t) = 0$ and D_{10R} is the euclidean ball of radius $10R$.

In particular, for $\varepsilon_2(n) := (100n)^{-n}\varepsilon(n)$ one gets

COROLLARY. *Let ϕ be a continuous map of the n-sphere $(n \geq 2)$ into a complete framed manifold. If the image of ϕ is contained in a distance ball of radius $\varepsilon_2(n)/\kappa$, then ϕ is homotopic to a constant.*

The construction of N can be outlined as follows. Let $\exp_p := \exp|T_pM$. The smallness assumption on $\kappa\varepsilon_1$ implies that the pullback $\tilde{\omega} := \exp_p^*\omega$ is C^0-close to dx on $B(0, 100\varepsilon_1)$, if linear coordinates x are chosen on T_pM such that dx and ω coincide at the origin. For $\rho \leq 100\varepsilon_1$ let $\Gamma_\rho := \exp_p^{-1}(p) \cap B(0, \rho)$. For $2.1\rho \leq 100\varepsilon_1$, each element x in Γ_ρ defines an imbedding $\gamma_x : B(0, \rho) \to B(0, 2.1\rho)$, uniquely determined by the conditions $\gamma_x(0) = x$ and $\exp \circ \gamma_x = \gamma_x$. Since $\gamma_x^*\tilde{\omega} = \tilde{\omega}$ and $\tilde{\omega}$ is close to dx, γ_x is C^1-close to the euclidean translation by x.

If $\exp(x_1) = \exp(x_2)$ for points x_1 and x_2 in $B(0, \rho)$, then by extending the locally defined map $(\exp_p |U(x_2))^{-1} \circ \exp_p |U(x_1)$ (where $U(x_i)$ is a small neighborhood of x_i) one obtains an x in $\Gamma_{2.1\rho}$ such that $\gamma_x(x_1) = x_2$.

By an inductive argument, one can find a radius R satisfying the required bounds such that for $\rho = 50R$, Γ_ρ is contained in an $o(\kappa \varepsilon_1)$-neighborhood of a linear subspace V of $T_p M$. There is a subset $\mathscr{O}$ of Γ_ρ such that $\#\mathscr{O} < c(n)$ and such that each x in $V \cap B(0, \rho)$ satisfies $\mathrm{dist}(x, \mathscr{O}) \leq \rho/100$. This means that $U := \exp B(0, \rho/3)$ is obtained from $B(0, \rho/3)$ by identifying points that are equivalent under a pseudogroup of almost-translations along the subspace V.

After rotating ω by $a \in SO(n)$, one can assume that V is given by $x^1 = \cdots = x^k = 0$. Define functions $\bar{f}^i$ $(i = 1, \ldots, k)$ on $B(0, \rho)$ by

$$\bar{f}^i(x) = \frac{\sum \mu(x') x^i(x')}{\sum \mu(x')}$$

where μ is a cutoff function supported in $B(0, \rho/3)$ such that $\mu = 1$ on $B(0, \rho/10)$ and the sum is taken over all points $x' \in B(0, \rho)$ such that $\exp(x') = \exp(x)$. The description of Γ_ρ implies that $\bar{f}^i$ is C^1-close x^i. By its definition, $\bar{f}^i$ descends to a smooth function f^i on $U \subseteq M$, and one defines $N = \{q \in U : f^i(q) = f^i(p)\}$.

Applying the method of Theorem B to the frame bundle of a Riemannian manifold carrying an almost flat connection, one obtains [12] a proof of the following generalization of the theorem of Gromov and Ruh.

THEOREM C. *There is a positive constant $\varepsilon(n)$ such that the following is true. Let (M, g) be a compact connected n-dimensional Riemannian manifold, d its diameter and ∇ a connection on TM compatible with the metric g. If the curvature and torsion tensors R and T of ∇ satisfy $(\|R\|_\infty + \|T\|_\infty^2)d^2 \leq \varepsilon(n)$, then M is diffeomorphic to an infranilmanifold $\Lambda \backslash G$.*

The term "diffeomorphic" can be replaced by "almost affinely isometric" with respect to (the quotients of) the canonical flat connection $\bar{D}$ and a suitable left invariant Riemannian metric on the universal covering G. It is known that the curvature tensor of the Levi-Cività connection of the metric g is bounded pointwise by $\|R^{LC}\| \leq \|R\| + 6\|T\|^2 + 3\|\nabla T\|$. No assumption on ∇T is made in Theorem C.

For the proof one can assume that $d \ll 1$ and apply the construction of Theorem B to the bundle P of orthonormal frames on M, choosing ε_1 large compared to d but small compared to the diameter of the fiber $O(n)$. Using the nilpotent structure on the soul N one proves that N is a finite regular covering space of M. By center of mass methods, N can be perturbed into a principal subbundle $Q \subseteq P$ close to the holonomy bundle of a flat connection with parallel torsion.

288 PATRICK GHANAAT

REFERENCES

1. P. Buser and H. Karcher, *Gromov's almost flat manifolds*, Astérisque, no. 81, Soc. Math. France, Paris, 1981.
2. J. Cheeger and M. Gromov, *Collapsing Riemannian manifolds while keeping their curvature bounded*. I, J. Differential Geom. **23** (1986), 309–346.
3. _____, *Collapsing Riemannian manifolds while keeping their curvature bounded*. II, J. Differential Geom. **32** (1990), 269–298.
4. J. Cheeger, K. Fukaya and M. Gromov, *Nilpotent structures and invariant metrics on collapsed manifolds*, Journal Amer. Math. Soc. **5** (1992), 327–372.
5. F. T. Farrell and W. C. Hsiang, *Topological characterization of flat and almost flat Riemannian manifolds M^n* ($n \neq 3, 4$), Amer. J. Math. **105** (1983), 641–672.
6. K. Fukaya, *Collapsing Riemannian manifolds to ones of lower dimensions*, J. Differential Geom. **25** (1987), 139–156.
7. _____, *Collapsing Riemannian manifolds to ones with lower dimension*.II, J. Math. Soc. Japan **41** (1989), 333–356.
8. _____, *Hausdorff convergence of Riemannian manifolds and its applications*, Adv. Stud. Pure Math., vol. 18-I, Kinokuniya, Tokyo, 1990, pp. 143–238.
9. _____, these proceedings.
10. P. Ghanaat, *A note on discrete uniform subgroups of Lie groups*, Geom. Dedicata **21** (1986), 13–17.
11. _____, *Almost Lie groups of type R^n* , J. Reine Angew. Math. **401** (1989), 60–81.
12. _____, *Geometric construction of holonomy coverings for almost flat manifolds*, J. Differential Geom. **34** (1991), 571–580.
13. P. Ghanaat, M. Min-Oo, and E. Ruh, *Local structure of Riemannian manifolds*, Indiana Univ. Math. J. **39** (1990), 1305–1312.
14. M. Gromov, *Almost flat manifolds*, J. Differential Geom. **13** (1978), 231–241.
15. _____, *Synthetic geometry in Riemannian manifolds*, Proc. Internat. Congr. of Math. (Helsinki, 1978), pp. 415–419.
16. _____, *Partial differential relations*, Springer-Verlag, 1986.
17. D. Hardorp, *All compact orientable three dimensional manifolds admit total foliations*, Mem. Amer. Math Soc. No. 233 (1980).
18. M. Hirsch, *On imbedding differentiable manifolds into euclidean space*, Ann. of Math. (2) **73** (1961), 566–571.
19. M. Min-Oo, *Almost symmetric spaces*, Astérisque, no. 163/164, Soc. Math. France, Paris, 1988, pp. 221–246.
20. K. Nomizu, *On the cohomology of compact homogeneous spaces of nilpotent Lie groups*, Ann. of Math. (2) **59** (1954), 531–538.
21. E. Ruh, *Almost flat manifolds*, J. Differential Geom. **17** (1982), 1–14.
22. _____, *Almost homogeneous spaces*, Astérisque, no. 132, Soc. Math. France, Paris, 1985, pp. 285–293.
23. _____, *Almost Lie groups*, Proc. Internat. Cong. Math. (Berkeley, 1986), Amer. Math. Soc., Providence, RI, pp. 561–564.
24. S. Sternberg, *Lectures on differential geometry (second edition)*, Chelsea, New York, 1983.
25. E. Stiefel, *Richtungsfelder und Fernparallelismus in n-dimensionalen Mannigfaltigkeiten*, Comment. Math. Helv. **8** (1935–36), 305–353.

UNIVERSITÄT BASEL, SWITZERLAND

Proceedings of Symposia in Pure Mathematics
Volume 54 (1993), Part 3

Constant Affine Mean Curvature Hypersurfaces of Decomposable Type

SALVADOR GIGENA

Introduction

Under the common name of "affine geometry," diverse geometrical theories have been developed through the years, mostly for hypersurfaces. Such are the cases of unimodular affine, centroaffine and relative geometries [1, 2, 8, 9, 14, 15, 20, 21, 24, 28, 29, 31]. In recent years the author succeeded in formulating a theory of invariants under the action of the full general affine group, also for hypersurfaces [12]. Compare to [17, 30].

The class of affine hyperspheres, in unimodular affine geometry, has received a good deal of attention and the classification of complete, locally strongly convex hyperspheres was achieved through the contributions of several people [2, 3, 4, 6, 10, 11, 16, 19, 22, 25, 26].

Another class of hypersurfaces dealt with in unimodular affine geometry has been that with constant mean curvature. Diverse authors have considered the problem, under the additional hypothesis that the hypersurface be the boundary of a convex body. The original result of Blaschke, "every ovaloid in 3-dimensional affine space with constant unimodular affine mean curvature is an ellipsoid," has been generalized to higher dimensions by using diverse methods [2, 15, 18, 21, 23]. Some of the results are still valid for compact surfaces and hypersurfaces with boundary (see [27] and references therein).

The so-called affine Bernstein problem consists in the classification of complete hypersurfaces with vanishing unimodular affine mean curvature, $H_{ua} = 0$. So far, this problem, posed by Chern in [7], has been solved by Calabi [5] for dimension $n = 2$: "every locally strongly convex surface in affine 3-space with $H_{ua} = 0$, which is, besides, both a global graph and metrically complete (i.e. with respect to the unimodular affine Berwald-Blaschke metric) must be an elliptical paraboloid."

1991 *Mathematics Subject Classification.* Primary 53A15; Secondary 34A34.

This paper is in final form and no version will be submitted for publication elsewhere.

In this article we consider a class of hypersurfaces with constant general affine mean curvature, $H_{ga} = constant$. It is, in a sense, very natural to consider this mean curvature, defined by averaging the trace of the third fundamental form, since the latter is the first real-valued symmetric 2-tensor that appears in the development of the general affine theory, as opposed to the first and second fundamental forms, and the general affine normal, which are all density-valued objects. Let us recall, incidentally, that the unimodular affine mean curvature is defined by averaging the contraction of the third fundamental form with the first fundamental form of unimodular affine geometry.

The class of hypersurfaces with constant general affine mean curvature that we study here consists of those which are expressible in the form of Monge, i.e., graphs over suitable subsets of a hyperplane, and such that the defining graph function can be decomposed into a sum of functions, each depending on only one of the ambient space coordinates. We prefer to use the term "decomposable" for such hypersurfaces, as opposed to "translation hypersurfaces" (schiebflächen" in German), because the latter represents a much wider class of hypersurfaces. Some people prefer to refer to the present "decomposable" class as "translation hypersurfaces with plane translation curves".

In this article we get a complete classification of the mentioned class of hypersurfaces, both from the local and global points of view. Working in dimension n greater or equal than 2, we shall obtain six original types of components. This number rises to twenty-two after suitable reflections. Every hypersurface in the class can be expressed in terms of these types.

The exposition of the present article is organized as follows: in §1 we summarize the general affine theory of hypersurfaces. In §2 we introduce the two hypotheses mentioned previously, and draw from them some important conclusions: first, that the nonlinear partial differential equation of the fourth order, that characterizes hypersurfaces with constant general affine mean curvature, reduces to a nonlinear ordinary differential equation to be satisfied by each of the components, and secondly, that such an ordinary differential equation, also of order four, reduces to a second-order, and then to a first-order ordinary differential equation. This last is the classifying differential equation for the class of hypersurfaces considered, and it depends on two constants: one coming from the hypothesis on the mean curvature, the other from the first integration that reduces the problem from second to first order. Three of the original types obtained are expressible as explicit solutions and can be integrated further in a simple fashion. For the remaining cases it is more feasible to integrate first the corresponding inverse functions: the qualitative analysis of these occupies §3. In §4 we integrate further the direct functions and get the full classification of the class considered. We compare, in §5, the mentioned class with the class of (proper and improper) affine hyperspheres: it turns out, in a sense, that very few are actually at the same time affine hyperspheres. Finally, some conclusions with regards to hypersurfaces with constant unimodular affine mean curvature are drawn in §6.

In summary, the purpose of this paper is two-fold: first, to show and analyze a good many new examples of hypersurfaces in affine geometry, and second, to prepare the way for using those examples in order to research more deeply the same topic and related partial differential equation problems. We conjecture that the methods developed and used in what follows shall be of help in formulating a solution to the Bernstein problem in higher dimensions.

1. General affine theory of hypersurfaces

In this section we deliver an abbreviated version of the general affine geometry of hypersurfaces. The interested reader can see further details in [12].

Let $X: M^n \to E^{n+1}$ be a codimension-one immersion of the real, oriented, n-dimensional, abstract differentiable manifold M, into the $(n+1)$-dimensional real vector space E. From the given immersion we can induce the tangent map as a section in a suitable bundle over the manifold; i.e.,

$$dX = DX \in \Gamma(\mathrm{Hom}(TM, E)) = \Gamma(E \otimes T^*M),$$

where TM is the tangent bundle of M, T^*M its cotangent bundle, and D represents the ambient space covariant derivative. In local coordinates $(t^1, t^2, \ldots, t^n)$ we can write

$$DX = D_i X \otimes dt^i = X_i \otimes dt^i \quad \text{(Summation convention)}.$$

Similarly, the Hessian form $D^2 X$ is given locally by

$$D^2 X = D_i(D_j X) \otimes (dt^i\, dt^j) = X_{ij} \otimes (dt^i\, dt^j).$$

For reasons of notation we choose a nonzero exterior $(n+1)$-form in E, or determinant function, its action being denoted by square brackets $[\ ,\ ,\ldots,\ ]$. This is commonly done in the case of unimodular affine geometry and it represents an essential object in the proper development of that theory. We expressly point out here, however, that in the present case it will only play the role of an auxiliary tool. All of the geometrical objects and properties to be described shall be independent of such a choice.

Now, if we consider the form

$$[D^2 X, (DX)^n] = [X_{ij}, X_1, \ldots, X_n](dt^i\, dt^j) \otimes (dt^1 \wedge \cdots \wedge dt^n),$$

and assume it to be *nondegenerate*, we can simultaneously normalize its scalar components, as well as its line bundle part, in order to introduce a pseudo-Riemannian structure invariant under the action of the full general affine group. This can be done in a unique way and leads to the construction of the first fundamental form of the geometry.

In what follows we proceed to use the method of moving frames, letting small latin letters denote indices ranging from 1 to $n = \dim(M)$, i.e. $1 \leq i, j, k, \ldots \leq n$, while small greek letters, when used as indices, range from 1 to $n+1 = \dim(E)$: $1 \leq \alpha, \beta, \gamma, \ldots, \leq n+1$.

Let $(f_1, \ldots, f_n)$ be a positively oriented frame field, locally defined on an open subset U of M, with corresponding dual coframe $(\sigma^1, \ldots, \sigma^n)$. We introduce a general affine frame field $(X, (e_1, \ldots, e_{n+1}))$ on $X(U)$ by putting $e_i = dX(f_i)$, and prescribing, besides, that e_{n+1} be a nonzero differentiable vector field, transversal, at each point, to the image hypersurface $X(U)$, i.e., $[e_1, e_2, \ldots, e_{n+1}] \neq 0$. If $(\omega^1, \ldots, \omega^{n+1})$ is the coframe of one forms, dual to $(e_1, \ldots, e_{n+1})$, we obtain readily the relationships

$$\sigma^i = X^*(\omega^i); \qquad \omega^{n+1}|_{X(U)} \equiv 0.$$

Hence, the tangent map dX can be written

(1.1) $$dX = \sigma^i e_i.$$

Similarly, we introduce the matrix (σ_α^β) of one forms by means of the equalities

(1.2) $$de_\alpha = \sigma_\alpha^\beta e_\beta.$$

Exterior differentiation, as applied to (1.1) and (1.2), provides us with the further relationships

(1.3) $$d\sigma^i = \sigma^j \wedge \sigma_j^i, \qquad \sigma^i \wedge \sigma_i^{n+1} = 0.$$

(1.4) $$d\sigma_\alpha^\beta = \sigma_\alpha^\gamma \wedge \sigma_\gamma^\beta$$

The two last numbered equations are the *Maurer-Cartan structural equations* of the immersion X with respect to invariance under the action of the general affine group $\mathrm{AGL}(n+1, R)$, as acting on the vector space E. They represent the *integrability conditions* in the language of moving frames.

From the last of equations (1.3) we obtain, by Cartan's lemma, the existence of differentiable functions h_{ij} such that

(1.5) $$\sigma_i^{n+1} = h_{ij}\sigma^j, \qquad h_{ij} = h_{ji},$$

with $H = \det(h_{ij}) \neq 0$, by the nondegeneracy assumed for X.

Then, by analyzing the variations occurring in these local objects when a change of frames is performed, it can be concluded that the symmetric bilinear form

$$\mathbf{I}_{\mathrm{ga}} = (g_{ij}\sigma^i\sigma^j) \otimes (\sigma^1 \wedge \cdots \wedge \sigma^n)^{-2/n}, \qquad g_{ij} = |H|^{-1/n}h_{ij},$$

defined on the tangent bundle TM, with values in the line bundle of scalar densities of weight $-2/n$, $[\Lambda^n(M)]^{-2/n}$, is an invariant of the immersion X under the action of the full general affine group $\mathrm{AGL}(n+1, R)$ as acting on E. Moreover, the real numbers $-2/n$, $-1/n$, are the unique ones to satisfy that defining property. $\mathbf{I}_{\mathrm{ga}}$ is the so-called *first fundamental form* of the geometry.

The nondegeneracy assumed for X also implies the existence, at each point on the image hypersurface $X(M)$, of a unique general affine invariant transversal straight line, called the *affine normal line* and characterized in our notation by

$$(1.6) \qquad \sigma_{n+1}^{n+1} + \frac{1}{n+2} d \log |H| - \frac{2}{n+2} d \log[e_1, \ldots, e_{n+1}] = 0.$$

If $X_N(M)$ is the normal bundle, constructed as the disjoint union of all affine normal lines, the local expression of the general affine normal N_{ga}, as a section in the bundle $X_N(M) \otimes [\Lambda^n(M)]^{2/n}$ is given by

$$(1.7) \qquad N_{ga} := |H|^{1/n} e_{n+1} \otimes (\sigma^1 \wedge \cdots \wedge \sigma^n)^{2/n}.$$

This construction, together with the definition of the second fundamental form, can also be seen alternatively as follows: the ambient space covariant derivative D induces on M many connections, in fact one for each choice of a distribution of straight lines transversal to $X(M)$, by projecting on the image tangent space and pulling back to M. All of these connections are $\mathrm{AGL}(n+1, R)$-invariant. Let us call any of those by ∇, and consider the first covariant derivative of I_{ga}, whose expression in our notation is given by the cubic form $\nabla I_{ga} = g_{ijk}(\sigma^i \sigma^j \sigma^k) \otimes (\sigma^1 \wedge \cdots \wedge \sigma^n)^{-2/n}$. By contracting this with (the inverse of) I_{ga} we introduce the Tschebyschev covector field $\tau = g^{ij} g_{ijk} \sigma^k$. Then, the general affine straight line at each point of $X(M)$ is uniquely determined by the uniform vanishing of this form, $\tau \equiv 0$. This relation, also called the *apolarity condition*, determines at once two more geometrical objects: a distinguished connection on M, called the *affine normal connection*, and a cubic (Fubini-Pick) form, as a section in the bundle $[T^*M]^3 \otimes [\Lambda^n(M)]^{-2/n}$, with local expression given by

$$(1.8) \qquad II_{ga} = g_{ijk}(\sigma^i \sigma^j \sigma^k) \otimes (\sigma^1 \wedge \cdots \wedge \sigma^n)^{-2/n},$$

which we call the *second fundamental form* of the geometry. Its identical annihilation characterizes quadric hypersurfaces.

A second remarkable connection can be induced on M from the first fundamental form I_{ga} (Levi-Civita type). However, the conditions of free-torsion and parallel translation of I_{ga} are not enough to provide uniqueness of the construction. This happens because I_{ga} is a density-valued object, and in covariantly differentiating such objects an extra term appears containing the trace of the connection matrix. That connecting object, which we call the *trace connection form*, plays an important role in the theory. We summarize these comments by stating that "there exists a unique affine connection $\tilde{\nabla}$ on M such that (a) $\tilde{\nabla} I_{ga} = 0$, (b) $\tilde{\nabla}$ is torsion-free, (c) the trace of $\tilde{\nabla}$ equals the trace of ∇, i.e. $\mathrm{tr}(\tilde{\nabla}) = \mathrm{tr}(\nabla)$, where ∇ is the normal connection."

The set of three objects consisting of the first fundamental form I_{ga}, the second fundamental form II_{ga}, and the trace connections form $\mathrm{tr}(\tilde{\nabla}) = \mathrm{tr}(\nabla)$ characterizes the immersed hypersurface $X(M)$ up to general affine motions of E.

We introduce next the *third fundamental form*: by exterior differentiating in equation (1.6), and also using (1.4), it follows that

$$(1.9) \qquad d\sigma_{n+1}^{n+1} = \sigma_{n+1}^{\alpha} \wedge \sigma_{\alpha}^{n+1} = \sigma_{n+1}^{k} \wedge \sigma_{k}^{n+1} = 0.$$

Cartan's lemma implies that

$$(1.10) \qquad \sigma_{n+1}^{i} = L^{ij}\sigma_{i}^{n+1}, \qquad L^{ij} = L^{ji}.$$

It follows that the fourth-order object represented locally by the expression

$$(1.11) \qquad \mathrm{III}_{\mathrm{ga}} = B_{ij}\sigma^{i}\sigma^{j},$$

where $B_{ij} := L^{pq}h_{pi}h_{qj}$, is an invariant of the geometry. Let us observe, moreover, that it is the first non-density-valued invariant object that appears in the development of the theory. Its paramount importance is strengthened, too, by the fact that, being a symmetric 2-tensor, it can be diagonalized, thus introducing the *general affine principal curvatures*. All of the symmetric functions constructed from these principal curvatures are invariants of the geometry. In particular, the *general affine mean curvature* is defined as the trace of $\mathrm{III}_{\mathrm{ga}}$ divided by n:

$$H_{\mathrm{ga}} = (1/n)\,\mathrm{tr}(\mathrm{III}_{\mathrm{ga}}).$$

We observe, on the other hand, that $\mathrm{III}_{\mathrm{ga}}$ can be expressed in terms of the other objects introduced previously. In fact, if we define $B_j^i := g^{ik}B_{kj}$, which are the scalar components of a section in a suitable bundle, we obtain that

$$B_j^i = \frac{1}{n-1}g^{pq}R_{pqj}^i,$$

where R_{pqj}^i are the scalar components of the curvature tensor of the affine normal connection.

To finish this section we ennumerate a complete set of integrability conditions where, besides the notations already introduced, we shall also use the contracted form of the second fundamental with respect to the first, i.e., $A_{jk}^i = (1/2)g^{ip}g_{pjk}$, the factor $1/2$ due to technical reasons. By a comma, we indicate covariant differentiation with respect to the Levi-Civita type connection, while a semicolon represents the one corresponding to the affine normal connection.

It is to be emphasized here, that these conditions represent the essential requirement that is needed in the formulation, and proof, of the so-called Fundamental Existence Theorem for the general affine theory of hypersurfaces, in general dimensions greater or equal than two. (See [12] for full details.)

Integrability conditions. (1) The scalar components g_{ij}, g_{ijk} are symmetric in all of their indices.

(2) The matrix (g_{ij}) is unimodular; i.e., $|\det(g_{ij})| = 1$.

(3) II_{ga} is apolar with respect to I_{ga}; i.e., $g^{ij}g_{ijk} = 0$, for every $k = 1, \ldots, n$.

(4) The trace connection form $\text{tr}(\nabla) = \text{tr}(\tilde{\nabla})$ is exact.

(5) It holds a "Mainardi-Codazzi" type condition. Namely,

$$g_{ijk,q} - g_{ijq,k} = B_{iq}g_{jk} + B_{jq}g_{ik} - B_{ik}g_{jq} - B_{jk}g_{iq}.$$

(6) It holds a "Gauss" type condition, expressing the curvature tensor of the Levi-Civita type connection:

$$\tilde{R}^{j}_{irq} = (1/2)(B^{j}_{q}g_{ir} - B^{j}_{r}g_{iq} + B_{ir}\delta^{j}_{q} - B_{iq}\delta^{j}_{r}) - (A^{s}_{iq}A^{j}_{sr} - A^{s}_{ir}A^{j}_{sq}).$$

(7) The affine normal covariant derivative of the third fundamental form satisfies

$$B^{k}_{j;q} = B^{k}_{q;j}.$$

This last is needed only for the case $n = 2$, since for $n \geq 3$ that last equation is a direct consequence of the Bianchi identity, as applied to the curvature tensor of the affine normal connection.

2. Hypersurfaces of decomposable type

We make in this section the main two hypotheses of the paper, namely

$(\mathscr{H}_1)$ We assume, first of all, that the hypersurface is projectable onto (part of) a hyperplane and that it can be expressed, with respect to a suitable affine coordinate system $(t^1, t^2, \ldots, t^n, t^{n+1})$ of the vector space E, in the form of Monge:

$$X(t^1, \ldots, t^n) = (t^1, \ldots, t^n, f(t^1, \ldots, t^n)),$$

with f enough differentiable. More precisely, we shall need in this work up to fourth-order derivatives, $\mathscr{C}^4$.

$(\mathscr{H}_2)$ We assume, furthermore, that the graph function f can be decomposed into a sum of n terms, each of them depending on only one of the independent variables $t^1, \ldots, t^n$:

$$f(t^1, \ldots, t^n) = f^1(t^1) + f^2(t^2) + \cdots + f^n(t^n).$$

By using the first hypothesis $(\mathscr{H}_1)$ we can write equation (1.1) as

(2.1) $$dX = \sigma^i e_i,$$

where now $e_i = \partial X/\partial t^i = (0, \ldots, 1, \ldots, f_i)$; $\sigma^i = dt^i$. (Here, and in what follows, subindices shall indicate partial derivatives, when applied to real- or vector-valued functions, i.e. $f_i = \partial f/\partial t^i$; $X_{ij} = \partial^2 X/\partial t^i \partial t^j$.) It follows that

(2.2) $$de_i = (0, \ldots, 0, f_{ij}\sigma^j).$$

Hence, if we choose the vector field e_{n+1} in the affine normal direction in such a way that the frame $(e_1, e_2, \ldots, e_{n+1})$ be positively oriented, write its components in the ambient space coordinate system as

$$e_{n+1} = (a^1, a^2, \ldots, a^{n+1})$$

and assume, without loss of generality, that the exterior $(n + 1)$-form (determinant function) $[\ ,\ ,\ldots,\ ]$ has been normalized with respect to that system, we obtain

$$(2.3) \qquad [e_1, e_2, \ldots, e_{n+1}] = \det \begin{pmatrix} 1 & 0 & \ldots & 0 & f_1 \\ 0 & 1 & 0 & \ldots & f_2 \\ \vdots & \vdots & \ddots & \vdots & \vdots \\ 0 & 0 & \ldots & 1 & f_n \\ a^1 & a^2 & \cdots & a^n & a^{n+1} \end{pmatrix}$$

$$= a^{n+1} - a^k f_k > 0.$$

Similarly, we calculate

$$(2.4) \qquad [e_1, \ldots, e_{j-1}, de_i, e_{j+1}, \ldots, e_{n+1}] = -a^j f_{ik}\sigma^k,$$

$$(2.5) \qquad [e_1, \ldots, e_n, de_i] = f_{ik}\sigma^k,$$

$$(2.6) \qquad [e_1, \ldots, e_n, de_{n+1}] = da^{n+1} - f_k da^k,$$

$$(2.7) \qquad \begin{aligned} &[e_1, \ldots, e_{j-1}, de_{n+1}, e_{j+1}, \ldots, e_{n+1}] \\ &= \left(a^{n+1} - \sum_{k \neq j} f_k a^k\right) da^j - a^j \left(da^{n+1} - \sum_{k \neq j} f_k da^k\right). \end{aligned}$$

From these equations and (2.1) we further obtain

$$(2.8) \qquad \sigma_i^j = \frac{[e_1, \ldots, e_{j-1}, de_i, e_{j+1}, \ldots, e_{n+1}]}{[e_1, \ldots, e_j, \ldots, e_{n+1}]} = -(a^{n+1} - a^k f_k)^{-1} a^j f_{iq}\sigma^q,$$

$$(2.9) \qquad \sigma_i^{n+1} = (a^{n+1} - a^k f_k)^{-1} f_{iq}\sigma^q,$$

$$(2.10) \qquad \sigma_{n+1}^{n+1} = (a^{n+1} - a^k f_k)^{-1}(da^{n+1} - f_q da^q),$$

$$(2.11) \qquad \begin{aligned} \sigma_{n+1}^j = (a^{n+1} - a^k f_k)^{-1} \Bigg[&\left(a^{n+1} - \sum_{k \neq j} f_k a^k\right) da^j \\ &- a^j \left(da^{n+1} - \sum_{k \neq j} f_k da^k\right)\Bigg]. \end{aligned}$$

Now, we can always assume that

$$(2.12) \qquad \sigma_{n+1}^{n+1} = 0.$$

In fact, if this were not the case, we substitute e_{n+1} by

$$\hat{e}_{n+1} = |H|^{1/(n+2)}[e_1, \ldots, e_{n+1}]^{-2/(n+2)} e_{n+1}.$$

This would keep the last vector field of the frame still contained in the affine normal line and, according to (1.6), with vanishing Pfaffian form $\hat{\sigma}_{n+1}^{n+1}$. Let us suppose, then, that e_{n+1} was already chosen to fulfill condition (2.12) above.

Then, it follows from (2.10) and (2.11) that

$$(2.13) \qquad \sigma_{n+1}^{j} = da^{j}.$$

Now, we apply the structural equations (1.4) to σ_i^{n+1}, and observe (1.5), (2.8), (2.9) and (2.12), to get

$$d\sigma_i^{n+1} = dh_{ij} \wedge \sigma^j = \sigma_i^k \wedge \sigma_k^{n+1} = -a^p h_{ij} h_{pk} \sigma^j \wedge \sigma^k.$$

From the second and fourth members of this it follows

$$\frac{\partial h_{ij}}{\partial t^k} \sigma^k \wedge \sigma^j = a^p h_{pk} h_{ij} \sigma^k \wedge \sigma^j,$$

which implies that

$$\frac{\partial h_{ij}}{\partial t^k} - a^p h_{pk} h_{ij} = \frac{\partial h_{ik}}{\partial t^j} - a^p h_{pj} h_{ik}.$$

Hence

$$(2.14) \qquad (n-1)a^p = h^{pk} h^{ij} \left(\frac{\partial h_{ij}}{\partial t^k} - \frac{\partial h_{ik}}{\partial t^j} \right),$$

where (h^{ij}) is the inverse matrix of (h_{ij}).

Now, since from (1.5) and (2.9) we have

$$(2.15) \qquad h_{ij} = (a^{n+1} - a^k f_k)^{-1} f_{ij},$$

it also follows that

$$|H| = |\det(h_{ij})| = (a^{n+1} - a^k f_k)^{-n} F, \quad \text{with } F := |\det(f_{ij})|.$$

This identity, together with (1.6), (2.3) and (2.12) furnish

$$(2.16) \qquad d\log(F) = (n+2)d\log(a^{n+1} - a^k f_k).$$

Therefore,

$$F = C(a^{n+1} - a^k f_k)^{n+2},$$

and we can assume, without loss of generality, to have the constant C normalized, i.e., $C = 1$, in order to write

$$(2.17) \qquad F = (a^{n+1} - a^k f_k)^{n+2},$$

and obtain in (2.15) that

$$(2.18) \qquad h_{ij} = F^{-1/(n+2)} f_{ij}.$$

This last equality, applied in (2.14), provides

$$(2.19) \qquad a^p = -\frac{1}{n+2} F^{-(n+1)/(n+2)} f^{kp} F_k.$$

If we now apply hypothesis $(\mathscr{H}_2)$, we obtain successively:

$$(2.20) \qquad f_i = \frac{\partial f}{\partial t^i} = (f^i)' = \frac{df^i}{dt^i},$$

for the first derivatives of f;

$$(2.21) \qquad f_{ij} = \begin{cases} (f^i)'', & \text{if } i = j, \\ 0, & \text{if } i \neq j, \end{cases}$$

for the second derivatives;

$$(2.22) \qquad F = |(f^1)'' \cdot (f^2)'' \cdots (f^n)''|,$$

for the absolute value of the Hessian matrix of f. Observe that, in each case, the symbol for ordinary differentiation applies to the corresponding independent variable, and recall, too, that we are assuming X to be nondegenerate. Hence, we further obtain

$$(2.23) \qquad F_k := \frac{\partial F}{\partial t^k} = F \frac{(f^k)'''}{(f^k)''},$$

and

$$(2.24) \qquad a^k = -\frac{1}{n+2} F^{1/(n+2)} \frac{(f^k)'''}{[(f^k)'']^2}.$$

We compute next the derivatives of the last to get

$$(2.25) \qquad \frac{\partial a^k}{\partial t^k} = -\frac{1}{(n+2)^2} F^{1/(n+2)} \left[(n+2) \frac{(f^k)^{iv}}{[(f^k)'']^2} - (2n+3) \frac{[(f^k)''']^2}{[(f^k)'']^3} \right],$$

and

$$(2.26) \qquad \frac{\partial a^k}{\partial t^j} = -\frac{1}{(n+2)^2} F^{1/(n+2)} \frac{(f^j)'''}{(f^j)''} \cdot \frac{(f^k)'''}{[(f^k)'']^2}, \quad \text{for } j \neq k.$$

Now, by applying to these expressions (1.10), (1.11) and (2.13), we obtain

$$(2.27) \qquad B_{kk} = -\frac{1}{(n+2)^2} \left[(n+2) \cdot \frac{(f^k)^{iv}}{(f^k)''} - (2n+3) \frac{[(f^k)''']^2}{[(f^k)'']^2} \right],$$

and

$$(2.28) \qquad B_{jk} = -\frac{1}{(n+2)^2} \frac{(f^j)'''}{(f^j)''} \frac{(f^k)'''}{(f^k)''}, \quad \text{if } j \neq k$$

for the matrix components of the third fundamental form III_{ga}.

Therefore, we come to the important conclusion that, for hypersurfaces of decomposable type, the affine mean curvature equation, which is a partial differential equation of order four, reduces in that case to an ordinary differential equation, also of fourth order, because of the splitting that occurs in the diagonal elements of the matrix representing the third fundamental form; i.e. $\text{tr}(\text{III}_{ga}) = \text{constant}$ if, and only if, $B_{kk} = \text{constant}$, for each of the entries in the diagonal of the matrix (B_{ij}). We state this result in the form of a lemma.

LEMMA (2.1). *Let* $X\colon M^n \to E^{n+1}$ *be a hypersurface of decomposable type. Then* X *is of constant general affine mean curvature,* $H_{ga} = constant$ *if, and only if, each of the elements in the diagonal of the matrix representing the third fundamental form is constant:* $B_{kk} = constant$.

Henceforth, we shall call

$$(2.29) \qquad (n+2)\frac{(f^k)^{iv}}{(f^k)''} - (2n+3)\frac{[(f^k)''']^2}{[(f^k)'']^2} = K = \text{constant}$$

the equation of constant general affine mean curvature for hypersurfaces of decomposable type and dimension equal to n.

An immediate observation that we can make is that, by substituting

$$(2.30) \qquad x = t^k, \qquad (f^k)'' = y = g(x)$$

into equation (2.29), this reduces to

$$(2.31) \qquad (n+2)\frac{y''}{y} - (2n+3)\frac{(y')^2}{y^2} = K.$$

Now, there is a standard method for finding the first integral of this non-linear ordinary differential equation. One writes
(2.32)

$$(y')^2 = z, \quad \text{and hence} \quad 2y'\frac{dy'}{dy} = \frac{dz}{dy}; \; y'' = \frac{dy'}{dx} = \frac{dy'}{dy}\frac{dy}{dx} = \frac{dy'}{dy}y',$$

and substitutes these identities into (2.31) to get

$$(2.33) \qquad \frac{n+2}{2}y\,dz - [(2n+3)z + Ky^2]dy = 0.$$

If we try to perform a first integration of this equation, under the additional condition for y to satisfy $y > 0$, that equation has $\mu = y^{-(5n+8)/(n+2)}$ as integrating factor. In fact, by multiplying in (2.33) by the given μ we obtain

$$(2.34) \quad \frac{n+2}{2}y^{-(4n+6)/(n+2)}dz - [(2n+3)zy^{-(5n+8)/(n+2)} + Ky^{-(3n+4)/(n+2)}] = 0,$$

which is an exact equation.

Its solution can be expressed by

$$(2.35) \qquad \frac{n+2}{2}y^{-(4n+6)/(n+2)}z + \frac{n+2}{2n+2}Ky^{-(2n+2)/(n+2)} = \frac{n+2}{2}C.$$

From this and (2.32) it follows that

$$(2.36) \qquad (y')^2 = y^2\left(Cy^{(2n+2)/(n+2)} - \frac{K}{n+1}\right).$$

Let us observe here that, for n odd, the same equation holds with $y < 0$. On the other hand, for n even and $y < 0$, the same no longer makes sense but, it is easy to see that in the later case, a solution can be found to be

$$(y')^2 = y^2\left(C(-y)^{(2n+2)/(n+2)} - \frac{K}{n+1}\right).$$

In order to avoid unnecessary complications in notation we shall deal, from now on, exclusively with solutions for which $y > 0$. However, it is clear, from the previous observation and from the context of what follows, that all of the conditions discussed, and properties obtained, are shared mutatis mutandis for the symmetrical hypothesis in which $y < 0$. For example, all graphs obtained as solutions with $y > 0$ can be symmetrized with respect to the x-axis furnishing new graphical solutions to the equation. It is obvious, too, that this symmetry extends to the integrals of these solutions and we, in fact, shall need to integrate twice more original solutions of the form $y = g(x)$, according to equations (2.30) and (2.31) above.

Another, in a sense similar, and also previous, observation that shall facilitate our exposition is that, when taking the square root in equation (2.36), the consideration of two separate cases for the different signs that appear, namely

$$(2.37) \qquad y' = \pm y \left(Cy^{(2n+2)/(n+2)} - \frac{K}{n+1} \right)^{1/2},$$

would amount only to a change in the orientation of the x-axis. In fact, in most cases we are going to see that it is more feasible to obtain, as first integration, x *as a function of* y. Therefore, we can drop the case of the $-$ sign in equation (2.37), and consider only the case of the $+$ sign. For each solution that we get under this hypothesis, there exists another one constructed just by symmetrizing the first with respect to the y-axis. This explains why, from now on, we shall consider only equation

$$(2.38) \qquad y' = y \left(Cy^{(2n+2)/(n+2)} - \frac{K}{n+1} \right)^{1/2}.$$

We collect these observations in a lemma.

LEMMA (2.2). *The analysis of solutions of the nonlinear second-order ordinary differential equation (2.31), including further integration, can be reduced to the analysis of positive solutions of the first-order nonlinear ordinary differential equation (2.38). There exist three remaining classes of solutions, whose corresponding properties, including further integration, can be obtained by reflection on the x- and y-axis.*

Except for some exceptional cases, for example when $C = 0$, further integration of equation (2.38) requires a qualitative analysis of the solutions. This is accomplished in the following sections.

3. Qualitative analysis for the inverse function

First we consider, in equation (2.38), the case $C = 0$. With this condition it follows that

$$y' = y \left(-\frac{K}{n+1} \right)^{1/2}$$

forcing K to be ≤ 0. The solution can be calculated explicitly, being convenient to separate both subcases:

Type I: $C = 0$, $K = 0$. Here, we have the solutions $y = \tilde{C}$, which after two more integrations will give for the corresponding component the solutions $f^k(t^k) = C_k(t^k)^2 + D_k t^k + E_k$ (k not summed). By an affine change of coordinates, we can even write

$$f^k(t^k) = (t^k)^2.$$

If all of the components are of this kind, or the corresponding ones obtained by reflection on the x-axis, we get all of the different types of paraboloids, which are very well known hypersurfaces in affine geometry.

Type II: $C = 0$, $K < 0$. We get in this case the solutions $y = \tilde{C} \exp(-\frac{K}{n+1})^{1/2} x$. After two more integrations, and normalization of the constants of integrations, one gets for the corresponding component of the hypersurface

$$f^k(t^k) = \exp\left[\left(-\frac{K}{n+1}\right)^{1/2} t^k\right].$$

Type III: $C > 0$, $K = 0$. Here it is again possible to integrate explicitly. After doing that, performing two more integrations and normalizing the constants one gets, for example, solutions of the form

$$f^k(t^k) = -(-t^k)^{n/(n+1)}.$$

For the remaining cases, where explicit solutions can no longer be computed in an immediate way, we are in pursuit of analyzing the local, as well as global, behaviour of y as a function of x: $y = g(x)$, when this represents a solution to equation (2.38). However, it is more convenient, first, to look at x as a function of y. For this purpose we consider (2.38) in its equivalent form

$$(3.1) \qquad \frac{dx}{dy} = \frac{1}{y\left(Cy^{(2n+2)/(n+2)} - \frac{K}{n+1}\right)^{1/2}}.$$

By fixing a point (x_0, y_0), with $y_0 > 0$, as initial condition, we can express a solution to the above by

$$(3.2) \qquad x = x_0 + \int_{y_0}^{y} \frac{dt}{t\left(Ct^{(2n+2)/(n+2)} - \frac{K}{n+1}\right)^{1/2}}.$$

We shall use the last expression in order to accomplish our goal of analyzing the behaviour of solutions for the remaining cases. Prior to that, we make a couple of observations that shall be of help:

First, the local behaviour shall depend mainly on the first two derivatives of x, with respect to y, which we now calculate and put together as

  (3.3)
$$\frac{dx}{dy} = y^{-1}\left(Cy^{(2n+2)/(n+2)} - \frac{K}{n+1}\right)^{-1/2},$$

$$\frac{d^2x}{dy^2} = \frac{-\frac{3}{2}Cy^{(2n+2)/(n+2)} + \frac{K}{n+1}}{y^2\left(Cy^{(2n+2)/(n+2)} - \frac{K}{n+1}\right)^{3/2}}.$$

Second, the global behaviour of the solutions shall also depend on the analysis in the neighborhood of singular points, as well as on the questions of convergence, or divergence, of the integral in (3.2).

Type IV: $C > 0$, $K > 0$. Since we must have $w = Cy^{(2n+2)/(n+2)} - \frac{K}{n+1} > 0$, we choose

$$y_0 > \bar{y}_0 := \left[\frac{KC^{-1}}{n+1}\right]^{(n+2)/(2n+2)},$$

and make the substitution $u = [\frac{(n+1)C}{K}]^{(n+2)/(2n+2)}t$, to write (3.2) in the form

(3.4) $\qquad x = x_0 + \left(\frac{n+1}{K}\right)^{1/2}\int_{z_0}^{z}\frac{du}{u(u^{(2n+2)/(n+2)} - 1)^{1/2}},$

with $z = [\frac{(n+1)C}{K}]^{(n+2)/(2n+2)}y$, and corresponding equality for z_0 in terms of y_0.

Now, the integral in (3.4) can be estimated. In fact, there exists u_0 $(> z_0)$ such that $u^{1+n/(n+2)} - 1 > u$, for all $u > u_0$. Therefore, since the integral $\int_{z_0}^{\infty} du/u^{3/2}$ is convergent, it follows that there exists a real number x_1 $(> x_0)$ such that

(3.5) $\qquad\qquad\qquad \lim_{y\to\infty} x = x_1.$

On the other hand, by writing $p = 1 + \frac{n}{n+2}$ and $v = u^p - 1$, for the integral in (3.4), we have

$$\int_{z_0}^{z}\frac{du}{u(u^p - 1)^{1/2}} = (1/p)\int_{v_0}^{v}\frac{dv}{(v+1)v^{1/2}}$$

with $v = z^p - 1$, $v_0 = z_0^p - 1$. For $1 < z < z_0$ it holds

$$0 < \int_{v}^{v_0}\frac{dv}{(v+1)v^{1/2}} < \int_{v}^{v_0}\frac{dv}{v^{1/2}}.$$

This proves that there exists $\bar{x}_1$ $(< x_0)$ such that

(3.6) $\qquad\qquad\qquad \lim_{y\to\bar{y}_0} x = \bar{x}_1.$

Finally, it is easy to see, from (3.3), that

$$(3.7) \qquad \frac{dx}{dy} > 0, \qquad \frac{d^2x}{dy^2} < 0,$$

for every $y > \overline{y}_0$.

The graph of x versus y is fully characterized by equations (3.5), (3.6) and (3.7) above.

Type V: $C > 0$, $K < 0$. Let us take as initial point (x_0, y_0), with $y_0 > 0$. Then we can write (3.2) in the form

$$x = x_0 + \frac{1}{C^{1/2}} \int_{y_0}^{y} \frac{dt}{t(t^p - a)^{1/2}},$$

where $a = KC^{-1}/(n+1)$ and $p = (2n+2)/(n+2)$. Thus, since $t^p - a > t^p$ for every $t > 0$, it follows easily that there exists x_2 such that

$$(3.8) \qquad \lim_{y \to \infty} x = x_2, \qquad x_2 > x_0.$$

On the other hand $t^p - a$ is bounded above by the value it takes in y_0, for $t < y_0$, and hence

$$(3.9) \qquad \lim_{y \to 0} x = -\infty.$$

From (3.3), one gets for every $y \in (0, \infty)$,

$$(3.10) \qquad \frac{dx}{dy} > 0, \qquad \frac{d^2x}{dy^2} < 0.$$

Type VI: $C < 0$, $K < 0$. Here we take (x_0, y_0) such that $y_0 \in (0, \overline{y}_0)$, with $\overline{y}_0 = [KC^{-1}/(n+1)]^{(n+2)/(2n+2)}$ being one of the singularities of (3.2). Then such equation can be written

$$x = x_0 + \frac{1}{(-C)^{1/2}} \int_{y_0}^{y} \frac{dt}{t(a - t^p)^{1/2}},$$

where, as in the previous case, $a = KC^{-1}/(n+1)$, $p = (2n+2)/(n+2)$.

We see that the integrand in the last expression is bounded below by $(a)^{-1/2}t^{-1}$. Hence, it follows that

$$(3.11) \qquad \lim_{y \to 0} x = -\infty.$$

On the other hand, by making the substitution $u = a - t^p$, we obtain for $y > y_0$, the estimate

$$\int_{y_0}^{y} \frac{dt}{t(a - t^p)^{1/2}} = -\frac{1}{p} \int_{a-y_0^p}^{a-y^p} \frac{du}{(a - u)u^{1/2}} < -\frac{1}{py_0^p} \int_{a-y_0^p}^{a-y^p} \frac{du}{u^{1/2}},$$

from which it follows that

$$(3.12) \qquad \lim_{y \to \overline{y}_0} x = x_3 > x_0.$$

Finally, by observing (3.3) for this case, we can see that, for $y \in (0, \overline{y}_0)$,

(3.13)
$$\frac{dx}{dy} > 0,$$

while the behaviour of the second derivative is given by

(3.14)
$$\frac{d^2x}{dy^2} < 0, \quad \text{in } (0, \check{y}_0), \qquad \frac{d^2x}{dy^2}(\check{y}_0) = 0;$$
$$\frac{d^2x}{dy^2} > 0 \quad \text{in } (\check{y}_0, \overline{y}_0),$$

where $\check{y}_0 = [(2/3)KC^{-1}/(n+1)]^{(n+2)/(2n+2)}$.

4. Integrating further the direct function

We have computed in the previous section explicit solutions for three of the types under consideration. From the analysis practiced on the remaining cases, which render the inverse functions of the solutions, $x = x(y)$, one gets the unified conclusion that this last is always invertible, allowing one to define y as a function of x: $y = g(x)$, for all possible cases. In order to obtain the components of the function defining the hypersurface we need, according to (2.30) and (2.31), to further integrate $y = g(x)$ two more times. This can be done for all of the types, as we are about to see in the current section. A fundamental tool for accomplishing that goal will be the ordinary differential equation (2.38), which we now rewrite as

(4.1)
$$\frac{dy}{dx} = g'(x) = g(x) \left(C[g(x)]^{(2n+2)/(n+2)} - \frac{K}{n+1} \right)^{1/2}.$$

We analyze separately the diverse remaining cases.

Type IV: $C > 0$, $K > 0$. It follows from (3.5), (3.6) and (3.7) that $y = g(x)$, with $g: (\overline{x}_1, x_1) \to R$, is characterized by

(4.2)
$$\lim_{x \to x_1} g(x) = +\infty.$$

(4.3)
$$\lim_{x \to \overline{x}_1} g(x) = \overline{y}_0 = \left[\frac{KC^{-1}}{n+1} \right]^{(n+2)/(2n+2)} > 0.$$

(4.4)
$$\frac{dg}{dx} > 0, \qquad \frac{d^2g}{dx^2} > 0,$$

for every $x \in (\overline{x}_1, x_1)$.

Now a typical integral of the function g can be written

(4.5)
$$G(x) = \check{y}_0 + \int_{\tilde{x}_0}^{x} g(t)\,dt,$$

for some point $(\tilde{x}_0, \tilde{y}_0)$, with $\tilde{x}_0 \in (\overline{x}_1, x_1)$.

By using (4.1) we can write $G(x)$ as

(4.6) $\qquad G(x) = \tilde{y}_0 + \left(\dfrac{n+1}{K}\right)^{1/2} \displaystyle\int_{\tilde{x}_0}^x \dfrac{g'(t)dt}{(a[g(t)]^{(2n+2)/(n+2)} - 1)^{1/2}}$,

with $a = \dfrac{(n+1)C}{K}$.

By integrating this last by parts one obtains

(4.7)
$$G(x) = \tilde{y}_0 + \left(\dfrac{n+1}{K}\right)^{1/2} \left[\dfrac{g(t)}{(a[g(t)]^{(2n+2)/(n+2)} - 1)^{1/2}}\right]_{x_0}^x$$
$$+ \left(\dfrac{n+1}{K}\right)^{1/2} \dfrac{a}{2} \int_{\tilde{x}_0}^x \dfrac{[g(t)]^{(2n+2)/(n+2)} g'(t)\, dt}{(a[g(t)]^{(2n+2)/(n+2)} - 1)^{3/2}}.$$

Now, since for $x > x_0$ the third term in the right-hand side of this identity is greater than 0, while the second one diverges to $+\infty$ as x converges to x_1, it follows that

(4.8) $\qquad\qquad\qquad \displaystyle\lim_{x \to x_1} G(x) = +\infty$.

On the other hand, from (4.3) and (4.5) one gets

(4.9) $\qquad\qquad\qquad \displaystyle\lim_{x \to \overline{x}_1} G(x) = \overline{G}_1 \in \mathbf{R}$.

Finally, by (4.2) through (4.5), it follows that

(4.10) $\qquad G'(x) = g(x) > 0, \qquad G''(x) = g'(x) > 0$,

for every $x \in (\overline{x}_1, x_1)$.

The last three numbered conditions characterize the graph of the function $G: (\overline{x}_1, x_1) \to \mathbf{R}$.

We need one further integration and write

(4.11) $\qquad\qquad\qquad F(x) = \hat{y}_0 + \displaystyle\int_{\hat{x}_0}^x G(t)\, dt$

with $\hat{x}_0 \in (\overline{x}_1, x_1)$.

It is immediate to obtain, from (4.8), (4.9) and (4.10), that the first two derivatives of F behave as follows:

(4.12)
$$F'(x) = G(X) \quad \text{ranges from } \overline{G}_1 \text{ to } +\infty,$$
$$F''(x) = G'(x) > 0, \quad \text{for every } x \in (\overline{x}_1, x_1).$$

It is also easy to see, on the other hand, that

(4.13) $\qquad\qquad\qquad \displaystyle\lim_{x \to \overline{x}_1} F(x) = \overline{F}_1 \in \mathbf{R}$.

To complete the study on the global behaviour of F we need to determine $\lim_{x \to x_1} F(x)$, and to accomplish this we have to analyze in more detail $G(x)$ near that limit point. Let us consider the integral part in equation (4.6). It is easy to see that there exists t_1 in the open interval $(\tilde{x}_0, x_1)$ such that for every $t \in (t_1, x_1)$ we have

(4.14) $[g(t)]^{1+1/(n+2)} < a[g(t)]^{(2n+2)/(n+2)} - 1$.

From this we obtain the estimate

$$\int_{t_1}^{x} \frac{g'(t)dt}{(a[g(t)]^{(2n+2)/(n+2)} - 1)^{1/2}} < \int_{t_1}^{x} \frac{g'(t)\,dt}{[g(t)]^{(n+3)/(2n+4)}}$$

(4.15)
$$= \frac{2n+4}{n+1}([g(x)]^{(n+1)/(2n+4)} - [g(t_1)]^{(n+1)/(2n+4)}),$$

for every x in the open interval (t_1, x_1). Next, we estimate the integral of $[g(t)]^p$, with $p = \frac{n+1}{2n+4}$. We can write, by using (4.1) and (4.14),

$$\int_{t_1}^{x} [g(t)]^p \, dt = \left(\frac{n+1}{K}\right)^{1/2} \int_{t_1}^{x} \frac{[g(t)]^{p-1} g'(t)\,dt}{(a[g(t)]^{(2n+2)/(n+2)} - 1)^{1/2}}$$

$$< -(n+2)\left(\frac{n+1}{K}\right)^{1/2}([g(x)]^{-1/(n+1)} - [g(t_1)]^{-1/(n+1)})$$

and since the last member has a finite limit as $x \to x_1$, it follows that

(4.16) $\lim_{x \to x_1} F(x) = F_1 \in \mathbf{R}$.

Therefore, $F: (\overline{x}_1, x_1) \to \mathbf{R}$ is characterized by conditions (4.12), (4.13) and (4.16). For example, the way its graph looks will depend on whether G_1 is < 0, $= 0$ or > 0.

Type V: $C > 0$, $K < 0$. According to (3.8), (3.9) and (3.10), we have for the direct function $y = g(x)$, $g: (-\infty, x_2) \to R$, the characterizing conditions

(4.17) $\lim_{x \to x_2} g(x) = +\infty$.

(4.18) $\lim_{x \to -\infty} g(x) = 0$.

(4.19) $\dfrac{dg}{dx} > 0, \qquad \dfrac{d^2 g}{dx^2} > 0$,

for every $x \in (-\infty, x_2)$.

Let us choose a point $(\tilde{x}_0, \tilde{y}_0)$, with $\tilde{x}_0 < x_2$. Then, by also using (4.1), a first integral of g can be written

(4.20) $G(x) := \tilde{y}_0 + \displaystyle\int_{\tilde{x}_0}^{x} g(t)\,dt = \tilde{y}_0 + \frac{1}{(C)^{1/2}} \int_{\tilde{x}_0}^{x} \frac{g'(t)dt}{([g(t)]^{(2n+2)/(n+2)} + b)^{1/2}}$,

with $b = -KC^{-1}/(n+1) > 0$. Next, by an argument quite similar to the one used in the previous case, that is, integrating by parts and observing that the term containing the integral is greater than zero while the other diverges to $+\infty$ as $x \to x_2$, it follows that

(4.21) $\lim_{x \to x_2} G(x) = +\infty$.

It is also immediate to check that

$$(4.22) \qquad G'(x) = g(x) > 0, \qquad G''(x) = g'(x) > 0,$$

for every x in the open interval $(-\infty, x_2)$.

Now, if we fix a real number r with $0 < r < \frac{2n+2}{n+2}$, we can find $t_2 \in (-\infty, x_2)$ such that

$$[g(t)]^r < [g(t)]^{(2n+2)/(n+2)} + b,$$

for every $t < t_2$. Hence, the integral in the last member of (4.20) can be estimated. In fact, for $x < t_2$ we have

$$(4.23) \quad 0 < \int_x^{t_2} \frac{g'(t)dt}{([g(t)]^{(2n+2)/(n+2)} + b)^{1/2}} < \int_x^{t_2} \frac{g'(t)\,dt}{[g(t)]^{r/2}} = \left[\frac{[g(t)]^{1-r/2}}{1 - \frac{r}{2}} \right]_x^{t_2},$$

and since this last has a positive finite limit as $x \to -\infty$, it follows that

$$(4.24) \qquad \lim_{x \to -\infty} G(x) = G_{-\infty} \in \mathbf{R}.$$

We proceed now to integrate the function G, by choosing a point $(\hat{x}_0, \hat{y}_0)$, with $\hat{x}_0 < x_2$, and define $F: (-\infty, x_2) \to \mathbf{R}$ by

$$(4.25) \qquad F(x) = \hat{y}_0 + \int_{\hat{x}_0}^x G(t)\, dt.$$

In order to analyze the behaviour of F we observe, first, that its second derivative is always positive; i.e.,

$$(4.26) \qquad \frac{d^2 F}{dx^2} = \frac{dG}{dx} = g(x) > 0,$$

while its first derivative $\frac{dF}{dx} = G(x)$ is always increasing and ranges from $G_{-\infty}$ to $+\infty$. It is fairly obvious to get conclusions as to the behaviour of F near $-\infty$ in the cases $G_{-\infty} > 0$, $G_{-\infty} < 0$. To see what happens in the remaining limiting case of $G_{-\infty} = 0$ we notice first that, in this case, we can write

$$G(x) = \int_{-\infty}^x g(t)\, dt.$$

Hence, from (4.23) we get the estimate

$$0 < G(x) < \frac{1}{1 - \frac{r}{2}} [g(x)]^{1-r/2}.$$

By integrating this, with $x < t_2$, and by applying (4.1) and (4.23), we obtain

$$0 < \int_x^{t_2} G(t)\, dt < \frac{1}{1 - \frac{r}{2}} \int_x^{t_2} [g(t)]^{1-r/2}\, dt < \frac{1}{1 - \frac{r}{2}} \int_x^{t_2} [g(t)]^{-r} g'(t)\, dt$$

$$= \frac{1}{1 - r} \frac{1}{1 - \frac{r}{2}} ([g(t_2)]^{1-r} - [g(x)]^{1-r}),$$

and since we can take $r < 1$, it follows that $\int_{-\infty}^{t_2} G(t)\,dt$ is finite. This allows us to conclude that

$$(4.27) \qquad \lim_{x \to -\infty} F(x) = \begin{cases} +\infty, & \text{if } G_{-\infty} < 0, \\ F_{-\infty} \in \mathbf{R}, & \text{if } G_{-\infty} = 0, \\ -\infty, & \text{if } G_{-\infty} > 0. \end{cases}$$

To finish the analysis of the function F for this case it remains only to study its behaviour near x_2. For this purpose we consider first the integral in the last member of (4.20). From there, since $b > 0$, we get, for $x > x_0$, the inequality

$$(4.28) \qquad 0 < \int_{x_0}^{x} \frac{g'(t)\,dt}{([g(t)]^{(2n+2)/(n+2)} + b)^{1/2}} < \int_{x_0}^{x} \frac{g'(t)\,dt}{[g(t)]^{(n+1)/(n+2)}}$$
$$= (n+2)([g(x)]^{1/(n+2)} - [g(x_0)]^{1/(n+2)}).$$

We estimate, next, the integral of the variable part in this last member, by also using again (4.1), to obtain

$$\int_{x_0}^{x} [g(t)]^{1/(n+2)}\,dt = \frac{1}{C^{1/2}} \int_{x_0}^{x} \frac{[g(t)]^{-(n+1)/(n+2)} g'(t)\,dt}{([g(t)]^{(2n+2)/(n+2)} + b)^{1/2}}$$
$$< \frac{1}{C^{1/2}} \int_{x_0}^{x} [g(t)]^{-(2n+2)/(n+2)} g'(t)\,dt$$

and since the last right-hand side converges to a positive finite limit as $x \to x_2$, we conclude finally that

$$(4.29) \qquad \lim_{x \to x_2} F(x) = F_2 \in \mathbf{R}.$$

Type VI: $C < 0$, $K < 0$. The direct function $y = g(x)$, $g: (-\infty, x_3) \to \mathbf{R}$, has the following properties

$$(4.30) \qquad \lim_{x \to -\infty} g(x) = 0.$$

$$(4.31) \qquad \lim_{x \to x_3} g(x) = \bar{y}_0 = \left[\frac{KC^{-1}}{n+1} \right]^{(n+2)/(2n+2)}.$$

$$(4.32) \qquad \frac{dg}{dx} > 0, \quad \text{for every } x \in (-\infty, x_3).$$

$$(4.33) \qquad \begin{aligned} \frac{d^2 g}{dx^2} > 0 \ \text{ in } (-\infty, \check{x}_0), \qquad & \frac{d^2 g}{dx^2}(\check{x}_0) = 0, \\ \frac{d^2 g}{dx^2} < 0, \quad \text{in } (\check{x}_0, x_3), & \end{aligned}$$

where $\check{x}_0 = g^{-1}(\check{y}_0)$ and $\check{y}_0 = [(2/3)KC^{-1}/(n+1)]^{(n+2)/(2n+2)}$.
These are easily obtained from (3.12), (3.13) and (3.14).

In order to integrate g, in the present case, we choose a point $(\tilde{x}_0, \tilde{y}_0)$, with $\tilde{x}_0 < x_3$, and define $G: (-\infty, x_3) \to \mathbf{R}$, by

$$(4.34) \quad G(x) := \tilde{y}_0 + \int_{\tilde{x}_0}^x g(t)\, dt = \tilde{y}_0 + \frac{1}{(-C)^{1/2}} \int_{\tilde{x}_0}^x \frac{g'(t)dt}{(a - [g(t)]^{(2n+2)/(n+2)})^{1/2}}$$

where the last equality is obtained from (4.1) by putting $a = KC^{-1}/(n+1)$.

Now, if we fix a real number $r \in (0, \frac{2n+2}{n+2})$, there exists $t_3 < x_3$ such that $[g(t)]^r < a - [g(t)]^{(2n+2)/(n+2)}$, for every $t < t_3$. Hence, we can write, for $x < t_3$, the estimate

$$(4.35) \quad 0 < \int_x^{t_3} \frac{g'(t)dt}{(a - [g(t)]^{(2n+2)/(n+2)})^{1/2}} < \int_x^{t_3} \frac{g'(t)\, dt}{[g(t)]^{r/2}},$$

and since the right-hand side has a finite limit as $x \to -\infty$, it follows that

$$(4.36) \quad \lim_{x \to -\infty} G(x) = G_{-\infty} \in \mathbf{R}.$$

On the other hand, it is obvious from (4.31) that there exists a real number G_3 such that

$$(4.37) \quad \lim_{x \to x_3} G(x) = G_3 \in \mathbf{R}.$$

From (4.30) through (4.32) we obtain the remaining characterizing conditions for G, namely

$$(4.38) \quad \begin{aligned} \frac{dG}{dx} &= g(x) > 0, \\ \frac{d^2 G}{dx^2} = \frac{dg}{dx} &> 0, \quad \text{for every } x \in (-\infty, x_3). \end{aligned}$$

We integrate once more by choosing $(\hat{x}_0, \hat{y}_0)$, with $\hat{x}_0 < x_3$, and defining

$$(4.39) \quad F(x) = \hat{y}_0 + \int_{\hat{x}_0}^x G(t)\, dt.$$

It is obvious, from (4.37), that

$$(4.40) \quad \lim_{x \to x_3} F(x) = F_3 \in \mathbf{R},$$

while (4.38) furnishes

$$(4.41) \quad \frac{d^2 F}{dx^2} = \frac{dG}{dx} > 0$$

and $dF/dx = G(x)$ ranges from $G_{-\infty}$ to G_3, according to (4.36) and (4.37). Now, $G_{-\infty}$ can be > 0, < 0, or $= 0$, with different behaviour for F near $-\infty$, in consistency with those three different possibilities. We analyze next the limit case where $G_{-\infty} = 0$, since the other two are easier to draw conclusions from. We can express, in that case, the function G as

$G(x) = \int_{-\infty}^{x} g(t)\,dt$, and by using (4.35) we can estimate it, for $x < t_3$, as follows:

$$(4.42) \qquad 0 < G(x) < \frac{1}{(-C)^{1/2}} \frac{1}{1 - \frac{r}{2}} ([g(x)]^{1-r/2}).$$

Hence we get, for $x < t_3$,

$$(4.43) \quad 0 < \int_{x}^{t_3} G(t)\,dt < \frac{1}{(-C)^{1/2}} \frac{2}{2-r} \cdot \frac{1}{1-r} ([g(t_3)]^{1-r} - [g(x)]^{1-r}),$$

by using (4.1), (4.34) and (4.35).

Since r can be chosen to be < 1, we conclude that $\int_{-\infty}^{t_3} G(t)\,dt$ is finite and, hence, $\lim_{x \to -\infty} F(x) = F_{-\infty}$ is also finite. Therefore,

$$(4.44) \qquad \lim_{x \to -\infty} F(x) = \begin{cases} +\infty, & \text{if } G_{-\infty} < 0, \\ F_{-\infty} \in \mathbf{R}, & \text{if } G_{-\infty} = 0, \\ -\infty, & \text{if } G_{-\infty} > 0. \end{cases}$$

This finishes the analysis of the function F in this case and, therefore, in all of possible cases.

THEOREM (4.1). *Let* $X: M^n \to E^{n+1}$ *be a hypersurface of decomposable type and constant general affine mean curvature. Then each of its components* $f^1, f^2, \ldots, f^n$, *must be of one of the original types* I *through* VI, *whose properties are enumerated below, or the corresponding three more kinds of types that are obtained from those original types by suitable reflections in the x- and y-axes. All of the solutions belonging to the original types share the common feature that their second derivatives,* $(f^k)'' = y > 0$, *satisfy, in each case, the classifying ordinary differential equation*

$$y' = y \left(Cy^{(2n+2)/(n+2)} - \frac{K}{n+1} \right)^{1/2}.$$

Type I: $C = 0$, $K = 0$. $f^k: \mathbf{R} \to \mathbf{R}$, given by

$$f^k(t^k) = (t^k)^2 \cdot \text{ (parabola)}.$$

Type II: $C = 0$, $K < 0$. $f^k: \mathbf{R} \to \mathbf{R}$, given by

$$f^k(t^k) = \exp\left[\left(-\frac{K}{n+1} \right)^{1/2} t^k \right] + C_k t^k + D_k \quad \text{(exponential type)}.$$

Type III: $C > 0$, $K = 0$. $f^k: (-\infty, 0) \to \mathbf{R}$, defined by

$$f^k(t^k) = -\frac{(n+2)^2}{n} C^{-(n+2)/(2n+2)} \left(-\frac{n+1}{n+2} t^k \right)^{n/(n+1)} + \overline{C}_k t^k + \overline{D}_k.$$

Type IV: $C > 0$, $K > 0$. f^k defined on a finite open interval, i.e. $f^k \colon (\bar{t}_1^k, t_1^k) \to \mathbf{R}$, such that

$$\lim_{t^k \to \bar{t}_1^k} f^k(t^k) = \bar{f}_1^k \in \mathbf{R}, \qquad \lim_{t^k \to t_1^k} f^k(t^k) = f_1^k \in \mathbf{R},$$

$$\lim_{t^k \to \bar{t}_1^k} (f^k)'(t^k) = (\bar{f}_1^k)' \in \mathbf{R}, \qquad \lim_{t^k \to t_1^k} (f^k)'(t^k) = +\infty,$$

$$\lim_{t^k \to \bar{t}_1^k} (f^k)''(t^k) = \left(\frac{K \cdot C^{-1}}{n+1}\right)^{(n+2)/(2n+2)}, \qquad \lim_{t^k \to t_1^k} (f^k)''(t^k) = +\infty,$$

$$(f^k)'''(t^k) > 0, \quad (f^k)^{\mathrm{iv}}(t^k) > 0, \quad \text{for every } t^k \in (\bar{t}_1^k, t_1^k).$$

Type V: $C > 0$, $K < 0$. f^k defined on a semi-infinite interval, $f^k \colon (-\infty, t_2^k)$ $\to \mathbf{R}$, such that

$$\lim_{t^k \to -\infty} f^k(t^k) = \begin{cases} +\infty, & \text{if } (f^k)'_{-\infty} < 0 \\ \bar{f}_{-\infty}^k \in \mathbf{R}, & \text{if } (f^k)'_{-\infty} = 0, \\ -\infty, & \text{if } (f^k)'_{-\infty} > 0 \end{cases}$$

$$\lim_{t^k \to t_2^k} f^k(t^k) = f_2^k \in \mathbf{R},$$

$$\lim_{t^k \to -\infty} (f^k)'(t^k) = (f^k)'_{-\infty} \in \mathbf{R}, \qquad \lim_{t^k \to t_2^k} (f^k)'(t^k) = +\infty,$$

$$\lim_{t^k \to -\infty} (f^k)''(t^k) = 0, \qquad \lim_{t^k \to t_2^k} (f^k)''(t^k) = +\infty,$$

$$(f^k)'''(t^k) > 0, \quad (f^k)^{\mathrm{iv}}(t^k) > 0, \quad \text{for every } t^k \in (-\infty, t_2^k).$$

Type VI: $C < 0$, $K < 0$. $f^k \colon (-\infty, t_3^k) \to \mathbf{R}$, with the following properties:

$$\lim_{t^k \to -\infty} f^k(t^k) = \begin{cases} +\infty, & \text{if } (f^k)'_{-\infty} < 0 \\ \bar{f}_{-\infty}^k \in \mathbf{R}, & \text{if } (f^k)'_{-\infty} = 0, \\ -\infty, & \text{if } (f^k)'_{-\infty} > 0 \end{cases}$$

$$\lim_{t^k \to t_3^k} f^k(t^k) = f_3^k \in \mathbf{R},$$

$$\lim_{t^k \to -\infty} (f^k)'(t^k) = (f^k)'_{-\infty} \in \mathbf{R}, \qquad \lim_{t^k \to t_3^k} (f^k)'(t^k) = (f_3^k)' \in \mathbf{R},$$

$$\lim_{t^k \to -\infty} (f^k)''(t^k) = 0, \qquad \lim_{t^k \to t_3^k} (f^k)''(t^k) = (f_3^k)'' := \left[\frac{K \cdot C^{-1}}{n+1}\right]^{(n+2)/(2n+2)},$$

$$(f^k)'''(t^k) > 0, \quad \text{for every } t^k \in (-\infty, t_3^k),$$

$$(f^k)^{\mathrm{iv}}(t^k) > 0 \quad \text{for } t^k \in (-\infty, t_0^k),$$

$$(f^k)^{\mathrm{iv}}(t_0^k) = 0, \quad (f^k)^{\mathrm{iv}}(t^k) < 0 \quad \text{for } t^k \in (t_0^k, t_3^k),$$

where

$$t_0^k = [(f^k)'']^{-1} \left(\frac{2}{3} \cdot \frac{K \cdot C^{-1}}{n+1} \right)^{(n+2)/(2n+2)}.$$

5. Affine hyperspheres

Affine hyperspheres have been the first important class of hypersurfaces that appeared in the historic development of the so-called "affine geometries." Many articles in the field have been devoted to the analysis of their properties and classification, since the pioneering works of Tzitzeica and Blaschke [2, 31].

In this section we shall determine which of the hypersurfaces of decomposable type and constant general affine mean curvature share the property of also being affine hyperspheres: proper or improper.

THEOREM (5.1). *There are no proper affine hyperspheres of decomposable type, regardless of whether the general affine mean curvature be constant or not.*

PROOF. Let us suppose that $X: M^n \to E^{n+1}$ is a proper affine hypersphere. Then there exists a constant vector $C \in E^{n+1}$ and a function $\phi: M \to \mathbf{R}$, such that

$$(5.1) \qquad X + \phi \cdot e_{n+1} = C,$$

where $e_{n+1} \neq 0$ is a transversal vector field in the direction of the affine normal line.

That condition is equivalent to

$$(5.2) \qquad dX + d\phi \cdot e_{n+1} + \phi \cdot de_{n+1} = 0.$$

We write now $e_{n+1} = (a^1, a^2, \ldots, a^{n+1})$, as in §2, and assume also that X is of decomposable type, so $dX = (dt^1, dt^2, \ldots, dt^n, (f^i)'dt^i)$.

By splitting (5.2) in its ambient space components we have

$$(5.3) \qquad \begin{aligned} dt^k + d\phi \cdot a^k + \phi \cdot da^k &= 0, \qquad k = 1, 2, \ldots, n. \\ \sum (f^i)'dt^i + d\phi \cdot a^{n+1} + \phi \cdot da^{n+1} &= 0. \end{aligned}$$

Let us recall that we can always assume $\sigma_{n+1}^{n+1} = 0$ (see equation (2.12)), which implies, by (2.10), that

$$(5.4) \qquad da^{n+1} = f_k da^k.$$

Hence, by using this last and (2.20), and by substituting the first n equations of (5.3) into the last, we obtain

$$(5.5) \qquad d\phi(a^{n+1} - (f^k)' \cdot a^k) = 0.$$

Now, if it were $d\phi = 0$, then $\phi = $ constant and (5.3) would also imply that $t^k + \phi \cdot a^k = C^k = $ constant for $k = 1, \ldots, n$. By using (2.26) we would obtain that at least $n - 1$ of the list of third derivatives $(f^1)'''$, $(f^2)''', \ldots, (f^n)'''$ would actually vanish.

But, if for some k, $(f^k)''' = 0$, then we would also have $a^k = 0$, by (2.24), and, hence $t^k = C^k$. Contradiction.

Therefore, $d\phi \neq 0$, and by (5.5) we should have

$$a^{n+1} = (f^k)' \cdot a^k = f_k a^k.$$

By differentiating this and applying (5.4), we would also have

$$a^k df_k = 0,$$

and since the Hessian matrix (f_{ij}) is nonsingular, the last would imply

$$a^1 = a^2 = \cdots = a^n = 0 = a^{n+1},$$

again a contradiction.

The theorem is proved.

THEOREM (5.2). *Let* $X: M^n \to E^{n+1}$ *be a hypersurface in the form of Monge,* $X(t^1, \ldots, t^n) = (t^1, \ldots, t^n, f(t^1, \ldots, t^n))$. *Then, if* $X(M)$ *is an improper affine hypersphere its third fundamental form vanishes identically and, hence,* $H_{\mathrm{ga}} = 0$. *Furthermore, if* X *is of decomposable type, then there are only two alternatives: either all of its components are of type* I, *in which case* X *is a paraboloid, or* $n - 1$ *of its components are of type* I, *and the remaining one is of type* III.

PROOF. If $X(M)$ is an improper affine hypersphere, then all of its affine normal lines are parallel. Hence, there exists a nonzero constant vector $A = (A^1, A^2, \ldots, A^n, A^{n+1}) \neq 0$, in E^{n+1}, and a differentiable function $\phi: M \to \mathbf{R}$, such that any vector field e_{n+1} in the affine normal direction satisfies

$$(5.6) \qquad e_{n+1} = (a^1, a^2, \ldots, a^n, a^{n+1}) = \phi \cdot A.$$

By using this, and by assuming again $\sigma_{n+1}^{n+1} = 0$, which also implies (5.4), i.e., $da^{n+1} = f_k da^k$, we obtain

$$(5.7) \qquad da^{n+1} = A^{n+1} d\phi = f_k da^k = f_k A^k d\phi.$$

The equality of second and fourth members of the last implies

$$(5.8) \qquad (A^{n+1} - f_k A^k) d\phi = 0.$$

Now, if $A^{n+1} = f_k A^k$, it follows that the n-dimensional vector $(A^1, A^2, \ldots, A^n)$ is in the kernel of the Hessian matrix (f_{ij}) and, since this is nonsingular, we obtain readily $A^1 = A^2 = \cdots = A^n = 0 = A^{n+1}$. Contradiction.

Hence we must have, from (5.8), $d\phi = 0$ which also implies $e_{n+1} =$ constant, by (5.6).

Therefore, by using (1.2), (1.10) and (1.11), we obtain

$$\mathrm{III}_{\mathrm{ga}} \equiv 0 .$$

Next, if we further assume $X(M)$ to be of decomposable type, (2.28) implies that

(5.9) $$\qquad (f^j)''' \cdot (f^k)''' = 0 , \quad \text{for } j \neq k .$$

Then, if $(f^j)''' = 0$ for every $j = 1, 2, \ldots , n$, it follows that all of the components $f^1, f^2, \ldots , f^n$ are of Type I: $X(M)$ is a paraboloid.

If, on the other hand, for some k_0, $(f^{k_0})''' \neq 0$, then (5.9) implies that the remaining components f^j, $j \neq k_0$, must satisfy $(f^j)''' = 0$: all of these, then, are of type I.

In that case, since we must also have $B_{k_0 k_0} = 0$, it follows from (2.27), (2.29) and Theorem (4.1), that f^{k_0} must be of Type III.

This finishes the proof.

In summary, we emphasize once more the important fact that was announced in the Introduction to this paper, namely, that actually very few of the hypersurfaces introduced and discussed here have, at the same time, the property of being affine hyperspheres.

6. Unimodular affine mean curvature

The methods deployed in past sections in order to classify those hypersurfaces of decomposable type with constant general affine mean curvature, can also be extended to study the class of hypersurfaces with constant unimodular affine mean curvature, $H_{\mathrm{ua}} = \text{constant}$. In particular, when that constant is zero, $H_{\mathrm{ua}} = 0$, one is considering the important class of affine minimal hypersurfaces [2, 7], alternatively called affine maximal [5].

If $\mathrm{I}_{\mathrm{ua}} = h_{ij} \sigma^i \sigma^j$ is the first fundamental form of unimodular affine geometry: the so-called Berwald-Blaschke metric of a given hypersurface $X(M)$ [1, 2, 5], then the eigenvalues of the third fundamental form $\mathrm{III}_{\mathrm{ga}}$ with respect to I_{ua} are the unimodular affine principal curvatures. Thus, the unimodular affine mean curvature is defined as the arithmetic mean of these eigenvalues, that is, the average of the contraction of $\mathrm{III}_{\mathrm{ga}}$ with (the inverse of) I_{ua}:

(6.1) $$\qquad H_{\mathrm{ua}} = \frac{1}{n} h^{ij} B_{ij} .$$

If the hypersurface is of decomposable type we obtain, by using (2.18), (2.21), (2.22) and (2.27),

$$H_{\mathrm{ua}} = \frac{1}{n} \cdot F^{1/(n+2)} \cdot \sum_{k=1}^{n} \frac{B_{kk}}{(f^k)''}$$

(6.2)
$$= -\frac{1}{n} \cdot \frac{1}{(n+2)^2} \cdot F^{1/(n+2)}$$
$$\cdot \sum_{k=1}^{n} \left[(n+2) \frac{(f^k)^{\mathrm{iv}}}{[(f^k)'']^2} - (2n+3) \frac{[(f^k)''']^2}{[(f^k)'']^3} \right] .$$

Thus, in working toward the classification of hypersurfaces of decomposable type and $H_{\mathrm{ua}} = \text{constant} \neq 0$, one has to be a bit more careful because of the presence in the last equation of the factor $F^{1/(n+2)}$. However, for affine maximal hypersurfaces, i.e. $H_{\mathrm{ua}} = 0$, there occurs again a splitting of this equation which is, then, equivalent to the set of equations

(6.3)
$$(n+2) \cdot \frac{(f^k)^{\mathrm{iv}}}{[(f^k)'']^2} - (2n+3) \cdot \frac{[(f^k)''']^2}{[(f^k)'']^3} = C_k = \text{constant},$$

with the condition $\sum C_k = 0$.

The corresponding reduction, as in equation (2.31), now becomes

(6.4)
$$(n+2) \cdot \frac{y''}{y^2} - (2n+3) \cdot \frac{(y')^2}{y^3} = C_k .$$

The procedure for solving this equation can follow the same pattern as before, producing several types of affine maximal, decomposable hypersurfaces. In particular, for $C_k = 0$, one obtains again the previously named Types I ($C = 0$, $K = 0$) and III ($C > 0$, $K = 0$). These are all, in a sense, Scherk's types of hypersurfaces. For example, in dimension $n = 2$, the graphs of

$$f(x, y) = x^2 - (y)^{2/3}$$

and

$$g(x, y) = -x^{2/3} - y^{2/3}$$

are affine minimal (maximal) surfaces of translation, with plane translation curves. Compare to [13].

Acknowledgments

I want to express here my gratitude to Edgar Acuña for installing T_EX in my PC. That made possible the appropriate typing of this article. I want to also thank Gabriele Castellini for helping me with some macros.

BIBLIOGRAPHY

1. L. Berwald, *Die Grundgleichungen der Hyperflächen in Euklidischen R_{n+1} gegenüber den inhaltstreuen Affinitäten*, Monats. Math. Phys. 32 (1922), 89–106.
2. W. Blaschke, *Vorlesungen über Differentialgeometrie*. II, Springer-Verlag, Berlin, 1923.
3. E. Calabi, *Improper affine hyperspheres of convex type and a generalization of a theorem by K. Jörgens*, Michigan Math. J. 5 (1958), 105–126.
4. ____, *Complete affine hyperspheres* I, Ist. Naz. Alta Mat. Sym. Mat. 10 (1972), 19–38.

5. ____, *Hypersurfaces with maximal affinely invariant area*, Amer. J. Math. **104** (1982), 91–126.

6. S.-Y. Cheng and S.-T. Yau, *Complete affine hypersurfaces, I. The completeness of affine metrics*, Comm. Pure Appl. Math. **39** (1986), 839–886.

7. S.-S. Chern, *Affine minimal hypersurfaces*, Minimal Submanifolds and Geodesics, Kaigai Publ., Tokyo, 1978.

8. J. Favard, *Cours de géométrie differentielle locale*, Gauthier-Villars, Paris, 1975.

9. H. Flanders, *Local theory of affine hypersurfaces*, J. Anal. Math. **15** (1965), 353–387.

10. S. Gigena, *Integral invariants of convex cones*, J. Differential Geom. **13** (1978), 191–222.

11. ____, *On a conjecture by E. Calabi*, Geom. Dedicata **11** (1981), 387–396.

12. ____, *General affine geometry of hypersurfaces. I*, to appear in Math. Notae (1992).

13. E. Glässner, *Ein affinanalogon zu den Scherkschen minimalflächen*, Arch. Math. (Basel) **27** (1977), 436–439.

14. H. W. Guggenheimer, *Differential geometry*, McGraw-Hill, New York, 1963.

15. C.-C. Hsiung and J. K. Shahin, *Affine differential geometry of closed hypersurfaces*, Proc. London Math. Soc. **17** (1967), 715–735.

16. K. Jörgens, *Über die Lösungen der Differentialgleichung $r.t - s^2 = 1$*, Math. Ann. **127** (1954), 130–134.

17. T. Miăilescu, *Geometrie différentielle affine générale des surfaces*, Acad. Roy. Bel. Cl. Sci. Mem. **35** (1965), fasc. 8 et dernier, 1–91.

18. S. Nakajima, *Über die Isoperimetrie der Ellipsoide und Eiflächen mit konstanter mittlerer Affinkrümung im $(n + 1)$-dimensionalen Raum*, Jap. J. Math. **2** (1927), 193–196.

19. A. V. Pogorelov, *On the improper affine hyperspheres*, Geom. Dedicata **1** (1972), 33–46.

20. E. Salkowski, *Affine Differentialgeometrie*, de Gruyter, Berlin and Leipzig, 1934.

21. L. A. Santaló, *Geometría diferencial afín y cuerpos convexos*, Math. Notae **16**, 20–42.

22. T. Sasaki, *Hyperbolic affine hyperspheres*, Nagoya Math. J. **77** (1980), 107–123.

23. ____, *On affine isoperimetric inequality for a strongly convex closed hypersurface in the unimodular affine space A^{n+1}*, Kumamoto J. Sci (Math.) **16** (1984), 23–38.

24. P. A. Schirokov and A. P. Schirokov, *Affine Differentialgeometrie*, Teubner, Leipzig, 1962.

25. R. Schneider, *Zur affinen Differentialgeometrie im Grossen. I*, Math. Z. **101** (1967), 375–406.

26. ____, *Zur affinen Differentialgeometrie im Grossen. II*, Math. Z. **102** (1967), 1–8.

27. A. Schwenk and U. Simon, *Hypersurfaces with constant equiaffine mean curvature*, Arch. Math. (Basel) **46** (1986), 85–90.

28. U. Simon, *Zur Entwicklung der affinen Flächentheorie nach Blaschke*, Gesammelte Werke Blaschke, Band 4, Thales Verlag, Essen, 1985.

29. M. Spivak, *A comprehensive introduction to differential geometry*. Vol. III, Publish or Perish Inc., Berkeley, CA, 1979.

30. L. K. Tutaev, *Lines and surfaces in three-dimensional affine space*, Oldbourne Press, London, 1965.

31. G. Tzitzeica, *Sur une nouvelle classe des surfaces*, Rend. Circ. Mat. Palermo **25** (1908), 180–187; **28** (1909), 210–216.

DEPARTMENTO DE MATEMÁTIC, FAC. DE CS. EXS., INGENIERÍA Y AGRIMENSURA, UNIVERSIDAD NACIONAL DE ROSARIO, AVDA. PELLEGRINI 250, 2000 ROSARIO, ARGENTINA

DEPARTMENTO DE MATEMÁTICA, FAC. DE CS. EXS., FÍSICAS Y NATURALES UNIVERSIDAD NACIONAL DE CÓRDOBA, AVDA. VELEZ SARSFIELD 299, 5000 CÓRDOBA, ARGENTINA

Proceedings of Symposia in Pure Mathematics
Volume **54** (1993), Part 3

Heat Equation Asymptotics

PETER B. GILKEY

ABSTRACT. We discuss the heat equation asymptotics $\operatorname{Tr}(e^{-tD})$ and $\operatorname{Tr}(Pe^{-tP^2})$ on a compact manifold M with boundary for a second-order operator D with scalar leading symbol and for a first-order operator P of Dirac type.

0. Introduction

Let M be a compact Riemannian manifold of dimension m; if $\partial M \neq \varnothing$, we impose Dirichlet or Neumann boundary conditions. Let $\Delta = \delta d$ be the geometer's Laplacian and let $\operatorname{spec}(M) = \{\lambda_\nu\}$ be the eigenvalues of Δ where $0 \leq \lambda_1 \leq \lambda_2 \cdots$. One wants to know the extent to which $\operatorname{spec}(M)$ determines the geometry and topology of M. The question for a domain in $\mathbf{R}^2$ with Dirichlet boundary conditions was originally posed by Kac [**Ka**] and has an attractive formulation due to Protter [**Pr**]: "Suppose a drum is being played in one room and a person with perfect pitch hears but cannot see the drum. Is it possible for her to deduce the precise shape of the drum just from hearing the fundamental tone and all the overtones?"

There is a vast literature on the subject; we refer to Bérard and Berger [**BB**] and Bérard [**Be**] for further references in addition to those cited in this paper.

There are many examples showing that neither the geometry nor the topology of M is determined by $\operatorname{spec}(M)$.

THEOREM 1.

1. (Milnor [**Mi**]) *There exist isospectral nonisometric flat tori of dimension* 16.

2. (Vigneras [**Vi**]) *There exist isospectral nonisometric Riemann surfaces.*

1991 *Mathematics Subject Classification.* Primary 58G25.
Research partially supported by the NSA, NSF, and IHES.
This paper is in final form and no version will be submitted for publication elsewhere.

3. (Vigneras [**Vi**]) *There exist isospectral manifolds with different fundamental groups if $m \geq 3$.*

4. (Ikeda [**Ik**]) *There exist isospectral nonisometric spherical space forms.*

5. (Urakawa [**Ur**]) *If $m \geq 5$, there exist regions in R^m which are isospectral but not isometric.*

REMARK. See also [**Su**] for a general method of constructing such examples.

These examples come in finite families. There are nontrivial isospectral deformations:

THEOREM 2.

1. (Gordon-Wilson [**GW**]) *There exists a nontrivial family of isospectral metrics which are not conformally equivalent.*

2. (Brooks-Gordon, [**BrGo**]) *There exists a nontrivial family of isospectral metrics which are conformally equivalent.*

In addition to these examples, there are some compactness results:

THEOREM 3.

1. (Osgood, Phillips, and Sarnak [**OPS**]) *Families of isospectral metrics on Riemann surfaces are compact modulo gauge equivalence.*

2. (Brooks, Chang, Perry, and Yang [**BPY, CY**]) *If $m = 3$, families of isospectral metrics within a conformal class are compact modulo gauge equivalence.*

3. (Brooks, Perry, and Petersen [**BPP**]) *Isospectral negative curvature manifolds contain only a finite number of topological types.*

1. Heat equation asymptotics for manifolds without boundary

Heat equation asymptotics are a fundamental tool. We refer to Gilkey [**Gi-2**] throughout this section for the necessary analytic details. Let

$$h(t) = \text{Tr}_{L^2} e^{-t\Delta} = \sum_\nu e^{-t\lambda_\nu}$$

be the trace of the heat kernel. As $t \to 0^+$, there is an asymptotic series

$$h(t) \sim \sum_n a_n(\Delta) t^{(n-m)}.$$

The $a_n(\Delta)$ are spectral invariants which are determined by the local geometry of M. Let italic indices i, j, etc. range from 1 through m and index a local orthonormal frame $\{e_1, ..., e_m\}$ for the tangent space $T(M)$. Let R_{ijkl} be the curvature tensor of the Levi-Civita connection with the sign convention $R_{1212} = -1$ on the standard sphere. We sum over repeated indices to define

$$\tau = -R_{ijij}, \quad \rho_{ij} = -R_{ikjk}, \quad \rho^2 = R_{ikjk} R_{iljl}, \quad R^2 = R_{ijkl} R_{ijkl}.$$

THEOREM 4 (Sakai [Sa], Gilkey [Gi-1]). *Let* $\partial M = \emptyset$. $a_n(\Delta) = 0$ *for n odd.*

1. $a_0(\Delta) = (4\pi)^{-m/2} \int_M 1$.
2. $a_2(\Delta) = 6^{-1}(4\pi)^{-m/2} \int_M \tau$.
3. $a_4(\Delta) = 360^{-1}(4\pi)^{-m/2} \int_M 5\tau^2 - 2\rho^2 + 2R^2$.
4. $a_6(\Delta) = 45360^{-1}(4\pi)^{-m/2} \int_M -142(\nabla\tau)^2 - 26(\nabla\rho)^2 - 7(\nabla R)^2 + 35\tau^3$
$-42\tau\rho^2 + 42\tau R^2 - 36\rho_{ij}\rho_{jk}\rho_{ki} - 20\rho_{ij}\rho_{kl}R_{ikjl} - 8\rho_{ij}R_{ikln}R_{jkln}$
$-24R_{ijkl}R_{ijnp}R_{klnp}$.

REMARK. Avramidi [AV] and Amsterdamski, Berkin, and O'Connor [ABC] have computed $a_8(D)$; a_8 has formidable combinatorial complexity.

Because $a_1(\Delta) = c \cdot \int_M \tau$, the Euler characteristic and hence the topological type of a Riemann surface are a spectral invariant if $m = 2$. There are many other results based on the heat equation invariants. For example:

THEOREM 5 (Tanno [Ta]). *If* $\mathrm{spec}(M) = \mathrm{spec}(S^m)$ *and* $m \leq 6$, *then* $M = S^m$.

We now consider more general operators. Let greek indices ν, μ, etc. range from 1 through m and index the coordinate frame for $T(M)$. Let $ds^2 = g_{\nu\mu} dx^\nu \circ dx^\mu$ be the metric tensor and let

$$\Gamma_{\nu\mu}{}^\sigma = \tfrac{1}{2} g^{\sigma\epsilon}(\partial_\nu g_{\epsilon\mu} + \partial_\mu g_{\epsilon\nu} - \partial_\epsilon g_{\nu\mu})$$

be the Christoffel symbols of the Levi-Civita connection. Let V be a smooth vector bundle over M and let D be a second-order partial differential operator on $C^\infty(V)$ with leading symbol given by the metric tensor. In local coordinates, D has the form

$$D = -(g^{\nu\mu} I \cdot \partial_\nu \partial_\mu + A^\nu \partial_\nu + B)$$

where A^ν and B are endomorphisms of V. We can work more invariantly. Let ∇ be a connection on V and let $E \in \mathrm{End}\, V$. Let

$$D(g, \nabla, E) = -\left(\sum_{\nu,\mu} g^{\nu\mu} \nabla_\nu \nabla_\mu + E\right)$$

on $C^\infty(V)$ be an operator with leading symbol given by the metric tensor.

THEOREM 6 (Gilkey [Gi-1]). *If* D *is a second-order operator on* $C^\infty(V)$ *with leading symbol given by the metric tensor, there exists a unique connection* ∇ *on* V *and a unique endomorphism* E *of* V *so* $D = D(g, \nabla, E)$. *If* ω_ν *is the connection 1-form of* ∇, *then*

$$\omega_\nu = \tfrac{1}{2} g_{\mu\nu}(A^\mu + g^{\sigma\epsilon}\Gamma_{\sigma\epsilon}{}^\mu),$$
$$E = B - g^{\mu\sigma}(\partial_\sigma \omega_\mu + \omega_\mu \omega_\sigma - \omega_\epsilon \Gamma_{\mu\sigma}{}^\epsilon).$$

EXAMPLE. Let $\Delta_p = d_{p-1}\delta_{p-1} + \delta_p d_p$ be the Laplacian on p forms. The associated connection is the Levi-Civita connection and $E(\Delta_p)$ is given by the Weitzenböck formulas. If $p = 0$, $E = 0$. If $p = 1$, $E(\Delta_p) = -\rho$.

EXAMPLE. Let Δ^s be the spin Laplacian. The associated connection is the spin connection. $E(\Delta^s) = -\frac{1}{4}\tau$.

The operator e^{-tD} maps $L^2(V)$ to $C^\infty(V)$. There is a smooth kernel function $K(t, x, y, D)$ so $e^{-tD}u(x) = \int K(t, x, y, D)u(y)\,dy$.

$$\text{Tr } K(t, x, x, D) \sim \sum_n a_n(D)(x)t^{(n-m)} \quad \text{as } t \to 0^+;$$

$$\text{Tr}_{L^2}(e^{-tD}) = \int_M \text{Tr } K(t, x, x, D) \sim \sum_n a_n(D)t^{n-m},$$

$$a_n(D) = \int_M a_n(D)(x).$$

The $a_n(D)(x)$ vanish if n is odd since $\partial M = \varnothing$. The $a_n(D)(x)$ are locally computable. Let Ω be the curvature of the connection ∇ on V.

THEOREM 7 (Gilkey [Gi-1]).

1. $a_0(D)(x) = (4\pi)^{-m/2}\,\text{Tr}(1)$.
2. $a_2(D)(x) = (4\pi)^{-m/2}6^{-1}\,\text{Tr}(6E + \tau)$.
3.

$$a_4(D)(x) = (4\pi)^{-m/2}360^{-1}\,\text{Tr}\{(60E_{;kk} + 60\tau E + 180E^2 + 30\Omega_{ij}\Omega_{ij}$$
$$+ 12\tau_{;kk} + 5\tau^2 - 2\rho^2 + 2R^2)\}.$$

REMARK. See [Gi-1] for $a_6(D)(x)$ and Avramidi [Av] for $a_8(D)(x)$.

This gives complete information concerning a_n for $n \leq 8$. Partial information about all the coefficients is also available.

THEOREM 8 (Gilkey [Gi-3]). *Let* $n \geq 6$ *and* $\partial M = \varnothing$. *Let*

$$c(n) = (4\pi)^{-m/2}/\{(-1)^n \cdot 2^{n+1} \cdot 1 \cdot 3 \cdot \ldots \cdot (2n + 1)\}.$$

Modulo cubic and higher degree terms which have fewer covariant derivatives,

$$a_{2n}(D) = c(n) \int_M \text{Tr}\left\{(n^2 - n - 1)|\nabla^{n-2}\tau|^2 + 2|\nabla^{n-2}\rho|^2\right.$$
$$+ 4(2n + 1)(n - 1)\nabla^{n-2}\tau \cdot \nabla^{n-2}E$$
$$+ 2(2n + 1)\nabla^{n-2}\Omega \cdot \nabla^{n-2}\Omega$$
$$\left. + 4(2n + 1)(2n - 1)\nabla^{n-2}E \cdot \nabla^{n-2}E + \cdots\right\}.$$

REMARK. See Osgood, Phillips and Sarnak [OPS] if $m = 2$ and $D = \Delta_0$.

2. Operators of Dirac type

In some cases, it is possible to take a square root of D within the category of differential operators; for example, the total form valued Laplacian has this property: $d\delta + \delta d = (d + \delta)^2$. This leads to additional spectral invariants. The Clifford algebra bundle $\text{Clif}(TM)$ is the natural setting for this discussion;

it is the universal complex unital algebra bundle generated by TM subject to the Clifford commutation relations

$$\partial_\nu * \partial_\mu + \partial_\mu * \partial_\nu = -2g_{\nu\mu}.$$

A $\mathrm{Clif}(M)$ module structure on V is a unital algebra morphism

$$\gamma : \mathrm{Clif}(M) \to \mathrm{End}(V).$$

If $\gamma^\nu = \gamma(dx^\nu)$, then $\gamma^\nu\gamma^\mu + \gamma^\mu\gamma^\nu = -2g^{\nu\mu}$. We may always choose a connection ∇ on V so $\nabla\gamma = 0$; fix such a connection henceforth. Let $\psi \in \mathrm{End}(V)$ and let

$$P = P(\nabla, \psi) = \gamma^\nu\nabla_\nu - \psi \quad \text{on } C^\infty(V)$$

be an operator of Dirac type; any operator of Dirac type can be expressed in this form. Let $D = P^2$ be the associated Laplacian.

Pe^{-tD} maps $L^2(V)$ to $C^\infty(V)$. Let $L(t, x, y, P) = P_x K(t, x, y, D)$ be the smooth kernel function so $Pe^{-tD}u(x) = \int L(t, x, y, P)u(y)\,dy$.

$$\mathrm{Tr}\, L(t, x, x, P) \sim \sum_n a_n(P)(x)t^{(n-m)} \quad \text{as } t \to 0^+;$$

$$\mathrm{Tr}_{L^2}(Pe^{-tP}) = \int_M \mathrm{Tr}\, K(t, x, x, P) \sim \sum_n a_n(P)t^{n-m},$$

$$a_n(P) = \int_M a_n(P)(x).$$

The $a_n(P)(x)$ vanish if n is even since $\partial M = \varnothing$. The $a_n(P)(x)$ are locally computable. Let $\{e_i\}$ be a local orthonormal frame for $T(M)$ and let $\gamma_i = \gamma(e_i)$. Let $\Psi = \gamma_i\psi\gamma_i$ and let $W_{ij} = \Omega_{ij} - \frac{1}{4}R_{ijkl}\gamma_k\gamma_l$.

THEOREM 9 (Branson and Gilkey [**BG-2**]).

1. $a_1(P)(x) = (4\pi)^{-m/2}(m-1)\mathrm{Tr}\{\psi\}.$
2.

$$a_3(P)(x)$$
$$= -12^{-1}(4\pi)^{-m/2}\,\mathrm{Tr}\,\{\{2(1-m)\psi_{:i} + 3(4-m)\psi\gamma_i\psi + 3\Psi\gamma_i\psi\}_{:i}$$
$$+ (m-3)\{\tau\psi + 6\gamma_i\gamma_jW_{ij}\psi + 6\psi\psi_{:i}\gamma_i$$
$$+ (4-m)\psi\psi\psi + 3\psi\psi\Psi\}\}.$$

3. Heat equation asymptotics for manifolds with boundary

If $\partial M \neq \varnothing$, we must impose boundary conditions. Let e_m be the inward unit normal. Let roman indices a, b, etc. range from 1 through $m-1$ and index an orthonormal frame for the tangent bundle of the boundary $T(\partial M)$. Let $\chi \in \mathrm{End}\, V|_{\partial M}$ with $\chi^2 = 1$. Let

$$\Pi_N = \tfrac{1}{2}(1 - \chi) \quad \text{and} \quad \Pi_D = \tfrac{1}{2}(1 + \chi)$$

be projection on the ± 1 eigenspaces of χ. Let $S \in \text{End } V_+|_{\partial M}$. Let $v \in C^\infty(V)$. Let

$$Bv = (\nabla_N + S)v_+|_{\partial M} \oplus v_-|_{\partial M}$$

define mixed Neumann and Dirichlet boundary conditions. Let $f \in C^\infty(M)$,

$$\text{Tr}_{L^2}(fe^{-tD_B}) \sim \sum_n t^{(n-m)} a_n(f, D, B) \quad \text{as } t \to 0^+.$$

Near ∂M, the heat kernel is a distribution; we recover this by considering the auxiliary parameter f. The usual heat equation asymptotics arise by setting $f = 1$ and need not vanish if n is odd.

Let $L_{ab} = (\nabla_{e_a} e_b, N)$ be the second fundamental form. Let ";" be multiple covariant differentiation with respect to the Levi-Civita connection ∇^{LC} of M and let ":" be multiple covariant differentiation tangentially with respect to the Levi-Civita connection ∇^{lc} of the boundary. Theorem 7 generalizes to this setting to become:

THEOREM 10 (Branson and Gilkey [**BG-1, BG-2**], Moss and Dowker [**MD**]).

1. $a_0(f, D) = (4\pi)^{-m/2} \int_M \text{Tr}(f)$.
2. $a_1(f, D) = 4^{-1}(4\pi)^{-(m-1)/2} \int_{\partial M} \text{Tr}(\chi f)$.
3.

$$a_2(f, D) = (4\pi)^{-m/2} 6^{-1} \left\{ \int_M \text{Tr}(6fE + f\tau) \right.$$
$$\left. + \int_{\partial M} \text{Tr}(2fL_{aa} + 3\chi f_{;m} + 12fS) \right\}$$

4.

$$a_3(f, D) = 4^{-1}(4\pi)^{-(m-1)/2} 96^{-1}$$
$$\times \left\{ \int_{\partial M} \text{Tr}\{f(96\chi E + 16\chi\tau + 8f\chi R_{amam} \right.$$
$$+ (13\Pi_N - 7\Pi_D)L_{aa}L_{bb}$$
$$+ (2\Pi_N + 10\Pi_D)L_{ab}L_{ab} + 96SL_{aa}$$
$$+ 192S^2 - 12\chi_{:a}\chi_{:a})$$
$$\left. + f_{;m}((6\Pi_N + 30\Pi_D)L_{aa} + 96S) + 24\chi f_{;mm}\} \right\}.$$

5.

$$a_4(f, D) = (4\pi)^{-m/2}360^{-1}$$

$$\times \left\{ \int_M \mathrm{Tr}\{f(60E_{;kk} + 60\tau E + 180E^2 + 30\Omega^2 + 12\tau_{;kk}\right.$$

$$+ 5\tau^2 - 2\rho^2 + 2R^2)\}$$

$$+ \int_{\partial M} \mathrm{Tr}\{f(240\Pi_N - 120\Pi_D)E_{;m} + (42\Pi_N - 18\Pi_D)\tau_{;m}$$

$$+ 24L_{aa:bb} + 0L_{ab:ab} + 120EL_{aa} + 20\tau L_{aa}$$

$$+ 4R_{amam}L_{bb} - 12R_{ambm}L_{ab} + 4R_{abcb}L_{ac} + 0\Omega_{im:i}$$

$$+ 21^{-1}\{(280\Pi_N + 40\Pi_D)L_{aa}L_{bb}L_{cc} + 0\chi_{:a}\Omega_{am}$$

$$+ (168\Pi_N - 264\Pi_D)L_{ab}L_{ab}L_{cc}$$

$$+ (224\Pi_N + 320\Pi_D)L_{ab}L_{bc}L_{ac}\}$$

$$+ 720SE + 120S\tau + 0SR_{amam} + 144SL_{aa}L_{bb}$$

$$+ 48SL_{ab}L_{ab} + 480S^2L_{aa} + 480S^3 + 120S_{:aa} + 60\chi\chi_{:a}\Omega_{am}$$

$$- 42\chi_{:a}\chi_{:a}L_{bb} + 6\chi_{:a}\chi_{:b}L_{ab} - 120\chi_{:a}\chi_{:a}S)$$

$$+ f_{;m}(180\chi E + 30\chi\tau + 0R_{amam} + (84\Pi_N - 180\Pi_D)/7 \cdot L_{aa}L_{bb}$$

$$\times (84\Pi_N + 60\Pi_D)/7 \cdot L_{ab}L_{ab} + 72SL_{aa}$$

$$+ 240S^2 - 18\chi_{:a}\chi_{:a})$$

$$+ f_{;mm}(24L_{aa} + 120S)$$

$$\left. \left. + 30\chi f_{;iim}\}\right\} \right\}$$

REMARK. See also Kennedy et al [KCD], Melmed [Me], Moss [Mo], and Smith [Sm].

Let P be an operator of Dirac type. Suppose $\gamma_m\chi + \chi\gamma_m = \gamma_a\chi - \chi\gamma_a = 0$ on ∂M; if m is odd, χ need not exist. Let $Bv = \Pi_-v|_{\partial M}$. We generalize Theorem 9 to manifolds with boundary:

THEOREM 11.

1. $a_0(f, P, \chi) = 0$.
2. $a_1(f, P, \chi) = (4\pi)^{-m/2}\int_M (m-1)f\,\mathrm{Tr}(\psi)$.
3. $a_2(f, P, \chi) = 4^{-1}(4\pi)^{-(m-1)/2}\int_{\partial M}(m-2)f\,\mathrm{Tr}(\psi\chi)$.

4.

$$a_3(f, P, \chi) = -12^{-1}(4\pi)^{-m/2}$$

$$\times \left\{ \int_M f \operatorname{Tr}\{2(1-m)\psi_{;i} + 3(4-m)\psi\gamma_i\psi + 3\Psi\gamma_i\psi\}_{;i} \right.$$

$$+ (m-3)f \operatorname{Tr}\{\tau\psi + 6\gamma_i\gamma_j W_{ij}\psi + 6\psi\psi_{;i}\gamma_i$$

$$+ (4-m)\psi\psi\psi + 3\psi\psi\Psi\}$$

$$+ \int_{\partial M} 6(2-m)f_{;m} \operatorname{Tr}\{\chi\psi\}$$

$$+ f \operatorname{Tr}\{(18-6m)\chi\psi_{;m} + (2-2m)\psi_{;m}$$

$$- 6\chi\gamma_m\gamma_a\psi_{;a} + 6(m-2)\chi\psi L_{aa}$$

$$+ 2(m-3)\psi L_{aa} + 6(3-m)\chi\gamma_m\psi\psi$$

$$+ 3\gamma_m\psi\Psi_T + 3(3-m)\chi\gamma_m\psi\chi\psi + 6\chi\gamma_a W_{am}\} \right\}$$

REMARK. It follows from the Atiyah, Patodi, Singer index theorem [APS] that $a_n(P) = 0$ if $n = m$; this can be checked directly from these formulas for $n \le 3$.

4. Properties of the heat equation invariants

We summarize below some of the properties of the $a_n(-)$.

THEOREM 12. *Let D be a second order operator with leading symbol given by the metric tensor and let B be a suitable boundary condition.*

1. $a_n(f, D, B)$ *is locally computable, i.e.,*

$$a_n(f, D, B) = \int_M f \cdot a_n(D)(x) + \sum_{\nu \le n} \int_{\partial M} \nabla^\nu_m f \cdot a_{n,\nu}(D, B)(y).$$

$a_n(D)(x)$ *and* $a_{n,\nu}(D, B)(y)$ *are universally computable in terms of the jets of the symbol of D.* $a_n(D)(x) = 0$ *if n is odd.*

2.

$$a_0(f, D, B) = (4\pi)^{-m/2} \int_M \operatorname{Tr}(f).$$

$$a_1(f, D, B) = \frac{1}{4}(4\pi)^{-(m-1)/2} \int_{\partial M} \operatorname{Tr}(\chi f).$$

3. $\frac{d}{d\epsilon}|_{\epsilon=0} a_n(1, D - \epsilon f, B) = a_{n-1}(f, D, B)$.

4. $\frac{d}{d\epsilon}|_{\epsilon=0} a_n(1, e^{-2\epsilon f}, B) = (m-n)a_n(f, D, B)$.

5. *Let* $M = M_1 \times M_2$ *and* $D = D_1 \otimes 1 + 1 \otimes D_2$. *Assume* $\partial M_2 = \emptyset$. *Let* $f = f_1 f_2$.

$$a_n(f, D, B) = \sum_{p+q=n} a_p(f_1, D_1, B) \cdot a_q(f_2, D_2).$$

6. *Let f be real. If (D,B) is self-adjoint or if (D,B) is real,* $a_n(f, D, B)$ *is real.*

THEOREM 13. *Let P be an operator of Dirac type and let χ define a suitable boundary condition. Let $D = P^2$.*

1. *$a_n(f, P, \chi)$ is locally computable, i.e.,*

$$a_n(f, P, \chi) = \int_M f \cdot a_n(P)(x) + \sum_{\nu \leq n} \int_{\partial M} \nabla_m^\nu f \cdot a_{n,\nu}(P, \chi)(y).$$

$a_n(P)(x)$ and $a_{n,\nu}(P, \chi)(y)$ are universally computable in terms of the jets of the symbol of D. $a_n(P)(x) = 0$ if n is even.

2. *Let $P(\epsilon) = P - \epsilon f$. Then*
   i. *$\frac{d}{d\epsilon}|_{\epsilon=0} a_n(1, P, \chi) = (m - n)a_{n-1}(f, D, B)$,*
   ii. *$\frac{d}{d\epsilon}|_{\epsilon=0} a_n(1, D, B) = 2a_{n-1}(f, P, \chi)$.*

3. *Let $P(\epsilon) = e^{-\epsilon f} P$. Then*
   i. *$\frac{d}{d\epsilon}|_{\epsilon=0} a_n(1, P, \chi) = (m - n)a_n(f, P, \chi)$,*
   ii. *$\frac{d}{d\epsilon}|_{\epsilon=0} a_n(1, D, B) = (m - n)a_{n-1}(f, D, B)$.*

4. *Let $M = M_1 \times M_2$ and let*

$$P = \begin{pmatrix} P_1 \otimes 1 & 1 \otimes P_2^* \\ 1 \otimes P_2 & -P_1 \otimes 1 \end{pmatrix}$$

Then $a_n(1, P) = a_{n-m_2}(P_1) \cdot \text{index}(P_2)$.

REMARK. The properties of Theorems 12 and 13 suffice to prove Theorems 7, 8, 9, 10, and 11. One could conjecture they suffice to compute the heat equation asymptotics in general, but this is not known.

REFERENCES

[ABC] P. Amsterdamski, A. Berkin, and D. O'Connor, b_8 'Hamidew' coefficient for a scalar field, Classical Quantum Gravity **6** (1989), 1981–1991.

[APS] M. F. Atiyah, V. K. Patodi, and I. M. Singer, Spectral asymmetry and Riemannian geometry I, Math. Proc. Cambridge Philos. Soc **77** (1975), 43–69.

[Av] I. G. Avramidi, The covariant technique for the calculation of the heat kernel asymptotic expansion, Phys. Lett. B **238** (1990), 92–97; see also articles in Theoret. Mat. Fiz. **79** (1989), 219 and also Yadernaya Fiz. **49** (1989), 1185. (Russian)

[Be] P. Bérard, Variétés Riemanniennes isospectrales non isométriques, Sém. Bourbaki, 41ème année, 1989–89, #705.

[BB] P. Bérard and M. Berger, Le spectre d'une variété Riemannienne en 1981, Lecture Notes in Math., vol. 1207, Springer-Verlag, 1986.

[BG-1] T. Branson and P. Gilkey, The asymptotics of the Laplacian on a manifold with boundary, Comm. Partial Differential Equations **15** (1990), 245–272.

[BG-2] _____, Residues of the eta function for an operator of Dirac type, J. Funct. Anal (to appear); see also Residues of the eta function for an operator of Dirac type with local boundary conditions, Int. J. Diff. Geo. and its applications (to appear).

[BGO] T. Branson, P. Gilkey, and B. Ørsted, Leading terms in the heat invariants, Proc. Amer. Math. Soc. (to appear).

[BPY] R. Brooks, P. Perry, and P. Yang, Isospectral sets of conformally equivalent metrics, Duke Math. J. **58** (1989), 131–150.

[BrGo] R. Brooks and C. Gordon, Isospectral families of conformally equivalent Riemannian metrics, Bull. Amer. Math. Soc. (to appear).

[BPP] R. Brooks, P. Perry, and P. Peterson, Finiteness of diffeomorphism types of isospectral manifolds, (to appear).

[CY] A. Chang and P. Yang, *Compactness of isospectral conformal metrics on* S^3, Comment. Math. Helv. **64** (1989), 363–374; *The conformal deformation equation and isospectral sets of conformal metrics*, Contemp. Math. **101** (1989), 165–178; *Isospectral conformal metrics on 3 manifolds*, J. Amer. Math. Soc. **3** (1990), 117–145.

[Gi-1] P. Gilkey, *The spectral geometry of a Riemannian manifold*, J. Differential Geom. **10** (1975), 601–618.

[Gi-2] ——, *Invariance theory, the heat equation, and the Atiyah-Singer index theorem*, Math. Lecture Ser., no. **11**, Publish or Perish Inc., Washington, DE, 1984.

[Gi-3] ——, *Leading terms in the asymptotics of the heat equation*, Contemp. Math. **73** (1988), 79–85.

[GW] C. Gordon and E. Wilson, *Isospectral deformations of compact solvmanifolds*, J. Differential Geom. **19** (1984), 241–256.

[Ik] A. Ikeda, *On spherical space forms which are isospectral but not isometric*, J. Math. Soc. Japan **35** (1983), 437–444.

[Ka] M. Kac, *Can one hear the shape of a drum?*, Amer. Math. Monthly **73** (1966), 1–23.

[KCD] G. Kennedy, R. Critchley, and J. S. Dowker, *Finite temperature field theory with boundaries: stress tensor and surface action renormalization*, Ann. Physics **125** (1980), 346–400.

[Me] J. Melmed, *Conformal invariance and the regularized one loop effective action*, J. Phys. A **21** (1989), L1131–1134.

[Mi] J. Milnor, *Eigenvalues of the Laplace operator on certain manifolds*, Proc. Nat. Acad. Sci. U.S.A. **51** (1964), 542.

[Mo] I. Moss, *Boundary terms in the heat kernel expansion*, Classical Quantum Gravity **6** (1989), 759–765.

[MD] I. Moss and J. S. Dowker, *The correct* B_4 *coefficient*, Phys. Lett. B **229** (1989), 261–263.

[OPS] B. Osgood, R. Phillips, and P. Sarnak, *Compact isospectral sets of surfaces*, J. Funct. Anal. **80** (1988), 212–234.

[Pr] M. Protter, SIAM Rev. **29** (1987), 185–197.

[Sa] T. Sakai, *On eigenvalues of Laplacian and curvature of Riemannian manifold*, Tôhuku Math. J. (2) **23** (1971), 589–603.

[Sm] L. Smith, *The asymptotics of the heat equation for a boundary value problem*, Invent. Math. **63** (1981), 467–493.

[Su] T. Sunada, *Riemannian coverings and isospectral manifolds*, Ann. of Math. (2) **121** (1985), 169–186.

[Ta] V. Tanno, *Eigenvalues of the Laplacian of Riemannian manifolds*, Tôhoku Math. J. **25** (1973), 391–403.

[Ur] H. Urakawa, *Bounded domains which are isospectral but not congruent*, Ann. Sci. Ecóle Norm. Sup. (4) **15** (1982), 441–456.

[Vi] M. Vigneras, *Variétés Riemanniennes isospectrals et non isométriques*, Ann. of Math. (2) **112** (1980), 21–32.

UNIVERSITY OF OREGON
E-mail address: gilkey@math.uoregon.edu

Proceedings of Symposia in Pure Mathematics
Volume **54** (1993), Part 3

Nonnegatively Curved Manifolds
Which Are Flat Outside a Compact Set

R. E. GREENE AND H. WU

About ten years ago, we proved a group of theorems about Riemannian manifolds which are reminiscent of Hartogs's extension theorem in several complex variables. The simplest of these (Theorem 1 of [GW4]) states

THEOREM A. *A complete noncompact Riemannian manifold which is simply connected at infinity and which has overall nonnegative sectional curvature but is flat outside a compact set must be isometric to Euclidean space.*

This result is obtained as a consequence of a general conclusion about complete manifolds with curvature zero outside a compact set [GW4, p. 736]:

PROPOSITION $(*)$. *If M is a complete noncompact Riemannian manifold such that, for some compact K, the sectional curvature is identically 0 on $M - K$ and such that M is simply connected at infinity, then there are compact sets K_1 in M and K_2 in $\mathbb{R}^n$, $n = $ dimension n, such that $M - K_1$ is isometric to $\mathbb{R}^n - K_2$.*

From this, Theorem A follows immediately: $M - K_1$ being isometric to $\mathbb{R}^n - K_2$ implies immediately that M has euclidean volume growth; i.e.,

$$\lim_{r \to +\infty} B(x, r) / \mathrm{vol}_0(r) = 1$$

for each $x \in M$, where $\mathrm{vol}_0(r) = $ the volume of the ball of radius r in $\mathbb{R}^n$. Thus if M has nonnegative sectional curvature everywhere in addition, the usual volume comparison shows that M is flat. This result, Theorem A, was stated explicitly for sectional curvature in [GW4] because sectional curvature is overall the subject in that paper. But Proposition $(*)$ implies *immediately* the Ricci curvature result:

1991 *Mathematics Subject Classification.* Primary 53C20.
This paper is in final form and no version of it will be submitted elsewhere..
Work of both authors partially supported by the National Science Foundation.

THEOREM A'. *If M is a complete noncompact Riemannian manifold, dimension n, simply connected at infinity, flat outside a compact set, and nonnegative Ricci curvature everywhere then M is flat.*

The transition from $(*)$ as proved in [GW4] to this result is immediate: One simply applies the volume growth estimate argument in [GW4] using the Ricci curvature volume comparison of Bishop (cf. [Bi]) to conclude flatness everywhere.

There is an alternative way using the Bochner technique to go from Proposition $(*)$ to Theorem A' which is of interest, though it is not so completely immediate: Choose a large (solid) cube C in $\mathbb{R}^n$ containing K_2 in its interior, where $M - K_1$ is isometric to $\mathbb{R}^n - K_2$, with K_1, K_2 compact via an isometry $I: M - K_1 \to \mathbb{R}^n - K_2$. The compact set $M - I^{-1}(\mathbb{R}^n - C)$, which contains K, can be made into a manifold by identifying opposite faces of the boundary = opposite faces $I_1^{-1}(\partial C)$. The resulting manifold inherits an obvious metric equal to the metric of M away from $I^{-1}(\partial C)$ and being the usual flat torus metric associated to C near ∂C. This manifold has nonnegative Ricci curvature. Moreover, by construction its first integral cohomology group has rank at least n, and hence its first Betti number is at least n. By the Bochner technique, it admits n parallel linearly independent vector fields and is hence flat. Since M is flat already outside K_1, M itself is flat everywhere.

Returning to [GW4], the key idea of the proof of Proposition $(*)$ and Theorem A was to make use of the structure theorem of van Heijenoort–Sacksteder [H and S] for complete hypersurfaces of nonnegative sectional curvature in Euclidean space. Theorem A, Proposition $(*)$ and their immediate consequences have since been reproved elsewhere and, on occasion, generalized in various directions, and the idea of using the van Heijenoort–Sacksteder structure theorem seems to have taken root (e.g. [SS] and [C]). Recently we realized that our original idea from [GW4] was in fact capable of providing detailed information even in more general circumstances, in particular without assuming simple connectivity conditions. The purpose of this paper is to investigate these more general situations. First we investigate the analogue of Proposition $(*)$ in this more general set-up.

Let M be an n-dimensional complete noncompact Riemannian manifold which is flat outside a compact set. By an observation in [GW2, p. 292], the Cheeger-Gromoll construction [CG1] in fact yields an exhaustion function $\tau: M \to \mathbb{R}$ which is convex outside a compact set. Denoting for each $t \in \mathbb{R}$ the sublevel set of τ by $M^t \equiv \{x \in M : \tau(x) \le t\}$, we see that each M^t is compact. We may assume that, for some $a > 0$, τ is convex in $M - M^a$ and that $M - M^a$ is flat. If we apply the Riemannian smoothing process of [GW1] to τ (to be completely precise, one needs to use a trivial variation of the argument in [GW1], noting that the argument does not depend on having a positive lower bound on the injectivity radius), then the flatness of $M - M^a$ implies that τ can be smoothed to a C^∞ convex function (and not just an *almost* convex function as happens in the nonflat case) on $M - M^b$ for any $b > a$. Now M is diffeomorphic to the interior of M^b, and the diffeomorphism is obtained by deforming along the gradient lines of τ (this was stated without proof in [GW3, p. 242], but the proof of a more general fact is given in [GS2]). Thus $M - M^b$ has only a finite number of components and in particular, the number of ends of M is finite. Let E be one of the components in $M - M^b$. Let $b < c$ and let c be a regular value of τ. Defining S to be the level set in E of $\tau = c$, S *is a compact oriented hypersurface.* Since τ is convex and E is flat, S *has nonnegative sectional curvature.* Now let the normal bundle of S be denoted by $\mathcal{N}(s)$ and let the zero secion of this bundle be denoted by S as usual. Then $\mathcal{N}(s) - S$ splits into two components. Let $\mathcal{N}^+(S)$ denote the component of vectors which are positive multiples of $\operatorname{grad} \tau|_S$. We shall prove the analogue in this general situation of Proposition $(*)$:

PROPOSITION $(\dagger)$. *If* exp *denotes the exponential map of* $\mathcal{N}(s)$, *then* exp: $\mathcal{N}^+(S) \to E - E^c$ *is a diffeomorphism, where* $E^c \equiv \{x \in E : \tau(x) \le c\}$.

This implies in particular that $E - E^c$ is diffeomorphic to $\mathbb{R} \times S$. Since the existence of some such diffeomorphism is a known fact [GS2], the interest in the proposition lies in the explicit description of this diffomorphism using geodesics normal to S. Coupled with the Cheeger-Gromoll structure theorem ([CG1] or [CG2]), this shows that *each end of* M *is covered by* $\mathbb{R}^k \times N$, *where* N *is an* $(n-k)$-*dimensional compact simply connected Riemannian manifold of nonnegative sectional curvature.*

If M has in addition everywhere nonnegative sectional curvature, then we can say more. The following structure theorem in effect extends Theorem A to the nonsimply connected case.

THEOREM B. *Let* M *be an* n-*dimensional complete noncompact Riemannian manifold of nonnegative sectional curvature. If* M *is flat outside a compact set, then either* (a) M *is flat, or* (b) *any soul of* M *is flat and has codimension* 2, *the universal covering of* M *splits isometrically as* $\mathbb{R}^{n-2} \times M_0$, *where* M_0 *is diffeomorphic to* $\mathbb{R}^2$ *and is flat outside a compact set but not flat everywhere, and finally the fundamental group of* M *is a Bieberbach group of rank* $n - 2$.

The fact that M has a flat soul under these hypotheses is a special case of a more general result of P. Petersen [P], proved via convergence techniques, that if its curvature goes to 0 at infinity then the soul of a complete nonnegative-curvature manifold is flat. Thus the emphasis here is on the particular technique we are using as well as the complete structural analysis we give of M itself.

Here is an example to show that case (b) occurs in a nontrivial way. Let M_0 be obtained from a 2-dimensional cone with the (singular) apex replaced by a smooth cap, in such a way that M_0 is a rotationally symmetric complete Riemannian manifold whose curvature is everywhere nonnegative but is zero outside a compact set. Let $\varphi : M_0 \to M_0$ be a rotation through an angle θ which is an irrational multiple of 2π. Let $M' = M_0 \times \mathbb{R}$ and for each integer k, let $\psi_k : M' \to M'$ be the isometry $\psi_k(p, s) = (\varphi^k(p), s + k)$. The manifold $M \equiv M'/\{\psi_k : k \in \mathbb{Z}\}$ is then flat outside a compact set but not flat everywhere, it is diffeomorphic to $\mathbb{R}^2 \times S^1$ (S^1 being the unit circle) but not isometric to a direct product, and finally it has $M_0 \times \mathbb{R}$ as its universal covering.

We do not know what is the proper analogue of this theorem for nonpositive sectional curvature, beyond the description in Proposition (†).

We now present the *proof of the Theorem* B, assuming Proposition (†). Let $\mathscr{S}$ be a soul of M. If codim $\mathscr{S} \leq 2$, we invoke Theorem 2 of [ESS] which says that for the case of codim $\mathscr{S} \leq 3$, as soon as it is known that the sectional curvature decays to zero at ∞, either M is flat, or the universal covering $\widetilde{M}$ splits isometrically as $M_0 \times \mathbb{R}^k$ ($n - k = \dim \mathscr{S}$), where M_0 is diffeomorphic to $\mathbb{R}^{n-k}$. Moreover $\pi_1(M)$ is a Bieberbach group of rank k and $\mathscr{S}$ must be a compact flat k-dimensional manifold. Under our hypothesis that M is flat outside a compact set, if codim $\mathscr{S} = 1$, then $\widetilde{M}$ is isometric to $\mathbb{R} \times \mathbb{R}^{n-1} = \mathbb{R}^n$ and we are done. If codim $\mathscr{S} = 2$, then $\widetilde{M}$ is isometric to $M_0 \times \mathbb{R}^{n-2}$, M_0 is diffeomorphic to $\mathbb{R}^2$ and $M \simeq M_0 \times \mathbb{R}^{n-2}/\pi_1(M)$, $\pi_1(M)$, $\pi_1(M)$ being a Bieberbach group of rank $(n-2)$. The latter implies that if M_0 has points of positive curvature diverging to ∞ so would M. Hence M_0 is flat outside a compact set.

It remains to deal with the general case of codim $\mathscr{S} \geq 3$. Let τ be a convex exhaustion function on M [CG1]. Let $M^s \equiv \{x \in M : \tau(x) \leq s\}$ for every $s \in R$, and as above, we may assume that for some $a > 0$, $M - M^a$ is flat. We may also assume that $\mathscr{S} \subset M^\circ$, i.e., $\min_M \tau = 0$. Introduce the notation: if $s \in \mathbb{R}$, $^sM \equiv \{x \in M : \tau(x) = s\}$. Then we claim

(a) For any $b > a$, ^bM is homeomorphic with the unit sphere bundle $\nu_1(\mathscr{S})$ of the normal bundle $\nu(\mathscr{S})$ of $\mathscr{S}$.

Let $\varepsilon > 0$ (the size of ε will be specified presently). We already observed above that ^bM is homeomorphic with $^\varepsilon M$ (see [GS1]). Moreover by [CG1], when δ is sufficiently small, the δ-disc bundle of the normal bundle $\nu(\mathscr{S})$ of $\mathscr{S}$ is homeomorphic with the δ-neighborhood $M^{0,\delta}$ of M^0; more precisely, $M^{0,\delta} \equiv \{x \in M : d(x, M^0) \leq \delta\}$. We will always take $0 < \delta < \varepsilon/2$. Now another construction in [CG1] shows that if ε is sufficiently small (the existence of such an ε being dependent on the compactness of M^0), then for each $q \in {}^\varepsilon M$, there exists a unique closest point $q' \in M^0$; in addition, the unique minimizing geodesic from q to q' will intersect the boundary $\partial M^{0,\delta}$ of $M^{0,\delta}$ at a unique point q^*, and the map $q \mapsto q^*$ gives a homeomorphism of $^\varepsilon M$ with $\partial M^{0,\delta}$. (To be exact, this is a trivial consequence of part (2) of

Lemma 2.4 in [CG1].) This ^bM is homeomorphic with $\partial M^{0,\delta}$, and hence with the δ-sphere bundle of $\nu(\mathscr{S})$. This proves (a).

Since we shall be dealing with $M - M^a$ exclusively, we shall smooth τ to a C^∞ convex function τ_1 on $M - M^a$ (see the remarks preceding Proposition (†)). Since we may take τ_1 to be an arbitrarily close approximation of τ, the level sets of τ_1 will be homeomorphic with the level sets of τ. The fact that this homeomorphism property will hold for the particular C^∞ approximations we are using is established in [GS2]. (In outline, the function τ is strictly decreasing along the integral curves of the gradient vector field for the approximation, so that one obtains a map for a given level set of the C^∞ function onto a (slightly) lower level set of the unsmoothed function and this map is shown to be a homeomorphism.) In the ensuing arguments, it is only the topology of the level sets of τ that interests us. Thus to minimize notation, we may as well assume that τ is itself C^∞. Then the convexity of τ implies that for each regular value b of τ on $M - M^a$, ^bM is a compact oriented hypersurface in $M - M^a$, whose second fundamental form relative to $\operatorname{grad} \tau|_{^bM}$ is positive semidefinite. We next claim

(b) $\pi_1(^bM) \simeq \pi_1(\mathscr{S}) \simeq \pi_1(M)$.

The second isomorphism is a consequence of the fact that M and $\mathscr{S}$ have the same homotopy type [CG1]. We now prove the first isomorphism. Since $\operatorname{codim} \mathscr{S} \geq 3$, the fibre of the fibration $\nu_1(\mathscr{S}) \to \mathscr{S}$ is a k-sphere with $k \geq 2$ and is hence simply connected. The homotopy sequence of the fibration then shows that $\pi_1(\mathscr{S}) \simeq \pi_1(\nu_1(\mathscr{S}))$. Assertion (b) now follows from (a).

As was remarked above in the discussion preceding Proposition (†), $M - M^b$ is diffeomorphic with $^bM \times \mathbb{R}$ (in fact, the proposition gives an explicit diffeomorphism). Thus $\pi_1(M - M^b) \simeq \pi_1(^bM)$. It follows from (b) that

(c) $\pi_1(M) \simeq \pi_1(M - M^b)$.

Now take the universal covering $\widetilde{M}$ of M and let $\pi : \widetilde{M} \to M$ be the covering map. Let $\widetilde{M}^b = \pi^{-1}(M^b)$. By (c), $\pi : \widetilde{M} - \widetilde{M}^b \to M - M^b$ is a universal covering space. Now take another regular value c of τ, $c > b$, and for simplicity, denote by S the level set ^cM of τ at c. Recall that S is a compact oriented $(n-1)$-dimensional manifold of nonnegative sectional curvature (by the convexity of τ and the flatness of $M - M^b$). Let $\widetilde{S} \equiv \pi^{-1}(S)$. Then $\widetilde{S}$ is a simply connected complete manifold of nonnegative sectional curvature which has a compact quotient, namely, S. By the structure theorem of Cheeger-Gromoll ([CG1] or [CG2]), $\widetilde{S}$ splits isometrically into $\mathbb{R}^{n-1-k} \times N$, where N is a k-dimensional simply connected *compact* manifold of nonnegative sectional curvature. We claim

(d) $k \geq 2$.

Suppose $k < 2$; then $k = 0$ because there is no compact simply connected manifold of dimension 1. So let $k = 0$. Then $\pi : \mathbb{R}^{n-1} \to S$ is a covering map. This means S is an oriented compact $(n-1)$-dimensional flat manifold; by standard Hodge theory, $\dim H^1(S, \mathbb{R}) = n - 1$. On the other hand, since $\mathscr{S}$ is a compact manifold of nonnegative sectional curvature, a classical theorem of Bochner says $\dim H^1(\mathscr{S}, \mathbb{R}) \leq n - 3$ $(= \dim \mathscr{S})$. So by (b), we have

$$
\begin{aligned}
(n - 1) &= \dim H^1(S, \mathbb{R}) \\
&= \operatorname{rank} \pi_1(S)/[\pi_1(S), \pi_1(S)] \\
&= \operatorname{rank} \pi_1(\mathscr{S})/[\pi_1(\mathscr{S}), \pi_1(\mathscr{S})] \\
&= \dim H^1(\mathscr{S}, \mathbb{R}) \leq n - 3,
\end{aligned}
$$

which is absurd. So (d) is true.

We make use of (d) in the following way. Since $\widetilde{M} - \widetilde{M}^b$ is flat and simply connected, a classical monodromy argument gives a local isometry $\Phi : \widetilde{M} - \widetilde{M}^b \to \mathbb{R}^n$. Then $\Phi(\widetilde{S})$ is an immersed complete hypersurface of nonnegative sectional curvature in $\mathbb{R}^n$. By the structure theorem of van Heijenoort–Sacksteder [H, S], $\Phi(\widetilde{S})$ must be imbedded and is in fact the boundary of a convex body. In particular, $\Phi(\widetilde{S})$ is an imbedded hypersurface in $\mathbb{R}^n$, $\widetilde{S}$ is complete and $\Phi : \widetilde{S} \to \Phi(\widetilde{S})$ is a local isometry. By a standard theorem in Riemannian geometry, $\Phi : \widetilde{S} \to \Phi(\widetilde{S})$ is a covering map. Now the boundaries of convex bodies in $\mathbb{R}^n$ are completely classified (see the argument on p. 3 of [Bu], for example), so we know that $\Phi(\widetilde{S})$ is isometric to either (i) $\mathbb{R}^{n-1-l} \times H_1$, where H_1 is the boundary of a *noncompact* convex body in $\mathbb{R}^{l+1}$ which contains no complete straight lines, or (ii) $\mathbb{R}^{n-1-l} \times H_2$, where H_2 is the boundary of a *compact* convex body in $\mathbb{R}^{l+1}$. Because of (d), case (i) is ruled out. In case (ii), H_2 is either two points (if $l = 0$), or a circle (if $l = 1$), or a simply connected sphere (if $l \geq 2$). The first two possiblities are ruled out again because of (d). So H_2 is simply connected and hence so is $\Phi(\widetilde{S})$. It follows that $\Phi : \widetilde{S} \to \Phi(\widetilde{S})$ is a global isometry. Let E' be the noncompact component of $\mathbb{R}^n - \Phi(\widetilde{S})$ and let E be $\widetilde{M} - \widetilde{M}^c$. We now claim

(e) $\Phi : E \to E'$ is a global isometry.

This is a straightforward argument and is given in [GW4, p. 737]. Briefly, let $\mathscr{N}^+(\widetilde{S})$ (resp. $\mathscr{N}^+(\Phi(\widetilde{S}))$) be those nonzero vectors of the normal bundle of $\widetilde{S}$ in $\widetilde{M}$ (resp. of $\Phi(\widetilde{S})$ in $\mathbb{R}^n$) pointing into E (resp. E'). Then the exponential map of the normal bundle of $\widetilde{S}$ defines by restriction a map

$$
\widetilde{\exp} : \mathscr{N}^+(\widetilde{S}) \to E.
$$

Similarly, the exponential map of the normal bundle of $\Phi(\widetilde{S})$ defines by restriction a map

$$
\exp : \mathscr{N}^+(\Phi(\widetilde{S})) \to E'.
$$

The map $\exp$ is easily seen to be globally diffeomorphic, and it follows also that so is $\widetilde{\exp}$. Since the local isometry $\Phi : E \to E'$ is a global isometry from $\widetilde{S}$ to $\Phi(\widetilde{S})$, one deduces immediately that $\Phi = \exp \circ d\Phi|_{\widetilde{S}} \circ \widetilde{\exp}^{-1}$. In particular, Φ is a global isometry, thereby proving (e).

The conclusion of the proof of the theorem is now easy. We use Φ to identify $\widetilde{M} - \widetilde{M}^c$ with E'. Let Σ_0 be the *round* sphere of radius r_0 in $\mathbb{R}^n$. We may assume that the convex hypersurface $\Phi(\widetilde{S})$ is $\mathbb{R}^{n-1-k} \times H$, where H is a compact convex hypersurface in $\mathbb{R}^{k+1}$ strictly contained in the sphere of radius $r_0/2$ in $\mathbb{R}^{k+1}$. In particular, the set $\Sigma_0' \overset{\text{def}}{=} \Sigma_0 \cap E'$ is nonempty. Fix some point $x_0 \in \widetilde{M}^c$ and let ρ be the distance function on $\widetilde{M}$ relative to x_0. Let $\theta = $ maximum of ρ on $\Phi^{-1}(\Sigma_0')$. If $B(s)$ denotes the closed geodesic ball of radius s around $x_0 \in \widetilde{M}$, then $\Phi^{-1}(\Sigma_0') \subset B(\theta)$. We claim

(f) If $s > \theta + r_0$, then $B(s)$ contains $\Phi^{-1}(\Sigma_{s-\theta-r_0}')$, where $\Sigma_{s-\theta-r_0}$ means the round sphere of radius $s - \theta - r_0$ in $\mathbb{R}^n$ concentric with Σ_0, and as usual, $\Sigma_{s-\theta-r_0}' = \Sigma_{s-\theta-r_0} \cap E'$.

Given $y \in \Sigma_{s-\theta-r_0}'$, clearly we can find a $y' \in \Sigma_0'$ such that the line segment $\overline{yy'}$ lies in E' (this elementary fact of course depends on the preceding arrangement that $\Phi(\widetilde{S})$ is inside $\mathbb{R}^{n-1-k} \times \{$ sphere of radius $r_0/2$ in $\mathbb{R}^{k+1}\}$). Then $|y - y'| \leq |y| + |y'| = (s - \theta - r_0) + r_0 = s - \theta$. Thus

$$
\begin{aligned}
\rho(\Phi^{-1}(y)) &= d(\Phi^{-1}(y), x_0) \\
&= d(\Phi^{-1}(y), \Phi^{-1}(y')) \\
&\quad + d(\Phi^{-1}(y'), x_0) \\
&= |y - y'| + d(\Phi^{-1}(y'), x_0) \\
&\leq (s - \theta) + \theta \\
&= s,
\end{aligned}
$$

thereby proving $\Phi^{-1}(y) \in B(s)$, and hence also (f).

We are now in a position to apply the elementary lemma (Lemma 1 of [GW4]) to the effect that: if $v_e(r)$ is the volume of the ball of radius r in $\mathbb{R}^n$ and $v(r)$ is the volume of the geodesic ball of radius r (around a given point) in a complete n-dimensional Riemannian manifold of nonnegative sectional curvature, then $v(r)/v_{e(r)} \leq 1$, and $\lim_{r \to \infty} \sup(v(r)/v_e(r)) = 1$ iff M is isometric to $\mathbb{R}^n$. In our situation, let $D(r) = B(r) - \widetilde{M}^c$. Then volume $D(r) \geq$ volume of the region between $\Sigma_{r-\theta-r_0}$ and $\Phi(\widetilde{S})$, using (f). Since $\Phi(\widetilde{S})$ is inside $\mathbb{R}^{n-1-k} \times \{$sphere of radius $r_0/2$ in $\mathbb{R}^{k+1}\}$, we see that volume

$$
D(r) \geq v_e(r - \theta - r_0) - \gamma_0(r - \theta)^{n-1-k},
$$

where γ_0 is a positive constant depending on r_0 and $k \geq 2$. Hence

$$1 \geq \frac{v(r)}{v_e(r)} \geq \frac{\text{volume } D(r)}{v_e(r)}$$

$$\geq \frac{v_e(r - \theta - r_0) - \gamma_0(r - \theta)^{n-1-k}}{v_e(r)}$$

$$= \frac{v_e(1)(r - \theta - r_0)^n - \gamma_0(r - \theta)^{n-1-k}}{v_e(1)r^n} \to 1.$$

This implies $\lim_{r\to\infty}(v(r)/v_e(r)) = 1$. Thus M is isometric to $\mathbb{R}^n$.   Q.E.D.

To conclude, we give the proof of Proposition (†). Since S has nonnegative sectional curvature and for each $v \in \mathcal{N}^+(S)$, the second fundamental form of S with respect to v is positive semidefinite, standard Jacobi field arguments show that $\exp : \mathcal{N}^+(S) \to E$ is an immersion. Moreover, if ζ is a geodesic starting from S such that $\dot\zeta(0)$ belongs to $\mathcal{N}^+(S)$, then the function $t \mapsto \tau(\zeta(t))$ is convex and has positive derivative at $t = 0$; consequently, $\tau(\zeta(t))$ is strictly increasing on $(0, \infty)$. In particular, $\tau(\zeta(t)) > c$ for all $t > 0$. This shows that $\exp : \mathcal{N}^+(s) \to E - E^c$ is well defined. We proceed to prove that this is actually a proper map; i.e., for any compact set $K \subset E - E^c$, $\exp^{-1}(K)$ is compact. Observe first that if ζ is as above and if ζ is parameterized by arc-length, then the function $h(t) \equiv \tau(\zeta(t))$ satisfies $h'(0) \geq A$ for some $A > 0$; this uses the fact that c is a regular value of τ. Since h is convex, $h(t) \geq At$ for all $t \geq 0$. Now K being compact, $\tau(K) \subset [B, C]$ for some positive constants B, C. Hence an elementary argument shows that, for all such geodesics ζ, $\zeta(t) \in K$ implies $t \in [B_1, C_1]$ for some other positive constants B_1, C_1 which depend on A, B, C. In terms of the exponential map, this clearly means $\exp^{-1}(K)$ is compact.

Thus $\exp : N^+(S) \to E - E^c$ is a proper immersion between manifolds of the same dimension. It is a standard fact that $\exp$ must be a covering map. The easiest way to show this is to equip $E - E^c$ with an arbitrary complete Riemannian metric G. Since $\exp$ is proper, $\exp^* G$ is also complete; and $\mathcal{N}^+(S) \to E - E^c$ is now a local isometry from a complete Riemannian manifold $N^+(S)$, and so must be a covering.

Now if $\bigcup_\varepsilon$ is an ε-neighborhood of S in $\mathcal{N}(S)$ (we identify S with the zero section in $\mathcal{N}(S)$), and if $\bigcup_\varepsilon^+ \equiv \bigcup_\varepsilon \cap \mathcal{N}^+(S)$, then $\exp^{-1}(\exp\bigcup_\varepsilon^+)$ in $\mathcal{N}^+(S)$ is just $\bigcup_\varepsilon^+$ itself; this uses the fact that the function h above is always strictly increasing. Moreover, if ε is small enough, $\exp : \bigcup_\varepsilon^+ \to \exp\bigcup_\varepsilon^+$ is diffeomorphic. Thus for each $x \in \exp\bigcup_\varepsilon^+$, $\exp^{-1}(x)$ consists of one point. It follows that $\exp : \mathcal{N}^+(S) \to E - E^c$ is a diffeomorphism.   Q.E.D.

REFERENCES

[Bi] R. Bishop and R. Crittenden, *Geometry of manifolds*, Academic Press, New York 1964.
[Bu] H. Busemann, *Convex surfaces*, Interscience, New York, 1958.
[C] C. Croke, *Simply connected manifolds with no conjugate points which are flat outside a compact set*, Proc. Amer. Math. Soc. **111** (1991), 297–298.

[CG1] J. Cheeger and D. Gromoll, *On the structure of complete manifolds of nonnegative curvature*, Ann. of Math. (2) **96** (1972), 413–443.

[CG2] _____, *The splitting theorem for manifolds of nonnegative Ricci curvature*, J. Differential Geom. **6** (1971), 119–128.

[ESS] J. Eschenburg, V. Schroeder, and M. Strake, *Curvature at infinity of open nonnegatively curved manifolds*, J. Differential Geom. **30** (1989), 155–166.

[GW1] R. E. Greene and H. Wu, *On the subharmonicity and plurisubharmonicity of geodesically convex functions*, Indiana Univ. Math. J. **22** (1973), 641–653.

[GW2] _____, *Integrals of subharmonic functions on manifolds of nonnegative curvature*, Invent. Math. **27** (1974), 265–298.

[GW3] _____, C^∞ *convex functions and manifolds of positive curvature*, Acta Math. **137** (1976), 209–245.

[GW4] _____, *Gap theorems for noncompact Riemannian manifolds*, Duke Math. J. **49** (1982), 731–756.

[GS1] R. E. Greene and K. Shiohama, *Convex functions on complete noncompact manifolds: topological structure*, Invent. Math. **63** (1981), 129–157.

[GS2] _____, *Convex functions on complete noncompact manifolds: differentiable structure*, Ann. Sci. École Norm. Sup. **14** (1981), 357–367.

[H] J. van Heijenoort, *On locally convex manifolds*, Comm. Pure Appl. Math. **5** (1952), 223–242.

[P] P. Petersen, V, *Rigidity of fibrations in manifolds of nonnegative curvature*, Preprint.

[S] R. Sacksteder, *On hypersurfaces with no negative sectional curvatures*, Amer. J. Math. **82** (1960), 609–630.

[SS] V. Schroeder and M. Strake, *Rigidity of convex domains in manifolds with nonnegative Ricci and sectional curvature*, Comment. Math. Helv. **64** (1989), 173–186.

UNIVERSITY OF CALIFORNIA, LOS ANGELES

UNIVERSITY OF CALIFORNIA, BERKELEY

Proceedings of Symposia in Pure Mathematics
Volume **54** (1993), Part 3

Spaces of Nonnegative Curvature

DETLEF GROMOLL

This is a somewhat enhanced version of a survey lecture on developments in the general area of positive curvature. When I gave an account of the same subject during the last Geometry Summer Institute at Stanford in 1973, the frontier of our knowledge had just been pushed forward substantially. The *Soul Theorem* provided a structure theory for complete manifolds of non-negative sectional curvature, reducing some basic questions to the compact case. For spaces with nonnegative Ricci curvature, at least the fundamental group, one of the first and most natural invariants in that class, had become fairly well understood—as a consequence of the *Splitting Theorem* and ideas centered about the concept of *polynomial growth* of a group. *Spin geometry*, notably index theory of the Dirac operator, began to play its crucial role in the study of positive scalar curvature, giving rise to surprising new obstructions, e.g., among exotic spheres.

Since then an explosion of interest in these exciting, mostly global aspects of riemannian geometry has led to remarkable further progress in almost all directions. It would be well beyond the scope of this article to describe or acknowledge all the contributions that have been important to the field more recently. We will rather try to give an overview by focussing on just a few central problems and results, the key techniques, and the improving state of the art of constructing examples. In no way is our discussion meant to be complete. The reader will find a wealth of more specialized information and references at various places throughout these Proceedings.

We consider the classes of riemannian spaces M^n (usually assumed to be smooth, complete, and without boundary) whose scalar, Ricci, or sectional curvatures scal, ric, or K are everywhere nonnegative. These increasingly restrictive curvature conditions correspond to underdetermined, well-posed, or overdetermined differential relations. In turn the study of spaces with

1991 *Mathematics Subject Classification.* Primary 53C20; Secondary 53C21.
This paper is in final form and no version of it will be submitted for publication elsewhere.

$K \geq 0$ often amounts to understanding the extent of their rigidity, a type of question notoriously difficult in geometry. The methods involved are largely metric; general results as well as examples remain quite scarce and intriguing. At the other end, the class scal ≥ 0 is very ample and highly nonrigid. While here a structure theory has advanced the most—primarily by analytic means—scalar curvature is a very weak invariant in higher dimensions and does not control much of the geometry in any metric sense. More recently the greatest activity has gravitated toward the class ric ≥ 0, which is also of considerable outside interest, say in algebraic geometry. Nonnegative Ricci curvature tends to capture some essential metric data of the underlying manifold, and yet it is often flexible enough and tractable by both geometric and analytic methods. Progress in this area has been stimulated a lot by the emergence of diverse large families of new examples.

There are two fundamental problems we want to discuss already now. The first one concerns the distinction between *nonnegative* and *positive* curvature. This distinction is not too severe in the classes scal ≥ 0 or ric ≥ 0, but very subtle for $K \geq 0$. The famous conjecture of H. Hopf earlier this century, that $S^2 \times S^2$ cannot carry a riemannian structure with everywhere positive sectional curvature, seems to remain essentially open, even for deformations of the standard metric. Despite various interesting efforts, only very little could be added to the discussion in [**Bg**] today. It has always been frustrating that, by contrast, the same question for $\mathbb{R}P^2 \times \mathbb{R}P^2$ is confirmed easily (with the classical Synge Lemma). It is not clear yet if there is *any* simply connected compact M^n with $K \geq 0$ which does not admit $K > 0$. In the complete noncompact case, all M^n not diffeomorphic to $\mathbb{R}^n$ are known to be such examples (by the Soul Theorem). The Hopf Conjecture marked the starting point for global geometry of nonnegative curvature. It may in fact turn out to be one of the most difficult problems to resolve, definitely in its ultimate form: how to characterize the difference between the classes $K \geq 0$ and $K > 0$, say from a topological point of view?

The classes ric ≥ 0 and ric > 0 do not coincide either. For example: $S^n \times S^1$ does not admit positive Ricci curvature, simply because the fundamental group is not finite (Myers's Theorem); the complex quartic in $\mathbb{C}P^3$ ($K3$ surface) is simply connected and does not admit scal > 0 (see §1), but it carries a Ricci flat Calabi-Yau metric; the complete manifolds $S^n \times \mathbb{R}$ cannot support ric > 0 (Splitting Theorem). The second example also shows that the condition scal > 0 is (somewhat) more restrictive than scal ≥ 0, in the compact case. However, there is only a distinction if the metric is already Ricci flat, which was observed by J. P. Bourguignon (see [**KW**]). Another reason why it seems that, for Ricci and scalar curvature, "nonnegative" and "positive" generally mean almost the same is the following: It is possible to deform a metric with ric ≥ 0 or scal ≥ 0 into a metric with ric > 0 or scal > 0, respectively, if the original metric had positive curvature at some point. This result of T. Aubin [Au1] can be proved in a direct geometric way by spreading positive curvature over the manifold using a sequence of natural deformations in annular regions of small convex balls (see also [Eh]). Local methods like this fail to produce $K > 0$ in the case $K \geq 0$, except when the points with zero curvatures are concentrated in a convex set.

The second basic question is the following: To what extent are the possible types of spaces in the classes scal ≥ 0, ric ≥ 0, $K \geq 0$ more and more restricted? Is there actually a difference between these curvature conditions, as one would expect in higher dimensions? This is surprisingly difficult to answer. Of course, spaces like the $K3$ surface provide Ricci flat (border line) examples that cannot support $K \geq 0$, since such a metric would have to be flat riemannian, by the above discussion. Much more to the point, there are now examples of manifolds M^n with ric > 0, for which it is known that they do not admit $K \geq 0$, compact or noncompact, simply connected, $n \geq 4$. In the compact case, one may take for M^n the connected sum of any sufficiently large number of copies of $S^k \times S^l$, $k + l = n$, and $k, l \geq 2$. These manifolds can be given metrics with ric > 0 by the Sha-Yang construction (see §3), and the existence of a metric with $K \geq 0$ is excluded by Gromov's a priori bound on the Betti numbers of such spaces in a given dimension (see §2). Unfortunately, the known bounds only give explicit examples with huge Betti numbers. In the noncompact case, analogous constructions are possible. But there are also other ways to produce topologically rather simple open manifolds with ric > 0, which do not have the homology of a compact manifold, and thus cannot support $K \geq 0$. Manifolds M^n admitting scal > 0, but not ric ≥ 0, abound if they are not required to be simply connected: Since ric ≥ 0 restricts the fundamental group very tightly (see §3), one can easily give examples, compact or noncompact, such as $M^n = N^2 \times S^{n-2}$ or $M^n = N^2 \times S^{n-2} \times \mathbb{R}$, where N^2 is a closed surface of genus ≥ 2 and the spherical factor is scaled to have large enough curvature. On the other hand, at this point there are no simply connected examples whatsoever, compact or not. The problem is that except for the fundamental group, we do not know any obstruction for ric ≥ 0 that is not already an obstruction for scal > 0.

Our discussion so far has only touched on the simplest orientational questions about nonnegatively curved spaces. The answers, as much as they can be given, already make use of many major results on the subject. We will now turn to more specific aspects of scal ≥ 0 in §1, $K \geq 0$ in §2, and ric ≥ 0 in §3—in this neither natural nor historical, but convenient order. Other recent surveys in and around the general area of this broad overview include [Ab, Bé, Ch, Gn, Gr, Gv, Me, Pe, RS, Sh].

1. Nonnegative scalar curvature

Comprehensive, up-to-date discussions of this class have been given in [RS], as well as in the monograph [LM] on spin geometry, which also provides a beautiful self-contained introduction to many crucial geometric techniques centered about index theory.

The condition scal ≥ 0 is considerably more significant for compact spaces M^n. Its relative weakness should be kept in mind, however. For example, if M carries a metric with scal > 0 then so does the product with an arbitrary compact manifold, after shrinking the factor M. We have already pointed out that we can assume scal > 0, as soon as M is not Ricci flat. An immediate question is then whether or not M^n also admits *constant* positive scalar curvature. The existence of such a metric in the conformal class of the original metric (Yamabe Problem) was finally established, primarily by T. Aubin [**Au2**] and R. Schoen [**Sch**], who dealt with the underlying nonlinear elliptic equation more or less analytically (see also [**Au, Kaz**], as well as [**Esc**] for the case of manifolds with boundary). It follows from this work and earlier results of J. Kazdan and F. Warner [**KW**] that precisely all somewhere positive functions can be realized as the scalar curvature of a metric conformal to a given metric with scal > 0. Constant negative scalar curvature is otherwise unobstructed in dimensions $n \geq 3$ and coexists with constant positive scalar curvature; simple explicit examples are left invariant metrics on S^3. Surprisingly, it seems that beyond its own merit, this "uniformization" theory for scalar curvature has not had many applications so far, even for 3-manifolds.

What are obstructions to scal > 0? The first result here was obtained rather late and has turned out to be very consequential: If M^n is a spin manifold, there is a canonical selfadjoint first-order elliptic operator D, the Dirac operator, defined on sections of the spinor bundle $S = S^+ \oplus S^-$ over M and interchanging sections of S^+ and S^-, i.e., positive and negative half-spinors. Sections in the kernel of D are called harmonic. In the note [**Li**], A. Lichnerowicz showed that harmonic spinors necessarily vanish if scal > 0; the analytic index of the Dirac operator from positive to negative half-spinors is therefore zero. It then follows from the Atiyah-Singer Index Theorem that an integral characteristic number of M, the $\hat{A}$-genus, must be zero in dimensions $n = 4k$. Since, for example, the $K3$ surface is spin and has nonzero $\hat{A}$-genus, it cannot carry scal > 0. The above vanishing result for harmonic spinors is proved by a variant of the powerful *Bochner Method*, which has many diverse applications in geometry. Roughly speaking, this technique associates with a suitable first-order operator a certain generalized Laplacian that can sometimes be expressed as the sum of a natural nonnegative second order operator and zero-order curvature terms (*Weitzenböck formulas*); if the last operator is positive, global solutions of the first-order homogeneous equation must vanish (see also [**Bé**] for an expository discussion). For the Dirac operator D, the Weitzenböck formula turns out to be $D^2 = \nabla^*\nabla + \text{scal}/4$, where D^2 is the *Dirac Laplacian* and $\nabla^*\nabla$ the *connection Laplacian*, which is clearly nonnegative.

These ideas were refined substantially by N. Hitchin [**Hi**], who exploited the fact that the real Dirac operator actually gives rise to a family of closely related Fredholm operators parametrized on $\mathbb{R}^n$ (invertible away from zero), which differ from each other by a Clifford multiplication. This family has an analytic index $\alpha(M^n)$ in the group $KR^{-n}(\text{point}) = KO^{-n}(\text{point})$, where $KO^{-n}(\text{pt}) = \mathbb{Z}$ if $n \equiv 0 \pmod 4$, $KO^{-n}(\text{pt}) = \mathbb{Z}_2$ if $n \equiv 1, 2 \pmod 8$, and $KO^{-n}(\text{pt}) = 0$ otherwise. The index theorem for such families now yields that $\alpha(M^n)$ is a topological invariant, the Atiyah-Milnor-Singer invariant of a spin manifold. In particular, $\alpha(M^{8k+1}) = $ dimension of harmonic spinors (mod 2) and $\alpha(M^{8k+2}) = $ dimension of positive harmonic half-spinors (mod 2) are subtle and very global 2-torsion invariants; they do not have a local curvature description like the $\widehat{A}$-genus in terms of Pontrjagin classes. The dimension of harmonic spinors usually depends on the metric. As before, $\alpha(M^{4k})$ is $\widehat{A}(M^{4k})$ or $\frac{1}{2}\widehat{A}(M^{4k})$. α is multiplicative on cartesian products and also additive on connected sums, when viewed as taking values in the graded ring $KO^{-*}(\text{pt})$. Being a KO-characteristic number, $\alpha(M^n)$ is a spin cobordism invariant, and thus there is a graded ring homomorphism $\alpha : \Omega_*^{\text{Spin}} \longrightarrow KO^{-*}(\text{pt})$ on spin cobordism, which is known to be surjective. In fact, by a deep result of F. Adams and J. Milnor in topology, $\alpha : \Gamma^n \cong \Theta_n \longrightarrow \mathbb{Z}_2$ is already surjective on the group of homotopy n-spheres for $8 < n \equiv 1, 2 \pmod 8$. Topologically, $\alpha(\Sigma) = 1$ iff the exotic sphere Σ does not bound a spin manifold (it always bounds an oriented manifold).

Again, if M^n is spin and scal > 0, then $\alpha(M) = 0$. This does not only show immediately that half of the exotic spheres in dimensions $8k + 1, 8k + 2$ (their number increases rapidly with k) cannot support any metric with positive scalar curvature. But moreover, for any spin manifold M^n in those dimensions, one of the two homeomorphic spaces, M or the connected sum $M \# \Sigma$, will not admit scal > 0, for any homotopy n-sphere Σ with $\alpha(\Sigma) = 1$. Actually, the other manifold will then carry such a metric, at least if M is simply connected, as we shall discuss in a moment.

Major progress toward a structure theory for simply connected spaces with scal > 0 was made by M. Gromov and H. B. Lawson [GL2], along quite different lines. They proved that in dimensions $n \geq 5$ a simply connected manifold M^n can be given a metric with scal > 0 if either M is not spin, or M is spin and the spin cobordism class of M contains some representative with scal > 0. The arguments depend on a basic surgery procedure, which was independently developed also by R. Schoen and S. T. Yau [SY2]: If M has scal > 0, then any manifold obtained from M by surgery in codimension ≥ 3 also admits a metric with scal > 0. In particular, if M_1^n, M_2^n support scal > 0, $n \geq 3$, then so does the connected sum $M_1 \# M_2$. The construction can be carried out in an elementary, direct geometric way; it is somewhat reminiscent of an earlier method of Lawson-Yau to produce metrics with scal > 0 on compact spin manifolds with nontrivial S^3-action. As a first surprising application, all simply connected compact M^n in dimensions $5 \leq n \leq 7$ admit scal > 0, since $\Omega_n^{\text{Spin}} = 0$ for those n. The spin cobordism classes that contain manifolds with scal > 0 form an ideal P_* in Ω_*^{Spin}, by the initial remark in this section. Since P_* is contained in the kernel of the KO-invariant α, it was a natural conjecture of Gromov-Lawson that $P_* = \ker \alpha$, or equivalently, a simply connected compact spin manifold M^n carries a metric with scal > 0 iff $\alpha(M^n) = 0$, $n \geq 5$. This has recently been confirmed in full generality by work of S. Stolz [St]. The proof circumvents previous difficulties arising from the lack of an explicit description for all the generators of the ring Ω_*^{Spin}. It is rather shown that for $n \geq 8$ and any class $[M^n]$ in $\ker \alpha$, an odd multiple of $[M^n]$ (that is enough) contains a bundle over some compact manifold, with fiber $\mathbb{HP}^2$ and structure group $PSp(3)$, which is easily seen to support scal > 0. In this form, the problem can be solved using methods from stable homotopy theory.

If M^n is not simply connected, the fundamental group π may further obstruct positive scalar curvature, notably when, in a certain sense, π is "large" compared to M^n. There are only partial results as yet. Pioneering work was done by R. Schoen and S. T. Yau [SY1], who used minimal surface methods to prove that the torus T^n cannot admit scal > 0, at least for $1 \leq n \leq 7$, but apparently also in higher dimensions. Another strong result along the same lines concerns the structure of compact 3-manifolds with scal > 0, which can only have the form of connected sums $M_1 \# \cdots \# M_k \# N_1 \# \cdots \# N_l$, where M_i has finite fundamental group and $N_j = S^1 \times S^2$; see also [SY] for further discussions. The main idea here is to study closed hypersurfaces in M^n of absolute minimal $(n-1)$-dimensional volume, in a nontrivial homology class. They must have positive total curvature for $n = 3$, and are therefore spherical, or they can be given scal > 0 themselves, by a suitable conformal change of the induced metric when $n > 3$ (setting the stage for a recursion). Although such techniques do not require the presence of a spin structure, they encounter difficulties for $n > 7$, because of regularity problems with minimal hypersurfaces.

It was a surprise when linear Dirac operator methods were introduced, by Gromov and Lawson [GL1], that not only provided alternate arguments for the last results, but also allowed ruling out scal > 0 for many other geometrically interesting M^n with large π. The crucial measure for π to be "large" is this: M^n is *enlargeable* if there are isometric coverings that admit contracting, compactly supported essential maps onto the euclidean unit sphere S^n, with arbitrarily small differentials. Enlargeability is a homotopy invariant of M^n. The class of enlargeable manifolds is quite big; it is closed under products and connected sums (with any manifold). In particular, it contains all manifolds with sectional curvature $K \leq 0$, as well as all solvmanifolds. The important theorem now is that an enlargeable spin manifold M^n cannot carry any metric with scal > 0. The spin restriction for M

is not too severe, since often a suitable cover is spin, even if M itself is not. For example, no compact manifold with $K \leq 0$ admits a metric with scal > 0. The new geometric ingredient in the proof of the theorem is to unwrap a twisted spinor bundle on coverings of the enlargeable manifold M^n. Assuming $n = 2k$ (which suffices), consider the twisted Dirac operator D_E on the bundle $S \otimes E$ over some (say compact) cover $\overline{M}$, where $E = f^*E_0$ is the pullback of a hermitian k-plane bundle E_0 over S^{2k} with $c_k \neq 0$, and $f : \overline{M} \longrightarrow S^{2k}$ has degree $\deg(f) \neq 0$. If f contracts enough, i.e., $|f_*|$ is sufficiently small, then the induced curvature of E will be small, and as before, $D_E^2 \approx \nabla^*\nabla + \text{scal}/4$ becomes positive for scal > 0. Hence, the analytic index of D_E on $S^+ \otimes E$ vanishes; but topologically, it is essentially $\deg(f)c_k(E_0)[S^{2k}] \neq 0$. It remains a challenging basic problem to decide if in general no manifold M^n of type $K(\pi, 1)$ can support scal > 0. Indications are, this might be true.

Finally, we mention two interesting miscellaneous results which can be proved by adaptations of the above methods in spin geometry. The space of all metrics with scal > 0 on a given compact spin manifold M^n tends to be topologically complicated (unless empty). It has infinitely many connected components whenever $n = 4k - 1 \geq 7$, for example, already in the case of the standard sphere $M = S^7$, even after dividing out by the natural action of the group $\text{Diff}(S^7)$. Moreover, the components need not be contractible. For complete open manifolds, the restriction scal > 0 seems very weak without additional conditions. However, there is a nice result: If M^k is a compact enlargeable manifold, then $M^k \times \mathbb{R}$ cannot carry any complete metric with scal > 0. This applies to $T^k \times \mathbb{R}$, whereas $T^k \times \mathbb{R}^2$ clearly carries such a (product) metric. For this and other related work, we also refer to [GL3].

2. Nonnegative sectional curvature

The single most important aspect of spaces M^n with $K \geq 0$ is the geometry of their geodesics. Second variation techniques originated here, leading to comparison theory and culminating in Toponogov's *Triangle Theorem*: Any geodesic triangle in M has angles not smaller than the corresponding angles of a euclidean triangle with the same sides. This result remains a marvel of both simplicity and power in geometry (for alternate versions, refinements, and comments see [Gr, Gv, Me, Sh]). In a very natural and useful way, the Triangle Theorem can also be taken to extend the concept $K \geq 0$, locally or globally, from "regular" smooth spaces to the more general setting of metric length spaces (which goes back to Alexandrov and Rinow). Such spaces have become increasingly important as "singular" limits of riemannian manifolds. So far the strongest applications of comparison theory in the class $K \geq 0$ are those that translate directly into basic properties of distance functions, primarily *convexity* and control of *critical points*. There are only a few other techniques available, of use in dealing with very specific problems: Bochner's Method (as a maximum principle, related to the second variation approach), integral formulas (like the Gauss-Bonnet Theorem), and other analytic methods (minimal surfaces, certain deformation equations). We will touch on some details later.

One reason why comparison is such an effective tool for $K \geq 0$ is that constant sectional curvature is well understood, and the simply connected models $\mathbb{R}^n$, S^n of space forms carry the most familiar elementary geometric structures. The corresponding situation is quite different in the class $\mathrm{ric} \geq 0$, and essentially meaningless for $\mathrm{scal} \geq 0$, as we saw in §1. Let us first consider compact manifolds M^n with $K \geq 0$. It is disturbing how little is still known here, even at the lowest topological level. Contrary to the case $\mathrm{scal} \geq 0$, at least the structure of the fundamental group π of M can be determined completely, modulo finite groups, although this is actually a theorem in the class $\mathrm{ric} \geq 0$ (see §3). Genuine results for $K \geq 0$ are so rare that they deserve mention. If one assumes $K > 0$, then π is finite (also true for $\mathrm{ric} > 0$), but in addition, by the Synge Lemma, M^n must be simply connected if orientable and n even, and orientable if n odd. This was among the very first classical applications of the second variation formula; yet no other specific obstruction to $K > 0$, of comparable strength, has been found since then. The long-standing problem of S. S. Chern, if abelian subgroups of π are necessarily cyclic, remains open. There is a weaker obstruction to $K \geq 0$, due to M. Gromov: The minimal number of generators for π is (explicitly) bounded in terms of n only. This follows, since geodesics from a point in the simply connected cover of M to the orbit of a minimal generator set with minimal total displacement make mutual angles $\geq \pi/3$, just by use of the Triangle Theorem. Of course, infinitely many nonisomorphic fundamental groups π arise, say for lens spaces M^3.

We now turn to compact simply connected manifolds M^n with $K > 0$. It was a second perceptive question of H. Hopf that started and stimulated all the developments in this area: Is there a "perturbation" of the basic theorem that M^n must be isometric to the euclidean sphere S^n if $K \equiv 1$? In fact, this question eventually led to the by now well established general principle that every rigidity result has some perturbation—especially when lower curvature bounds are involved. It took a very long time and considerable efforts to solve the original problem in a reasonably precise way. The celebrated classical pinching theorems of H. E. Rauch, M. Berger, and W. Klingenberg finally showed that M^n is homeomorphic to S^n if $4 > K \geq 1$, which is sharp in dimensions $n = 2k > 2$, since for example, $\mathbb{C}P^k$ satisfies $4 \geq K \geq 1$, as a symmetric space of rank 1. This was only the beginning of a rich series of results that became known as *Sphere Theorems*. An excellent account of this work has been given in [**Sh**], and we can limit our discussion to a few highlights and remarks. Many geometers have made contributions in the context of *differentiable* sphere theorems, which naturally require stronger curvature assumptions. The best quantitative result still seems to arise out of the pinching case [GKR]: If $1.31 > K \geq 1$ then M^n is diffeomorphic to S^n. There is an explicit metric on the Milnor sphere Σ^7 with $K \geq 0$, and $K > 0$ on an open dense subset of points (see [GM1]). However, we do not know if any exotic sphere can support a metric with $K > 0$ everywhere. There are many examples with ric > 0 (see §3). On the other hand, a large number of homotopy spheres do not even admit scal > 0 in higher dimensions (see §1); they are already differentiably quite "asymmetric". It is an intriguing problem to better understand exotic spheres from a geometric point of view.

The classical Sphere Theorem needed the upper curvature bound only to control conjugate points and in turn, to find a lower bound $r_0 > \pi/2$ for the *injectivity radius* of M^n (maximal size of balls with geodesic coordinates). In particular, the diameter then satisfies $\operatorname{diam}(M) > \pi/2$. The lower curvature bound was much more essential: it entered the above estimate for odd n (via Morse Theory), and also the final argument (via triangle comparison) that two balls of radius r_0 about points at maximal distance cover M. It is remarkable that the upper curvature bound can be replaced by the (substantially weaker) corresponding lower diameter bound, which leads to the Diameter Sphere Theorem of K. Grove and K. Shiohama [GS]: If $K \geq 1$ and $\operatorname{diam}(M) > \pi/2$ then M^n is homemorphic to S^n. Here it is not necessary to assume M is simply connected. Note that $\operatorname{diam}(M) \leq \pi$ by Myers's Theorem, and if $\operatorname{diam}(M) = \pi$ is maximal, then M^n is isometric to S^n, which was the earlier rigidity result of V. Toponogov. The Diameter Sphere Theorem was a milestone for two reasons. First, it suggested that upper bounds should not play much of a role for nonnegative curvature, as long as only topological questions are concerned, and uniform smooth local coordinates can be sacrificed. This seems to become another guiding principle supported by growing evidence. At the same time, there is a trend to use more and more *assumptions* on often complicated *global* data, rather than on easily accessible *local* invariants. It is not always clear when this is reasonable. A typical example is conditions on the injectivity radius.

The second novel aspect of the Diameter Sphere Theorem was the method of its proof. A homotopy version had actually been obtained before by M. Berger, again in part by means of Morse Theory on loops in M. Grove and Shiohama introduced a crude, but very effective Morse theoretical idea that applies directly to the distance function d_p from a point p, also where it is nondifferentiable: *Regular* points of d_p are precisely those $x \neq p$ such that for some point $q \neq x$, all geodesic triangles pxq are obtuse at x; i.e., $\sphericalangle pxq > \pi/2$. The remaining points of M form the closed set of *critical* points of d_p, which consists of p and a usually very small subset of the cut locus of p. The fact that a relative maximum of d_p is a critical point in this sense, was a well-known lemma of Berger. Regularity of d_p at x means, by definition, all incoming minimal geodesics from p have directions that lie in some proper half-space of the tangent space at x, and thus cannot average out to zero. This fact was exploited originally by showing that suitable smoothings cannot have critical points. Then Gromov approximated such a "nonsmooth" prospective gradient field of d_p by a smooth nonzero gradient-like vector field, in a more elementary way. In any case, since there is just one kind of critical point, the corresponding Morse theory basically consists only of an "Isotopy Lemma" for annular regions without critical points. But that is quite enough for a large number of impressive applications (see [**Ch, Gv, Me**]). Now a proof of the Diameter Sphere Theorem is not difficult: If p, q are points at maximal distance $d(p, q) > \pi/2$, and x is any other point, then triangles pxq are obtuse at x, by comparison with the unit sphere S^2. Thus d_p has only two critical points, and the above Morse theory allows to construct a homeomorphism $M^n \longrightarrow S^n$, more or less as in the proof of the smooth Reeb Theorem.

Like its predecessor, the Diameter Sphere Theorem is quantitatively sharp, and it has a delicate extension to the case $\operatorname{diam}(M) = \pi/2$ [**GG**], generalizing Berger's Rigidity Theorem in the simply connected case: M^n is topologically a sphere or isometric to a complex or quaternionic projective space, except possibly when M has the cohomology ring of the Cayley plane (see also [**Gv**]). A substantial part of this work reduces to the problem of characterizing the Hopf fibrations as the only riemannian fibrations of the standard spheres, up to congruence. It should be pointed out that riemannian fibrations seem to play an essential role in the geometry of manifolds with $K \geq 0$; see also the discussion of open manifolds.

The last rigidity results characterize the compact simply connected symmetric spaces with $K > 0$. Beyond early work of Cheeger, which assumed bounds on derivatives of curvature, perturbation theorems here were proved first in the pinching case, by M. Berger, and then also with lower diameter bounds, by O. Durumeric [**Du**]. All the arguments use convergence techniques for metrics in a now fairly typical way (see also [**Ab, Pe**]). This approach often works well for qualitative perturbations, but it also tends to require laborious proofs of known rigidity results under weak regularity assumptions—and it rarely allows giving effective estimates. From a different, more Lie-theoretical point of view, M. Min-Oo and E. Ruh [**MR**] developed a very interesting deformation technique that allowed differentiable comparison of manifolds with compact symmetric spaces (of any rank), without lower bounds on the injectivity radius. In a similar vein, Ruh obtained a first perturbation of Schur's Lemma, i.e., a sphere theorem for curvature that is almost constant only at each point of M^n, $n \geq 3$. It was quite a surprise when M. Micallef and J. D. Moore [**MM**] proved that the classical Sphere Theorem is already true under the pointwise pinching condition $4\lambda(p) > K \geq \lambda(p)$, where λ is some positive function on M, $n \geq 4$. The beautiful key fact is that conformal branched minimal surfaces S^2 in M^n can be considered critical points of the energy for an appropriate Morse Theory, and they must have index $\geq (n-3)/2$ under the curvature assumptions, which translate into a positivity condition for the second variation. For some additional facts, notably in dimension 4, see [**Cn**].

Beyond the last conclusions, there are hardly any quantitative results for a compact M^n with $K \geq 0$. The Bochner Method has some old and new applications, of value here in connection with harmonic 2-forms, primarily in low dimensions; [**Se**] is the latest reference. The Gallot-Meyer Theorem for manifolds with positive curvature operator (see [**GaM**]) takes a stronger form in the setting of [**MM**]. Estimates for Betti numbers in terms of harmonic forms appear to be difficult, even with upper curvature bounds; see [**Bé**]. A more modest approach is to look for geometric *finiteness* theorems; see [**Ab, Gv, Pe**]. A most remarkable result in this direction was proved by M. Gromov [**Gr1**]: The sum of the Betti numbers of M^n, with respect to coefficients in any field, are a priori bounded by a constant $C(n)$ which depends only on the dimension. However, there can be infinitely many homotopy types. The arguments make delicate use of the critical point theory for distance functions, triangle comparison, and ball coverings. Although the original estimates for $C(n)$ were improved considerably in [**Ab1**], they are still very far away from the expected value 2^n, which is exact for the torus T^n. Gromov's theorem is not specific for $K \geq 0$. It is an intriguing conjecture of Grove and Halperin [**GH**] that the condition $K \geq 0$ might imply that M^n is a rationally elliptic space in the sense of homotopy theory. If this were true, it would not only confirm the above bound 2^n, but also that $\chi_M \geq 0$ for the Euler number of M, which is a third classical conjecture of H. Hopf. From an example of Geroch, it is known that the last relation cannot be proved by use of the Gauss-Bonnet Theorem $(n \geq 6)$.

It is time to say a few words about the known examples of compact manifolds with $K \geq 0$. One way or another, they all come out of constructions, ultimately leading back to compact Lie groups. The basic tool is that riemannian fibrations do not decrease horizontal sectional curvatures, by the formula of O'Neill. But it is hard to find such fibrations, other than by isometric group actions. There are practically no gluing procedures for $K \geq 0$, but see [**Ch1**], where the first example was given which is not homeomorphic to a homogeneous space. Later, J.-H. Eschenburg found such a manifold even with $K > 0$, in dimension 6. A comprehensive reference for many of these questions is [**Es**]. Very recently, M. Kreck and S. Stolz found 7-dimensional homogeneous spaces with $K > 0$ that are homeomorphic, but not diffeomorphic, among the Wallach examples (see [**WZ**] for the reference). But although this family of spaces contains infinitely many homotopy types, they only differ by a very subtle torsion invariant of their homology. In particular for

$K > 0$, having some selection of examples is a phenomenon of certain low dimensions. It is a sobering fact that for $n > 24$, we know only *one* simply connected example if n odd—the sphere, and *four* examples altogether if n even. The last new positively curved space was discovered some time ago!

Now let us briefly look at complete open spaces M^n with $K \geq 0$. The structure theory in [CG1] reduced all topological questions to a nonnegatively curved compact (totally convex) submanifold S, a *soul*, and its normal bundle, which is diffeomorphic to M. Exactly which vector bundles arise is still very poorly understood. G. Walschap has described bundles over a flat torus that cannot admit $K \geq 0$. All bundles associated to the tangent bundle of S^n can be given metrics with $K \geq 0$ [Ch1], but there also seem to be some others, both in the stable and unstable range. Otherwise, the study of such spaces has risen to a higher level of beautiful and intricate geometry. There is much rigidity, but not to the extent we had originally expected. We can only refer to the main developments: A very nice basic survey is [Es1]. Various aspects of the souls S (often not unique) were considered by J.-W. Yim [Yi]. We had asked if M allows a riemannian fibration over a soul; this and related questions were addressed by G. Walschap [Wa] and M. Strake [Str]. V. A. Sharafudtinov [Shf] provided an essential distance-decreasing retraction onto a soul. V. Marenich [Ma] gave a solution of the long-standing conjecture that M must be diffeomorphic to $\mathbb{R}^n$ if K is positive at some point (for all planes), in the real analytic case.

Gap phenomena and asymptotic flatness were analyzed by R. E. Greene, H. Wu, and A. Kasue [GnW, Ka].

For other general finiteness theorems that may have some special meaning in the case of (almost) nonnegative sectional curvature, we refer to [Ab, Ch, Gn, Gv, Pe]. We should not conclude this section without mentioning two structure results which can be considered perturbations of rigidity aspects for $K \geq 0$, but involve additional ideas: In a given dimension n, normalize $\mathrm{diam}(M^n) \leq 1$. M. Gromov [Gr2] had shown that if M is *almost flat*, i.e., $|K|$ is small enough, then a finite cover of M is a nilmanifold, and vice versa. T. Yamaguchi [Ya] showed the existence of $\varepsilon > 0$ so that if $K > -\varepsilon$ is only almost nonnegative, then a finite cover of M still fibers over a torus T^k, where $k = b_1(M)$ is the first Betti number.

3. Nonnegative Ricci curvature

This area has received by far the most attention recently and is very much in flux. Our discussion will have to remain sketchy and somewhat arbitrary. A first obstacle in the study of spaces M^n with $\mathrm{ric} \geq 0$ is the still marginal understanding of *Einstein manifolds*, i.e., spaces with constant Ricci curvature $\mathrm{ric} \equiv \lambda$. The comprehensive reference [Be] also touches on some other topics of this section. Let us make just a few remarks. Only Ricci flat Einstein manifolds can be noncompact. It is not clear if ric $\equiv \lambda$ is more of a geometric or analytic condition in dimensions $n \geq 4$ (for $n \leq 3$, such M^n are space forms). Large classes of interesting examples have been constructed by diverse methods. For $\lambda > 0$, aside from many (locally) homogeneous spaces, there are infinitely many homotopy types of simply connected nonhomogeneous Einstein manifolds in almost all dimensions ≥ 7, arising as certain torus bundles [WZ]. In the case $\lambda = 0$, Yau's solution of the Calabi conjecture provides nonflat Kähler-Einstein metrics on compact Kähler manifolds with vanishing first Chern class, like hypersurfaces of degree $d = m + 2$ in $\mathbb{C}P^{m+1}$. Techniques centered about the complex Monge-Ampère equation can also be used to produce noncompact examples among complements of suitable divisors in projective manifolds [TY]. By a completely different and quite explicit method, Ricci flat spaces of infinite type were constructed in [AKL], using the Gibbons-Hawking metric as a building block. Very little seems to be known about obstructions to the existence of ric $\equiv \lambda$ on a given manifold with ric ≥ 0 (see [Be]).

How big is the general class ric ≥ 0? Diverse types of examples have emerged during the last two decades—after essentially none were known outside the small class $K \geq 0$ (see §2). This is an encouraging development, but many more types of examples are needed! We briefly summarize some progress in this direction. A versatile approach is to construct examples as quotients of riemannian fibrations $E \longrightarrow M$, where E usually has ric ≥ 0. *Warping* often improves the situation; it means changing the metric of E along the fibers by a conformal factor $e^{2\varphi}$, where φ is a function on M. For example, warping the fibration of the (punctured) plane by concentric circles describes surfaces of revolution. In the presence of two fibrations, *double warping* uses this technique simultaneously (or better, successively) and may yield stronger results. First applications of such ideas by J. Nash and A. Poor led to the existence of ric > 0 on certain fiber bundles, including more exotic spheres (see [GM]). Then P. Nabonnand and L. Bérard Bergery [BB] utilized double warping to prove that $M^n \times \mathbb{R}^k$ can be given a metric with ric > 0 if M is compact with ric ≥ 0 and $k \geq 3$. This result also provides alternate explicit examples of complete manifolds with ric > 0 that cannot admit $K \geq 0$, say when M^4 is the K3 surface. A remarkable refinement of these methods was developed by J.-P. Sha and D.-G. Yang [ShY1], who constructed complete metrics with ric > 0 on compact spaces, like connected sums $M_l^{p,q}$ of any l copies of $S^p \times S^q$ if $p, q \geq 2$, and also on noncompact manifolds, like $W_l^{p,q} = W_0^{p,q} \# M_l^{p,q}$, where $W_0^{p,q} = S^{p-1} \times (\mathbb{R}^{q+1} \backslash D^{q+1}) \cup D^p \times S^q$ and l can actually be infinite. The important conclusion is that, in any fixed dimension $n \geq 4$, there are compact manifolds with ric > 0 and arbitrarily large homology, as well as complete manifolds with ric > 0 and infinite topological type. Obviously, the arguments needed cannot be local in nature. Another theorem, inspired by similar ideas, is due to G. Wei [**We**]: If M^n is a complete nilmanifold, then $M^n \times \mathbb{R}^k$ carries a metric with ric > 0 for $k \gg n$. This is somewhat surprising because the Ricci curvature of any left invariant metric on a (nonabelian) nilpotent Lie group always changes sign. We will mention an interesting corollary soon. For some more constructions along these lines, see [**Ch1, An1, Ot, An2**].

Kähler geometry is a different source of spaces with ric ≥ 0. If M^k is compact Kähler with $c_1 > 0$ (or ≥ 0), then by Yau's work, there is a Kähler metric essentially realizing c_1 as the Ricci curvature, so ric > 0 (or ≥ 0); [**SP**] is still a very useful general reference. For complete intersections $V^k(d_1, \ldots, d_r)$ in $\mathbb{C}P^{k+r}$, one has $c_1 = (k + r + 1 - d)\omega$, where $d = \sum d_i$ and ω is the restriction of the Kähler form. Therefore, $c_1 > 0$ for $d < k + r + 1$. The number of such examples is relatively small in a given dimension, their topology fairly complicated: For hypersurfaces, the Euler number is of magnitude d^{n+1}. There are also geometrically very interesting complete examples with ric ≥ 0 among the quasi-projective manifolds studied by Tian and Yau in [**TY**]. The complement of a smooth hypersurface of degree $\leq k$ in $\mathbb{C}P^k$ admits a complete Kähler metric with positive Ricci curvature. Other much larger classes of spaces with ric > 0, both compact and noncompact, arise in a real algebraic setting and can be obtained by direct geometric methods (see [**GM2**]): Let f be any multihomogeneous polynomial on $\mathbb{R}^m$, with isolated singularity at the origin, and F the polynomial on $\mathbb{R}^{m+p+q}$ with $F(z, x, y) = f(z) + |x|^2 - |y|^2$. Then, the intersection V_0 of $F = 0$ with a (small) sphere S admits ric > 0 if $p, q \gg \deg f$. Moreover, the intersection V_- of $F < 0$ and S can be given a complete metric with ric > 0 if also $p - q$ is large. Perhaps the simplest interesting case is $f(z) = (\text{Re } z^{l+1})$ on $\mathbb{R}^2 = \mathbb{C}$ and $l \geq 1$. Topologically the V_- were the first and leanest examples of simply connected spaces with ric > 0 that do not have the homotopy type of a compact manifold, and therefore cannot admit $K \geq 0$. The V_0 are exactly the special compact Sha-Yang examples described above. Of course, in a fixed dimension (which has to be bigger than 4), only a finite number of them seem to carry ric > 0 by this approach.

Before turning to some results on ric ≥ 0 we should briefly talk about the basic techniques that are available so far. Again, variational methods and comparison theory along geodesics play a central role, traditionally mostly in an *average* sense, by the very definition of Ricci curvature as an average of sectional curvatures in a given direction: Whereas $K \geq 0$ controls the hessian of distance functions, and then distances themselves, ric ≥ 0 controls their Laplacian, and then the volume of metric balls. For example, in geodesic coordinates, distances for $K \geq 0$ and volumes for ric ≥ 0 are not greater than in euclidean space, by theorems of Rauch and Bishop. The concavity of distances to convex boundaries (including infinity) is weakened to superharmonicity. More quantitatively, Gromov recognized the importance of a relative volume comparison (some aspects of which had been used earlier by Cheeger): If $\mathrm{vol}_r(p)$ is the volume of the metric ball $B_r(p)$ in M^n about a point p, then $\mathrm{vol}_r/\mathrm{vol}_{r_0} \leq (r/r_0)^n$, whenever $r \geq r_0$. This inequality has remarkably strong applications. In particular, it gives bounds on the size of ball coverings of various types. Basic linear comparison for the Laplacian of the distance function d_p from p yields $\Delta d \leq (n-1)/d$, in an *upper barrier* sense at cut points. This means that d_p is almost superharmonic far away from p, which is the key to the Splitting Theorem. A more careful analysis of this situation has led to a new result in [AG] that actually allows one to estimate some *distance* relations in triangles, only with the assumption ric ≥ 0: Introducing the length *excess* function E with respect to points p, q by $E(x) = xp + xq - pq \geq 0$, one can show that $E \leq 4\left(\frac{h}{s-h}\right)^{1/n-1} h$, where s is the shorter of the two sides xp and xq, and h is the height of the triangle pxq, i.e., the distance from x to a minimal geodesic connecting p and q (cf. also [Ch]). Of course, this excess estimate is inherently weaker than triangle comparison in the case $K \geq 0$; on the other hand, it improves the triangle inequality $E \leq 2h$ substantially for *thin* triangles $(h \ll s)$. Other general techniques for dealing with ric ≥ 0 have also included Bochner's Method, but its use is very limited. It would appear that nonlinear analytic methods should be powerful tools when dealing with Ricci curvature, as they are in Kähler geometry. But this has not yet materialized, except perhaps in dimensions ≤ 4 (minimal surfaces, evolution equations, gauge theory).

What are the general results on ric ≥ 0? Our knowledge is not that extensive. The fundamental group π of M^n is probably understood best. The classical argument of Myers that $\mathrm{diam}(M^n) \leq \pi$ if ric $\geq n - 1$, and thus π finite for any compact M with ric > 0, is still the best theorem. Any finite group π arises as fundamental group of such a manifold M, in a stable sense, since π can be imbedded as a subgroup of some $U(k) \subset SU(k+1)$. Bochner's Method can be applied to show that $b_1(M) \leq n$ if M^n compact and only ric ≥ 0, by looking at harmonic 1-forms, which have to be parallel. But as usual, metric techniques are stronger: All lines in the universal cover split off as a euclidean factor, and that forces π to be essentially a crystallographic group (see [CG2, EH]). If M^n is not compact and ric ≥ 0, it follows from [Mi], by use of volume comparison, that any finitely generated subgroup of π has polynomial growth of order $\leq n$. M. Anderson [An3] proved that $b_1(M^n) \leq n - 3$ if ric > 0, and this is optimal, by the above result of Bérard Bergery. It is a major open problem whether or not π has to be finitely generated, without additional conditions. For example, is $\pi = \mathbb{Q}$ possible? If π is finitely generated, then polynomial growth implies that π must be nilpotent, up to finite index, by a theorem of Gromov. Conversely, any finitely generated, torsion-free nilpotent group can be realized as the fundamental group of a complete manifold with ric > 0, according to the above result of Wei.

In low dimensions $n \leq 4$, some special methods lead to strong results. R. S. Hamilton proved that any metric on a compact manifold M^3 with ric $>$ 0 can be deformed into a constant curvature metric. This is accomplished by means of an evolution equation $\partial g / \partial t = -2\mathrm{ric}(g)$, after reparametrizing to unit volume (see [**Ha, Be**]). In fact, metrics with $K > 0$ remain in their class under the deformation. Already in dimension 4, this technique seems to require positivity of the curvature operator. However, Sha and Yang [**ShY2**] were able to use their above construction to show that a simply connected compact manifold M^4 admits ric > 0 exactly when it admits scal > 0, up to a homeomorphism. Motivated by the Soul Theorem, Schoen and Yau [**SY2**] proved that any complete open manifold M^3 with ric > 0 must be diffeomorphic to $\mathbb{R}^3$. Here minimal surfaces play an essential new role in the arguments.

Further results on ric ≥ 0 in the compact case have been almost exclusively directed toward sphere theorems. It is clear that here Ricci curvature assumptions are not enough. If ric $\geq n - 1$ and $\mathrm{diam}(M^n) = \pi$, M^n is isometric to the standard sphere. This is Cheng's Diameter Theorem. There are many different proofs, both analytic and geometric, and numerous perturbation results have been obtained (see [**Sh**]). We just mention two very recent theorems that seem quite optimal, at least qualitatively: M. Anderson [**An4**] proves that M^n is diffeomorphic to S^n if $C \geq \mathrm{ric} \geq n - 1$ and the volume of M^n is close enough to the volume of the unit sphere, in terms of the upper bound C, which is needed to get lower bounds on the size of harmonic coordinates. If one drops the upper bound for the curvature, then G. Perelman [**Pm**] has shown that M must be a homotopy sphere.

In the noncompact case, the Sha-Yang examples $W_\infty^{p,q}$ make it clear that not even the weakest topological aspects of the Soul Theorem generalize to ric ≥ 0 without further restrictions. Here natural conditions involve volume growth vol_r and diameter growth diam_r with respect to a fixed reference point p. It is an interesting fact, due to Calabi and Yau, that the volume vol_r grows at least linearly. Roughly speaking, diam_r is the diameter of the distance sphere of radius r centered about p, measured *outside* the ball $B_{r/2}(p)$. It follows by a packing argument that, for ric ≥ 0, diameter growth is always of order $o(r)$, i.e., at most linear. We mention two theorems in this context: It was shown in [**AG**] that if ric ≥ 0 and the diameter growth is of (small) order $o(r^{1/n})$, then M is of finite topological type, provided the sectional curvature is bounded from below. There are already many such examples with *bounded* diameter near infinity; the diameter growth of $W_\infty^{p,q}$ can be at least of order $o(r^{2/3})$. See [**Shn**] for a generalization. At the other extreme, Perelman has also proved that if ric ≥ 0 and the volume growth is of (close to maximal) order $\mathrm{vol}_r / cr^n \geq 1 - \varepsilon$ for some sufficiently small $\varepsilon > 0$ and $c = \mathrm{vol}_1$ in $\mathbb{R}^n$, then M is contractible. It is known that necessarily $\varepsilon < 1/2$. The last three results all make use of the excess estimate.

Finiteness results under reasonable geometric restrictions are of great interest in the ample class of spaces M^n with (almost) nonnegative Ricci curvature. Here the basic conditions are upper diameter and lower volume bounds. Work of M. Anderson and J. Cheeger so far gives smooth finiteness under additional conditions, either by bounding the injectivity radius from below, or by assuming also an upper bound for the Ricci curvature as well as for the (global) $L^{n/2}$ norm of the full curvature tensor. But these results do not specifically make use of *nonnegative* Ricci curvature; see also the remarks and references at the end of §2. There we had mentioned Yamaguchi's fibration theorem for almost nonnegative sectional curvature. This result extends to almost nonnegative Ricci curvature, in terms of a bound on $|K|$; but it fails to hold without any additional assumption, by examples of Anderson. Finally, an interesting perturbation of the Splitting Theorem for ric ≥ 0 has been obtained by P. Li and L.-F. Tam analytically, and by M. Cai, and also independently by Z.-d. Liu, in a geometric way (using ball packings): If ric ≥ 0 outside some compact set, then M^n has only finitely many ends.

REFERENCES

[Ab] U. Abresch, *Endlichkeitssätze in der Riemannschen Geometrie*, Jubiläumstagung Bremen 1990: 100 Jahre DMV, Teubner, Stuttgart, 1992, 152–176.

[Ab1] _____, *Lower curvature bounds, Toponogov's Theorem, and bounded topology. II*, Ann. Sci. École Norm. Sup. (4) **20** (1987), 475–502.

[AG] U. Abresch and D. Gromoll, *On complete manifolds with nonnegative Ricci curvature*, J. Amer. Math. Soc. **3** (1990), 355–374.

[An1] M. Anderson, *Metrics of positive Ricci curvature with large diameter*, Manuscripta Math. **68** (1990), 405–415.

[An2] _____, *Short geodesics and gravitational instantons*, J. Differential Geom. **31** (1990), 265–275.

[An3] _____, *On the topology of complete manifolds of nonnegative Ricci curvature*, Topology **29** (1990), 41–55.

[An4] _____, *Convergence and rigidity of manifolds under Ricci curvature bounds*, Invent. Math. **102** (1990), 429–445.

[AKL] M. Anderson, P. Kronheimer, and C. LeBrun, *Complete Ricci-flat Kähler manifolds of infinite topological type*, Comm. Math. Phys. **125** (1989), 637–642.

[Au] T. Aubin, *Nonlinear analysis on manifolds: Monge-Ampère equations*, Springer-Verlag, New York, 1982.

[Au1] _____, *Métriques riemanniennes et courbure*, J. Differential Geom. **4** (1970), 383–424.

[Au2] _____, *Le problème de Yamabe concernant le courbure scalaire*, C. R. Acad. Sci. Paris Sér. I Math. **280** (1975), 721–724.

[Bé] P. E. Bérard, *From vanishing theorems to estimating theorems: the Bochner technique revisited*, Bull. Amer. Math. Soc. (N.S.) **19** (1988), 371–406.

[BB] L. Bérard Bergery, *Quelques exemples de variétés riemanniennes complètes non compactes à courbure de Ricci positive*, C. R. Acad. Sci. Paris Sér. I Math. **302** (1986), 159–161.

[Be] A. L. Besse, *Einstein manifolds*, Springer-Verlag, New York, 1986.

[Bg] J.P. Bourguignon, *Some constructions related to H. Hopf's Conjecture on product manifolds*, Proc. Sympos. Pure Math., vol. 27, part 1, Amer. Math. Soc., Providence, RI, 1975, pp. 33–37.

[Ch] J. Cheeger, *Critical points of distance functions and applications to geometry*, Lecture Notes in Math., Springer-Verlag, **1504** (1991), 1–38.

[Ch1]   ——, *Some examples of manifolds of nonnegative curvature*, J. Differential Geom. **8** (1973), 223–228.

[CG1]   J. Cheeger and D. Gromoll, *On the structure of complete manifolds of nonnegative curvature*, Ann. of Math. (2) **96** (1972), 413–443.

[CG2]   ——, *The splitting theorem for manifolds of nonnegative Ricci curvature*, J. Differential Geom. **6** (1971), 119–129.

[Cn]   H. Chen, *Pointwise $\frac{1}{4}$-pinched 4-manifolds*, Ann. Global Anal. Geom. **9** (1991), 161–176.

[Du]   O. Durumeric, *A generalization of Berger's almost $\frac{1}{4}$-pinched manifolds theorem*, J. Differential Geom. **26** (1987), 101–139.

[Eh]   P. E. Ehrlich, *Metric deformations and curvature. I: Local convex deformations*, Geom. Dedicata **5** (1976), 1–29.

[Es]   J.-H. Eschenburg, *Freie isometrische Aktionen auf kompakten Liegruppen mit positiv gekrümmten Orbiträumen*, Schriftenreihe Math. Inst. Univ. Münster, Ser. 2, vol. 32, Univ. Münster, Münster, 1984.

[Es1]   ——, *Open manifolds with nonnegative curvature*, lecture notes (1990).

[EH]   J.-H. Eschenburg and E. Heintze, *An elementary proof of the Cheeger-Gromoll Splitting Theorem*, Ann. Glob. Anal. Geom. **2** (1984), 141–151.

[Esc]   J. F. Escobar, *The Yamabe problem on manifolds with boundary*, J. Differential Geom. **35** (1992), 21–84.

[GaM]   S. Gallot and D. Meyer, *Opérateur de courbure et laplacien des formes différentielles d'une variété riemannienne*, J. Math. Pures Appl. **54** (1975), 259–284.

[Gn]   R. E. Greene, *Some recent developments in riemannian geometry*, Contemp. Math. no. 101, Amer. Math. Soc., Providence, RI, 1989, 1–30.

[GnW]   R. E. Greene and H. Wu, *Gap theorems for noncompact riemannian manifolds*, Duke Math. J. **49** (1982), 731–756.

[GG]   D. Gromoll and K. Grove, *A generalization of Berger's rigidity theorem for positively curved manifolds*, Ann. Sci. École Norm. Sup. (4) **20** (1987), 227–239.

[GM1]   D. Gromoll and W. T. Meyer, *An exotic sphere with nonnegative sectional curvature*, Ann. of Math. (2) **100** (1974), 401–408.

[GM2]   ——, *Examples of complete manifolds with positive Ricci curvature*, J. Differential Geom. **21** (1985), 195–211.

[Gr]   M. Gromov, *Sign and geometric meaning of curvature*, lecture notes, Inst. Hautes Ètudes Sci, 1990.

[Gr1]   ——, *Curvature, diameter, and Betti numbers*, Comment Math. Helv. **56** (1981), 179–195.

[Gr2]   ——, *Almost flat manifolds*, J. Differential Geom. **13** (1978), 235–241.

[GL1]   M. Gromov and H. B. Lawson, *Spin and scalar curvature in the presence of a fundamental group*. I, Ann. of Math. (2) **111** (1980), 209–230.

[GL2]   ——, *The classification of simply connected manifolds of positive scalar curvature*, Ann. of Math. (2) **111** (1980), 423–434.

[GL3]   ——*Positive scalar curvature and the Dirac operator on complete riemannian manifolds*, Publ. Math. I. H. E. S. **58** (1983), 295–408.

[Gv]   K. Grove, *Critical point theory for distance functions*, these Proceedings.

[GH]   K. Grove and S. Halperin, *Contributions of rational homotopy theory to global problems in geometry*, Inst. Hautes Ètudes Sci. Publ. Math. **56** (1983), 379–385.

[GKR]   K. Grove, H. Karcher, and E. Ruh, *Jacobi fields and Finsler metrics with applications to differentiable pinching problems*, Math. Ann. **211** (1974), 7–21.

[GS]   K. Grove and K. Shiohama, *A generalized sphere theorem*, Ann. of Math. (2) **106** (1977), 201–211.

[Ha]   R. S. Hamilton, *Four-manifolds with positive curvature operator*, J. Differential Geom. **24** (1986), 153–179.

[Hi]   N. Hitchin, *Harmonic spinors*, Adv. in Math. **14** (1974), 1–55.

[Ka]   A. Kasue, *A convergence theorem for riemannian manifolds and some applications*, Nagoya Math. J. **114** (1989), 21–51.

[Kaz] J. L. Kazdan, *Prescribing the Curvature of a riemannian manifold*, CBMS Regional Conf. Ser. in Math., no. 57, Amer. Math. Soc., Providence, RI, 1985.

[KW] J. L. Kazdan and F. W. Warner, *Prescribing curvatures*, Proc. Symp. Pure Math., vol. 27, Part 2, Amer. Math. Soc., Providence, RI, 1975, pp. 309–319.

[Li] A. Lichnerowicz, *Spineurs harmoniques*, C. R. Acad. Sci. Paris Sér. A-B **257** (1963), 7–9.

[LM] H. B. Lawson and M.-L. Michelsohn, *Spin geometry*, Princeton Univ. Press, Princeton NJ, 1989.

[Ma] V. Marenich, *Structure of open manifolds of nonnegative curvature*, Dokl. Acad. SSSR **305**:6 (1989), 1311–1314.

[Me] W. T. Meyer, *Toponogov's Theorem and applications*, lecture notes, College on Differential Geometry, Trieste, 1989.

[MM] M. J. Micallef and J. D. Moore, *Minimal 2-spheres and the topology of manifolds with positive curvature on totally isotropic 2-planes*, Ann. of Math. (2) **127** (1988), 199–227.

[Mi] J. Milnor, *A note on curvature and fundamental group*, J. Differential Geom. **2** (1968), 1–7.

[MR] M. Min-Oo and E. Ruh, *Comparison theorems for compact symmetric spaces*, Ann. Sci. École Norm. Sup. (4) **12** (1979), 335–353.

[Ot] Y. Otsu, *On manifolds of positive Ricci curvature with large diameter*, Math. Z. **206** (1991), 255–264.

[Pe] P. Petersen V, *Gromov-Hausdorff convergence of metric spaces*, Amer. Math. Soc., PSPUM **54** (Part 3), pp. 489–504.

[Pm] G. Perelman, *Manifolds of positive Ricci curvature with almost maximal volume*, Preprint (1992).

[RS] J. Rosenberg and S. Stolz, *Manifolds of positive scalar curvature*, Proc. Workshop on Applications of Algebraic Topology to Geometry and Analysis, Math. Sci. Res. Inst. Berkeley, CA, 1990 (to appear).

[Sch] R. Schoen, *Conformal deformation of a riemannian metric to constant scalar curvature*, J. Differential Geom. **20** (1984), 479–495.

[SY] R. Schoen and S. T. Yau, *The structure of manifolds with positive scalar curvature*, Directions in partial differential equations, Academic Press, 1987, pp. 235–242.

[SY1] _____, *Existence of incompressible minimal surfaces and the topology of three dimensional manifolds with nonnegative scalar curvature*, Ann. of Math. (2) **110** (1979), 127–142.

[SY2] _____, *The structure of manifolds with positive scalar curvature*, Manuscripta Math. **28** (1979), 159–183.

[SY3] _____, *Complete three dimensional manifolds with positive Ricci curvature and scalar curvature*, Ann. of Math. Stud., vol. 102, Princeton Univ. Press, Princeton, NJ, 1982, pp. 209–228.

[Se] W. Seaman, *Harmonic two-forms in four dimensions*, Proc. Amer. Math. Soc. **112** (1991), 545–548.

[SP] Séminaire Palaiseau 1978, *Première classe de Chern et courbure de Ricci: preuve de la conjecture de Calabi*, Soc. Math. de France, Astérisque **58** (1978).

[ShY1] J.-P. Sha and D.-G. Yang, *Positive Ricci curvature on the connected sums of $S^n \times S^m$*, J. Differential Geom. **33** (1990), 127–138.

[ShY2] _____, *Positive Ricci curvature on compact simply connected 4-manifolds*, Amer. Math. Soc., PSPUM, **54** (Part 3) 529–538.

[Shf] V. A. Sharafudtinov, *Convex sets in a manifold of nonnegative curvature*, Mat. Zametki **26** (1979), 129–136.

[Shn] Z. Shen, *On complete manifolds of nonnegative kth-Ricci curvature*, Trans. Amer. Math. Soc. (to appear).

[Sh] K. Shiohama, *Recent developments in sphere theorems*, Amer. Math. Soc., PSPUM, **54** (Part 3) 551–576.

[St] S. Stolz, *Simply connected manifolds of positive scalar curvature*, Bull. Amer. Math. Soc. (N.S.) **23** (1990), 427–432. The detailed version will appear in Ann. of Math.

[Str] M. Strake, *A splitting theorem for open nonnegatively curved manifolds*, Manuscripta Math. **61** (1988), 315–325.

[TY]   G. Tian and S. T. Yau, *Complete Kähler manifolds with zero Ricci curvature*. I, J. Amer. Math. Soc. **3** (1990), 579–609.

[Wa]   G. Walschap, *Nonnegatively curved manifolds with souls of codimension* 2, J. Differential Geom. **27** (1988), 525–537.

[WZ]   M. Y. Wang and W. Ziller, *Einstein metrics on principal torus bundles*, J. Differential Geom. **31** (1990), 215–248.

[We]   G. Wei, *Examples of complete manifolds of positive Ricci curvature with nilpotent isometry groups*, Bull. Amer. Math. Soc. (N.S.) **19** (1988), 311–313.

[Ya]   T. Yamaguchi, *Collapsing and pinching under a lower curvature bound*, Ann. of Math. (2) **133** (1991), 317–357.

[Yi]   J.-W. Yim, *Space of souls in a complete open manifold of nonnegative curvature*, J. Differential Geom. **32** (1990), 429–456.

STATE UNIVERSITY OF NEW YORK AT STONY BROOK

Proceedings of Symposia in Pure Mathematics
Volume 54 (1993), Part 3

Critical Point Theory for Distance Functions

KARSTEN GROVE

Introduction

One of the fundamental themes in riemannian geometry is to relate properties of a riemannian manifold as a *metric space* to differential *topological* properties of it as a smooth *manifold*.

It is well known that the topology of a smooth manifold is intimately related to the smooth functions it supports via Morse theory (cf. [47]). Applying this fact to functions associated with the manifold as a metric space, such as distance functions, would seem to provide a natural bridge between geometry and topology. The only problem with such a program is that distance functions generally are non smooth, and much less Morse functions. Nonetheless, there is a notion of *regular/critical points* for distance functions, which is *equivalent* to the usual notion for smooth functions (cf. 1.1). The importance of this idea, which was conceived in [37], lies primarily in the observation that some basic principles for smooth functions remain valid for distance functions. In particular, the level set of a regular value is a (topological) submanifold, and the region between two regular levels is (topologically) a product if it contains no critical points (cf. 1.7, 1.8 and 1.14). Moreover, we show that with these techniques a complete Lusternik-Schnirelman theory for distance functions is available (cf. 1.16–1.20).

Although the theory has applications in curvature free settings, (cf., e.g., [14, 15, 16, 17, 18, 28, 29, 39, 41, 42, 71], it becomes particularly powerful when used in conjunction with Toponogov's comparison theorem, i.e., when a lower curvature bound is present. However, in contrast to earlier work in global riemannian geometry (cf. [20, 9]), which was based on getting (good) a priori estimates for the injectivity radius, an upper bound for the curvature is irrelevant for this theory. Rather than trying to cover a majority of applications, we will focus on four simple principles, the *convexity, regularity,*

1991 *Mathematics Subject Classification.* Primary 53C20, 53C23; Secondary 57R70.
Supported in part by a grant from the National Science Foundation.
This paper is in final form and no version of it will be submitted for publication elsewhere.

criticality, and *shrinking principles* (cf. 1.3, 2.5, 2.8 and 2.10). The utility of these principles is then illustrated in different types of problems, namely *recognition*, *structure*, and *finiteness problems* (cf. §§3, 4 and 5).

There are two natural ways in which to extend this theory. One of them is to consider more than one distance function and develop a similar "calculus", in particular an implicit function theorem. Although this has not yet been done systematically, applications of this sort of idea may be found in, e.g., [49, 66, 67]. The other extension is to distance functions on Aleksandrov spaces, i.e. inner metric spaces with a lower curvature bound in the sense of distance comparison (cf. [56]). This extension is actually related to the first. In fact, an "inverse function theorem" for n distance functions on an n-dimensional Aleksandrov space, together with an inverse induction argument, forms the basis for getting a suitable theory for one function in this setting (cf. forthcoming work of Perelman [1] announced in [6]). For lack of space and focus, we confine our discussion here to the simplest case of one distance function on a complete riemannian manifold. Strong applications are to be expected, however, when these ideas are used in conjunction with ideas of Gromov-Hausdorff convergence (cf. [30, 33, 34, 36]), and critical point theory for distance functions on Aleksandrov (limit) spaces.

This paper was written during a visit to Aarhus University. It is a pleasure to thank the department for its hospitality and SNF for its support. Parts of the presentation have been directly influenced by questions and comments from W. Browder, M. Böckstedt, I. Madsen, P. Petersen, J. Tornehave and B. Williams. Also, our discussion of Lusternik-Schnirelman theory owes much to the beautiful expositions given by Palais in [50] and [51]. An apology is due for omitting much of the beautiful work involving ideas related to those discussed here. In an attempt to make up for this, at least in part, we have included an expanded list of references. At the same time we also refer to the treatments of critical point theory given in the lecture notes by Cheeger [7] and Meyer [46]. For basic results and tools from riemannian geometry that will be used freely we refer to, e.g., [9, 20, 27, 40, 43].

0. Preliminaries

Throughout M will denote a smooth connected n-dimensional manifold with riemannian metric g. The *distance function*, dist : $M \times M \to \mathbb{R}$, is defined by

$$\text{dist}(p, q) = \inf \text{length}(c),$$

where the infimum is taken over all piecewise C^1-curves $c : [0, 1] \to M$, with $c(0) = p$, $c(1) = q$ and

$$\text{length}(c) = \int_0^1 \sqrt{g(\dot{c}(t), \dot{c}(t))}\, dt.$$

---

[1] G. Perelman, *A. D. Alexandrovs spaces with curvature bounded from below*. II, preprint.

Ascoli's theorem implies that if (M, dist) is a complete metric space, then any $p, q \in M$ can be joined by a *segment* in M, i.e., a shortest path parametrized proportional to arc length. Curves that are everywhere locally segments are called *geodesics*. Every geodesic is a smooth curve uniquely determined by any one of its velocity vectors. We assume throughout that (M, dist) is a complete metric space, or equivalently, by the Hopf-Rinow theorem, that every geodesic extends indefinitely in either direction.

For every tangent vector $v \in T_p M \subset TM$ at $p \in M$, let $c_v : \mathbb{R} \to M$ denote the unique geodesic determined by $\dot{c}_v(0) = v$. The *exponential map*, $\exp : TM \to M$, is then defined by

$$\exp(v) = c_v(1),$$

and its restriction to $T_p M$ will be denoted by $\exp_p$. A vector $v \in T_p M$ belongs to the *tangent cut locus*, $\text{tancut}(p) \subset T_p M$, if and only if $c_v : [0, 1] \to M$ is a segment, but $c_v : [0, 1+\varepsilon] \to M$ is not for any $\varepsilon > 0$. The *cut locus* $\text{cut}(p) \subset M$ of p in M is then by definition $\text{cut}(p) = \exp_p(\text{tancut}(p))$. If the *segment domain*, $\text{seg}(p) \subset T_p M$, is the star-shaped subset of $T_p M$ bounded by $\text{tancut}(p)$, then $\exp_p : \text{seg}(p) \to M$ is surjective and its restriction to $\text{intseg}(p) = \text{seg}(p) - \text{tancut}(p)$ provides a diffeomorphism with $M - \text{cut}(p)$. In particular, the distance function from p, $\text{dist}_p : M \to \mathbb{R}$, is smooth when restricted to $M - \text{cut}(p) \cup p$. Moreover, according to the so-called Gauss lemma, its gradient is the unit radial vector field on this set; i.e., at $q = \exp_p(v) \in M - \text{cut}(p) \cup p$ we have $\text{grad}(\text{dist}_p) = \dot{c}_v(1)/\|\dot{c}_v(1)\| \in T_q M$. However, dist_p is clearly not differentiable at points $q \in \text{cut}(p)$ that are joined to p by more than one segment. Nevertheless, it turns out that if these segments do not spread out too much as seen from q, then dist_p behaves as a smooth function near q with $r = \text{dist}(p, q)$ as regular value (cf. §1).

More generally we need to consider closed subsets $A \subset M$. In this case

$$\text{dist}(A, q) = \min_{p \in A} \text{dist}(p, q)$$

for all $q \in M$. The open and closed r-neighborhoods of A will be denoted by

$$B_A(r) = \{x \in M \mid \text{dist}(A, x) < r\}$$

and

$$D_A(r) = \{x \in M \mid \text{dist}(A, x) \le r\}$$

respectively. Similarly we set

$$S_A(r) = \{x \in M \mid \text{dist}(A, x) = r\}.$$

As usual, if A is compact, its *diameter* is defined by

$$\text{diam}\, A = \max_{p \in A} \max_{q \in A} \text{dist}(p, q)$$

whereas its *radius* is

$$\text{rad}\, A = \min_{p \in A} \max_{q \in A} \text{dist}(p, q).$$

Clearly $\operatorname{rad} A \leq \operatorname{diam} A \leq 2 \operatorname{rad} A$ and $A \subset D_p(\operatorname{diam} A)$ for all $p \in A$, whereas $A \subset D_p(\operatorname{rad} A)$ for some $p \in A$.

1. Lusternik-Schnirelman theory

In this section we will discuss the notion of critical points for distance functions, and show how all of the classical Lusternik-Schnirelman theory for smooth functions on a manifold carries over to this case.

Fix a closed subset A of a complete connected Riemannian manifold M. The distance function from A will be denoted by dist_A, i.e.,

$$\operatorname{dist}_A : M \to \mathbb{R}, \qquad p \to \operatorname{dist}(p, A)$$

for all $p \in M$. As for smooth functions, where the Taylor expansion can be applied, we say that $p \in M$ is a *regular point* for dist_A, or simply for A, if and only if there is a unit vector $v \in S_p \subset T_p M$ and a $c > 0$ such that

$$(1.1) \qquad\qquad \operatorname{dist}_A(c_v(t)) \geq \operatorname{dist}_A(c_v(0)) + c \cdot t$$

for all sufficiently small $t > 0$. If $p \in M$ is not a regular point for dist_A it is called *critical*.

A simple argument based on standard local distance comparison shows that definition (1.1) above is equivalent to the following more commonly used characterization. A point $p \in M$ is *regular* for A if and only if there is a $v \in S_p$ such that

$$(1.1)' \qquad\qquad \angle(v, \dot{c}(0)) > \pi/2$$

for any segment c from p to A. Similarly, $p \in M$ is a *critical point* for dist_A if and only if every vector $u \in S_p$ makes an angle $\leq \pi/2$ to some segment from p to A.

Thus, whether a point p is regular or critical for A depends entirely on the geometry of segments from p to A, seen from p. More precisely, if $S_{pA} \subset S_p$ denotes the *set of directions* for segments from p to A, then p is *regular* if and only if S_{pA} is contained in an open hemisphere of S_p, and p is *critical* if and only if S_{pA} is a *weak $\pi/2$-net* in S_p; i.e., any $u \in S_p$ has distance at most $\pi/2$ to some point of S_{pA}.

REMARK 1.2. In a Riemannian manifold it is of course equivalent to say that there is a $v \in S_p$ so that S_{pA} is contained in the complement of the closed hemisphere centered at v, and to say that there is a $u \in S_p$ such that S_{pA} is contained in the open hemisphere centered at u. This is not the case, however, in singular spaces like orbit and limit spaces, where the theory works as well with the definition $(1.1')$ above.

EXAMPLE 1.3 (Convexity principle). From the work of Cheeger and Gromoll in [10], it is well known that any convex set $C \subset M$ has the structure of a topological manifold with (possibly empty) boundary, ∂C. Moreover, the *interior* $\operatorname{int} C = C - \partial C$ is smooth and totally geodesic. If $\partial C \neq \varnothing$, every $p \in \partial C$ has a *supporting half-space*; i.e., there is a $v \in S_p$ so that any segment from p to an interior point of C makes an obtuse angle with v. In particular, if $A \subset \text{int } C$, then any $p \in \partial C$ is a regular point for A.

EXAMPLE 1.4 (Cut locus). A point $p \in M - A$ is called a *cut point* for A if and only if no segment from p to A can be extended as a segment to A beyond p. The set of all cut points for A, cut(A), is called the cut locus of A. It is clear from (1.1) that any critical point for A is in cut(A). The converse, however, is usually not true. This, in fact, is one of the main reasons for the introduction and utility of the concept.

A simple first variation argument yields the following basic

LEMMA 1.5. *Let $\alpha : [0, a] \to M$ be a differentiable curve parametrized by arc length. Suppose there is a $\theta \in [0, \pi/2)$ such that for any $t \in [0, a]$*

$$\angle(\dot{\alpha}(t), v) \geq \pi - \theta$$

for some $v \in S_{\alpha(t)A}$. Then

$$\text{dist}_A(\alpha(t)) \geq \text{dist}_A(\alpha(0)) + t \cdot \cos\theta,$$

and $\alpha(t)$ is a regular point for each $t \in [0, a]$.

This lemma also shows that if α is a differentiable curve parametrized by arc length, then $S_{\alpha(t)A}$ is completely contained in one of the closed half-spaces in $T_{\alpha(t)}M$ determined by $\dot{\alpha}(t)^\perp$, unless $\alpha(t)$ is a *local maximum* point for $\text{dist}_A \circ \alpha$. So far, however, a reasonable notion of *index* of a critical point is known only in special cases.

In the most important applications of Lemma 1.5, α is a curve where $\angle(\dot{\alpha}(t), v) \geq \pi - \theta$ for all $v \in S_{\alpha(t)A}$. Indeed, suppose X is a unit vector field on some open set $U \subset M$ and $\angle(X_p, S_{pA}) \geq \pi - \theta$, $0 \leq \theta < \pi/2$, for all $p \in U$. Then any integral curve α of X will satisfy the conditions in Lemma 1.5. When allowing θ to depend on $p \in U$, such a *gradient-like vector field* exists on all of $U = M - \text{crit}(A)$, where $\text{crit}(A)$ is the closed subset of M consisting of all the critical points for dist_A. This is a direct consequence of $(1.1)'$ and a partition of unity argument. The existence of a fixed θ is of course guaranteed on any compact subset of U.

The first immediate consequence of this discussion is the following important observation, which predates the general idea of critical points as presented here.

PROPOSITION 1.6 (Berger Lemma). *Any local maximum point for dist_A is critical.*

In analogy to the case of smooth functions, regular level sets of dist_A have the following structure.

PROPOSITION 1.7 (Implicit Function Theorem). *Let $r > 0$ be a regular value for dist_A; i.e., $\text{dist}_A^{-1}(r) \cap \text{crit}(A) = \varnothing$. Then $\text{dist}_A^{-1}(r) = S_A(r)$ is a* *topological $(n-1)$-dimensional manifold, locally flatly embedded in M (cf. also Remark 1.14).*

PROOF. Fix a point $p \in \text{dist}_A^{-1}(r)$. Let X be a unit vector field defined in an open neighborhood U of p and satisfying $\angle(X, S_{qA}) \geq \pi - \theta$ for all $q \in U$ and some fixed positive $\theta < \pi/2$. Choose a local hypersurface H through p and transversal to X. For $\varepsilon > 0$ sufficiently small, the flow of X defines a diffeomorphism $\Phi : H \times (-\varepsilon, \varepsilon) \to W \subset M$, where W is an open neighborhood of $p \in M$. By Lemma 1.5

$$\text{dist}_A(\Phi(p, \varepsilon)) \geq \text{dist}_A(\Phi(p, 0)) + \varepsilon \cos \theta = r + \varepsilon \cos \theta$$

and similarly $\text{dist}_A(\Phi(p, -\varepsilon)) \leq r - \varepsilon \cos \theta$. Therefore, for H small enough each integral curve $\Phi(q, t)$, $q \in H$, $t \in (-\varepsilon, \varepsilon)$, intersects $\text{dist}_A^{-1}(r)$ in exactly one point, $\Phi(q, f(q))$, and $f : H \to (-\varepsilon, \varepsilon)$ defined this way is obviously continuous. For $\varepsilon > 0$ sufficiently small, the map $H \times (-\varepsilon, \varepsilon) \to M$, $(q, t) \to \Phi(q, f(q)+t)$ defines the desired submanifold chart for dist_A^{-1} near p. □

As in Morse and Lusternik-Schnirelman theory one has the following key result.

PROPOSITION 1.8 (Isotopy Lemma). *Let $A \subset M$ be a compact subset of M, and suppose $[r_1, r_2] \subset R_+$ contains only regular values for dist_A. Then all the levels $\text{dist}_A^{-1}(r)$, $r \in [r_1, r_2]$, are homeomorphic, and the annulus*

$$R(r_1, r_2) = D_A(r_2) - B_A(r_1) = \{q \in M | r_1 \leq \text{dist}_A(q) \leq r_2\}$$

is homeomorphic to $\text{dist}_A^{-1}(r_1) \times [r_1, r_2]$ (cf. also Remark 1.14).

PROOF. Compactness of A implies that the sets $D_A(r) = \{q \in M \mid \text{dist}_A(q) \leq r\}$, $r \geq 0$, are compact. In particular, if X is a gradient-like vector field on $M - \text{crit}(A)$, then $\angle(X_p, S_{pA}) \geq \pi - \theta$, for a fixed $0 \leq \theta < \pi/2$ and all p in $R(r_1, r_2)$. It now suffices to invoke Lemma 1.5 to complete the proof. □

The proof of Proposition 1.8 also applies to the case $r_2 = \infty$, i.e.,

COROLLARY 1.9 (Finite-Type Lemma). *Let M be a complete noncompact riemannian manifold and $A \subset M$ a compact subset. If dist_A has no critical points in $M - B_A(r)$, then $D_A(r)$ is a compact manifold with boundary $S_A(r)$ and M is diffeomorphic to the interior $B_A(r)$ of $D_A(r)$.*

In the other extreme case where $r_1 = 0$, we have

COROLLARY 1.10 (Soul Lemma). *Let $A \subset M$ be a compact submanifold in M without boundary. If there are no critical points in $D_A(r) - A$, then $B_A(r)$ is diffeomorphic to the normal bundle of A in M.*

PROOF. First pick $\varepsilon > 0$ so that $D_A(\varepsilon) \cap \text{cut}(A) = \varnothing$. In particular, there is a unique segment from each $p \in \partial D_A(\varepsilon) = \text{dist}_A^{-1}(\varepsilon)$ to A. Using this it is easy to construct a gradient like vector field X on $D_A(r) - A$ which is radial near A. The desired diffeomorphism takes each normal segment emanating from A to the corresponding integral curve of X.   □

As a trivial combination of Corollaries 1.9 and 1.10 we get

THEOREM 1.11 (Disc Theorem). *Let M be a complete noncompact riemannian n-manifold. If there is a $p \in M$ so that dist_p has no critical points (other than p), then M is diffeomorphic to $\mathbb{R}^n$.*

The compact version of this is contained in

THEOREM 1.12 (Sphere Theorem). *Let M be a closed riemannian n-manifold. If there is a $p \in M$ so that dist_p has only one critical point (other than p), then M is homeomorphic to S^n.*

PROOF. By assumption, there is only one point q at maximal distance from p, say $d(p, q) = r_0$. Pick $\varepsilon > 0$ smaller than the injectivity radii at p and at q. From the isotopy lemma it then follows that $D_p(r)$ is homeomorphic (in fact diffeomorphic, cf. Remark 1.14), to D^n for any $0 < r < r_0$. If therefore $M - B_p(r) \subset B_q(\varepsilon)$ for some r, our claim is a consequence of Proposition 1.7 and the Generalized Schoenflies theorem (cf., e.g., [57]). Now assume on the contrary, that $M - B_p(r) \not\subset B_q(\varepsilon)$ for any $r < r_0$; i.e., there is a sequence of points $x_n \in M$ with $\operatorname{dist}(p, x_n) \to r_0$ and $\operatorname{dist}(q, x_n) \geq \varepsilon$ for all n. Since M is compact we find an accumulation point $x \in M$ with $\operatorname{dist}(p, x) = r_0$ and $\operatorname{dist}(x, q) \geq \varepsilon$. This contradicts the assumption that only q was critical for dist_p.   □

It should be mentioned that any twisted (exotic) sphere has a riemannian metric which satisfies the hypotheses in Theorem 1.12. This follows from a general construction due to Weinstein (cf. [4, p. 231]).

In the absence of a good notion of index for critical points, there is nothing to predict the change in topology when crossing a critical level. Rather than pursuing Morse theory any further, we proceed to show that Lusternik-Schnirelman theory is valid for distance functions. The key to this is the following

LEMMA 1.13 (Deformation Lemma). *Let M be a complete riemannian manifold and $A \subset M$ a compact subset. Suppose $r > 0$ is an isolated critical value of dist_A. For every open neighborhood U of $\operatorname{crit}(A) \cap \operatorname{dist}_A^{-1}(r)$ there is an $\varepsilon > 0$, such that $D_A(r + \varepsilon) - U$ can be isotoped into $D_A(r - \varepsilon)$.*

PROOF. As in the proof of Proposition 1.7 we see that for each $p \in \operatorname{dist}_A^{-1}(r) - U$, there is an $\varepsilon_p > 0$ and a neighborhood U_p of p in M, such that $U_p \cap R(r - \varepsilon_p, r + \varepsilon_p)$ is homeomorphic to $(U_p \cap \operatorname{dist}_A^{-1}(r)) \times [r - \varepsilon_p, r + \varepsilon_p]$. By compactness, cover $\operatorname{dist}_A^{-1}(r) - U$ by finitely many sets $U_{p_i} \cap \operatorname{dist}_A^{-1}(r)$, $i = 1, \ldots, l$. With $\varepsilon = \min_i \varepsilon_{p_i}$, and $W = \bigcup_i U_{p_i}$ clearly $W \cap R(r - \varepsilon, r + \varepsilon)$ is homeomorphic to $(W \cap \operatorname{dist}_A^{-1}(r)) \times [r - \varepsilon, r + \varepsilon]$ and $W \cup U \supset \operatorname{dist}_A^{-1}(r)$.

By possibly choosing a smaller ε we can assume $R(r - \varepsilon, r + \varepsilon) \subset W \cup U$. In particular $R(r - \varepsilon, r + \varepsilon) - U \subset W \cap R(r - \varepsilon, r + \varepsilon)$ and the proof is completed. $\square$

REMARK 1.14. A simple modification of the argument given above shows that the isotopy in Lemma 1.13 can be chosen globally on M and fixing everything outside an arbitrarily small neighborhood of $W \cap R(r - \varepsilon, r + \varepsilon)$. Note also, that if r in Lemma 1.13 is a regular value, then U can be chosen empty. By Proposition 1.8 and smoothing theory (cf. [44]) it then follows that $\text{dist}_A^{-1}(r)$ has a smooth structure.

The deformation lemma helps in locating critical points other than the obvious minimum and maximum points. The method for this is referred to as the *minimax principle*. Let $\mathscr{F}$ be a family of subsets of M. Define the minimax of dist_A over $\mathscr{F}$ by

$$(1.15) \qquad \text{Minmax}(\text{dist}_A, \mathscr{F}) = \inf_{F \in \mathscr{F}} \sup\{\text{dist}_A(p) | p \in F\}$$

or equivalently

$$(1.15)' \qquad \text{Minmax}(\text{dist}_A, \mathscr{F}) = \inf\{r \in \mathbb{R}_+ | \exists F \in \mathscr{F} \text{ with } F \subset D_A(r)\}.$$

A family $\mathscr{F}$ is called *isotopy invariant* if every isotopy of M takes any subset of M from $\mathscr{F}$ to a subset of M from $\mathscr{F}$. From Lemma 1.13 (and Remark 1.14) we now get immediately

THEOREM 1.16 (Minimax Principle). *Suppose $A \subset M$ is compact and that $\mathscr{F}$ is an isotopy invariant family of subsets in M. Then $\text{Minmax}(\text{dist}_A, \mathscr{F})$ is a critical value of dist_A.*

There are many interesting examples of isotopy invariant families. We mention here only a few of the most important ones.

EXAMPLES 1.17 (Isotopy Invariant Families).

(i) Let S be any topological space and $[S, M]$ the set of homotopy classes of maps from S to M. For fixed $f : S \to M$, the family $\mathscr{F}_{[f]} = \{g(S) \subset M | g \in [f]\}$ is clearly an isotopy invariant family.

(a) If $S = M$ is closed we have $\text{Minmax}(\text{dist}_A, \mathscr{F}_{[\text{id}]}) = \text{Max dist}_A$.

(b) If $S = \{\text{point}\}$ then $\text{Minmax}(\text{dist}_A, \mathscr{F}_{[\cdot]}) = \text{Min dist}_A = 0$.

(c) If $S = S^k$, the minimax principle associates to each element $[f] \in \pi_k(M)$ of the kth *homotopy group* of M a critical value, $\text{Minmax}(\text{dist}_A, \mathscr{F}_{[f]})$ of dist_A.

(ii) Let $H_k(M, R)$ be the kth singular *homology module* of M with coefficients in a ring R. For each k-cycle z let $\mathscr{F}_{[z]} = \{\text{carrier of } w | w \text{ a } k\text{-cycle with } [w] = [z] \in H_k(M)\}$. Here the carrier of a singular k-chain, $c = \sum n_\alpha \sigma_\alpha$, $\sigma_\alpha : \Delta_k \to M$ is simply $\bigcup_\alpha \sigma_\alpha(\Delta_k) \subset M$.

(a) If $[M] \in H_n(M)$ is a fundamental class, clearly

$$\text{Minmax}(\text{dist}_A \mathscr{F}_{[M]}) = \text{Max dist}_A.$$

(b) If $[\cdot] \in H_0(M)$, then $\text{Minmax}(\text{dist}_A, \mathscr{F}_{[\cdot]}) = \text{Min dist}_A = 0$.

(iii) Recall that a subset $X \subset M$ has *Lusternik-Schnirelman category*, $\text{cat}(X; M) = m$, if it can be covered by m (but not fewer) closed subsets of M, each of which is contractible to a point inside M. For each $m \leq \text{cat}(M) := \text{cat}(M; M)$ let $\mathscr{F}_m$ be the family of subsets $X \subset M$ with $\text{cat}(X; M) \geq m$. By the minimax principle

$$c_A(m) = \text{Minmax}(\text{dist}_A, \mathscr{F}_m)$$

is a critical value for each $m \leq \text{cat}(M)$.

Since obviously

$$\text{cat}(X; M) \leq \text{cat}(Y; M) \quad \text{if } X \subset Y$$

we can also write

$$c_A(m) = \inf\{r \in \mathbb{R}_+ \mid \text{cat}(D_A(r); M) \geq m\}.$$

From this or $\mathscr{F}_{m+1} \subset \mathscr{F}_m$ we get

$$0 = c_A(1) \leq \cdots \leq c_A(m) \leq c_A(m+1) \leq \cdots \leq c_A(\text{cat}(M)) \leq \text{Max dist}_A.$$

In the last example it can of course happen that equality occurs. For example if $M = \text{RP}^2$ with constant curvature 1 and $A = \{p\}$ then $0 = c_{\{p\}}(1) < c_{\{p\}}(2) = c_{\{p\}}(3) = \pi/2$ (if instead $A = \text{RP}^1 \subset \text{RP}^2$ then $0 = c_{\text{RP}^1}(1) = c_{\text{RP}^1}(2) < c_{\text{RP}^1}(3) = \pi/2$). However, if equality does occur, one gets the following remarkable compensation.

THEOREM 1.18 (Main Theorem of L.-S. Theory). *Let M be a complete riemannian manifold and $A \subset M$ a compact subset. For each $m \leq \text{cat}(M)$,*

$$c_A(m) = \inf\{r \in \mathbb{R}_+ \mid \text{cat}(D_A(r); M) \geq m\}$$

is a critical value of dist_A *and*

$$0 = c_A(1) \leq \cdots \leq c_A(m) \leq \cdots \leq c_A(m+k) \leq \cdots \leq c_A(\text{cat } M).$$

If moreover, $c_A(m) = c_A(m+k) = c$, *then* $\text{cat}(\text{crit}_A(c); M) \geq k+1$ *and in particular* $\dim \text{crit}_A(c) \geq k$.

PROOF. It remains to consider the case $c = c_A(m) = c_A(m+k)$. From this and the definition of the c_A's we have

$$\text{cat}(D_A(c - \varepsilon); M) \leq m - 1 \quad \text{and} \quad \text{cat}(D_A(c + \varepsilon); M) \geq m + k$$

for any $\varepsilon > 0$. Now let U be a neighborhood of $\text{crit}_A(c)$ and choose $\varepsilon > 0$ as in Lemma 1.13. From trivial properties of $\text{cat}(\cdot, M)$ we then get

$$\text{cat}(U; M) \geq \text{cat}(D_A(c + \varepsilon) \cup U; M) - \text{cat}(D_A(c + \varepsilon) - U; M)$$
$$\geq \text{cat}(D_A(c + \varepsilon); M) - \text{cat}(D_A(c - \varepsilon); M) \geq k + 1$$

for every $U \supset \text{crit}_A(c)$. The desired inequality then follows once we have seen that there is a U with $\text{cat}(U; M) = \text{cat}(\text{crit}_A(c); M) = l$. For this let $\text{crit}_A(c) \subset F_1 \cup \cdots \cup F_l$, where each F_i, $i = 1, \ldots, l$, is closed and there are homotopies $\varphi_i : F_i \times [0, 1] \to M$ with $\varphi_i(p, 0) = p$ and $\varphi_i(p, 1) = p_i \in M$, for all $p \in F_i$, $i = 1, \dots, l$. By the homotopy extension property of M we may assume that each φ_i is defined on all of $M \times [0, 1]$. For each $i = 1, \dots, l$ let $\mathscr{O}_i$ be a neighborhood of p_i with $\bar{\mathscr{O}}_i$ contractible, and U_i an open neighborhood of F_i with $\bar{U}_i \subset \varphi_i(\cdot, 1)^{-1}(\mathscr{O}_i)$. Then $U = U_1 \cup \cdots \cup U_l \supset \mathrm{crit}_A(c)$ has category l because $\mathrm{cat}(U_i; M) = \mathrm{cat}(\bar{U}_i; M) \le \mathrm{cat}(\varphi_i(\bar{U}_i; 1); M) \le \mathrm{cat}(\mathscr{O}_i; M) = \mathrm{cat}(\bar{\mathscr{O}}_i; M) = 1$.

The claim $\dim \mathrm{crit}_A(c) \ge k$ now follows since $\mathrm{cat}(X; M) \le \dim X + 1$ for any closed subset $X \subset M$ (cf., e.g., [50]). $\square$

For any compact subset $A \subset M$ we see in particular that dist_A has at least $\mathrm{cat}(M)$ critical points.

EXAMPLE 1.19. Let M be the real projective plane RP^2 with riemannian metric so that $\widetilde{M}$ is an ellipsoid in $\mathbb{R}^3$ with three different axes. If $A = \{p\}$ is the point in M corresponding to the pair at maximal distance in $\widetilde{M}$, then clearly $\mathrm{dist}_A : M \to \mathbb{R}$ has exactly $3 = \mathrm{cat}(\mathrm{RP}^2)$ critical points (including of course $A = p$ itself).

A lower bound for $\mathrm{cat}(M)$ is provided by the so-called *cuplength* of M. Here $\mathrm{cuplong}(M)$ is the largest integer l such that for some field F there are l cohomology classes $\omega_1, \dots, \omega_l \in H^*(M; F)$ each of positive degree and $\omega_1 \cup \cdots \cup \omega_l \ne 0$. The following comparison between $\mathrm{cat}(M)$ and $\mathrm{cuplong}(M)$ is proved for metric and path connected spaces M in [5].

THEOREM 1.20. *For any riemannian manifold M, $\mathrm{cuplong}(M) + 1 \le \mathrm{cat}(M) \le \dim M + 1$.*

This concludes our general discussion of critical point theory for distance functions. In the next sections we will see how to use this in conjunction with comparison theory.

2. Comparison theory and critical points

The utility of critical point theory, as discussed in §1, has been particularly apparent so far, in the presence of a lower (sectional) curvature bound. Before we attempt to isolate a few essential ideas behind this, we recall the basic results from comparison theory that are used.

Following [56] we let S_k^n denote the simply connected n-dimensional *space form* of constant curvature k. Points in S_k^n will be written as $\bar{p}, \bar{q}$, etc. rather than p, q, etc., which will continue to denote points in a general manifold M.

There are several equivalent formulations of the basic *distance comparison* theorem, usually referred to as Toponogov's triangle comparison theorem. Here are three of them.

THEOREM 2.1 (Toponogov). *Let M be a complete riemannian manifold with sectional curvature, $\sec M$, satisfying $\sec M \ge k$. The following equiva-*

lent statements hold:

(Δ) *For every geodesic triangle* (c_0, c_1, c_2) *in* M *with minimal sides, there is a triangle* $(\bar{c}_0, \bar{c}_1, \bar{c}_2)$ *in* S_k^2 *with* Length$(\bar{c}_i)$ = Length(c_i), i = 0, 1, 2, *and for corresponding angles* $\theta_i \geq \bar{\theta}_i$, $i = 0, 1, 2$.

(Λ) *Let* $(c_0, c_1; \theta)$ *be any geodesic hinge in* M *with minimal sides, and* $(\bar{c}_0, \bar{c}_i, \theta)$ *the corresponding hinge in* S_k^2. *Then for the hinge endpoints* (p_1, p_0) *and* $(\bar{p}_1, \bar{p}_0)$, dist$(p_1, p_2) \leq$ dist$(\bar{p}_1, \bar{p}_0)$.

(T) *Consider any pair* $(c_0; p_0)$, *where* c_0 *is a minimal geodesic in* M *and* $p_0 \in M$. *Let* $(\bar{c}_0; \bar{p}_0)$ *be the corresponding pair in* S_k^2; *i.e., the distances from* $\bar{p}_0$ *to the endpoints of* $\bar{c}_0$ *are the same as from* p_0 *to the endpoints of* c_0. *Then* dist$(p_0, q) \geq$ dist$(\bar{p}_0, \bar{q})$ *for any* $q \in c_0$ *and corresponding* $\bar{q} \in \bar{c}_0$.

In each of these statements c_0 *does not need to be minimal, only* Length(c_0) $\leq \pi/\sqrt{k}$ *if* $k > 0$. *In this case, however, the angle comparison in* (Δ) *holds only for the angles adjacent to* c_0.

There are important rigidity companions to (T) *and* (Λ) *above in cases of equality*:

[Λ] *Suppose* $0 < \theta < \pi$ *and* dist(p_1, p_0) = dist$(\bar{p}_1, \bar{p}_0)$. *Then* (c_0, c_1) *spans a surface in* M *isometric to the unique triangular surface in* S_k^2 *spanned by* $(\bar{c}_0, \bar{c}_1)$, *and with totally geodesic interior*.

[T] *Assume* $p_0 \notin c_0$ *and* dist(p_0, q) = dist$(\bar{p}_0, \bar{q})$ *for some interior* $q \in c_0$. *Then every minimal geodesic* c_q *from* p_0 *to* q *spans together with* c_0 *a unique surface isometric to the triangular surface in* S_k^2 *spanned by* $\bar{c}_0$ *and* $\bar{p}_0$, *and with totally geodesic interior*.

One advantage with the T-version of Theorem 2.1 is that it makes sense in more general inner metric spaces where angles are not a priori defined (cf. [56]). If S is such a space with curv $S \geq k$, i.e., local distance comparison á la Theorem 2.1 holds in S, then indeed global distance comparison holds as well (see [6] [2]). The corresponding rigidity results have been proved and applied recently in [30].

It is sometimes useful to interpret (Λ) in terms of the exponential maps as in [34]: If $p \in M$ and rad(p) = max dist$_p$ we endow $D_0(\text{rad}(p)) \subset T_p M$ with the constant curvature k metric obtained from the euclidean metric by a radial conformal change (when $k > 0$ and rad$(p) = \pi/\sqrt{k}$ we interpret $D_0(\text{rad}(p))$ as S_k^n). In this way we view the segment domain seg(p) as a subset of S_k^n. Clearly Theorem 2.1 (Λ) is equivalent to

$$(2.1)' \qquad \exp_p : \text{seg}(p) \to M \quad \text{is distance nonincreasing.}$$

This is the basis also for volume comparison of various important *metrically defined subsets* of M. We mention only two examples of this, both of which are special cases of one general result from [33] (cf. also [11]).

---

[2] See also C. Plaut, *Spaces with Wald Curvature Bounded Below*. I, II, preprints.

EXAMPLE 2.2 (Half-spaces). Fix $p \in M$ and a closed subset $Q \subset M$. Let $\bar{p} = \exp_p^{-1}(p)$ and $\bar{Q} = \exp_p^{-1}(Q)$ in $\operatorname{seg}(p) \subset S_k^n$. For the *half-spaces* $H(p, Q) = \{x \in M \mid \operatorname{dist}(x, p) \leq \operatorname{dist}(x, Q)\}$ and $H(\bar{p}, \bar{Q})$ in M and S_k^n respectively we then have

$$\operatorname{vol} H(p, Q) \leq \operatorname{vol} H(\bar{p}, \bar{Q}).$$

EXAMPLE 2.3 (Swiss cheeses). With p and Q as in Example 2.2 fix $R > 0$ and an arbitrary function $r : Q \to \mathbb{R}_+$. By definition the *Swiss cheese* $K = K((Q, r); (p, R))$ in $D_p(R)$ relative to r is the set $K = D_p(R) - \bigcup_{q \in Q} B_q(r(q))$. Then

$$\operatorname{vol} K((Q, r); (p, R)) \leq \operatorname{vol} K((\bar{Q}, \bar{r}); (\bar{p}, R))$$

where $\bar{r} = r \circ \exp_p : \bar{Q} \to \mathbb{R}_+$.

The volume estimates given in Examples 2.2 and 2.3 *do not hold* under the weaker curvature assumption $\operatorname{Ric} M \geq (n-1)k$. This, however, is sufficient for the following simple extension of the so-called Bishop-Gromov volume comparison theorem (cf., e.g., [7]).

THEOREM 2.4 (Relative Volume Comparison). *Let M be a complete riemannian n-manifold with $\operatorname{Ric} M \geq (n-1)k$ and suppose $Q \subset M$ is compact. If we set $v_k^n(R) = \operatorname{vol} D_{\bar{p}}(R)$ in S_k^n, then*

$$R \to \operatorname{vol} D_Q(R)/v_k^n(R)$$

is a nonincreasing function.

We are now ready to present some principles frequently used in the detection of critical or regular points.

Throughout we fix a complete riemannian n-manifold M with $\operatorname{sec} M \geq k$.

In the *regularity principle* we consider two points $p, q \in M$ and fix $\bar{p}, \bar{q} \in S_k^2$ with $\operatorname{dist}(p, q) = \operatorname{dist}(\bar{p}, \bar{q})$. Except for the single case where $k > 0, d(p, q) = \pi/\sqrt{k}$ and in particular $M \equiv S_k^n$, there is a continuous map $T : M \to S_k^2$ defined by the requirement $\operatorname{dist}(p, x) = \operatorname{dist}(\bar{p}, T(x))$ and $\operatorname{dist}(q, x) = \operatorname{dist}(\bar{q}, T(x))$ for all $x \in M$. T is unique up to reflection in the segment $\bar{p}\bar{q}$ from $\bar{p}$ to $\bar{q}$ in S_k^2. If

$$\operatorname{reg}(\bar{p}, \bar{q}) = \{\bar{x} \in S_k^2 \mid \angle(\bar{p}, \bar{x}, \bar{q}) > \pi/2\}$$

(cf. Figure 2.6) we have

LEMMA 2.5 (Regularity Principle). *For any $p, q \in M$ the set*

$$\operatorname{reg}(p, q) := T^{-1}(\operatorname{reg}(\bar{p}, \bar{q}))$$

consists of regular points for p as well as for q.

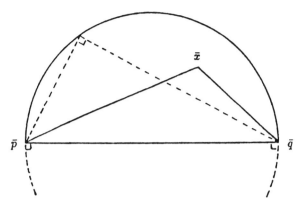

FIGURE 2.6. $\mathrm{reg}(\bar{p}, \bar{q})$ for $k = 0$

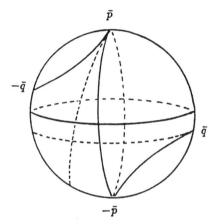

FIGURE 2.7. $\mathrm{reg}\,\bar{p}, \bar{q})$ for $k > 0$ and $\mathrm{dist}(\bar{p}, \bar{q}) > \pi/2\sqrt{k}$

PROOF. Let $x \in T^{-1}(\mathrm{reg}(\bar{p}, \bar{q}))$. Then by Theorem 2.1 (Δ), the angle at x between any two segments from x to p and from x to q is obtuse. □

In the special case where $k > 0$ and $\mathrm{dist}(\bar{p}, \bar{q}) = \pi/2\sqrt{k}$, $\mathrm{reg}(\bar{p}, \bar{q}) \subset S_k^2$ has two connected components, one of which is $B(\bar{p}, \pi/2\sqrt{k}) \cap B(\bar{q}, \pi/2\sqrt{k})$. If $\mathrm{dist}(\bar{p}, \bar{q}) = D > \pi/2\sqrt{k}$ then $S_k^2 - \overline{\mathrm{reg}(\bar{p}, \bar{q})} = \mathrm{reg}(-\bar{p}, \bar{q}) \cup \mathrm{reg}(-\bar{q}, \bar{p})$, and in particular $\mathrm{seg}(\bar{p}, \bar{q}) \supset B_{\bar{p}}(D) \cap B_{\bar{q}}(D)$ (cf. Figure 2.7).

In the *criticality principle* we consider points $p, q \in M$ where $q \in \mathrm{crit}(p)$. Fix corresponding points $\bar{p}, \bar{q} \in S_k^2$ and for each $r > \mathrm{dist}(\bar{p}, \bar{q}) = \mathrm{dist}(p, q) = d$ let $\bar{r} \in S_k^2$ be the unique point (up to reflection in segment $\bar{p}\bar{q}$) with $\mathrm{dist}(\bar{p}, \bar{r}) = r$ and $\angle(\bar{p}, \bar{q}, \bar{r}) = \pi/2$ (cf. Figure 2.9 on next page). If $\theta_k(r, d) = \angle(\bar{r}, \bar{p}, \bar{q})$ then $\theta_k(r, d)$ is increasing in r and decreasing in d (if $k > 0$ we assume here $d < r < \pi/2\sqrt{k}$). With this notation we have

LEMMA 2.8 (Criticality Principle). *Let $p \in M$ and suppose $q \in \mathrm{crit}(p)$ with $\mathrm{dist}(p, q) = d$. Then for all $x \in M$ with $\mathrm{dist}(p, x) = r > d$, the angle between any pair of segments from p to q and from p to x is at least $\theta_k(r, d)$.*

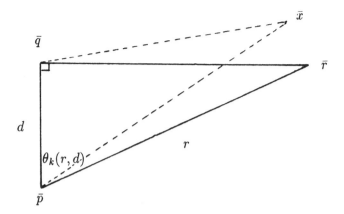

FIGURE 2.9

PROOF. Let c_0 be a segment from p to q, c_1 a segment from p to x and c_2 a segment from q to x. If $\angle(\dot{c}_0(0), \dot{c}_1(0)) < \theta_k(r, d)$ so is the comparison angle in the triangle $(\bar{c}_0, \bar{c}_1, \bar{c}_2)$ by Theorem 2.1 (Δ). By definition of $\theta_k(r, d)$, therefore, $\angle(-\dot{\bar{c}}_0(d), \dot{\bar{c}}_2(0)) > \pi/2$. This means, however, that also $\angle(-\dot{c}_0(d), \dot{c}_2(0)) > \pi/2$ by Theorem 2.1 (Δ). Since c_0 was arbitrary, this contradicts the assumption $q \in \mathrm{crit}(p)$. Thus $\angle(\dot{c}_0(0), \dot{c}_1(0)) \geq \theta_k(r, d)$ as claimed. □

The *shrinking principle* below is based on $(2.1)'$ as applied in the volume estimates of Examples 2.2 and 2.3.

LEMMA 2.10 (Shrinking Principle). *Let* $p \in M$ *and suppose* $q \in \mathrm{crit}(p)$ *with* $\mathrm{dist}(p, q) = d$. *Then*

(i) $\mathrm{vol}\, H(q ; p) \leq \mathrm{vol}\, H(\bar{q}, \{\bar{p}_0, \bar{p}_1\})$, *where* $H(\bar{q}, \{\bar{p}_0, \bar{p}_2\}) \subset D_{\bar{q}}(\mathrm{rad}(q))$ $\subset S_k^n$, $\bar{q}$ *is the midpoint of segment* $\bar{p}_0 \bar{p}_1$ *and* $\mathrm{dist}(\bar{q}, \bar{p}_i) = d$.

(ii) $\mathrm{vol}(M - B_p(r)) \leq \mathrm{vol}(D_{\bar{q}}(\mathrm{rad}(q)) - (B_{\bar{p}_0}(r) \cup B_{\bar{p}_1}(r)))$ *for any* $r > 0$.

PROOF. From Examples 2.2 and 2.3 we have

$$\mathrm{vol}\, H(q ; p) \leq \mathrm{vol}\, H(\bar{q}, \exp_q^{-1}(p))$$

and $\mathrm{vol}(M - B_p(r)) \leq \mathrm{vol}\, K((\exp_q^{-1}(p), r); (\bar{q}, \mathrm{rad}(q)))$. It remains to see that if we replace $\exp_q^{-1}(p)$ by $\{\bar{p}_0, \bar{p}_1\}$ chosen as in Lemma 2.10, then we get subsets in S_k^n with (possibly) even bigger volume. For this observe that $S_{qp} \subset T_q M$ forms a weak $\pi/2$-net in the unit sphere $S_q \subset T_q M$ because $q \in \mathrm{crit}(p)$. From Theorem 2.4 we conclude that $\mathrm{vol}\, D_{S_{qp}}(\theta)/v_1^{n-1}(\theta) \geq$ $\mathrm{vol}\, S_{qp}(\pi/2)/v_1^{n-1}(\pi/2) = \mathrm{vol}\, S_1^{n-1}/v_1^{n-1}(\pi/2) = 2$ for all $\theta \leq \pi/2$, where $D_{S_{qp}}(\theta) \subset S_q$ is the θ-neighborhood of S_{qp} in $S_q \equiv S_1^{n-1}$. Thus $\mathrm{vol}\, D_{S_{qp}}(\theta)$ $\geq \mathrm{vol}\, D_{\{v_0, v_1\}}(\theta)$, where $v_0, v_1 \in S_q$ is any antipodal pair. By integrating the reverse inequality for the corresponding complements we derive our claim. □

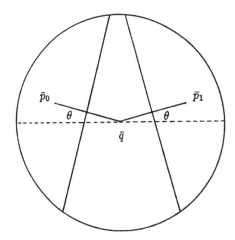

FIGURE 2.11. $H(\bar{q}; \{\bar{p}_0, \bar{p}_1\}) \subset D_{\bar{q}}(R)$

REMARK 2.12. The shrinking lemma has a natural generalization, where instead of assuming q to be a critical point for p, one assumes that S_{qp} is a weak $\frac{1}{2}\pi + \theta$ net in S_q for some $0 \le \theta < \pi/2$. In this case the points $\bar{p}_0, \bar{p}_1$ must be chosen so that $\angle(\bar{p}_0\bar{q}\bar{p}_1) = \pi - 2\theta$, where the corresponding $\{v_0, v_1\}$ form a weak $\frac{1}{2}\pi + \theta$ net in S_q.

In the subsequent sections we will apply these principles to different types of problems in riemannian geometry.

3. Recognition theorems

In this section we give examples of situations where critical point theory is used to determine the type of the manifold. The simplest and first such application was given in [36]:

THEOREM 3.1 (Diameter Sphere Theorem). *Any complete riemannian manifold M with sec $M \ge 1$ and diam $M > \pi/2$ is a twisted sphere.*

PROOF. Choose $p, q \in M$ with $\mathrm{dist}(p, q) = \mathrm{diam}\, M$. By the regularity principle in Lemma 2.5 all points in $M - \{p, q\}$ are regular for p as well as for q (cf. also Figure 2.7). The conclusion then follows from Theorem 1.12. □

We point out that in the special case of Theorem 3.1, one does not have to appeal to the generalized Schoenflies Theorem as we did in Theorem 1.12. It suffices to observe that a gradient-like vector field X can be constructed on $M - \{p, q\}$ which is radial near p and q. We also note that the following question remains open.

PROBLEM 3.2. Is there a $d = d(n) < \pi$ so that any complete riemannian n-manifold M with sec $M \ge 1$ and diam $M \ge d$ is diffeomorphic with S^n?

If in this problem we replace the diameter by the radius, an affirmative answer has been given in [65]. The idea here is to show that when sec $M \ge 1$ and rad $M \sim \pi$, then M is metrically close to S_1^n in the Gromov-Hausdorff, and hence Lipschitz, sense (cf. [71]).

Except for the question about existence of different differentiable structures on manifolds M as in Theorem 3.1, it is well known that the diameter sphere theorem is optimal. However, when $\operatorname{diam} M > \pi/2$ is replaced by $\operatorname{diam} M \geq \pi/2$ one has the following essentially complete metric classification proved in [21] and [22].

THEOREM 3.3 (Diameter rigidity). *Let M be a complete riemannian manifold with $\sec M \geq 1$ and $\operatorname{diam} M = \pi/2$. Then, either M is a twisted sphere or else*

(i) *If $\pi_1(M) = \Gamma \neq \{1\}$, M is isometric to*
    (a) *the unique $\mathbb{Z}_2$-quotient of a complex odd dimensional projective space, or*
    (b) *S_1^n/Γ, where $\Gamma \to O(n+1)$ is reducible.*
(ii) *If $\pi_1(M) = \{1\}$, M is isometric to a projective space, except possibly if M has the cohomology ring of the Cayley plane, $C_a P^2$.*

In this theorem, one would also have rigidity in the exceptional case where $H^*(M) \cong H^*(C_a P^2)$, provided the following holds

CONJECTURE 3.4. *Any riemannian submersion $S_1^{15} \to M^8$ is congruent to the Hopf map $S_1^{15} \to S_4^8$.*

The proof of Theorem 3.3 is rather long and intricate. Here we only give an outline in order to show how critical point theory enters:

As in Theorem 3.1 we begin by choosing points $p, q \in M$ with $\operatorname{dist}(p, q) = \operatorname{diam} M = \pi/2$. Then

$$A = \{x \in M \mid \operatorname{dist}_p(x) = \pi/2\} \quad \text{and} \quad \bar{A} = \{x \in M \mid \operatorname{dist}_A(x) = \pi/2\}$$

are both nonempty, and totally π-convex by Theorem 2.1 (T); i.e., any geodesic in M of length $< \pi$ and with endpoints in A (resp. $\bar{A}$) is entirely contained in A (resp. $\bar{A}$).

The regularity principle of Lemma 2.5 shows that all points in $M - (A \cup \bar{A})$ are regular for dist_A as well for $\operatorname{dist}_{\bar{A}}$. In particular, for any $\varepsilon > 0$, $M - B_A(\varepsilon) \cup B_{\bar{A}}(\varepsilon)$ is homeomorphic to $\operatorname{dist}_A^{-1}(\varepsilon) \times [0, 1]$. Moreover, from the structure of convex sets (cf. Example 1.3) A (resp. $\bar{A}$) is either a closed smooth totally geodesic submanifold of M or else a compact topological submanifold with $\partial A \neq \varnothing$ and smooth totally geodesic interior. In the first case, $B_A(\varepsilon)$ is diffeomorphic to the normal bundle of A in M by Corollary 1.10. If on the other hand $\partial A \neq \varnothing$, then $\operatorname{dist}_{\partial A} : A \to \mathbb{R}$ is strictly concave since $\sec M > 0$ (cf. [10]). In particular there is a unique point $s \in \operatorname{int} A$ at maximal distance from ∂A and all points in A are s-regular as explained in Example 1.3. Thus for $\varepsilon > 0$ sufficiently small, all points in $D_A(\varepsilon)$ are regular for dist_s and therefore $D_A(\varepsilon)$ is diffeomorphic to the $\dim M$ dimensional euclidean disc by Corollary 1.10. All in all we conclude that there are totally geodesic smooth submanifolds of M (possibly points), so that M is the union of their tubular neighborhoods.

Observe in particular, that if $\partial A \neq \varnothing$, and $\partial \bar{A} \neq \varnothing$ then M is a twisted sphere.

Now suppose M is not a sphere. We claim that $\partial A = \partial \bar{A} = \varnothing$; i.e., A and $\bar{A}$ are smooth totally geodesic submanifolds of M (one of them possibly a point). Indeed, if say $\partial \bar{A} \neq \varnothing$ pick $\bar{p} \in \bar{A}$ arbitrarily and let c be a segment from $\bar{p}$ to A. By the rigidity comparison theorem, Theorem 2.1 [Λ], it follows that any normal vector $u \in T_q A^{\perp}$ which is obtained from $-\dot{c}(\pi/2) \in T_{c(\pi/2)} A^{\perp}$ by parallel translation along a curve in A defines a segment c_u from $q \in A$ to $\bar{p}$. Moreover, this set $U \subset TA^{\perp}$ of normal vectors to A is a smooth closed submanifold of the unit normal bundle $SA^{\perp}$ to A with fiber $F \subset U$ being the orbit of $-\dot{c}(\pi/2)$ in $S_{c(\pi/2)} A^{\perp}$ under the normal holonomy group. By assumption $SA^{\perp} \simeq \text{dist}_A^{-1}(\varepsilon) \simeq \text{dist}_{\bar{A}}^{-1}(\varepsilon)$ is diffeomorphic to the $\dim M - 1$ dimensional sphere. If therefore $SA^{\perp} - U \neq \varnothing$, the (normal) projection $U \to A$ is homotopic to a constant map $U \to \text{pt} \in A$. Using the homotopy lifting property of $U \to A$ we get a homotopy $h : U \times [0, 1] \to U$ where $h_0 = \text{id}_U$ and $h_1(U) \subset F$. Since U and $F \subset U$ are closed manifolds this is impossible and hence $U = SA^{\perp}$. This on the other hand implies that every normal vector to A defines a segment from A to $\bar{p} \in \bar{A}$. Since $\bar{p} \in \bar{A}$ was chosen arbitrarily this yields the desired contradiction.

An elaboration of the argument just given shows that $\text{cut}(A) = \bar{A}$ and $\text{cut}(\bar{A}) = A$. Moreover, for each $\bar{p} \in \bar{A}$ the map

$$S_p \bar{A}^{\perp} \to A, \qquad \bar{v} \to \exp_{\bar{p}}(\pi/2 \cdot \bar{v})$$

is a riemannian submersion from the euclidean unit normal sphere $S_p \bar{A}^{\perp}$ onto A. An essentially complete metric classification of such fibrations was given in [22]. The only case left is when $S_p \bar{A}^{\perp} \equiv S_1^{15}$ and A is a simply connected 8-dimensional manifold $\simeq S^8$ (cf. Conjecture 3.4).

The remaining part of the proof of Theorem 3.3 is separated into the cases (i) $\pi_1(M) = \Gamma \neq \{1\}$ and (ii) $\pi_1(M) = \{1\}$. The topological decomposition of M obtained via critical point theory is used together with Morse theory for geodesics to show that when $\pi_1(M)$ is trivial so are $\pi_1(A)$ and $\pi_1(\bar{A})$. In this case, the classification of fibrations $S_p \bar{A}^{\perp} \to A$ gives in particular that A and $\bar{A}$ are rank-1 symmetric spaces and then that M itself is such a space. When $\pi_1(M) \neq \{1\}$, one considers the universal cover $\widetilde{M}$ of M. Clearly sec $\widetilde{M} \geq 1$ and diam $\widetilde{M} \geq \pi/2$. In view of the classification given already for simply connected manifolds the remaining case of interest is when diam $\widetilde{M} > \pi/2$. In particular, $\widetilde{M}$ is a topological sphere by Theorem 3.1, which is decomposed similarly using the lifts $\widetilde{A}, \widetilde{\bar{A}}$ of $A, \bar{A}$. By a second variation argument of Synge type we have in general $\dim A + \dim \bar{A} \leq \dim M - 1$. However, when $\widetilde{M}$ is a sphere a simple transversality argument then implies $\dim \widetilde{A} + \dim \widetilde{\bar{A}} = \dim \widetilde{M} - 1$. The riemannian fibrations constructed above are then local isometries. It is now fairly easy to show that $\widetilde{A}, \widetilde{\widetilde{A}}$ is an orthogonal pair of totally geodesic subspheres in the unit sphere $\widetilde{M}$. This concludes the outline of Theorem 3.3. □

Before leaving the class of manifolds M with sec $M \geq 1$ and diam $M \geq \pi/2$ we like to point out some interesting volume problems related to critical point theory. The first is a natural analogue of a classical area problem due to A. D. Aleksandrov (cf. [34]).

CONJECTURE 3.5. *Let M be a closed riemannian n-manifold with sec $M \geq 1$ and diam $M = d > \pi/2$. Then $\mathrm{vol}\, M < 2v_1^n(d/2)$, and this estimate is optimal. Note that $\mathrm{vol}\, X = 2\, v_1^n(d/2)$ for the singular spherical space X obtained by gluing two copies of $D_1^n(d/2) \subset S_1^n$ together along their boundary.*

If in Conjecture 3.5 we replace the diameter by the radius an optimal estimate has been found in [34]. There it was also proved that if sec $M \geq 1$ and rad $M > \pi/2$, then every dist_p, $p \in M$, has exactly two critical points (including p itself). This then gives a lower bound for the Filling Radius, and hence the volume of M (cf. [24]). The optimal lower bound, however, is not known. The following was proposed in [34].

CONJECTURE 3.6. *If M is a closed riemannian n-manifold with sec $M \geq 1$ and rad $M = r > \pi/2$, then $\mathrm{vol}\, M \geq \mathrm{vol}\, S_{(\pi/r)^2}^n$; i.e., the volume of M is at least the volume of the constant curvature n-sphere with diameter r.*

We close this section with a recognition theorem for exotic spheres. Following the terminology of [3], the *excess function* associated with $p, q \in M$ is given by

$$(3.7) \qquad \mathrm{exc}_{p,q}(x) = \mathrm{dist}(p, x) + \mathrm{dist}(x, q) - \mathrm{dist}(p, q)$$

for all $x \in M$. Based on this we define the *excess* of M, $\mathrm{exc}\, M$, as in [34] by

$$(3.8) \qquad \mathrm{exc}\, M = \min_{p,q} \max_x \mathrm{exc}_{p,q}(x).$$

Other excess type invariants have been introduced in [48], [59] and [30]. Observe that $\mathrm{exc}\, M = 0$ if and only if there are points $p, q \in M$ such that $\mathrm{cut}(p) = \{q\}$ and $\mathrm{cut}(q) = \{p\}$. In particular M is a twisted sphere. Conversely, any twisted sphere has a riemannian metric with excess $= 0$ (cf. [4, p. 231]). However, it is easy to see that small excess has no topological significance in general: Simply take any manifold M and concentrate all of its topology in a tiny metric ball whose complement is the complement of a tiny ball in the unit sphere. In this simple construction there is of course no curvature control. The following problem posed in [35] appears to be significant.

PROBLEM 3.9. For fixed $k \in \mathbb{R}$ and $D \in \mathbb{R}_+$ describe all closed riemannian n-manifolds with sec $M \geq k$, diam $\leq D$ and $\mathrm{exc}\, M$ arbitrarily small.

By excluding the possibility of collapse one has the following answer from [35].

THEOREM 3.10 (Exotic Sphere Theorem). *Given an integer $n \geq 2$, a real k and $v, D > 0$. There is an $\varepsilon = \varepsilon(n, k, D, v)$ such that any closed riemannian n-manifold M with $\sec M \geq k$, $\operatorname{diam} M \leq D$ and $\operatorname{vol} M \geq v$ is a homotopy sphere whenever $\operatorname{exc} M \leq \varepsilon$.*

PROOF. Pick $p, q \in M$ with $\min \operatorname{exc}_{p,q} = \operatorname{exc} M$. A simple application of the regularity principle of Lemma 2.5 shows that for every $\delta > 0$ there is an $\varepsilon = \varepsilon(\delta, k, D)$ so that $M - B_p(\delta) \cup B_q(\delta)$ consists of regular points for p as well as for q, whenever $\operatorname{exc} M \leq \varepsilon$. In view of the isotopy Lemma 1.8, therefore, it suffices to show that $B_p(\delta)$ and $B_q(\delta)$ are contractible to points inside M: Indeed, in this case $M = X_1 \cup X_2$, where $X_1 = M - B_q(\delta)$ and $X_2 = M - B_p(\delta)$ are contractible in M. Then $H^*(M, X_i) \to H^*(M)$ is surjective for $i = 1, 2$ and therefore $H^*(M)$ has trivial cup product structure for any coefficient ring. Using $\mathbb{Z}_2$ as coefficient field, it follows from Poincaré duality that, in particular, $H_1(M; \mathbb{Z}_2) = 0$. From the Mayer-Victoris sequence

$$0 \to H_0(X_1 \cap X_2; \mathbb{Z}_2) \to H_0(X_1; \mathbb{Z}_2) \oplus H_0(X_2; \mathbb{Z}_2) \to H_0(M; \mathbb{Z}_2) \to 0$$

we conclude that $X_1 \cap X_2$ is connected. This in turn implies that $\pi_1(M) = \{1\}$ by Van Kampen's theorem. Now any simply connected homology sphere is a homotopy sphere by theorems of Hurewicz and Whitehead. The proof of Theorem 3.10 is therefore complete once the following basic *local geometric contractibility* result has been established. $\square$

THEOREM 3.11 (LGC-Lemma). *Given an integer $n \geq 2$, a real k and $D, v > 0$. There is a $\delta = \delta(n, k, D, v)$ such all points $(p, q) \in M \times M$ with $\operatorname{dist}(p, q) \leq \delta$ are regular for the diagonal $\Delta(M) \subset M \times M$. In particular, for each $p \in M$, $D_p(\delta)$ is contractible in M to a point.*

PROOF. Assume $(p, q) \in M \times M$ is Δ-critical and $\operatorname{dist}(p, q) = d$. Then clearly p is q-critical and q is p-critical. By the shrinking principle of Lemma 2.10,

$$\operatorname{vol} M = \operatorname{vol} H(p; q) + \operatorname{vol} H(q; p) \leq 2 \operatorname{vol} H(\bar{q}; \{\bar{p}_0, \bar{p}_1\}),$$

where

$$H(\bar{q}; \{\bar{p}_0, \bar{p}_1\}) = \{\bar{x} \in D_q(D) \mid \operatorname{dist}(\bar{q}, \bar{x}) \leq \operatorname{dist}(\bar{x}, \{\bar{p}_0, \bar{p}_1\})\} \subset S_k^n$$

and $\bar{q}$ is the midpoint of $\bar{p}_0 \bar{p}_1$ and $\operatorname{dist}(\bar{q}, \bar{p}_i) = d$ (cf. Figure 2.11 with $\theta = 0$). Since obviously $\operatorname{vol} H(\bar{q}; \{\bar{p}_0, \bar{p}_1\}) \to 0$ and $d \to 0$ this proves the first claim. The deformation retraction defined near $\Delta \subset M \times M$ by following the integral curves of a gradient-like vector field for $\operatorname{dist}_\Delta$ will also provide the desired deformation of $D_p(\delta)$ via the embedding $D_p(\delta) \subset \{p\} \times D_p(\delta) \subset M \times M$. $\square$

REMARK 3.12 (Contractibility functions). By appealing to the more general θ-version in Remark 2.12 of the shrinking lemma, one gets an important sharpening of the LGC-lemma above: There are $\delta = \delta(n, k, D, v) > 0$

and $\theta = \theta(n, k, D, v) > 0$ such that any point $(p, q) \in M \times M$ with $\operatorname{dist}(p, q) \leq \delta$ is θ-*regular*; i.e., $S_{(p, q)\Delta}$ is contained in a $\frac{1}{2}\pi - \theta$ ball in $S_{(p, q)}M \times M$. From Lemma 1.5 we then find an $R = R(n, k, D, v)$ so that any r-ball in M with $r \leq \delta$ is contractible inside the concentric ball of radius $R \cdot r$.

The local geometric contractibility control described above is crucial for example in the derivation of homotopy and homeomorphism finiteness results (cf. §5).

REMARK 3.13. An application of the relative volume comparison theorem, Theorem 2.4, shows that any n-manifold M with $\operatorname{Ric} M \geq (n - 1)$ and $\operatorname{diam} M \geq \pi - \varepsilon$ has small excess. Theorem 3.10 therefore has a diameter Ricci curvature sphere theorem as corollary (cf. [35]).

4. Structure theorems

It is too optimistic to expect a recognition type solution to the excess problem mentioned in Problem 3.9. Rather, one would hope to at least be able to find restrictions for the structure of such manifolds. The applications of critical point theory given in this section are to problems of that kind.

Although all manifolds M in this section will be complete and noncompact, the idea in the following observation due to Gromov plays a key role in the Betti- number finiteness theorem of §5.

THEOREM 4.1. *Let M be a noncompact riemannian manifold with $\sec M \geq 0$. Then M is diffeomorphic to the interior of a compact manifold with boundary.*

PROOF. We claim that for any $p \in M$, there is an $r > 0$ so that dist_p has no critical points in $M - B_p(r)$. In fact, if this is not the case let q_i, $i = 1, 2$, be a sequence of points in $\operatorname{crit}(p)$ with say $\operatorname{dist}(p, q_{n+1}) \geq 2 \operatorname{dist}(p, q_n)$. By the criticality principle of Lemma 2.8 the angle between any segment from p to q_i and from p to q_j, is bounded below by $\theta_0(1, 2)$ $(= \theta_0(d, 2d)$ for any $d > 0$) independent of i, j. This is of course impossible since $S_p \subset T_p M$ is compact. The claim now follows from the finite type Lemma 1.9. □

The following much stronger structure theorem is due to Cheeger and Gromoll [10].

THEOREM 4.2 (Soul Theorem). *Any complete noncompact riemannian manifold M with $\sec M \geq 0$ is diffeomorphic to the normal bundle of some compact totally geodesic submanifold $S \subset M$.*

The dominating feature in the proof of this result is convexity. Here is an outline:

Choose $p \in M$ arbitrarily and let $c : [0, \infty] \to M$ be a ray in M; i.e., $c|[s, t]$ is a segment for any s, t. The existence of such a ray is a simple consequence of the Hopf-Rinow theorem. For each $t \geq 0$, $H_t(c) =$

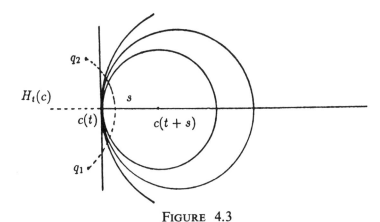

FIGURE 4.3

$M - \bigcup_s B_{c(t+s)}(s)$ is called the half space associated with $c : [t, \infty) \to M$ (cf. Figure 4.3).

Let $c_0 : [0, l] \to M$ be a geodesic in M with $c_0(0) = q_1$, $c_0(l) = q_2 \in H_t(c)$. If $c_0(r) \in B_{c(t+s_0)}(s_0)$ for some s_0 then $\text{dist}(c_0(r), c(t+s)) \le s - \varepsilon$ for some $\varepsilon > 0$ and all $s \ge s_0$ by the triangle inequality. Since $\text{dist}(p_i, c(t+s)) \ge s$ for all $s \ge 0$ this contradicts Toponogov's theorem, Theorem 2.1 (T). Thus $H_t(c)$ is totally convex for all $t \ge 0$.

Now consider the compact totally convex subsets $C_t = \bigcap_c H_t(c)$, where the intersection is taken over all rays c emanating from p. Clearly $\bigcup_{t \ge 0} C_t = M$, and for each $t > 0$, $\partial C_t \ne \varnothing$.

If $C \subset M$ is a convex set with $\partial C \ne \varnothing$ then $\text{dist}_{\partial C} : C \to \mathbb{R}$ is a concave function because $\sec M \ge 0$ [10]. In particular $C_a = \{x \in C \mid \text{dist}_{\partial C}(x) \ge a\}$ is convex for all $a \le a_0 = \max \text{dist}_{\partial C}$. Moreover $\dim C^{a_0} < \dim C$. This construction terminates in finitely many steps when one arrives at a convex set $S \subset C$ without boundary.

By applying the above construction to one of the sets C_t, $t > 0$, we get a totally geodesic, closed submanifold $S \subset C_t \subset M$. Such an S is called a soul of M. It is now easy to see that each $x \in M - S$ lies on the boundary of some convex set $C \supset S$. By the convexity principle of Example 1.3, x is a regular point for $\text{dist}_S : M \to \mathbb{R}$. The soul theorem is then a consequence of Corollaries 1.9 and 1.10. $\quad \square$

It is not known if the converse to Theorem 4.2 holds, i.e.,

PROBLEM 4.4. Does the total space E of any vector bundle $E \to M$ over a closed riemannian manifold M with $\sec M \ge 0$ carry a complete metric with $\sec E \ge 0$?[3]

This problem remains unsolved even when $M = S^n$ (cf. [10, 54, 55]).

In trying to extend some of these ideas to complete noncompact manifolds M with $\sec M \ge k$, $k < 0$, we reinterpret some of the above constructions

---

[3] A counterexample has recently been found by M.Özaydin and G. Walschap, *Vector bundles with no soul*, preprint.

for $k = 0$. Specifically we observe that $\text{dist}_p : M \to \mathbb{R}$ has no critical points outside the compact set $C_t = \bigcap_c H_t(c)$ for any $t \geq 0$. In fact any point $x \in M - C_t$ belongs to $\bigcup_s B_{c(t+s)}(s)$ for some ray $c : [0, \infty) \to M$ with $c(0) = p$. Now $\bigcup_s B_{c(t+s)}(s) \subset \bigcup_s \text{reg}(c(t), c(t + 2s))$ by Theorem 2.1 and the regularity principle of Lemma 2.5 applies.

With this in mind suppose M is a complete noncompact riemannian manifold with $\sec M \geq -k$, $k > 0$, and finitely many ends. For fixed $p \in M$ and $r > 0$ let $R(p, r) = \{c(r) | c : [0, \infty) \to M \text{ ray}, c(0) = p\} \subset \text{dist}_p^{-1}(r)$. Following [7] we define the essential diameter of ends at distance r from p, $\text{esdi}(p, r)$, as

(4.5) $\text{esdi}(p, r) = \sup \text{diam} \Sigma_r$

where the supremum is taken over all connected components Σ_r of $\text{dist}_p^{-1}(r)$ with $\Sigma_r \cap R(p, r) \neq \emptyset$ (corresponding to the boundary of unbounded components of $M - D(p, r)$). The *essential end diameter* of M is then

(4.6) $\text{esdi}_\infty(M) = \inf_p \limsup_{r \to \infty} \text{esdi}(p, r)$.

Of course $\text{esdi}_\infty(M) = \infty$ in general. In view of the discussion of $\sec M \geq 0$ above, however, the following result of Shen [59] can be interpreted as a generalization of Theorem 4.1.

THEOREM 4.7 (Bounded End Theorem). *Suppose M is a complete noncompact riemannian manifold with finitely many ends and $\sec M \geq -k$, $k > 0$. If $\text{esdi}_\infty(M) < \frac{1}{\sqrt{k}} \log(\frac{3+\sqrt{5}}{2})$, then M is diffeomorphic to the interior of a compact manifold with boundary.*

PROOF. By assumption there exist a $p \in M$ and an $R > 0$ so that $\text{diam} \Sigma_r < \frac{1}{\sqrt{k}} \log(\frac{3+\sqrt{5}}{2})$ for all connected components $\Sigma_r \subset \text{dist}_p^{-1}(r)$ with $\Sigma_r \cap R(p, r) \neq \emptyset, r \geq R$. In particular $\text{dist}(x, R(p, r)) < \frac{1}{\sqrt{k}} \log(\frac{3+\sqrt{5}}{2})$ for all x in such Σ's. An easy application of the regularity principle of Lemma 2.5 (cf. Figure 4.8) then shows that all points of $\Sigma_r \subset \text{dist}_p^{-1}(r)$ with $\Sigma_r \cap R(p, r) \neq \emptyset, r \geq R$, are regular points for p. Now if U is one of the finitely many unbounded connected components of $M - D_p(R)$, there is a ray $c : [0, \infty) \to M$ emanating from p with $c(r) \in U$ for any $r > R$. The isotopy Lemma 1.8 then implies that all $\Sigma_r, r \geq R$, determined by $c(r) \in \Sigma_r$ are homeomorphic and in fact U is homeomorphic to $\Sigma_R \times (R, \infty)$. From this and the fact that $M - D_p(R)$ has at most finitely many bounded components, we conclude that there is an $R_1 \geq R$ so that all points in $M - B_p(R_1)$ are regular points for dist_p.   □

When the essential end diameter (4.6) is infinite (or violates the assumption in Theorem 4.7), the regularity argument used above does not apply directly. However, existence of p-critical points in $M - B_p(r)$ for any r restricts suitable excess-invariants of M severely. This was first observed in [3] (cf. also [59] and [7]). Abresh and Gromoll also made the fundamental

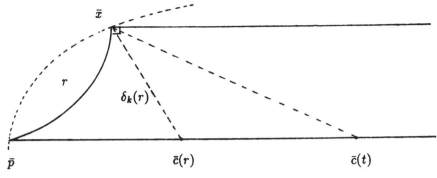

FIGURE 4.8. In S_k^2,

$$\text{reg}(\bar{p}, \bar{c}) = \bigcup_t \text{reg}(\bar{p}, \bar{c}(t)) = \bigcup_r \text{dist}_{\bar{p}}^{-1}(r) \cap B_{c(r)}(\delta_k(r))$$

$$\text{where } \delta_k(r) = \frac{1}{\sqrt{k}} \cosh^{-1}\left(\frac{\cosh^3 \sqrt{k}r - \sinh^3 \sqrt{k}r}{\cosh \sqrt{k}r}\right).$$

discovery that excess functions can be controlled in terms of Ricci curvature. If the end diameter growth function (4.5) is not too big, this excess control then violates the restrictions stemming from the existence of critical points in $M - B_p(r)$ for all r. More precisely [3],

THEOREM 4.9. *Let M be a complete noncompact riemannian n-manifold with* $\sec M \geq -k$, $k > 0$, *and end diameter growth satisfying*

$$\inf_p \limsup_{r \to \infty} \text{esdi}(p, r) \cdot r^{-1/n} < \frac{1}{8} k^{-(n-1)/2n}.$$

Then M is diffeomorphic to the interior of a compact manifold with boundary if also $\text{Ric } M \geq 0$.

For other similar results involving intermediate curvatures and volume restrictions see [59] and [62].

It should also be mentioned here that examples due to Sha and Yang [58] show that the diameter growth assumption in Theorem 4.9 is necessary. For more details about all of this we refer to the survey article by Gromoll [19] in these proceedings.

REMARK 4.10. In the framework of this survey, there are so far no general structure theorems for natural classes of closed manifolds, e.g., positively or non negatively curved manifolds. There is reasonable hope, however, that results of this type may emerge via our increased understanding of the Gromov-Hausdorff topology on such natural classes of closed manifolds. For more about this we refer to the surveys by Petersen [52], and Fukaya [13].

5. Finiteness theorems

It is not unreasonable to view a finiteness theorem as a first step in a recognition process, or for that matter as an approximation to a structure theorem.

In the results we will discuss here, the following simple packing lemma plays a crucial role.

LEMMA 5.1 (Packing Lemma). *Fix an integer $n \geq 2$, and $k \in \mathbb{R}$. For every $\varepsilon, D > 0$ there is an $N = N(n, k, \varepsilon, D)$ such that any D-ball in a riemannian n-manifold M with $\mathrm{Ric}\, M \geq k(n - 1)$ can be covered by $\leq N$ ε-balls.*

PROOF. Let $\{x_i, \ldots, x_N\}$ be the centers of a maximal set of disjoint $\varepsilon/2$-balls in the given D-ball in M. By maximality, the corresponding ε-balls provide a cover. Moreover, the relative volume comparison Lemma 2.4 yields

$$N \leq v_k^n(D)/v_k^n(\varepsilon/2) = N(n, k, \varepsilon, D). \quad \square$$

It is useful to know that when an ε-cover is chosen as efficiently as in the proof of Lemma 5.1, the number of ε-balls with nonempty intersection is also a priori bounded independent of ε. This is another simple consequence of Theorem 2.4 (cf. e.g. [31]). We now have all the ingredients used in the proof of the following result (cf. [31]).

THEOREM 5.2 (Homotopy Finiteness). *Given an integer $n \geq 2$, a real k and positive D, v, there are at most finitely many homotopy types among closed riemannian n-manifolds M satisfying $\sec M \geq k$, $\mathrm{diam}\, M \leq D$ and $\mathrm{vol}\, M \geq v$.*

PROOF. By the local geometric contractibility Lemma 3.11 (3.12), there are constants $r = r(n, k, D, v) > 0$ and $L = (n, k, D, v) > 0$, such that any pair $(p_1, p_2) \in M \times M$ with $\mathrm{dist}(p_1, p_2) = \varepsilon \leq r$ is joined by a path $\sigma_{p_1 p_2} : [0, 1] \to M$ depending continuously on (p_1, p_2) and with $\mathrm{Length}(\sigma_{p_1 p_2}) \leq L \cdot \varepsilon$. This allows us to think of $\sigma_{p_1 p_2}(t)$ as the "center of mass" of the points p_1, p_2 with "weights" $1 - t$ and t respectively. For any integer $l \geq 2$ it is clear that by iterating this construction, we can assign to any ordered $(l+1)$-tuple $(p_0, \ldots, p_l)$ in M a map $\sigma_{p_0, \ldots, p_l} : \Delta_l \to M$ of the standard l-simplex, "spanning" $(p_0, \ldots, p_l)$ and depending continuously on $(p_0, \ldots, p_l)$, whenever $\mathrm{dist}(p_i, p_j) \leq \varepsilon = \varepsilon(l, r, L) \leq r$, $i, j = 0, \ldots, l$.

As pointed out after Lemma 5.1 there is an $l = l(n, k, D)$ such that in any "efficient" cover of M by ε-balls at most l such balls meet, independent of ε! With this l, choose $\varepsilon = \varepsilon(l, r, L) = \varepsilon(n, k, D, v)$ as above and fix for each closed riemannian n-manifold M with $\sec M \geq k$, $\mathrm{diam}\, M \leq D$ and $\mathrm{vol}\, M \geq v$ an "efficient" cover by $\varepsilon/2$-balls, $B(p_1, \varepsilon/2), \ldots, B(p_m, \varepsilon/2)$. According to the packing Lemma 5.1, $m \leq N = N(n, k, D)$.

For each $m \in \{1, \ldots, N\}$ consider all the manifolds M that are covered by exactly $m\varepsilon/2$-balls as above, and where the corresponding nerves have isomorphic 1-skeleta; i.e., ball number i intersects ball number j in all M or ball number i does not intersect ball number j in any M. This obviously divides our class of manifolds M into finitely many subclasses. From the above choices and constructions it now follows that by choosing a partition of unity subordinate to each fixed $\varepsilon/2$-cover we get continuous maps f, f^1 between any two manifolds M, M^1 from the same subclass. Moreover, by construction $\mathrm{dist}(f^1 \circ f, \mathrm{id}_M) \leq C \cdot \varepsilon$, for some $C = C(L, l) = C(n, k, D, v)$. Making sure that also $C \cdot \varepsilon \leq r$ then shows that our maps f, f^1 are homotopy equivalences. This completes the proof.

The conclusion in Theorem 5.2 can actually be sharpened considerably (cf. [36] and the announcement in [6] of recent work of Perelman). This relies heavily on Gromov-Hausdorff convergence techniques and geometric topology, and goes beyond the main topic discussed here (see [52]).

THEOREM 5.3 (Topological Finiteness). *For fixed n, k, D and v as above the class of closed riemannian n-manifolds M with $\sec M \geq k$, $\mathrm{diam}\, M \leq D$ and $\mathrm{vol}\, M \geq v$ contains at most finitely many homeomorphic types.*

Since any closed topological manifold of dimension $n \neq 4$ has at most finitely many differentiable structures [44], the above theorem yields actually *finiteness of diffeomorphism types* when $n \neq 4$. Except for $n = 4$, Theorem 5.3 therefore generalizes Cheeger's finiteness theorem [8].

QUESTION 5.4. Are there $k \in \mathbb{R}$, D, $v > 0$, and infinitely many diffeomorphism types of closed riemannian 4-manifolds M satisfying $\sec M \geq k$, $\mathrm{diam}\, M \leq D$ and $\mathrm{vol}\, M \geq v$?

The volume assumption in 5.2 (5.3) prevents the phenomenon called collapsing. It is remarkable that even without this assumption, one has the following general finiteness theorem due to Gromov [23].

THEOREM 5.4 (Betti Number Theorem). *Given an integer $n \geq 2$, a real k and positive D, there is a $C = C(n, k, D)$ so that for every field F of coefficients, $\dim H * (M; F) \leq C$ when M is any closed riemannian n-manifold satisfying $\sec M \geq k$ and $\mathrm{diam}\, M \leq D$.*

Note that since $\sec M \geq 0$ is a scalings invariant property, it follows that in this theorem the diameter assumption is superfluous when $k = 0$.

Besides the original proof of this theorem in [23] (cf. also [1, 2]), detailed proofs have been given in lecture notes by Cheeger [7] and Meyer [46]. Here, therefore, we will only attempt to elaborate on the main ideas and strategies of the proof:

First observe that if for some reason one could control the homology of all ε-balls for some a priori ε, then the packing lemma combined with a Mayer-Victoris type argument would give control on the homology of all M^n with $\sec M \geq k$ and $\mathrm{diam}\, M \leq D$. For fixed k, however, there is a small $\varepsilon = \varepsilon(k)$ so that all comparison arguments applied to such ε-balls are essentially the same as comparison arguments used for $k = 0$. This then "reduces" the proof to the case $\sec M \geq 0$ and no diameter assumption. Now, however, one needs to be able to control the homology of all balls B. It is important that all of the above ideas work if one replaces $\dim H(B)$ (which could be

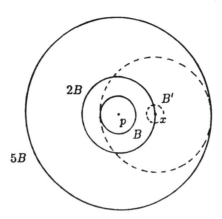

FIGURE 5.5

infinite) by the finite number

$$\mathrm{cont}\, B = \mathrm{rank}\left(\underset{*}{i} : \underset{*}{H(B)} \to \underset{*}{H(5B)} \right),$$

called the *content* of B. Here $5B$ is concentric with B and has 5 times larger radius. The only significance of 5 is that it is a fixed number > 1.

In order to estimate $\mathrm{cont}(B)$, one makes the following simple but basic observation: If B can be isotoped inside $5B$ to a ball B' with $5B' \subset 5B$, then

$$\mathrm{cont}\, B \le \mathrm{cont}\, B'.$$

In this case we say that B can be *compressed* to B'. If $\mathrm{rad}\, B' \le \frac{1}{2}\,\mathrm{rad}\, B$ we say that B is *compressible*. Thus, B is *incompressible* if it does not compress to any B' with $\mathrm{rad}\, B' \le \frac{1}{2}\,\mathrm{rad}\, B$.

Obviously, if a ball B can be compressed sufficiently many times, so as to end up in a contractible ball (e.g., a ball with radius smaller than $\mathrm{inj}(M)$) its content is 1.

If on the other hand $B_p(r)$ is incompressible then by the isotopy Lemma 1.8 any point $x \in 2B$ has a critical value in the interval $[\frac{1}{2}r, r + \mathrm{dist}(x, p)]$ (cf. Figure 5.5).

Now cover $B_p(r)$ efficiently with balls of radius say $r/10$ as in the packing Lemma 5.1. If all of these balls have content 1 we get an a priori estimate for $\mathrm{cont}\, B_p(r)$. Otherwise we find a $q \in B_p(\frac{11}{10}r)$ so that all $x' \in 2B_q(\frac{1}{10}r')$, $r' \le r$ have critical values in the interval $[\frac{1}{20}r', \frac{1}{10}r' + \mathrm{dist}(x', q)]$ as well as in the interval $[\frac{1}{2}r, r + d(x', p)]$ from before. By the criticality principle of Lemma 2.8, this process must terminate after a finite a priori number of steps where all involved balls have content 1. The a priori control on the number of such balls needed then gives via the Mayer-Victoris type argument alluded to above the desired conclusion.

We point out that the constructive proofs discussed above yield explicit estimates, although probably very far from being sharp. The following conjecture has been formulated by Gromov.

CONJECTURE 5.6. *For any complete riemannian n-manifold M with* $\sec M \geq 0$, $\dim \underset{*}{H}(M\,;\,F) \leq \dim \underset{*}{H}(T^n\,;\,F) = 2^n$.

Another natural question is whether the product structure of $\underset{*}{H}(M\,;\,F)$ is controlled for manifolds M in the class $\sec M \geq k$ and $\operatorname{diam} M \leq D$. In particular

QUESTION 5.7. Given an integer $n \geq 2$, $k \in \mathbb{R}$ and $D > 0$, are there only finitely many rational homotopy types among closed simply connected riemannian n-manifolds M, satisfying $\sec M \geq k$ and $\operatorname{diam} M \leq D$?

REFERENCES

1. U. Abresch, *Lower curvature bounds, Toponogov's theorem, and bounded topology.* I, Ann. Sci. École Norm. Sup. (4) **18** (1985), 563–633.
2. ____, *Lower curvature bounds, Toponogov's theorem, and bounded topology.* II, Ann. Sci. École Norm. Sup. (4) **20** (1987), 475–502.
3. U. Abresch and D. Gromoll, *On complete manifolds with non-negative Ricci curvature*, J. Amer. Math. Soc. **3** (1990), 355–374.
4. A. L. Besse, *Manifolds all of whose geodesics are closed*, Ergeb. Math. Grenzgeb.(3), vol. 93, Springer-Verlag, 1978.
5. J. Bernstein and T. Ganea, *Homotopical nilpotency*, Illinois J. Math. **5** (1961), 99–130.
6. Y. Burago, M. Gromov, and G. Perelman, *A. D. Aleksandrov's spaces with curvature bounded from below.* I, Uspechi Mat. Nauk (to appear).
7. J. Cheeger, *Critical points of distance functions and applications to geometry*, Geometric Topology: Recent developements, Montecatini Termé, 1990 (Ed. P.de Bartolomeis, F. Tricerri), Springer Lecture notes, **1504** (1991), 1–38.
8. ____, *Finiteness theorems for Riemannian manifolds*, Amer. J. Math. **92** (1970), 61–74.
9. J. Cheeger and D. G. Ebin, *Comparison theorems in Riemannian geometry*, North Holland Math Library **9** (1975), Amsterdam.
10. J. Cheeger and D. Gromoll, *On the structure of complete manifolds of non-negative curvature*, Ann. of Math. **96** (1972), 413–443.
11. O. Durumeric, *Manifolds of almost half of the maximum volume*, Proc. Amer. Math. Soc. **104** (1988), 277–283.
12. J.-H. Eschenburg, *Diameter, volume and topology for positive Ricci curvature*, J. Differential Geom. **33** (1991), 743–747.
13. K. Fukaya, *Hausdorff convergence of Riemannian manifolds and its applications*, Adv. Stud. Pure Math., vol. 18-I, Kinokuniya-North Holland, 1990, pp. 143–238.
14. R. Greene and P. Petersen V, *Little topology, big volume*, Duke Math. J. (to appear).
15. R. Greene and K. Shiohama, *Convex functions on complete non compact manifolds: topological structure*, Invent. Math. **63** (1981), 129–157.
16. ____, *Convex functions on complete non compact manifolds: differentiable structure*, Ann. Sci. École Norm. Sup. (4) **14** (1981), 357–367.
17. R. Greene and H. Wu, C^∞ *convex function and manifolds of positive curvature*, Acta Math. **137** (1976), 209–245.
18. ____, *Integrals of subharmonic functions on manifolds of nonnegative curvature*, Invent. Math. **27** (1974), 265–298.
19. D. Gromoll, *Spaces of nonnegative curvature*, PSPUM **54** (1993) 337–357.
20. D. Gromoll, W. T. Meyer, and W. Klingenberg, *Riemannsche Geometrie im Grossen*, Lecture Notes in Math., vol. **55**, Springer-Verlag, 1968.
21. D. Gromoll and K. Grove, *A generalization of Berger's rigidity theorem for positively curved manifolds*, Ann. Sci. École Norm. Sup. (4) **20** (1987), 227–239.
22. ____, *The low-dimensional metric foliations of euclidean spheres*, J. Differential Geom. **28** (1988), 143–156.
23. M. Gromov, *Curvature, Diameter and Betti numbers*, Comment Math. Helv. **56** (1981), 179–195.

24. ——, *Filling Riemannian manifolds*, J. Differential Geom. **18** (1983), 1–148.
25. ——, *Manifolds of negative curvature*, J. Differential Geom. **13** (1978), 223–230.
26. M. Gromov, J. Lafontaine, and P. Pansu, *Structures métrique pour les variétés riemanniennes*, Textes Math., vol. 1, CEDIC/Fernand Nathan, 1981.
27. K. Grove, *Metric differential geometry*, Differential Geometry, (Proc. Nordic Summer School, Lyngby 1985), (V. L. Hansen, ed.), Lecture Notes in Math., vol. 1263, Springer-Verlag, 1987, 171–227.
28. K. Grove and S. Halperin, *Dupin hypersurfaces, group actions and the double mapping cylinder*, J. Differential Geom. **26** (1987), 429–459.
29. ——, *Elliptic isometries, condition* (C) *and proper maps*, Arch. Math. (Basel) **56** (1991), 288–299.
30. K. Grove and S. Markvorsen, *Metric invariants for the riemannian recognition program, via Aleksandrov geometry*, Preprint.
31. K. Grove and P. Petersen V, *Bounding homotopy types by geometry*, Ann. of Math. (2) **128** (1988), 195–206.
32. ——, *Homotopy types of positively curved manifolds with large volume*, Amer. J. Math. **110** (1988), 1183–1188.
33. ——, *Manifolds near the boundary of existence*, J. Differential Geom. **33** (1991), 379–394.
34. ——, *Volume comparison à la Aleksandrov*, Acta Math. (to appear).
35. ——, *A pinching theorem for homotopy spheres*, J. Amer. Math. Soc. **3** (1990), 671–677.
36. K. Grove, P. Petersen V, and J.-Y Wu, *Geometric finiteness theorems via controlled topology*, Invent. Math. **99** (1990), 205–213; erratum **104** (1991) 221–222.
37. K. Grove and K. Shiohama, *A generalized sphere theorem*, Ann. of Math. (2) **106** (1977), 201–211.
38. W.-Y. Hsiang and B. Kleiner, *On the topology of positively curved 4-manifolds with symmetry*, J. Differential Geom. **29** (1989), 615–621.
39. J. Itoh and T. Sakai, *On the quasiconvexity and contractibility of domains in* $\mathbb{R}^n$, Math. J. Okayama Univ. **27** (1985), 221–227.
40. H. Karcher, *Riemannian comparison constructions*, Global Differential Geometry, MAA Stud. Math., vol. 27, Math. Assoc. of America, Washington, D.C., 1989.
41. M. Katz, *Diameter-extremal subsets of spheres*, Discrete Comput. Geom. **4** (1989), 117–137.
42. ——, *On neighborhoods of the Kuratowski imbedding beyond the first extremum of the diameter functional*, Preprint.
43. W. Klingenberg, *Riemannian geometry*, de Gruyter Studies in Math., vol. 1, de Gruyter, Berlin, 1982.
44. R. C. Kirby and L. C. Siebenmann, *Foundational essays on topological manifolds, smoothings, and triangulations*, Ann. of Math. Stud., **88**, Princeton Univ. Press, Princeton, NJ, 1977.
45. Lusternik and Schnirelmann, *Methodes topologiques dans les problèmes variationells*, Hermann, Paris, 1934.
46. W. Meyer, *Toponogov's theorem with applications*, College on Differential Geometry, Trieste, Lecture Notes.
47. J. Milnor, *Morse theory*, Ann. of Math. Stud., vol. 51, Princeton Univ. Press, Princeton, NJ, 1963.
48. Y. Otsu, *On manifolds of small excess*, Amer. J. Math. (to appear).
49. Y. Otsu, K. Shiohama, and T. Yamaguchil, *A new version of differentiable sphere theorem*, Invent. Math. **98** (1989), 219–228.
50. R. Palais, *Lusternik-Schnirelman theory on Banach manifolds*, Topology **5** (1966), 115–132.
51. ——, *Critical point theory and the minimax principle*, Proc. Sympos. in Pure Math., vol. 15, Amer. Math. Soc., Providence, RI, 1970, pp. 185–212.
52. P. Petersen V, *Gromov-Hausdorff convergence of metric spaces*, these proceedings.
53. ——, *Small excess and Ricci curvature*, J. Geom. Anal. **1** (1991), 383–387.
54. A. Rigas, *Geodesic spheres as generators of the homotopy groups of* O, BO, J. Differential Geom. **13** (1979), 527–545.

55. _____, *Some bundles of non-negative curvature*, Math. Ann. **232** (1978), 187–193.
56. W. Rinow, *Die innere Geometrie der metrischen Räume*, Springer-Verlag, 1961.
57. T. B. Rushing, *Topological embeddings*, Pure Appl. Math., vol. 52, 1974.
58. J.-P. Sha and D. G. Yang, *Examples of manifolds of positive Ricci curvature*, J. Differential Geom. **29** (1989), 95–103.
59. Z. Shen, *On complete manifolds of nonnegative kth Ricci curvature*, Trans. Amer. Math. Soc. (to appear).
60. _____, *Scalar invariants and ridigity theorems for complete riemannian manifolds*, Preprint.
61. _____, *On the upper estimates of the Betti numbers of long riemannian manifolds*, Preprint.
62. Z. Shen and G. Wei, *Volume growth and finite topological type*, Indiana Univ. Math. J. **40** (1991), 551–565.
63. K. Shiohama, *A sphere theorem for manifolds of positive Ricci curvature*, Trans. Amer. Math. Soc. **275** (1983), 811–819.
64. _____, *Recent developments in sphere theorems*, these proceedings.
65. K. Shiohama and T. Yamaguchi, *Positively curved manifolds with restricted diameters*, Geometry of Manifolds, Perspect. Math., Academic Press, 1989, pp. 345–350.
66. T. Shioya, *On the excess of open manifolds*, Preprint.
67. _____, *Spitting theorems for nonnegatively curved open manifolds with large ideal boundary*, Preprint.
68. F. Wilhelm, *On the Filling radius of positively curved manifolds*, Invent. Math. **107** (1992), 653–668.
69. J.-Y. Wu, *Applications of critical point theory of distance functions to geometry*, Geom. Dedicata **40** (1991), 213–222.
70. T. Yamaguchi, *Collapsing and pinching under a lower curvature bound*, Ann. of Math. (2) **133** (1991), 317–357.
71. _____, *On the structure of locally convex filtrations on complete manifolds*, J. Math. Soc. Japan **40** (1988), 221–234.
72. _____, *Homotopy type finiteness theorems for certain precompact families of Riemannian manifolds*, Proc. Amer. Math. Soc. **102** (1988), 660–666.

UNIVERSITY OF MARYLAND

Proceedings of Symposia in Pure Mathematics
Volume **54** (1993), Part 3

Spectral Geometry in Higher Ranks:
Closed Geodesics and Flat Tori

DAVID GURARIE

ABSTRACT. After a brief survey of the classical "shape problems" of spectral geometry: *shape of the metric* and *shape of the potential* (for Schrödinger operators), we outline some recent progress in special cases: the n-sphere Schrödinger theory, and operators on higher rank symmetric spaces. We explore the role of *length spectrum* of M (the length of all closed path/geodesics), and the related *Radon transform of V*, and then indicate how new geometric objects, *geodesically flat tori*, naturally arise in higher ranks and play the role of closed geodesics here.

In general terms spectral geometry is concerned with the relation between *spectral data* (eigenvalues, eigenfunctions) of differential operators on manifolds (Laplacians, Schrödinger operators), and the underlying *geometry/dynamics* on M. One usually distinguishes two types of spectral problems: the *direct problem*, that asks to determine spectral data (its asymptotics) through geometry data, and the *inverse problem*, that asks to recover geometry (metric, potential, shape of the boundary, etc.) from a suitable spectral data.

Let us remark that the relations between spectra and geometry have deep physical meaning and significance. In the terminology of quantum theory manifold M represents a *configuration space* of the *classical mechanical system*, whereas operators on M (differential and other) give *quantum observables*. The correspondence "geometry" → "spectra" manifests relationship of certain quantum characteristic (e.g. energy) to the underlying classical dynamics, and is often called the *Correspondence Principle* in quantum theory.

The first known example of geometric data, that comes from spectral asymptotics, appeared in the celebrated Weyl (volume-counting) principle [We]. It describes large eigenvalue asymptotics of Laplacians on manifolds or domains with boundary [Ka, Hö, Se, Me, Iv]. The *counting function*

1991 *Mathematics Subject Classification.* Primary 58F19, 58G25, 58G35, 22E46.
This paper is in final form and no version will be submitted for publication elsewhere.

$N(\lambda) = \#\{\text{eigen's } \lambda_k(\Delta) \le \lambda\}$ admits the following asymptotics,

(1) $$N(\lambda) \sim C_0 |M| \lambda^{n/2} + C_1 |\partial M| \lambda^{(n-1)/2}, \quad \text{as } \lambda \to \infty,$$

where $|M|$, $|\partial M|$ denote the volume of M and the area of ∂M, constants C_0, C_1 depend only on the $\dim M = n$. Formula (2) was established for fairly general classes of elliptic (subelliptic) differential/pseudodifferential operators on manifolds and domains.[1] It yields the simplest geometric data (volume $|M|$, and area $|\partial M|$), from asymptotics of $\{\lambda_k\}$. Formula (2) could be further refined by replacing $N(\lambda)$ with certain its averages, like *zeta* or *theta-functions* of Δ,

(2) $$Z(s) = \text{tr}(\Delta^{-s/2}) = \sum_\lambda \lambda^{-s/2}; \qquad \Theta(t) = \text{tr}(e^{-t\Delta}) = \sum_\lambda e^{-\lambda t}.$$

The large-λ asymptotics of $N(\lambda)$ could be linked to the small-t asymptotics of $\Theta(t)$, and to the residues of $Z(s)$, via Abelian/Tauberian theorems. But averaged functions usually behave much better than $N(\lambda)$, as they admit asymptotic expansions to all orders! The foremost is the heat-expansion of Munakshisundaram-Pleijel [**MP**], elaborated and extended by many other authors (see [**McS, Gi**]),

(3) $$\Theta(t) \sim t^{-n/2}\{b_0 + b_1 t^{1/2} + b_2 t + \cdots\}.$$

Coefficients $\{b_j\}$ called *heat-invariants* (as they obviously represent spectral invariants of Δ_M!), carry some important geometric-topological data. The first two are simply related to the Weyl coefficients,

$$b_0 = \Gamma(\tfrac{n}{2}+1)C_0|M|; \qquad b_1 = \Gamma(\tfrac{n}{2})C_1|\partial M|,$$

while the third was found by McKean-Singer [**McS**],

$$b_2 = \frac{1}{3}\left(\int_M K\,dV - \int_{\partial M} J\,dS\right),$$

where $K = \text{tr}(\text{Ric})$ denotes the scalar curvature of M, while $J = \text{tr}$ (2nd fundamental form) denotes the mean curvature of the boundary. For 2-D surfaces invariant b_2 yields the Euler characteristics of M ($1 -$ "genus M") by the Gauss-Bonet theorem. So the "topological shape" of M is audible (can be discerned from the spectrum of Δ_M)!

Higher heat-invariants are given by the universal polynomial expressions in curvature and its covariant derivatives, integrated over M. But the computations become increasingly hard; only few $\{b_j\}$ have been found explicitly

---

[1] Here it takes the form

(4) $$N(\lambda) \sim \text{vol}\{(x;\xi) \in T^*(M): \text{"symbol}_H(x;\xi)\text{"} \le \lambda\}.$$

The cotangent bundle $T^*(M)$ represents the classical-mechanical phase-space of the system, function $h(x;\xi) = \text{symbol}(H)$, becomes the classical hamiltonian on $T^*(M)$, while the operator $(\psi\,\text{do})H = h(x;i\nabla)$ on M gives the corresponding "quantum hamiltonian." In such form (4) asserts that the number $N(\lambda)$ of quantum states (eigenfunctions of H) below the energy level λ, is roughly equal to the "number" (phase-space volume) of classical states $\{(x;\xi): \text{of energy } h(x;\xi) \le \lambda\}$, whence comes the name "(phase) volume counting principle"!

[Gil, Sm]. One does not know if the entire set $\{b_j\}$ completely characterizes the *isospectral class* of Riemannian metrics on M.

1. Periodic orbits and "shape of the metric"

The Weyl/heat invariants for Laplacians on Riemannian manifolds provide some useful geometric data, but fall short of determining metric g on M. So one needs to look beyond heat-invariants to approach the inverse metric problem. One such class of geometric data are *closed geodesics*[2] $\{\gamma\}$ on M. Their role in spectral theory was extensively analyzed both in the physical **[Gut, BB, BT, Vo, AA]**, and the mathematical literature **[Ch, DG, Ra, Do, DKV]**, to name a few. The physical heuristics behind the "eigenvalues" $\leftrightarrow$ "closed path" relations reflects once again the basic Correspondence Principle between the quantum hamiltonian (differential operator, ψdo) $H = h(x; \nabla)$ on Hilbert space $L^2(M)$, and the underlying classical dynamical system, geodesic/bicharacteristic flow on the classical phase-space $T^*(M)$, given by symbol of H.

The general principle states: *spectrum of* Δ_M determines the *length spectrum* of M (length of all closed path: $l(\gamma) = |\gamma|$), as well as other related geometric data: Morse indices $\sigma(\gamma)$, Poincaré numbers (measuring the "twist of the flow" along γ), etc. Specifically one could show (cf. **[DG]**) *for any stable closed path* $l = l(\gamma)$ *there exists an almost arithmetic sequence of eigenvalues in* spec $\sqrt{\Delta}$:

$$(4) \qquad \lambda_k \simeq \frac{2\pi}{l}k + d; \qquad d = \sum m_j \theta_j + \nu(\gamma).$$

Here $\{\theta_j\}$ are Poincaré angles and ν the Morse index along γ. One way to establish (4) is via the so called quasi-mode construction **[Co3]**, i.e., embedding of space $L^2(\gamma) \to L^2(M)$, that "almost intertwines" operators $(d^2/ds^2)|_\gamma$ with the Laplacian Δ_M.

Another way to explain how closed paths enter spectral asymptotics of differential operators (and in many cases to rigorously prove it **[DG]**) exploits the so-called *wave-kernel* of Δ, $U_t = e^{it\sqrt{\Delta}}$—the fundamental solution (Green's function) of the wave equation:

$$u_{tt} + \Delta u = \delta(t)\delta(x; y) \quad \text{on } M \times \mathbb{R}.$$

It turns out that

$$\chi(t) = \operatorname{tr} U_t = \sum e^{it\sqrt{\lambda_k}},$$

understood as a distribution[3] on $\mathbb{R}$, has singularities located at precisely

---

[2]More generally one looks at closed (periodic) *bicharacteristics*, trajectories of the hamiltonian flow of the operator H on the phase-space $T^*(M)$, or the cosphere bundle $S^*(M)$.

[3]Obviously, unitary operators U_t do not have "trace" in the usual sense. But integrating U_t against nice (test) functions f on $\mathbb{R}$, the resulting "smoothed out" operators $U_f = \int f(t)U_t \, dt$, become of trace class on $L^2(M)$. So distribution χ can be defined via pairing to testing $\{f\}$, $\langle \chi; f \rangle = \operatorname{tr} U_f$.

the length spectrum $\{\ell(\gamma)\}$ of M, including the "big singularity" at $t = 0$, corresponding the "zero-length" (one-point) path. Precisely, $\chi = \chi_0 + \sum \chi_\gamma$, the sum of terms representing contributions of different closed paths $\{\gamma\}$. Furthermore, the main singularity χ_0 has its Fourier transform $\hat{\chi}_0$ asymptotically expanded as

$$\hat{\chi}_0(\mu) \sim c_0 \mu^{n-1} + c_1 \mu^{n-2} + \cdots, \quad \text{as } \mu \to \infty,$$

with coefficients $\{c_0; c_1; \dots\}$ related to the Weyl heat invariants (2)–(4). All other (nonzeros) terms yield

$$(5) \qquad (\chi - \chi_0)(t) \sim \frac{1}{2\pi} \sum_\gamma \sum_{m=1}^{\infty} \frac{l(\gamma) e^{i(\pi/2)\sigma(\gamma)}}{(t - l(\gamma^m))\sqrt{\det(I - P(\gamma)^m)}}.$$

Here γ denotes a primitive closed classical path (geodesic), γ^m the mth iterate of γ, $\sigma(\gamma)$ the Morse index of γ, and $P(\gamma)$ the Poincaré map along γ. Let us remark that singular terms of the type $[t - l(\gamma^m)]^{-1}$ in (5) correspond to an isolated closed path γ. In more degenerate cases (e.g., orbits of a given period forming a d-parameter family) more singular distributions of order $\frac{d+1}{2}$ appear in place of $(t - l)^{-1}$.

Formula (5) provides a far-reaching generalization of the classical *Poisson summation formula* for the wave-kernel $e^{it\sqrt{\Delta}}$ on the n-torus $M = \mathbb{T}^n$, as well as its noncommutative analog, the celebrated *Selberg-trace formula* on hyperbolic spaces $M = \mathbb{H}\backslash\Gamma$, Poincaré half-plane modulo a discrete (Fuschian) subgroup Γ of $\mathrm{SL}(2; \mathbb{R})$, acting on $\mathbb{H}$ by fractional linear transformations [**Sel, Mc**]. The remarkable feature of the classical (Poisson, Selberg) cases, however, is that the asymptotic relation (5) becomes exact.

This additional set of geometric data furnished by the wave method, namely, the length spectrum and Poincaré numbers, allows one to approach the inverse spectral problem for Riemannian metrics. Two basic questions arise here:

(i) *Uniqueness*: is metric $ds^2 = g_{ij} dx^i dx^j$ uniquely determined by the spectrum of the corresponding Laplace-Beltrami operator Δ_g;

(ii) *Isospectral classes and rigidity*: describe metrics $\{g\}$ with identical spectrum of Δ_g; in particular, prove (or disprove) that the only isospectral deformations of metric g are given by isometries of M (*Rigidity Hypothesis*).

There are several examples of continuous nontrivial isospectral deformations of the metric (nonuniqueness!) [**GW**], but those are mostly in high dimensions. It is not clear to what extent these examples are "generic" in any sense. In contrast many low dimension cases come very close to resolving the uniqueness and rigidity problems. For instance, among hyperbolic 2-D surfaces of constant negative curvature $M = \mathbb{H}\backslash\Gamma$, generic M is uniquely determined by $\mathrm{spec}(\Delta_M)$ [**Vo**]. At the same time there exist exceptional finite isospectral families of surfaces in any genus [**Vi**]. For nonconstant curvature and higher-D hyperbolic manifolds a weaker form of uniqueness holds, the so-called infinitesimal rigidity established in [**GK**]. The best general results to date in dimensions 2 and 3 establish *compactness of isospectral metrics in* C^∞-*topology* [**OPS, CY**]. They exploit yet another interesting spectral invariant: *determinant of the Laplacian.*

The next question, that comes to mind in connection with the length spectrum (especially in higher dimensions), is *whether other geometric objects could have "spectral content" and play the role of closed geodesics?* One known example is given by *invariant Lagrangians* $\Lambda \subset T^*(M)$-maximal isotropic submanifolds of the symplectic space $T^*(M)$, invariant under the hamiltonian (geodesic) flow [**MF, Ke, KR, Vo**]. Based on such a Lagrangian, one can construct approximate eigenfunctions, or spectral projections of H. The method usually requires "large families of Lagrangians," rather than individual terms, for instance foliation of the phase-space $T^*(M)$ into invariant tori $\{\Lambda(\alpha_1, \ldots, \alpha_n)\}$ of an integrable hamiltonian. Continuous parameters $\{\alpha_1, \ldots, \alpha_n\}$ could then be "quantized" (discretized), according to the EBK (Einstein-Brilluoin-Keller) rules to produce approximate sequences of eigenvalues $\lambda(\alpha_1 \cdots \alpha_n)$,

$$(6) \qquad A_j(\alpha) = \oint_{\gamma_j(\alpha)} p \cdot dq = \pi(m_j + \mathrm{ind}(\gamma_j)).$$

Here $\{\gamma_j : 1 \le j \le n\}$ denote a system of fundamental path (homotopy classes) on Λ, and $\mathrm{ind}(\gamma)$ refers to its Maslow index. Equations (6) yield a discrete set of $\{\alpha\text{'s}\}$, which in turn give "quantized energy levels," eigenvalues, $\lambda = \lambda(\alpha)$ of an appropriate operator H. The requirement of Lagrangian foliation, however, essentially limits the method to completely integrable hamiltonians! In this regard closed path are far less restrictive.

In this paper we want to propose another candidate, that occupies an intermediate place between closed path and Lagrangians, namely, the *geodesically flat tori* embedded in M. An early attempt along these lines is due to Voros [**Vo1**], where he studied a semiclassical eigenvalue problem for nonintegrable (nonseparable) hamiltonians. Voros considered Lagrangian foliations $\{\Lambda\}$ in the phase-space $T^*(M)$ of intermediate dimensions: $1 \le \dim \Lambda \le n$, and established an analog of the EBK-quantization rules (6). Our work will be similar in spirit, but with a different emphasis: flat tori appear as geometric structures on manifold M itself, rather than the phase-space $T^*(M)$. In fact, we shall concentrate on special classes on manifold, where flat tori naturally arise. These are *higher rank* symmetric, or locally-symmetric, spaces of two types:

(i) *compact symmetric spaces* $M = K\backslash G$, quotients of compact semisimple Lie groups G modulo certain subgroups K (see [**He**]), discussed in §3;

(ii) *higher rank compact hyperbolic spaces* $M = \Gamma\backslash G/K$, i.e., quotients of hyperbolic symmetric spaces $\mathbb{H} = G/K$ (complex semisimple Lie group G, over a maximal compact subgroup $K \subset G$), modulo a discrete cocompact subgroup $\Gamma \subset G$ (so the fundamental region $\Gamma \backslash \mathbb{H} \simeq M$ becomes compact [**Va**]).

In the first case (Laplacians on compact symmetric spaces) $\mathrm{spec}(\Delta_M)$ is well known. Eigenvalues are labeled by certain lattice points k (so-called *highest weights*),

$$\lambda_k = (k + \rho)^2 - \rho^2, \quad \text{where } \rho = \tfrac{1}{2} \sum \text{"positive roots"},$$

The lattice $\{k\}$ is naturally identified with the dual lattice Λ of a flat primitive torus $\gamma \simeq \mathbb{T}^r \subset M$, while each $k \in \Lambda$ defines a k-fold cover of γ. Parameter ρ could be linked to multiplicities of $\{\lambda_k\}$, and the dimension of space $\mathscr{O}_r$, of all flat r-tori in M. This clearly indicates connection between "flat tori" and the spectrum of Δ_M.

For hyperbolic spaces the eigenvalues of Δ_M are hard to compute in any closed form. However, one should be able in principle to uncover their combined contribution by a suitable version of the "Selberg trace-formula", along the lines of (4)–(5). Such results are well known in many rank-1 cases [**Sel, Hej, Mc, Vo**]. There are also some higher-rank versions of the "Selberg-trace," that seem to indicate a contribution of "flat tori" to $\mathrm{spec}(\Delta_M)$ [**HST**]. But to our knowledge this part has not yet been carried out explicitly. To understand these connections better and to extend them to other manifolds would require further developement of the wave-method (cf. [**DKV**]).

In the next sections (§§2–3) we shall turn to the *potential (Schrödinger)* problem in both "rank-one" and "higher-ranks," and demonstrate the effects of geometry: *geodesics, flat tori* and the related *Radon transforms*, on spectral asymptotics of $H = \Delta + V$. In some cases of rank-one (§2) it leads to a complete solution of the inverse problem and rigidity. In higher-ranks (§3) we are only beginning to develop the appropriate tools.

2. Schrödinger operators on manifolds and the n-sphere theory

Potential V of the Schrödinger operator $H = \Delta + V$ on M can be regarded as a regular (relatively "small") perturbation of the Laplacian. So one expects the closed path $\{\gamma\}$ of §1 to enter spectral asymptotics of H in the form of integrals: $\oint_\gamma V\, ds$. In other words the role of the *length/period spectrum* of the metric problem will be played now by the *Radon transform* of V,

$$(7) \qquad\qquad \widetilde{V}(\gamma) \overset{\text{def}}{=} \oint_\gamma V\, ds.$$

A heuristic explanation of (7) is provided by the Kac-Feynman path-integral representation of the heat-kernel of H,

$$(8) \qquad\qquad e^{t(\Delta - V)}(x\,;x) = \int d\mathscr{B}_t[\gamma] e^{-\oint_\gamma V\, ds}.$$

Integration in (8) extends over of all closed path $\gamma = \gamma_t(x)$, connecting $\{x\}$ to itself in time t, and $d\mathscr{B}_t[\gamma]$ denotes the corresponding Wiener (Brownian) measure [**Mo, AA**]. Formal application of the "stationary phase/steepest descend" method to the path-integral (8) would keep the contribution of *stationary path* (*geodesics*) only, which explains the role of (7). Such considerations suggest a 2-step approach to the inverse potential problem:

(i) derivation of the Radon transform $\tilde{V}$, or certain "functionals of $\tilde{V}$," from spectral asymptotics of H;

(ii) the proper Radon-inversion procedure.

Before we proceed to multi-D problems let us briefly review the classical case of regular Strum-Liouville (S-L) operators: $H = -\partial^2 + V(x)$ on $[0; 1]$, with a suitable boundary condition (2-point; periodic; Floquet, etc.). S-L operators are well known to have simple (multiplicity free) spectrum $\{\lambda_k\}$. The asymptotics of $\{\lambda_k\}$ were found by Borg [**Bo**] (see also [**Le**]),

$$(9) \qquad \lambda_k = (\pi k)^2 + b_0 + k^{-1} b_1 + \cdots, \quad \text{as } k \to \infty,$$

where $b_0 = \int V \, dx$, while other $\{b_j\}$ are expressed in terms of the Fourier coefficients of V. It turns out that the correspondence $V \to \mathrm{spec}(H_V)$ for S-L operators is highly nonunique: *there exist large* (∞-D) *isospectral classes of potentials*, both in the periodic/Floquet [**La, No**], and in the "2-point boundary" case [**IMT, PT**]. So a unique determination of V would require an additional infinite set of data, like the "KdV-flow parameters," "norming constants," etc. These results were extended also to some singular S-L problems [**MT, GR, Gur1**].

Compared to the S-L theory multidimensional Schrödinger operators are much less studied and understood. They exhibit quite different qualitative features. For instance, it is believed that multi-D Schrödinger operators are *spectrally rigid*, in the sense that the only possible isospectral deformations are given by natural geometric symmetries (isometries), rather than hidden "KdV-type symmetries."

The *rigidity hypothesis* proposed by V. Guillemin was confirmed in a number of examples. The foremost is the case of negatively curved manifolds [**GK**], where under some additional assumptions, it was shown that V *is uniquely determined by* $\mathrm{spec}(H_V)$, i.e., $\mathrm{Iso}(V) = \{V\}$.[4] In this respect the potential problem on hyperbolic spaces proved to be more manageable than the metric problem. The main tool of [**GK**] was to show that the Radon transform of V is determined by spectral asymptotics of H_V. This data combined with some specific features of hyperbolic geometry (a discrete and rapidly proliferating with length family of closed geodesics) proved to be sufficient to recover V.

Another well-studied case is the flat torus $\mathbb{T}^n$. Its symmetries consist of translations and reflections. It was shown that *generic* (*real analytic*) *potentials*[5] V *on* $\mathbb{T}^n$ are spectrally rigid [**ERT, MN**]. Here once again the Radon transform and certain 1-D reduced S-L operators played an important role.

---

[4]Negatively curved manifolds have typically no continuous internal (geometric) symmetries.

[5]The assumption of real analyticity seems to be more of technical, than substantial nature. But genericity could not be dropped, as one can easily construct special potentials, like $V = V_1(x_1) + \cdots + V_n(x_n)$, the sum of 1-variable functions, with "large" (infinite-D) isospectral classes, obtained by the KdV-flow hierarchies applied to 1-D constituents $\{V_j(x_j)\}$ of V.

The case of the positively curved (and highly symmetric) n-sphere turned out to be the most difficult. So far only partial results were obtained [**Wei, Gui, Co, Ur, Wi, Gur2–4**]. We shall briefly outline the state of the n-sphere theory. The n-sphere Laplacian has regular distributed and highly degenerate spectrum:

$$\{\lambda_k = k(k+n-1) = (k+\rho)^2 - \rho^2 : k = 0, 1, \ldots\}, \qquad \rho = \frac{n-1}{2},$$

the multiplicity of λ_k rapidly increasing with k, $d_k = \mathscr{O}(k^{n-1})$. The degeneracy results from the underlying rotational symmetry $SO(n+1)$, since each eigensubspace carries an irreducible representation π^k of $SO(n+1)$. Adding potential V breaks the rotational symmetry, so $\mathrm{spec}(H_V)$ splits into clusters of simple (or less degenerate) eigenvalues $\Lambda_k = \{\lambda_{km} = \lambda_k + \mu_{km} : m = 1, \ldots, d_k\}$. The clusters are asymptotically well localized: the cluster size,

$$|\Lambda_k| = \begin{cases} \mathscr{O}(1), & \text{for even/generic } V, \\ \mathscr{O}(k^{-2}), & \text{for odd } V, \end{cases}$$

while the distance between neighboring Λ_k increases in proportion to k [**Gui2**]. So it seems natural to look for spectral invariants, associated with the distribution of spectral shifts within clusters.

To find such invariants Weinstein [**Wei**] introduced a sequence of *cluster-distribution measures*,

$$d\nu_k = \frac{1}{d_k} \sum_m \delta(x - \mu_{km}).$$

He proved that sequence $\{d\nu_k\}$ converges to a continuous measure $\beta_0(\lambda)\, d\lambda$ on $\mathbb{R}$, whose density is equal to the distribution function of (not surprisingly!) the *Radon transform* $\tilde{V}$, $d\nu_k \to \beta_0\, d\lambda$, where distribution β_0 paired to any test function f on $\mathbb{R}$, gives

$$(10) \qquad \langle f; \beta_0 \rangle = \int_{\mathscr{O}} f \circ \tilde{V}\, dS,$$

integration over the space $\mathscr{O}$ of all closed geodesics (great circles)[6] on S^n (Figure 1). Moreover, by analogy with (10) sequence $\{d\nu_k\}$ can be expanded in powers of k^{-1},

$$(11) \qquad d\nu_k \sim \beta_0 + k^{-1}\beta_1 + \cdots$$

whose coefficients $\{\beta_0, \beta_1, \ldots\}$ are certain (Schwartz) distributions on $\mathbb{R}$ depending on V. Weinstein named $\{\beta_k\}$ *band-invariants*, and calculated

---

[6]Let us remark that the space of closed geodesics $\mathscr{O}$ on S^n is very different from the hyperbolic or flat (torus) cases. Instead of a discrete sequence of isolated path $\{\gamma_m\}$ (hyperbolic space), or a discrete union of cells $\mathscr{O}_l = \{\gamma : |\gamma| = l\} \simeq \mathbb{T}^{n-1}$, $l = (\sum m_j^2)^{1/2}$ (torus), the spherical space $\mathscr{O}$ consists of a single "fat" cell $\mathscr{O} \simeq S^*(S^n)/\mathbb{T}$, of $\dim = 2(n-1)$, which has many other nice structures: homogeneous space of $SO(n+1)$, complex projective variety [**Gui, Ur**]. In particular the entire geodesic flow is periodic of period 2π, so S^n belongs to the class of the so-called Zoll manifold.

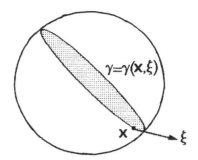

FIGURE 1. Closed geodesics on S^n are great circles $\gamma = \gamma(x; \xi)$ through point x in the direction ξ.

β_0 = distribution function of the Radon transform of V (10). The higher band-invariants turned out much harder to calculate explicitly; only two of them $\{\beta_1; \beta_2\}$ have been found so far [**Ur1-2, Gur4**].

The Weistein result suggested a link to the inverse problem, determination of V (modulo rigid motions) from $\text{spec}(H_V)$, if one could only "invert" the map $\beta_0 \to V$. However, the main difficulty comes here, since the "*distribution density* of $\tilde{V}$" does not determine $\tilde{V}$ itself (even for 1-variable functions!), whereas the S^n-Radon transform in general depends on $2(n-1)$ parameters. Of course, if one knew $\tilde{V}$ (rather than the distribution of its values), the inverse problem would be easily solved by the Radon inversion. Nevertheless, Guillemin [**Gui1**] successfully applied β_0 along with the classical heat-invariants to prove spectral rigidity for special classes of potentials: low-degree spherical harmonics on S^2. In our recent works [**Gur2-4**] we studied a different class, so-called *zonal (axisymmetric) potentials* V, i.e., functions $\{V\}$ invariant under the group $\text{SO}(n)$ of rotations about the "north pole," so $V = V(\theta)$ depends only on the angle θ between point on the sphere and the vertical axis (see Figure 2 on next page). The corresponding Schrödinger operators possess an additional symmetry: on S^2 they commute with z-component of the angular momentum operator $\mathcal{M} = i\partial_\theta$, while on S^n the role of $\mathcal{M}$ being played by the entire Lie algebra $\text{so}(n)$. In this setup one can study the *joint spectrum* of H and $\mathcal{M}$. In other words spectral shifts $\{\mu_{km}\}$ of the kth cluster acquire an additional (bigraded) structure, with index m labeling the angular momentum of the kth eigenfunction ψ_{km},

(12) $$H[\psi_{km}] = (\lambda_k + \mu_{km})\psi_{km}, \qquad \mathcal{M}[\psi_{km}] = m\psi_{km}.$$

In this context we made a significant improvement over the Weinstein's result (10), by replacing asymptotics of cluster-distribution measures $\{d\nu_k\}$ by asymptotics of individual spectral shifts $\{\mu_{km}\}$. The main result of [**Gur4**] states

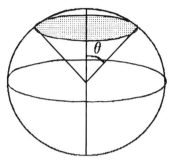

FIGURE 2. Zonal axisymmetric potentials depend only
on the angle between the point on the sphere and the
vertical axis.

THEOREM 1. *Spectral shifts* $\{\mu_{km}\}$ *of the joint* H, $\mathcal{M}$-*eigenvalue problem*
(12) *admit an asymptotic expansion, similar to* (11)

$$(13) \qquad \mu_{km} \sim a\left(\frac{m}{k}\right) + k^{-1}b\left(\frac{m}{k}\right) + k^{-2}c\left(\frac{m}{k}\right) + \cdots,$$

with function-coefficients $a(x)$, $b(x)$, $c(x)$, *depending on* $V(x)$. *Coefficients*
$\{a, b, c\}$ *are compuated explicitly in terms of certain transforms on* $[0; 1]$-
functions: the reduced Radon transform[7] $\mathfrak{R}$,

$$\mathfrak{R}: f(x) \to \tilde{f}(r) = \frac{2}{\pi}\int_0^{\sqrt{1-r^2}} f(x)\frac{dx}{\sqrt{1-r^2-x^2}};$$

the Gegenbauer-Legendre operator (*zonally reduced Laplacian*), *and the Euler
operator* $\mathfrak{E} = x\frac{d}{dx}$, *applied to the even and odd parts of potential*: $V = V_{\mathrm{ev}} + V_{\mathrm{od}}$.

In special cases of even [**Gur2**] and odd [**Gur3**] zonal potentials V on S^2,
we get

$$\mu_{km} = \begin{cases} \tilde{V}\left(\frac{m}{k}\right) + \mathscr{O}(k^{-1}), & \text{even } V, \\ k^{-2}U\left(\frac{m}{k}\right) + \mathscr{O}(k^{-3}), & \text{odd } V, \end{cases}$$

where function $U(x)$ is obtained from V_{od}^2 by transform

$$\frac{1}{4}\left(\mathfrak{R} - \frac{r^2}{1-r^2}\mathfrak{R}\mathfrak{E}\right).$$

As a corollary of Theorem 1 we obtain a unique and explicit solution
of the inverse problem for the joint $(H, \mathcal{M})$-spectrum, and establish *local
spectral rigidity* for generic zonal potential on S^n, which provides a partial
answer to the Guillemin's rigidity hypothesis. We have also derived from

---

[7]Let us remark that zonal functions $\{f(x)\}$, as well as their Radon transforms $\{\tilde{f}(r)\}$,
depend on a single variable. Variable $x \in [-1; 1]$ of f runs along the symmetry axis, while
variable r of $\tilde{f}$ measures the x-coordinate of the "north pole" of the geodesics (great circle)
γ, i.e., $\tilde{f}(\gamma) = \tilde{f}(r(\gamma))$ (Figure 2).

(13) some explicit formulae of the first three Weinstein band-invariants for zonal potentials.

To give the reader the feeling of the techniques and ideas employed in the n-sphere potential theory, let us briefly outline the derivation of the Weinstein's band-invariant β_0. It is convenient to replace differential operators H and Δ with square roots, ψdo's

$$A = \sqrt{\Delta + \rho^2} - \rho, \quad \text{and} \quad \sqrt{H} = A + B, \quad \rho = \frac{n-1}{2},$$

where ψdo$B \approx \frac{1}{2}A^{-1}V - \frac{1}{4}A^{-2}[A;V] + \cdots$ (see [**Gur4**]). The key idea of [**Wei**] was to average perturbation B of A and to get an operator $\overline{B}$ that would commute with A; hence both A and $\overline{B}$ could be simultaneously diagonalized.

The Weinstein averaging procedure consisted in conjugating perturbation B with the unitary group generated by A $\{e^{itA}\}$, and exploiting 2π-periodicity of $U_t = e^{itA}$, (spec $A = \{0, 1, 2, \ldots\}$!). So one takes conjugates,

$$B(t) = e^{-itA}Be^{itA},$$

and average them over the period of A,

(14)
$$\overline{B} = \frac{1}{2\pi}\int_0^{2\pi} B(t)\,dt.$$

The resulting operator $\overline{B}$ commutes with A, and one can show that $A + B$ and $A + \overline{B}$ are "almost unitarily equivalent." Namely,

$$U^{-1}(A + B)U = A + \overline{B} + R,$$

with the remainder R estimated in terms of A,

$$|R| = (RR^*)^{1/2} \leq \text{Const}\,A^{-3}.$$

As a consequence, spectral shifts $\{\mu_{km}\}$ of $A + B$ can be approximated by the shifts $\{\overline{\mu}_{km}\}$ of $A + \overline{B}$ ("eigenvalues of $\overline{B}$"),

$$|\mu_{km} - \overline{\mu}_{km}| \leq \text{Const}\,k^{-3}.$$

The asymptotic distribution of the latter is then computed using the standard symbolic calculus on S^n (Szegö limit theorem [**Gui**]), which yields the Weinstein invariant β_0. Namely, pairing a test function f on $\mathbb{R}$ against cluster-distribution measures $\{d\nu_k\}$, and passing to the limit $k \to \infty$, we get

$$\langle f; d\nu_k \rangle = \frac{1}{d_k}\sum_m f(\mu_{km}) \approx \frac{\text{tr}[P_k f(\overline{B})P_k]}{\text{rk}\,P_k}\int_{S^*(S^n)} (f \circ \text{symbol}\,\overline{B})\,dS.$$

By the standard symbolic calculus on S^n symbol of the conjugate operator (ψdo) $B(t)$ is obtained by composing $\sigma_B = V(x)/2|\xi|$ with the hamiltonian (geodesic) flow, $\exp(t\Xi)$, of "symbol A" $= |\xi|$. Here Ξ denotes the hamiltonian vector field of "symbol A". This yields symbol of the average operator $\overline{B}$,

$$\sigma_{\overline{B}} = \frac{1}{2\pi} \int_0^{2\pi} \sigma_B \circ \exp(t\Xi)\, dt,$$

and the latter is nothing but the Radon transform $\widetilde{V}(\gamma)$, evaluated on a closed path (great circle) $\gamma = \gamma_{x\xi}$, passing through $x \in S^n$ in the direction $\xi \in T_x^*$. Thus we get the first band-invariant β_0 in terms of the Radon transform $\widetilde{V}$.

Derivation of higher band-invariants [Ur], and the proof of Theorem 1 [Gur4] are more involved. The procedure exploits more elaborate *averaging* (cf. [Gui, Ur, Gur2-4]); a form of *complete symbolic calculus* on S^n, developed by A. Uribe [Ur] (based on the representation theory of $SO(n)$, the *Wick-Berezin symbols* and *coherent states* on S^n); and the author's *zonal reduction* (see [Gur4] for details).

Let us remark that the techniques of the n-sphere Schrödinger theory rely heavily on *spherical symmetries*, and the 2π-periodicity of both the classical (geodesic flow), and the quantum operator $A = \sqrt{-\Delta}$ (*periodic (integral) spectrum!*). Both features, symmetry and periodicity, are present in any rank-1 symmetric space, so most of the above results extend to rank-1 spaces, as well as more general *Zoll manifolds* [Wei, Ku].

However, the higher rank spaces pose a very different situation. They lack *periodicity*, both on quantum and classical level: eigenvalues of $\sqrt{\Delta}$ are no more integers (as demonstrated by simple torus $\mathbb{T}^r$), and the geodesic flow $\exp t\Xi$ is no more periodic! Although one still has a fair number of closed paths in $T^*(M)$, the dynamics is more likely to be dominated by aperiodic γ (irrational slopes!), which densely "fill in" geodesically flat tori $\gamma \simeq \mathbb{T}^r$, where $r = \operatorname{rank} M$.

The aperiodicity of A and of the flow $\{\exp(t\Xi)\}$ requires a modification of the basic Weinstein averaging method. Namely, the 2π-periodic averages (14) should be replaced by *ergodic averages*,

$$\lim \left\{ \overline{B}_T = \frac{1}{T} \int_0^T B(t)\, dt \right\}, \quad \text{as } T \to \infty.$$

In paper [Gur5] we developed the multiparameter averaging procedure and applied it to anharmonic oscillators on $\mathbb{R}^n$. In the next section we shall adopt this method to Schrödinger operators on symmetric spaces.

3. Schrödinger operators on higher rank symmetric spaces

We recall that symmetric space $M = K\backslash G$ represents a quotient of a compact semisimple Lie group G, modulo a subgroup K, which stabilizes an involutive automorphism θ of G, $K = \{u \in G : \theta(u) = u\}$, $\theta^2 = 1$. Such an automorphism θ splits the Lie algebra $\mathfrak{G}$ of G into the direct (orthogonal) sum of the subalgebra $\mathfrak{K}$ of K, and a subspace $\mathfrak{P}$, identified with the tangent space of M at $x_0 = \{K\}$. Furthermore, $\theta\mathfrak{K} \simeq I$, $\theta\mathfrak{P} \simeq -I$, so the resulting Lie brackets between the $\mathfrak{K}$ and $\mathfrak{P}$ components take the form: $[\mathfrak{K}; \mathfrak{K}] \subset \mathfrak{K}$; $[\mathfrak{K}; \mathfrak{P}] \subset \mathfrak{P}$; $[\mathfrak{P}; \mathfrak{P}] \subset \mathfrak{K}$.

Space $\mathfrak{P}$ contains a maximal abelian (*Cartan*) subalgebras $\mathfrak{T} \simeq \mathbb{R}^r$, whose dimension $r = \dim \mathfrak{T}$ is called the rank of M. The image of $\mathfrak{T}$ under the exponential map $\mathfrak{G} \to G$, forms a maximal geodesically flat torus $\gamma = \exp \mathfrak{T} \simeq \mathbb{T}^r$ in M. The space $\mathscr{O}_r$ of all flat r-tori $\{\gamma\}$ is itself a smooth manifold, whose dimension depends on $\dim G$ and its rank. Group G acts upon $\mathscr{O}_r$, turning it into a homogeneous space $K_0\backslash G$, where K_0 is the stabilizer of $\mathfrak{T}$ in G.

Symmetric spaces were completely classified by Cartan (see [**He**; **Gur6**, Chapter 5]). We shall mention a few examples of so-called *irreducible globally symmetric compact spaces*:

(i) $\mathrm{SU}(n)/\mathrm{SO}(n)$ $\theta: X \to \overline{X}$ (complex conjugation) $r = n - 1$

(ii) $\mathrm{SU}(2n)/\mathrm{Sp}(n)$ $\theta(X) = J_n \overline{X} J_n^{-1}$, where J_n is $r = n - 1$

                   symplectic matrix $\begin{pmatrix} & I \\ -I & \end{pmatrix}$ in $\mathbb{C}^{2n}$

(iii) $\mathrm{SU}(p+q)/$ $\theta(X) = I_{pq} \overline{X} I_{pq}$, where I_{pq} is $r = \min(p; q)$

      $\mathrm{S}(\mathrm{U}(p) \times \mathrm{U}(q))$ the matrix of the indefinite

                   $(p; q)$-form:

$$\sum_1^p x_j y_j - \sum_{p+1}^{p+1} x_j y_j,$$

                   in $\mathbb{C}^{p+q}$;

(iv) $\mathrm{SO}(p+q)/\mathrm{SO}(p)$ $\theta(X) = I_{pq} X I_{pq}$, same I_{pq} as $r = \min(p; q)$

      $\times \mathrm{SO}(q)$ above in $\mathbb{R}^{p+1}$;

(v) $\mathrm{SO}(2n)/\mathrm{U}(n)$ $\theta(X) = J_n X J_n^{-1}$ $r = [\frac{n}{2}]$

(vi) $\mathrm{Sp}(n)/\mathrm{U}(n)$ $\theta(X) = \overline{X} = J_n X J_n^{-1}$ $r = n$

Other (more general) symmetric spaces can be decomposed into products of irreducible components $M_1 \times M_2 \times \cdots$. The simplest example is the product of 2-spheres $(S^2 \times S^2)/\mathbb{Z}_2$, a quotient $\mathrm{SO}(4)/\mathrm{SO}(2) \times \mathrm{SO}(2)$.

Spectra of Laplacians on symmetric spaces are well known:

$$\lambda_k = (k + \rho)^2 - \rho^2,$$

where k varies over the so-called restricted *weight lattice* $\Lambda = \{k = \sum k_j \alpha_j;\ k_j \in \mathbb{Z}_+\}$ in $\mathfrak{T}$, i.e. the lattice spanned by the *basic positive roots* $\{\alpha_j\}$ of $\mathfrak{G}$, that lie in $\mathfrak{T}$. Weight $\rho = \frac{1}{2}\sum \alpha$, the half sum of all positive roots taken with their multiplicities. We shall illustrate the foregoing with 2 examples:

(1) *The n-sphere*: $S^n \simeq \mathrm{SO}(n+1)/\mathrm{SO}(n)$. Here subalgebra $\mathfrak{K} \simeq \mathrm{so}(n)$, subspace $\mathfrak{P}$ consists of matrices:

$$X = X_b = \begin{pmatrix} 0 & b \\ -^\mathrm{T}b & 0 \end{pmatrix}$$

with columnar vector $b \in \mathbb{R}^n$; isotropy subgroup $K = \mathrm{SO}(n)$ acts on $\mathfrak{P}$ by rotations: $u^{-1}X_b u = X_{u(b)}$. Maximal abelian subalgebras $\mathfrak{T} \subset \mathfrak{P}$ are 1-dimensional (rank 1!), and the exponential image of $\mathfrak{T}$ becomes a closed geodesics (great circle γ) in S^n. Lattice Λ consists of integers k, and the basic positive root $\{1\}$ has multiplicity $n - 1$; hence $\rho = \frac{n-1}{2}$.

The kth eigen of the Laplacian Δ_{S^n}, $\lambda_k = k(k+n-1) = (k+\rho)^2 - \rho^2$, and the eigensubspace $\mathscr{H}_k$ (spherical harmonics of degree k) coincides with a π^k-irreducible component of the regular representation R of $G = \mathrm{SO}(n+1)$ on S^n of weight $k \simeq (k; 0; 0; \dots)$.

(2) *Space* $\mathrm{SU}(n)/\mathrm{SO}(n)$. Here $\mathfrak{P}$ is realized by real symmetric matrices (any $\mathrm{su}(n)$ element $Z = X + iY$, $X \in \mathrm{so}(n)$-real orthogonal, $Y \in \mathrm{Sym}_n$). Group $K = \mathrm{SO}(n)$ acts on $\mathfrak{P}$ by conjugation: $X \to u^{-1}Xu$. Cartan subgroup $\mathfrak{T} \subset \mathfrak{P}$ consists of diagonal matrices, $\mathfrak{T} \simeq \mathbb{R}^{n-1}$ (rank $= n - 1$!), and $\exp \mathfrak{T}$ spans a geodesically flat torus $\gamma \simeq \mathbb{T}^{n-1}$ in M. Positive roots in $\mathfrak{T} = \mathfrak{H} \cap \mathfrak{P} \simeq \mathbb{R}^{n-1}$ are of the form

$$H_{ij} = \mathrm{diag}(0; \dots 1; \dots -1; \dots) \quad \text{on the } i\text{th and } j\text{th place},$$

the basis consists of $H_j = \mathrm{diag}(\dots 1; -1; \dots)$; while weights are made of all integer combinations of the basic roots:

$$k = \sum_j k_j H_j = \mathrm{diag}(k_1; k_2 - k_1; \dots; -k_{n-1}).$$

The half sum of positive roots $\frac{1}{2}\rho = \frac{1}{2}(n - 1; n - 3; \dots; -n + 1)$, while the inner product (Killing form),

$$\langle H; H' \rangle = \mathrm{tr}(\mathrm{ad}_H \, \mathrm{ad}_{H'}) = \sum_{i<j}(h_i - h_j)(h'_i - h'_j).$$

So the eigenvalues of Δ,

$$\lambda_k = \sum_{i<j}(k_{ij} + \rho_{ij})^2 - \rho_{ij}^2,$$

where $k_{ij} = (k_i - k_{i-1}) - (k_j - k_{j-1})$; $\rho_{ij} = \rho_i - \rho_j = 2(j - i)$.

Turning to Schrödinger operators on symmetric spaces, our first goal is to find the proper analog of the band-invariant β_0. As above (§2) the spectrum of $H = \Delta + V$ consists of clusters $\{\lambda_k + \mu_{km}\}$. We observe that the spectrum of operator H_V is still made of clusters $\Lambda_\alpha = \{\lambda_\alpha + \mu_{\alpha m} : 1 \le m \le d(\alpha)\}$, that result from splitting of multiple eigenvalues of Δ. But this time label α varies over the lattice points $\alpha \in \Gamma_+$ in the Weyl chamber of $\mathfrak{T}$. Additional complication may arises here as clusters $\{\Lambda_\alpha\}$, corresponding to different weights $\{\alpha\}$, may overlap (see Figure 3).

We assume clusters could be asymptotically separated (at large α) into

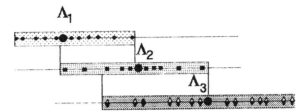

FIGURE 3. For higher rank spaces spectral clusters $\{\Lambda_\alpha\}$ of Schrödinger operators H_V may overlap, since eigenvalues $\{\lambda_\alpha\}$ of Δ do not separate asymptotically.

disjoint subsets[8] of $\mathbb{R}$, and define a sequence of cluster-distribution measures,

$$(15) \qquad d\nu_\alpha = \frac{1}{d(\alpha)} \sum \delta(\lambda - \mu_{\alpha m}).$$

We are interested in the asymptotic distribution of measures $\{d\nu_\alpha\}$, when $\alpha \to \infty$. Guillemin [**Gui3**] studied a similar problem for asymptotic distribution of "normalized" weight diagrams: $\Sigma_\alpha = \{-\alpha \le \beta \le \alpha\}$ of irreducible representations[9] π^α. He introduced a sequence of discrete distribution-measures for normalized weight-diagrams $\Sigma' = \{\frac{\beta}{|\alpha|} : \beta \in \Sigma_\alpha\}$,

$$(16) \qquad d\nu_\alpha = \frac{1}{d(\alpha)} \sum_{\beta \in \Sigma} \delta\left(\lambda - \frac{\beta}{|\alpha|}\right), \quad \text{on } \mathfrak{A}, \alpha \in \Gamma_+$$

and proved the following

SZEGÖ-TYPE THEOREM (GUILLEMIN). *Consider a sequence of normalized weights* $\alpha_k/|\alpha_k| \to \alpha_0$ *in the Cartan subalgebra* $\mathfrak{H} \subset \mathfrak{G}$, *the corresponding representation spaces* $\mathscr{V}_k = \mathscr{V}(\alpha_k) \subset L^2(G)$ *with highest weights* $\{\alpha_k\}$, *and orthogonal projections* $P_k: L^2 \to \mathscr{V}_k$. *Then*

(i) *Sequence of measures* $d\nu_k$ (16) *converges to a continuous distribution* β_0, *supported on the weight-diagram of* α.

(ii) *Distribution* β_0 *coincides with a projection of the natural G-invariant measure* $d_\mathscr{O}$ *on the co-adjoint orbit* $\mathscr{O}_0 = \mathscr{O}(\alpha_0) = \{\mathrm{ad}_g(\alpha_0): g \in G\}$, *through* $\alpha_0 \in \mathfrak{H}$,

$$d\nu_{\alpha_k} \to d\beta(\lambda) = \mathrm{proj}_\mathfrak{H}[d_\mathscr{O}(\cdots)], \quad \text{as } k \to \infty,$$

where $\mathrm{proj}_\mathfrak{H}$ *stands for the natural projection:* $\mathfrak{G} \to \mathfrak{H}$, *determined by the Killing form on Lie algebra* $\mathfrak{G}$.

---

[8]Such splitting could be always accomplished for small potentials V, or adiabatically evolving families $\{V(x; \varepsilon)$, e.g. $\varepsilon V(x)$, with small parameter $\varepsilon\}$.

[9]We recall that any irreducible π^α is determined by a finite set of its weights $\{\beta\}$ (linear functionals on $\mathfrak{H}$), obtained by restricting representation-operators $\{\pi_H^\alpha\}$ on all Cartan elements $\{H \in \mathfrak{H}\}$, $\pi^\alpha|\mathfrak{H} \simeq \bigoplus_\beta \langle\beta|H\rangle$. One of them α (highest) uniquely determines π^α; all other $\{\beta\}$ in the weight diagram Λ_α are squeezed between $-\alpha$ and α, and transformed one into the other by Weyl elements, a finite group generated by all (reflectional) symmetries of the root system of the pair in $\mathfrak{H} \subset \mathfrak{G}$.

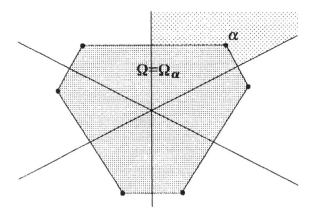

FIGURE 4. This shows a weight diagram of any positive weight α in the Weyl chamber (lightly shaded sector) of Lie algebra $\mathfrak{G} = \mathrm{su}(3)$. The same figure describes restricted weights of symmetric space $\mathrm{SU}(3)/\mathrm{SO}(3)$. The G-orbit of element α in $\mathrm{su}(3)$ projects down onto its weight-diagram (the dark shaded region $\Omega = \Omega_\alpha$), the convex hull of α, reflected by all elements of the Weyl group $\mathbf{W}$ of $\mathrm{su}(3)$.

(iii) *For any* 0*th order pseudo-differential operator* B *on* G, *sequence*

$$\frac{\mathrm{tr}(P_k B P_k)}{d(\alpha)} \to \int_{\mathscr{O}(\alpha)} \mathrm{symb}(B)d_{\mathscr{O}}(x\,;\xi).$$

Let us notice that a G-orbit $\mathscr{O} \subset \mathfrak{G}$, passing through α, projects onto the region in $\mathfrak{H}$, bounded by the weight-diagram Λ_α (shaded region Ω in Figure 4), and gives certain density on Ω. A similar result holds for symmetric spaces $M = G/K$, but this time one takes restricted weights: $\alpha \in \mathfrak{A} \subset \mathfrak{P} \cap \mathfrak{H}$, and a G-orbit $\mathscr{O}$ in the cotangent bundle $T^*(M)$, naturally embedded in $\mathfrak{G}$.

The following simple example illustrates the statement of the Szego-type theorem.

EXAMPLE. Take rank-1 symmetric space $S^2 \simeq \mathrm{SO}(3)/\mathrm{SO}(2)$. Then $\alpha = k$, an integer, the weight-diagram $\{\beta = m\}$ coincides with the interval: $-k \leq m \leq k$, and measures,

$$d\nu_k(x) = \frac{1}{2k+1} \sum_{-k}^{k} \delta\left(x - \frac{m}{k}\right) \to dx = \mathrm{proj}\{dS|_{\mathscr{O}}\},$$

where orbit $\mathscr{O} \simeq S^2 \subset T^*(S^2)$.

We want to establish an analogous result for Schrödinger operators on M. Here a generalized Radon transform of potential V will replace the invariant measure $d\mathscr{O}$. Precisely, we shall consider,

$$\tilde{V}(\gamma) = \int_\gamma V\, d^r S\,;$$

integration over all flat r-tori γ in $\mathscr{M}$, and prove

THEOREM 2. *Let the sequence of normalized weights* $\{\alpha_k\} \subset \mathfrak{A}$ *converge to a point* $\alpha_k/|\alpha_k| \to \alpha_0 \in \mathfrak{A}$. *Then the sequence of cluster-distribution measures*

$$d\nu_{\alpha_k} \to \beta(\lambda)\,d\lambda, \quad as\ k \to \infty.$$

The resulting continuous density $\beta(\lambda) = \beta_0$ *(band-invariant) is equal to the distribution function of the generalized Radon transform* $\widetilde{V}(\gamma)$ *restricted on the* α_0-*orbit*[10] $\mathscr{O}(\alpha_0) \subset T^*(M)$.

The proof follows the basic steps of [**Gur5**], as outlined in §2. However, due to aperiodicity of the problem (spec$A$ and $\exp(t\Xi)$), the Weinstein averages (16) need to be replaced by *ergodic averages* [**Gur5**],

$$\overline{B}(T) = \frac{1}{T}\int_0^T B(t)\,dt, \quad as\ T \to \infty.$$

Let us observe that operators $\{\overline{B}(T)\}$ almost commute with A at large T,

$$[A\,; \overline{B}(T)] = \frac{1}{T}\int iB'(t)\,dt = \frac{1}{T}(B(t) - B(0)) \to 0, \quad as\ T \to \infty.$$

We need to show that "averaged operators" $\{A + \overline{B}(T)\}$ are almost unitarily equivalent to $A + B$, as $T \to \infty$. The equivalence is implemented by antisymmetric operators,

$$Q(T) = \frac{i}{T}\int_0^T \int_0^t B(s)\,ds\,dt.$$

Indeed, A and Q obey the commutation relation

$$[A\,; Q(T)] = \overline{B}(T) - B,$$

which yields via the standard algebra (cf. [**Wei**]) the commutation relation for unitary operators $U = e^Q$,

$$Ae^Q - e^Q A = \overline{B}e^Q - e^Q B + R.$$

The remainder $R = R(T)$, a ψdo of order -3 on M, is bounded by

(17) $$|R| = (RR^*)^{1/2} \leq \mathrm{Const}\,A^{-3},$$

the estimate[11] being independent of T! Consequently, spectral shifts $\{\mu_{km}\}$ in the kth cluster of $A+B$ can be approximated by "average spectral shifts" $\{\mu_{km}(T)\}$ of $A + \overline{B}(T)$,

$$|\mu_{km} - \mu_{km}(T)| \leq \mathrm{Const}\,|k|^{-3}.$$

---

[10]Let us remark an important difference between rank-one and higher ranks: the former have a single G-orbit to cover the entire co-sphere bundle $S^*(M)$, while the latter have continuous families of such orbits.

[11]The "inequality" (17) refers to comparison between two positive selfadjoint operators.

To find large k-asymptotics of cluster-distribution measures $\{d\nu_k\}$, we approximate them by "average measures" $\{d\nu_k^T\}$, and apply the Szegö limit theorem to the latter,

(18)
$$\langle f|d\nu_k\rangle \simeq \langle f|d\nu_k^T\rangle = \frac{\mathrm{tr}(P_k f \circ \overline{B}(T)P_k)}{\mathrm{tr}\, P_k}$$
$$\to \lim_{T\to\infty} \frac{1}{T}\int_0^T dt \left(\int_{S^*(M)} f \circ \sigma_{\overline{B}(T)}(x;\xi)\, dS\right).$$

As before, P_k denotes the kth spectral projection of operator A (or Δ), f is an arbitrary test-function on $\mathbb{R}$, $\sigma_{\overline{B}}$ stands for the principal symbol of the ψdo $\overline{B}$ on $T^*(M)$. Inner integration in (18) extends over the unit cosphere bundle of M. Remembering that $\sigma_{\overline{B}}(x;\xi) = V \circ \exp t\Xi(x;\xi)$, and implementing ergodic averaging in (18), we obtain the band-invariant β_0 in terms of the *generalized Radon transform* of V,

(19) $\tilde{V}(\gamma) = \displaystyle\int_\gamma V\, d^r S$ (integral of V over the geodesically flat γ).

Indeed, almost all directions $(x;\xi)$ in $S^*(M)$ are ergodic with respect to $\exp(t\Xi)$! The rest of the argument is completed using Guillemin's (Szego-type) theorem.

As a consequence of Theorem 2 we could pass from cluster-distributions over $\{\Lambda_\alpha\}$ to "superclusters" $\bigcup_\alpha \Lambda_\alpha$, made up of all weights $\{\alpha\}$, of the same eigenvalue $\lambda = \lambda_\alpha$ of the Laplacian.

COROLLARY. *The limiting distribution (band-invariant) β_0 of spectral clusters $\{\mu_{km}\}$ of the Schrödinger operator $H = \Delta + V$ on a symmetric space M is equal to the distribution density of the generalized Radon transform of V; i.e.,*

$$\langle f|d\nu_k\rangle \to \int_{\mathscr{O}_r} (f \circ \tilde{V})(\gamma)\, dS(\gamma),$$

for any nice test-function f on $\mathbb{R}$.

COMMENTS. Theorem 2 represents a first step in the program of extending the n-sphere spectral theory to higher rank spaces. Next steps include derivation of higher band-invariants $\{\beta_j\}$ (cf. [Ur]), and explicit solution of the inverse problem for *zonal K-invariant*) *potentials* on $M = K\backslash G$, along the lines of [Gur2-4]. Specifically, we shall mention 2 problems:

1. To study "spectral contribution" of flat tori in the metric problem (Laplacians on Riemannian $\mathscr{M}$). We conjecture that any torus γ yields an "approximate lattice" $(\Lambda = \Lambda(\gamma))$ of eigenvalues:

$$\lambda_k \simeq (k+\rho)^2 - \rho, \qquad k \in \Lambda(\gamma),\ \rho(?)\ \text{are "Poincaré numbers/Jacobian,"}$$

by analogy with the well-known "almost arithmetic sequence" $\sqrt{\lambda_k} \simeq \frac{2\pi}{L}k + \ldots$, for a closed path (geodesic) of the length/period L. Furthermore, there should be an analog of the wave-trace formula (5) in higher ranks. The natural examples of such spaces are locally symmetric hyperbolic spaces $\mathcal{M} = \mathbb{H}/\Gamma$ (hyperbolic manifolds $\mathbb{H} = G/K$, modulo discrete $\Gamma \subset G$). Here trace formula (5) should be closely connected to a higher-rank version of "Selberg trace formula." Some partial results along these lines are due to Voros [Vo].

2. Extend our results on the "potential problem" to higher band-invariants along the lines of [Ur, Gur2-4]. This could include among the study of zonal potentials, inverse problems and rigidity on compact symmetric spaces. This work is in progress now.

Acknowledgment

We would like to acknowledge stimulating discussions with A. Uribe on the subject of the paper, who pointed to us to a number of important references.

References

[AA] S. Albeverio and T. Arede, *The relation between quantum mechanics and classical mechanics: a survey of some mathematical aspects*, Chaotic Behavior in Quantum Systems (G. Casati, ed.), Plenum Press, New York, 1985, pp. 37–76.

[BT] M. Berry and M. Tabor, *Closed orbits and the regular bound spectrum*, Proc. Roy. Soc. London Ser. A. **349** (1976), 101–123.

[Bo] G. Borg, *Umkerhrung der Sturm-Liouvillischen Eigebnvertanfgabe Bestimung der differentialgleichung die Eigenverte*, Acta Math. **78** (1946), 1–96.

[Ch] J. Chazarain, *Formule de Poisson pour les varietes Riemanniens*, Invent. Math. **24** (1974), 65–82.

[CY] A. Chang and P. Yang, *Isospectral conformal metrics on 3-manifolds*, Preprint (1989).

[Co1] Y. Colin deVerdiere, *Spectre conjoint d'operatuers pseudodifferentials qui commutent*, Math. Z. **171** (1980), 51–73.

[Co2] ___, *Sue les longuers des trajectoires periodic d'un billiard*, Sem. Sud-Rhodanien de Geometrie **111** (1983), 122–140.

[Co3] ___, *Quasi-modes sur les varietes riemanniennes*, Invent. Math., **43** (1977), 15–52.

[Do] H. Donnelly, *On the wave equation asymptotics of a compact negatively curved surface*, Invent. Math. **45** (1978), 115–137.

[DG] J. Duistermaat and V. Guillemin, *The spectrum of positive elliptic operators and periodic bicharacteristics*, Invent. Math. **29** (1975), 39–79.

[DKV1] J. Duistermaat, J. Kolk and V. Varadarajan, *Spectra of compact locally symmetric manifolds of negative curvature*, Invent. Math. **52** (1979), 27–93.

[DKV2] ___, *Functions, flows and oscillatory integrals on flag manifolds and conjugacy classes of real semisimple Lie groups*, Compostio. Math. **49** (1983), 309–398.

[ERT] G. Eskin, J. Ralston and E. Trubowitz, *The multidimensional inverse spectral problem with a periodic potential*, Contemp. Math., no. 27, Amer. Math. Soc., Providence, RI, 1984.

[Gi1] P. B. Gilkey, *Invariance theory, the heat equation and Atiyah-Singer index theorem*, Publish or Perish, Boston, 1985.

[Gi2] ___, *The spectral geometry of a Riemannian manifold*, J. Differential Geom. **10** (1975), 601–618.

[GW] C. Gordon and E. Wilson, *Isospectral deformations of compact solvemanifolds*, J. Differential Geom. **19** (1984), 241–256.

[GR] J. C. Guillot and J. V. Ralston, *Inverse spectral theory for a singular Sturm-Liouville problems on* [0; 1], J. Differential Equations **76** (1988), 353–373.

[Gui1] V. Guillemin, *Spectral theory on S^2: some open questions*, Adv. in Math. **42** (1981), 283–290.

[Gui2] ___, *Some spectral results for the Laplacian on the n-sphere*, Adv. in Math. **27** (1978), 273–286.

[Gui3] ____, *A Szego-type theorem for symmetric spaces*, Seminar on Micro-local Anal., Ann. of Math. Stud., vol. 93, Princeton Univ. Press, Princeton, NJ, 1979.

[Gui4] ____, *Symplectic spinors and partial differential equations*, CNRS Colloq. Internat. Geom. Symplect. Phys. Math., 1974, pp. 217-252.

[GK] V. Guillemin and D. Kazdan, *Some inverse spectral results for negatively curved manifolds*, Proc. Sympos. Pure Math., vol. 36, Amer. Math. Soc., Providence, RI, 1980, pp. 153-180.

[Gut1] M. Gutzwiller, *Periodic orbits and classical quantization conditions*, J. Math. Phys. **12** (1971), 343-358.

[Gut2] M. Gutzwiller, *Chaos in classical and quantum mechanics*, Springer, 1990

[Gur1] D. Gurarie, *Asymptotic inverse spectral problem for anharmonic oscillators*, Comm. Math. Phys. **112** (1987), 491-502.

[Gur2] ____, *Inverse spectral problem for the 2-sphere Schrödinger operators with zonal potentials*, Lett. Math. Phys. **16** (1988), 313-323.

[Gur3] ____, *Two-sphere Schrödinger operators with odd potentials*, Inverse Problems 6, (1990), 371-378.

[Gur4] ____, *Zonal Schrödinger operators on the n-sphere: Inverse spectral problem and rigidity*, Comm. Math. Phys. **131** (1990), 571-603.

[Gur5] ____, *Multiparameter averaging and inverse spectral problem for anharmonic oscillators in $\mathbb{R}^n$*, Comput. Math. Appl. **22** (1991), 81-92.

[Gur6] ____, *Symmetries and Laplacians: Introduction to harmonic analysis, group representations and applications*, North Holland Math. Stud. **174** North-Holland, 1992.

[He] S. Helgason, *Differential geometry and symmetric spaces*, Academic Press, 1962.

[Hej] D. A. Hejhal, *The Selberg trace formula for* PSL(2 ; $\mathbb{R}$) , Lecture Notes in Math., vol. 548, Springer-Verlag, 1976.

[HST] D. Hejhal, P. Sarnak, and A. Terras (Editors), *The Selberg trace formula and related topics*, Contemp. Math., no. 53, Amer. Math. Soc., Providence, RI, 1986.

[Hö1] L. Hörmander, *The spectral function of an elliptic operator*, Acta Math. **121** (1968), 193-218.

[Hö2] ____, *On the asymptotic distribution of eigenvalues of pseudodifferential operators*, Ark. Math. **17** (1979), 297-313.

[IMT] E. Isaacson, H. P. McKean, and E. Trubowitz, *The inverse Sturm-Liouville problem*. I, II, Comm. Pure Appl. Math. **36** (1983), 767-783; **37** (1984), 1-11.

[Iv] V. Ivrii, *Second term of the spectral asymptotic expansion of the Laplace-Beltrami operator on manifolds with boundary*, Functional Anal. Appl. **14** (1980), 98-106.

[Ka] M. Kac, *Can one hear the shape of the drum?* Amer. Math. Monthly **73** (1966), 1-23.

[Ke] J. B. Keller, *Corrected Bohr-Sommerfeld quantum conditions for nonseparable systems*, Ann. of Phys. **4** (1985), 180-188.

[KR] J. Keller and Rubinow, *Asymptotic solution of eigenvalue problems*, Ann. of Phys. **9** (1960), 24-75.

[Ku1] R. Kubovara, *Band asymptotics of eigenvalues for the Zoll manifold*, Lett. Math. Phys. **16** (1988).

[Ku2] ____, *Spectrum of the laplacian on vector bundles over $C_{2\pi}$-manifolds*, J. Differential Geom. **27** (1988), 241-258.

[La] P. Lax, *Periodic solutions of the KdV problem*, Comm. Pure Appl. Math. **28** (1975), 141-188.

[Le] B. M. Levitan, *The inverse Sturm-Liouville problem*, "Nauka", Moscow, 1984. (Russian)

[MF] V. Maslov and M. Fedoryuk, *Quasiclassical approximations for the equations of quantum mechanics*, "Nauka", Moscow, 1976. (Russian)

[McS] H. P. McKean and I. M. Singer, *Curvature and the eigenvalues of the Laplacian*, J. Differential Geom. **1** (1967), 43-69.

[MT] H. P. McKean and E. Trubowitz, *The spectral class of the quantum mechanical harmonic oscillator*, Comm. Math. Phys. **82** (1982), 471-495.

[Mc] H. P. McKean, *Selberg trace formula as applied to a compact Riemann surface*, Comm. Pure Appl. Math. **25** (1972), 225-246.

[Me] R. Melrose, *Weyl's conjecture for manifolds with concave boundary*, Proc. Sympos. Pure Math., vol. 36, Amer. Math. Soc., Providence, RI, 1980, pp. 254-274.

[Mo] S. Molchanov, *Brownian motion and Riemannian geometry*, Russian Math. Survey, (1976).

[MN] S. Molchanov and M. Novitski, *On spectral invariants of Schrödinger operators on the torus*, Math. Phys., Funct. Anal., "Naukova Dumka", Kiev, 1986, pp. 34–39.

[MP] S. Munakshisundaram and A. Pleijel, *Some properties of the eigenfunctions of the Laplace operator on Riemannian manifolds*, Canad. J. Math. **1** (1949).

[No] S. P. Novikov, *The periodic KdV problem*, Functional Anal. Appl. **8** (1974), 236–246.

[OPS1] B. Osgood, R. Phillips and P. Sarnak, *Extremals of determinants of Laplacians*, J. Funct. Anal. **80** (1988), 148–211.

[OPS2] ____, *Compact isospectral sets of surfaces*, J. Funct. Anal. **80** (1988), 212–234.

[PT] J. Pöschel and E. Trubowitz, *Inverse spectral theory*, Pure Appl. Math., vol. 130, Academic Press, 1987.

[Se] R. Seeley, *An estimate near the boundary for the spectral function of the Laplace operator*, Amer. J. Math. **102** (1980), 869–902.

[Sel] A. Selberg, *Harmonic analysis and discontinuous groups in weakly symmetric Riemannian spaces with applications to Dirichlet series*, J. Indian Math. Soc. (N.S.) **20** (1956), 47–87.

[Sm] L. Smith, *The asymptotics of the heat equation for a boundary value problem*, Invent. Math. **63** (1981), 467–493.

[Ur1] A. Uribe, *A symbol calculus for a class of pseudodifferential operators on S^n*, J. Funct. Anal. **59** (1984), 535–556.

[Ur2] ____, *Band invariants and closed trajectories on S^n*, Adv. in Math. **58**, (1985), 285–299.

[Vo1] A. Voros, *The WKB-Maslov method for nonseparable systems*, CNRS Colloq. Internat. Geom. Symplect. Phys. Math. **237** (1974), 277–287.

[Vo2] ____, *Unstable periodic orbits and semiclassical quantization*, J. Phys. A (1989).

[Va] V. S. Varadarajan, *The eigenvalue problem on negatively curved compact locally symmetric manifolds*, Contemp. Math., no. 53, Amer. Math. Soc., Providence, RI, 1986, pp. 449–461.

[Vi] M. Vigneras, *Varietes riemanniennes isospectrales et non isometriques*, Ann. of Math. (2) **112** (1980), 21–32.

[Wei] A. Weinstein, *Asymptotics of eigenvalue clusters for the Laplacian plus a potential*, Duke Math. J. **44** (1977), 883–892.

[We1] H. Weyl, *Über die asymptotische verteilung der Eigenwerte*, Gott. Nach. (1911), 110–117.

[We2] ____, *Ramifications, old and new, of the eigenvalue problem*, Bull. Amer. Math. Soc. **56** (1950), 115–139.

[Wi] H. Widom, *Szego's theorem and a complete symbolic calculus for pseudo-differential operators*, Seminar on Singularities of Solutions of Linear Differential Equations, Ann. of Math. Stud., vol. 91, Princeton Univ. Press, Princeton, NJ, 1979.

CASE WESTERN RESERVE UNIVERSITY

Proceedings of Symposia in Pure Mathematics
Volume **54** (1993), Part 3

Ellipticity of Local Isometric Embeddings

CHONG-KYU HAN

Introduction

Let Ω be an open neighborhood of the origin in R^n and let $g = g_{ij}(x)$ be a Riemanian metric on Ω. We consider a system of partial differential equations:

$$(1) \qquad \sum_{\alpha=1}^{n+1} \frac{\partial u^\alpha}{\partial x_i} \frac{\partial u^\alpha}{\partial x_j} = g_{ij}, \qquad i, j = 1, \dots, n.$$

for a system of real-valued functions $u = (u^1, \dots, u^N)$. A C^1 solution of (1) is an isometric immersion, thus locally an isometric embedding of (Ω, g) into R^N. Since $g_{ij} = g_{ji}$, the number of equations in (1) is $n(n+1)/2$. The Cartan-Janet theorem states that a real analytic Riemannian manifold of dimension n always has an analytic local isometric embedding into $R^{n(n+1)/2}$; that is, if g_{ij} are real analytic the system (1) with $N = n(n+1)/2$ has a real analytic solution, in a possibly smaller neighborhood of the origin.

The existence of solutions of (1) in nonanalytic cases has been studied by Nash [10] and many other authors including Greene [3], Jacobowitz [8], Bryant-Griffiths-Yang [2] and very recently by Günther (cf. [14]).

A solution of (1) is not unique. If u is a solution of (1), then a composition $\Phi \circ u$ with any rigid motion Φ of R^N is another. If these are the only possibilities, u is said to be *rigid*. More precisely, an isometric immersion $u : (\Omega, g) \to R^N$ is *locally rigid* at a point $O \in \Omega$ if for any open neighborhood $U_1 \subseteq \Omega$ of O there exists an open set U_2 with $O \in U_2 \subseteq U_1$ having the following property : if u' is any isometric immersion of U_2 into R^N, there

---

1991 *Mathematics Subject Classification.* Primary 53C42.
Research supported by grant P90004 of POSTECH Research Fund.
This paper is in final form and no reversion will be submitted for publication elsewhere.

exists a unique isometry Φ of R^N such that $u' = \Phi \circ u$. Global rigidity is defined in the obvious manner (cf. [13]).

We are concerned with the problem of determining whether (1) is equivalent to a system of elliptic equations.

The notion of ellipticity of isometric embeddings was first introduced by N. Tanaka [13]. He defined an isometric immersion of (Ω, g) into R^N to be elliptic if the second fundamental form corresponding to each normal has at least two nonzero eigenvalues of the same sign. Associated with an isometric embedding u he also defined a linear partial differential operator L, which is the operator of infinitesimal isometric deformations of u. Then u is elliptic if and only if L is elliptic as a linear partial differential operator. In the case $n = 2$ and $N = 3$, L is elliptic if the gaussian curvature $K > 0$, and hyperbolic if $K < 0$. He also defined in [9] an isometric immersion to be of *finite type* if the associated operator L has a symbol whose third prolongation vanishes.

If an isometric immersion u is of finite type, then u is elliptic. If u is of finite type and u' is an isometric immersion sufficiently close to u in the C^2 topology, then there exists a unique rigid motion Φ of R^N such that $u' = \Phi \circ u$ [13]. If u is an elliptic isometric immersion of a compact Riemannian manifold M into R^N and is infinitesimally rigid, then u is globally rigid [9].

In this note we present results on the prolongation of the isometric embedding equation (1) to an elliptic system. The motivation for this problem was the question of regularity of isometric embeddings and the analogous problems in CR geometry (cf. [4]).

A key observation is that the equation of Gauss can be obtained by differentiating (1) twice and eliminating the third-order terms through an algebraic process.

Computing the principal symbol of the linearized Gauss equation (cf. [5]), we can prove

THEOREM 1. *Let* Ω *be an open subset in* R^n, $n \geq 2$, *and let* $g = (g_{ij})$ *be a* C^∞ *Riemannian metric on* Ω. *Suppose that* $u : \Omega \to R^N$ *is a* C^2 *solution of* (1) *which is elliptic in Tanaka's sense. Then* (1) *can be prolonged to a system of second order which is elliptic at* u.

We also prove

THEOREM 2. *Let* (Ω, g) *be as in Theorem 1. Let* $n \geq 2$ *and* $N = n + 1$. *Suppose that all the sectional curvatures of* (Ω, g) *are positive. Then* (1) *can be prolonged to an elliptic system of second order.*

If $u : \Omega \to R^N$ is a rigid immersion, then an infinitesimal deformation of u must come from an infinitesimal isometry of R^N. Using this fact we can prove

THEOREM 3. *Let* (Ω, g) *be as in Theorem* 1. *Suppose that* $u : \Omega \to R^N$ *is a* C^3 *solution of* (1) *which is rigid at* $O \in M$. *Then* (1) *can be prolonged to a system of third order which is elliptic at* u.

We present the idea of the proofs below.

1. Invariants of the imbedding equation

Consider the following operations on the right side of (1):

(1) multiplication by a C^∞ function,
(2) sum,
(3) product,
(4) inversion for nonzero expressions,
(5) $\sqrt[n]{}$,
(6) partial differentiation.

By an invariant we shall mean the expression obtained by a composite of operations (1)–(6) on the g_{ij}'s. Such an expression is certainly invariant under changes of isometric embedding u.

The set of all invariants can be graded as follows: let I_k be the set of invariants involving the partial derivatives of g_{ij} of order k and not those of order $\geq k+1$. The set I_k will be called the set of invariants of order k. For each invariant a of order k we get a partial differential equation

$$(2) \qquad L_a(x, D^\alpha u : |\alpha| \leq k+1) = a(x),$$

where L_a is the expression obtained from the left sides of (1) by the same process that gave $a(x)$ from the right sides of (1).

DEFINITION 1.1. *The function* $a \in I_k$ *is said to be a proper invariant of order* k *if* L_a *in* (2) *involves only the partial derivatives up to order* k.

As an example we will show that the Gaussian curvature is a proper invariant of (1) in the case $n = 2$ and $N = 3$.

PROPOSITION 1.2. *Let* Ω *be an open set in* $\mathbb{R}^2$ *and* $g = (g_{ij})$, $i, j = 1, 2$, *be a* C^∞ *Riemannian metric on* Ω. *Writing* E, F *and* G *for* g_{11}, g_{12} *and* g_{22}, *respectively*,

$$(3) \qquad \frac{1}{H} \left\{ \left(\frac{FE_2 - EG_1}{2HE} \right)_1 + \left(\frac{2EF_1 - FE_1 - EE_2}{2HE} \right)_2 \right\},$$

where $H = (EG - F^2)^{1/2}$ *and the subscripts* 1 *and* 2 *denote the partial derivatives, is a proper invariant of order* 2.

PROOF. (3) is the Gaussian curvature of the metric, which involves the second derivatives of g_{ij}'s. We will show that all the third order derivatives of u cancel out as the left side of (1) passes through the same operations as in (3). All the third order derivatives appear in the following:

$$
\frac{1}{H}\frac{1}{2HE}\left(FE_{21} - EG_{11} + 2EF_{12} - FE_{12} - EE_{22}\right)
$$

$$
= \frac{1}{2H^2}\left(-G_{11} + 2F_{12} - E_{22}\right)
$$

$$
= \frac{1}{2H^2}\{-(2u_1^1 u_{122}^1 + 2u_1^2 u_{122}^2 + 2u_1^3 u_{122}^3)
$$

$$
+2(u_{112}^1 u_2^1 + u_1^1 u_{212}^1 + u_{112}^2 u_2^2 + u_1^2 u_{122}^2 + u_{112}^3 u_2^3 + u_1^3 u_{122}^3)
$$

$$
-(2u_2^1 u_{211}^1 + 2u_2^2 u_{211}^2 + 2u_2^3 u_{211}^3)\}
$$

$$
= 0,
$$

which proves the proposition. □

2. Prolongation to elliptic systems

By a prolongation of a given system of PDE's we mean a new system obtained by differentiating the original equations and combining them by the operations (1)–(6) of §1. A merit of finding proper invariants is that the properties of solutions that are due to the lower order terms of (1) may be revealed by eliminating the principal part of the prolongations. Ellipticity of a system of PDE's is determined by the principal part. Although (1) is not elliptic, under certain conditions on some proper invariants one can prolong it to an elliptic system of order ≥ 2.

For instance, a new system of PDE's consisting of (1) differentiated once together with the new equations given by the sectional curvatures, which are proper invariants of order 2, forms an elliptic system of order 2. We will sketch the proof of Theorem 2 for the case $n = 2$. Namely, we prove

PROPOSITION 2.1. *Let Ω be an open neighborhood of the origin of $\mathbb{R}^2$ and let $g = (g_{ij})$, $i, j = 1, 2$, be a C^∞ Riemann metric of Ω. Suppose that the curvature of g is positive. Then the imbedding equation (1) can be prolonged to a second order elliptic system.*

PROOF. To find the prolongation $L_K(x, D^\alpha u) = K(x)$ of (1) associated with the Gaussian curvature $K(x)$ we make use of the Gauss equation instead of computing directly. The summation convention will be used for repeated Greek indices. Without loss of generality, we assume that

$$
\frac{\partial u^i}{\partial x_j}(0) = \delta_{ij} \text{ (Kronecker delta)},
$$

and

$$
\frac{\partial u^3}{\partial x_i}(0) = 0, \qquad i, j = 1, 2.
$$

Let (e_1, e_2) be an orthonormal frame such that $e_j(0) = \partial/\partial x_j$, $j = 1, 2$.

Let η^1, η^2, η^3 be functions of the arguments $(e_i u^j)$ so that the matrix

$$M \equiv \begin{bmatrix} e_1 u^1 & e_1 u^2 & e_1 u^3 \\ e_2 u^1 & e_2 u^2 & e_2 u^3 \\ \eta^1 & \eta^2 & \eta^3 \end{bmatrix}$$

belongs to $SO(3)$. Then

(4) $$\eta^j = (e_j u^3)B^j + (e_\lambda u^3)\zeta^\lambda, \qquad j = 1, 2,$$

and

$$\eta^3 = (e_1 u^1)(e_2 u^2)B^j + \zeta,$$

where B^j is a C^∞ function in $(x, D^\alpha u^k : |\alpha| \le 1, k \ne 3)$, $B^j = 1$ at $(0, D^\alpha u^k(0))$, and each ζ^λ, ζ are C^∞ functions in $(x, D^\alpha u^k : |\alpha| \le 1, k \ne 3)$ which vanish at $(0, D^\alpha u(0))$.

For each $i, j = 1, 2, 3$, let $A_{ij} = \sum_{\lambda=1}^3 (e_i \eta^\lambda)(e_j \eta^\lambda)$. Substituting (4) for η^λ we get

$$A_{ij} = (e_i e_j u^3)B_j(e_j u^j) + (e_\lambda e_\mu u^\nu)\zeta_\nu^{\lambda\mu}$$

where each $\zeta_\nu^{\lambda\mu}$ are C^∞ functions in $(x, D^\alpha u^k : |\alpha| \le 1)$ and $\zeta_\nu^{\lambda\mu} = 0$ at $(0, D^\alpha u(0))$. Then the Gaussian curvature is $K(x) = A_{11}A_{22} - (A_{12})^2$. In terms of coordinates (x_1, x_2) the above equation becomes

$$\begin{aligned} G(x, D^\alpha u) &\equiv (u_{11}^3 B_1 u_1^1 + u_{\lambda\mu}^\nu \zeta_\nu^{\lambda\mu} + \zeta_1)(u_{22}^3 B_2 u_2^2 + u_{\lambda\mu}^\nu \zeta'^{\lambda\mu}_\nu + \zeta_2) \\ &\quad - (u_{12}^3 B_2 u_2^2 + u_{\lambda\mu}^\nu \zeta''^{\lambda\mu}_\nu + \zeta_{12})^2 - K(x) \\ &= 0 \end{aligned}$$

where B's are 1, ζ's are 0 at $(0, D^\alpha u(0))$.

On the other hand, differentiating (1), we get

$$H_{ij}(x, D^\alpha u) \equiv u_{ii}^1 u_j^1 + u_i^1 u_{ij}^1 + u_{ii}^2 u_j^2 + u_i^2 u_{ij}^2 + u_{ii}^3 u_j^3 + u_i^3 u_{ij}^3 = 0.$$

Let $\tilde{H}_{ij}$ and $\tilde{G}$ be the linearizations at u of H_{ij} and G, respectively. Namely,

$$\tilde{H}_{ij}w = \sum_{|\alpha|\le 2} \frac{\partial H_{ij}}{\partial(D^\alpha u^k)} D^\alpha w^k,$$

$$\tilde{G}w = \sum_{|\alpha|\le 2} \frac{\partial G}{\partial(D^\alpha u^k)} D^\alpha w^k, \quad \text{where } w = (w^1, w^2, w^3).$$

Then the principal symbol $\sigma(x, \xi)$, $x \in \Omega$, $\xi \in \mathbb{R}^3$, of the system of linear partial differential operators $\tilde{H}_{11}$, $\tilde{H}_{12}$, $\tilde{H}_{21}$, $\tilde{H}_{22}$, $\tilde{G}$ is a 5×3 matrix. An easy computation shows that

$$\sigma(0, \xi) = \begin{bmatrix} 2(\xi_1)^2 & 0 & 0 \\ \xi_1\xi_2 & (\xi_1)^2 & 0 \\ (\xi_2)^2 & \xi_1\xi_2 & 0 \\ 0 & 2(\xi_2)^2 & 0 \\ 0 & 0 & * \end{bmatrix},$$

where $*$ is equal to $u^3_{22}(0)(\xi_1)^2 + u^3_{11}(0)(\xi_2)^2 - 2u^3_{12}(0)\xi_1\xi_2$. Since $K(0) = u^3_{11}(0)u^3_{22}(0) - (u^3_{12}(0))^2 > 0$, $*$ never vanishes unless $\xi = 0$. Thus we see that $\sigma(0, \xi)$ is of rank 3, if $\xi \neq 0$, which completes the proof. $\square$

REFERENCES

1. E. Berger, R. Bryant, and P. Griffiths, *The Gauss equations and rigidity of isometric imbeddings*, Duke Math. J. **50** (1983), 803–892.
2. R. Bryant, P. Griffiths, and D. Yang, *Characteristics and existence of isometric imbeddings*, Duke Math. J. **50** (1983), 893–994.
3. R. Greene, *Isometric embedding of Riemannian and pseudo-Riemannian manifolds*, Mem. Amer. Math. Soc. No. 97 (1970).
4. C. K. Han, *Regularity of isometric immersions of positively curved Riemannian manifolds and its analogy with CR geometry*, J. Differential Geom. **28** (1988), 477–484.
5. _____, *Regularity of certain rigid isometric immersions of n-dimensional Riemannian manifolds into* $\mathbb{R}^{n+1}$, Michigan Math. J. **36** (1989), 245–250.
6. _____, *Regularity of mappings of G-structures of Frobenius type*, Proc. Amer. Math. Soc. **105**(1989), 127–137.
7. _____, *The Theorema Egregium of Gauss from a viewpoint of partial differential equations*, Mathematical Heritage of C. F. Gauss (G. M. Rassias, ed.) (1991), 326–333.
8. H. Jacobowitz, *Local isometric embeddings*, Ann. of Math. Stud., vol. 102, Princeton Univ. Press, Princeton, NJ, 1982, pp. 381–393.
9. E. Kaneda and N. Tanaka, *Rigidity for isometric embeddings*, J. Math. Kyoto Univ. **18** (1978), 1–70.
10. J. Nash, *The embedding problem for Riemannian manifolds*, Ann. of Math. (2) **63** (1956), 20–63.
11. _____, C^1 *isometric embeddings*, Ann. of Math. (2) **60** (1954), 383–396.
12. S. Sternberg, *Lectures on differential geometry*, Chelsea, New York, 1983.
13. N. Tanaka, *Rigidity for elliptic isometric imbeddings*, Nagoya Math. J. **51** (1973), 137–160.
14. M. Günther, *Isometric embeddings of Riemannian manifolds*, Proc. Int. Cong. Math. Math. Soc. Japan (1991), 1137–1143.

POHANG INSTITUTE OF SCIENCE AND TECHNOLOGY, SOUTH KOREA

Proceedings of Symposia in Pure Mathematics
Volume **54** (1993), Part 3

Curve Flows on Surfaces and Intersections of Curves

JOEL HASS AND PETER SCOTT

ABSTRACT. New techniques from evolution equations hold promise as a tool in low-dimensional topology. Here a new curve shortening flow is introduced and used to study the intersections of curves on surfaces.

0. Introduction

Various methods of evolving a curve in a manifold have been used over the years, mainly with the problem of establishing the existence of simple closed geodesics in mind. For this purpose, a curve evolution process should (a) not increase the number of self-intersections of a curve, (b) exist for all time or until a curve collapses to a point and (c) shorten curves sufficiently fast so that curves which exist for all time converge to a geodesic. Birkhoff originated what is now known as the *Birkhoff curve flow*, where midpoints of polygonal approximations to a curve are successively connected by geodesic segments [**Bi**]. This type of shortening has the advantage that (b) and (c) are easy to establish, but (a) is not automatic and seems difficult to arrange. Many variations of this process have since been described, among them [**B**, **K**, **J**, **L-S**]. The Birkhoff process is a rough approximation of the evolution undergone by a curve on a surface which is moving so as to shrink its length as fast as possible. Such a deformation corresponds to the curve moving in the direction of its curvature at each point, at a speed proportional to the magnitude of the curvature. A process of this type, which we will refer to as the *curvature flow*, has recently been studied with considerable success [**G**, **G-H**, **Gr1**]. All three conditions can be shown to hold for an embedded curve on a surface evolving by this process. For nonembedded curves, singularities may develop during the curvature flow. These have been studied recently by Angenent [**A**] who has developed a technique to flow through a singularity,

1991 *Mathematics Subject Classification*. Primary 57M25; Secondary 53A10.

The first author was partially supported by National Science Foundation grant DMS-8823009 and the Alfred P. Sloan Foundation; and the second author by National Science Foundation grant DMS-8702519.

The detailed version of this paper will appear elsewhere.

but from the point of view of the applications we develop here, some key questions remain unanswered. See [Gr2] for a discussion of some of the phenomena that can occur for the curvature flow of a singular curve.

In this paper, we describe a new curve shortening flow which is easily shown to satisfy (a), (b) and (c). Our curve flow is in some sense a modification of the Birkhoff process which picks out its piecewise-geodesic structure from the surface rather than from the curve. This flow, called the *disk flow*, is developed in §1. As with the curvature flow, we can use this flow to study the evolution of families of curves. This is carried out in §2, where we establish the existence of a simply geodesic on an arbitrary Riemannian 2-sphere. In §3, we use this flow to solve a purely topological problem concerning curves on two dimensional surfaces. Turaev posed the problem in the following form:

QUESTION 1 [T]. Let s_0 and s_1 be homotopic curves on a surface, each with k double points. Is there a homotopy s_t from s_0 to s_1 with the property that each curve s_t in the homotopy has at most k double points?

The curvature flow can also be used to solve this question in the case where k is the minimal number of double points in the homotopy class of s_0, a fact also observed by M. Shepard [S]. However, when there are excess numbers of double points, the curvature flow does not suffice to solve the problem, whereas the disk flow does.

The solution of this problem is, we feel, an early example of a solution of a topological question by flow type methods. Some concluding remarks are made in §4. Full details will appear in [H-S].

1. A curve shortening flow

We give a simple construction of a curve flow that takes a finite collection of curves on a surface and homotops it in such a way that

1. the number of self-intersection points of each curve is nonincreasing,

2. the number of intersection points between each pair of curves is non-increasing,

3. either a curve disappears in a finite time or it eventually lies arbitrarily close to a geodesic.

This flow seems to have all the benefits of the curvature flow for geometrical applications, but has the advantages of being easy to construct and understand, particularly for singular curves. A precise definition of our flow can be found in [H-S], but we give here a rough description. We will explore the properties and convergence of this flow, and show that, if suitably defined, the number of intersection points is nonincreasing. We first state a preliminary result.

LEMMA 2. *A finite collection of piecewise smooth arcs in a convex disk, with no two arcs having a common boundary point, can be homotoped (rel boundary) to geodesic arcs so that the number of self-intersection points of each arc and the number of self-intersection points between each pair of arcs is nonincreasing during the homotopy.*

At first glance this may seem as hard to solve as Question 1, but the fact that everything takes place in the disk allows one to use induction arguments which make this case much easier.

Now let $\{\gamma\}$ be a piecewise-smooth transversely immersed curve on a Riemannian surface F. Cover F with convex disks $D_1, D_2, \ldots, D_n$ of radius r sufficiently small so that there is a unique geodesic connecting any two points in a disk, this geodesic is contained in the disk, and such that the disks of radius $r/2$ with the same centers also cover F. Define the disks $D_{n+1}, D_{n+2}, \ldots$ so that $D_{n+i} = D_i$. Perturb the disks if necessary so that no two arcs of $\gamma \cap D_i$ have a common boundary point. Homotop each arc of $\gamma \cap D_1$ to the unique geodesic arc with the same endpoints, using the homotopy defined in Question 1. Repeat for $D_2, D_3, \ldots, D_n$. Cycle through the disks again, repeating for each D_i indefinitely, so that γ_t for $t \in [i-1, i]$, $i = 1, 2, \ldots$, is the result of performing the straightening process of Lemma 2 in the disk D_i on $\gamma_{i-1} \cap D_i$. Although it is not necessarily true that the homotopy decreases lengths, it is true that the lengths at integral time periods are nonincreasing. Note that we could choose the values $1 - 1/(i+1)$ and $1 - 1/i$ for the time interval if we want t to take values in $[0, 1]$. If two arcs of $\gamma_j \cap D_i$ have a common boundary point, we first shrink D_i, by a factor of at most $1 - 1/4^i$, so that this no longer holds, and then perform the homotopy of Lemma 2. If γ_{i-1} is completely contained in D_i then we set γ_i to be the empty set.

If γ_0 has transverse self-intersection, small shrinkings of D_i will always suffice to ensure that no two arcs have a common endpoint, since straightened arcs intersect transversely. This shrinking process is necessary in order to avoid a situation where two arcs have common boundary points, and their straightenings overlap along an arc. If we worked with an embedded curve, or counted only transverse intersection points, then we would not need to shrink disks. Figure 1 illustrates two stages in the disk flow of a curve.

A drawback of the flow is that it is not canonical, since the shrinking and the choice of homotopy in Lemma 2 are not uniquely defined. Nonetheless we can establish that it has the following useful properties.

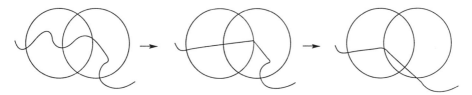

Figure 1

THEOREM 3. *The disk flow applied to a connected curve* γ *has the following properties*:

(1) *The number of self-intersection points of* γ_t *is nonincreasing with* t, $t \in [0, \infty)$.

(2) *Either* γ_t *disappears in finite time, or a subsequence of* $\{\gamma_i\}$ *converges to a geodesic as* $i \to \infty$. *In the second case, if* U *is an open neighborhood of the set of geodesics homotopic to* γ *then there is a* $T > 0$ *such that* γ_t *lies in* U *for* $t > T$.

(3) *If a sequence* $\{\gamma_j\}$ *converges to a geodesic* γ_∞ *as* $i \to \infty$ *then* $\mathrm{length}(\gamma_i) \to \mathrm{length}(\gamma_\infty)$ *as* $i \to \infty$.

(4) *The number of intersection points between* γ_t *and a second flowing curve* γ_t' *is nonincreasing with* t, $t \in [0, \infty)$.

(5) *For each integer* i, $\mathrm{length}(\gamma_{i+1}) \le \mathrm{length}(\gamma_i)$, *with equality if and only if* γ_i *is a geodesic.*

Note that it is always true that a subsequence of $\{\gamma_i\}$ converges to a geodesic, but the entire sequence may possibly oscillate between different geodesics. This same phenomenon is encountered in all of the curve flows. The sequence $\{\gamma_i\}$ does converge smoothly to a geodesic except possibly when there are an infinite number of nonparallel geodesics of uniformly bounded length, each homotopic to γ, where two geodesics are *parallel* if they cobound a flat annulus. The existence of infinitely many such geodesics is nongeneric, and in particular does not hold for metrics of nonpositive curvature.

2. Flowing families of curves

We can extend the disk flow to a family of curves parametrized by the unit interval by applying separately to each one the flow defined in §1. It may not be true now that the number of intersection points between two distinct curves γ_t and γ_t' is nonincreasing with t, as the straightening process of Lemma 1 only applies to finitely many curves. However, it is immediate that the number of intersection points between two curves is non-increasing at integral values of t. Moreover embedded curves stay embedded for all $t > 0$. Thus the chief difficulty of the Birkhoff curve flow does not arise. Instead, it becomes more difficult to extend the flow continuously to families of curves. A problem arises in case a curve in the family is tangent to the boundary of a disk D_i. We then need to fill in a 'gap' in order to maintain a continuous family of curves.

LEMMA 4. *Given a* 1-*parameter family, we can deform the disks* D_i *slightly so that no curve in the family has more than one interior tangency point to* ∂D_i.

The gap is then filled as illustrated in Figure 2.

The new family is again parametrized by the unit interval, after replacing a curve with an interior tangency with a closed interval of curves which fill the

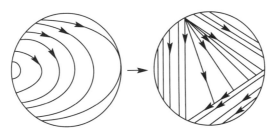

FIGURE 2. An example of a straightened 1-parameter family of curves

gap. This defines a map $f_1 : I \to I$ with $f_1(s_1) = s_0$, where s_1 is the index of any curve which originated with γ_{s_0}. Repeat for $D_2, D_3, \ldots$. A sequence of parameters $S = (s_0, s_1, s_2, \ldots)$ and maps $f_i : I \to I$ with $f_i(s_i) = s_{i-1}$ determines the evolution of a given curve at integral times. The parameter space $(s_0, s_1, s_2, \ldots)$ with the inverse limit topology is homeomorphic to an interval. So we will have a 1-parameter family of curves.

We now show how to establish the existence of a geodesic on a Riemannian 2-sphere, a special case of the Lusternik-Schnirelman theorem.

THEOREM 5. *A Riemannian 2-sphere contains a simple geodesic.*

PROOF. Consider a family of embedded curves, parametrized by I and starting and ending with a point, which together define a degree one map of the 2-sphere. Pick $\delta > 0$ so that any embedded curve of length at most δ lies in one of the disks $\{D_i\}$. Then, if all curves in the family have length less than δ, we can extend the map of the sphere to a continuous map of the ball to the sphere, a contradiction to the assumption that the family represents a degree one map of the 2-sphere. So the length of a maximal length curve in such family can never be less than δ. Apply the disk curve flow to the family described above. We can shrink the curve D_i slightly if necessary, to ensure that $\{\partial D_i\}$ is tangent to any curve in the family in at most one point, and to finitely many curves in the family. The set of parameters of curves that flow to a point and disappear is open, since such a curve eventually lies inside a single disk D_i, and thus so do a 1-parameter family of nearby curves. Let $S = (s_0, s_1, s_2, \ldots)$ be the smallest parameter, ordered lexicographically, which does not disappear. Then γ_S has a subsequence converging to a geodesic.

The existence of three geodesics is much more technical and is discussed in [H-S]. Additional complications arise since a curve in a generic k-parameter families of curves can be tangent to ∂D_i at k points.

3. Double points and homotopies of curves

In this section we apply the disk flow to solve the problem of Turaev mentioned in the introduction. In counting double points it is required that the curve s_t be self-transverse, though not necessarily in general position.

This property is satisfied by curves in a generic homotopy. A k-tuple point is then counted with multiplicity, counting as $k(k-1)/2$ double points. Turaev's question is answered affirmatively through the following two results.

THEOREM 6. *Let s_0 and s_1 be homotopic curves in general position on a surface, each minimizing the number of double points in their common homotopy class. Then there is a homotopy s_t from s_0 to s_1 such that s_t is self-transverse for all t and the number of double points of s_t is constant.*

THEOREM 7. *Let s_0 be a curve in general position on a surface which does not minimize the number of double points in its homotopy class. Then there is a homotopy s_t from s_0 to a curve s_1 which has minimal self-intersection such that the number of double points of the curve s_t is nonincreasing with t. s_t is a regular homotopy except for a finite number of times when a small loop in the curve disappears (see Figure 3).*

COROLLARY 8. *Let s_0 and s_1 be homotopic curves on a surface, each with k double points. There is a homotopy s_t from s_0 to s_1 with the property that each curve s_t has at most k double points.*

PROOF. Applying Theorem 7 to s_0, we can homotop it to a curve s_0' which has minimal self-intersection, without increasing the number of double points. Similarly, we can homotop s_1 to a curve s_1' which has minimal self-intersection without increasing the number of double points. Theorem 6 implies that s_0' and s_1' can be homotoped to one another without increasing the number of double points. Combining the three homotopies proves the corollary.

EXAMPLE 9. The corresponding result for nonconnected curves is false, as illustrated by the pair of homotopic two component curves in Figure 4.

The proofs of Theorems 6 and 7 show that the obstruction to generalizing the result to nonconnected curves is completely illustrated in this example. If

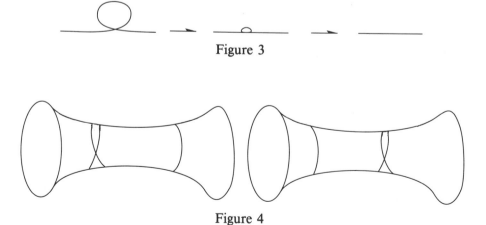

Figure 3

Figure 4

no pair of components are powers of parallel curves, then the corresponding result still holds. The idea of the proofs is to look first at the case where the curves each represent primitive elements of the fundamental group of the surface, which we endow with a hyperbolic metric. There is then a unique hyperbolic geodesic which in the homotopy class of the curve, and this can be flowed to without ever introducing any double curves on the way. The nonprimitive case entails a detailed analysis of the behavior of curves on the annulus. Details are given in [H-S].

4. Concluding remarks

An interesting open problem is whether the disk curve flow converges to the curvature flow if an appropriate limiting procedure is used in which the disks have radius shrinking to zero. If so, this could yield a new proof of the main results of Grayson on the curvature flow in [Gr1].

Finding generalizations of these flows to higher dimensions is a promising area for future investigations. Flows of surfaces in 3-manifolds, for example, are a natural way to investigate diffeomorphisms of 3-manifolds.

REFERENCES

[A] S. Angenent, *Parabolic equations for curves on surfaces*. I, Ann. of Math. (2) **132** (1990), 451–484.

[B] W. Ballman, *Doppelpunkte freie geschlossene Geodatische auf kompacten flachen*, Math. Z. **161** (1978), 41–46.

[Bi] G. D. Birkhoff, *Dynamical systems with two degrees of freedom*, Trans. Amer. Math. Soc. **18** (1917), 199–300.

[G] M. Gage, *Curve shortening makes convex curves circular*, Invent. Math. **76** (1984), 357–364.

[G-H] M. Gage and R. S. Hamilton, *The heat equation shrinking convex plane curves*, J. Differential Geom. **23** (1986), 69–96.

[Gr1] M. Grayson, *Shortening embedded curves*, Ann. of Math. (2) **129** (1989) 71–111.

[Gr2] ____, *The shape of a figure-eight curve under the curve shortening flow*, Invent. Math. **96** (1989), 177–180.

[H-S] J. Hass and G. P. Scott, *Shortening curves on surfaces*, Preprint.

[J] J. Jost, *A nonparametric proof of the theorem of Lusternik and Schnirelman*, Arch. Math. **53** (1989) 497–509.

[K] W. Klingenberg, *Lectures on closed geodesics*, Grundlehren Math. Wiss., vol. 230, Springer, 1978.

[L-S] L. Lusternik and L. Schnirelman, *Sur le probleme de trois geodesiques fermees sur les surface de genre* 0 , C. R. Acad. Sci. Paris **189** (1929), 269–271.

[S] M. Shepard, *Topology of shortest curves on surfaces*, Ph.D. thesis, U. C. Berkeley, 1991.

[T] V. Turaev, *Problem list from workshop on low dimensional topology*, Problem 10, Luminy, 1989 (preprint).

UNIVERSITY OF CALIFORNIA AT DAVIS
E-mail address: hass@ucdmath.ucdavis.edu, jhass@ucdavis.bitnet

UNIVERSITY OF MICHIGAN
E-mail address: peter_scott@ub.cc.umich.edu

Proceedings of Symposia in Pure Mathematics
Volume 54 (1993), Part 3

Curve Shortening, Equivariant Morse Theory, and Closed Geodesics on the 2-Sphere

NANCY HINGSTON

We present two easy and rather surprising results. The first has to do with curve shortening and the second with equivariant Morse theory on the free loop space of a compact surface. We will indicate how these results may relate to the problem of finding closed geodesics on a Riemannian manifold homeomorphic to S^2.

Background

The following is known:

(1) (Moser [**15**], Klingenberg-Takens [**12**], Hingston [**8**]) For a "generic" metric on S^2, there are infinitely many closed geodesics. The proof uses a local result involving the "Birkhoff-Lewis fixed point theorem" in the non-hyperbolic case (Moser, Klingenberg-Takens) and equivarant Morse theory in the hyperbolic case (Hingston).

(2) (Lusternik-Schnirelmann [**13**], Ballmann [**2**], Grayson [**7**], Jost [**10**]) For *any* metric on S^2, there are three simple closed geodesics. Simple means the curves have no self-intersections; this proof will be outlined below.

(3) (Bangert [**3**]) A metric on S^2 always gives infinitely many closed geodesics unless the above three "Lusternik-Schnirelmann geodesics" are all degenerate. (Degenerate means degenerate in the Morse theory sense, as critical points of the energy function.) The proof extends the use of Birkhoff's annulus mappings [**5**] from the convex to the "nonwaisted" case and uses a result of Bangert and Klingenberg [**4**] in the "waisted" case.

However, it is unknown, e.g., whether, even close to the standard metric, there are always at least four closed geodesics on S^2.[1] One would like to show

---

1991 *Mathematics Subject Classification.* Primary 58E10.

The final version of this paper will be submitted for publication elsewhere.

[1] John Franks (preprint) has proved, using Birkhoff's annulus mapping, that a metric on S^2 gives infinitely many closed geodesics in the case not included in Bangert'sresult. Thus there are infinitely many for *any* metric.

that there are always infinitely many. Ziller's paper on the Katok examples
[17] describes some startling cases of Finsler metrics on S^2 with only finitely
many closed geodesics.

Another instructive example is that of the ellipsoid

$$\frac{x^2}{a^2} + \frac{y^2}{b^2} + \frac{z^2}{c^2} = 1.$$

An ellipsoid always has infinitely many closed geodesics since the geodesic
flow is completely integrable [11], and always has at least three simple ones
by (2) above. Close to the standard metric $a = b = c = 1$, these three are
given by the intersections of the ellipsoid with the three coordinates planes.
But (this was known already to Morse [14]!) if we look at ellipsoids with axes
$a < b < c$, as we approach the standard metric the *length of the fourth shortest
closed geodesic goes to infinity*. Moreover, close to the standard metric there
will be only three simple closed geodesics. This example is important for two
reasons. First, it shows that, while (1) above may not be optimal (because
of the word generic), the gap between (1) and (2) is in some sense real and
not just an accident of proof; the fourth closed geodesic in general is not
simple and is much longer and harder to find than the first three. Second,
it tells us something about the equivariant topology of the free loop space
of S^2 which is relevant to the degenerate case mentioned in (3) and to any
attempt to prove the existence of infinitely many closed geodesics on S^2
using topology: It says that the *equivariant topology of the free loop space of
S^2 looks like that of an "idealized ellipsoid" with exactly three (degenerate)
closed geodesics*.

Using the equivariant topology of the free loop space

$$\Lambda S^2 = \text{Maps}(S^1, S^2)$$

to predict the existence of closed geodesics on S^2 is an idea which goes
back to Morse [14]. The energy function $E: \Lambda \to \mathbb{R}$ has as its critical
points the closed geodesics; thus Morse theory allows one to use the topology of Λ (which is, of course, independent of the metric) to predict critical
points, i.e., closed geodesics. An example of how this works will be given
below. The main difficulty with this approach is that, given a closed geodesic
$\gamma: [0, 1] \to S^2$, its iterates $\gamma^m: [0, 1] \to S^2$, $\gamma^m(t) = \gamma(mt)$ (as well as
its arclength reparametrizations), all appear as distinct critical points in Λ
although they are of course "geometrically" the same. It was clear already
to Morse that, given these difficulties, the topology of Λ for a sphere was
"too weak" to predict the existence of many geometrically distinct closed
geodesics; thus he looked at the *equivariant* topology of Λ. The $O(2)$ action
on S^1 induces a natural action on Λ (by reparametrization) which leaves
the function E invariant and thus we get extra information by using the
equivariant topology of Λ, i.e., the data given by the $O(2)$-space Λ. The
moral of the ellipsoid example is that *topology alone is insufficient to prove*

the existence of infinitely many closed geodesics on S^2. The fact that the length of the fourth closed geodesic goes to infinity means that it is "topologically invisible"; no algebraic computation (e.g., computing the rank of some group) on Λ can by itself rule out the possibility of a degenerate metric with exactly three closed geodesics. Some geometry must be used. (It is worth noting, however, that one is guaranteed the existence of infinitely many closed geodesics if the lengths of any two of the Lusternik-Schnirelmann geodesics coincide; thus there is hope for a combination of topology and geometry to succeed in proving the existence of infinitely many closed geodesics for any metric. We are quite confident that such a method gives at least four. It is also important to note that the Katok examples mentioned above show that one *must* use the full $O(2)$-action to prove the existence of infinitely many; the $SO(2)$-action alone will not do.)

Curve shortening

We next sketch the proof of (2): We start with the three-parameter family of "circles" on S^2 with the standard metric. By circle we mean the intersection of S^2 with any plane in $\mathbb{R}^3$. Given a metric g on S^2 and a homeomorphism $(S^2, \text{std.}) \rightarrow (S^2, g)$, we get a three-parameter family of simple curves on (S^2, g). Next continuously shorten these curves by some process which

(a) only "stops" at closed geodesics (or point-curves) and

(b) keeps the curves simple.

The three-parameter family contains a two-parameter and a one-parameter family; together these represent three "subordinated" homology classes in ΛS^2 which Lusternik-Schnirelmann theory guarantees will "get stuck" on three distinct closed geodesics. (This is a typical example of how the topology of Λ leads to closed geodesics.)

This beautiful idea was outlined in a two-page paper by Lusternik and Schnirelmann in 1929 [13]. The difficulty is in finding an appropriate curve-shortening process. The details of the Lusternik-Schnirelmann process were finally worked out by Ballmann in 1978 [2]. The papers by Grayson and Jost describe two other such processes. We are interested in the process used by Grayson, which has also been studied by Gage and Hamilton [6]. It is a natural process with many beautiful and amazing properties. A curve evolves by its curvature vector (see Figure 1 on next page): The properties (a) and (b) are not difficult to verify; the difficulty lies in showing that, starting with an embedded curve, the flow is defined for all time.

Our first result is the following

OBSERVATION. Take any metric g on S^2. Take any homeomorphism

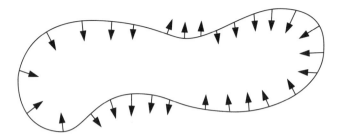

FIGURE 1

$(S^2, \text{std.}) \to (S^2, g)$ and, starting with the images of the standard circles on (S^2, g), flow by the curvature flow. Then for all time,

(i) Given distinct points P, Q, $R \in S^2$, there is a unique "circle" through P, Q and R.

(ii) Given distinct points P, $Q \in S^2$ and a vector V at P, there is a unique "circle" through P and Q which is tangent to V at P.

We find this very surprising. The proof is quite easy using established properties of the flow. Existence follows from existence at $t = 0$ and a degree argument. Uniqueness comes as follows: Initially, two distinct circles intersect twice transversely, once tangentially or not at all. A result of Angenent [1] insures that, as the two curves evolve, "intersections only decrease" (See Figure 2.):
This is really a variation on the property that simple curves stay simple. As an example of how this can be used to relate the topology of Λ to geometry, we have the

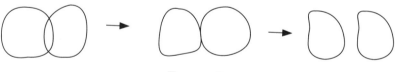

FIGURE 2

COROLLARY. *Given a metric on S^2, normalize so that all three Lusternik-Schnirelmann goedesics have length $\leq 2\pi$. Then for any $n \geq 0$, any class in $H_*(\Lambda S^2)$ which can be represented in the standard metric using curves of length $\leq 2n\pi$ can be represented in the given metric using curves of length $\leq 2n\pi$.*

SKETCH OF PROOF. Given any $\varepsilon > 0$ there is a continuous deformation from the set of curves of length $\leq 2n\pi$ in the standard metric to the set of curves consisting of n standard "circle pieces" of length $\leq 2\pi - \varepsilon/n$ and one piece of length $\leq \varepsilon$. The pieces can then be deformed, end points fixed, to give n "circle" pieces of length $\leq 2\pi$ in the given metric (since the normalization ensures that all "circles" have length $\leq 2\pi$) and one piece of length $\leq D\varepsilon$, where D is the dilation of a homeomorphism $(S^2, \text{std.}) \to (S^2, g)$.

REMARK. Unfortunately, the above construction is not equivariant.

Equivariant Morse theory on the free loop space of a surface

Let M be a compact Riemannian manifold with all closed geodesics nondegenerate. "Ordinary" Morse theory gives as a model for ΛM a CW complex with one λ-cell D_c^λ (summit) and one $(\lambda+1)$-cell $S^1 \times D_c^\lambda/\mathbb{Z}_m$ (tunnel) for each (oriented) closed geodesic c of index λ, multiplicity m (multiplicity m means c has stability group $\mathbb{Z}_m \subset O(2)$). However, in general there is no way to construct this CW complex equivariantly (i.e., with equivariant attaching maps) with respect to the $SO(2)$- or $O(2)$-action. Thus the model CW complex carries the homotopy type, but not the $O(2)$-homotopy type, of Λ. Rademacher [16] has constructed the *equivariant Morse complex* using the strata of D_c^λ by the $\mathbb{Z}_m$-action as cells. This is a "$(\mathbb{Z}_N, O(2))$-CW complex" carrying the $O(2)$-homotopy type of ΛM. However, it has many more cells than the "ordinary" Morse complex. The disadvantage of this is the following: One would like Morse theory to give a "map"

$$\text{Topology of } \Lambda \to \text{Closed geodesics}$$

in such a way that different "topological data", e.g., different equivariant homology classes, yield different closed geodesics (at least different in $\Lambda/O(2)$ if not necessarily geometrically distinct). But, roughly speaking, if a geodesic c requires attaching n cells (for different strata), we should expect n different "data" to yield the same closed geodesic c.

THEOREM. *Let M be a compact surface with all closed geodesics nondegenerate. Then "we can put an $O(2)$-action on the ordinary Morse complex." Specifically, suppose all closed geodesics with energy $\leq a$ have multiplicity dividing N. Then the equivariant Morse complex $\mathcal{M}^a$ has the structure of an (ordinary) CW complex $\widehat{\mathcal{M}}^a$ with cells*

    (i) $\omega_N^k D_c^\lambda$, $T\omega_N^k D_c^\lambda$ (*summits*),

    (ii) $\omega_N^k \delta_N D_c^\lambda$, $T\omega_N^k \delta_N D_c^\lambda$ (*tunnels*) *for each unoriented closed geodesic* c *with* $E(c) \leq a$, $0 \leq k < \frac{N}{m}$, $m =$ *multiplicity* (c) . *The attaching maps remain equivariant.*

Here

$$\omega_N = e^{2\pi i/N} \text{ generates } \mathbb{Z}_N \subset O(2),$$
$$T \text{ generates } \mathbb{Z}_2 \subset O(2),$$
$$\delta_N = \left\{ e^{2\pi it} \mid 0 \leq t \leq \frac{1}{N} \right\} \subset SO(2).$$

In other words, the underlying $O(2)$-topological space $\mathcal{M}^a$ can be given the structure of an ordinary CW complex *with an* $O(2)$-*action and with only top-dimensional cells.*

COROLLARY. *We can play "summits and tunnels". Let* $\mathbb{D}_N \subset O(2)$ *be the dihedral group of order* $2N$, *generated by* T *and* ω_N. *Put*

$$M_k = \mathbb{Z}[\mathbb{D}_N]\text{-module spanned by}$$
$$\begin{cases} D_c^\lambda, & \text{index }(c) = \lambda = k, \\ \delta_N D_c^\lambda, & \text{index }(c) = \lambda = k-1 \end{cases}$$

with relations

$$(1 - \omega_m)D_c^\lambda = 0 \quad \text{if } m = \text{mult}(c),$$
$$(1 - \omega_m)\delta_N D_c^\lambda = 0 \quad \text{if } m = \text{mult}(c).$$

There is a boundary map $d \colon M_k \to M_{k-1}$ *commuting with the* $\mathbb{D}_N$-*action*;

$$d^2 = \delta_N^2 = 0, \qquad d\delta_N + \delta_N d = (1 - \omega_N),$$
$$\delta_N T + \omega_N T \delta_N = 0, \qquad \delta_N \omega_N - \omega_N \delta_N = 0.$$

$(M, d, \delta_N, \mathbb{D}_N)$ *computes all ordinary and equivariant homology of* Λ. *All homology of* Λ: *ordinary, relative (e.g., sets of subordinated homology classes), equivariant and relative equivariant can be "represented" in the complex* M.

For example $\mathbb{D}_N$ or $O(2)$-homology is represented in a triple complex with entries in M and maps given by d, δ_N, $\mathbb{D}_N$.

One can even "mod out by tunnels": Let

$$\delta = (1 + \omega_N + \cdots + \omega_N^{N-1})\delta_N \qquad (\text{" } = SO(2)\text{"}).$$

Then

$$\text{Ker } \delta = \text{Im}(1 - \omega_N) + \text{Im } \delta_N,$$

and

$$M_k / \text{Ker } \delta \simeq \mathbb{Z}[\mathscr{G}^k / SO(2)],$$

where $\mathscr{G}^k$ is the set of closed geodesics of index k. Since $d \colon \text{Ker } \delta \to \text{Ker } \delta$, d descends to a map

$$d \colon M / \text{Ker } \delta \to M / \text{Ker } \delta;$$

this seems to compute the rational homology of $\Lambda/SO(2)$. (It is not clear if it computes anything over $\mathbb{Z}$.) One can also divide out by the full $O(2)$-action by modding out by the kernel of $\delta(1 + T)$.

The lemma follows from the fact that for every pair c, c' of closed geodesics in the same component of ΛM we have, for $i = 0$ and $i = 1$,

$$\text{index}(c') < \text{index}(c) - i \Rightarrow \text{index}((c')^m) < \text{index}(c^m) - i \forall m.$$

Application. Take a metric on S^2 with all three Lusternik-Schnirelmann geodesics nondegenerate. If there are only three "short" closed geodesics, then there are infinitely many closed geodesics.

This is much weaker than Bangert's result (3). The proof is completely different and uses the $O(2)$-homology of ΛS^2. This $O(2)$-homology can be computed by extending the argument in [9], where we compute the $SO(2)$-homology using Morse theory and an appropriate Finsler metric on S^2.

REFERENCES

1. S. Angenent, *Parabolic equations for curves on surfaces. II. Intersections, blow up and generalized solutions*, Preprint, Univ. of Wisconsin-Madison, 1989.
2. W. Ballmann, *Doppelpunktfreie geschlossene Geodätische auf kompakten Flächen*, Math. Z. **161** (1978), 41–46.
3. V. Bangert, *Geodätische Linien auf Riemannschen Mannigfaltigkeiten*, Jahresber. Deutsch. Math.-Verein. **87** (1985), 39–66.
4. V. Bangert and W. Klingenberg, *Homology generated by iterated closed geodesics*, Topology **22** (1983), 379–388.
5. G. D. Birkhoff, *Dynamical systems*, Amer. Math. Soc. Colloq. Pub., vol. 9, Amer. Math. Soc., Providence, RI, 1927.
6. M. Gage and R. S. Hamilton, *The heat equation shrinking convex plane curves*, J. Differential Geom. **23** (1986), 69–96.
7. M. Grayson, *Shortening embedded curves*, Ann. of Math. **129** (1989), 71–111.
8. N. Hingston, *Equivariant Morse theory and closed geodesics*, J. Diff. Geom. 19 (1984), 85–116.
9. ____, *Equivariant homology of the free loop space of S^2*, preprint, 1987.
10. J. Jost, *A nonparametric proof of the theorem of Lusternik and Schnirelmann*, Sonderdruck aus Arch. Math., **53** (1989), 497–509.
11. W. Klingenberg, *Lectures on closed geodesics*, Springer-Verlag, 1978.
12. W. Klingenberg and F. Takens, *Generic properties of geodesic flows*, Math. Ann. 197 (1972), 323–334.
13. L. Lusternik and L. Schnirelmann, *Sur le problème de trois géodésiques fermées sur les surfaces de genre O*, C. R. Acad. Sci. Paris **189** (1929), 269–271.
14. M. Morse, *The calculus of variations in the large*, Colloq. Publ. vol. 18, Amer. Math. Soc., Providence, RI, 1934.
15. J. Moser, *Proof of a generalized form of a fixed point theorem due to G. D. Birkhoff*, Lecture Notes in Math. vol. 597, Springer-Verlag, 1977, 464–494.
16. H. B. Rademacher, *On the equivariant Morse chain complex on the space of closed curves*, Math. Z. **201** (1989), 279–302.
17. W. Ziller, *Geometry of the Katok examples*, Ergodic Theory Dynamical Systems 3 (1982), 135–157.

TRENTON STATE COLLEGE

Proceedings of Symposia in Pure Mathematics
Volume **54** (1993), Part 3

Morse Theory for Geodesics

DIANE KALISH

ABSTRACT. This paper gives a brief account of the Morse index theorem and its subsequent generalizations to various settings. Also given is an outline of a proof in the two endmanifold case on both Riemannian and Lorentzian manifolds which calculates the index of the energy function using focal points and a second fundamental form at one submanifold. The simplicity of this formulation allows for an easy computation of the homotopy type of some specific path spaces joining two submanifolds.

Introduction

A remarkable theorem shows that the critical points of a smooth function defined on a manifold M determine the homotopy type of M to be the same as a CW-complex with a cell of dimension λ for each critical point of index λ. The task of formulating an analogous theory in the infinite-dimensional context involves defining appropriate concepts such as critical point, derivative, etc. These ideas were considered and developed by Marston Morse (see [11]). The following section describes variations and extensions of Morse's theory. Section 2 will briefly outline the author's proof of the Morse index theorem for the space of paths connecting two submanifolds so that the above-mentioned theorem calculating homotopy type can be easily extended to the two endmanifold case. A few such calculations are given. The advantage of this proof of the Morse index theorem over previous ones is in the simplicity of the calculation of the index of geodesic. This calculation is then extended to the Lorentzian setting.

1991 *Mathematics Subject Classification.* Primary 58E05; Secondary 58B05, 53C22, 53C50, 55P35.

This paper is in final form and no version will be submitted for publication elsewhere.

Research for this paper was partially supported by the School of Science and Mathematics of William Paterson College under the Governor's Challenge Grant.

I wish to thank Dean Simpson for his support, which encouraged me to attend the AMS Summer Institute on Differential Geometry, 1992.

1

It was one of Morse's goals to formulate a theory relating critical points of functions defined on an infinite-dimensional manifold to the topology of the manifold. In [11] he considered the space of piecewise smooth paths $\Omega(M; p, q)$ connecting two points p and q on a finite-dimensional Riemann ian manifold M together with the length functional defined on $\Omega(M; p, q)$. "Critical points" are now geodesics joining p and q, while the "tangent space" at a point is the set of piecewise smooth vector fields along a path in $\Omega(M; p, q)$. The hessian called the Morse index form is the second variation representing two vector fields along a geodesic. Morse obtains that the index of a geodesic γ joining p and q is the number of conjugate points between p and q counted with multiplicities. Another proof for the Morse index theorem can be found in Milnor's book [10], in which he uses the energy function and obtains a far less complicated proof.

Extensions of the Morse index theorem involve path spaces connecting a submanifold P to a point q or a submanifold P to a submanifold Q, denoted respectively by $\Omega(M; P, q)$ and $\Omega(M; P, Q)$. A proof for $\Omega(M; P, q)$ can be found in [3], where P-focal points are employed in place of conjugate points, and a similar result follows. However, in the traditional extensions such as in [1, 4] to the two endmanifold case, the index of the extremal has not been computed so simply from the behavior of Jacobi fields along the geodesic. Ambrose defined P–Q conjugate points which involves translating boundary conditions at Q along the geodesic [1]. Bolton uses positive and negative P–Q focal points which also involves a translation of a boundary condition [4]. The author's formulation uses focal points as in the manifold point case together with a boundary condition at one end, thereby achieving a simpler calculation of the index and thus allowing for the computation of the homotopy type of $\Omega(M; P, Q)$ as in [8, 9].

Uhlenbeck calculates the homotopy type of a globally hyperbolic manifold by considering null geodesics from a point to an infinite time-like line in [12], while Beem and Ehrlich [2] determined index in the two-point case by generalizing Riemannian results to the Lorentzian setting. Ehrlich and Kim have recently written a paper [5] adapting the author's proof to the situation of a null geodesic segment connecting two spacelike submanifolds, by constructing a quotient bundle as employed in general relativity. An alternate simpler proof omitting this construction is offered in §2. At present, Hingston and the author have completed a proof for finding the index of an extremal in $\Omega(M; P, Q)$ in the case that Q is at a P-focal point (see [7]).

An interesting application of the Morse index theorem to medicine can be found in a paper by Greensite entitled "Topological foundations of electrocardiology" [6], in which Morse theory is used to derive geometric properties of the heart muscle.

2

Let M be a Riemannian manifold with submanifolds P and Q of codimension greater than zero. The energy function E is defined on the space Ω of continuous piecewise C^∞ paths joining P and Q. A path $\gamma \in \Omega$, $\gamma: [0, T] \to M$, is a critical point of E when γ is a geodesic orthogonal at its endpoints to P and Q. We denote by H the space of piecewise C^∞ vector fields orthogonal to γ with initial and final vectors in $P_{\gamma(0)}$ and $Q_{\gamma(T)}$. A symmetric bilinear functional is defined by

$$I(X, Y) = \int_0^T \langle RX - \ddot{X}, Y \rangle \, dt + \sum_i \langle \dot{X}(p_i^-) - \dot{X}(p_i^+), Y(p_i) \rangle$$

$$+ \langle \dot{X}(t) - S_T X(t), Y(t) \rangle \big|_0^T$$

where the summation is evaluated at points of discontinuity of $\dot{X}$, and S_0 and S_T are the second fundamental forms of P and Q with respect to $\dot{\gamma}(0)$ and $\dot{\gamma}(T)$. A P-Jacobi field J is a Jacobi field along γ such that $J(0) \in P_{\gamma(0)}$ and $\dot{J}(0) - S_0 J_{(0)} \perp P_{\gamma(0)}$. P-focal points are where P-Jacobi fields vanish and the dimension of this space is the multiplicity of the P-focal point.

We consider the case where $\gamma(T)$ is not a P-focal point and define a symmetric bilinear map A on the subspace spanned by P-Jacobi fields with nonvanishing final value in $Q_{\gamma(T)}$ by

$$A(J_1, J_2) = \langle \dot{J}_1(T) - S_T J_1(T), J_2(T) \rangle.$$

Statement of extended Morse index theorem.

THEOREM. *The index of I is equal to the sum of the number of P-focal points counted with multiplicities and the index of A. (Assume $\gamma(T)$ is not a P-focal point.)*

Since $I(X, X)$ can be thought of as a second derivative of E in the direction X, the above index of I tells us the number of linearly independent directions γ can be pushed to yield decreased energy.

The proof is indicated in Figures 1 and 2 (see next page), noting that the vector fields drawn give index (see [8] for details).

At a P-focal point τ, we add a multiple of a bump function to the vanishing P-Jacobi field J as shown in Figure 1.

At T, we choose those P-Jacobi fields K such that $A(K) < 0$ (See Figure 2).

We can then write $H = (B \oplus B_+^c) \oplus (B_-^c)$, where B_-^c is the span of all the pictured vector fields V and K. The subspace B consists of vector fields in H which vanish at T and are functional combinations of P-Jacobi fields and B_+^c is the space spanned by P-Jacobi fields K such that $A(K) \geq 0$.

We then show that

$$I(B \oplus B_+^c) \geq 0, \quad I(B_-^c) < 0, \quad \text{and} \quad I(B \oplus B_+^c, B_-^c) = 0,$$

which proves the theorem.

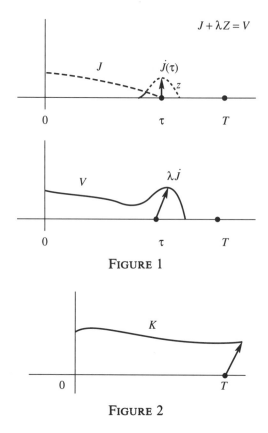

FIGURE 1

FIGURE 2

Statement of theorem relating homotopy type to critical points. Following Milnor's proof in [**10**] and making the necessary modifications for path spaces joining submanifolds, we can prove the following theorem.

THEOREM. *Let M be a complete Riemannian manifold with P and Q as two closed and bounded submanifolds of M which are not conjugate along any geodesic orthogonal to P and Q. Then Ω or Ω^* has the homotopy type of a countable CW-complex which contains one cell of dimension λ for each geodesic orthogonal to P and Q of index λ (see [**9**] for details).*

In the above statement Ω^* is the space of continuous paths joining P and Q. P and Q are said to be conjugate along an extremal geodesic if there exists a nonzero P-Jacobi field which is also a Q-Jacobi field, i.e., a vector field in the null space of I. In order to obtain a finite-dimensional faithful model for the infinite-dimensional path spaces we are considering, we define a space which consists of broken geodesic segments from P to Q such that the segment starting (ending) at $P(Q)$ is orthogonal to $P(Q)$ and with restrictions on the energy.

Examples.

EXAMPLE 1. The homotopy type of the path space joining two circles on S^2 (see Figure 3).

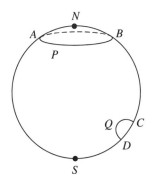

FIGURE 3

THE GEODESICS ORTHOGONAL TO P AND Q:

Clockwise	Counterclockwise
$\gamma_1 : BC$	$\gamma_5 : AD$
$\gamma_2 : BD$	$\gamma_6 : AC$
$\gamma_3 : AC$	$\gamma_7 : BD$
$\gamma_4 : AD$	$\gamma_8 : BC$

We obtain the other geodesics by adding $2\pi n$ $(n = 1, 2, 3, \ldots)$ onto each of the eight geodesics listed.

$J(t) = (\sin t)E(t)$ can serve as a P-Jacobi field for the geodesics. To find the index at each geodesic we compute $A(J) = \langle \dot{J}(T) - S_T J(T), J(T) \rangle$ and count the number of times $J(t)$ passes through the north or south poles depending on which geodesic we are considering. For example, at γ_4 the index of A is 1 and γ_4 passes through N, which is a P-focal point for γ_4. We therefore obtain the index at γ_4 as 2. Our final answer for the homotopy type is

$$e^0 \cup e^1 \cup e^1 \cup e^2 \cup e^1 \cup e^2 \cup e^2 \cup e^3 \cup e^2 \cup e^3 \cup \cdots \text{etc.}$$

EXAMPLE 2. The path space joining S^n to S^n in R^{n+1} (see Figure 4).

(Dotted line in Figure 4 indicates variation through geodesics which gives a P-Jacobi field. The origin is the only possible P-focal point.)

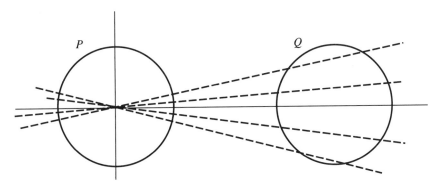

FIGURE 4

Using similar methods we get that Ω has the homotopy type of the CW-complex

$$e^0 \cup e^n \cup e^n \cup e^{2n}.$$

For $n > 1$ we can compute the integral homology groups.

Statement of the focal index theorem for null geodesics.

THEOREM. *Let $\beta : [0, T] \to M$ be a null geodesic segment in a Lorentzian manifold of dimension $n \geq 3$. Let P and Q be two spacelike submanifolds of M of dimensions $\leq n - 2$, with β orthogonal to P and Q at $\beta(0)$ and $\beta(T)$. Assuming $\beta(T)$ is not a P-focal point, the index of I at β is equal to the number of P-focal points counted with multiplicities plus the index of A.*

Ehrilich and Kim [5] have given a proof for this theorem by adapting the techniques in [8] to the Lorentzian setting by constructing a quotient bundle, as suggested by general relativity theory. One reason the quotient bundle is employed in the Lorentzian setting is that there is an infinite-dimensional space of vector fields which are not Jacobi fields and which are in the nullity of I. However, this fact plays no role in the proof of the index theorem, so the methods used in [8] carry over to the Lorentzian setting.

Note that in the Lorentzian setting the index is defined as the dimension of the largest subspace on which I is *positive* definite, so that index tells us which directions *increase* energy.

The proof carries over directly with the following two items to consider.

1. We let $J_1(t) = t\dot{\beta}(t)$. We can then find $n - 2$ P-Jacobi fields which together with $J_1(t)$ span the space of P-Jacobi fields.

2. We define B and B^c as before:

$$B = \left\{ x \in H : x = \sum_{i=1}^{n-1} f_i J_i \text{ and } x(T) = 0 \right\}.$$

However, in order to show $H = B \oplus B^c$ we need to prove the following lemma.

LEMMA. *If $J(t)$ is a P-Jacobi field such that $J(t_0) = 0$ for $t_0 > 0$ and $\dot{J}(t_0) = c\dot{\beta}(t_0)$, then $c = 0$ and $J = 0$.*

PROOF. Since $J(t)$ is a P-Jacobi field, it satisfies a second-order linear differential equation with initial conditions $J(t_0) = 0$ and $\dot{J}(t_0) = c\dot{\beta}(t_0)$. By the uniqueness of solutions of such equations with conditions at t_0, we have

$$J(t) = c(t - t_0)\dot{\beta}(t).$$

But since J is a P-Jacobi field, $J(0) = ct_0\dot{\beta}(0) \in TP_{\gamma(0)}$, which implies that $c = 0$, since vectors in $TP_{\gamma(0)}$ are spacelike.

REFERENCES

1. W. Ambrose, *The index theorem in Riemannian geometry*, Ann. of Math. (2) **73** (1961), 49–86.
2. J. Beem and P. Ehrlich, *Global Lorentzian geometry*, Marcel Dekker, New York, 1981.
3. R. Bishop, and R. Crittenden, *Geometry of manifolds*, Pure and Appl. Math. vol. 15, Academic Press, 1964.
4. J. Bolton, *The Morse index theorem in the case of two variable endpoints*, J. Differential Geom. **12** (1977), 567–581.
5. P. Ehrlich, and S.-B. Kim, *A focal index theorem for null geodesics*, J. Geom. Phys. **6** (1989), no. 4, 657–670.
6. F. Greensite, *Topological foundations of electrocardiology*, Adv. Appl. Math. **6** (1985), 230–258.
7. N. Hingston and D. Kalish, *The extended Morse index theorem*, Proc. Amer. Math. Soc. (to appear).
8. D. Kalish, *The Morse index theorem where the ends are submanifolds*, Trans. Amer. Math. Soc. **308** (1988), 341–348.
9. ____, *Homotopy type of path spaces*, Proc. Amer. Math. Soc. (to appear).
10. J. Milnor, *Morse theory*, Ann. of Math. Stud., no. 51, Princeton Univ. Press, Princeton, NJ, 1973, p. 95.
11. M. Morse, *The calculus of variations in the large*, Amer. Math. Soc. Colloq. Publ. vol. 18, Amer. Math. Soc., Providence, RI, 1934.
12. K. Uhlenbeck, *A Morse theory for geodesics on a Lorentz manifold*, Topology **14** (1975), 69–90.

WILLIAM PATERSON COLLEGE

Proceedings of Symposia in Pure Mathematics
Volume **54** (1993), Part 3

On Some Metrics on $S^2 \times S^2$

MASATAKE KURANISHI

One of the unsolved problems in differential geometry is the question of the existence of metric with positive sectional curvature on $S^2 \times S^2$. The oldest reference to the problem seems to be in S. S. Chern [7]. The conjecture asserting that such metric does not exist is often referred to as the H. Hopf conjecture.

Only a few restraints on the topology of a compact manifold with positive sectional curvature are known: (1) If the dimension is even, its fundamental group is either $\mathbf{Z}_2$ or trivial (cf. [14]). (2) If the dimension is odd, it is orientable and its universal covering space is still compact (cf. [13]).

For compact 3-manifolds R. S. Hamilton showed in [10] that a manifold of positive Ricci curvature also admits a metric of constant positive curvature. Since the manifolds of constant curvature have been completely classified by J. A. Wolf [18], we have a complete answer on the existence of a metric of positive Ricci curvature on a compact 3-manifold. In fact, it is diffeomorphic to a quotient of one of spaces S^3, $S^2 \times \mathbf{R}^1$, $\mathbf{R}^3$ by a group of fixed point free isometries in the standard metrics.

For compact 4-manifolds with positive curvature operators R. S. Hamilton [11] showed that they are diffeomorphic to S^4 or the real projective space. He also showed that a compact 4-manifold with nonnegative curvature operator is diffeomorphic to a quotient of one of the above spaces or $S^3 \times \mathbf{R}^1$ or $S^2 \times S^2$ or $S^2 \times \mathbf{R}^2$ or $\mathbf{R}^4$ by a finite group of fixed point free isometries in the standard metric. For an alternate proof see [5].

A compact manifold M with positive sectional curvature must have some differential geometric properties. For example, it cannot have two disjoint totally geodesic closed submanifolds such that the sum of the dimensions is at least that of M. This is because by the Frankel's construction we can find tangential planes of zero sectional curvature using such submanifolds

1991 *Mathematics Subject Classification.* Primary 53-02.
Work partially supported by the National Science Foundation under grant DMS-87-04209.
This paper is in final form and no version of it will be submitted for publication elsewhere.

(cf. [9]). Note that on $S^2 \times S^2$ with the standard product metric we have such a pair of submanifolds S^2. W. Hsiang and B. Kleiner showed that only orientable compact 4-manifolds with positive sectional curvature which admits nontrivial circle actions are S^4 or complex projective space (cf. [12]). A. Weinstein showed that any metric on $S^2 \times S^2$ induced by an immersion in $\mathbf{R}^6$ cannot be of positive sectional curvature (cf. [17]). This is a corollary of his general theorem on the topology of a compact manifold of dimension n isometrically immersed in the euclidean space of dimension $n + 2$.

It seems that the next thing to do is to study deformations of a metric with nonnegative curvature and try to see if we can improve it. In the case of a product manifold with a product metric, where each of the factor metrics is of positive sectional curvature, the set of tangential planes with zero sectional curvature is the set of planes generated by tangent vectors from each of the factor manifolds. Such planes are called mixed planes. Let a deformation of a product metric as above be given by a parameter, say t, starting at $t = 0$. Then the deformation is called nonnegative, or stationary, at first order if for all mixed planes the derivative of the sectional curvature by t at $t = 0$ is nonnegative, or zero, respectively. M. Berger showed in [1] that every nonnegative variation at first order is stationary at first order. Therefore we have to study higher order variations. This was done by J. P. Bourguignon, A. Deschamps and P. Sentenac (cf. [2, 3, 4]). Actually their result is rather striking. They noted first that a deformation of a metric by applying diffeomorphisms should be regarded as a trivial deformation. On the other hand D. G. Ebin proved (cf. [8]) that the action of the group of diffeomorphisms on metrics admits a slice at a metric. Therefore they study deformations where metrics are in the slice. In the case the product metric has no infinitesimal isometry, they showed that, as formal deformations, any nonnegative deformation is actually a product deformation of the deformations of each of the factor metrics. If there are infinitesimal isometries, the situation is complicated. However, they found that they can use an isometry to construct a deformation such that the sectional curvature of any mixed plane is an increasing function of the parameter. As they note, this does not imply that we can make the sectional curvature positive. Indeed, in the case of a deformation by a parameter t starting at $t = 0$, let us consider a family of tangential planes parameterized by a small real parameter r such that r is also a distance to the set of mixed planes. If the sectional curvature of the plane is given by $ct^2 - rt$ (mod t^3) with a constant $c > 0$ for example, then the sectional curvature of the mixed plane at $r = 0$ is an increasing function of t but the sectional curvature of the plane at the parameter $r \neq 0$ is negative for sufficiently small $t > 0$. In fact this is what happens in many examples. Their construction of the deformations using isometries which improve the situation at mixed planes is related to the construction of J. Cheeger [6]. He used isometries to construct a new metric. Namely, let a compact group G

act by isometries on a compact manifold M with a metric g. Let b be a bi-invariant metric on G. Then G also acts by isometries on $G \times M$ with the metric $b \times g$, where the action by an element y in G sends (x, m) to (xy^{-1}, ym). Since the quotient of $G \times M$ by the action of G is diffeomorphic to M, we have a metric $\bar{g}$ on M such that the projection of $G \times M$ to M is a riemannian submersion. Since the sectional curvature of a horizontal plane can only increase by taking the projection thanks to the O'Neill's formula (cf. [15]), we have a possibly improved new metric $\bar{g}$ as far as the positivity of sectional curvature is concerned.

In an unpublished manuscript [16], J. Wason developed an idea to construct a new metric on $S^2 \times S^2$ using submersion of $S^3 \times S^3$ by Hopf maps. This led to a construction of a metric on $S^2 \times S^2$ which very much looks like Cheeger's metric obtained by applying his construction on the standard product metric on $S^2 \times S^2$. This is a metric of nonnegative sectional curvature. The set of planes of zero sectional curvature is a subset of the set of mixed planes. In fact, at any point $(\underline{x}, \underline{y})$ in $S^2 \times S^2$ where $\underline{x} \neq \underline{y}$ or $\underline{x} \neq -\underline{y}$, there is exactly one plane of zero sectional plane. It is tempting to see what happens when we deform this metric. We have not yet completely analyzed the situation. This is a progress report on the calculation.

Let us first describe the metric on $S^2 \times S^2$. We can write down a metric more easily if the manifold we are working on is parallelizable. Namely, if $\underline{\omega}^j$ is a base of 1-forms, a metric $\underline{g}$ is written down as

$$(1) \qquad\qquad \underline{g} = g_{jk} \underline{\omega}^j \underline{\omega}^k,$$

where $g = (g_{jk})$ is a symmetric matrix-valued function. For a general manifold M, we may construct a bundle $\widetilde{M}$ over M such that $\widetilde{M}$ is parallelizable. For a metric $\mathbf{g}$ on M we may lift it to a metric $\underline{g}$ in such a way that the projection is a submersion. By means of the O'Neill's formula we can then calculate the sectional curvature using $\underline{g}$ and the horizontal lift of the tangential planes. This is what we are going to do.

S^2 is not parallelizable. However, S^3 is, and there is the Hopf map $S^3 \to S^2$. Therefore we may use $S^3 \times S^3$ as the above $\widetilde{M}$. However, we find it more convenient to use its quotient by a $\mathbf{Z}^2$ action on the second factor. Namely, the frame bundle $F(S_a)$ of S^3 with the standard metric of radius $1/a$. We consider the metric such that the Levi-Civita connection forms ω^j, ω^k_j are orthonormal. This metric depends on a. To make the above construction more explicit, let S_a be the sphere of radius $1/a$ centered at origin in the field Q of quaternions with the standard metric. Note first that $F(S_a)$ has a global section. Namely, when we denote by $1, \mathbf{i}, \mathbf{j}, \mathbf{k}$ the standard base of Q, the frame formed by $\mathbf{i}, \mathbf{j}$, and $\mathbf{k}$ may be regarded as a positively oriented orthonormal frame to S_1 at the point 1. For a point q in S_a, we may move the frame to q by multiplying q from the left and by adjusting the length by multiplying the scalar a. Denote by σ the section.

We now have a global chart:

(2) $$S_a \times O^+(3) \ni q \times h \mapsto \sigma(q)R_h \in F(S_a)$$

where R_h denotes the right action by h. We call it the standard chart.

We next define an operation of the two-dimensional torus on $F(S_a)$. For $t \in \mathbf{R}$ set

(3) $$\gamma(t) = \cos t + (\sin t)\mathbf{k}$$

and denote by ϕ_t the diffeomorphism of S_a obtained by multiplying $\gamma(t)$ from the left. Since ϕ_t is an isometry of S_a, it induces a bundle map

(4) $$\Phi_t = (\phi_t)_* : F(S_a) \to F(S_a)$$

over ϕ_t. Since σ is obtained by multiplication of $\mathbf{i}, \mathbf{j}, \mathbf{k}$ from the right, we see that the section σ is mapped onto itself by Φ_t. Therefore the expression of Φ_t in the standard chart is given by

(5) $$(q, h) \mapsto (\gamma(t)q, h).$$

Denote by h_s the one-parameter group of rotation in $\mathbf{R}^3$ around the third standard axis with the velocity 1. Then we have a bundle map

(6) $$\Psi_s : F(S_a) \ni u \mapsto R_{h_s} u \in F(S_a).$$

In terms of the standard chart Ψ_s is given by

(7) $$(q, h) \mapsto (q, hh_s).$$

Hence Φ_t and Ψ_s commute. We thus have an operation of $T^2 = \mathbf{R}^2/(2\pi\mathbf{Z}^2)$. To see that $F(S_a)/T^2$ is diffeomorphic to $S^2 \times S^2$, note first that we may identify $\mathbf{R}^3$ with the subspace Q_0 of Q generated by $\mathbf{i}, \mathbf{j}, \mathbf{k}$. Namely, we identify (x^1, x^2, x^3) with $x^1\mathbf{i} + x^2\mathbf{j} + x^3\mathbf{k}$. Nonzero q in Q acts as an orthogonal transformation by

(8) $$\mu(q^{-1}) : \mathbf{R}^3 = Q_0 \ni q_0 \mapsto q^{-1}q_0 q \in Q_0 = \mathbf{R}^3.$$

Denote by e_1, e_2, e_3 the standard base of $\mathbf{R}^3$. We now define a smooth map ρ of $F(S_a)$ to $S^2 \times S^2$ by sending (q, h) to $(\underline{x}, \underline{y})$, where

(9) $$\underline{x} = he_3, \quad \underline{y} = \mu(q^{-1})\mathbf{k}.$$

We then see easily that ρ induces a diffeomorphism of $F(S_a)/T^2$ with $S^2 \times S^2$.

Denote by ω^j, ω^j_k $(j, k = 1, 2, 3)$ the Levi-Civita connection forms on $F(S_a)$. Set

(10) $$\omega^{\bar{1}} = \omega^2_3, \quad \omega^{\bar{2}} = \omega^3_1, \quad \omega^{\bar{3}} = \omega^1_2.$$

Our metric on $F(S_a)$ is

(11) $$\underline{g}_0 = \sum_{j=1,2,3} (\omega^j \omega^j + \omega^{\bar{j}} \omega^{\bar{j}}).$$

We see easily that the action of T^2 on $F(S_a)$ is an isometry. Therefore ρ may be considered as a submersion of $\underline{g}_0$ to a metric $\mathbf{g}_0$ on $S^2 \times S^2$.

In order to calculate the sectional curvature of a metric as $\underline{g}_0$, it will be advantageous to write down an expression of the sectional curvature of a metric given as in (1) on a manifold M of dimension m. We are going to apply the formula to the case of a submersion. Hence we are interested in horizontal tangential planes. The horizontal condition for a vector X obviously depends on g. However, when we denote by $\underline{\omega}$ the $\mathbf{R}^m$-valued 1-form $(\underline{\omega}^1, \ldots, \underline{\omega}^m)$, the expression of the horizontal condition in terms of

$$(12) \qquad \tilde{\xi} = g\underline{\omega}(X)$$

is independent of g because the vertical vector fields are independent of $\underline{g}$. Therefore, it is advantageous for us to use the chart $\tilde{\xi}$ as far as it does not complicate the formula.

For general $\xi = (\xi^1, \ldots, \xi^m)$ and $\eta = (\eta^1, \ldots, \eta^m)$ in $\mathbf{R}^m$, we denote by $\langle \xi, \eta \rangle$ the standard inner product. For a symmetric matrix $a = (a_{st})$, $s, t = 1, \ldots, m$, we set

$$(13) \qquad \langle \xi, \eta \rangle_a = \sum a_{st} \xi^s \eta^t.$$

For tangent vectors X, Y to M we denote by $\langle X, Y \rangle_g$ the inner product. Hence

$$(14) \qquad \langle X, Y \rangle_{\underline{g}} = \langle \underline{\omega}(X), \underline{\omega}(Y) \rangle_g.$$

Denote by E_k the base of tangent vectors dual to $\underline{\omega}^j$. We set

$$(15) \qquad \mathscr{L}_g(\xi)\tilde{\eta} = -\sum \xi^k (E_k g^{-1})\tilde{\eta}.$$

We also introduce a symmetric form $\mathscr{L}(\tilde{\eta}, \tilde{\xi})$ by

$$(16) \qquad \langle \mathscr{L}_g(\theta)\tilde{\eta}, \tilde{\xi} \rangle = \langle \theta, \mathscr{L}_g(\tilde{\eta}, \tilde{\xi}) \rangle.$$

We need a linear map $\beta \colon \mathbf{R}^m \to \mathrm{Hom}(\mathbf{R}^m, \mathbf{R}^m)$ depending on points in M, which is given by the formula

$$(17) \qquad d\underline{\omega}(X, Y) = \beta(\underline{\omega}(X))\underline{\omega}(Y)$$

for any tangent vectors X, Y at any point in M. We may further require that

$$(18) \qquad \beta(\xi)\eta + \beta(\eta)\xi = 0$$

for all ξ, η in $\mathbf{R}^m$.

As usual we define $R(X, Y)$ by

$$(19) \qquad \nabla_X \nabla_Y - \nabla_Y \nabla_X = \nabla_{[X, Y]} + R(X, Y).$$

We set

$$(20) \qquad K(X, Y)_{\underline{g}} = \langle R(X, Y)Y, X \rangle_{\underline{g}}.$$

What we need is an expression of $K(X, Y)_{\underline{g}}$. After some calculation we find that

$$(21) \qquad K(X, Y)_{\underline{g}} = K_{\langle 0 \rangle} + K_{\langle 1 \rangle} + K_{\langle 2 \rangle} + K_{\langle 3 \rangle} + K_{\langle 4 \rangle} + K_{\langle 5 \rangle} + K_{\langle 6 \rangle},$$

where

$$K_{\langle 0 \rangle} = K_{01} + K_{02} \quad \text{with}$$

$$(22) \qquad K_{01} = \frac{1}{4}\|g^{-1/2}(\beta^{tr}(\xi)\tilde{\eta} - \beta^{tr}(\eta)\tilde{\xi}) - g^{1/2}\beta(\xi)\eta\|^2 - \|g^{1/2}\beta(\xi)\eta\|^2,$$

$$K_{02} = \langle \beta^{tr}(\xi)\tilde{\eta}, \beta^{tr}(\eta)\tilde{\xi} \rangle_{g^{-1}} - \langle \beta^{tr}(\xi)\tilde{\xi}, \beta^{tr}(\eta)\tilde{\eta} \rangle_{g^{-1}}.$$

$$(23) \quad K_{\langle 1 \rangle} = -\frac{3}{2}\langle \beta(\xi)\eta, \mathscr{L}_g(\xi)\tilde{\eta} - \mathscr{L}_g(\eta)\tilde{\xi} \rangle_g - \frac{3}{4}\|\mathscr{L}_g(\xi)\tilde{\eta} - \mathscr{L}_g(\eta)\tilde{\xi}\|^2_g,$$

$$K_{\langle 2 \rangle} = \frac{1}{2}\langle \beta^{tr}(\xi)\tilde{\eta} + \beta^{tr}(\eta)\tilde{\xi}, \mathscr{L}_g(\xi)\tilde{\eta} + \mathscr{L}_g(\eta)\tilde{\xi} - g^{-1}\mathscr{L}_g(\tilde{\xi}, \tilde{\eta}) \rangle$$

$$(24) \qquad - \langle \beta^{tr}(\xi)\tilde{\xi}, \mathscr{L}_g(\eta)\tilde{\eta} - \frac{1}{2}g^{-1}\mathscr{L}_g(\tilde{\eta}, \tilde{\eta}) \rangle$$

$$\qquad - \langle \beta^{tr}(\eta)\tilde{\eta}, \mathscr{L}_g(\xi)\tilde{\xi} - \frac{1}{2}g^{-1}\mathscr{L}_g(\tilde{\xi}, \tilde{\xi}) \rangle,$$

To write down $K_{\langle 3 \rangle}$ we set

$$(25) \qquad \Delta(\eta, \xi) = \eta^j \xi^k E_k E_j.$$

Then

$$(26) \qquad K_{\langle 3 \rangle} = \frac{1}{2}(\langle (\Delta(\xi, \xi)g^{-1})\tilde{\eta}, \tilde{\eta} \rangle + \langle (\Delta(\eta, \eta)g^{-1})\tilde{\xi}, \tilde{\xi} \rangle$$

$$- \langle (\Delta(\eta, \xi)g^{-1} + \Delta(\xi, \eta)g^{-1})\tilde{\xi}, \tilde{\eta} \rangle),$$

$$(27) \qquad K_{\langle 4 \rangle} = \frac{1}{2}(\langle \mathscr{L}_g(\xi)\tilde{\xi}, \mathscr{L}_g(\tilde{\eta}, \tilde{\eta}) \rangle + \langle \mathscr{L}_g(\eta)\tilde{\eta}, \mathscr{L}_g(\tilde{\xi}, \tilde{\xi}) \rangle$$

$$- \langle \mathscr{L}_g(\xi)\tilde{\eta} + \mathscr{L}_g(\eta)\tilde{\xi}, \mathscr{L}_g(\tilde{\xi}, \tilde{\eta}) \rangle),$$

$$(28) \qquad K_{\langle 5 \rangle} = \frac{1}{4}(\|\mathscr{L}_g(\tilde{\xi}, \tilde{\eta})\|_{g^{-1}})^2 - \langle \mathscr{L}_g(\tilde{\xi}, \tilde{\xi}), \mathscr{L}_g(\tilde{\eta}, \tilde{\eta}) \rangle_{g^{-1}}.$$

$$(29) \qquad K_{\langle 6 \rangle} = \langle [X, \beta](\eta)\xi, \eta \rangle_g + \langle [Y, \beta](\xi)\eta, \xi \rangle_g.$$

We now go back to $F(S_a)$ and calculate the sectional curvature using the above formula. We are interested in metrics on $F(S_a)$ invariant under the operation of T^2. To treat such metrics it is more convenient to use

$$(30) \qquad \underline{\omega}^j = h_k^j \omega^k, \qquad \underline{\omega}^{\overline{j}} = h_k^j \omega^{\overline{k}}$$

instead of ω^α where we denote by α the indexes $1, 2, 3, \overline{1}, \overline{2}, \overline{3}$. This is because we have the following:

(31) PROPOSITION. *A metric $\underline{g}$ given on $F(S_a)$ by*

$$(32) \qquad \underline{g} = g_{\alpha\beta}\underline{\omega}^\alpha \underline{\omega}^\beta$$

is invariant under the operation of T^2 if and only if $g_{\alpha\beta}$ is invariant under the operation of T^2.

Note that the metric $\underline{g}_0$ is given by

$$(33) \qquad \underline{g}_0 = \sum \underline{\omega}^\alpha \underline{\omega}^\alpha.$$

For $j, k, l = 1, 2, 3$ denote by $\varepsilon(j, k, l)$ the signature of the permutation $(1, 2, 3) \to (j, k, l)$. The structure equation of S_a takes the form:

$$(34) \qquad d\underline{\omega}^j = -a\varepsilon(j, k, l)\underline{\omega}^k \wedge \underline{\omega}^l,$$

$$(35) \qquad d\underline{\omega}^{\bar{j}} = \varepsilon(j, k, l)\left(-\frac{1}{2}\underline{\omega}^{\bar{k}} \wedge \underline{\omega}^{\bar{l}} + \frac{1}{2}a^2\underline{\omega}^k \wedge \underline{\omega}^l - a\underline{\omega}^k \wedge \underline{\omega}^{\bar{l}}\right).$$

To write down β, set $\underline{\omega}' = (\underline{\omega}^1, \underline{\omega}^2, \underline{\omega}^3)$, $\underline{\omega}'' = (\underline{\omega}^{\bar{1}}, \underline{\omega}^{\bar{2}}, \underline{\omega}^{\bar{3}})$, and

$$(36) \qquad \xi' = \underline{\omega}'(X), \quad \xi'' = \underline{\omega}''(X), \quad \eta' = \underline{\omega}'(Y), \quad \eta'' = \underline{\omega}''(Y).$$

We define

$$(37) \qquad \tilde{\Gamma} = \tilde{\xi}' \times \tilde{\eta}'' + \tilde{\xi}'' \times \tilde{\eta}', \quad \tilde{\Pi}' = \tilde{\xi}' \times \tilde{\eta}', \quad \tilde{\Pi}'' = \tilde{\xi}'' \times \tilde{\eta}'',$$

where $\times$ denotes the cross product. We then find by (34)–(35)

$$(38) \qquad \beta(\tilde{\xi})\tilde{\eta} = (-2a\tilde{\Pi}', a^2\tilde{\Pi}' - \tilde{\Pi}'' - a\tilde{\Gamma}).$$

Hence β is of constant coefficients. Therefore we see that

$$(39) \qquad K_{(6)} = 0.$$

For the metric $\underline{g}_0$, g is the identity matrix and $\tilde{\xi} = \xi$. Therefore $K(X, Y)_g = K_{(0)}$ can be calculated by using (38). Applying Jacobi's identity for the cross product several times to the term K_{02}, we find that

$$K(X, Y)_{\underline{g}_0} = A(\tilde{\xi}, \tilde{\eta}) \qquad \text{where}$$

$$(40) \qquad A(\tilde{\xi}, \tilde{\eta}) = \frac{1}{4}\|a^2\tilde{\Gamma}\|^2 + \frac{1}{4}\|\tilde{\Pi}'' + (3 - 2a^2)a^2\tilde{\Pi}'\|^2 + (1 - a^2)^3\|a\tilde{\Pi}'\|^2.$$

Therefore the sectional curvature is ≥ 0, provided $0 < a \leq 1$. In the following we always assume that

$$(41) \qquad 0 < a < 1.$$

We then find that a plane is of zero sectional curvature if and only if it is generated by X and Y such that

$$(42) \qquad \tilde{\xi} = (\theta, 0) \quad \text{and} \quad \tilde{\eta} = (0, \theta)$$

for a θ in $\mathbf{R}^3$. On the other hand we see by the definition of the operation of T^2 in (3)–(8) that the horizontal condition for a vector X is given by

$$(43) \qquad \langle \tilde{\xi}' - a\tilde{\xi}'', \underline{y} \rangle = \langle \tilde{\xi}'', \underline{x} \rangle = 0.$$

Hence a horizontal plane is of zero sectional curvature if and only if it is generated by X, Y such that

$$(44) \qquad \tilde{\xi} = (w, 0) \quad \text{and} \quad \tilde{\eta} = (0, w) \quad \text{with} \quad \langle w, \underline{x} \rangle = \langle w, \underline{y} \rangle = 0.$$

Therefore we find that over a point $(\underline{x}, \underline{y}) \in S^2 \times S^2$ with $\underline{z} = \underline{x} \times \underline{y} \neq 0$ there is exactly one horizontal plane. This plane is generated by X, Y where

$$(45) \qquad\qquad \tilde{\xi} = (\underline{z}, 0), \qquad \tilde{\eta} = (0, \underline{z}).$$

For a tangential plane to $S^2 \times S^2$ generated by $\mathbf{X}$ and $\mathbf{Y}$, we take their horizontal lift X, Y. Then $K(\mathbf{X}, \mathbf{Y})_{\mathbf{g}_0}$ is obtained by adding to $K(X, Y)_{\underline{g}_0}$ the O'Neill term:

$$(46) \qquad\qquad \frac{3}{4} \|\rho_V [X, Y]\|_{\underline{g}_0}^2$$

where ρ_V denotes the projection to the vertical tangential vector space. However, it turns out that this term depends only on $\tilde{\Gamma}, \tilde{\Pi}', \tilde{\Pi}''$. Therefore the planes of zero sectional curvature of $\mathbf{g}_0$ are the projections of the horizontal planes of zero sectional curvature of $\underline{g}_0$. Using the expression (9) of the projection we find that the above unique plane (to $S^2 \times S^2$) of zero sectional curvature is generated by $(\underline{x} \times \underline{z}, 0)$ and $(0, \underline{y} \times \underline{z})$.

At this point it is of some interest to note that the diagonal $\{(\underline{x}, \underline{x})\}$ and the antidiagonal $\{(-\underline{x}), \underline{x})\}$ are totally geodesic with respect to the metric $\mathbf{g}_0$ and the set of planes of zero sectional curvatures is exactly the set constructed in the proof of Frankel's theorem (cf. [9]).

We next consider deformations $\underline{g}$ of $\underline{g}_0$ which is invariant under the operation of T^2 and depending on a small ε. We call a horizontal plane critical when it is generated by X, Y such that $A(\tilde{\xi}, \tilde{\eta}) = 0$ (cf. (40)). A critical horizontal plane is also characterized by the formula (44).

We denote by $\mathscr{C}_{\underline{g}}$ the set of horizontal critical planes. By (44) we see that $\underline{g}\omega(\mathscr{C}_{\underline{g}})$ is independent of $\underline{g}$ and may be identified with $\underline{\omega}(\mathscr{C}_{\underline{g}_0})$. Let $\mathscr{U}_{\underline{g}}$ be a neighborhood of $\mathscr{C}_{\underline{g}}$ in the manifold of horizontal planes. We consider the case when $\underline{g}$ is a family of deformations of $\underline{g}_0$ depending on a small parameter ε. Assume that $\underline{g}\underline{\omega}(\mathscr{U}_{\underline{g}})$ is independent of ε. Then the following is obvious.

(48) PROPOSITION. *Under the assumption as above there is $\varepsilon_0 > 0$ such that for any ε with $|\varepsilon| < \varepsilon_0$ and for any horizontal plane Σ outside of $\mathscr{U}_{\underline{g}}$ the sectional curvature of Σ is > 0.*

In order to see if we can make the sectional curvature positive by a small deformation of $\underline{g}_0$ we find by the above that it is enough to consider only tangential planes in a small neighborhood of $\mathscr{C}_{\underline{g}}$. A convenient such neighborhood we use is the following: For $\delta > 0$ we denote by $\mathscr{U}_{\delta}$ the set of horizontal planes generated by X, Y such that

(49)
$$\tilde{\xi} = (w + w_1, w_2), \qquad \tilde{\eta} = (w_3, w)$$
$$\text{with } \|w\| = 1, \quad \langle w_1, w \rangle = \langle w_2, w \rangle = 0, \quad \|w_j\| \leq \delta \|w\|$$

$(j = 1, 2, 3)$. In the following we only consider planes in $\mathcal{U}_\delta$. We set

(50)
$$g^{-1} = I + \dot{f}, \qquad g = I + \dot{g}.$$

Hence $\underline{g}$ is given when we give $\dot{f}$. We find that

(51)
$$\xi = \tilde{\xi} + \dot{\xi}, \quad \text{where } \dot{\xi} = \dot{f}\tilde{\xi}.$$

Since $\dot{f}$ is a symmetric matrix, we can write

(52)
$$\dot{\xi}' = \dot{f}_1 \tilde{\xi}' + \dot{f}_3{}^{tr} \tilde{\xi}'', \qquad \dot{\xi}'' = \dot{f}_3 \tilde{\xi}' + \dot{f}_2 \tilde{\xi}'$$

where $\dot{f}_1$ and $\dot{f}_2$ are symmetric 3×3 matrix-valued functions and $\dot{f}_3$ is 3×3 matrix-valued. We consider the case

(53)
$$\dot{f}_3^{tr} = \dot{f}_3.$$

We will be concerned in this paper with deformations where $\dot{f}_j$ are constant matrices depending on small ε. Therefore we see by (23)–(26) that $K_{(l)} = 0$ for $l > 1$. We only have to calculate $K_{(0)}$. It is more convenient for calculation to rewrite the formula (22) of K_{01}. Note first that, since $g(I + \dot{f}) = I$,

(54)
$$\dot{g} = -g\dot{f}.$$

It then follows that $g = I + \dot{f} - (I + g)\dot{f}$. Therefore

(55)
$$g = g^{-1} - (I + g)\dot{f}.$$

Hence we see by (22) that

$$K_{01} = \frac{1}{4} \|\beta^{tr}(\xi)\tilde{\eta} - \beta^{tr}(\eta)\tilde{\xi} - g\beta(\xi)\eta\|_{g^{-1}}^2 - \|\beta(\xi)\eta\|_{g^{-1}}^2 + \|\beta(\xi)\eta\|_{(I+g)\dot{f}}^2.$$

Let us set

(56)
$$P = \beta(\xi)\eta,$$
$$Q = \beta^{tr}(\xi)\tilde{\eta} - \beta^{tr}(\eta)\tilde{\xi} - \beta(\xi)\eta.$$

In view of (54) it then follows that

(57)
$$K_{01} = K_{01P} + K_{01R}$$
$$\text{where } K_{01P} = \frac{1}{4}\|Q\|_{g^{-1}}^2 - \|P\|_{g^{-1}}^2,$$
$$K_{01R} = \frac{1}{2}\langle Q, P \rangle_{\dot{f}} + \frac{1}{4}\|\dot{g}P\|_{g^{-1}}^2 + \|P\|_{(I+g)\dot{f}}^2.$$

Noting by (54) that $\dot{g}g^{-1}\dot{g} = g\dot{f}^2$ and $(I + g)\dot{f} = (2I + \dot{g})\dot{f} = 2\dot{f} - g\dot{f}^2$, we find that

(58)
$$K_{01R} = \frac{1}{2}\langle Q, P \rangle_{\dot{f}} + 2\|P\|_{\dot{f}}^2 - \frac{3}{4}\langle P, g\dot{f}^2 P \rangle.$$

Note that K_{01} and K_{02} are expressed in the metric g^{-1}. For $K_{01P} + K_{02}$ we repeat the calculation similar to the one which led to the formula (40). Note also that each term in K_{01R} contains at least one dot.

For the calculation it is convenient to set (cf. (51))

$$(59) \qquad \dot{\xi}^{\#} = \dot{\xi}'' - a\dot{\xi}', \qquad \dot{\xi}^{\flat} = \dot{\xi}'' + a\dot{\xi}' .$$

and

$$(60) \qquad \begin{aligned} \Phi' &= \dot{\xi}^{\#} \times \tilde{\eta}' - \dot{\eta}^{\#} \times \tilde{\xi}', & \Phi'' &= \dot{\xi}^{\#} \times \tilde{\eta}'' - \dot{\eta}^{\#} \times \tilde{\xi}'', \\ \Theta' &= \dot{\xi}^{\flat} \times \tilde{\eta}' - \dot{\eta}^{\flat} \times \tilde{\xi}', & \Theta'' &= \dot{\xi}^{\flat} \times \tilde{\eta}'' - \dot{\eta}^{\flat} \times \tilde{\xi}''. \end{aligned}$$

We also set

$$(61) \qquad \widehat{\Pi}' = a\widetilde{\Pi}' - \Phi', \quad \widehat{\Pi}'' = \widetilde{\Pi}'', \quad \widehat{\Gamma} = a^2\widetilde{\Gamma} - \Phi',$$

and

$$(62) \qquad \dot{\Gamma} = \dot{\xi}' \times \dot{\eta}'' + \dot{\xi}'' \times \dot{\eta}', \quad \dot{\Pi}' = \dot{\xi}' \times \dot{\eta}', \quad \dot{\Pi}'' = \dot{\xi}'' \times \dot{\eta}''.$$

Then we see by (38) that with

$$(63) \qquad \begin{aligned} S' &= \Phi' + \Theta', & S'' &= a^{-1}\Phi' + \Theta'', \\ T' &= 2a\dot{\Pi}', & T'' &= -a^2\dot{\Pi}' + \dot{\Pi}'' + a\dot{\Gamma}, \end{aligned}$$

we have the formula

$$(64) \qquad \begin{aligned} P' &= -2\widehat{\Pi}' - S' - T', \\ P'' &= a\widehat{\Pi}' - \widehat{\Pi}'' - a^{-1}\widehat{\Gamma} - S'' - T''. \end{aligned}$$

We also find by (38) that

$$(65) \qquad \begin{aligned} \beta^{tr}(\xi)'\tilde{\eta} &= 2a\widetilde{\Pi}' + a\widetilde{\Pi}'' - a^2\dot{\xi}' \times \tilde{\eta}'', \\ \beta^{tr}(\xi)''\tilde{\eta} &= \widetilde{\Pi}'' + a\dot{\xi}' \times \tilde{\eta}''. \end{aligned}$$

It then follows that

$$(66) \qquad \begin{aligned} Q' &= 6\widehat{\Pi}' + 2a\widehat{\Pi}'' - \widehat{\Gamma} + 2S' + (\Phi' + a\Phi'') + T', \\ Q'' &= -a\widehat{\Pi}' + 3\widehat{\Pi}'' + 2a^{-1}\widehat{\Gamma} + 2S'' + T''. \end{aligned}$$

It is convenient to denote by E any terms such that

$$(67) \qquad |E| \leq \varepsilon(\|\widehat{\Gamma}\|^2 + \|\widehat{\Pi}'\|^2 + \|\widehat{\Pi}''\|^2) + \varepsilon^2\delta\|w\|^2 + \varepsilon^3\|w\|^2 .$$

It is hoped that E-terms are absorbed in the principal positive term.

After a considerable calculation we find that

$$(68)$$
$$K_{(0)} = \widehat{A} + B + C + E,$$

$$\text{where } \widehat{A} = \frac{1}{4}\|a^2\widehat{\Gamma}\|^2 + \frac{1}{4}\|\widehat{\Pi}'' + (3 - 2a^2)a^2\widehat{\Pi}'\|^2 + (1 - a^2)^3\|a\widehat{\Pi}'\|^2 ,$$

$$B = \left\langle 2\widehat{\Pi}' - \frac{1}{2}\widehat{\Gamma}, \Phi' + a\Phi'' \right\rangle ,$$

$$C = \frac{1}{4}\|\Phi' + a\Phi''\|^2 .$$

The above formula shows that in order to have the positivity $(\mathrm{mod}\ \varepsilon^3)$ for small deformations we have to consider the case

(69) $$\Phi' + a\Phi'' \equiv 0 \quad (\mathrm{mod}\ \delta\varepsilon,\ \varepsilon^2).$$

We have to examine the above condition. Note by (59) and (52) that

(70)
$$\xi^\# = -F_1\tilde{\xi}' + F_2\tilde{\xi}'', \qquad \dot{\eta}^\# = -F_1\tilde{\eta}' + F_2\tilde{\eta}'',$$
$$\text{where } F_1 = a\dot{f}_1 - \dot{f}_3,\ F_2 = \dot{f}_2 - a\dot{f}_3.$$

Hence we find by (60) that

(71)
$$\Phi' = -F_2 w \times w + \Phi'_1$$
$$\text{where } \Phi'_1 = -F_1 w \times w_3 + F_1 w_3 \times w - F_1 w_1 \times w_3 + F_1 w_3 \times w_1$$
$$- F_2 w \times w_1 + F_2 w_2 \times w_3,$$

(72)
$$\Phi'' = -F_1 w \times w + \Phi''_1$$
$$\text{with } \Phi''_1 = -F_2 w \times w_2 + F_2 w_2 \times w - F_1 w_1 \times w + F_1 w_3 \times w_2.$$

Therefore

(73) $$\Phi' + a\Phi'' \equiv -Fw \times w \quad (\mathrm{mod}\ \varepsilon\delta) \quad \text{where } F + F_2 + aF_1.$$

Therefore we see by (70) the condition (69) is satisfied when

(74) $$a^2\dot{f}_1 + \dot{f}_2 - 2a\dot{f}_3 = 0 \quad (\mathrm{mod}\ \varepsilon^2).$$

Therefore under the above condition

(75) $$K_{\langle 0 \rangle} = \widehat{A} + E.$$

Our situation does not seem encouraging. When we compare (40) with (75) we notice that the situation has not improved going from $\underline{g}_0$ to $\underline{g}$. Actually there are tangential planes in $\mathscr{U}_\delta$ at which $\widehat{A} = 0$. Is it possible that we can get the positivity when we calculate the coefficient of ε^l for $l \geq 3$? Or is there a structure in the formula which makes it impossible to obtain the positivity? These are tantalizing questions which remain to be answered.

REFERENCES

1. M. Berger, *Trois remarque sur les varieties riemanniennes a coubure positive*, C. R. Acad, Sci. Paris Sér. A-B **263** (1966), A76–A78.
2. J. P. Bourguignon, *Some constructions related to H. Hopf's conjecture on product manifolds*, Proc. Sympos. Pure Math., vol. 25, Amer. Math. Soc., Providence, RI, 1975, pp. 33–37.
3. J. P. Bourguignon, A. Deschamps, and P. Sentenac, *Conjecture de H. Hopf sur les produits de varietes*, Ann. Sci. École Norm. Sup. (4) **5** (1972), fasc. 2.
4. ____, *Quelques variations particulieres d'un produit de metriques*, Ann. Sci. École Norm. Sup. (4) **6** (1973), fasc. 1.
5. Jianguo Ca. *Some 4-manifolds of nonnegative curvature*, Mimeographed manuscript, Cornell University.
6. J. Cheeger, *Some example of manifolds of non-negative curvature*, J. Differential Geom. **8** (1972), 623–628.

7. S. S. Chern, *The geometry of G-structures*, Bull. Amer. Math. Soc. **72** (1966), 167–219.

8. D. G. Ebin, *The manifolds of Riemannian metrics*, Proc. Sympos. Pure Math., vol. 15, Amer. Math. Soc., Providence, RI, 1970, pp. 11–40.

9. T. Frankel, *Manifolds with positive sectional curvature*, Pacific J. Math. **11** (1961), 165–1174.

10. R. S. Hamilton, *Three-manifolds with positive Ricci curvature*, J. Differential Geom. **17** (1982), 255–306.

11. ____, *Four-manifolds with positive curvature operator*, J. Differential Geom. **24** (1986), 153–179.

12. W. Hsiang and B. Kleiner, *On the topology of positively curved 4-manifolds with symmetry*, J. Differential Geom. **29** (1989), 615–621.

13. S. Kobayashi, *Transformation groups in differential geometry*, Springer-Verlag, 1972.

14. S. Kobayashi and N. Nomizu, *Foundation of differential geometry*. Vols. I, II, Interscience, 1963, 1969.

15. B. O'Neill, *The fundamental equations of a submersion*, Michigan Math. J. **13** (1966), 451–469.

16. J. Wason, *A metric of positive sectional curvature on $S^2 \times S^2$*, unpublished.

17. A. Weinstein, *Positively curved n-manifolds in $\mathbf{R}^{n+2}$*, J. Differential Geom. **4** (1970), 1–4.

18. J. A. Wolf, *Spaces of constant curvature*, McGraw-Hill, New York, 1967, p. 408.

COLUMBIA UNIVERSITY

Proceedings of Symposia in Pure Mathematics
Volume **54** (1993), Part 3

Curve-straightening

ANDERS LINNÉR

ABSTRACT. The flow in the negative direction of the gradient vector field associated with the functional total squared (geodesic) curvature $\int_\gamma k^2 \, ds$ is the so-called curve-straightening flow. This paper will survey the main results concerning this flow as they have appeared during the last decade. The curve-straightening flow will be compared with the curve-shortening flow and other related flows. The most significant difference is that if the initial curve is in a trivial homotopy class the latter flows tend to shrink such curve to a point. On the other hand the curve-straightening flow will carry any initial curve to some nontrivial critical curve provided the functional is modified to be $\int_\gamma k^2 + \lambda \, ds$ for some positive real number λ. The set of critical curves contains geodesics as well as the so-called elastic curves. If λ is sufficiently small then on the standard sphere all nongeodesic critical curves are unstable. It follows that if the flow is used as a means of finding closed geodesics we expect the curve-straightening flow to be more useful than the other flows, in particular if the underlying manifold is sphere-like. Note that the curve-straightening flow exhibits a much different geometric behavior than the curve-shortening flow and examples of this will be given for planar curves.

0. Introduction

Curve-straightening is loosely speaking the flow generated when the trajectories of the gradient of the total squared (geodesic) curvature functional $\int_\gamma k^2 \, ds$ is followed in the negative direction.

This paper will give a survey of what is known about the so-called curve-straightening flow up to present time. The paper will also compare the curve-straightening flow with other flows such as the curve-shortening flow and the bibliography has been prepared accordingly. Another theme of the paper is to describe the advantages of the curve-straightening flow as a means of finding closed geodesics. In conjunction with this some significant open problems will be discussed.

1991 *Mathematics Subject Classification.* Primary 58F25; Secondary 58E10, 53C21.

Key words and phrases. Closed geodesics, Closed elastic curves, Curve-straightening, Total squared curvature, Steepest descent, Gradient flow.

This paper is in final form and no version will be submitted for publication elsewhere.

One of the motivations for considering this flow is a desire to be able to find (or construct) closed geodesics. If this problem is approached by variational methods a functional has to be selected and consequently a domain of this functional is also given. A minimal requirement on such functional is that the geodesics should be critical. The length functional $\int_\gamma ds$ and the energy functional $\int_\gamma ds^2$ are two examples of such functionals. Note that the length functional is independent of the parametrization of the curve but the energy functional is not. Geometrically natural functionals, such as length, tend to be invariant under coordinate transformation and the same can be said about functionals that physicists extremize to define basic physical laws. Below we will describe one of the principal tools in dealing with such functionals: the (generalized) indicatrix.

First we outline the steps of a general approach to the problem of constructing closed geodesics using variational methods. A first observation is that the periodic boundary conditions require a search not only for an extremal but also for the correct boundary values. In this paper all methods mentioned take care of the boundary conditions right away by considering flows defined only on spaces of closed curves. Note that we consider a curve closed only if both endpoints are the same and the directions are also the same at the endpoints. The curves of interest may be found by minimizing functionals on these spaces of closed curves. If such space of curves is given a Hilbert manifold structure then the most efficient way of minimizing the functional is to follow the gradient trajectory in the negative direction. For this approach to be successful a number of steps are needed. Let Ω be some space representing all "geometrically distinct" closed curves in some given manifold. We are deliberately vague here but we will be more precise as we move on. Here is the general program:

1. Select a functional $F: \Omega \to R$ such that the critical points include the closed geodesics. Note that a point $\hat{x}$ is critical if the derivative $DF(\hat{x})$ vanish.

2. Describe and maybe classify the set of critical points. There may exist critical points other than the closed geodesics.

3. Supply Ω with a Hilbert manifold structure $\langle \, , \, \rangle$ so that the gradient may be defined by $\langle \nabla F(x), v \rangle = DF(x)v$ for all (x, v) in $T\Omega$. Show that the trajectories of the negative gradient vector field converge to critical points. This may be done by verifying the Palais-Smale condition but it is not an easy task. Recall that the Palais-Smale condition (Condition C) says that if $\{x_n\}$ is any sequence in Ω for which $|F(x_n)|$ is bounded and for which $\|\nabla F(x_n)\| \to 0$ then there is a convergent subsequence $\{x_{n_k}\}$ (see [1, p. 181]).

4. Determine which critical points are stable (i.e., local minima). Initial curves sufficiently close would then converge to such stable critical point.

5. Analyze the behavior of the flow along the trajectories. In particular

answer questions about various properties such as embeddedness and convexity.

6. Develop algorithms that can be used to effectively solve the flow equations.

7. Implement the algorithms.

A number of these steps have been completed for the energy functional. Most of this work has been unified in Klingenberg's books [2, 3] and we refer to their respective biliographies for references. The energy functional has only geodesics and constant curves as critical points. In a trivial homotopy class curves tend to shrink to points if they flow along trajectories of the negative energy gradient. When searching for closed geodesics on sphere-like surfaces this behavior is undesirable.

Now the obvious question is if it is possible to choose a different functional that excludes constant curves from its domain of definition and still has closed geodesics as critical points but nevertheless satisfies the convergence properties in step 3. During the 1980s a good candidate has emerged and it is quite natural in the following sense. Consider a finite-dimensional possibly nonlinear system of equations. One possible way of solving such a system is to consider the real-valued function consisting of the sum of the squares of the component functions and then minimize this function. If it has a global minimum zero then this must be a solution and by computing the gradient of the function we may search for the minimum by steepest descent. If we let k be the geodesic curvature and we want to find curves such that $k = 0$ then the infinite-dimensional analog would be to consider the total squared curvature functional $\int_\gamma k^2\, ds$ and flow in the negative gradient direction. Note that the total squared curvature functional is independent of parametrizations of the curve and this causes a problem in step 3 since such functional always violates Condition C. For a discussion of this with the length and the energy functional as examples see [1, pp. 243–252].

The use of indicatrices as a means of dealing with this problem is a central idea in curve-straightening. What it amounts to is that instead of considering the functional as defined on a space of curves one restricts the attention to curves parametrized proportional to arclength. By choosing a frame at the initial point of the curve and then parallel translating this frame along the curve one may regard the unit tangent vector field along the curve as a map into a sphere. This map together with the length of the curve and an element of the orthonormal frame bundle representing the initial point of the curve and some specific choice of initial frame constitutes the indicatrix. The indicatrix simplifies the computation of the geodesic curvature and it consequently simplifies functionals defined in terms of the geodesic curvature. On the other hand it is difficult to recognize if a given indicatrix corresponds to a closed curve. It turns out that despite this, the space of indicatrices representing closed curves may be given a Hilbert manifold structure.

To illustrate these concepts consider the standard two-sphere in three-dimensional Euclidean space. Next consider the plane given by $z = h$. If $h \in (-1, 1)$ then the plane cuts the sphere along a circle and we will give the indicatrix in this case. Suppose we choose the initial point straight above the positive x-axis and follow the circle in the positive direction about the z-axis. Next choose an initial frame so that its first axis is parallel to the initial direction of the circle and the second axis so that with the outward normal we get a coordinate system of right-hand orientation. As this frame is parallel translated around the circle it will rotate in the negative direction about the outward normal. The indicatrix in this case consists of a map into a circle, the length $L = 2\pi\sqrt{1 - h^2}$, the initial point and the initial frame. If we let $\theta\colon [0, 1] \to R$ represent the map into the circle then $\theta(s) = 2\pi h s$ for $s \in [0, 1]$. The signed geodesic curvature is the constant $\frac{1}{L}\theta'(s) = h/\sqrt{1 - h^2}$ and the total squared curvature is $2\pi h^2/\sqrt{1 - h^2}$. Note that this quantity increases without bound as the circle shrinks to a point when h approaches 1 or -1. The paper is organized as follows:

  0. Introduction
  1. Curve-straightening and the general program, a progress report.
  2. Curve-straightening and its relation to curve-shortening and other flows.

1. Curve-straightening and the general program, a progress report

1.0. General remarks. Let M be a smooth (C^∞) n-dimensional Riemannian manifold and let γ be a smooth curve in M (Sobolev H^2 is enough for 1.3 below).

1.1. Choice of functional. The functional we choose is given by

$$F^\lambda(\gamma) = \int_\gamma k^2 + \lambda\, ds;$$

here k is the geodesic curvature along the curve γ and λ is some nonnegative real number. Later in the paper when we use indicatrices we will carefully state the domain of the functional. The need for the parameter λ will be explained in 1.3 below.

1.2. Description and classification of the critical points. Let T denote the unit tangent vector field along γ. Let R be the Riemann curvature tensor and ∇ the Riemannian connection on M. Critical points of F^λ satisfy

$$2(\nabla_T)^3 T + \nabla_T(3k^2 - \lambda)T + 2R(\nabla_T T, T)T = 0.$$

Geodesics $(\nabla_T T = 0, k = 0)$ are clearly always critical. The nongeodesic critical points are the so-called elastic curves. In manifolds of constant sectional curvature this equation has been integrated and the critical points subsequently classified (see [4] and for a more general approach to the integration [5, 6, 7], for curves in R^3 see [8] and for curves in the hyperbolic plane see [9]).

1.3. The gradient flow and its convergence. To be able to use gradient flow methods it is necessary to turn the space of all curves in M into a Riemannian manifold. It turns out that by restricting attention to curves parametrized proportional to arclength and their corresponding indicatrices a well-defined gradient flow may be defined on the space of indicatrices. If S denotes all Sobolev H^1 maps into the standard sphere S^{n-1}, let R^+ be the positive reals and OM the orthonormal frame bundle. The domain of our functional is given to be $\Omega \subset S \times R^+ \times OM$, the set of indicatrices representing closed curves. In [10] Ω is given a Hilbert manifold structure. The fundamental result now is that the Palais-Smale condition is satisfied for the total squared curvature functional when defined on Ω. For curves of fixed length in R^2 and R^3 condition C is established in [11]. For curves of variable length and positive λ or for curves of fixed length and any λ condition C is established in arbitrary Riemannian manifolds in [10]. An indication why condition C fails when $\lambda = 0$ is given by considering for increasing p the sequence of p-fold great circles on the standard sphere. This sequence has no convergent subsequence but it does not provide an example if $\lambda > 0$ since the functional in this case is unbounded. A significant open problem is to understand the convergence (or divergence) of the gradient flow even when $\lambda = 0$. Note that if $\lambda = 0$ then all closed geodesics are automatically stable since they are global minima.

1.4. Stability. Nongeodesic elastic curves have been shown unstable in two-dimensional manifolds for $\lambda = 0$ and constant nonnegative Gaussian curvature G in [4] and in the sphere of constant curvature G for $0 \leq \lambda \leq \frac{8}{7}G$ in [10]. The stability of the critical points in the nonconstant curvature case is still a significant open problem. The second variation formula may be derived using the methods in [4] (see [12] for the expression along closed geodesics) but it is quite complicated and difficult to analyze.

In the plane we let the rotation number η be the number of times the unit tangent vector along the curve rotates around the circle in the positive direction. When $\int_\gamma k^2 \, ds$ is restricted to closed curves of fixed length it turns out that for each nonzero η only the η-folded circle is critical. When $\eta = 0$ there is however an infinite number of critical points. There is a particular figure eight defined in terms of elliptic functions giving the global minimum. The remaining critical points are multiple covers of the same figure eight scaled suitably. For instance the double covered figure eight is half the size and so on and so forth. All these multiple covers are unstable critical points (see [11]).

1.5. The behavior along the trajectories. In the plane rotationally symmetric convex curves remain convex during the flow (see for instance [11] or for more details [13]). It was conjectured in [11] that this was true for nonsymmetric curves as well but counterexamples were found in [13]. Explicit examples of embedded curves flowing to nonembedded curves are also given in [13]. As will be seen in §2 this is in sharp contrast with the curve-shortening flow.

1.6. Effective algorithms. For rotationally symmetric curves of rotation number $\eta = 1$ the curve-straightening flow has an explicit solution (see [13]). For arbitrary planar curves the gradient vector field is known explicitly in terms of integrals of the indicatrix (see [13]). The gradient vector field is not known explicitly for any other manifolds, including the standard sphere. A complete algorithm for the flow in Euclidean submanifolds is presented in [12] where the main contribution is to show how to reduce the problem of computing the gradient vector field to a problem of solving systems of ordinary differential equations and systems of linear equations.

1.7. Implementations. There is an implementation of the curve-straightening flow for curves of fixed length in the plane. The most spectacular animated computer graphics example produced shows an entire trajectory starting close to the unstable double covered figure eight (see 1.4 above) and ending at the global minimum, the single covered figure eight. For a series of snap shots see [14].

2. Curve-straightening and its relation to curve-shortening and other flows

2.0. General remarks. In this section we will relate curve-straightening to some other flows that have been studied in conjunction with geodesic problems. It should be understood that the flows below are not necessarily gradient flows. Let us mention that the curve-shortening flow may be regarded as the L^2-gradient of the length functional (see the introduction in [13] for instance). The method these flows try to improve is the so-called Birkhoff curve-shortening process (see [15]).

2.1. Curve-shortening. In a series of papers Gage, Hamilton and Grayson investigated the so-called curve-shortening flow which is when a curve $\gamma: S^1 \to M$ evolves according to

$$\frac{\partial \gamma}{\partial t} = kN;$$

here t is the flow parameter, k is the curvature of the curve and N is the normal to the curve. Unlike curve-straightening this flow preserves convexity (see [16]–[18]) and embedded curves stay embedded (see [19, 20]). For a nonembedded example see [21]. Curve-shortening will tend to shrink homotopically trivial curves down to points. If the flow is modified and restrictions are put on the initial curve this can be avoided (see [22, 23]).

Despite the fact that the evolution equation above is seemingly simpler than the evolution equation for curve-straightening (see [13]) it is surprisingly difficult to give an effective algorithm for the curve-shortening flow (see [24]–[26]). In attempt to be complete we also list [27, 28].

2.2. Miscellaneous flows. Here we like to mention the so-called normalized curve-shortening flow (see [29]). This flow gives insight into the asymptotic behavior of the curve-shortening flow. A classification of curves that evolve by scaling is given. The only embedded such curve is the circle. More about the asymptotic behavior can be found in [30]–[32]. There is also the so-called polygonal curve flow which is closer to the Birkhoff process than the other flows (see [33, 34]). It has also been shown to preserve embeddedness.

2.3. Curve-straightening and nonclosed curves. The total elastic energy in a thin beam or wire is given by the total squared curvature functional $\int_\gamma k^2\, ds$. It is therefore of interest to minimize the functional over a space containing curves satisfying an arbitrary combination of boundary condition. Flow methods may still be used as in [14] but a different considerably less general approach is presented in [35].

REFERENCES

1. R. S. Palais and C. Terng, *Critical point theory and submanifold geometry*, Lecture Notes in Math., vol. 1353, Springer-Verlag, 1988.
2. W. Klingenberg, *Lectures on closed geodesics*, Springer-Verlag, 1978.
3. ____, *Riemannian geometry*, de Gruyter, Berlin, New York 1982.
4. J. Langer and D. A. Singer, *The total squared curvature of closed curves*, J. Differential Geom. **20** (1984), 1–22.
5. R. Bryant and P. Griffiths, *Reduction for constrained variational problems and $\int k^2/2\, ds$*, Amer. J. Math. **108** (1982), 525–570.
6. P. Griffiths, *Exterior differential systems and the calculus of variations*, Birkhäuser, 1982.
7. V. Jurdjevic, *Singular optimal problems*, Perspectives in control theory (Sielpia 1988) Progr. Systems Control Theory, Birkhäuser, Boston **2** (1990), 75–88.
8. J. Langer and D. A. Singer, *Knotted elastic curves in R^3*, J. London Math. Soc. **30** (1984), 512–520.
9. ____, *Curves in the hyperbolic plane and mean curvature of tori in 3-space*, Bull. London Math. Soc. **16** (1984), 531–534.
10. ____, *Curve-straightening in Riemannian manifolds*, Ann. Global Anal. Geom. **5** (1987), 133–150.
11. ____, *Curve-straightening and a minimax argument for closed elastic curves*, Topology **24** (1985), 75–88.
12. A. Linnér, *Curve-straightening in closed Euclidean submanifolds*, Commun. Math. Phys. **138** (1991), 33–49.
13. ____, *Some properties of the curve straightening flow in the plane*, Trans. Amer. Math. Soc. **314** (1989), 605–617.
14. ____, *Steepest descent as a tool to find critical points of $\int k^2\, ds$ restricted to planar curves satisfying arbitrary boundary conditions*, Proc. MSRI Workshop Differential Geom., Springer-Verlag Geometric Analysis and Computer Graphics, MSRI Publications, Vol. 17, Springer (1991), 127–138.
15. G. D. Birkhoff, *Dynamical systems with two degrees of freedom*, Trans. Amer. Math. Soc. **18** (1917), 199–300.
16. M. Gage, *An isoperimetric inequality with applications to curve shortening*, Duke Math. J. **50** (1983), 1225–1229.
17. ____, *Curve shortening makes convex curves circular*, Invent. Math. **76** (1984), 357–364.
18. M. Gage and R. Hamilton, *The shrinking of convex plane curves by the heat equation*, J. Differential Geom. **23** (1986), 69–96.
19. M. Grayson, *The heat equation shrinks embedded plane curves to round points*, J. Differential Geom. **26** (1987), 285–314.

20. ____, *Shortening embedded curves*, Ann of Math. (2) **129** (1989), 71–111.

21. ____, *The shape of a figure-eight curve under the curve shortening flow*, Invent. Math. **96** (1989), 177–180.

22. M. Gage, *On an area-preserving evolution equation for plane curves*, Contemp. Math., no. 51, Amer. Math. Soc., Providence, RI, 1986, 51–62 .

23. ____, *Curve shortening on surfaces*, Ann. Sci. École Norm. Sup. (4) **23** (1990) no. 2, 229–256.

24. J. A. Sethian, *Curvature and the evolution of fronts*, Comm. Math. Phys. **101** (1985), 487–499.

25. S. Osher and J. A. Sethian, *Fronts propagating with curvature dependent speed: Algorithms based on Hamilton-Jacobi formulations*, J. Comput. Phys. **79** (1988), 12–49.

26. J. A. Sethian, *Numerical algorithms for propagating interfaces: Hamilton-Jacobi equations and conservtoin laws* J. Differential Geom. **31** (1990), 131–161.

27. C. Epstein and M. Gage, *The curve shortening flow*, Wave Motion: Theory, Modelling, and Computation, Proc. Conf. in honor of the 60th birthday of Peter D. Lax, Math. Sci. Res. Inst. Publ., Springer-Verlag, 1987.

28. C. Epstein and M. Weinstein, *A stable manifold theorem for the curve shortening equation*, Comm. Pure Appl. Math. **40** (1987), 119–139.

29. U. Abresch and J. Langer, *The normalized curve shortening flow and homothetic solutions*, J. Differential Geom. **23** (1986), 175–196.

30. S. Angenent, *Parabolic equations for curves on surfaces. I. Curves with p-integrable curvature*, Ann. of Math. **132** (1990), 451–483.

31. ____, *Parabolic equations for curves on surfaces. II. Intersections, blowup and generalized solutions*, Preprint.Ann. of Math. **133** (1991), 171–215.

32. ____, *On the formation of singularities in the curve shortening flow*, J. Differential Geom. **33** (1991), 601–633.

33. J. Hass and G. P. Scott, *Intersections of curves on surfaces*, Israel J. Math. **51** (1985), 90–120.

34. ____, *Shortening curves on surfaces*, Preprint (1990).

35. B. K. P. Horn, *The curve of least energy*, ACM Trans. Math. Software **9** (1983), 441–460.

NORTHERN ILLINOIS UNIVERSITY

Proceedings of Symposia in Pure Mathematics
Volume 54 (1993), Part 3

Ball Covering Property and
Nonnegative Ricci Curvature Outside a Compact Set

ZHONG-DONG LIU

ABSTRACT. Let M be a complete open Riemannian manifold with nonnegative Ricci curvature outside a compact set B. We show that the usual ball covering property holds.

1. Introduction

On a complete open Riemannian manifold with nonnegative Ricci curvature, the volume comparison of Bishop and Gromov [**BC, GLP**] can be used to show that the geometric growth of the ends of the manifold is well controlled. Let us name a few properties that will be discussed here:

(A) There are a bounded number of ends (Cheeger-Gromoll's splitting theorem implies that there are at most two ends).

(B) Every end has at most linear diameter growth [**AG**].

(C) There is a *ball covering property*: for a fixed point p and $0 < \mu < 1$, $B_p(r)$ can be covered by a bounded number of balls of radius μr; this bound, called the packing number, is independent of r.

When the global condition $\mathrm{Ric}_M \geq 0$ is relaxed to $\mathrm{Ric}_{M-B} \geq 0$, where B is a compact set, the relative volume comparison is weakened. However, as we observed in [**L**], the ball covering property (C) implies (A) and (B) with no additional assumptions on curvatures. Thus, it is natural to examine the ball covering property in this case. The question of whether the ball covering property is true when $\mathrm{Ric}_{M-B} \geq 0$ was first raised by P. Li and L. Tam in [**LT1**].

In [**L**], we assumed an arbitrary lower bound on the sectional curvature in addition to the assumption that $\mathrm{Ric}_{M-B} \geq 0$ and proved the ball covering property. At the AMS Summer Institute on Differential Geometry held in U.C.L.A., July 1990, P. Li announced that he and L. Tam have proved (A),

1991 *Mathematics Subject Classification*. Primary 53C20.
Key words and phrases. Ricci curvature, end, volume comparison, geodesic ball.
This paper is in final form and no version of it will be submitted for publication elsewhere.

(without assuming a lower bound on the sectional curvature); i.e., a complete Riemannian manifold with nonnegative Ricci curvature outside a compact set has finitely many ends (see [LT2]). Their approach is analytic in nature. At about the same time, M. Cai [C] independently proved this same result by purely geometric means.

Cai's argument suggested to us that our original approach in [L] could be slightly modified by using the triangle inequality argument in place of Toponogov's theorem. This is carried out here, establishing the ball covering property assuming only $\text{Ric}_{M-B} \geq 0$. Furthermore, the covering number can be estimated in terms of the dimension, the lower bound on the Ricci curvature, and the radius of the ball $B_p(D)$ that contains the compact set B. (Ball covering is not treated in either [C] or [LT2].)

We state and prove here a general version of the ball covering property.

THEOREM 1. *Let M^n be a complete Riemannian manifold with nonnegative Ricci curvature outside a compact set B. Assume that $\text{Ric}_B \geq (n-1)H$ and $B \subset B_{p_0}(D_0)$. Then for any $\mu > 0$, there exists $N = N(n, HD_0^2, \mu) > 0$ such that for any $r > 0$, if S is a subset satisfying*

$$S \subset \overline{B}_{p_0}(r),$$

we can find $p_1, \ldots, p_k \in S$, $k \leq N$, with

$$\bigcup_{j=1}^{k} B_{p_j}(\mu \cdot r) \supset S.$$

2. Proof of Theorem 1

First, let us recall the well-known relative volume comparisons.

Let M be an n-dimensional Riemannian manifold. Let $B_p(r)$ denote the geodesic ball of radius r at $p \in M$. Put $A_p(R, r) := B_p(R) - \overline{B_p(r)}$, $R > r > 0$. $V_p(r) := \text{vol}(B_p(r))$, $V_p(R, r) := \text{vol}(A_p(R, r))$. We use $V^H(r)$ to denote the volume of a ball of radius r in the space form of constant curvature H of the same dimension. $V^H(R, r) := V^H(R) - V^H(r)$. We have (see [GLP]) the following:

LEMMA 1. *If on $B_p(R)$, $\text{Ric} \geq (n-1)H$, and $0 < r < R$, then*

$$V_p(R)/V_p(r) \leq V^H(R)/V^H(r),$$

$$V_p(R, r)/V_p(R) \leq V^H(R, r)/V^H(R),$$

and

$$V_p(R, r)/V_p(r) \leq V^H(R, r)/V^H(r).$$

There is also a relative volume comparison for star-shaped sets. A star-shaped set S_p at p is a set containing p such that whenever $x \in S_p$ is not on the cut-locus of p, any point on the minimal geodesic joining p and x is also in S_p. We then have [CGT, §4, Remark 4.1]

LEMMA 2. *If S_p is a star-shaped set and $\mathrm{Ric}|S_p \geq (n-1)H$, then for $R > r > 0$ we have*

$$\mathrm{Vol}(S_p \cap B_p(R))/\mathrm{Vol}(S_p \cap B_p(r)) \leq V^H(R)/V^H(r).$$

The following lemma is proved in [L].

LEMMA 3. *Assume that $\mathrm{Ric}|B_p(R) \geq 0$. Given $m > 1$, let*

$$\delta = (1 - 1/(2m))^{1/n}.$$

If $W \subset B_p(R)$ has

$$\mathrm{Vol}(W) \geq 1/m \cdot \mathrm{Vol}(B_p(R)) \ ,$$

then there exists $q \in W$ such that $\mathrm{dist}(q, p) \leq \delta R$. Hence

$$B_q((1 - \delta)R) \subset B_p(R).$$

PROOF OF THEOREM 1. We assume that all geodesics are parametrized by arclength.

We can assume that $H < 0$. As in [L], we multiply the metric on M by $\sqrt{-H}$ and work with the rescaled metric. Write $D = \sqrt{-H} \cdot D_0$. Then $B \subset B_{p_0}(D)$ and $\mathrm{Ric}_B \geq -(n-1)$.

Let $\mu > 0$ be given. We may assume that $\mu \leq 2$. Otherwise the theorem is trivial. We divide S into the union of S_1 and S_2, where

$$S_1 = S \cap B_{p_0}(\mu r/2), \qquad S_2 = S - S_1.$$

If S_1 is not empty, it can be covered by just one $B_p(\mu r)$ with p in S_1. So we only have to estimate the covering number for S_2, which is contained in $\overline{B}_{p_0}(r) - B_{p_0}(\mu r/2)$. Also, it suffices to count the number of balls of radius $\mu r/4$ needed to cover S_2. Let us denote $t := \mu/4$.

First, we assume that $t \cdot r > 2D$. Note that for any $q \in S_2$,

$$B_q(tr) \cap \overline{B}_{p_0}(2D) = \emptyset.$$

Write $\partial B_{p_0}(2D)$ as the union of subsets $\{U_1, ..., U_m\}$ such that $\forall x, y \in U_a$, $d_M(x, y) \leq 2D$. This can be done as follows. Take a maximal set of points $\{q_1, ..., q_m\} \subset \partial B_{p_0}(2D)$ such that $\mathrm{dist}(q_a, q_b) \geq D$, $a \neq b$. Then

$$\bigcup B_{q_a}(D) \supset \partial B_{p_0}(2D);$$

$$B_{p_a}(D/2) \cap B_{p_b}(D/2) = \emptyset, \qquad a \neq b.$$

Suppose $B_{p_s}(D/2)$ has the smallest volume among all $B_{p_a}(D/2)$. Since $\bigcup_{a=1}^m B_{p_a}(D/2) \subset B_{p_s}(5D)$, relative volume comparison (Lemma 1) implies that

(1) $m \le V^{-1}(5D)/V^{-1}(D/2)$.

Note that the right-hand side of (1) depends only on n and D. We define

$$U_a = B_{q_a}(D) \cap \partial B_{p_0}(2D), \qquad a = 1, \ldots, m.$$

Let M_r be the subset of M consisting of all points on any minimal geodesic emanating from p_0 that is no shorter than r. Note that $M_r \supset M - B_{p_0}(r)$, and M_r is star-shaped at p_0.

We now divide M_{2D} into m cones K_a by defining K_a to be the subset consisting of all points on any minimal geodesic emanating from p_0 that intersects U_a.

REMARK. By the triangle inequality, if

$$d(x_i, p_0) > 2D, \qquad x_i \in K_a, \, i = 1, 2,$$

then any minimal geodesic connecting x_1 and x_2 will not pass through $B_{p_0}(D)$.

Indeed, let γ_i be a minimal geodesic from p_0 to x_i with $\gamma_i(2D) \in U_a$, $i = 1, 2$. Then the broken geodesic from x_1 to $\gamma_1(2D)$ to $\gamma_2(2D)$ to x_2 has length $\le d(x_1, p_0) + d(x_2, p_0) - 2D$. On the other hand, if a minimal geodesic connecting x_1 and x_2 intersects $B_{p_0}(D)$, then it would have a length $> d(x_1, p_0) + d(x_2, p_0) - 2D$, which is a contradiction. (It is this observation, inspired by our reading of [C], that allows us to eliminate the use of Toponogov's theorem and hence, the requirement of an arbitrary lower bound on the sectional curvature imposed in [L].)

Now we can estimate the covering number just as in [L]. For the convenience of the reader, we will repeat some of the constructions.

Take a maximal set of points $\{p_1, \ldots, p_k\}$ in S_2 such that

$$\text{dist}(p_i, p_j) \ge tr, \qquad i \ne j.$$

Then

(2) $$\bigcup B_{p_i}(tr) \supset S_2;$$

(3) $$B_{p_i}(tr/2) \cap B_{p_j}(tr/2) = \varnothing, \qquad i \ne j.$$

We then divide the balls $B_{p_i}(tr/2)$ into m families as follows: for each ball $B_{p_i}(tr/2)$, look at $\text{Vol}(B_{p_i}(tr/2) \cap K_a)$, $a = 1, \ldots, m$. Fix an a_i such that $\text{Vol}(B_{p_i}(tr/2) \cap K_{a_i})$ is maximal. Then

(4) $$\text{Vol}(B_{p_i}(tr/2) \cap K_{a_i}) \ge \frac{1}{m} \text{Vol}(B_{p_i}(tr/2)).$$

We denote

$$B_{p_i}(tr/2) \cap K_{a_i} = B_{p_i}^{L,a_i},$$

or simply $B_{p_i}^L$, and place the ball $B_{p_i}(tr/2)$ in the a_ith family. Fix an K_a. Suppose B_p^L has the smallest volume among all $B_{p_i}^{L,a}$ in this cone. By Lemma 3 we can find a $q \in B_p^L$ such that

$$B_q((1-\delta)tr/2) \subset B_p(tr/2).$$

Let W_q be the star-shaped set such that $y \in W_q$ if and only if there is a point x belonging to either $B_q((1-\delta)tr/2)$ or one of $B_{p_i}^{L,a}$ in the cone K_a and there is a minimal geodesic γ connecting q and x which passes y. By the previous remark,

$$B_q((1-\delta)tr/2) \subset W_q \subset (M - B_{p_0}(D)) \cap B_q((2+t)r).$$

The same argument in [L] yields that the number of balls in the ath family is bounded by

$$m \cdot 2^n (2+t)^n t^{-n} (1-\delta)^{-n};$$

hence the total number of balls

$$(5) \qquad\qquad k \leq m^2 2^n (2+t)^n t^{-n} (1-\delta)^{-n},$$

where δ is defined in Lemma 3. Since m depends only on n and D, the right hand side of (5) is a function of n, D, and μ. (Recall that $t = \mu/4$.)

If $tr \leq 2D$, that is $\mu r \leq 8D$, as in the proof of (1), we can bound k by

$$\max_{0 < \mu r \leq 8D} V^{-1}((2+\mu)r)/V^{-1}(\mu r/2),$$

which again depends only on μ, n and D and not on r. Theorem 1 is proved.

3. Applications

As mentioned in [L], the following two corollaries are immediate consequences of ball covering property.

COROLLARY 1 [C, LT2]. *Under the same assumption as in Theorem 1, the number of ends is finite and bounded by some $N_1 = N_1(n, D_0^2 H)$.*

COROLLARY 2. *Under the same assumption as in Theorem 1, the diameter growth of each end is at most linear.*

Let $b_i(p, r)$ denote the rank of $i_*: H_i(B_p(r), \mathbb{R}) \to H_i(M, \mathbb{R})$. Using Gromov's isotopy lemma [G] and ball covering property, Z. Shen proved the following theorem on topological growth of geodesic balls.

THEOREM 2 [Sh2]. *If Ricci curvature of M is nonnegative and sectional curvature $K_M \geq -1$, then there is a constant $C(n)$ such that*

$$\sum_{0 \leq i \leq n} b_i(p, r) \leq C(n)(1 + r)^n, \qquad r > 0.$$

The only use of nonnegative Ricci curvature in his proof is to apply the ball covering property. Since this property can be generalized to the case of nonnegative Ricci curvature outside a compact set, we have

THEOREM 3. *If* $\mathrm{Ric}_{M-B} \geq 0$, $K_M \geq -1$, *and* $B \subset B_{p_0}(D)$, *then there is a* $C(n, D)$, *such that*

$$\sum_{0 \leq i \leq n} b_i(p, r) \leq C(n, D)(1 + r)^n, \qquad r > 0.$$

Acknowledgment

I would like to thank Professor J. Cheeger for his guidance and encouragement. I would also like to thank Professor D. Gromoll for his interest in this work and several helpful discussions.

REFERENCES

[AG] U. Abresch and D. Gromoll, *On complete manifolds with nonnegative Ricci curvature*, J. Amer. Math. Soc. vol. 3 (1990), 355–374.

[BC] R. Bishop and R. Critenden, *Geometry of manifolds*, Academic Press, 1964.

[C]   M. Cai, *Ends of Riemannian manifolds with nonnegative Ricci curvature outside a compact set*, Bull Amer. Math. Soc. 24 (1991), 371–377.

[CGT] J. Cheeger, M. Gromov and M. Taylor, *Finite propagation speed, kernel estimates for functions of the Laplace operator, and the geometry of complete Riemannian manifolds*, J. Differential Geom. 17 (1982), 15–53.

[G]   M. Gromov, *Curvature, diameter and Betti numbers*, Comment. Math. Helv. 56 (1981), 179–195.

[GLP] M. Gromov, J. Lafontaine, and P. Pansu, *Metrique pour les varietes Riemanniennes*, Nathan, Paris, 1982.

[LT1] P. Li and L. Tam, *Positive harmonic functions on complete manifolds with non-negative curvature outside a compact Set*, Ann. of Math. (2) 125 (1987), 171–207.

[LT2] _____, *Harmonic functions and the structure of complete manifolds*, J. Differential Geom. 35 (1992), 359–383.

[L]   Z. Liu, *Ball Covering on manifolds with nonnegative Ricci curvature near infinity*, Proc. Amer. Math. Soc. 115 (1992), 211–219.

[Sh1] Z. Shen, *Finiteness and vanishing theorems for complete open Riemannian manifolds*, Bull. Amer. Math. Soc. (N.S.) 21 (1989), 241–244.

[Sh2] Z. Shen, *Finite topological type and vanishing theorems for Riemannian manifolds* (Ph.D. Thesis), SUNY Stony Brook, 1990.

STATE UNIVERSITY OF NEW YORK AT STONY BROOK

Proceedings of Symposia in Pure Mathematics
Volume **54** (1993), Part 3

General Conjugate Loci are not Closed

CHRISTOPHE M. MARGERIN

1. Introduction

It is an elementary remark in basic Riemannian geometry that the cut locus of a point is a closed set. See, for example, Kobayashi's "On Conjugate & Cut Loci" in *Studies in Global Geometry & Analysis* edited by Chern (M.A.A. Studies in Mathematics, Vol. 4, 1967). Although there is no specific reason of any kind for the conjugate locus to be closed, this seems to be taken for granted by many—see, for example, Cheeger and Ebin's textbook *Comparison Theorems in Riemannian Geometry* (1975), on p. 94, when proving Proposition 5.4: "the set of singular values of the exponential map is closed; an accumulation point of conjugate points is a conjugate point." If this had been true it could have been of some help in the understanding of the structure of distance ball in a singular space. Unhappily we prove here:

1. in general—even when the sectional curvature is positive, and the manifold compact (in which case *each* "caustic" is compact)—a sequence of conjugate points to a fixed point P may converge to a point which is not conjugate to P;

2. in general—even for a compact manifold of nonnegative curvature—the set of first conjugate points to a given point P may not be closed: its closure may even contain points which are not conjugate to P.

The only purpose of this is to improve the common intuition of how badly behaved a conjugate locus may be.

2. Caustics accumulation towards nonconjugate points

We show, for any smooth function $f: [-1, +1] \to \mathbf{R}^+$, even, constant and equal to some irrational number α on $[-a, +a]$, $a \in (0, 1)$, monotonically increasing on $(a, 1]$ and equal to 1 at 1 (see Figure 1 on next page), the metric defined by

1991 *Mathematics Subject Classification*. Primary 53C22, 53C20, 51M09, 51M15.
This paper is in final form and no version of it will be submitted for publication elsewhere.

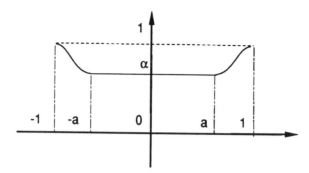

FIGURE 1. A generic function f considered in the construction.

$$(1) \qquad g = f(\cos r)^2 \, dr^2 + \sin^2 r \, d\theta^2,$$

where (r, θ) is the usual polar coordinate system of the round sphere $\mathbf{S}^2$, is a smooth metric on $\mathbf{S}^2$, for which the conjugate locus of any point on the equator $r = \frac{\pi}{2}$, intersects the equator in a dense countable subset of the equator; in particular the conjugate set of any such point cannot be closed. This construction may be trivially suspended into higher dimensions, since the geodesics of such a metric are well known (see Darboux's "Leçons sur la théorie des surfaces" (1894), or Besse's "Manifolds all of whose geodesics are closed") and characterized by the following system of equations:

$$(2) \qquad \frac{d\theta}{ds} = \varepsilon_1 \frac{\sin i}{\sin^2 r} \quad \text{(Clairaut's first integral)},$$

$$(3) \qquad \frac{d\theta}{dr} = \frac{\varepsilon_1 \varepsilon \sin i f(\cos r)}{\sin r (\sin^2 r - \sin^2 i)^{1/2}},$$

$$(4) \qquad \frac{ds}{dr} = \frac{\varepsilon \sin r f(\cos r)}{(\sin^2 r - \sin^2 i)^{1/2}},$$

where

- ε_1 is $+$ or $-$ according to the orientation of the geodesic,
- ε switches from $+$ to $-$ and vice versa each time the r-function reaches $\pi - i$ or i.

It is not difficult to get the following identity for the angle which separates two consecutive intersections of the geodesic culminating at $r = i$ or $\pi - i$ with the equator

$$(5) \qquad \Theta(i) = \varepsilon_1 \varepsilon \int_i^{\pi - i} \frac{\sin i}{\sin r} \frac{f(\cos r)}{(\sin^2 r - \sin^2 i)^{1/2}} \, dr.$$

One has the picture given in Figure 2. When noticing $\Theta(i) = \pi$ for the "round" sphere $(f \equiv 1)$, this provides cheap counterexamples (singular—or noncomplete when removing the two isolated singularities): take f to be constant and equal to some irrational number α (see Figure 3).

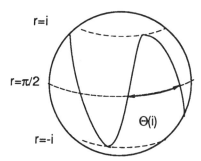

FIGURE 2. Definition of the angle $\Theta(i)$.

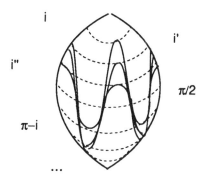

FIGURE 3. An easy singular or noncomplete counterexample.

The purpose of the construction is to "smooth" the singularities without introducing new conjugate points on the equator. To do so—i.e., to get a handle on the behavior of the conjugate locus—we explicitly solve the Jacobi equations. This is possible because of the S^1-isometric action which provides us with one more Jacobi field $(\frac{\partial}{\partial\theta})$ than the usual $\frac{d\gamma}{ds}(s)$ and $s\cdot\frac{d\gamma}{ds}(s)$. Writing

$$\text{(6)} \qquad \frac{\partial}{\partial\theta} = \sin i \frac{d\gamma}{ds}(s) + \varepsilon(\sin^2 r - \sin^2 i)^{1/2} \cdot \vec{n}$$

and projecting the Jacobi equation on the normal vector $\vec{n}$, one reduces the problem to a second-order O.D.E. $y'' + yK = 0$ (where K is the curvature at the point under consideration), which, in turn, reduces to a first-order O.D.E. on each geodesic segment $[\pi - i, i]$, since $(\sin^2 r - \sin^2 i)^{1/2}$ is known as a solution according to (6) for the Killing vector field $\frac{\partial}{\partial\theta}$: the relative length function, ℓ, defined by

$$\text{(7)} \qquad y = \tilde{y}_0 l, \quad \text{where } \tilde{y}_0 = (\sin^2 r - \sin^2 i)^{1/2},$$

satisfies the following differential equation:

$$\text{(8)} \qquad \frac{dl}{dr} = \frac{\varepsilon \sin r f(\cos r)}{(\sin^2 r - \sin^2 i)^{3/2}};$$

a general Jacobi field normal to the geodesic segment $[\pi - i, i]$, vanishing at $r = \frac{\pi}{2}$ and $\theta = 0$, is characterized by the following equation:

$$(9) \qquad Y_0(r) = \left(\int_{\pi/2}^r \frac{\varepsilon \sin \rho f(\cos \rho)}{(\sin^2 \rho - \sin^2 i)^{3/2}} \, d\rho \right) (\sin^2 r - \sin^2 i)^{1/2} \cdot \vec{n}$$

$$(10) \qquad = -\varepsilon \left(\frac{\cos r f(\cos r)}{\cos^2 i (\sin^2 r - \sin^2 i)^{1/2}} + \int_{\pi/2}^r \frac{\cos \rho \sin \rho f'(\cos \rho)}{\cos^2 i (\sin^2 \rho - \sin^2 i)^{1/2}} \, d\rho \right) \cdot \vec{n}$$

where $\varepsilon = +1$.

The next technical problem consists in solving the equation through a singularity ($r = i$ or $r = \pi - i$). I proved (cf. Margerin, "De l'hypothétique fermeture du lieu conjugué"—Preprint M878.0289 de l'école polytechnique)

LEMMA 1. *One has the following characterization of the only nontrivial Jacobi field along the kth geodesic arc $[i, \pi - i]$:*

$$(11) \qquad Y_k = Y_0(r) + (-1)^k k (\sin^2 r - \sin^2 i)^{1/2} P(i) \cdot \vec{n}$$

where

$$(12) \qquad P(i) = \int_i^{\pi - i} \frac{\cos \rho \sin \rho f'(\cos \rho)}{\cos^2 i (\sin^2 \rho - \sin^2 i)} \, d\rho,$$

is the period associated to the geodesic culminating at $r = i$ and $r = \pi - i$.

PROOF. The proof is by induction on k. Equation (8) gives the expression of the Jacobi field along the first geodesic segment $[\pi - i, i]$, which is called $Y_0(r)$ when it is looked at as a function of r. On the following segment $r \in [i, \pi - i]$, $l_1(r)$ is determined by
  1. equation (8),
  2. initial conditions

$$a. \quad \ell_1(i)\tilde{y}_1(i) = \ell_0(i)\tilde{y}_0(i),$$

$$b. \quad \frac{d}{ds}(\ell_1(r)\tilde{y}_1(r))|_{r=i} = \frac{d}{ds}(\ell_0(r)\tilde{y}_0(r))|_{r=i}.$$

From (8), one deduces the existence of a constant A such that

$$(13) \qquad (Y_1(r), \vec{n}) = -(\sin^2 r - \sin^2 i)^{1/2}(A - \ell_0(r))$$

$$(14) \qquad = (Y_0(r), \vec{n}) - A(\sin^2 r - \sin^2 i)^{1/2}.$$

Using the following identities

$$(15) \qquad \begin{aligned} \frac{d(Y_1, \vec{n})}{ds} &= -\frac{d(Y_0, \vec{n})}{ds} \frac{(\sin^2 r - \sin^2 i)^{1/2}}{\sin r f(\cos r)} \\ &\quad + A \frac{\sin r \cos r}{(\sin^2 r - \sin^2 i)^{1/2}} \frac{(\sin^2 r - \sin^2 i)^{1/2}}{\sin r f(\cos r)}, \end{aligned}$$

and

(16)
$$\frac{d(Y_0, \vec{n})}{ds} = \frac{d(Y_0, \vec{n})}{dr} \frac{(\sin^2 r - \sin^2 i)^{1/2}}{\sin r f(\cos r)},$$

the fact that Y_1 and Y_0 fit well together at $r = i$ (a Jacobi field is a smooth vector field along the geodesic) translates into the following characterization of A:

(17)
$$\frac{A \cos i}{f(\cos i)} = 2 \left(\frac{d}{dr} (\ell_0(r)(\sin^2 r - \sin^2 i)^{1/2}) \frac{(\sin^2 r - \sin^2 i)^{1/2}}{\sin r f(\cos r)} \right)_{r=i}$$

(18)
$$= -2 \frac{\cos i}{f(\cos i)} \int_{\pi/2}^{i} \frac{\cos \rho \sin \rho f'(\cos \rho)}{\cos^2 i (\sin^2 \rho - \sin^2 i)^{1/2}} d\rho.$$

The lemma for Y_1 is now a straight consequence of identity (14), and of the previous computation of the period A.

For a general "melding" at $r = i$, equation (8) leads to the identity

(19)
$$(Y_{2k+1}(r), \vec{n}) = (Y_{2k}(r), \vec{n}) - B(\sin^2 r - \sin^2 i)^{1/2},$$

from which it is deduced by differentiation

(20)
$$\frac{d(Y_{2k+1}, \vec{n})}{ds} = - \frac{(d Y_{2k}, \vec{n})}{dr} \frac{(\sin^2 r - \sin^2 i)^{1/2}}{\sin r f(\cos r)}$$
$$+ \frac{B \sin r \cos r}{(\sin^2 r - \sin^2 i)^{1/2}} \frac{(\sin^2 r - \sin^2 i)^{1/2}}{\sin r f(\cos r)}.$$

Using

(21)
$$\frac{d(Y_{2k}, \vec{n})}{ds} = \frac{d(Y_{2k}, \vec{n})}{dr} \frac{(\sin^2 r - \sin^2 i)^{1/2}}{\sin r f(\cos r)},$$

and the induction assumption at stage $2k$, which reads
(22)
$$\frac{d(Y_{2k}, \vec{n})}{ds} = \frac{(\sin^2 r - \sin^2 i)^{1/2}}{\sin r f(\cos r)} \left(\frac{d(Y_0, \vec{n})}{dr} + 2k \frac{\sin r \cos r}{(\sin^2 r - \sin^2 i)^{1/2}} \right.$$
$$\left. \times \int_i^{\pi-i} \frac{\cos \rho \sin \rho f'(\cos \rho)}{\cos^2 i (\sin^2 \rho - \sin^2 i)^{1/2}} d\rho \right),$$

leads to the following evaluation of the constant B

(23)
$$B = -2 \int_{\pi/2}^{i} \frac{\cos \rho \sin \rho f'(\cos \rho)}{\cos^2 i (\sin^2 \rho - \sin^2 i)^{1/2}}$$
$$+ 4k \int_i^{\pi-i} \frac{\cos \rho \sin \rho f'(\cos \rho)}{\cos^2 i (\sin^2 \rho - \sin^2 i)^{1/2}} d\rho,$$

which, up to the induction assumption, is what is claimed in the lemma. The same proof—mutatis mutandis—gives the announced expression for $Y_{2k+2}(r)$, considering a matching-up at $r = \pi - i$.

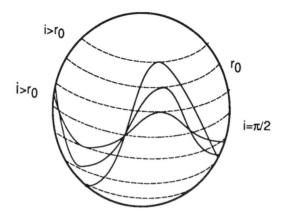

FIGURE 4. None of the geodesic culminating at $r < r_0$ has a point on the equator conjugated to P.

To end the proof of the first claim of the paper, we remark that

$$(24) \qquad Y_0\left(\frac{\pi}{2}\right) = 0,$$

and that

$$(25) \qquad \left(Y_{2k}\left(\frac{\pi}{2}\right), \vec{n}\right) = 2(-1)^k k(1 - \sin^2 i)^{1/2}$$
$$\int_i^{\pi-i} \frac{\cos\rho\sin\rho f'(\cos\rho)}{\cos^2 i(\sin^2\rho - \sin^2 i)}\,d\rho,$$

which is different from zero as soon as k is not zero and f is a monotonic function on the interval $(0, \cos i)$ that is not constant—i.e., when $|\cos i| > a$, as shown in Figure 4.

Let me stress the curvature of such a rotationally symmetric metric is given by the following expression:

$$(26) \qquad K(r) = \frac{1}{f^3(\cos r)}(f(\cos r) - \cos r f'(\cos r)),$$

in such a way it may be achieved to be positive by choosing α close to 1 and a close to 0.

3. Closure to the first conjugate locus need not be made of conjugate points

As already mentioned, there is no hope for counterexample with curvature bounded from below by a positive constant. We will construct many examples as soon as the curvature is nonnegative, even on a compact manifold—here $\mathbf{S}^2$.

To do so, let us choose the system of coordinates on $\mathbf{S}^2$ as shown in Figure 5, and introduce the following metrics

$$(27) \qquad g(u, \theta) = du^2 + a^2(u)\,d\theta^2,$$

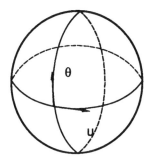

FIGURE 5. The (u, θ) system of coordinate on $\mathbf{S}^2$; $u \in [-\alpha, \alpha]$, $\alpha \in \mathbf{R}_+^*$, and $\theta \in [0, 2\pi]$.

which are known (same references as before) to be smooth, provided $a > 0$, except at the end-points $\pm\alpha$, where $a(\alpha) = a(-\alpha) = 0$, $a'(\alpha) = -a'(-\alpha) = -1$, and $a^{2p}(\alpha) = a^{2p}(-\alpha) = 0$ for all p in $\mathbf{N}^*$. We will restrict ourselves to even functions a, and convex (for the curvature, which is given by $K = a''/a$, to be nonnegative). For such a metric the equator is shown to be contained in the closure of the first conjugate locus of any (fixed) point of the equator, although there are only countably many conjugate points to that fixed point on the equator—at least as soon as the function a is analytic in u.

3.1. The first step is very similar to the previous section: it amounts to an explicit formula for the only nontrivial Jacobi field along a geodesic. Because of the $\mathbf{S}^1$-symmetry of the considered metrics, it reduces to a first order O.D.E. on each geodesic segment $[-u_1, u_1]$. The only technical and not obvious result is the "period" computation.

LEMMA 2. *The only nontrivial normal Jacobi field is characterized by the following expression*:

(28) $$Y_k(u) = Y_0(u) - (-1)^k k B (a^2(u) - a^2(u_1))^{1/2} P' \cdot \vec{n}$$

where the period P' is given by the expression

(29)
$$P'(u_1) = 2 \left(\int_{u_1/2}^{u_1} \frac{a''}{a'^2} \frac{1}{(a^2 - a^2(u_1))^{1/2}} \, du - \frac{1}{a'(u_1/2)} \frac{1}{(a^2(u_1/2) - a^2(u_1))^{1/2}} \right.$$
$$\left. - \int_0^{u_1/2} \frac{a(u)}{(a^2(u) - a^2(u_1))^{3/2}} du \right)$$

PROOF. For any metric of the previous type—see (27)—the geodesics are characterized by the equation

(30) $$\frac{d\theta}{ds} = \frac{a(u_1)}{a^2(u)}$$

where $a(u_1)$ is the radius of the shortest "parallel" reached by the geodesic which is considered; for shortness sake, I will say this geodesic "culminates" in u_1. In the (u, θ)-coordinates the tangent and normal vectors are characterized by the following equations:

$$(31) \qquad \dot{\gamma}(u, \theta) = \begin{pmatrix} \dfrac{du}{ds} \\ \dfrac{d\theta}{ds} \end{pmatrix}$$

and, respectively,

$$(32) \qquad \vec{n}(u, \theta) = \begin{pmatrix} -a\dfrac{d\theta}{ds} \\ \dfrac{1}{a}\dfrac{du}{ds} \end{pmatrix}.$$

The Killing vector field

$$(33) \qquad \frac{\partial}{\partial \theta} = \begin{pmatrix} 0 \\ 1 \end{pmatrix} = a^2 \frac{d\theta}{ds} \cdot \dot{\gamma}(s) + a\frac{du}{ds} \cdot \vec{n}(s)$$

is then a Jacobi field; in particular $\tilde{y}(u) = a(u)\frac{du}{ds}$ satisfies the (scalar) Jacobi equation

$$(34) \qquad y'' + Ky = 0.$$

In order to get the general solution, we introduce a relative "length function" ℓ defined by the identity $y = \ell \tilde{y}$, the same way we did in the first part. This function then satisfies a first-order O.D.E.

$$(35) \qquad \frac{d\ell}{ds} = \frac{A}{a(u)^2} \frac{1}{(\frac{du}{ds})^2}$$

that we get by integration of the following translation of (34):

$$(36) \qquad \frac{\ell''}{\ell'} = -\frac{2\tilde{y}'}{\tilde{y}}.$$

Thanks to the trivial identity

$$(37) \qquad \frac{du}{ds} = \varepsilon \left(1 - \frac{a^2(u_1)}{a^2(u)} \right)^{1/2},$$

the previous differential equation rewrites

$$(38) \qquad \frac{d\ell}{ds} = \frac{A}{(a^2(u) - a^2(u_1))^{1/2}} \quad \text{and} \quad \frac{d\ell}{du} = \frac{\varepsilon A a(u)}{(a^2(u) - a^2(u_1))^{3/2}},$$

from which we deduce the following expression of a general normal Jacobi field that vanishes at the origin of the geodesic:

$$(39) \qquad Y(u) = \varepsilon \ell(u)(a^2(u) - a^2(u_1))^{1/2} \cdot \vec{n};$$

in particular, on the first geodesic segment $[-u_1, u_1]$,

$$(40) \qquad Y_0(u) = \int_0^u \frac{a(u)\,du}{(a^2(u) - a^2(u_1))^{1/2}} (a^2(u) - a^2(u_1))^{1/2},$$

which I wish to write
—if $u > 0$

$$
(41) \qquad
\begin{aligned}
Y_0(u) = {} & -\frac{1}{a'(u)} + (a^2(u) - a^2(u_1))^{1/2} \\
& \cdot \Bigg(-\int_{u_1/2}^u \frac{1}{(a^2(u) - a^2(u_1))^{1/2}} \frac{a''}{a'^2}\,du \\
& \quad + \frac{1}{a'(u_1/2)} \frac{1}{(a^2(u_1/2) - a^2(u_1))^{1/2}} \\
& \quad + \int_0^{u_1/2} \frac{a(u)}{(a^2(u) - a^2(u_1))^{3/2}}\,du \Bigg) ;
\end{aligned}
$$

—if $u < 0$

$$
(42) \qquad
\begin{aligned}
Y_0(u) = {} & -\frac{1}{a'(u)} + (a^2(u) - a^2(u_1))^{1/2} \\
& \cdot \Bigg(-\int_{-u_1/2}^u \frac{1}{(a^2(u) - a^2(u_1))^{1/2}} \frac{a''}{a'^2}\,du \\
& \quad + \frac{1}{a'(-u_1/2)} \frac{1}{(a^2(u_1/2) - a^2(u_1))^{1/2}} \\
& \quad + \int_0^{u_1/2} \frac{a(u)}{(a^2(u) - a^2(u_1))^{3/2}}\,du \Bigg)
\end{aligned}
$$

According to equation (38), the expression of a general Jacobi field along the kth geodesic arc can be deduced from the expression on the $(k-1)$th segment, up to a constant $B(k)$, which we earlier called "period"

$$(43) \qquad Y_1(u) = Y_0(u) + B(a^2(u) - a^2(u_1))^{1/2}.$$

Because a Jacobi field is smooth, the expressions for Y_0 and Y_1 must fit at $u = u_1$ up to order 2 (the Jacobi equation is a second-order differential equation!)

$$
(44) \qquad
\begin{aligned}
\frac{dY_1}{ds}\bigg|_{u=u_1} &= \frac{dY_1}{du} \cdot \frac{du}{ds}\bigg|_{u=u_1} = \frac{dY_0}{ds}\bigg|_{u=u_1} \\
&= \left\{ -\left(\frac{dY_0}{du} + \frac{Baa'}{(a^2(u) - a^2(u_1))^{1/2}} \right) \frac{(a^2 - a^2(u_1))^{1/2}}{a} \right\}_{u=u_1}.
\end{aligned}
$$

Since the differentiation of (41) reads

$$\frac{dY_0}{ds}\bigg|_{u=u_1} = \left(-\int_{u_1/2}^{u_1} \frac{1}{(a^2 - a^2(u_1))^{1/2}} \frac{a''}{a'^2} du \right.$$

(45)
$$+ \frac{1}{a'(u_1/2)} \frac{1}{(a^2(u_1/2) - a^2(u_1))^{1/2}}$$

$$+ \left. \int_0^{u_1/2} \frac{a(u)}{(a^2(u) - a^2(u_1))^{3/2}} du \right) a'(u_1),$$

this leads to the following evaluation of the constant B:

(46) $$B = 2 \left(\int_{u_1/2}^{u_1} \frac{a''}{a'^2} \frac{1}{(a^2 - a^2(u_1))^{1/2}} du - \frac{1}{a'(u_1/2)} \frac{1}{(a^2(u_1/2) - a^2(u_1))^{1/2}} \right.$$

$$\left. - \int_0^{u_1/2} \frac{a(u)}{(a^2(u) - a^2(u_1))^{3/2}} du \right),$$

which achieves the proof of the lemma for $k = 1$.

As for the general case, equation (38) reads

(47) $$Y_k = Y_{k-1} + \alpha(a^2 - a^2(u_1))^{1/2}.$$

Let us assume, to start with, k is *odd*. Then Y_{k-1} and Y_k will have to match together at $u = u_1$, which, at order one, reads

(48) $$\frac{dY_k}{ds} = -\frac{dY_{k-1}}{ds} - \alpha\alpha', \quad \text{i.e., } \alpha = -\frac{2}{a'(u_1)} \frac{dY_{k-1}}{ds}\bigg|_{u_1}.$$

Differentiating the induction assumption (at order $k - 1$), we get

(49) $$\alpha = -\frac{2}{a'(u_1)} \left(\frac{d}{ds}Y_0\bigg|_{u_1} - (k-1)Ba'(u_1) \right) = (2k-1)B;$$

that is, with the help of (47) and the induction assumption at stage $k - 1$

(50) $$Y_k(u) = Y_{k-1} + (2k-1)B(a^2 - a^2(u_1))^{1/2}$$

(51) $$= Y_0(u) + kB(a^2 - a^2(u_1))^{1/2}.$$

We deal with the case k is even in exactly the same way—mutatis mutandis. This ends the proof of the lemma.

3.2. We now want to prove the period function P' has at most countably many zeros on $[0, \alpha]$. To do so we remark

LEMMA 3. *The period function* $P': u_1 \mapsto P'(u_1)$ *is real analytic on any closed subinterval of* $(0, \alpha)$, *provided the function* a *is analytic. In particular the period function has at most countably many zeros on* $[0, \alpha]$.

PROOF.

3.2.1. As soon as a is analytic, $a(u_1/2) \neq a(u_1)$, and $a'(u_1/2) \neq 0$—which amounts to saying $u_1 \neq 0$, since a is decreasing on $[0, \alpha]$—the following function of u_1,

$$
(52) \qquad \frac{1}{a'(u_1/2)} \frac{1}{(a^2(u_1/2) - a^2(u_1))^{1/2}} \, ,
$$

is clearly analytic in a neighborhood of u_1.

3.2.2. If a is analytic, if $u_1 \notin \{\alpha, 0\}$, then $a(u_1) > 0$, $a'(u_1) < 0$ and

$$
(53) \qquad \frac{a''}{a'^2(a(u) + a(u_1))}
$$

is analytic on $[u_1/2, u_1]$. Around any such point u_1 the analytic function a has the following power series expansion (with positive convergence radius)

$$
(54) \qquad a(u) = \sum_N a_n(u_1 - u)^n .
$$

Since $a_1 = -a'(u_1) > 0$ (provided $u_1 > 0$), we can write

$$
(55) \quad \frac{a''}{a'^2} \frac{1}{(a^2 - a^2(u_1))^{1/2}} = \frac{a''}{a'^2} \frac{1}{(a + a(u_1))^{1/2}} \frac{1}{(a - a(u_1))^{1/2}} = \frac{F(u, u_1)}{(u_1 - u)^{1/2}}
$$

where F is an analytic function in both its arguments. The following change of variable,

$$
(56) \qquad
\begin{aligned}
\int_{u_1/2}^{u_1} \frac{F(u, u_1)}{(u_1 - u)^{1/2}} du &= \int_{\sqrt{u_1/2}}^{\sqrt{u_1}} 2F(u_1 - v^2, u_1) dv \\
&= \int_{\sqrt{u_1/2}}^{\sqrt{u_1}} G(v, u_1) dv ,
\end{aligned}
$$

where $G(v, u_1)$ is analytic in both its arguments, makes explicit the analyticity of

$$
(57) \qquad u_1 \mapsto \int_{u_1/2}^{u_1} \frac{a''}{a'^2} \frac{1}{(a^2 - a^2(u_1))^{1/2}} \, .
$$

3.2.3. As for the last term

$$
(58) \qquad \int_0^{u_1/2} \frac{a(u)}{(a^2(u) - a^2(u_1))^{3/2}} \, ,
$$

this is clearly analytic in the u_1-variable on $(0, \alpha]$.

From the three previous remarks it should be clear the period map is analytic on $(0, \alpha)$.

3.3. To get the equator $\{u = 0\}$ is contained in the closure of the conjugate locus to P, we note $(Y_0, \vec{n}) > 0$ for $u > 0$ and $u < u_1$, which leads to the following lower bound for the θ-coordinate of the first conjugate point:

$$(59) \qquad \int_0^{u_1} \frac{d\theta}{ds}\,ds = \int_0^{u_1} \frac{a(u_1)}{a(u)}\,\frac{du}{(a^2 - a^2(u_1))^{1/2}}$$

$$(60) \qquad = \int_0^{u_1} \frac{a(u_1)}{-a'(u)a^2(u)}\,\frac{-a'(u)a(u)}{(a^2 - a^2(u_1))^{1/2}}\,du$$

$$(61) \qquad \geq \left(-\frac{a(u_1)}{a(0)a'(u_1)}\right)(a^2(0) - a^2(u_1))^{1/2}.$$

Since $a'(0) = 0 = a''(0) = a'''(0) = a^{(2p+1)}(0)$ for every $p \in \mathbf{N}$, we can write

$$(62) \qquad a(0) - a(u) = u^{4+2p} f(u),$$

where f is a smooth function and $f(0) > 0$, so that we obtain the following lower bound for the θ-coordinate of the first conjugate point:

$$(63) \qquad \begin{aligned} &\int_0^{u_1} \frac{a(u_1)}{a(u)}\,\frac{du}{(a^2(u) - a^2(u_1))^{1/2}} \\ &\geq \frac{a(u_1)}{a^2(0)}\,\frac{u_1^{2+p} f'^{1/2}(u_1)(a(0) + a(u_1))^{1/2}}{(4 + 2p)f(u_1)u_1^{3+2p} + f'(u_1)u_1^{4+2p}}. \end{aligned}$$

From this we deduce

$$(64) \qquad \lim_{u_1 \to 0} \int_0^{u_1} \frac{a(u_1)}{a(u)}\,\frac{du}{(a^2 - a^2(u_1))^{1/2}} = +\infty \qquad \text{(see Figure 6).}$$

The equation of the geodesic running out of the point P on the equator,

$$(65) \qquad \frac{d\theta}{ds} = \frac{a(u_1)}{a^2(u)},$$

leads to the following characterization of the initial tangent vector

$$(66) \qquad \left(\begin{array}{c} \dfrac{du}{ds} \\[2mm] \dfrac{d\theta}{ds} \end{array}\right)\Bigg|_{s=0} = \left(\begin{array}{c} 1 - \dfrac{a^2(u_1)}{a^2(0)} \\[2mm] \dfrac{a(u_1)}{a^2(0)} \end{array}\right);$$

because of the function a being decreasing on $[0, \alpha]$ and continuous, we deduce from the previous characterization that the map

$(u_1 \mapsto$ initial tangent vector (at P) to the geodesic culminating at $u = u_1)$

is a homeomorphism onto its image, $\mathbf{S}^1$ minus two points.

Since the cut distance by conjugacy, θ_c, is a continuous function of the initial tangent vector of the considered geodesic, it is also a continuous function of u_1, the culminating height of the geodesic. We have the following obvious translation of the fact that θ_c is a well-defined function out of zero:

$$(67) \qquad \forall \theta_0 \in [0, 2\pi]\ \forall k \in \mathbf{N}\ \exists p \in \mathbf{N} \quad \theta_c\left(\frac{1}{k}\right) < \theta_0 + 2p\pi,$$

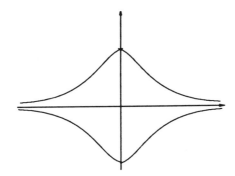

FIGURE 6. The first critical set in $T_P M$.

which, combined with the previous lower bound for the cut distance by conjugacy

(68) $$\exists \alpha < \frac{1}{k} \quad \theta_c(\alpha) > \theta_0 + 2p\pi ,$$

and the continuity of the function θ_c, leads to the assertion

(69) $$\forall k \in \mathbf{N} \ \exists \alpha_k \ 0 < \alpha_k < \frac{1}{k} \quad \theta_c(\alpha_k) = \theta_0 \pmod{2\pi}.$$

It can be remarked the u-coordinate of any point of the geodesic culminating in α_k satisfies the following bounds:

(70) $$|u| \leq \alpha_k ;$$

doing so, we have just proved any point $(\theta_0, u = 0)$ of the equator is the accumulation point of a sequence of first conjugate points to P, a given point of the equator, along geodesics culminating at lower and lower heights (see Figure 7). This ends the proof of the claims we started with.

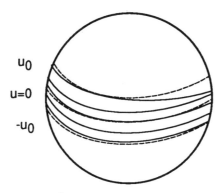

FIGURE 7. This $(\mathbf{S}^2, g)$ is flat along the equator $u = 0$: the smaller is u_0 the more the geodesic culminating at u_0 spins around the equator before it hits its first conjugate point.

REFERENCES

1. A. Besse, *Manifolds all of whose geodesics are closed*, Ergeb. Math. Grenzgeb., vol. 93, Springer, 1978.
2. J. Cheeger and D. Ebin, *Comparison theorems in Riemannian geometry*, North-Holland, 1975.
3. G. Darboux, *Leçons sur la théorie générale des surfaces*. Vols. 1, 2, 3, 4, Gauthier-Villars, Paris, 1894–1905.
4. S. Kobayashi, *On conjugate and cut loci*, "Studies in Global Geometry and Analysis" (S. S. Chern, ed.), M.A.A. Stud. Math., vol. 4, 1967.

CENTRE DE MATHÉMATIQUES, ÉCOLE POLYTECHNIQUE, FRANCE

Proceedings of Symposia in Pure Mathematics
Volume **54** (1993), Part 3

Collapsing of Riemannian Manifolds and Their Excesses

YUKIO OTSU

1. Introduction

A sequence of n-dimensional Riemannian manifolds $\{M_i\}_{i=1,2,\ldots}$ is said to collapse to a space N if $\{M_i\}$ converges to N with respect to the Hausdorff distance d_H [**G**] and if the Hausdorff dimension of N is less than n. When N is a Riemannian manifold and $\{M_i\}$ is contained in a certain class of Riemannian manifolds, the phenomena are studied in [**F**, **Y2**]. In [**Y2**] Yamaguchi showed that there exists $\varepsilon = \varepsilon(n, i_0) > 0$ such that if M is an n-dimensional complete Riemannian manifold with sectional curvature $K_M \geq -1$, N is a complete Riemannian manifold with dimension $\dim N \leq n$, $|K_N| \leq 1$ and injectivity radius $\operatorname{inj}(N) \geq i_0 > 0$, and if they satisfy $d_H(M, N) \leq \varepsilon$, then there exists a fibration $f: M \to N$ such that f is an almost Riemannian submersion and the first Betti number of the fiber is not greater than its dimension. The purpose of this paper is to announce the results on the descriptions of the collapsing phenomena of this type without referring to the manifold N, and certain finiteness and pinching theorems.

Let $X = (X, d)$ be a metric space and p and q be two points of X. The excess function $e_{pq}: X \to \mathbf{R}$ is defined by

$$e_{pq}(x) := d(p, x) + d(q, x) - d(p, q)$$

for $x \in X$. The radius of X is defined by

$$\operatorname{rad}(X) := \inf_{p \in X} \sup_{q \in X} d(p, q).$$

Let X be a length space and $d \in (0, \operatorname{rad}(X)]$. We define

$$e^d(X) := \sup_{p \in X} \sup_{x \in B_d(p)} \inf_{q \in S_d(p)} e_{pq}(x),$$

1991 *Mathematics Subject Classification.* Primary 53C20, 53C22.
The detailed version of this paper has been submitted for publication elsewhere [**O2**].

where $B_r(p)$ (or $B_r(p, X)$) is the metric r-ball around p and $S_r(p)$ is the metric r-sphere around p.

REMARK. Note that e_{pq} was defined in [AG] (see also [GP1]) and $\text{rad}(X)$ was defined in [SY].

EXAMPLE. (1) If $X = S^n(1)$, then $e^d(X) = 0$ for $d \leq \text{rad}(X) = \pi$.

(2) If $X = [0, 1]$, then $e^d(X) = d$ for $d \leq \text{rad}(X) = 1/2$.

(3) If $X = \mathbf{R}^n$, then $e^d(X) = 0$ for any $d > 0$.

(4) If X is a flat circular cone in $\mathbf{R}^3$ with cone angle $\theta > \pi/6$ then $e^d(X) = d(1/\sin(\frac{\pi}{2}\sin\theta) - 1)$ for any $d > 0$. Hence if θ is close to $\frac{\pi}{2}$, $e^d(X)$ becomes small.

(5) If X is isometric to $S^1(1) \times S^1(\varepsilon)$ for $\varepsilon > 0$, then

$$e^d(X) = (\sqrt{d^2 - \varepsilon^2} + \varepsilon - d)\pi < \varepsilon\pi$$

for $d \leq \pi$.

We indicate here the following two suggestive properties of e^d: If Riemannian manifolds M and N satisfy $d_H(M, N) \leq \varepsilon$, then $|e^d(M) - e^d(N)| \leq 5\varepsilon$ for any $d < \text{rad}(M) - 3\varepsilon$. If N satisfies $\text{inj}(N) \geq i_0 > 0$, then $\text{rad}(N) \geq i_0$ and $e^d(N) = 0$ for $d \leq i_0$. Thus if a sequence of Riemannian manifolds $\{M_i\}_{i=1,2,\ldots}$ collapses to a Riemannian manifold N with $\text{inj}(N) \geq i_0 > 0$, then $e^d(M_i) \to 0$ as $i \to \infty$ for $d \leq i_0$. Hence we are naturally led to the question of whether we can characterize them from that point of view. Our main theorem is an affirmative answer to it.

To state the theorem precisely, we present the following definitions: Let $X = (X, d_X)$ and $Y = (Y, d_Y)$ be two metric spaces and $L > 0$, $\varepsilon > 0$. A map $f: X \to Y$ is an (ε, L)-Hausdorff approximation if it satisfies

$$|d_X(p, q) - d_Y(f(p), f(q))| < \varepsilon$$

for any $p, q \in X$ with $d(p, q) < L$ and

$$B_\varepsilon(f(X); Y) = Y.$$

Let M and N be Riemannian manifolds and $f: M \to N$ be a differentiable map. For $\tau \in (0, 1)$ the map f is a τ-almost Riemannian submersion if f satisfies

$$\left|\frac{|df(\xi)|}{|\xi|} - 1\right| \leq \tau$$

for any tangent vector ξ orthogonal to the fibers of f.

With these definitions we can state our main theorem.

THEOREM 1. *Given $n \in \mathbf{N}$, $k > 0$, $d > 0$ and $\tau \in (0, 1)$. There exist $\delta = \delta(n, k, d; \tau) > 0$, $i_0 = i_0(n, k, d; \tau) > 0$ and $\Delta = \Delta(n, k, d, \tau) > 0$ such that if an n-dimensional complete Riemannian manifold M satisfies*

$$K_M \geq -k^2, \quad \text{rad}(M) \geq d \quad and \quad e^d(M) \leq \delta,$$

then there exist a complete Riemannian manifold N and a fibration $f \colon M \to N$ with

$$1 \leq \dim N \leq n, \quad \mathrm{inj}(N) \geq i_0, \quad |K_N| \leq \Delta$$

and f being a $(\tau L, L)$-Hausdorff approximation for any $L > i_0$ and a τ-almost Riemannian submersion, where $\lim_{\tau \to 0} \delta = 0$, $\lim_{\tau \to 0} i_0 = 0$ and $\lim_{\tau \to 0} \Delta = \infty$. Furthermore there is $\tau_0 > 0$ such that if $\tau < \tau_0$, then the first Betti number of the fiber is not greater than $n - \dim N$.

For any $n \in \mathbf{N}$ and $k, D, d, \delta > 0$ we define $m_\delta := m_\delta(n, k, D, d) = \{M \colon n\text{-dimensional complete Riemannian manifolds with } K_M \geq -k^2$, diameter $d(M) \leq D$, $\mathrm{rad}(M) \geq d$ and $e^d(M) \leq \delta\}$. Note that m_δ contains the class of n-dimensional Riemannian manifolds M with $K_M \geq -k^2$, $d(M) \leq D$, $\mathrm{inj}(M) \geq i_0 > 0$. Using Theorem 1 and the Cheeger finiteness theorem [C] we have some finiteness theorem.

COROLLARY. *For any $n \in \mathbf{N}$ and $k, D, d > 0$ there exists $\delta > 0$ such that*

(1) for any $M \in m_\delta$ there exist a Riemannian manifold $N = N_M$ and a fibration $f \colon M \to N$ such that $1 \leq \dim N \leq n$ and the first Betti number of the fiber is not greater than $n - \dim N$;

(2) the number of diffeomorphism classes of $\{N_M \colon M \in m_\delta\}$ is finite.

REMARK. The finiteness theorem has close relation with the Gromov convergence theorem and precompactness theorem ([G, K, P, GW], etc.). Our theorem is considered as an extension of the convergence theorem to collapsing case. Note that the convergence theorem is recently extended in several ways [A1, AC, GPW] in noncollapsing categories.

As applications of Theorem 1 we have certain pinching theorems. Let us define

$$E(X) := \sup_{p \in X} \inf_{q \in X} \sup_{x \in X} e_{pq}(x).$$

Note that a complete Riemannian manifold M is isometric to the standard sphere if and only if $E(M) = 0$. Since $E(M) \geq e^d(M)$ for any $d \leq \mathrm{rad}(M) - E(M)/2$, we can apply Theorem 1 to the pinching problem for $E(M)$.

THEOREM 2. *Given $n \in \mathbf{N}$, $k > 0$ and $d > 0$, there exists $\delta = \delta(n, k, d) > 0$ such that if an n-dimensional complete Riemannian manifold M satisfies*

$$K_M \geq -k^2, \quad d(M) \leq d \quad \text{and} \quad E(M) \leq \delta$$

then M is homeomorphic to a fiber bundle over $S^m(1)$ $(0 \leq m \leq n)$ with fiber F whose first Betti number is not greater than $n - m$, where $S^0(1)$ denotes $\{1\}$.

Next we apply Theorem 1 to a pinching problem in the noncollapsing case. In [OSY] we proved that there is $\varepsilon > 0$ such that if $K_M \geq 1$ and

$\mathrm{vol}(M) \geq \mathrm{vol}(S^n(1)) - \varepsilon$, then M is diffeomorphic to $S^n(1)$. This result was extended replacing the necessity conditions in two ways; $\mathrm{Ric}_M \geq n - 1$, $K_M \geq -k^2$ and $\mathrm{vol}(M) \geq \mathrm{vol}(S^n(1)) - \varepsilon$ for small $\varepsilon > 0$ [Y1] or $K_M \geq 1$ and $\mathrm{rad}(M) \geq \pi - \varepsilon'$ for small $\varepsilon' > 0$ [SY]. Since $\mathrm{vol}(M) \geq \mathrm{vol}(S^n(1)) - \varepsilon$ and $K_M \geq 1$ imply $\mathrm{rad}(M) \geq \pi - \varepsilon'$, the second condition is weaker than the previous one. We can extend these results as follows.

THEOREM 3. *Given* $n \in \mathbf{N}$, $k > 0$ *and* $\nu_0 > 0$. *There exists* $\varepsilon = \varepsilon(n, k, \nu_0) > 0$ *such that if an* n-*dimensional complete Riemannian manifold* M *satisfies*

$$\mathrm{Ric}_M \geq n - 1, \quad K_M \geq -k^2, \quad \mathrm{vol}(M) \geq \nu_0 \quad and \quad \mathrm{rad}(M) \geq \pi - \varepsilon$$

then M *is diffeomorphic to* $S^n(1)$.

REMARK. Relevant works are [GS, S, I, K, A1, A2, O1], etc.

2. Outline of the proof

We give the outline of the proof of Theorem 1. First we investigate the local structure of M. Our main technical tool is only the Toponogov comparison theorem.

LEMMA 1. *For any* ρ, $\rho' > 0$ *there exists* $\delta_1 = \delta_1(k, d, \rho; \rho') > 0$ *such that if* $e^d(M) < \delta_1$, *then for any* $p, x \in M$ *with* $d(p, x) > \rho$, *there exist* $u, u' \in U_p M$ *such that* $\gamma(t) := \exp_p tu$ *and* $\gamma'(t) := \exp_p tu'$ *are minimal geodesics on* $[0, d/2]$ *with*

$$\sphericalangle(u, v) < \rho' \quad and \quad \sphericalangle(-u', v) < \rho',$$

where $v \in U_p M$ *is the initial vector of any minimal geodesic from* p *to* x.

LEMMA 2. *There exist* $\alpha_1 > 0$, $r_1 = r_1(k, d) > 0$ *and* $\delta_2 = \delta_2(k, d; r) > 0$ *for any* $0 < r < r_1$ *such that if* $e^d(M) < \delta_2$, *then for any* $p \in M$ *and* $u, v \in U_p M$ *with* $\gamma(t) := \exp_p tu : [0, d/2] \to M$ *and* $\sigma(s) := \exp_p sv : [0, a] \to M$ *being minimal geodesics for* $0 < a < r$ *it follows*

$$|d(\gamma(t), \sigma(s)) - |tu - sv|| < r^{1+\alpha_1}$$

for any $0 < s \leq a$ *and* $0 < t \leq r$.

PROOF. It follows from the Toponogov theorem that

$$d(\gamma(t), \sigma(s))^2 < (1 + O(r^2))^2 |tu - sv|^2 < |tu - sv|^2 + O(r^4)$$

for $0 < s \leq a$ and $0 < t \leq r$ if r is sufficiently small.

In Lemma 1 we take $\rho = d/2$ and $\rho' = r$. If $e^d(M) < \delta_2 := \delta_1(k, d; d/2, r)$ there is a minimal geodesic $\gamma'(t) = \exp_p tu' : [0, d/2] \to M$ with $\sphericalangle(u, u') > \pi - r$. It follows from the Toponogov theorem that

$$d(\gamma'(d/2), \sigma(s)) \leq \frac{d}{2} - s(u', v) + O(r^2)$$

$$\leq \frac{d}{2} + s(u, v) + O(r^2)$$

for $s < a$, where $(\ ,\ )$ is the induced metric on $T_p M$. We have

$$d(\gamma(d/2), \sigma(s)) + d(\gamma'(d/2), \sigma(s)) \geq d - \delta_2.$$

Since we may assume $\delta_2 \leq r^2$ we have

$$d(\gamma(d/2), \sigma(s)) \geq \frac{d}{2} - s(u, v) - O(r^2).$$

It follows from Alexandrov convexity (the Toponogov theorem) that

$$d(\gamma(t), \sigma(s))^2 \geq |tu - sv|^2 - O(r^3).$$

Thus we have

$$|d(\gamma(t), \sigma(s))^2 - |tu - sv|^2| < O(r^3) < r^{2(1+\alpha_1)}$$

for $\alpha_1 < 0.5$ if r is sufficiently small. $\quad\square$

Since the order of $r^{1+\alpha_1}$ is greater than 1, we may consider that it is arbitrarily small in comparison with r. Note that if we take small r, then δ_2 becomes small; that is, the condition on $e^d(M)$ also becomes stronger. From the above lemmas we can see the local triviality of M.

LEMMA 3. *There exist* $\alpha_2 \in (0, \alpha_1)$, $r_2 = r_2(n, k, d) > 0$ *and* $\delta(r) = \delta(k, d; r) > 0$ *for any* $r \in (0, r_2)$ *such that if* $e^d(M) < \delta(r)$, *then there exists* $m \in \{1, \ldots, n\}$ *depending on* M *which satisfies*

$$d_H(B_r(p, M), B_r(0, \mathbf{R}^m)) < r^{1+\alpha_2}$$

for any $p \in M$.

PROOF. Fix $c > 10$. Let $\varepsilon_0 := r^{\alpha_1}$ and

$$\delta(r) := \min\{\delta_1(k, d, cr\varepsilon_0; \varepsilon_0), \delta_2(k, d; r)\}$$

for $r < r_1$. Assume that $e^d(M) < \delta(r)$. Let p be a fixed point of M and let

$$A := \{u \in U_p M : \gamma(t) := \exp_p tu : [0, d/2] \to M \text{ is a minimal geodesic}\},$$

$$A_r := \{tu \in T_p M : u \in A \text{ and } 0 \leq t \leq r\}.$$

Then we have $d_H(A_r, B_r(p)) \leq 3cr\varepsilon_0$, since $B_{cr\varepsilon_0}(\exp_p A_r) \subset B_r(p)$ and $d_H(A_r, \exp_p A_r) \leq 2cr\varepsilon_0$ from Lemmas 1 and 2.

The following claim, which is proved using Lemmas 1 and 2 repeatedly, is essential in the proof.

CLAIM. *There are* $c_1 > 2c$ *and* $c_2 \in (0, 0.2c_1)$ *which only depend on* c *such that if* $u, v \in A$ *satisfy*

$$c_1\varepsilon_0 < \sphericalangle(u, v) < \pi - c_1\varepsilon_0,$$

then there is a 2-dimensional linear subspace $T \subset T_p M$ *such that*

$$u \in T, \quad v \in B_{c_2\varepsilon_0}(T \cap U_p M) \quad \text{and} \quad T \cap U_p M \subset B_{c_2\varepsilon_0}(A).$$

Since $\mathrm{rad}(M) > d$, there is $u \in A$. Since from Lemma 1 there is $u' \in A$ with $\sphericalangle(-u, u') \le \varepsilon_0$, for the subspace T_1 spanned by u we have

$$T_1 \cap U_p M \subset B_{\varepsilon_0}(A).$$

If $A \subset B_{c_1 \varepsilon_0}(T_1 \cap U_p M)$, m is equal to 1. If $A \not\subset B_{c_1 \varepsilon_0}(T_1 \cap U_p M)$, from Lemma 2 we have a 2-dimensional subspace $T_2 \supset T_1$ with

$$T_2 \cap U_p M \subset B_{c_2 \varepsilon_0}(A).$$

Suppose that there is an l-dimensional linear subspace $T_l \subset T_p M$ such that

$$T_l \cap U_p M =: S_{l-1} \subset B_{\varepsilon_l}(A) \quad \text{and} \quad A \not\subset B_{4\varepsilon_l}(S_{l-1})$$

for a certain $\varepsilon_l \ge c_2 \varepsilon_0$. Then from the claim we can find $w \in A$ with $\sphericalangle(w, S_{l-1}) > \pi/2 - 2c_2\varepsilon_0$. Let T_{l+1} be the subspace spanned by w and T_l and let $S_l := T_{l+1} \cap U_p M$. Then it also follows from the claim that

$$S_l \subset B_{\varepsilon_{l+1}}(A),$$

where ε_{l+1} ($> \varepsilon_l$) only depends on ε_l. Hence we find a sequence $\varepsilon_1 < \varepsilon_2 < \cdots < \varepsilon_n$ such that there is an m-dimensional linear subspace T_m with

$$T_m \cap U_p M \subset B_{\varepsilon_m}(A) \quad \text{and} \quad A \subset B_{4\varepsilon_m}(T_m \cap U_p M),$$

which implies that $d_H(B_r(p, M), B_r(0, T_m)) < O(r^{1+\alpha_1})$.

Since this property is open, it does not depend on the choice of $p \in M$. $\square$

Take α_2, r_2 and $\delta(r)$ in Lemma 3. Let $r < r_2/4$. Then for a certain m there is a $(4r)^{1+\alpha_2}$-Hausdorff approximation $f_p \colon B_{4r}(p, M) \to B_{4r}(0, R^m)$ for any $p \in M$ with $f_p(p) = 0$. In the following we construct a smooth approximation of f_p by almost linear functions. Let $\mathbf{e}_1, \ldots, \mathbf{e}_m$ be the standard basis of $\mathbf{R}^m$ and $\mathbf{e}_{m+1} := -\mathbf{e}_1, \ldots, \mathbf{e}_{2m} := -\mathbf{e}_m$. We can choose base points $q_1, \ldots, q_{2m}$ in $B_{4r}(p, M)$ such that

$$|f_p(q_\alpha) - 3r\mathbf{e}_\alpha| < (4r)^{1+\alpha_2}$$

for $\alpha = 1, \ldots, 2m$. For sufficiently small $\varepsilon' < r^{1+\alpha_2}/10$ we define smooth functions λ_α by

$$\lambda_\alpha(q) = \int_{y \in B_{\varepsilon'}(q_\alpha)} d(q, y) \, d\mu_\alpha(y)$$

for $q \in M$, where $d\mu_\alpha$ is the volume form on $B_{\varepsilon'}(q_\alpha)$ normalized by its volume. Then λ_α is a C^1-function (see [F]). Let

$$\chi_\alpha(q) := \frac{(\lambda_{m+\alpha}^2(q) - \lambda_\alpha^2(q))}{12r}$$

for $i = 1, \ldots, m$ [JK]. Let us define a C^1 map $\tilde{f}_p \colon B_{4r}(p) \to \mathbf{R}^m$ by

$$\tilde{f}_p(q) := (\chi_1(q), \ldots, \chi_m(q)).$$

Then we can prove that $\tilde{f}_p\colon B_{4r}(p, M) \to B_{4r+r^{1+\alpha_3}}(0, \mathbf{R}^m)$ is a $r^{1+\alpha_3}$-Hausdorff approximation and $\tilde{f}_p|B_{0.9r}(p, M)\colon B_{0.9r}(p, M) \to B_r(0, \mathbf{R}^m)$ is a r^{α_3}-almost Riemannian submersion for a certain constant $\alpha_3 \in (0, \alpha_2)$ and $r < r_3$ because we can compare the geometry of $B_{4r}(p, M)$ with that of $\mathbf{R}^m$. Note that the arguments here have some similarity with the one in [OSY].

Next we construct N using the above information. For $r < r_3$ let $P_{0,2r} = \{p_i \in M : i = 1, \dots\}$ be a $0.2r$-net in M. Let us denote

$$E_i := \mathbf{R}^m \supset U_i := B_r(0, E_i) \supset V_i := B_{0.4r}(0, E_i)$$

for $i \in I := \{i; p_i \in P_{0,2r}\}$. From Lemma 4 we find a $r^{1+\alpha_3}$-Hausdorff approximation map $\tilde{f}_i := \tilde{f}_{p_i}\colon B_{4r}(p_i, M) \to B_{4r+r^{1+\alpha_3}}(0, E_i)$ for any $i \in I$ such that $\tilde{f}_i\colon B_{0.9r}(p_i, M) \to U_i$ is a r^{α_3}-almost Riemannian submersion.

We construct coordinate transformations of N so that they preserve the information of $\tilde{f}_i$. For $i, j \in I$ with $d(p_i, p_j) \le 0.8r$ we can find an affine map $F_{ji}\colon U_i \to E_j$ such that

$$|F_{ji}(\tilde{f}_i(q)) - \tilde{f}_j(q)| < r^{1+\alpha_4} \quad \text{and} \quad \|dF_{ji} \circ d\tilde{f}_i - d\tilde{f}_j\| < r^{1+\alpha_4},$$

for $\alpha_4 \in (0, \alpha_3)$ and $r < r_4 = r_4(n, k, d)$.

Unfortunately these F_{ji} do not satisfy the chain rule for i and j. Thus we deform them one by one so as to satisfy the chain rule. Let $\phi\colon [0, \infty) \to [0, \infty)$ be a C^∞-function such that

$$\begin{aligned} \phi(t) &= 1 \quad \text{for } t \le \tfrac{1}{2}, \\ \phi'(t) &< 0 \quad \text{for } \tfrac{1}{2} < t < 1, \\ \phi(t) &= 0 \quad \text{for } t \ge 1. \end{aligned}$$

Let $\psi_i\colon U_i \to \mathbf{R}$ be the function defined by $\psi_i(x) := \phi(|x|/0.8r)$. First for $i = 1$ and $j = 2, \dots$ we define $\widetilde{F}_{j1} := F_{j1}$ and $\widetilde{F}_{1j} := \widetilde{F}_{j1}^{-1}$. For $i = 2$ we define $\widetilde{F}_{j2}\colon U_2 \to E_j$ by

$$\widetilde{F}_{j2}(x) := \psi_1 \circ \widetilde{F}_{12}(x)\widetilde{F}_{j1} \circ \widetilde{F}_{12}(x) + (1 - \psi_1 \circ \widetilde{F}_{12}(x))F_{j2}(x)$$

for $j = 3, \dots$. Then we have

$$\widetilde{F}_{j2}(x) = \widetilde{F}_{j1} \circ \widetilde{F}_{12}(x)$$

for $x \in \widetilde{F}_{21}(V_1)$. We define $\widetilde{F}_{2j} := \widetilde{F}_{j2}^{-1}$. Note that we need to check that $\widetilde{F}_{2j}$ is a well-defined C^∞ map.

Inductively we define $\widetilde{F}_{ji}\colon U_i \to E_j$ for $i < j$ by

$$\widetilde{F}_{ji}(x) := (1 - \psi_i) \sum_{l=1}^{i-1} \frac{\Psi_l}{\sum_{l=1}^{i-1}\Psi_l} \widetilde{F}_{jl} \circ \widetilde{F}_{li} + \Psi_i F_{ji}$$

for $x \in U_i$ with $\sum_{l=1}^{i-1}\Psi_l(x) \ne 0$ and

$$\widetilde{F}_{ji}(x) := \widetilde{F}_{jl} \circ \widetilde{F}_{li}(x)$$

for $x \in U_i$ with $\sum_{l=1}^{i-1} \Psi_l(x) = 0$, and for a certain l with $x \notin \widetilde{F}_{il}(V_l)$, where

$$\Psi_1 := \psi_1 \circ \widetilde{F}_{1i}(x)(1 - \psi_2 \circ \widetilde{F}_{2i}(x)) \cdots (1 - \psi_{i-1} \circ \widetilde{F}_{i-1i}(x)),$$

$$\cdots$$

$$\Psi_{i-1} := (1 - \psi_1 \circ \widetilde{F}_{1i}(x)) \cdots (1 - \psi_{i-2} \circ \widetilde{F}_{i-2i}(x))\psi_{i-1} \circ \widetilde{F}_{i-1i}(x),$$

$$\Psi_i := (1 - \psi_1 \circ \widetilde{F}_{1i}(x)) \cdots (1 - \psi_{i-1} \circ \widetilde{F}_{i-1i}(x)).$$

Note that the number of $j \in I$ with $\Psi_j(x) \neq 0$ on U_i is at most 10^m. We define $\widetilde{F}_{ij} = \widetilde{F}_{ji}^{-1}$ for $i < j$. By choosing ϕ properly we can prove that they are well-defined C^∞ maps and

$$(1) \qquad\qquad \widetilde{F}_{ji}(x) := \widetilde{F}_{jl} \circ \widetilde{F}_{li}(x)$$

for $l < i < j$ with $x \in \widetilde{F}_{ij}(V_j) \cap \widetilde{F}_{il}(V_l)$ and

$$(2) \qquad\qquad \widetilde{F}_{ij} \circ \widetilde{F}_{ji} = \mathrm{id} \quad \text{on } \widetilde{F}_{ij}(V_j).$$

They also satisfy

$$(3) \qquad\qquad |\widetilde{F}_{ji}(\tilde{f}_i(q)) - \tilde{f}_j(q)| \leq r^{1+\alpha_5},$$

$$\|d\widetilde{F}_{ji} \circ d\tilde{f}_i - d\tilde{f}_j\| \leq r^{\alpha_5} < 0.1$$

for any $q \in M$ and $i < j$ with $q \in \tilde{f}_i^{-1}(U_i) \cap \tilde{f}_j^{-1}(U_j)$ for a certain constant $\alpha_5 \in (0, \alpha_4)$ and sufficiently small $r > 0$.

From (1) and (2) we can interpret $\{\widetilde{F}_{ji}|V_i \cap \widetilde{F}_{ij}(V_j)\}$ as coordinate transformations of a coordinate system of a manifold, which is the desired manifold N. The Riemannian metric on N and fibration $f: M \to N$ are similarly constructed so that f satisfies the conditions in Theorem 1 because of (3).

REFERENCES

[A1] M. Anderson, *Convergence and rigidity of manifolds under Ricci curvature bounds*, Invent. Math. **102** (1990), 429–445.

[A2] ___, *Metrics of positive Ricci curvature with large diameter*, Manuscripta Math. **68** (1990), 405–415.

[AG] U. Abresch and D. Gromoll, *On complete manifolds with nonnegative Ricci curvature*, J. Amer. Math. Soc. **3** (1990), 355–374.

[AC] M. Anderson and J. Cheeger, *C^α compactness for manifolds with Ricci curvature and injectivity radius bounded below*, Preprint.

[B] A. Besse, *Manifolds all of whose geodesics are closed*, Springer-Verlag, 1978.

[C] J. Cheeger, *Finiteness theorems for Riemannian manifolds*, Amer. J. Math. **92** (1970), 61–74.

[CE] J. Cheeger and D. Ebin, *Comparison theorems in Riemannian geometry*, North-Holland Math. Library, vol. 9, North-Holland, 1975.

[CG] J. Cheeger and M. Gromov, *Collapsing Riemannian manifold while keeping their curvature bound*. I, J. Differential Geom. **23** (1986), 309–346; II, **32** (1990), 355–374.

[F] K. Fukaya, *Collapsing Riemannian manifolds to ones of lower dimensions*, J. Differential Geom. **25** (1987), 139–156.

[G] M. Gromov, *Structure metrique pour les varietes riemanniennes* (J. Lafontaine and P. Pansu, eds.), Cedic/Fernand Nathan, Paris, 1981.

[GP1] K. Grove and P. Petersen, *On the excess of metric spaces and manifolds*, Preprint.

[GP2] ____, *Volume comparison a La Aleksandrov*, to appear in Acta Math.

[GPW] K. Grove, P. Petersen and J. Wu, *Geometric finiteness theorems via controlled topology*, Invent. Math. **99** (1990), 205–213.

[GS] K. Grove and K. Shiohama, *A generalized sphere theorem*, Ann. of Math. (2) **106** (1977), 201–211.

[GW] R. Greene and H. Wu, *Lipschitz convergence of Riemannian manifolds*, Pacific J. Math. **131** (1988), 119–141.

[I] Y. Itokawa, *The topology of certain Riemannian manifolds with positive Ricci curvature*, J. Differential Geom. **18** (1983), 151–155.

[JK] J. Jost and H. Karcher, *Geometrische Methoden zur Gewinnung von a-priori Schranker fur harmonische Abbildungen*, Manuscripta Math. **40** (1982), 27–77.

[K] A. Katsuda, *Gromov's convergence theorem and its application*, Nagoya Math. J. **100** (1985), 11–48.

[O1] Y. Otsu, *On manifolds of positive Ricci curvature with large diameter*, Math. Z. **206** (1991), 255–264.

[O2] ____, *On manifolds of small excess*, to appear in Amer. J. Math.

[OSY] Y. Otsu, K. Shiohama, and T. Yamaguchi, *A new version of differentiable sphere theorem*, Invent. Math. **98** (1989), 219–228.

[P] S. Peters, *Convergence of Riemannian manifolds*, Compositio Math. **62** (1987), 3–16.

[S] K. Shiohama, *A sphere theorem for manifolds of positive Ricci curvature*, Trans. Amer. Math. Soc. **273** (1983), 811–819.

[Sy] T. Shioya, *Splitting theorems for nonnegatively curved open manifolds with large ideal boundary*, to appear in Math. Z.

[SY] K. Shiohama and T. Yamaguchi, *Positively curved manifolds and restricted diameters: Geometry of manifolds*, Perspect. Math., vol. 8, Academic Press, 1989, pp. 345–350.

[Y1] T. Yamaguchi, *Lipschitz convergence of manifolds of positive Ricci curvature with large volume*, Math. Ann. **284** (1989), 423–436.

[Y2] ____, *Collapsing and pinching under a lower curvature bound*, Ann. of Math. **133** (1991), 317–357.

UNIVERSITY OF TOKYO, JAPAN

Proceedings of Symposia in Pure Mathematics
Volume **54** (1993), Part 3

Gromov-Hausdorff Convergence of Metric Spaces

PETER PETERSEN V

1. Introduction and basic notation

In this section survey article we wish to explain some relationships between topology, the Gromov-Hausdorff distance, and finiteness theorems.

Around 1980 Gromov (see [**14**]) extended the classical Hausdorff distance to what we shall call the Gromov-Hausdorff distance between abstract metric spaces. Two metric spaces which are close in this metric generally do not have to be topologically alike. In these notes we shall try to find natural topological and geometric assumptions, which ensure that close spaces also are topologically alike. This in turn can be used to understand how geometric constraints on metric spaces give rise to topological constraints. Such problems are in geometry often referred to as finiteness questions.

In the language of analysis one can think of the Gromov-Hausdorff distance as a very weak metric. The conditions we impose are then regularity conditions, which let us conclude some kind of stronger convergence from weak convergence.

In these notes we will only deal with Gromov-Hausdorff convergence of compact spaces in a very general context, and with a few geometric applications. For further applications the reader could consult [**14, 11, 12, 13, 33, 17, 18**].

Unless otherwise specified all spaces under consideration will be separable metric spaces. The metric on a space X is denoted by $d(\cdot, \cdot)$. If $A \subset X$ then int A and $\bar{A}$ denote, respectively, the interior, and the closure of A in X. The distance $d(A, B)$ between two subsets in X is

$$d(A, B) = \inf\{d(a, b) : a \in A, \ b \in B\}.$$

This distance is not to be confused with the Hausdorff distance between A and B in X. For $\varepsilon > 0$ and $A \subset X$ the closed ε-tube around A is

1991 *Mathematics Subject Classification.* Primary 53C20.
Supported in part by the National Science Foundation and the Alfred P. Sloan Foundation.
This paper is in final form and no version of it will appear elsewhere.

$D(A, \varepsilon) = \{x \in X : d(x, A) \leq \varepsilon\}$, and the open ε-tube around A is $B(A, \varepsilon) = \{x \in X : d(x, A) < \varepsilon\}$. When A is a point these tubes are of course the closed (resp. open) metric ε-ball around the specified point.

If A is a set (usually finite or countable) then $\mathbb{R}^A$ is the vector space with basis A. Thus $\mathbb{R}^A = \{f : A \to \mathbb{R} : f(a) = 0$ for all but finitely many $a \in A\}$. We will always think of this vector space as being equipped with the euclidean metric

$$d(f, g) = \left(\sum_{a \in A} (f(a) - g(a))^2 \right)^{1/2}.$$

If $a_0, \ldots, a_n \in A$ then the n-dimensional simplex

$$\left\{ \sum \lambda_i a_i : 0 \leq \lambda_i \leq 1, \ \sum \lambda_i = 1 \right\}$$

spanned by $a_0, \ldots, a_n$ is denoted by $[a_0, \ldots, a_n]$. A (locally finite) simplical complex (see [30]) can always be thought of as a subset of $\mathbb{R}^A$ (A is the set of vertices in the complex) which is the union of simplices of the above type. This also yields a natural piecewise linear metric on such complexes.

2. The distance between metric spaces

In this section we will introduce the Gromov-Hausdorff distance and show how to establish compactness for classes of metric spaces.

The classical Hausdorff distance between two subsets $A, B \subset Z$ in a metric space Z is

$$d_H^Z(A, B) = \inf\{\varepsilon : A \subset B(B, \varepsilon), \ B \subset B(A, \varepsilon)\}.$$

This metric is only a pseudometric since $d_H^Z(A, B) = 0$ only implies that $\bar{A} = \bar{B}$. However, we do get a metric if we restrict attention to closed subsets of Z. Over the years many people have studied this concept (see e.g. [2, 3]) and much of what they did is only now being rediscovered by geometers that are interested in using the (Gromov-) Hausdorff distance in geometry.

Around 1980 Gromov brought new life to this notion in his landmark paper on groups of polynomial growth. By de-emphasizing the role of the ambient space he came up with what we shall call the Gromov-Hausdorff distance between abstract metric spaces. Denote by $\mathcal{M}$ the class of all (isometry classes of) compact metric spaces.

If $X, Y \in \mathcal{M}$ define $d_{GH}(X, Y) = \inf\{\varepsilon :$ there is $Z \in \mathcal{M}$ and isometric embeddings $i : X \to Z, j : Y \to Z$ with $d_H^Z(i(X), j(Y)) < \varepsilon\}$. (When we say isometric embedding we mean a distance preserving map, thus a sphere with its usual Riemannian structure is not isometrically embedded in euclidean space). It is not hard to see that if $d_{GH}(X, Y) = 0$ then X and Y are isometric. The Gromov-Hausdorff distance therefore defines a metric on the space $\mathcal{M}$ of all isometry classes of compact metric spaces.

The importance of this concept lies not so much in the fact that we have a distance function, but in that we have a way of measuring when metric spaces look alike. The following equivalent definition might therefore seem more natural:

$d_{GH}(X, Y) \leq \varepsilon$ if there is a metric on $X \amalg Y$ extending the metrics on X and Y and such that $X \subset D(Y, \varepsilon)$, $Y \subset D(X, \varepsilon)$. The ambient space Z has now disappeared, and the question is merely to find metrics on the disjoint union of X and Y such that X and Y look like they are close to each other. Using this definition also makes it easier to prove that d_{GH} satisfies the triangle inequality.

A sequence of compact metric spaces $\{X_i\}$ is said to converge to a compact metric space X provided $d_{GH}(X_i, X) \to 0$ as $i \to \infty$.

It is usually very hard to compute the Gromov-Hausdorff distance exactly, but often one can give upper bounds.

EXAMPLE 1. Let X, Y be compact metric spaces with diameter diam X, diam $Y \leq D$. Define a distance on $X \amalg Y$ by $d(x, y) = D/2$ if $x \in X$, $y \in Y$. One easily checks that this is in fact a metric and that $X \subset D(Y, D/2)$, $Y \subset D(X, D/2)$. Hence $d_{GH}(X, Y) \leq D/2$.

EXAMPLE 2. Let X, Y be compact metric spaces. Suppose there are $\varepsilon > 0$, $x_1, \ldots, x_n \in X$ and $y_1, \ldots, y_n \in Y$ with $|d(x_i, x_j) - d(y_i, y_j)| < \varepsilon$ for all $i, j = 1, \ldots, n$, $D(\{x_i\}, \varepsilon) = X$, and $D(\{y_i\}, \varepsilon) = Y$. Define $d(x, y) = \min\{d(x, x_i) + \varepsilon + d(y_i, y) : i = 1, \ldots, n\}$ for $x \in X$, $y \in Y$. This gives a metric on $X \amalg Y$ with $X \subset D(Y, 2\varepsilon)$ and $Y \subset D(X, 2\varepsilon)$. In particular $d_{GH}(X, Y) \leq 2\varepsilon$.

Example 2 makes it possible to give very good estimates for d_{GH} and also to check in concrete examples that a given sequence converges.

EXAMPLE 3. Consider the unit sphere $S^3 \subset \mathbb{C}^2$ with S^1 acting on S^3 in the canonical way. The cyclic group $\mathbb{Z}_n$ of order n naturally lies on S^1. Denote by L_n the lenspace $\mathbb{Z}_n \backslash S^3$. As $n \to \infty$ $\mathbb{Z}_n$ fills up S^1 and $L_n \to S^3/S^1 = S^2 \subset \mathbb{R}^3$ with curvature $= 4$. One can also shrink the fiber S^1 on S^3 and thereby get a sequence of metrics g_ε on S^3 such that $(S^3, g_\varepsilon) \to S^2$. The metrics g_ε on S^3 are called Berger spheres. This phenomenon is called a collapse because the dimension of the limit is smaller than the dimension of the elements in the sequence.

EXAMPLE 4. In the unit cube $[0, 1]^3 \subset \mathbb{R}^3$ consider the grid $X_n = \{(x, y, z) \in [0, 1]^3 :$ at least two coordinates is a rational number with n in the denominator$\}$, and the jungle gym $Y_n = \partial D(X_n, 2^{-n}) \subset \mathbb{R}^3$. ($Y_n$ is a surface of very high genus). As $n \to \infty$ both X_n and Y_n fill up $[0, 1]^3$ so $X_n \to [0, 1]^3$, $Y_n \to [0, 1]^3$. This phenomenon is called an explosion, since the limit space has larger dimension than the elements in the sequence.

Both Examples 3 and 4 show that although all the spaces under consideration are nice there does not have to be any relation between the topology of Hausdorff close spaces. The next example shows how a sequence of nice spaces give rise to an awful limit space.

EXAMPLE 5. Let $D = D(0, 2) \subset \mathbb{R}^2$ and $D_n = D - B(0, \frac{1}{4}) - B(\frac{1}{2}, \frac{1}{8}) - \cdots - B(2^{1/n}, 2^{1/n+2})$. As $n \to \infty$ D_n converges to $D_\infty = \bigcap_{n=1}^\infty D_n \subset D$ which is not an ANR. If we double D_n we will get a sequence of surfaces of genus n converging to a space which is not an ANR.

In **[14]** it is shown that every Cauchy sequence in $\mathcal{M}$ has a subsequence which is isometrically embedded in some compact metric space, such that it is also a Cauchy sequence in this space. Thus many results on Hausdorff convergence will carry over to Gromov-Hausdorff convergence. Here, however, we will develop the theory without using this fact.

The rest of this section will be devoted to studying criteria for convergence and (pre)compactness for families of metric spaces.

PROPOSITION 6. $(\mathcal{M}, d_{Gh})$ *is complete.*

PROOF. Let $\{X_i\}$ be a Cauchy sequence. It suffices to show that some subsequence converges, so we can without loss of generality assume that $d_{\text{GH}}(X_i, X_{i+1}) < 2^{-i}$ for all $i = 1, 2, \ldots$. Then choose metrics $d^{i\,i+1}$ on $X_i \amalg X_{i+1}$ such that $d_{\text{H}}^{i\,i+1}(X_i, X_{i+1}) < 2^{-i}$. With these choices we can construct metrics d^{ij} on $X_i \amalg X_j$, where $i < j$ as follows

$$d^{ij}(x, y) = \inf\left\{\sum_{k=i}^{j-1} d^{k\,k+1}(x_k, x_{k+1}): x_k \in X_k \quad \text{and} \quad x_i = x, \, x_j = u\right\}.$$

These metrics clearly satisfy

$$d^{ik}(x_i, x_k) \leq d^{ij}(x_i, x_j) + d^{jk}(x_j, x_k)$$

if $i \leq j \leq k$ and $x_i \in X_i$, $x_j \in X_j$, $x_k \in X_k$. Therefore

$$d_{\text{H}}^{ij}(X_i, X_j) \leq \sum_{k=i}^{j-1} d_{\text{H}}^{k\,k+1}(X_k, X_{k+1}) \leq 2^{-i+1} \quad \text{if } i \leq j.$$

Let $\widehat{X} = \{(x_j): x_j \in X_j$ and $d_{(x_i, x_j)}^{ij} \to 0$ as $i, j \to \infty\}$. There is a pseudo metric on $\widehat{X}$ defined by $d((x_j), (y_j)) = \lim_{j \to \infty} d(x_j, y_j)$. We contend that the metric space X, obtained from $\widehat{X}$ by identifying points which have zero distance, is the limit of $\{X_i\}$.

Construct a metric d^i on $X_i \amalg X$ by $d^i(y, (x_j)) = \limsup_{j \to \infty} d^{ij}(y, x_j)$, where $y \in X_i$ and (x_i) represents an element in X. This is easily seen to give a well defined metric on $X \amalg X_i$. We claim that $d_{\text{H}}^i(X, X_i) < 2^{-i+2}$. Let (x_j) represent an element in X. Choose $n \geq i$ such that $d^n(x_n, (x_j)) < 2^{-i}$, and then $y \in X_i$ with $d^{in}(y, x_n) \leq 2^{-i+1}$. Thus

$$d^i(y, (x_j)) = \limsup_{j \to \infty} d^{ij}(y, (x_j))$$

$$\leq \limsup_{j \to \infty} d^{in}(y, x_n) + d^{nj}(x_n, x_j)$$

$$\leq d^{in}(y, x_n) + d^n(x_n, (x_j)) \leq 2^{-i} + 2^{-i+1} \leq 2^{-i+2}.$$

Conversely suppose $y \in X_i$. We can then successively find $x_j \in X_j$, $j \geq i$ and $y = x_i$ and $d^{jj+1}(x_j, x_{j+1}) < 2^{-j}$. The sequence (x_j) then defines an element in X and by construction $d^i(y, (x_i)) = \limsup_{j \to \infty} d^{ij}(y, x_j) \leq \lim_{j \to \infty} \sum_{k=i}^{j-1} 2^{-k} = 2^{-i+1}$. $\square$

For a compact metric space X define $\mathrm{Cov}(X, \varepsilon) = \min\{n: X \text{ is covered}$ by n closed ε-balls$\}$ and $\mathrm{Cap}(X, \varepsilon) = \max\{n: X \text{ contains } n \text{ disjoint } \frac{\varepsilon}{2} -$ balls$\}$. It is clear that $\mathrm{Cov}(X, \varepsilon) \leq \mathrm{Cap}(X, \varepsilon)$. If X and Y are two compact metric spaces with $d_{\mathrm{GH}}(X, Y) < \delta$ then one can easily show:

$$\mathrm{Cov}(X, \varepsilon) \geq \mathrm{Cov}(Y, \varepsilon + 2\delta)$$
$$\mathrm{Cap}(X, \varepsilon) \geq \mathrm{Cap}(Y, \varepsilon + 2\delta).$$

Using this we can formulate and prove the most important compactness result.

THE GROMOV COMPACTNESS THEOREM 7. *Let* $\mathscr{C} \subset \mathscr{M}$ *be a class of compact metric spaces. The following are equivalent.*

(i) $\mathscr{C}$ *is precompact i.e. every sequence in* $\mathscr{C}$ *contains a subsequence which converges in* $\mathscr{M}$.

(ii) *There exists a function* $N(\varepsilon): (0, \beta) \to (0, \infty)$ *such that*: $\mathrm{Cap}(X, \varepsilon) \leq N(\varepsilon)$ *for all* $\varepsilon \in (0, \beta)$ *and* $X \in \mathscr{C}$.

(iii) *There exists a function* $N(\varepsilon): (0, \beta) \to (0, \infty)$ *such that* $\mathrm{Cov}(X, \varepsilon) \leq N(\varepsilon)$ *for all* $\varepsilon \in (0, \beta)$ *and* $X \in \mathscr{C}$.

PROOF. (i) $\Rightarrow$ (ii) Fix $\varepsilon > 0$. By precompactness there are $X_1, \ldots, X_{n(\varepsilon)} \in \mathscr{C}$ such that for every $Y \in \mathscr{C}$ $d_{\mathrm{GH}}(Y, X_i) < \varepsilon/4$ for some $i = 1, \ldots, n(\varepsilon)$. Then $\mathrm{Cap}(Y, \varepsilon) \leq \mathrm{Cap}(X_i, \varepsilon - 2\frac{\varepsilon}{4}) = \mathrm{Cap}(X_i, \varepsilon/2)$. So we can use $N(\varepsilon) = \max\{\mathrm{Cap}(X_i, \varepsilon/2): i = 1, \ldots, n(\varepsilon)\}$

(ii) $\Rightarrow$ (iii) is trivial.

(iii) $\Rightarrow$ (i) It suffices to show that for every sequence $\{X_i\}$ in $\mathscr{C}$ and every $\varepsilon > 0$ there is a subsequence $\{X_k'\}$ where $d_{\mathrm{GH}}(X_k, X_l) < \varepsilon$ for all elements in $\{X_k'\}$.

Every X_i is covered by at most $N(\varepsilon/2)$ $\varepsilon/2$-balls. Hence there is a subsequence $\{X_k'\}$ where each X_k' is covered by exactly N $\varepsilon/2$-balls for some $N \leq N(\varepsilon/2)$. Denote by x_k^α, $\alpha = 1, \ldots, N$, the centers of these balls in X_k'. For each k consider the matrix of numbers $\{d(x_k^\alpha, x_k^\beta)\}_{\alpha, \beta = 1, \ldots, N}$. All these numbers are bounded by $\mathrm{diam} \, X_k' \leq \varepsilon \cdot N \leq \varepsilon \cdot N(\varepsilon/2)$. The Dirichlet selection principle then assures us that there is a subsequence $\{X_k''\}$ of $\{X_k'\}$ such that $|d(x_l^\alpha, x_l^\beta) - d(x_m^\alpha, x_m^\beta)| < \varepsilon/2$ for all l, m. By Example 2 we then have $d_{\mathrm{GH}}(X_l'', X_m'') < \varepsilon$ for all l, m. $\square$

COROLLARY 8. *Denote by* $\mathscr{C}(N(\varepsilon))$ *the class of compact metric spaces* X *with* $\mathrm{Cov}(X, \varepsilon) \leq N(\varepsilon)$ *for all* $\varepsilon \in (0, \beta)$. *The class* $\mathscr{C}(N(\varepsilon))$ *is compact.*

PROOF. From Theorem 7 we know that $\mathscr{C}(N(\varepsilon))$ is precompact, so it remains to show it is closed. Let $\{X_k\}$ be a sequence in $\mathscr{C}(N(\varepsilon))$ converging to X in $\mathcal{M}$. We can suppose X is constructed as in the proof of Proposition 6. Fix $\varepsilon > 0$ and choose finite sets $A_i \subset X_i$ such that $|A_i| \leq N(\varepsilon)$ and $D(A_i, \varepsilon) = X_i$. Using Theorem 7 the family $\{A_i\}$ is precompact and will therefore contain a convergent subsequence $\{A'_j\}$. The limit space A can obviously be thought of as a subset of X and satisfies $|A| \leq N(\varepsilon)$. It is not hard to check that also $D(A, \varepsilon) = X$ so $\mathrm{Cov}(X, \varepsilon) \leq N(\varepsilon)$. $\square$

Note that to check whether or not a collection $\mathcal{C} \subset \mathcal{M}$ is precompact one does not need to use the Gromov-Hausdorff distance itself.

EXAMPLE 9. Denote by $\mathcal{C}(n, i_0, V)$ the collection of closed connected Riemannian n-manifolds with injectivity radius $\geq i_0 > 0$ and volume $\leq V < \infty$. We claim that there is a number $C > 1$ depending on n, i_0 and V only, such that $\mathrm{Cov}(M, \varepsilon) \leq C \cdot \varepsilon^{-n}$ for all $\varepsilon \leq i_0/2$, $M \in \mathcal{C}(n, i_0, V)$.

Given $M \in \mathcal{C}(n, i_0, V)$ and $\varepsilon \leq i_0/2$, choose a maximal set of points $x_1, \ldots, x_N \in M$ with $d(x_i, x_j) \geq \varepsilon$. Then $B(x_i, \varepsilon)$, $i = 1, \ldots, N$, covers M, so $\mathrm{Cov}(M, \varepsilon) \leq N$. In [6] it is shown that vol $B(x_i, \varepsilon/2) \geq c(n)(\frac{\varepsilon}{2})^n$ for each i and some constant $c(n)$ depending only on n. Thus $N \cdot \min\{\mathrm{vol}\, B(x_i, \varepsilon/2): i = 1, \ldots, N\} \leq \mathrm{vol}\, M \leq V$, because the balls $B(x_i, \varepsilon/2)$ are disjoint. In particular $N \leq V \cdot c(n) \cdot 2^n \cdot \varepsilon^{-n}$.

EXAMPLE 10. Denote by $\mathcal{C}(n, k, D)$ the collection of closed connected Riemannian n-manifolds with diameter $\leq D$ and ricci curvature $\geq (n-1)k$. Again we can find $C = C(n, k, D)$ such that $\mathrm{Cov}(M, \varepsilon) \leq C \cdot \varepsilon^{-n}$ for all $\varepsilon \leq D$, $M \in \mathcal{C}(n, k, D)$.

To see this choose $x_1, \ldots, x_N$ as on Example 9. By the relative volume comparison theorem (see [15]).

$$\frac{\mathrm{vol}\, M}{\mathrm{vol}\, B(x_i, \varepsilon/2)} \leq \frac{\int_0^D (\sin h(\sqrt{|k|}t))^{n-1} dt}{\int_0^{\varepsilon/2} (\sin h(\sqrt{|k|}t))^{n-1} dt} \leq C\varepsilon^{-n}$$

for all $\varepsilon \leq D$ and some $C = C(n, k, D)$. Then it follows as before that $N \leq C \cdot \varepsilon^{-n}$.

3. Dimension theory

In this section we introduce the concepts of covering dimension and Hausdorff dimension. Understanding these notions in a metric context makes it possible to understand when collapsing and explosions do not occur (see Examples 2.3 and 2.4).

The covering dimension of a space X is said to be $\leq n$ if for every open covering $\{V_\alpha\}$ there is a refinement $\{V_\beta\}$ of order $\leq n+1$ (i.e., no more than $n+1$ V_β's have nonempty intersection). The covering dimension, denoted dim X, is then the smallest integer n such that X has covering dimension $\leq n$.

EXAMPLE 1. If X is an n-dimensional simplex or an n-dimensional manifold, then it is easy to see that X has covering dimension $\leq n$. In fact dim $X = n$, but the proof will be deferred to §4.

EXAMPLE 2. If $A \subset X$ then dim $A \le$ dim X.

Let $\{V_\alpha\}$ be an open locally finite (i.e., finite order) covering of X. Associated to this covering we have the nerve N. N is a simplicial complex whose vertices are $\{V_\alpha\}$ and where $[V_{\alpha_0}, \ldots, V_{\alpha_n}]$ is an n-simplex iff $V_{\alpha_0} \cap \cdots \cap V_{\alpha_n} \ne \varnothing$. The order of $\{V_\alpha\}$ is the same as the number of vertices in the highest dimensional simplex in N. With a metric on X we can find a partition of unity $\{\varphi_\alpha\}$ such that $\{x : \varphi_\alpha(x) > 0\} = V_\alpha$ and $\sum_\alpha \varphi_\alpha = 1$ (e.g. $\varphi_\alpha(x) = d(x, X - V_\alpha) \cdot (\sum_\beta d(x, X - V_\beta))^{-1})$. In this way we get a map $\Phi : X \to N$ by declaring $\Phi(x)$ to have barycentric coordinates $\{\varphi_\alpha(x)\}$. The map Φ is clearly a $\{V_\alpha\}$ map (i.e., for every $p \in N$ $\Phi^{-1}(p) \subset V_\alpha$ for some α).

Using this construction it is not hard to show

THEOREM 3 [1]. dim $X \le n$ iff for every covering $\{V_\alpha\}$ there exists a $\{V_\alpha\}$ map $f: X \to P$ to a simplicial complex P of dimension $\le n$.

EXAMPLE 4. From this theorem we immediately get dim $(X \times Y) \le$ dim X + dim X.

When X is a compact space it obviously suffices to consider finite coverings and finite complexes. This simplification makes it easier to compute the covering dimension.

A covering $\{V_\alpha\}$ of X is said to be a D-covering if diam $V_\alpha < D$ for all α. A map $f: X \to Y$ is a D-map provided diam $f^{-1}(y) < D$ for all $y \in Y$. D-maps $f : X \to Y$ between compact spaces are easily seen to satisfy the following inverse continuity condition: for all $\varepsilon > 0$ there is $\delta > 0$ such that $d(f(x_1), f(x_2)) < \delta \Rightarrow d(x_1, x_2) < D + \varepsilon$. With this is in mind we can strengthen Theorem 3 for compact spaces X.

THEOREM 5. The following are equivalent:
(i) dim $X \le n$.
(ii) For every $D > 0$ there is a D-map $f: X \to P$ to a finite simplicial complex of dimension $\le n$.
(iii) For every $D > 0$ there is a finite open D-covering $\{V_\alpha\}$ of X of order $\le n + 1$.
(iv) For every $D > 0$ there is a finite closed D-covering $\{F_\alpha\}$ of X of order $\le n + 1$.

COROLLARY 6. Let X be a compact subset of an n-dimensional complex P. If dim $X = n$ then X has nonempty interior in some n-simplex in P.

PROOF. It suffices to check the case where P is an n-simplex itself. For each $D > 0$ construct a triangulation of P where all simplices have diameter $\le \frac{1}{2}D$. If X has no interior points, then int $\Delta^n - X \ne 0$ for all n-simplices Δ on this triangulation. $X \cap \Delta$ can therefore be radially projected onto the $(n-1)$ face $\partial \Delta^n$ keeping $X \cap \partial \Delta^n$ fixed. This gives us a map $X \to (n-1)$ skeleton of the triangulation. Preimages of points lie in stars of $(n-1)$

simplices and therefore have diameter $\leq D$. Hence dim $X \leq n - 1$. $\square$

Urysohn (see ([**31**]) used the characterization of covering dimension in Theorem 5 to define the k-spread or k-diameter $\operatorname{diam}_k X$ of a compact metric space X. This quantity is supposed to measure how much X deviates from being a k-dimensional space.

PROPOSITION-DEFINITION 7. *Let X be a compact metric space. The following quantities are equal and by definition* $\operatorname{diam}_k X$:

(i) $\inf\{D:$ *there is a finite open D-covering of order* $\leq k + 1\}$,

(ii) $\inf\{D:$ *there is a finite closed D-covering of order* $\leq k + 1\}$,

(iii) $\inf\{D:$ *there is a D-map* $f: X \to P$ *to a finite k-dimensional complex*}.

Clearly $\operatorname{diam}_0 X \geq \operatorname{diam}_1 X \geq \cdots$, and $\operatorname{diam}_n X = 0$ if dim $X \leq n$. Furthermore $\operatorname{diam}_0 X = \max\{\operatorname{diam} A: A$ connected component of $X\}$. Urysohn originally introduced these quantities to study when a decreasing family of compact metric spaces $X_1 \supset X_2 \supset X_3 \supset \cdots$ all of dimension $= n$ has the property that dim $\bigcap_i X_i = n$. He showed that this is true provided $\operatorname{diam}_{n-1} X_i > \delta > 0$ for some δ and all $i \geq 1$. Here $\bigcap_i X_i$ is of course the Hausdorff limit of the sequence $\{X_i\}$ inside X_1. One might therefore expect a generalization to the situation where $\{X_i\}$ is merely a sequence of spaces converging to X in the (Gromov-) Hausdorff metric.

THEOREM 8. *Suppose $\{X_i\}$ is a sequence of compact metric spaces converging to X in the Gromov-Hausdorff metric. Then*

(i) $\lim \operatorname{diam} X_i = \operatorname{diam} X$.

(ii) $\limsup \operatorname{diam}_k X_i \leq \operatorname{diam}_k X$.

PROOF. For each i choose a metric d^i on $X \amalg X_i$ such that $d_H^i(X, X_i) \leq 2 d_{GH}(X, X_i)$.

(i) If $d(p, q) = \operatorname{diam} X$, then we can choose $p_i, q_i \in X_i$ with $d^i(p, p_i) \leq 2 d_{GH}(X, X_i)$, $d^i(q, q_i) \leq 2 d_{GH}(X, X_i)$. Thus diam $X_i \geq$ diam $X - 4d(X, X_i)$. Conversely one can also see that diam $X \geq \operatorname{diam} X_i - 4d_{GH}(X, X_i)$.

(ii) Let $\{F_\alpha\}$ be a finite closed D-covering of X of order $\leq k + 1$. We can find $\delta > 0$ with the property that if $F_{\alpha_1} \cap \cdots \cap F_{\alpha_l} = \varnothing$ then $d(F_{\alpha_1}, F_{\alpha_2} \cap \cdots \cap F_{\alpha_l}) > \delta$. For each i define $F_\alpha^i = \{x \in X: d^i(x, F_\alpha) \leq 2 d_{GH}(X, X_i)\}$. Clearly $\{F_\alpha^i\}$ is a closed finite $D + 4 d_{GH}(X, X_i)$-covering of X_i. Now suppose $F_{\alpha_1}^i \cap \cdots \cap F_{\alpha_{k+2}}^i \neq \varnothing$. We can then find $x_{\alpha_j} \in F_{\alpha_j}$ with $d^i(x_{\alpha_j}, y) \leq 2 d_{GH}(X, X_i)$, for some $y \in F_{\alpha_1}^i \cap \cdots \cap F_{\alpha_{k+2}}^i$. Then $d(x_{\alpha_j}, x_{\alpha_l}) \leq 4 d_{GH}(X, X_i)$ for all $j, l, \ldots, k + 2$. The map $F: F_{\alpha_1} \times \cdots \times F_{\alpha_{k+2}}) \to \mathbb{R}$ defined by $F(x_1, \ldots, x_{k+2}) = \max\{d(x_i, x_j): i, j = 1, \ldots, k + 2\}$ must have a minimum value $\delta > 0$, depending only on our covering $\{F_\alpha\}$ and k, because the order of this covering is $k + 1$. In particular we see that the existence of the above x_{α_j}, $j = 1, \ldots, k+2$, implies that $4d_{\mathrm{GH}}(X, X_i) \geq \delta$. For large i the covering (F_α^i) must therefore also have order $\leq k+1$. □

In the next section we will get effective ways of bounding the k-diameters from below. This will then make it possible to prevent collapse in certain cases.

The other dimension concept we need is the Hausdorff dimension. As opposed to covering dimension it is a metric rather than a topological concept, and it may take nonintegral values. For a compact metric space X we can define it as

$$\mathrm{H.\,dim}\ X = \lim_{\varepsilon \to 0} \sup \frac{\log \mathrm{Cov}(X, \varepsilon)}{-\log \varepsilon}.$$

So if $\mathrm{Cov}(X, \varepsilon) \leq C \cdot \varepsilon^{-n}$ we have $\mathrm{H.\,dim}\ X \leq n$.

EXAMPLE 9. If $X \subset \mathbb{R}^n$ is compact and has interior points, then there are constants c_1, c_2 depending on X such that $c_1 \varepsilon^{-n} \leq \mathrm{Cov}(X, \varepsilon) \leq c_2 \varepsilon^{-n}$ for all $\varepsilon \leq 1$. Hence $\mathrm{H.\,dim}\ X = n$.

EXAMPLE 10. Various Cantor sets and snowflake curves have nonintegral Hausdorff dimension.

EXAMPLE 11. If X is a finite simplicial complex with its piecewise linear metric, then it follows from Example 9 that $\mathrm{H.\,dim}\ X = n$.

THEOREM 12 [25]. *Let X be a compact metric space with $\dim X = n$. Then there is a constant $c_1 > 0$ with $c_1 \varepsilon^{-n} \leq \mathrm{Cov}(X, \varepsilon)$. In particular $\mathrm{H.\,dim}\ X \geq \dim X$.*

PROOF. Choose $\delta \ll \mathrm{diam}_{n-1} X$ and a finite open δ-covering $\{V_\alpha\}$ of X of order $\leq n+1$. The nerve map $\Phi: X \to N$ defined as above is easily seen to be a Lipschitz map. Hence $\mathrm{Cov}(X, \varepsilon) \geq \mathrm{Cov}(\Phi(X) \subset N, L\varepsilon)$, for some $L \geq 1$. Since $\delta \ll \mathrm{diam}_{n-1} X$, the image $\Phi(X)$ must have dimension n. Then the result follows from Corollary 6 and Example 11. □

The Hausdorff dimension can now be used to control explosions. Assume for instance $X_i \to X$ in $\mathscr{M}$ and that $\mathrm{Cov}(X_i, \varepsilon) \leq C\varepsilon^{-n}$ for all $\varepsilon \leq 1$, $i = 1, 2, \ldots$. Then also X satisfies $\mathrm{Cov}(X, \varepsilon) \leq C\varepsilon^{-n}$, $\varepsilon \leq 1$. Hence $\mathrm{H.\,dim}\ X \leq n$.

EXAMPLE 13. If $X_i \in \mathscr{C}(n, k, D)$ as in Example 2.10 and $X = \lim X_i$ then $\dim X \leq n$. So we have no explosions happening in this case. Collapse, however, may occur as in Example 2.3.

EXAMPLE 14. Suppose $X_i \in \mathscr{C}(n, i_0, V)$ as in Example 2.9 and that $X = \lim X_i$. Again we see that $\dim X \leq n$. In the next section the $(n-1)$-diameter will be used to show that $\dim X = n$. Thus neither collapse nor explosions can occur in this class.

4. Local contractibility and topological control

In the previous section we tried to understand how the dimension is affected under convergence of spaces. Here we will study how more subtle global invariants such as homotopy type and homeomorphism type fit in with Gromov-Hausdorff convergence.

A contractibility function $\rho(\varepsilon) : [0, \ r) \to [0, \ \infty]$ is a function which satisfies: (a) $\rho(0) = 0$, (b) $\rho(\varepsilon) \geq \varepsilon$, (c) $\rho(\varepsilon) \to 0$ as $\varepsilon \to 0$, (d) ρ is nondecreasing. One can without loss of generality assume ρ is concave (see [3]), but we will not need this fact.

A metric space X is said to be LGC(ρ) for some contractibility function ρ provided that for every $\varepsilon \in [0, \ r]$ and $x \in X$ the ball $B(x, \ \varepsilon)$ is contractible inside $B(x, \ \rho(\varepsilon))$. More generally we say that X is LGC$(n, \ \rho)$ for some integer $n = 0, 1, 2, \ldots$ if for every $k = 0, 1, \ldots, n$, $\varepsilon \in [0, \ r)$, $x \in X$ and continuous map $f : S^k \to B(x, \ \varepsilon)$ there is an extension:

$$
\begin{array}{ccc}
S^k & \longrightarrow & B(x, \ \varepsilon) \\[2pt]
\cap & & \cap \\[2pt]
D^{k+1} & \dashrightarrow & B(x, \ \rho(\varepsilon))
\end{array}
$$

where $S^k = \partial D^{k+1}$ is the k-sphere bounding the $k+1$ disk D^{k+1}. Here LGC stands for locally geometrically contractible. The words locally contractible are self-explanatory; geometrically refers to the fact that we require certain size balls to have the contractibility properties.

EXAMPLE 1. If X is a compact metric space and an ANR then X is LGC(ρ) for some function ρ. Conversely we shall see that if X is LGC$(n, \ \rho)$ and $\dim X \leq n$ then X is an ANR (see [20]).

EXAMPLE 2. To geometers it might seem more natural to have $\rho(\varepsilon) = \varepsilon$: $[0, \ r) \to [0, \ r)$, so that the notion parallels that of injectivity radius. This, however, is too restrictive and does not give any advantages. A flat cone for instance is not LGC(ρ) if $\rho(\varepsilon) = \varepsilon$ no matter how small r is.

Another notion we need is that of ε-continuity or ε-almost continuity. A map $f : X \to Y$ between compact metric spaces is ε-continuous provided there is an $\delta > 0$ such that $d(x_1, \ x_2) < \delta$ implies $d(f(x_1), \ f(x_2)) < \varepsilon$. A map $f : X \to Y$ is thus (uniformly) continuous iff f is ε-continuous for all $\varepsilon > 0$.

Most everything in this section is about when one can make ε-continuous maps into continuous maps by a small pertubation.

Define recursively $\rho_0(\varepsilon) = \rho(\varepsilon)$ and $\rho_i(\varepsilon) = \rho(\varepsilon + \rho_{i-1}(\varepsilon))$, $i = 1, 2, \ldots$. With this we can now state and prove the

MAIN OBSTRUCTION RESULT 3. *Let X, Y be compact metric spaces where Y is LGC$(n-1, \ \rho)$ and $A \subset X$ a closed subset with $\dim(X - A) \leq n$. Suppose we have a continuous map $f : A \to Y$ and some possibly discontinuous extension $g : X \to Y$ with*

(i) $g|A = f$,

(ii) *g is continuous at A (i.e., for all $\varepsilon > 0$ there is $\delta > 0$ such that, for all $x \in A$, $g(B(x, \ \varepsilon)) \subset B(g(x), \ \delta)$, where $B(x, \ \delta) \subset X$),*

(iii) g is ε-continuous.

If $\rho_{n-1}(\varepsilon) + \varepsilon \leq r$ there exists a continuous extension $\bar{f}: X \to Y$ of $f: A \to Y$ with $d(g(x), \bar{f}(x)) \leq \varepsilon + \rho_n(\varepsilon)$ for all $x \in X$.

PROOF. Choose $\delta > 0$ such that $d(x_1, x_2) < \delta$ implies $d(g(x_1), g(x_2)) < \varepsilon$. Then consider the canonical covering $\{B(x, r(x))\}_{x \in X-A}$ of $X - A$, where $r(x) = \min\{\frac{1}{10} \cdot \delta, \frac{1}{10} \cdot d(x, A)\}$. Since $\dim(X - A) \leq n$ (and X, A are compact), we can find a (countable) open refinement $\{U_\alpha\}$ of order $\leq n + 1$. By construction diam $U_\alpha \leq \min\{\frac{1}{2} \cdot \delta, \frac{1}{2}d(U_\alpha, A)\}$. Denote by N the associated nerve and $\Phi: X - A \to N$ the nerve map. It is not hard to topologize (see [20]) $N \cup A$ so that the natural map

$$i: X \to N \cup A(i|A = \mathrm{id}_A \quad \text{and} \quad i|X - A = \Phi)$$

is continuous. The construction of $\bar{f}$ is now done by induction on the skeleta $N^0 \cup A, N^1 \cup A, \ldots, N^n \cup A = N \cup A$.

For a vertex U_α in N^0 define $\bar{f}(U_\alpha) = g(x)$ for some $x \in U_\alpha$. Then $\bar{f}: N^0 \cup A \to Y$ is continuous because of (ii) and satisfies $d(\bar{f}(U_\alpha), \bar{f}(U_\beta)) \leq \varepsilon$ if $U_\alpha \cap U_\beta \neq \varnothing$.

Suppose now we have a continuous extension $\bar{f}: N^k \cup A \to Y$ such that $\bar{f}|N^0 \cup A =$ above map, and if $U_{\alpha_0} \cap \cdots \cap U_{\alpha_k} \neq \varnothing$ then $\bar{f}([U_{\alpha_0}, \ldots, U_{\alpha_k}]) \subset B(f(U_{\alpha_i}), \rho_k(\varepsilon))$ for some $i = 0, \ldots, k$. If $U_{\alpha_0} \cap \cdots \cap U_{\alpha_{k+1}} \neq \varnothing$ then $\bar{f}(\partial[U_{\alpha_0}, \ldots, U_{\alpha_{n+1}}]) \subset B(\bar{f}(U_{\alpha_i}), \eta + \rho_k(\eta))$ for all $i = 0, \ldots, k+1$, where $\eta \leq \varepsilon$ and since $\bar{f}: N^k \cup A \to Y$ is continuous we can choose $\eta \to 0$ as diam$(U_\alpha) \to 0$. Thus we can extend $\bar{f}$ over $[U_{\alpha_0}, \ldots, U_{\alpha_{n+1}}]$ inside $B(\bar{f}(U_{\alpha_i}), \rho(\eta + \rho_k(\eta))) = B(\bar{f}(U_{\alpha_i}), \rho_{k+1}(\eta))$ for some $i = 0, \ldots, k+1$ as long as $\varepsilon + \rho_k(\varepsilon) < r$. Since η was chosen smaller and smaller the closer U_{α_i} is to A, it is not hard to see that $\bar{f}: N^{k+1} \cup A \to Y$ is continuous.

In this way we get a continuous map $\bar{f}: N \cup A \to Y$ such that

$$\bar{f}([U_{\alpha_0}, \ldots, U_{\alpha_n}]) \subset B(\bar{f}(U_{\alpha_i}), \rho_n(\varepsilon))$$

for some $i = 0, \ldots, k$ wherever $U_{\alpha_0} \cap \cdots \cap U_{\alpha_k} \neq \varnothing$. Now if $x \in X$ we can find U_β such that $i(x)$ is in a simplex Δ with U_β as vertex and $\bar{f}(\Delta) \subset B(\bar{f}(U_\beta), \rho_n(\varepsilon))$. Hence $d(\bar{f} \circ i(x), g(x)) \leq d(\bar{f} \circ i(x), \bar{f}(U_\beta)) + d(\bar{f}(U_\beta), g(x)) \leq \rho_n(\varepsilon) + \varepsilon$. $\square$

This theorem will be used to prove most of our subsequent results.

EXAMPLE 4. Suppose X is compact, dim $X \leq n$ and X is LGC(n, ρ). Using Theorem 3 one can easily see that X is LGC$(\bar{\rho})$ where $\bar{\rho}(\varepsilon) = 2\varepsilon + \rho_{n+1}(\varepsilon): [0, \bar{r}) \to [0, \infty)$. We can then apply Theorem 3 again to see that X is an ANR in the category of finite-dimensional spaces. Now any finite-dimensional space can be embedded in a euclidian space (see [21]). Thus X must be an ENR and in particular an ANR.

EXAMPLE 5. Assume X, Y are compact metric spaces with $d_{\mathrm{GH}}(X, Y) < \varepsilon$ and Y is LGC$(n-1, \rho)$. Fix a metric on $X \amalg Y$ such that the Hausdorff distance between X and Y is $< \varepsilon$. If $f : T \to X$ is a continuous map, with dim $T \leq n$, then we can find $g : T \to Y$ with the property that $d(f(t), g(t)) < \varepsilon$ for all $t \in T$. In particular $g : T \to X$ is 2ε-continuous. Hence if $2\varepsilon + \rho_n(2\varepsilon) < r$ we can find a continuous map $\bar{f} : T \to Y$, using Theorem 3, with

$$d(\bar{f}, g) < 3\varepsilon + \rho_n(2\varepsilon) \quad \text{and} \quad d(f, \bar{f}) \leq 3\varepsilon + \rho_n(2\varepsilon).$$

COROLLARY 6. *Let X, Y be compact metric spaces, such that dim $X \leq n$ and Y is $\mathrm{LGC}(n, \rho)$. If f, $g : X \to Y$ are continuous maps with $d(f, g) < \varepsilon$ and $\varepsilon + \rho_n(\varepsilon) < r$, then f and g are $2\varepsilon + \rho_{n+1}(\varepsilon)$ controlled homotopy equivalent.*

PROOF. Use Theorem 3 with $X \times [0, 1]$, $A = X \times \{0, 1\}$ and $h : A \to Y$ is f on $X \times \{0\}$ and g on $X \times \{1\}$. By defining $\hat{h}$ on $X \times [0, 1]$ as f on $X \times [0, \frac{1}{2}]$ and g on $X \times (\frac{1}{2}, 1]$, we get an ε-continuous extension of h satisfying (i)–(iii) of Theorem 3. Thus we get a continuous map $H : X \times I \to Y$ which is a homotopy from f to g, with

$$d(H(x, t), f(x)) \leq 2\varepsilon + \rho_{n+1}(\varepsilon),$$
$$d(H(x, t), g(x)) \leq 2\varepsilon + \rho_{n+1}(\varepsilon) \quad \text{for all } t, x. \quad \square$$

COROLLARY 7. *Let X be a compact metric space such that* dim $X \leq n$, X *is* $\mathrm{LGC}(n, \rho)$ *and* $H_k(X, G) \neq 0$ *for some abelian group* G. *Then* $\mathrm{diam}_{k-1} X \geq \delta$, *as long as* $(2\delta + \rho_k(\delta) + \rho_n(2\delta + \rho_k(\delta))) < r$.

PROOF. Let $f : X \to C$ be a continuous ε-map onto a simplicial complex of dimension $\leq k - 1$. We can then find an ε-continuous inverse $g : C \to X$ with the property that $d(x, g \circ f(x)) < \varepsilon$ for all $x \in X$. Theorem 3 then insures us a continuous map $\bar{g} : C \to X$ with $d(x, \bar{g} \circ f(x)) \leq 2\varepsilon + \rho_k(\varepsilon)$ provided $\varepsilon + \rho_{k-1}(\varepsilon) < r$. Then Corollary 6 implies that id $: X \to X$ and $\bar{g} \circ f : X \to X$ are homotopic provided $(2\varepsilon + \rho_k(\varepsilon) + \rho_n(2\varepsilon + \rho_k(\varepsilon))) < r$. Thus we get a commutative diagram

$$
\begin{array}{ccc}
H_k(X) & \overset{\mathrm{id}}{\to} & H_k(X) \\
& \searrow \quad \swarrow & \\
& H_k(C) = 0. &
\end{array}
$$

which is a contradiction. $\square$

Corollary 7 shows in particular that any closed n-manifold has $\mathrm{diam}_{n-1} > 0$ and must therefore have dimension n.

There are also relative versions of Corollary 7 where one assumes $H_k(X, A, G) \neq \varnothing$ and that both X and A and $\mathrm{LGC}(n, \rho)$ to get estimates on $\mathrm{diam}_{k-1} X$.

We now turn our attention to how these results can be used in connection with the Gromov-Hausdorff distance.

THEOREM 8. *Let X, Y be compact metric spaces which are $\mathrm{LGC}(n, \rho)$ and have dimension $\leq n$. If*

$$d_{\mathrm{GH}}(X, Y) < \varepsilon, \quad \text{where } (6\varepsilon + 2\rho_n(\varepsilon) + \rho_n(6\varepsilon + 2\rho_n(\varepsilon))) < r,$$

then X and Y are $(12\varepsilon + 4\rho_n(\varepsilon) + \rho_{n+1}(6\varepsilon + 2\rho_n(\varepsilon)))$ controlled homotopy equivalent.

PROOF. First choose a metric on $X \amalg Y$ in which the Hausdorff distance between X and Y is $< \varepsilon$. Use Example 5 on id: $A \to X$ and id: $Y \to Y$ to get continuous maps $f: X \to Y$ and $g: Y \to X$ with $d(x, g(x)) < 3\varepsilon + \rho_n(2\varepsilon)$ and $d(y, f(y)) < 3\varepsilon + \rho_n(2\varepsilon)$. Then $d(x, g \circ f(x)) < 6\varepsilon + 2\rho_n(\varepsilon)$ and $d(y_1 f \circ g(y)) < 6\varepsilon + 2\rho_n(\varepsilon)$. The result is then an immediate consequence of Corollary 6. □

THEOREM 9 (see [3]). *Let $\{X_i\}$ be a sequence of compact metric spaces converging to X in the Gromov-Hausdorff distance, if all X_i's are $\mathrm{LGC}(n, \rho)$, then X is also $\mathrm{LGC}(n, \rho)$.*

PROOF. The notation is as in the proof of Propositon 2.6. Fix $p \in X$, $\varepsilon < r$ and a map $f: S^k \to B(p, \varepsilon) \subset X$, where $k \in \{0, 1, \dots, n\}$. Choose $\delta < \varepsilon$ such that $f(S^k) \subset B(p, \delta)$ and a sequence $p_i \in X_i$ representing $p \in X$. Using Example 5 we get continuous maps $f_i : S^k \to X_i$ with $d^i(f(t)), f_i(t)) < 2\varepsilon_i$ where $\varepsilon_i \searrow 0$ and $d_{\mathrm{GH}}(X, X_i) < \varepsilon_i$. We can then find N so that $f_i(S^k) \subset B(p_i, \delta)$ for all $i \geq N$.

Clearly each $f_i: S^k \to B(p_i, \delta) \subset X_i$ admits an extension $D^{k+1} \to B(p_i, \rho(\delta)) \subset X_i$, but this sequence of extensions does not necessarily represent a map into X. We therefore need to modify these extensions.

Choose an extension $\bar{f}_N: D^{k+1} \to B(p_N, \rho(\delta)) \subset X_N$ of f_N. Then construct a $2\varepsilon_N$-continuous extension $g_{N+1}: D^{k+1} \to X_{N+1}$ as in Example 5, such that $g_{N+1} = f_{N+1}$ on a neighborhood of S^k. By applying Theorem 3 we thus get a continuous extension $\bar{f}_{N+1}: D^{k+1} \to X_{N+1}$ of f_{N+1} with $d^{N, N+1}(\bar{f}_N(t), \bar{f}_{N+1}(t)) \leq 4\varepsilon_N + \rho_k(2\varepsilon_N)$; in particular $\bar{f}_{N+1}(D^{k+1}) \subset B(\rho_{N+1}, \rho(\delta) + 6\varepsilon_N + \rho_k(2\varepsilon_N))$.

Provided $\sum_{j=N}^{\infty} 6\varepsilon_j + \rho_k(2\varepsilon_j)$ is very small we can then inductively construct extensions $\bar{f}_i: D^{k+1} \to B(p_i, \rho(\delta) + \sum_{j=N}^{i} 6\varepsilon_j + \rho_k(2\varepsilon_j))$ of $f_i: S^k \to B(p_i, \delta)$, satisfying $d^{ik}(\bar{f}_i(t), \bar{f}_k(t)) \leq \sum_{j=i}^{k} 4\varepsilon_j + \rho_k(2\varepsilon_j)$. These extensions $\bar{f}_i / i = N, \dots, \infty$ must then represent a continuous extension $\bar{f}: D^{k+1} \to B(p, \rho(\delta) + \sum_{j+N}^{\infty} 6\varepsilon_j + \rho_k(2\varepsilon_j))$ of $f: S^k \to B(p, \delta) \subset X$.

After possibly passing to a subsequence of $\{X_i\}$ we can suppose that $\sum_{j=N}^{\infty} 6\varepsilon_j + \rho_k(2\varepsilon_j) \leq \min\{\varepsilon - \delta, r, \rho(\varepsilon) - \rho(\delta)\}$. The induction procedure will then work and give the desired extension. □

Using Theorems 8 and 9 and 2.8 we immediately get

COROLLARY 10. *Fix a covering function* $N(\varepsilon) \colon [0, \alpha] \to [0, \infty)$ *with* $\limsup_{\varepsilon \to 0} \varepsilon^n N(\varepsilon) < \infty$ *and a contractibility function* $\rho(\varepsilon) \colon [0, r) \to [0, \infty)$. *The class* $\mathscr{C}(N(\varepsilon), \rho)$ *of compact metric spaces* X, *such that* X *is* $\mathrm{LGC}(\rho)$ *and* $\mathrm{Cov}(X, \varepsilon) \le N(\varepsilon)$ *for all* $\varepsilon \in [0, \alpha]$, *is compact and contains only finitely many homotopy types.*

REMARK. Since it is possible to define the simple homotopy type of an ANR (see [5]), it is natural to ask whether one can bound the number of simple homotopy types in Theorem 10. Suppose X_i is a sequence in $\mathscr{C}(N(\varepsilon), \rho)$ of mutually nonsimply homotopic spaces. By compactness we can assume that X_i converges to a space X in $\mathscr{C}(N(\varepsilon), \rho)$.

In [8$_1$] it is shown that there is an $\varepsilon > 0$ depending on X, such that any ε-map $f \colon X \to Y$ into an ANR is homotopy equivalent to a simple homotopy equivalence between X and Y. Now for all sufficiently large i we are guaranteed to have an ε-map $f_i \colon X \to X_i$. Thus all the X_i's are simple homotopy equivalent for large i. This, however, contradicts our hypothesis, so $\mathscr{C}(N(\varepsilon), \rho)$ contains only finitely many simple homotopy types (see also [10] for a more general statement).

To control the homeomorphism types in $\mathscr{C}(N(\varepsilon), \rho)$ one needs to make further assumptions on the spaces. A natural choice seems to be the subclass $\mathscr{M}(N(\varepsilon), n, \rho) \subset \mathscr{C}(N(\varepsilon), \rho)$ of metric spaces which are in addition closed topological n-manifolds. Even for this class, however, several complications occur due to the fact that $\mathscr{M}(N(\varepsilon), n, \rho)$ might not be compact.

EXAMPLE 11 (see [23]). Given a manifold say S^3, one can find a sequence of metrics d_i on S^3, such that all the spaces (S^3, d_i) are $\mathrm{LGC}(\rho)$ for some ρ but (S^3, d_i) converges to a space X which is not a manifold. X could for instance be an identification space $f \colon S^3 \to X$, where some infinitely knotted curve in S^3 is mapped to a point and f is a homeomorphism on the complement of the curve. One can even have a sequence of n-manifolds $\{M_i\}$ which are all $\mathrm{LGC}(\rho)$ converging to an infinite-dimensional space.

Nevertheless it is still possible to prove

THEOREM 12 (see [19]). *Given* $N(\varepsilon)$, $n \ne 3$, *and* ρ *as in Corollary 10. The class* $\mathscr{M}(N(\varepsilon), n, \rho)$ *contains only finitely many homeomorphism types.*

The proof relies on deep results in geometric topology (see [26–29, 7, 9]) and will be omitted here. When $n = 3$ the theorem is more or less equivalent to the Poincaré conjecture.

From the results we have obtained so far we get two geometric corollaries.

THEOREM 13 (see [19]). *The class* $\mathscr{C}(n, i_0, V)$ *from Example 2.9 contains only finitely many simple homotopy types when* $n = 3$ *and only finitely many homeomorphism types when* $n \ne 3$.

THEOREM 14 (see [19]). *The class of Riemannian* n-*manifolds with sectional curvature* $\ge k$, *volume* $\ge v$ *and diameter* $\le D$, *for arbitrary* $k \in$

$\mathbb{R}$, v, $D > 0$ *contains only finitely many simple homotopy types when* $n = 3$ *and only finitely many homeomorphism types when* $n \neq 3$.

PROOF. In [16] it is shown that all elements in this class are $\mathrm{LGC}(\rho)$ for some function ρ depending only on n, k, v and D. This together with Example 2.10 gives the hypothesis needed to apply Corollary 10 and Theorem 12. □

REMARK. With the help of [22] one can even show that the classes in Theorems 13 and 14 contain only finitely many diffeomorphisms types, when $n \geq 5$.

REMARK. Theorems 13 and 14 generalize Cheeger's finiteness theorem (see [4] and [24]) at least when $n \geq 5$.

REMARK. One would hope that there are more geometric proofs of Theorems 13 and 14 which could also settle the problems in dimension 3 and 4.

REFERENCES

1. P. Aleksandroff, *Über Den Allgremeinen-Dimensionsbegriff und seine Beziehungen zur elementaren geometrischen Anschauung*, Math. Ann. **98** (1928), 617–635.
2. E. B. Begle, *Regular convergence*, Duke Math. J. **11** (1944), 441–450.
3. K. Borsuk, *On some metrizations of the hyperspace of compact sets*, Fund. Math. **41** (1955), 168–201.
4. J. Cheeger, *Finiteness theorems for Riemannian manifolds*, Amer. J. Math. **96** (1970), 61–74.
5. M. Cohen, *A course in simple homotopy theory*, Springer-Verlag, 1973.
6. C. Croke, *Some isoperimetric inequalities and eigenvalue estimates*, Ann. Sci. École Norm. Sup. (4) **13** (1980), 419–435.
7. R. D. Edwards, *The topology of manifolds and cell-like maps*, Proc. Internat. Congr. Math., Academia Scientiarum Fennica (Helsinki, 1978), 1980, pp. 111–127.
8. S. Ferry, *The homeomorphism group of a compact Hilbert cube manifold is an ANR*, Ann. of Math. (2) **106** (1977), 101–119.
9. ____, *Homotoping ε-maps to homeomorphisms*, Amer. J. Math. **101** (1979), 567–582.
10. ____, *An extension of simple-homotopy theory*, Abstract Amer. Math. Soc. **11** (1990), no. 3, 857-55-44, 254.
11. K. Fukaya, *A boundary of the set of Riemannian manifolds with bounded curvatures and diameters*, J. Differential Geom. **28** (1988), 1–21.
12. ____, *Collapsing of Riemannian manifolds and eigenvalues of Laplace operator*, Invent. Math. **25** (1987), 517–547.
13. K. Fukaya and T. Yamaguchi, *Almost nonpositively curved manifolds*, Amer. Math. Journal Ann. of Math.
14. M. Gromov, *Groups of polynomial growth and expanding maps*, Inst. Hautes Études Sci. Publ. Math. **53** (1981), 183–215.
15. K. Grove, *Metric differential geometry*, Differential Geometry (V. L. Hansen, ed.), Lecture Notes in Math., vol. 1263, Springer-Verlag, 1987.
16. K. Grove and P. Petersen V, *Bounding homotopy types by geometry*, Ann. of Math. (2) **128** (1988) 195-206.
17. ____, *Manifolds near the boundary of existence*, J. Differential Geom. **33** (1991), 379–394.
18. ____, *Volume comparison á la Aleksandrov*, Acta Math. (to appear).
19. K. Grove, P. Petersen V, and T.-Y. Wu, *Geometric finiteness theorems via controlled topology*, Invent. Math. **99** (1990), 205–213.
20. S. T. Hu, *Theory of retracts*, Wayne State Univ. Press, Detroit, 1965.

21. W. Hurewicz and H. Wallman, *Dimension theory,* Princeton Univ. Press, Princeton, NJ, 1941.
22. P. Kirby and L. Siebermann, *Foundational essays on topological manifolds, smoothings and triangulations,* Ann. of Math. Stud. vol. 88, Princeton Univ. Press, Princeton, NJ, 1977.
23. T. Moore, *Gromov-Hausdorff convergence to non-manifolds,* Preprint.
24. S. Peters, *Cheeger's finiteness theorem for diffeomorphism classes of Riemannian manifolds,* J. Reine Angew. Math. **394** (1984), 77–82.
25. L. Pontriagin and L. Schnirelmann, *Sur une propiété métrique de la dimension,* Ann. of Math. **33** (1932), 156–162.
26. F. Quinn, *Resolutions of homology manifolds, and the topological characterization of manifolds,* Invent. Math. **72** (1983), 167–284.
27. ____, *An obstruction to the resolution of homotology manifolds,* Michigan Math. J. **34** (1987), 285–291.
28. ____, *Ends of maps.* I, Ann. of Math. (2) **110** (1979), 275–331.
29. ____, *Ends of maps* III: *dimensions 4 and 5,* J. Differential Geom. **17** (1982), 503–521.
30. E. H. Spanier, *Algebraic topology,* Springer-Verlag, 1966.
31. P. Urysohn, *Mémoir sur les multiplicit'es Cantoriennes,* Fund. Math. **8** (1926), 225–359.
32. T. Yamaguchi, *Homotopy type finiteness theorems for certain precompact families of Riemannian manifolds,* Proc. Amer. Math. Soc.
33. ____, *Collapsing and pinching under a lower curvature bound,* Ann. of Math. **133** (1991), 317–357.

UNIVERSITY OF CALIFORNIA, LOS ANGELES

Proceedings of Symposia in Pure Mathematics
Volume 54 (1993), Part 3

Cartan Connections

ERNST A. RUH

1. Introduction

In the classical theory of connections, due to Levi-Civita, the tangent space of a manifold M is considered to be a vector space, and a connection is a law of linear parallel displacement of vectors along curves. In the theory of connections due to Cartan, the tangent space is considered to be an affine space on which the group of affine transformations acts transitively. A connection in this theory is a law of affine parallel displacement of points along curves. The classical development of the manifold M is then obtained by restricting the parallel displacement to the point field defined by the origin in each tangent space.

The point of view of Cartan allows for an important generalization. The tangent space need not be an affine space; a homogeneous space $F = G/K$ of the correct dimension will do. Just as the tangent vector space, the tangent homogeneous space is viewed as an approximation to the manifold M. The theory of Cartan connections can be understood as an infinitesimal version of Klein's Erlangen Program, where the emphasis is on the automorphism group of a geometry. In general, of course, the automorphism group of a manifold is not transitive. However it is still possible to approximate the manifold at every point with one (or several) homogeneous spaces which can serve as a model for the manifold. The quality of the approximation is measured by the Cartan curvature. In particular, vanishing Cartan curvature means that the manifold is locally homogeneous.

The theory of Cartan connections was rigorously established by Ehresmann [3], and further clarified by Kobayashi [17]. To define it we need a principal bundle P over the manifold M under consideration as well as a homogeneous space $F = G/K$. The total space P is a generalization of

1991 *Mathematics Subject Classification.* Primary 53C20.
Supported by EPF-Lausanne and NSF Grant DMS-8601282.
This paper is in final form and no version of it will be submitted for publication elsewhere.

the automorphism group G; in particular both spaces are of the same dimension. The tangent homogeneous space bundle is the associated bundle $B = P \times_\rho F$, where ρ is the isotopy representation of K in $\mathrm{Diff}(F)$, obtained by letting an element $k \in K$ operate as the diffeomorphism of F induced by the left translation of k on G. The fact that the fibers of B approximate the manifold M is expressed by the "condition de soudure"; i.e., it is postulated that tangent bundle TM is the associated bundle of P via the infinitesimal isotropy representation of K on $\mathbb{R}^n = T_0 F$. A Cartan connection is a generalization of the Maurer-Cartan form on G with values in the Lie algebra $\mathfrak{J}$ of G to a $\mathfrak{J}$-valued 1-form on P with certain properties. The main difference is that the Maurer-Cartan equation may no longer hold. The excess, $\Omega = d\omega + [\omega, \omega]$, is called the Cartan curvature.

A Cartan connection ω on P induces a unique connection (in the usual sense) $\tilde{\omega}$ on the bundle $\tilde{P}$ obtained from P by enlarging the fiber from K to G. This connection $\tilde{\omega}$ defines a parallel translation in $\tilde{P}$ as well as in any bundle associated to $\tilde{P}$. In case of the bundle B with fiber $F = G/K$ and structure group enlarged from K to G, the parallel translation is the homogeneous development.

The existence of a Cartan connection on a manifold provides additional structure to the manifold and may impose topological restrictions. The Cartan curvature measures how well the tangent homogeneous space approximates a manifold with this additional structure. Many of the well-known comparison theorems in Riemannian geometry are equivalent to the statement that a certain Cartan connection with small curvature can be replaced by a Cartan connection with vanishing curvature. A successful strategy to obtain such a flat Cartan connection is to set up an evolution equation for connections with the given connection as initial condition. We will explore this strategy later on. No general condition for success is known. It appears reasonable to expect that if the automorphism group of the model space is semisimple, then a Cartan connection with small curvature can be deformed into a flat connection. An example of Gromov and Thurston [14] shows that this is not the case, or at least, that the question of how small is small needs to be clarified. In case the automorphism group of the model is not semisimple it appears to be unreasonable to expect that an almost flat Cartan connection can be deformed into a flat one. However, in the example of almost flat manifolds, compare Gromov [12] and Ruh [28]; this can be done as long as we allow the model homogeneous space to change from $\mathbb{R}^n$ to a nilpotent Lie group during the deformation (compare also [8], [29]).

To accommodate the change of the model in the theory of Cartan connections it is convenient to generalize the concept of Cartan connection and curvature. We no longer assume that a Cartan connection form ω is Lie-algebra valued. We only require that the connection form takes values in a vector space with skew product. There is, essentially, no penalty to this generalization because $\Omega = 0$ implies that the skew product satisfies the

Jacobi equation and hence defines a Lie algebra. There are technical difficulties resulting from this generalization; in particular, the Bianchi equation is no longer satisfied exactly.

Another generalization of the concept of Cartan connections may prove to be useful. In the theory of connections, as well as Cartan connections, it is assumed that the connection form restricted to the fiber K of the principal bundle P is essentially the Maurer-Cartan form. This may be too restrictive an assumption in view of the fact that there are sequences of compact Riemannian manifolds topologically distinct from homogeneous spaces whose curvatures converge to the curvature of a homogeneous space (compare [4]). Therefore, it is hopeless in this case to try to deform an almost flat Cartan connection into a flat one. Still, the sequence is constructed from a Lie group and it might be possible to recover the Maurer-Cartan form of this Lie group from an almost flat Cartan connection. So far, this possibility has not been explored systematically.

The author would like to thank Professor P. Buser and the Swiss Federal Institute of Technology for their support and hospitality.

2. Definitions and examples

Let $F = G/K$, where G is a Lie group and $K \subset G$ a closed subgroup, be a homogeneous space. The coset K is called the origin $o \in F$. Left translation on G by $g \in G$ induces a diffeomorphism of F. We assume that the action of G on F is effective, i.e., that the only element of g which acts as identity on F is the identity $e \in G$. The restriction of this action to $K \subset G$ is called the isotropy representation. By definition, the isotropy representation fixes $o \in F$. Its differential at $o \in F$ is called the linear isotropy representation. In the present exposition we assume that the linear isotropy is a faithful representation of K on $\mathbb{R}^n = T_0 F$. This restriction rules out projective and conformal Cartan connections. For Cartan connections on higher-order jet bundles we refer to [18].

Let $P \to^\pi M$ denote a principal bundle with structure group K as above over the manifold M. Let ρ denote a representation of K as diffeomorphisms of a manifold F, or more specifically, a representation as linear maps of a vector space. By definition, the associated bundle to P via ρ is $P \times_\rho F$, where $P \times_\rho F$ is the set of equivalence classes of $P \times F$ defined by the relation $(pa, f) \sim (p, \rho(a)f)$. Let B denote the associated bundle to P via the isotropy representation of K on F. B, or P, is said to satisfy the "condition de soudure" if the tangent bundle TM of M is associated to P via the infinitesimal isotropy representation of K on $T_0 F = \mathbb{R}^n$, where $F = G/K$.

Let $\mathfrak{J}$ denote the Lie algebra of the Lie group G.

DEFINITION. A $\mathfrak{J}$-valued 1-form $\omega \colon TP \to \mathfrak{J}$ on P is called a *Cartan connection* of type (G, K) if it satisfies the following three conditions, and P satisfies the "condition de soudure."

(2.1) $\omega(A^*) = A$, where A^* is the fundamental vector field on P corresponding to the 1-parameter group defined by the right action of $\exp tA$, $A \in \mathfrak{k}$, the Lie algebra of K, on P.

(2.2) $(R_a)^*\omega = \mathrm{ad}(a^{-1})\omega$, where R_a denotes the right action of $a \in K$ on P and $\mathrm{ad} = \mathrm{ad}_K$ is the restriction of the adjoint representation of G on $\mathfrak{J}$ to K.

(2.3) If $\omega(X) = 0$ then X is the zero vector field on P.

The model for Cartan connections is the Maurer-Cartan form $\omega \colon TG \to \mathfrak{J}$, defined by $\omega(X_a) = (L_{a^{-1}})X_a \in T_eG = \mathfrak{J}$, on the principal bundle $G \to G/K = F$ which satisfies the "condition de soudure." The Maurer-Cartan equation, $d\omega + [\omega, \omega] = 0$, is the motivation for the definition of Cartan curvature.

(2.4) DEFINITION. The $\mathfrak{J}$-valued 2-form, $\Omega = d\omega + [\omega, \omega]$ is the *Cartan curvature* of the Cartan connection form ω defined above.

An important fact, already known to Lie, is that $\Omega = 0$ is the integrability condition for a local diffeomorphism $\phi \colon P \to G$, where G is the simply connected Lie group with Lie algebra $\mathfrak{J}$, and ϕ^* pulls back the Maurer-Cartan form $\overline{\omega}$ on G to ω on P, i.e., $\phi^*\overline{\omega} = \omega$. In another words, ω defines the structure of a local Lie group on P. The graph of ϕ is obtained as the integral manifold of the distribution $\{(X, \overline{X}), \omega(X) = \overline{\omega}(\overline{X})\}$ on $P \times G$ which passes through $(e, \overline{e})$, where $e \in P$ is arbitrary and $\overline{e}$ is the identity in G. Since $\Omega = \overline{\Omega} = 0$ the integrability condition is satisfied.

It is possible to think of Cartan connections as a special case of connections in the usual sense. To do this we enlarge the structure group K of the bundle P to G; i.e., we define the principal bundle $\widetilde{P}$ associate to P via the inclusion $K \subset G$. By extending the property (2.2) of Cartan connections from K to G we obtain a unique extension $\tilde{\omega}$ of the $\mathfrak{J}$-valued 1-form ω on P to a $\mathfrak{J}$-valued 1-form on $\widetilde{P}$. The property (2.1) is then satisfied for $A \in \mathfrak{J}$ and therefore $\tilde{\omega}$ is a connection form. Obviously, the restriction of $\tilde{\omega}$ to P satisfies (3.3). This property will be essential for the definition of the development of the base space M along curves. A connection form $\tilde{\omega}$ on $\widetilde{P}$ with the above properties is also called a Cartan connection.

Development. A Cartan connection ω on the K-principal bundle P induces a connection $\tilde{\omega}$ on $\widetilde{P} = P \times_i G$, where $i \colon K \to G$ is the inclusion, as above. This connection defines a parallel displacement in any bundle associated to $\widetilde{P}$; in particular in $B = \widetilde{P} \times_\rho G/K$, where $\rho \colon G \to \mathrm{Diff}(F)$, $F = G/K$, is the representation induced by left translation on G.

Let $x_o \in M$, the base space of P, denote a preferred point and $\tau \colon [0, 1] \to M$ a curve with $\tau(0) = x_o$. The *development of the curve* τ into the fiber F_{x_o} of B over x_o is the curve $\tilde{\tau} \colon [0, 1] \to F_{x_o}$, where $\tilde{\tau}(t)$ is defined by parallel translation of $o \in F_x$, $x = \tau(t)$, along the curve τ restricted to $[o, t]$, from $\tau(t) = x$ to $\tau(o) = x_o$. In case the Cartan connection Ω vanishes, then $\widetilde{\Omega}$ vanishes as well, and parallel translation depends only on the homotopy class of a curve. We obtain a local diffeomorphism

(2.5) $\phi \colon M \to F_{x_o}$,

where $\phi(x)$ is the development of the endpoint of any curve (in a neighborhood of x_o) with initial point x_o and endpoint x. Because of (2.3), the differential of ϕ is nondegenerate. If M is simply connected, then ϕ is globally defined and is a covering map.

EXAMPLE 0. Let $K = \mathrm{GL}(n, \mathbb{R})$, $G = \mathrm{Aff}(n) = \{\begin{pmatrix} a & x \\ 0 & 1 \end{pmatrix}, a \in \mathrm{GL}(n, \mathbb{R})$, $x \in \mathbb{R}^n)\}$ be the affine group. $F = G/K = A^n$ is the affine n-space. Let P be the $\mathrm{GL}(n, \mathbb{R})$-principal bundle of linear frames on F, and let $\eta \colon TP \to \mathfrak{Jl}(n, \mathbb{R})$ denote the standard connection form with vanishing torsion and curvature defined by the affine structure of F. Let θ be the identity gauge transformation $\theta \colon TF \to TF$. We identify θ with the canonical 1-form $\theta \colon TP \to \mathbb{R}^n$ defined by $\theta(X) = u^{-1}\pi(X)$ where a linear frame $u \in P$ is viewed as a linear isomorphism $u \colon \mathbb{R}^n \to T_{\pi(u)}F$ as usual, and the tensor θ on F is identified with the corresponding tensorial 1-form on P. We define

$$\omega = \theta + \eta \colon TP \to \mathbb{R}^n \oplus \mathfrak{Jl}(n, \mathbb{R}) = \mathfrak{aff}(n),$$

where $\mathfrak{aff}(n)$ is the Lie algebra of the affine group $\mathrm{Aff}(n)$. Since the torsion form, $d\theta + [\eta, \theta] + [\theta, \eta]$, as well as the curvature form, $d\eta + [\eta, \eta]$, vanishes, we obtain

$$\Omega = d\omega + [\omega, \omega] = (d\theta + [\eta, \theta] + [\theta, \eta]) + (d\eta + [\eta, \eta]) = 0.$$

Of course, this calculation is not really necessary because we know that θ and η are simply the projections to $\mathbb{R}^n$ and $\mathfrak{Jl}(n, \mathbb{R})$ respectively, of the Maurer-Cartan form ω; and $\Omega = 0$ is the Maurer-Cartan equation. The reason for our emphasis on the total space P of the principal bundle of frames over F and not the Lie group $\mathrm{Aff}(n)$ which it equals, is that in the general cases to be considered later, the principal bundle is the given data and the corresponding group G may or may not exist.

The development map (2.5) in this case is $\phi \colon A^n \to A^n$. It identifies the base space A^n of the bundle B with the fiber A^n over the point $o \in A^n = \mathrm{Aff}(n)/\mathrm{GL}(n, \mathbb{R})$.

EXAMPLE 1. Let $K = \mathrm{SO}(n)$, $G = \mathrm{SO}(n + 1)$, and $F = G/K = S^n$ is the standard sphere. Let P be the $\mathrm{SO}(n)$-principal bundle of orthogonal frames over $F = S^n$ with respect to the standard metric on S^n. As above, $\theta \colon TP \to \mathbb{R}^n$, defined by $\theta = u^{-1}\pi$ is the canonical 1-form, and $\eta \colon TP \to \mathfrak{so}(n)$ is the standard connection form, in this case the Levi-Civita connection of S^n. We identify $x \in \mathbb{R}^n$ with $\begin{pmatrix} 0 & x \\ -x & 0 \end{pmatrix} \in \mathfrak{so}(n + 1)$ and define

$$\omega = \theta + \eta \colon TP \to \mathbb{R}^n \oplus \mathfrak{so}(n) = \mathfrak{so}(n + 1).$$

The $\mathfrak{so}(n + 1)$-valued 1-form ω satisfies the conditions (2.1), (2.2), (2.3); in addition, P satisfies "la condition de soudure," and ω is a Cartan connection. Again, $\Omega = 0$ easily verified. We obtain

$$\Omega = d\omega + [\omega, \omega] = (d\theta + [\eta, \theta] + [\theta, \eta]) + (d\eta = [\eta, \eta] + [\theta, \theta]) = 0,$$

since the first expression, the torsion form vanishes, and the vanishing of the second expression is equivalent to the fact that the sectional curvature of the standard metric on S^n is equal to one. As in the example above, this calculation is not necessary. The vanishing of Ω is equivalent to the Maurer-Cartan equation for the Maurer-Cartan form ω on $P = \mathrm{SO}(n+1)$. Again we emphasize the principal bundle P over the Riemannian manifold $M = S^n$ and not the Lie group $\mathrm{SO}(n+1)$ because, for a general manifold M, the principal bundle P still exists while the group of automorphisms of M in general is trivial. In the standard situation at hand, the canonical 1-form θ and the Levi-Civita connection form η are simply the projections of the Maurer-Cartan form, $\omega \colon T\mathrm{SO}(n+1) \to \mathfrak{so}(n+1)$, to the summands $\mathbb{R}^n$ and $\mathfrak{so}(n)$, respectively.

The development map (2.5) in this case is $\phi \colon S^n \to S^n$. It is an isometry of the base space S^n of the fiber bundle B with the fiber S^n over the point $o \in S^n = \mathrm{SO}(n+1)/\mathrm{SO}(n)$. In this special case, the development map ϕ is essentially the same as the Gauss map.

EXAMPLE 2. Let G be a semisimple Lie group of noncompact type and K a maximal compact subgroup. $F = G/K$ is a Riemannian symmetric space of noncompact type. Let $\Gamma \subset G$ be a discrete subgroup with $M = \Gamma \backslash F$ a locally symmetric space. We set $P = \Gamma \backslash G$ and note that $P \to^\pi M$ is a K-principal bundle over M. The Maurer-Cartan form on G provides a Cartan connection ω with vanishing curvature on P. As in the previous examples, the projections of ω to the summands of $\mathfrak{J} = \mathfrak{k} \oplus \mathfrak{p}$, the Cartan decomposition of the Lie algebra $\mathfrak{J}$ of G, are the canonical 1-form, and the Levi-Civita connection form, respectively.

The development map (2.5) in this case,

$$\phi \colon M \to F = G/K,$$

is only defined locally because M is not simply connected.

Next, we review briefly a method for obtaining vanishing theorems for the cohomology groups $H^*(\Gamma, \rho)$ of the fundamental group Γ of the manifold M with values in an irreducible representation ρ of $\Gamma \subset G$ (compare [19]). The reason for this is that this method can be generalized to yield an important part of the proof of certain comparison theorems with symmetric spaces of noncompact type as models. For these geometric applications, outlined in the next section, only the adjoint representation will play a role. Even though the method applies to irreducible representations in general we restrict our attention to the adjoint representation ad of G on $\mathfrak{J}$.

Let $E_{\mathfrak{J}}$ be the vector bundle $E_{\mathfrak{J}} = \tilde{P} \times_{\mathrm{ad}} \mathfrak{J}$, where $\tilde{P}$ is obtained by extending the structure group K of P to G. It is well known (compare [19]), that $H^*(\Gamma, \mathrm{ad})$ is isomorphic to the space of harmonic forms on M with values in $E_{\mathfrak{J}}$. Note that the Maurer-Cartan form ω extends to a flat connection form $\tilde{\omega}$ on $\tilde{P}$ which in turn defines a flat covariant derivative D in the bundle $E_{\mathfrak{J}}$. This covariant derivative, and the metric on $E_{\mathfrak{J}}$ provided by the Killing form on $\mathfrak{J}$ after sign change in the $\mathfrak{k}$-direction, is used to define the exterior derivative d^D and its adjoint δ^D. As usual the Laplacian is $\Delta^D = d^D\delta^D + \delta^D d^D$.

The bundle $E_{\mathfrak{J}} = P \times_{\mathrm{ad}_K} \mathfrak{J} = TM \oplus E_{\mathfrak{k}}$, where $E_{\mathfrak{k}} = P \times_{\mathrm{ad}} \mathfrak{k}$, and $TM = P \times_{\mathrm{ad}_K} \mathbb{R}^n$ is the tangent bundle. Since the Levi-Civita connection form is obtained by the projection $TP \to^\omega \mathfrak{J} = \mathfrak{k} \oplus \mathfrak{p} \to \mathfrak{k}$, we obtain the following formula for the covariant derivative D on $E_{\mathfrak{J}}$ induced by ω.

$$(2.6) \qquad D_X s = \nabla_X s + [X, s],$$

where s is a section in $E_{\mathfrak{J}}$, $[\ ,\ ]$ is the Lie bracket on the fibers of $E_{\mathfrak{J}}$ obtained from the Lie bracket in $\mathfrak{J}$, ∇ is the levi-Civita connection on TM extended to $E_{\mathfrak{J}}$, and X is a vector field on M identified with the corresponding section in $E_{\mathfrak{J}}$. Since D is flat, the usual formula for the exterior derivative d applies to $E_{\mathfrak{J}}$-valued p-forms α.

$$(2.7) \qquad \begin{aligned} (d^D\alpha)(X_0, \ldots, X_p) &= \sum_{i=0}^{p} (-1)^i D_{X_i}\alpha(X_0, \ldots, \widehat{X}_i, \ldots, X_p) \\ &= \sum_{0 \le i < j \le p} (-1)^{i+j}\alpha([X_i, X_j], X_0, \ldots, \widehat{X}_i, \ldots, \widehat{X}_j, \ldots, X_p). \end{aligned}$$

The decomposition (2.6) produces the following decomposition of d^D:

$$d^D = d^{\nabla} + d^{\mathfrak{J}}, \quad \text{where } d^{\nabla} \text{ is obtained from (2.7) by replacing } D \text{ by } \nabla,$$

and

$$(2.8) \qquad (d^{\mathfrak{J}}\alpha)(X_0, \ldots, X_p) = \sum_{i=0}^{p} (-1)^i [X_i, \alpha(X_0, \ldots, \widehat{X}_i, \ldots, X_p)],$$

where X_i is identified with the corresponding section in $E_{\mathfrak{J}}$ and $[\ ,\ ]$ is the Lie bracket in the fiber $\mathfrak{J}$ of $E_{\mathfrak{J}}$.

The adjoint δ^D of d^D is $\delta^D = \delta^{\nabla} + \delta^{\mathfrak{J}}$, where δ^{∇} is the adjoint of d^{∇} with respect to the metrics on M and $E_{\mathfrak{J}}$ above, and

$$(2.9) \qquad (\delta^{\mathfrak{J}}\alpha)(X_2, \ldots, X_p) = \sum_{k=1}^{n} [e_k, \alpha(e_k, X_2, \ldots, X_p)],$$

where $\{e_k\}$ is an orthonormal basis in $TM \subset E_{\mathfrak{J}}$. The usual minus sign does not appear because bracketing with e_k is selfadjoint.

The Laplacian, $\Delta^D = d^D\delta^D + \delta^D d^D$, for $E_{\mathfrak{J}}$-valued differential forms on M is $\Delta^D = \Delta^{\nabla} + \Delta^{\mathfrak{J}}$, where $\Delta^{\nabla} = d^{\nabla}\delta^{\nabla} + \delta^{\nabla}d^{\nabla}$, and

$$(2.10) \qquad \begin{aligned} (\Delta^{\mathfrak{J}}\alpha)(X_1, \ldots, X_p) &= \sum_{k=1}^{n} [e_k, \alpha[e_k, \alpha(X_1, \ldots, X_p)]] \\ &+ \sum_{k=1}^{n}\sum_{i=1}^{P} (-1)^{i+1}[[X_i, e_k]\alpha(e_k, X_1, \ldots, \widehat{X}_i, \ldots, X_p)] \end{aligned}$$

(compare [23]). It is a nontrivial and very useful fact that the sum of mixed terms in $\Delta^D = \Delta^\nabla + \Delta^3$ vanishes.

The strict positivity of Δ^D on p-forms is equivalent to the vanishing of the cohomology $H^p(\Gamma, \mathrm{ad})$. to prove this it is sufficient to prove that Δ^3 is strictly positive, since Δ^∇ is nonnegative. In particular, to prove the classical vanishing theorem for $H^1(\Gamma, \mathrm{ad})$ due to A. Weil which is valid for all noncompact semisimple Lie groups without $\mathrm{SL}(2, \mathbb{R})$ factor, it is sufficient to prove that Δ^3 is strictly positive on 1-forms.

For the intended geometrical application we need to consider E_3-valued 2-forms on M. The reason is that the curvature form Ω is an E_3-valued 2-form. The calculations will be complicated by the fact that, unless $\Omega = 0$, the covariant derivative D is no longer flat and $d^D d^D$ may not vanish. In the standard model we obtain, compare [20, 23].

(2.11) PROPOSITION. *Let G/K be an irreducible symmetric space of noncompact type, dimension > 6, and rank > 1. The operator Δ^3 is strictly positive on E_3-valued 2-forms. In particular, $H^2(\Gamma, \mathrm{ad})$ vanishes for any cocompact discrete subgroup $\Gamma \subset G$.*

3. Applications of Cartan connections

3.1. **Sphere pinching.** This application is related to Example 1, where $K = \mathrm{SO}(n)$, $G = \mathrm{SO}(n+1)$, and $F = S^n = G/K$. Here we will not review the optimal results obtained so far (see [15, 30]), but outline the first instance where Cartan connections appeared (implicitly) in the context of comparison theorems.

(3.1) THEOREM. *Let M be a compact Riemannian manifold of dimension n and sectional curvature K, and $0.98 < K \le 1$. Then, M is diffeomorphic to $\Gamma \backslash S^n$, where $\Gamma \subset \mathrm{O}(n+1)$ is a finite subgroup.*

REMARK. If M is simply connected, the pinching constant is better but even in this case the optimal constant is not known.

Next, we outline the proof in [27] for the special case where M is simply connected. Let P denote the principal bundle of orthonormal frames over M. The construction of Example 1 yields a Cartan connection form.

$$\omega = \theta + \eta \colon TP \to \mathbb{R}^n \oplus \mathfrak{so}(n) = \mathfrak{so}(n+1).$$

For the torsion free Levi-Civita, connection, the corresponding Cartan curvature reduces to $\Omega = d\eta + [\eta, \eta] + [\theta, \theta]$, which is small because, by assumption, the sectional curvature is close to 1. For a good choice of norms, see [15]. Enlarging the structure group $SO(n)$ of P from $\mathrm{SO}(n)$ to $\mathrm{SO}(n+1)$ we obtain $\tilde{P} = P \times_i \mathrm{SO}(n+1)$, where $i \colon \mathrm{SO}(n) \to \mathrm{SO}(n+1)$ is the inclusion. The Cartan connection ω induces a connection $\tilde{\omega}$ on $\tilde{P}$ with small curvature $\tilde{\Omega}$. In [27] it is shown that for $\|\tilde{\Omega}\|$ sufficiently small, a connection

$\overline{\omega}$ near $\tilde{\omega}$ with vanishing curvature can be constructed. Since the condition (2.3) for Cartan connections is an open condition, $\overline{\omega}$ is also a Cartan connection, provided $\|\widetilde{\Omega}\|$ is small enough, and the development map (2.5) yields a diffeomorphism $\phi \colon M \to S^n$.

3.2. **Comparison theorems for noncompact symmetric spaces.** This application is related to Example 2, where G is a semisimple Lie group of noncompact type and $K \subset G$ is a maximal compact subgroup. The method outlined below works only for compact Riemannian manifolds whose curvature, roughly speaking, is close to the curvature of one of the symmetric spaces of Proposition 2.11. These are, with a few exceptions, the symmetric spaces of noncompact type and rank > 1. For more details we refer to [20, 23].

Let M be a compact manifold and P a K-principal bundle over M for which the "condition de soudure" is satisfied with respect to the isotropy representation of K on $F = G/K$. In particular, this assumption is satisfied for a reduction of the bundle of frames over M to the subgroup K represented in $\mathbb{R}^n$ by the linear isotropy. In this case, we can define a Cartan connection by $\omega = \theta + \eta$, where θ is the canonical 1-form $\theta = u^{-1}\pi$ on the bundle of frames restricted to P, and η is a connection on P. For this application the Levi-Civita connection form is too special. The reason is that is the Levi-Civita connection restricts to P, and rank $G/K > 1$, then M is already locally symmetric (compare [1, 31]).

We will use a deformation technique to prove the following

(3.2) THEOREM. *Let G/K be an irreducible Riemannian symmetric space of noncompact type and rank > 1, and $\omega \colon TP \to \mathfrak{J}$ a Cartan connection of type (G, K) on a K-principal bundle P over a compact manifold M. For any $n > 6$ and $d > 0$ there exists $\varepsilon = \varepsilon(n, d) > 0$ such that if the Cartan curvature Ω satisfies $\sup|\Omega| < \varepsilon$, and the diameter of M, defined in terms of ω, is bounded by d, then M, $\dim M = n$, is diffeomorphic to a compact quotient of $F = G/K$ by a discrete group of isometries $\Gamma \subset G$.*

REMARK. The theorem is also true for the 3-dimensional hyperbolic space as a model, since by Poincaré duality and the theorem of A. Weil, $H^2(M, \mathrm{ad})$ vanishes. As in the theorem above it is not known whether a pinching theorem exists where the curvature bound does not depend on the diameter.

OUTLINE OF PROOF [20]. In the following evolution equation ω_0 denotes the Cartan connection form $\omega = \omega_0 \colon TP \to \mathfrak{J}$ of the theorem. The definitions of Example 2 for D, d^D, δ^D, Δ^D apply to the present more general case as well. The only difference, of course, is that the Cartan curvature no longer vanishes. We define a flow of Cartan connections by the equation

(3.3) $\dot{\omega} = -\delta^D\Omega$,   initial condition $\omega(0) = \omega_0$.

This equation is not parabolic but the integrability condition of [16, Theorem 5.1] of Hamilton is satisfied thanks to the Bianchi equation $d^D\Omega = 0$.

Therefore, the equation has at least a short-time solution.

To prove that for a suitable choice of $\varepsilon(n, d)$ the equation (3.3) has a solution for the time interval $[0, \infty)$ we derive an equation for the flow of Cartan curvatures. Making use of the Bianchi equation, $d^D\Omega = 0$, and $\dot{\Omega} = d^D\dot{\omega}$, we have

$$(3.4) \qquad\qquad \dot{\Omega} = -\Delta^D\Omega.$$

The cartan curvature Ω is a section of the bundle $E_{ad} = P \times_{ad_K} \mathfrak{J}$. The formula $\Delta^D = \Delta^\nabla + \Delta^\mathfrak{J}$, except for an additional term due to the fact that the torsion of ∇ no longer vanishes, applies as in Example 2. The extra term is bounded by the norm of the Cartan curvature. By the assumptions of the theorem $\Delta^\mathfrak{J}$ is strictly positive on $E_\mathfrak{J}$-valued 2 forms. In addition, Δ^∇ is nonnegative on $E_\mathfrak{J}$-valued forms with respect to the L^2-norm provided by ω. Therefore, we obtain a differential inequality for the flow of the L^2-norm of Ω, $\|\Omega\|_{L^2} = (\int_M \langle \Omega, \Omega \rangle)^{1/2}$.

$$(3.5) \qquad\qquad \frac{d}{dt}\|\Omega\|_{L^2} \le -c\|\Omega\|_{L^2},$$

where the constant has to be chosen somewhat smaller than the constant expressing the strict positively of $\Delta^\mathfrak{J}$ in the proof of Proposition (2.11). The reason is the small additional term in Δ^D due to the nonvanishing torsion has to be accommodated.

To prove that the flow (3.3) exists for $t \in [0, \infty)$ it suffices to prove that $\sup|\Omega|$ stays small, e.g., bounded by 10ε during the evolution in which $\|\Omega\|_{L^2}$ decays exponentially. This is usually done by an adaptation of the iteration technique of Moser [25] to Riemannian manifolds. The essential ingredient is an estimate of the Sobolev constant (see [6]). The limiting Cartan connection $\overline{\omega} = \lim_{t\to\infty} \omega$ is flat and this proves the theorem.

3.3. Almost Lie groups of compact semisimple type.
In this application the model space is a compact semisimple Lie group G, i.e., $K = \{e\}$. Let P be a compact manifold and $\omega: TP \to \mathfrak{J}$ a Cartan connection. The Killing form on the Lie algebra $\mathfrak{J}$ of G induces a Riemannian metric on P. It is used to define the norm $|\Omega|$ of the Cartan curvature Ω at every point of P. The following result was obtained in Min-Oo and Ruh [22].

(3.6) THEOREM. *Let $\omega: TP \to \mathfrak{J}$ be a Cartan connection with P compact and $\mathfrak{J}$ a compact semisimple Lie algebra. There is $\varepsilon = \varepsilon(\mathfrak{J}) > 0$ such that $\sup|\Omega| < \varepsilon$ implies that P is diffeomorphic to $\Gamma\backslash G$ with G the simply connected Lie group with Lie algebra $\mathfrak{J}$, and $\Gamma \subset G$ is a finite subgroup.*

This theorem is motivated by the comparison theorems with compact symmetric spaces obtained in [22] as corollaries.

OUTLINE OF PROOF. We give an outline of a new proof which fits the general pattern of using an evolution equation to deform the original structure into a standard one. This method proved to be elegant and powerful (compare [16]). One of the advantages of this method is that it unifies the various proofs of comparison theorems. The essential calculations in this proof are the same as in [22] and will not be reproduced here.

Let $E = P \times \mathfrak{J}$ be the trivial bundle over P. The Cartan connection $\omega: TP \to \mathfrak{J}$ defines a covariant derivative D on E. The formula for D

$$(3.7) \qquad D_X s = Xs + [X, s]$$

is the same as in (2.6) except that ∇_x is replaced by the derivative in direction X since the bundle is trivial, and the Lie bracket is that of the compact semisimple Lie algebra $\mathfrak{J}$. As before, s is a section in E. The curvature R^D of D is $R^D(X, Y)s = [\Omega(X, Y), s]$.

The exterior derivative d^D induced by D as well as its adjoint δ^D is given by the usual formulas, i.e.,

$$(3.8) \qquad (d^D \alpha)(X_0, \ldots, X_p) = \sum_{i=0}^{p} (-1)^i D_{X_i} \alpha(X_0, \ldots, \widehat{X}_i, \ldots, (X_p)$$
$$+ \sum_{0 \le i < j \le p} (-1)^{i+j} \alpha([X_i, X_j], X_0, \ldots, \widehat{X}_i, \ldots, \widehat{X}_j, \ldots, X_p),$$

$$(3.9) \qquad (\delta^D \alpha)(X_2, \ldots, X_p) = -\sum_{k=1}^{m} (D_{e_i} \alpha)(e_i, X_2, \ldots, X_p).$$

We define $\Delta^D = d^D \delta^D + \delta^D d^D$ as in the previous example and set

$$(3.10) \qquad \dot{\omega} = -\delta^D \Omega, \qquad \text{initial condition } \omega_0 = \omega(0).$$

Again, (3.10) is not parabolic but the condition of [16, Theorem 5.1] applies and the equation has a short-time solution. The Bianchi equation, $d^D \Omega = 0$, and $\dot{\Omega} = d^D \dot{\omega}$ imply

$$(3.11) \qquad \dot{\Omega} = -\Delta^D \Omega.$$

In order to prove that the solution of (3.10) exists for all times we prove that $\sup |\Omega|$ decreases exponentially. The limiting Cartan connection $\overline{\omega} = \lim_{t \to \infty} \omega$ is flat and proves the theorem. We expect exponential decay of $\sup |\Omega|$ because the second cohomology of any compact semisimple Lie group vanishes, since D is nearly flat and the local geometry of P does not differ much from the local geometry of G. To prove that this expectation is correct, we utilize the estimate [22, 4.18] for a certain operator Δ' on E-valued 2-forms. The estimate shows that, for any point where $|\alpha|$ is maximal, $\langle \Delta' \alpha, \alpha \rangle > c|\alpha|^2$ with $c > 0$. The operator Δ' differs from Δ^D used here by a first-order operator L_1 whose coefficients are controlled by $|\Omega|$. In [22] Δ' was chosen over Δ^D because it satisfies a strict maximum principle. In the present proof Δ^D is more convenient. Since Δ^D is a small lower-order perturbation of Δ', the solution of $\dot{\Omega} = -\Delta^D\Omega$ decreases at nearly the same exponential rate as a solution of the unperturbed equation.

Of course, there are many other applications of Cartan connections to geometry. Here the emphasis is on comparison theorems where the positivity of a certain elliptic operator on 2-forms plays a key role. The positivity of the Laplacian on 1-forms with values in the adjoint representation is essential for the rigidity theorem of A. Weil (see [26]). A more recent application of the positivity on 1-forms, Min-Oo [21], is on scalar curvature rigidity of asymptotically hyperbolic spin manifolds.

4. Generalized Cartan connections

In this section we no longer assume that the connection form takes its values in a Lie algebra. We replace the Lie algebra by a vector space with skew product and do not assume that the Jacobi equation is satisfied. The reason for this generalization is that it may not be clear initially which Lie algebra should be chosen, and the space of Lie algebra structures is more complicated than the space of skew products. In the end, when the connection is deformed into a flat connection with respect to a suitably deformed skew product, it turns out that this resulting skew product satisfies the Jacobi identity after all and therefore defines a Lie algebra.

Let P be a differentiable manifold of dimension m and let $[\ ,\ ]: \mathbb{R}^m \times \mathbb{R}^m \to \mathbb{R}^m$ be a skew product, i.e., skew symmetric and bilinear.

(4.1) DEFINITION. $\omega: TP \to \mathbb{R}^m$ is a generalized Cartan connection if $\dim P = m$, and if $\omega(X) = 0$ implies $X = 0$.

(4.2) DEFINITION. $\Omega = d\omega + [\omega, \omega]$, where $[\ ,\ ]$ is a skew product on $\mathbb{R}^m$ is the generalized Cartan curvature of ω.

The following proposition explains why the above generalizations make sense.

(4.3) PROPOSITION. *If the generalized Cartan curvature vanishes, then the skew product used in its definition satisfies the Jacobi identity and hence defines a Lie algebra.*

PROOF. Let X and Y be parallel vector fields; i.e., $\omega(X)$ and $\omega(Y)$ are constant vectors in $\mathbb{R}^m$. We have

$$0 = \Omega(X, Y) = X\omega(Y) - Y\omega(X) - \omega([X, Y]) + [\omega(X), \omega(Y)]$$
$$= -\omega([X, Y]) + [\omega(X), \omega(Y)],$$

where the first bracket on the right-hand side is the vector field bracket on P, and the second is the skew product on $\mathbb{R}^m$. It follows that $[X, Y]$ is also a parallel vector field, and therefore the vector space of parallel vector fields is closed under the vector field bracket. Since the vector field bracket satisfies the Jacobi identity, and ω is an isomorphism of the Lie algebra of vector fields with $\mathbb{R}^m$ and skew product, the proposition is proved.

The first application is to parallelizations with small torsion. The norms are defined in terms of the Riemannian metric on P induced by a metric on $\mathbb{R}^m$ by the connection form $\omega\colon TP \to \mathbb{R}^m$. In the following theorem of Ghanaat [8], the initial skew product vanishes and $\Omega = d\omega$. A generalized Cartan connection form is simply the dual form of a parallelization of P.

(4.4) THEOREM. *Let $\omega\colon TP \to \mathbb{R}^m$ be the dual 1-form of a parallelization of the compact manifold P. There is $\varepsilon = \varepsilon(m) > 0$ such that $\sup|d\omega|d < \varepsilon$, where d is the diameter of P, implies that P is diffeomorphic to N/L with N a simply connected nilpotent Lie group and $L \subset N$ a lattice.*

The above theorem is the converse of a simple fact observed by Gromov [12]: Given $\varepsilon > 0$ and any lattice L in any simply connected nilpotent Lie group N, the metric in the Lie algebra of N can be adjusted such that the Maurer-Cartan forms satisfies $|d\omega|d < \varepsilon$.

OUTLINE OF PROOF. The outline presented here fits the general pattern of using an evolution equation to deform the original structure into a standard structure. In this case we define a flow which starts with the generalized Cartan connection with small curvature and ends up with a flat Cartan connection. The resulting Lie algebra is automatically nilpotent because there are no other Lie groups with small quotients by a theorem of Zassenhaus (compare [7]).

Formally, the evolution equation is the same as in the previous applications. We set

(4.5) $$\dot{\omega} = -\delta^D\Omega, \qquad \text{initial condition } \omega(0) = \omega_0,$$

where the skew product used in the definition of the generalized Cartan curvature at time t will be specified later. Again, (4.5) is not parabolic but the condition of [16, Theorem 5.1] applies and the equation has a short-time solution. Since we deform through generalized Cartan connections, the Bianchi equation, $\delta^D\Omega = 0$, is no longer satisfied. An estimate $|\delta^D\Omega| < c|\Omega|$, with c a small constant controlled by the $\varepsilon(m)$ of the theorem, still holds. This leads to a correction of the evolution equations (3.4), (3.11) for the Cartan curvature. We obtain

(4.6) $$\dot{\Omega} = -(\Delta_0 + L_1)\Omega,$$

where Δ_0 is the operator defined by letting the Laplacian operate on the coefficient functions of Ω with respect to the parallelization defined by ω, and L_1 is a small first-order perturbation.

In order to obtain exponential decay for $\|\Omega\|_{L^2}$ we fix the skew product on $\mathbb{R}^m$ such that Ω is perpendicular to the kernel of $\Delta_0 + L_1^*$, where L_1^* is the adjoint of L_1. To prove that this can be done with a change in the skew product on $\mathbb{R}^m$ we have to estimate the first nonzero eigenvalue of Δ_0 and prove that it is larger than a suitable norm of L_1.

The eigenvalue estimates for the Laplacian are usually done in terms of the curvature of the Riemannian metric. Since the sectional curvatures involve first derivatives of Ω, they are not under control of the assumptions of the theorem. In order to estimate the first eigenvalue of the Laplacian in terms of the diameter of P and a bound on Ω, an estimate of Buser is adapted to this situation (see [2]). Modulo the above modifications the proof proceeds as in 3.2 and 3.3.

The next application is a local version of Theorem 4.4; (see also Fukaya [5] for related results. The assumptions of the local version require a parallelization of the manifold P. The main examples are reductions of the principal bundle of linear frames on a manifold to a subgroup $H \subset \mathrm{GL}(n, \mathbb{R})$.

(4.7) THEOREM (Ghanaat, Min-Oo and Ruh [10]). *Let $\omega: TP \to \mathbb{R}^m$ be the dual 1-form of a parallelization of the manifold P. Assume that the metric on $\mathbb{R}^m$ is scaled such that $|d\omega| \leq 1$. There is a universal radius $\rho = \rho(m) > 0$ such that any ball B_ρ with radius ρ in P is contained in a neighborhood diffeomorphic to $N^q \times D^g$ with N^p a compact nilmanifold of dimension p and D^q a disk of complementary dimension q.*

The proof is a reduction to Theorem 4.4 by means of a geometric argument inspired by some of the pictures in Gromov [11].

Recently, Ghanaat [9] obtained a new proof of the following theorem of Gromov [12] in the stronger version of [28]. The new proof is independent of [12, 28], and relies on Theorem 4.7 instead.

(4.8) THEOREM. *There is $\varepsilon = \varepsilon(m) > 0$ such that any compact Riemannian manifold, whose sectional curvature K and diameter d satisfy $|K|d^2 \leq \varepsilon$, is diffeomorphic to N/Γ with N a simply connected nilpotent Lie group and Γ a finite extension of a lattice $L \subset N$.*

For further results and conjectures on this subject the reader is referred to a recent paper by Gromov [13] on stability and pinching.

REFERENCES

1. M. Berger, *Sur les groupes d'homologie des variétés à connexion affine et des variétés riemanniennes*, Bull. Soc. Math. France **83c** (1955), 279–330.
2. P. Buser, *Eine untere Schranke für λ_1 auf Mannigfaltigkeiten mit fast negativer Krümmung*, Arch. Math. (Basel) **30** (1978), 528–531.
3. C. Ehresmann, *Les connexions infinitésimales dans un fibré différentiable*, Colloque de Topologie, Bruxelles, 1950.
4. J. Eschenburg, *New examples of manifolds with strictly positive curvature*, Invent. Math. **66** (1982), 469–480.
5. K. Fukaya, *Hausdorff convergence of Riemannian manifolds and its application*, Recent Topics in Differential and Analytic Geometry, ed. T. Ochiai, Kinokuniya, Tokyo, 1990.
6. S. Gallot, *Isoperimetric inequalities based on integral norms of Ricci curvature*, Astérisque, no. 157–158, Soc. Math. France, Paris, 1988, pp. 191–212.
7. P. Ghanaat, *A note on discrete uniform subgroups of Lie groups*, Geom. Dedicata **21** (1986), 13–17.
8. _____, *Almost Lie groups of type $\mathbb{R}^n$*, J. Reine Angew. Math. **401** (1989), 60–81.
9. _____, *Geometric construction of holonomy coverings for almost flat manifolds*, J. Differential Geom. **34** (1991), 571–580.

10. P. Ghanaat, M. Min-Oo, and E. Ruh, *Local structure of Riemannian manifolds*, Indiana Univ. Math. J. **39** (1990), 1305–1312.

11. M. Gromov, *Structures métriques pour les variétés riemanniennes*, CEDIC/Ferdinand Nathan, 1981.

12. ____, *Almost flat manifolds*, J. Differential Geom. **13** (1978), 231–241.

13. ____, *Stability and pinching*, Preprint, Inst. Hautes Études Sci., 1990.

14. M. Gromov and W. Thurston, *Pinching constants for hyperbolic manifolds*, Invent. Math. **89** (1987), 1–12.

15. K. Grove, H. Karcher, and E. Ruh, *Jacobi fields and Finsler metrics on compact Lie groups with an application to differentiable pinching problems*, Math. Ann. **211** (1974), 7–21.

16. R. Hamilton, *Three-manifolds with positive Ricci curvature*, J. Differential Geom. **17** (1982), 255–306.

17. S. Kobayashi, *On connections of Cartan*, Canad. J. Math. **8** (1956), 145–156.

18. ____, *Transformation groups in differential geometry*, Erg. Math. Grenzgeb., 70, Springer-Verlag, 1972.

19. Y. Matsushima and S. Murakami, *On vector bundle valued harmonic forms and automorphic forms on symmetric Riemannian manifolds*, Ann. of Math. (2) **78** (1963), 365–416.

20. M. Min-Oo, *Almost symmetric spaces*, Astérique, no. 163–164, Soc. Math. France, Paris, pp. 221–246.

21. ____, *Scalar curvature rigidity of asymptotically hyperbolic spin manifolds*, Math. Ann. **285** (1989), 527–539.

22. M. Min-Oo and E. Ruh, *Comparison theorems for compact symmetric spaces*, Ann. Sci. École Norm. Sup. **12** (1979), 335–353.

23. ____, *Vanishing theorems and almost symmetric spaces of non-compact type*, Math. Ann. **257** (1981), 419–433.

24. ____, *Curvature deformations*, Curvature and Topology of Riemannian Manifolds (Proc. Katata), Lecture Notes in Math. vol. 1201, Springer-Verlag, 1986.

25. J. Moser, *A Harnack inequality for parabolic differential equations*, Comm. Pure Appl. Math. **17** (1964), 101–134.

26. M. Raghunathan, *Discrete subgroups of Lie groups*, Erg. Math. Grenzgeb., 68, Springer-Verlag, 1972.

27. E. Ruh, *Curvature and differentiable structure on spheres*, Comment. Math. Helv. **46** (1971), 127–136.

28. ____, *Almost flat manifolds*, J. Differential Geom. **17** (1982), 1–14.

29. ____, *Almost homogeneous spaces*, Astérique, no. 132, Soc. Math. France, Paris, 1985, pp. 285–293.

30. ____, *Almost Lie groups*, Proc. Internat. Congr. Math. (Berkeley, 1986), Vol. 1, Amer. Math. Soc., Providence, RI, 1987, pp. 561–564.

31. J. Simons, *On the transitivity of holonomy systems*, Ann. of Math. (2) **76** (1962), 213–234.

OHIO STATE UNIVERSITY

Proceedings of Symposia in Pure Mathematics
Volume **54** (1993), Part 3

Eigenvalue Estimates for the Laplacian with Lower Order Terms on a Compact Riemannian Manifold

ALBERTO G. SETTI

1. Introduction

Let M be an n-dimensional, complete Riemannian manifold without boundary and let $w > 0$ be a given smooth function on M that will be referred to as a weight function.

We consider on M the differential operator

$$L = -\Delta - \nabla(\log w)$$

which is essentially selfadjoint and positive on $C_c^\infty(M) \subset L^2(M, w\, dV)$, dV being the standard Riemannian measure on M. L is associated to the quadratic form

$$Q(u, v) = \int_M \langle \nabla u, \nabla v \rangle w\, dV, \qquad u, v \in C_c^\infty(M),$$

obtained by replacing the measure dV with the weighted measure $w\, dV$ in the standard Dirichlet form. In this sense L arises as a natural generalization of the Laplacian.

The operator L has already been considered in [**Bk1, BkE, D, Dl, DlS**], which mainly deal with properties of the diffusion semigroup generated by L, and in [**Bk2**], where the Riesz transform generated by L is investigated. In some of the above-mentioned papers the study of L is motivated by the fact that it is the finite-dimensional analogue of operators on infinite-dimensional product manifolds ("infinite volume limits") that arise in statistical mechanics [**DlS**].

We consider the case of a compact manifold M, so that L has purely discrete spectrum $0 = \lambda_0 < \lambda_1 \leq \lambda_2 \leq \cdots \nearrow \infty$, and we are interested in

1991 *Mathematics Subject Classification.* Primary 58G25; Secondary 35P15.

The final version of this paper will appear elsewhere.

estimates for the eigenvalues of L, and in particular for the first nonzero eigenvalue λ_1, a problem also considered in [DI] and [DIS].

Extensive activity has been devoted to the analogous problem for the Laplacian. Among the results concerning lower bounds for the first nonzero eigenvalue of Δ, we mention the classical Lichnerowicz theorem [L, Cl], which gives a sharp bound for λ_1^Δ in terms of the dimension and of a (strictly positive) lower bound for the Ricci curvature, and estimates due to Li and Yau which depend on dimension, a (not necessarily positive) lower bound for the Ricci curvature and the diameter of M [Li1, Li2, LY1]. As for bounds from above we quote Cheng's theorem [Ch], which provides an upper bound for all the eigenvalues of Δ in terms of dimension, upper bound for the Ricci curvature and diameter, and which is especially accurate in the case of λ_1^Δ. Bounds for the eigenvalues of L will necessarily depend on the interplay between the geometry of M (mainly via the curvature) and the behavior of the weight function w.

The purpose of this report is to describe some estimates that extend to L the results for the Laplacian on a compact manifold mentioned above. In these generalizations the interaction between the geometry of M and the behavior of w is taken into account by means of two "modified Ricci curvature" defined by

$$R_w = \text{Ric} - w^{-1} \text{Hess} \, w \, ,$$
$$S_w = \text{Ric} - \text{Hess}(\log w) = R_w + w^{-2} \, dw \otimes dw \, .$$

S_w is the tensor associated to Bakry's "carré du camp itéré" Γ_2 [BK2, BKE] and is, in a certain sense, more natural than R_w, which, nevertheless, appears to be overall more useful.

Only indications of the proofs will be given. More details are contained in [Se].

2. Lower bounds for λ_1

I will present three kinds of results: one of perturbative nature, which was shown to me by J-D. Deuschel, one that depends on an extension of the BLW formula and generalizes Lichnerowicz's theorem and two bounds that depend upon the use of the technique of gradient estimates a lá Li and Yau.

PERTURBATIVE BOUND. *If $\lambda_1^{-\Delta}$ denotes the first nonzero eigenvalue of the Laplacian on M, then*

$$\lambda_1 \geq \frac{\min w}{\max w} \lambda_1^{-\Delta} \, .$$

The quantity $\min w / \max w$ can be estimated by integrating $|\nabla(\log w)|$ along the minimizing geodesics joining the points where w attains respectively its minimum and its maximum, thus yielding

$$\lambda_1 \geq \lambda_1^{-\Delta} \exp \left\{ - \left(\sup_M |\nabla(\log w)| \right) d \right\} \, ,$$

where d denotes the diameter of M. Together with known estimates for $\lambda_1^{-\Delta}$ the method gives bounds for λ_1 in terms of the Ricci curvature, the diameter of M and of $\sup_M\{|\nabla(\log w)|)d\}$. The dependency upon the last quantity reflects the perturbative nature of the method employed.

The extension of Lichnerowicz's theorem depend on a generalization of the BLW formula already obtained in an equivalent but somewhat different form by Bakry [**Bk2, BkE**]. In our notation it is stated as follows:

GENERALIZED BLW. *if $u \in C^3(M)$ then*

(GBLW)

$$-\frac{1}{2}L(|\nabla u|^2) = |\operatorname{Hess} u|^2 - \langle \nabla u, \nabla(Lu)\rangle + R_w(\nabla u, \nabla u) + w^{-2}\langle \nabla w, \nabla u\rangle$$

$$= |\operatorname{Hess} u|^2 - \langle \nabla u, \nabla(Lu)\rangle + S_w(\nabla u, \nabla u).$$

The proof follows by simply using the definition of L in the classical BLW formula

$$\frac{1}{2}\Delta(|\nabla u|^2) = |\operatorname{Hess} u|^2 + \langle \nabla u, \nabla(\Delta u)\rangle + \operatorname{Ric}(\nabla u, \nabla u).$$

Applying (GLBW) to an eigenfunction of L belonging to the eigenvalue λ_1 one gets

GENERALIZED LICHNEROWICZ INEQUALITY. *Let M be a compact, n-dimensional Riemannian manifold with weight function w, and let λ_1 be the first nonzero eigenvalue of L. Assume that*

$$R_w \geq \kappa \quad \text{with } \kappa > 0.$$

Then

$$\lambda_1 \geq \frac{(n+1)}{n}\kappa.$$

Other bounds for λ_1 that depend on estimates on S_w have been obtained using similar techniques by Deuschel-Stroock [**DLS**].

Now we come to extension of the technique of gradient estimates. These techniques, whose use goes back to Payne-Weinberger [**PW**] and Cheng-Yau [**CY**], have then been successfully employed by Li, Yau and others [**Li1, Li2, LY1**] to obtain lower bounds for $\lambda_1^{-\Delta}$. In a different form, they have been used by Li and Yau [**LY2**] to obtain Harnack inequalities and estimates for the heat kernel. The fact that they can be used in the case of the operator L is a further indication of their great flexibility.

THEOREM 3. *Let u be an eigenfunction of L with eigenvalue $\lambda > 0$. Assume that $R_w \geq \kappa$. Then $\forall \beta \geq 1$ we have*

$$|\nabla u|^2 \leq 2\left\{(n+1)\frac{\lambda\beta}{\beta-1} - n\kappa\right\}(\beta \sup|u| - u)^2.$$

As in the original proof of Li and Yau [**Li1, LY1**] the gradient estimates follow by defining a function F by

$$F(x) = \frac{|\nabla u|^2}{(\beta - u)^2},$$

and by applying the maximum principle and (GBLW) at the point where F attains it its maximum.

Letting u be an eigenfunction of λ_1, and integrating $|\nabla u|/(\beta \sup |u| - u)$ along the shortest geodesic from the nodal set of u to the point x where u attains its maximum, and using the first gradient estimate one finds

THEOREM 4. *Assume that* $R_w \geq \kappa$ *and let* d *denote the diameter of* M. *Then the first nonzero eigenvalue* λ_1 *of* L *satisfies*

$$\lambda_1 \geq \frac{1}{(n+1)d^2} \exp\left\{ -1 - \sqrt{1 - 2n\kappa d^2} \right\} .$$

The estimate above clearly holds only for $2n\kappa \leq d^{-2}$. For a manifold with nonnegative R_w, however, one obtains a sharper bound using a different gradient estimate.

SECOND GRADIENT ESTIMATE. *Let* u *be an eigenfunction of* L *belonging to a nonzero eigenvalue* λ *assume that* $S_w \geq \kappa$. *Then* $\forall \alpha \geq 0$ *and* $\beta^2 \geq \sup(\alpha + u)^2$,

$$|\nabla u|^2 \leq \max_{x \in M} \left\{ \frac{\lambda(\beta^2 - (\alpha + u)\alpha) - (\beta^2 - (\alpha + u)^2)\kappa}{\beta^2} \right\} (\beta^2 - (\alpha + u)^2).$$

Notice that, since $S_w = R_w + w^{-2} dw \otimes dw \geq R_w$, the estimate certainly holds with the same constant if we assume instead that $R_w \geq \kappa$.

To prove the second gradient estimate one applies to the function

$$G(x) = \frac{|\nabla u|^2}{\beta^2 - (\alpha + u)^2}$$

the same kind of reasoning used to prove the first gradient estimate. Notice that, assuming that $\sup u \geq |\inf u|$, and setting $\alpha = 0$ and $\beta = \sup u$, the estimate becomes

$$\frac{|\nabla u|}{\sqrt{(\sup u)^2 - u^2}} \leq (\lambda_1 + \max\{-\kappa, 0\})^{1/2} .$$

Integrating the inequality along the minimizing geodesic joining the point where u achieves its supremum to the point where it attains its infimum, and using Li's argument [Li2], one has the following:

THEOREM 5. *Assume that* $S_w \geq \kappa$. *Then, denoting by* d *the diameter of* M, *we have*

$$\lambda_1 \geq \frac{\pi^2}{2d^2} - \max\{-\kappa, 0\} .$$

3. Estimates from above

In this section, essentially extending ideas of S. Y. Cheng [Ch] we present bounds from above for the eigenvalues of L. First some notation: For $p \in M$

and $\xi \in ST_p M$, let $c(\xi)$ be the distance along the geodesic $\gamma_\xi = \exp_p(t\xi)$ from p to the cut locus of p. Also let $\sqrt{g}(r, \xi)$ be the area element in geodesic spherical coordinates at p, so that the Riemannian volume element of M is given by $dV = \sqrt{g}(r, \xi)\, dr\, d\xi$. To simplify the notation we also let

$$
S_\kappa(r) = \begin{cases} (\sqrt{-\kappa})^{-1} \sinh(\sqrt{-\kappa}\, r) & \text{for } \kappa > 0, \\ r & \text{for } \kappa = 0, \\ (\sqrt{\kappa})^{-1} \sin(\sqrt{\kappa}\, r) & \text{for } \kappa > 0, \end{cases}
$$

$$
C_\kappa(r) = \begin{cases} \cosh(\sqrt{-\kappa}\, r) & \text{for } \kappa < 0, \\ 1 & \text{for } \kappa = 0, \\ \cos(\sqrt{\kappa}\, r) & \text{for } \kappa > 0. \end{cases}
$$

Then we have the following generalizations of Bishop's comparison theorem.

THEOREM 6. *Let M be a Riemannian manifold with weight function w. Assume that*

$$
R_w = (\mathrm{Ric} - w^{-1} \mathrm{Hess}\, w) \geq \alpha.
$$

Then, $\forall \xi \in ST_p M$ and $\forall r \in (0, c(\xi))$,

$$
(*) \qquad (w\sqrt{g})^{-1} \frac{\partial(w\sqrt{g})}{\partial r}(r, \xi) \leq n \frac{C_\beta(r)}{S_\beta(r)},
$$

where $\beta = \alpha/n$. Moreover, if $\alpha > 0$, $c(\xi) \leq \pi/\sqrt{\beta}$.

As to the relevance of the estimate $(*)$, note that bounds on $(w\sqrt{g})^{-1} \times \partial(w\sqrt{g})/\partial r)(r, \xi)$ play the same role for L as bounds on $(\sqrt{g})^{-1} \times \partial(\sqrt{g})/\partial r)(r, \xi)$ do for the Laplacian. This can be easily seen by noting that the quadratic form associated to L can be written in polar geodesic coordinates as

$$
\int_{ST_p M} \int_0^{c(\xi)} |\nabla f|^2 w\sqrt{g}(r, \xi)\, dr\, d\xi.
$$

Observe moreover that if $\alpha > 0$ we can conclude, as in the Bonnet-Myer theorem, that M is compact and that its diameter is bounded above by $\pi/\sqrt{\beta}$.

The proof of the theorem is done by defining ϕ to be the function $(w\sqrt{g})^{-1} \partial(w\sqrt{g})/\partial r)(r, \xi)$. The assumption on R_w implies that ϕ satisfies the differential inequality

$$
\phi' + \frac{1}{n}\phi^2 + n\beta \leq 0, \quad \text{in } (0, c(\xi)),
$$

while $(r) = n(C_\beta/S_\beta)(r)$ satisfies the corresponding differential equation. Then one proceeds as in the proof of Bishop's comparison theorem (cf. [C1, p. 73]).

The theorem above allows us to generalize Cheng's theorem as follows:

THEOREM 7. *Let $p \in M$ and $R > 0$, and consider the generalized Dirichlet problem for L on the geodesic ball in M centered at p with radius R, $B_p(R)$. If*

$$(\mathrm{Ric} - w^{-1} \mathrm{Hess}\, w) \geq \alpha$$

in $B_p(R)$, then

$$\lambda_0(B_p(R)) \leq \lambda^{-\Delta}(B_\beta^{n+1}(R)),$$

where $\lambda_0(B_p(R))$ is the bottom of the spectrum of L on $B_p](R)$ and $\lambda^{-\Delta}(B_\beta^{n+1})$ is the smallest Dirichlet eigenvalue for $-\Delta$ on the disk of radius R in $\mathbf{M}_\beta^{n+1}$, the $(n+1)$-dimensional space with constant curvature $\beta = \alpha/n$.

This in turn yields the following estimates from above for the eigenvalues of L on M:

COROLLARY 8. *Let M be a compact manifold with weight function w and assume that*

$$R_w = (\mathrm{Ric} - w^{-1} \mathrm{Hess}\, w) \geq \alpha.$$

Let $0 = \lambda_0 < \lambda_1 \leq \lambda_2 \leq \cdots \leq \lambda_m \leq \cdots$ be the sequence of the eigenvalues of L, where each eigenvalue is repeated according to its multiplicity. Then for every m we have

$$\lambda_m \leq \lambda^{-\Delta}\left(B_\beta^{n+1}\left(\frac{d}{2m}\right)\right),$$

where d is the diameter of M and β and $\lambda^{-\Delta}(B_\beta^{n+1}(R))$ are defined above.

Using the estimate for $\lambda^{-\Delta}(B_\beta^n(R))$ obtained by Cheng, we also have

COROLLARY 9. *Let M be as above. Then*
(i) *if $\alpha > 0$,*

$$\lambda_m \leq \frac{C(n)m^2}{d^2};$$

(ii) *if $\alpha < 0$,*

$$\lambda_m \leq C(n)|\alpha| + \frac{C'(n)m^2}{d^2}.$$

Acknowledgments

It is a pleasure to thank Prof. R. S. Strichartz for his help and many illuminating discussions. I am also grateful to Prof. J-D. Deuschel for several stimulating conversations. Financial support from the Alfred P. Sloan Foundation in the form of a dissertation fellowship is also gratefully acknowledged.

REFERENCES

[Bk1] D. Bakry, *Un critere de non-explosion pour certaines diffusions sur une variété riemanniene complete*, C. R. Acad. Sci. Paris Sér. I Math. **303** (1986), 23–26.

[Bk2] ___, *Étude des transformations de Riesz dans les variétés riemannienne à courboure de Ricci minorée*, Séminaire de Probabilités XXI, Lecture Notes in Math., vol. 1247, Springer-Verlag, 1987, pp. 137–172.

[BkE] D. Bakry and M. Emery, *Diffusion hypercontractives*, Séminaire de Probabilités XIX, Lecture Notes in Math., vol. 1123, Springer-Verlag, 1985, pp. 179–206.

[Ch] S. Y. Cheng, *Eigenvalue comparison theorems and its geometrical applications*, Math. Z. **143** (1975), 289–297.

[Cl] I. Chavel, *Eigenvalues in Riemannian geometry*, Academic Press, Orlando, 1984.

[CY] S. Y. Cheng and S. T. Yau, *Differential equations in Riemannian geometry and their geometric applications*, Comm. Pure Appl. Math. **28** (1975), 333–354.

[D] E. B. Davies, *Heat kernel bounds for second order elliptic operators on Riemannian manifolds*, Amer. J. Math. **109** (1987), 545–570.

[D1] J-D. Deuschel, *On estimating the hypercontractive constant of a diffusion process on a compact manifold*, preprint.

[DlS] J-D. Deuschel and D. Stroock, *Hypercontractivity and spectral gap of symmetric diffusions with applications to the stochastic Ising model*, J. Funct. Anal. **92** (1990), 30–48.

[L] A. Lichnerowicz, *Gëométrie des groups de transformations*, Dunod, Paris, 1958.

[Li1] P. Li, *A lower bound for the first eigenvalue of the Laplacian on a compact Riemannian manifold*, Indiana, Univ. Math. J. **28** (1979), 1013–1013.

[Li2] ___, *Poincaré inequalities on Riemannian manifolds*, Seminar in Differential Geometry, Ann. of Math. Stud., vol. 102, Princeton Univ. Press, Princeton, NJ, 1982, pp. 73–84.

[LY1] P. Li and S. T. Yau, *Estimates of eigenvalues of a compact Riemannian manifold*, Proc. Symp. Pure Math., vol. 36, Amer. Math. Soc., Providence, RI, 1980, pp. 205–240.

[LY2] ___, *On the parabolic kernel of the Schrödinger operator*, Acta Math. **156** (1986), 153–201.

[PW] L. Payne and H. Weinberger, *The optimal Poincaré inequality for convex domains*, Arch. Rational Mech. Anal. **5** (1960), 282–292.

[Se] A. Setti, *Eigenvalue estimates for the weighted Laplacian on a Riemannian manifold*, Preprint.

CORNELL UNIVERSITY

Proceedings of Symposia in Pure Mathematics
Volume **54** (1993), Part 3

Positive Ricci Curvature
on Compact Simply Connected 4-Manifolds

JI-PING SHA AND DAGANG YANG

0

The purpose of this note is to show the following:

THEOREM. *A closed simply connected 4-manifold is homeomorphic to a manifold with positive Ricci curvature if and only if it is homeomorphic to one with positive scalar curvature.*

A basic problem in Riemannian geometry is to distinguish manifolds with curvature of a definite sign by their topology. It has been confirmed recently that the class of simply connected manifolds supporting positive Ricci curvature is much larger than the class supporting nonnegative sectional curvature [6, 13]. On the other hand, not much is known about the topological obstructions for a manifold to have positive Ricci curvature. So far the only known obstructions are the fundamental group [12] and the $\hat{A}$-genus which even prevents a closed spin manifold from having positive scalar curvature [11]. There is not any known example of a simply connected manifold with positive scalar curvature which does not admit any metric of positive Ricci curvature. The theorem above shows that, modulo the differentiable structure, such an example for closed manifolds cannot exist in dimension 4.

REMARK. While exotic differentiable structures do exist in dimension 4, their geometrical properties are rarely understood. All the manifolds with positive Ricci curvature constructed in this paper are with standard smooth structures. We thank Professor S. Kwasik for pointing out this to us.

REMARK. It is well known that a metric with positive Ricci curvature on a closed 3-manifold can be deformed via the Ricci flow to a metric with positive constant sectional curvature, due to the work of R. Hamilton [8]. It

1991 *Mathematics Subject Classification.* Primary 53C20.
Partially supported by National Science Foundation grants.
This paper is in final form and no version will be submitted for publication elsewhere.

has been thought that Ricci flow might deform a metric with positive Ricci curvature on a closed 4-manifold to an Einstein metric as well. However, the theorem above shows that, if the Ricci flow does deform the metric to an Einstein one, singularities may occur. For example, the connected sum of k copies of CP^2 supports a metric with positive Ricci curvature according to our theorem, while it admits no Einstein metric for $k \geq 4$ (cf. [3]).

A closed spin 4-manifold with nonzero signature has nonzero $\widehat{A}$-genus. Thus it carries no metric with positive scalar curvature, and therefore carries no metric with positive Ricci curvature. In a previous paper [14], the authors have shown the existence of metrics with positive Ricci curvature on all closed simply connected 4-manifolds with zero signature. Therefore, S^4, the connected sum of k copies of $S^2 \times S^2$, and the connected sum of k copies of $CP^2 \# (-CP^2)$, $k = 1, 2, \ldots,$ all carry metrics with positive Ricci curvature. The existence of metrices with positive Ricci curvature on the connected sum of an arbitrary number of copies of $S^2 \times S^2$ has also been obtained independently in [1] by M. Anderson. The bulk of this paper is to construct metrics with positive Ricci curvature on the connected sum $((k+l)CP^2) \# (k(-CP^2))$, for all $k \geq 0$ and $l \geq 1$. The theorem then follows from the homeomorphism classification of smooth closed simply connected 4-manifolds due to M. Freedman and S. Donaldson (cf. [10]).

The idea of the construction is fairly simple. One begins with a gravitational multi-instanton, i.e., a complete asymptotically flat Kähler Einstein 4-manifold Y with zero scalar curvature, as constructed by G. Gibbons, S. Hawking and N. Hitchin [5, 9]. After a conformal deformation similar to the one from the flat $\mathbf{R}^4$ to the standard S^4, one obtains a noncomplete metric on Y with nonnegative Ricci curvature. A further deformation then gives a metric with constant positive sectional curvature away from a small compact subset of Y while keeping the Ricci curvature nonnegative. It is then clear that one can complete the above deformed metric on Y by adding a 2-cycle on it to obtain a smooth metric with positive Ricci curvature on lCP^2. In particular, this implies that lCP^2 and $l(S^2 \times S^2)$ (the double of Y), $l = 1, 2, \ldots,$ carry metrics with positive Ricci curvature everywhere and constant sectional curvature 1 away from a subset of arbitrarily small volume. It is easy to see that the resulting manifold of k $(1, 2)$-surgeries (sec §1) on lCP^2 is the connected sum $(lCP^2) \# (k(S^2 \times S^2))$, which is homeomorphic to $((l + k)CP^2) \# (k(-CP^2))$ by the topological classification of smooth closed simply connected 4-manifolds. However, unlike the construction of a metric with positive scalar curvature, it can be preserved by geometrically local codimension ≥ 3 surgeries [7]; in order to perform the k $(1, 2)$-surgeries geometrically while keeping the Ricci curvature positive, a substantial deformation of the metric above on lCP^2 is necessary. We will outline the construction in the next section and leave the details to §§2 and 3.

1

One way to visualize the connected sum of CP^2's is as follows.

Let $[z_1 : z_2 : z_3]$ be the homogeneous coordinates of CP^2, where $|z_2|^2 + |z_2|^2 + |z_3|^2 = 1$, and let S^1 act on CP^2 by $e^{i\theta} \cdot [z_1 : z_2 : z_3] = [z_1 : z_2 : e^{i\theta}z_3]$. This action is free except at the point $[0 : 0 : 1]$ and the complex line $z_3 = 0$ which are the fixed points. For convenience, let us present the CP^2 by the picture in Figure 1, where the thick line and points represent the fixed points.

Let $B_1 = \{[z_1 : z_2 : z_3] \mid |z_1| > \frac{1}{4}\}$ and $B_2 = \{[z_1 : z_2 : z_3] \mid |z_2| > \frac{1}{4}\}$ be two disjoint balls around the points $[1 : 0 : 0]$ and $[0 : 1 : 0]$ respectively. Figure 2 represents $CP^2 \setminus (B_1 \cup B_2)$. Then the connected sum lCP^2 can be represented by the picture in Figure 3, where the identification along the vertical lines is made by $[\frac{1}{4} : z_2; z_3] \sim [\overline{z}_2 : \frac{1}{4} : z_3]$. Then obviously, S^1 acts on lCP^2 freely except at l points $p_1, \dots, p_l$ and at an embedded 2-sphere S which are fixed. It is easy to see that the normal bundle of S in lCP^2, viewed as a complex line bundle, is with Chern class $-l$.

We now introduce some notations. Let X be a tubular neighborhood of S in lCP^2 with $\partial X \approx S^3/\mathbf{Z}_l$ and let $Y = (lCP^2) \setminus \overline{X}$. We will choose k disjoint S^1's in $X \setminus S$ and replace k disjoint tubular neighborhoods of them by k $S^2 \times D^2$'s respectively, where the replacement is made by identifying the boundary of each tubular neighborhood with the boundary of the respective

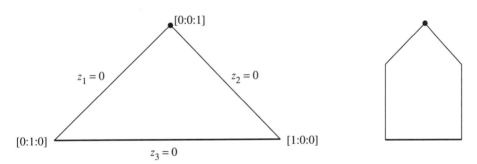

Figure 1   Figure 2

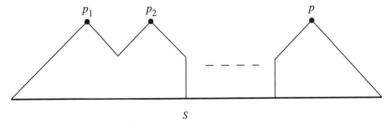

Figure 3

$S^2 \times D^2$. For simplicity, each such replacement is called a $(1, 2)$-surgery. The resulting manifold by these k $(1, 2)$-surgeries on X is denoted by Z. Then $M \equiv \overline{Y} \cup_{\partial Y = \partial Z} Z$ is easily seen to be $(l\mathbf{CP}^2) \# (k(S^2 \times S^2))$ which is homeomorphic to $((l + k)\mathbf{CP}^2) \# (k(-\mathbf{CP}^2))$ by the classification.

The construction of a metric with positive Ricci curvature on M is roughly as follows. We first construct a metric with Ric > 0 on X such that a subset of X is isometric to $S_\varepsilon^1 \times D_\delta^3$ where S_ε^1 is a circle of length $2\pi\varepsilon$ and D_δ^3 is a geodesic ball of radius δ in the euclidean S^3. Furthermore, $\delta > 0$ is fixed and $\varepsilon > 0$ can be arbitrarily small. Then by the result in [14], for sufficiently small ε, one can do k $(1, 2)$-surgeries in this subset to get a metric with Ric > 0 on Z. We next construct a metric with Ric > 0 on Y so that near ∂Y it coincides with the metric near ∂X. Then together they give the desired metric on M.

2

Let us now be more specific. We start with S^4 and parametrize it in $\mathbf{R} \times \mathbf{C}^2$ by

$$(\sin \tau \sin \varphi, \cos \varphi e^{i\alpha}, \cos \tau \sin \varphi e^{i\beta}), \qquad -\frac{\pi}{2} \leq \tau \leq \frac{\pi}{2}, \ 0 \leq \varphi \leq \frac{\pi}{2}$$

(i.e., view S^4 as a join $S^1 * S^2$). Define a metric on S^4 by

$$(1) \qquad d\varphi^2 + \varepsilon^2 \cos^2 h(\varphi) \, d\alpha^2 + \sin^2 \varphi (dr^2 + \cos^2 \tau \, d\beta^2)$$

where ε is a positive number which can be arbitrarily small and h is a smooth function on $[0, \frac{\pi}{2}]$. The function h will be specified later. At this point let us only point out that it should satisfy the following conditions.

$$h(0) = 0 \quad \text{and} \quad h \text{ is odd at } 0;$$
$$h\left(\frac{\pi}{2}\right) = \frac{\pi}{2}; \qquad h' > 0,$$
$$h'(\varphi) \equiv \frac{1}{\varepsilon} \quad \text{when } \varphi \text{ is near } \frac{\pi}{2}; \qquad h'' \geq 0.$$

It can easily be verified that, with any h satisfying the conditions above, (1) is a well-defined smooth metric on S^4 and is with positive sectional curvature. Notice also that (1) is invariant under the S^1-action: $e^{i\theta}(x, z_1, z_2) = (x, e^{i\theta} z_1, e^{i\theta} z_2)$.

We now change the parameters for S^4 to

$$(\sin t, \cos t \cos \psi e^{i\alpha}, \cos t \sin \psi e^{i\beta}), \qquad -\frac{\pi}{2} \leq t \leq \frac{\pi}{2}, \ 0 \leq \psi \leq \frac{\pi}{2}$$

Then the metric (1) can be written as

$$(2) \quad dt^2 + \cos^2 t \, d\psi^2 + \varepsilon^2 \cos^2 h(\cos^{-1}(\cos t \cos \psi)) \, d\alpha^2 + \cos^2 t \sin^2 \psi \, d\beta^2.$$

Next consider the metric

$$(3) \qquad d\psi^2 + \varepsilon^2 \cos h(\psi) \, d\alpha^2 + \sin^2 \psi \, d\beta^2$$

on S^3 which is parametrized in $\mathbf{C}^2$ by

$$(\cos \psi e^{i\alpha}, \sin \psi e^{i\beta}), \qquad 0 \leq \psi \leq \frac{\pi}{2}.$$

Obviously (3) is well defined and with positive sectional curvature. It is invariant under the free S^1-action $e^{i\theta} \cdot (z_1, z_2) = (e^{i\theta}z_1, e^{i\theta}z_2)$. Scale the metric (3) on the S^1-orbit direction by

$$\frac{c^2}{c^2 l^2 - \sin^2 \psi - \varepsilon^2 \cos^2 h(\psi)}$$

while keeping it unchanged on the orthogonal complement, where $c \geq 2$ is some constant. Denote S^3 with this modified metric, which is also S^1-invariant, by S^3_*. Clearly when c is large enough, S^3_* is also with positive sectional curvature.

Let S^2_c be the 2-sphere in $\mathbf{R}^3 \approx \mathbf{R} \times \mathbf{C}$ of radius c. The metric on S^2_c under the parameterization

$$\left(\sin \frac{t}{c}, \cos \frac{t}{c} e^{i\theta}\right), \qquad -\frac{c\pi}{2} \leq t \leq \frac{c\pi}{2}$$

is given by

$$(4) \qquad dt^2 + c^2 \cos^2 \frac{t}{c} d\theta^2.$$

(4) is invariant under the S^1-action $e^{i\theta} \cdot (x, z) = (x, e^{ik\theta}z)$.

Let S^1 act on S^3_* and S^2_c as above; then S^1 acts on $S^2_c \times S^3_*$ freely. Therefore the induced metric on the quotient $S^2_c \times S^3_*/S^1$ is with nonnegative sectional curvature. It is easy to see that the Ricci curvature is positive everywhere. Consider the region in $S^2_c \times S^3_*/S^1$ corresponding to $-\frac{c\pi}{2} < t < \frac{c\pi}{2}$. The metric on this region can be lifted to a metric on $(-\frac{c\pi}{2}, \frac{c\pi}{2}) \times S^3$ which is written as

$$(5) \qquad dt^2 + d\psi^2 + a_1(t, \psi) d\alpha^2 + 2a_2(t, \psi) d\alpha d\beta + a_3(t, \psi) d\beta^2$$

where a_1, a_2, a_3 are some functions of t and ψ. They can be written out explicitly. Since we do not need them, they are omitted. It is easy to see that the S^3 corresponding to $t = 0$ is a totally geodesic hypersurface and by the construction of S^3_* the induced metric on it is exactly (3).

The S^4 with metric (2) also has a totally geodesic hypersurface S^3 corresponding to $t = 0$ with the induced metric (3). By essentially the same argument as that in Proposition 2.1 of [4] (cf. [4] for details), one can show that there is a metric with Ric > 0,

$$(6) \qquad dt^2 + b_1(t) d\psi^2 + b_2(t, \psi) d\alpha^2 + 2b_2(t, \psi) d\alpha d\beta + b_4(t, \psi) d\beta^2$$

on $(-\frac{c\pi}{2}, \frac{\pi}{2}) \times S^3$, where b_1, b_2, b_3 are some functions of t or t and ψ, such that for some $-\frac{c\pi}{2} < t_1 < t_2 < 0$, (6) coincides with (5) on the region $-\frac{c\pi}{2} < t < t_1$ and (6) coincides with (2) on the region $t_2 < t < \frac{\pi}{2}$.

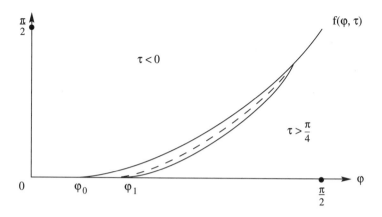

Figure 4

The same argument also shows that one can deform the metric (2) nearby the point $(1, 0, 0)$ on S^4 while keeping Ric > 0 such that for any $t_3 > 0$, there is t_4 with $t_3 < t_4 < \frac{\pi}{2}$ so that (6) coincides with (2) on the region $t_2 < t < t_3$ and (6) becomes

$$(7) \qquad dt^2 + \cos^2 t\, d\psi^2 + \cos^2 t \cos^2 \psi\, d\alpha^2 + \cos^2 t \sin^2 \psi\, d\beta^2,$$

i.e., the metric of constant sectional curvature 1 on S^4 near the point $(1, 0, 0)$, on the region $t_4 \le t < \frac{\pi}{2}$.

Now let us take the metric (6) on $(-\frac{c\pi}{2}, t_4] \times S^3$. This metric obviously descends to a metric with Ric > 0 on $(-\frac{c\pi}{2}, t_4] \times S^3/\mathbf{Z}_l \approx \overline{X} \setminus S$ which extends to a well-defined metric on $\overline{X}$. Then ∂X corresponds to $t = t_4$.

The metric so obtained is essentially what we want. In order to have a region in X is isometric to $S^1_\varepsilon \times D^3_\delta$ (see §1), we need to modify the function h in (1) a little. First we can let h satisfy that $h(\varphi) = 0$ for $\varphi < \varphi_0$, where φ_0 is some positive number $< |t_2|$. Note that t_2 may depend on ε. We then replace h by a function f of φ and τ, such that $f(\varphi, \tau) = h(\varphi)$ for $\tau < 0$; $f(\varphi, \tau) = \overline{h}(\varphi)$ for $\tau > \frac{\pi}{4}$, where $\overline{h}$ is a function which satisfies those conditions for h and furthermore, $\overline{h}(\varphi) = 0$ for $\varphi < \varphi_1$, where φ_1 is a positive number which does not depend on ε. See Figure 4.

It is easy to see that the metric

$$(8) \qquad d\varphi^2 + \varepsilon^2 \cos^2 f(\varphi, \tau)\, d\alpha^2 + \sin^2 \varphi (d\tau^2 + \cos^2 \tau\, d\beta^2)$$

is also a well-defined metric on S^4 and that replacing (1) by (8) does not affect the deformation done above.

One computes that, with respect to the orthogonal basis

$$e_1 = \frac{\partial}{\partial \varphi}, \quad e_2 = \frac{1}{\sin \varphi}\frac{\partial}{\partial \tau}, \quad e_3 = \frac{1}{\varepsilon \cos f}\frac{\partial}{\partial \alpha}, \quad e_4 = \frac{1}{\sin \varphi \cos \tau}\frac{\partial}{\partial \beta},$$

the components of the Ricci tensor are

$$R_{11} = 2 + f^2_{\varphi} + \frac{\sin f}{\cos f} f_{\varphi\varphi},$$

$$R_{22} = 2 + \frac{\cos \varphi \sin f}{\sin \varphi \cos f} f_{\varphi} + \frac{1}{\sin^2 \varphi} \left(f^2_{\tau} + \frac{\sin f}{\cos f} f_{\tau\tau} \right),$$

$$R_{33} = f^2_{\varphi} + \frac{2 \cos \varphi \sin f}{\sin \varphi \cos f} f_{\varphi} + \frac{\sin f}{\cos f} f_{\varphi\varphi}$$
$$+ \frac{1}{\sin^2 \varphi} \left(f^2_{\tau} + \frac{\sin \tau \sin f}{\cos \tau \cos f} f_{\tau} + \frac{\sin f}{\cos f} f_{\tau\tau} \right),$$

$$R_{44} = 2 + \frac{\cos \varphi \sin f}{\sin \varphi \cos f} f_{\varphi} - \frac{1}{\sin^2 \varphi} \frac{\sin \tau \sin f}{\cos \tau \cos f} f_{\tau},$$

$$R_{12} = \frac{1}{\sin \varphi} \left(f_{\varphi} f_{\tau} + \frac{\sin f}{\cos f} f_{\tau\tau} - \frac{\cos \varphi \sin f}{\sin \varphi \cos f} f_{\tau} \right),$$

$$R_{13} = R_{14} = R_{23} = R_{24} = R_{34} = 0.$$

It is not hard to see that one can choose h and f properly so that the Ricci curvature of (8) is nonnegative.

Putting this modification into (6), one gets a region in $(\frac{-c\pi}{2}, \frac{\pi}{2}) \times S^3$ (actually in $(0, t_3) \times S^3$) which is isometric to $S^1_{\varepsilon} \times D^3_{\delta_1}$, where $\delta_1 > 0$ does not depend on ε. Descending to X, it is easily seen that there is a region isometric to $S^1_{\varepsilon} \times D^3_{\delta}$ for arbitrarily small ε and some δ which does not depend on ε. This completes the construction of the desired metric on X.

3

We now proceed to set up the metric on Y. There is a family of Ricci flat metrics on Y, using the graviational multi-instantons constructed by Gibbons, Hawking and Hitchin. We briefly recall one description of this metric as follows (cf. [2]).

Choose l points in $\mathbf{R}^3$ which, by abusing notation slightly, are also denoted by $p_1, \dots, p_l$. There is a smooth map $\pi: Y \to \mathbf{R}^3$ such that $\pi^{-1}(p_j)$ is the point p_j in Y (see §1) for all j, but $\pi^{-1}(p) \approx S^1$ for $p \in \mathbf{R}^3 \setminus \{p_j\}$. Let $Y_0 = Y \setminus \{p_j\}$; then the restriction $\pi_0: Y_0 \to \mathbf{R}^3 \setminus \{p_j\}$ is a principal S^1 bundle whose Chern class is -1 when restricted to a small sphere around any p_j. Define a smooth function $V: \mathbf{R}^3 \setminus \{p_j\} \to \mathbf{R}$ by

$$V(p) = \frac{1}{2} \sum_{j=1}^{l} \frac{1}{\|p - p_j\|}.$$

Then $\frac{1}{2\pi} * dV$ is a closed 2-form and represents the Chern class of $\pi_0: Y_0 \to \mathbf{R}^3 \setminus \{p_j\}$ in deRham cohomology, where $*$ is the Hodge $*$ operator of $\mathbf{R}^3$. There is therefore a connection ω on $\pi_0: Y_0 \to \mathbf{R}^3 \setminus \{p_j\}$ with curvature

$*dV$; i.e., $d\omega = \pi_0^*(dV)$. The form ω is unique up to gauge transformations, since $\mathbf{R}^3 \setminus \{p_j\}$ is simply connected. The Gibbons-Hawking-Hitchin
metric on Y_0 is given by

$$(9) \qquad g = \frac{1}{\pi_0^*V}\omega \odot \omega + \pi_0^*V\pi_0^*\,ds^2$$

where ds^2 is the euclidean metric on $\mathbf{R}^3$. Metric (9) extends smoothly to
a metric on Y and is Ricci-flat and asymptotically flat. It is observed by
Anderson [1] that one can bend (9) near ∂Y, by deforming it conformally,
to get a metric with positive Ricci curvature on, for example, the double of
Y, which is homeomorphic to $l(S^2 \times S^2)$.

We also propose to bend the metric (9) conformally. The situation here
is more delicate. We need to deform the metric (9), while keeping $\mathrm{Ric} \geq 0$,
so that near ∂Y it coincides globally with the metric cross ∂X, which after
being lifted to universal covering is the boundary of a geodesic ball of radius
$\frac{\pi}{2} - t_4$ on S^4 (see (7), §2). Notice that $\frac{\pi}{2} - t_4$ could be arbitrarily small.

To begin with, let us consider Gibbons-Hawking-Hitchin metric in the
special case when $l = 1$. In this case Y is diffeomorphic to $\mathbf{R}^4$ and $\tilde{\pi}: \mathbf{R}^4 \to$
$\mathbf{R}^3$ is the standard Hopf projection. Let $r(p) = \|p\|$ for $p \in \mathbf{R}^3$ or $\mathbf{R}^4$. Then
the Gibbons-Hawking-Hitchin metric

$$\tilde{g} = 2r\tilde{\omega} \odot \tilde{\omega} + \frac{1}{2r}\tilde{\pi}_0^*\,ds^2$$

is isometric to the euclidean metric on $\mathbf{R}^4$, where $\tilde{\omega}$ is the connection on the
restriction $\tilde{\pi}_0: \mathbf{R}^4 \setminus \{0\} \to \mathbf{R}^3 \setminus \{0\}$ with curvature $*d\frac{1}{2r}$. With this metric,
$\|p\|_{\tilde{g}} = (2r(p))^{1/2}$. It is well known that for any $a > 0$, $(\frac{a}{2} + \frac{r}{a})^{-2}\tilde{g}$ is with
constant curvature 1. In fact $(\frac{a}{2} + \frac{r}{a})^{-2}\tilde{g}$ is the pull back metric from the
stereographic projection shown in Figure 5.

We fix an a such that the ball of radius $6^{1/2}l^{-1/4}$ around 0 in $\mathbf{R}^4_{\tilde{g}}$ is

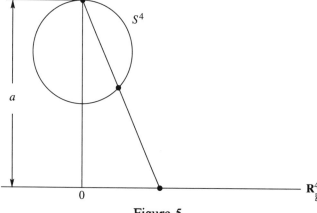

Figure 5

projected onto by the geodesic ball of radius $\frac{\pi}{2} - t_4$ around the south pole of the S^4.

Now we go back to the metric (9) in general case. Let $\rho = \max_j \|p_j\|$; one can assume $\rho < 1$, and let $B(b) = \{p \in \mathbf{R}^3 | r(p) < b\}$. Choose a smooth convex function $u(r)$ (i.e., $u'' \geq 0$) such that $u(r) = \text{constant} > 0$ for $r \leq 1$ and $u(r) = \frac{a}{2} + \frac{r}{a}$ for $r \geq 2$. Fix such a function. One computes directly that, when ρ is small enough, the Ricci tensor of $u^{-2}g$ is nonnegative.

Notice that the closed 2-form $\frac{1}{2\pi} * d\frac{l}{2r}$ represents the Chern class of the restriction $\overline{\pi}: Y \setminus \pi^{-1}(B(1)) \to \mathbf{R}^3 \setminus B(1)$. There is therefore a connection $\overline{\omega}$ on $\overline{\pi}: Y \setminus \pi^{-1}(B(1)) \to \mathbf{R}^3 \setminus B(1)$ such that $d\overline{\omega} = *d\frac{l}{2r}$. Define a metric on $Y \setminus \pi^{-1}(B(1))$ by

$$\overline{g} = \frac{2r}{l}\overline{\omega} \odot \overline{\omega} + \frac{l}{2r} ds^2.$$

There is a l-fold covering $\sigma: \mathbf{R}^4 \setminus \tilde{\pi}^{-1}(B(1)) \to Y \setminus \pi^{-1}(B(1))$. Obviously, one may assume that $\tilde{\omega} = \frac{1}{l}\sigma^*\overline{\omega}$ on $\mathbf{R}^4 \setminus \tilde{\pi}^{-1}(B(1))$. Then

$$\sigma^*\overline{g} = l\tilde{g}.$$

Therefore $u^{-2}\overline{g}$ is with constant sectional curvature $\frac{1}{l}$ on $Y \setminus \pi^{-1}(B(2))$.

Choose a nondecreasing function $\gamma(r)$ such that $\gamma(r) = 0$ for $r \leq 2$ and $\gamma(r) = 1$ for $r \geq 3$. Define a metric on Y by

$$\hat{g} = (1 - \gamma)u^{-2}g + \gamma u^{-2}\overline{g}.$$

Then $\hat{g}$ is with Ric ≥ 0 on $\pi^{-1}(B(2))$ and with constant sectional curvature $\frac{1}{l}$ on $\pi^{-1}(\mathbf{R}^3 \setminus B(3))$.

Notice that when $\rho \to 0$, $\pi_0^* V \to \frac{l}{2r}$ uniformly on $\pi^{-1}(B(3) \setminus B(2))$ in C^∞ topology. One then also has that $\omega \to \overline{\omega}$ uniformly ($\overline{\omega} - \omega$ can actually be written out explicitly by the transgression formula). Therefore $\hat{g} \to u^{-2}\overline{g}$ uniformly on $\pi^{-1}(B(3) \setminus B(2))$ as $\rho \to 0$. Thus when ρ is small enough, $\hat{g}$ is also with positive sectional curvature on $\pi^{-1}(B(3) \setminus B(2))$.

One now simply views $\pi^{-1}(B(3))$ as Y and put the metric $\frac{1}{l}\hat{g}$ on it. By our choice of a, this completes the construction.

REFERENCES

1. M. Anderson, *Short geodesics and gravitational instantons*, J. Differential Geom. **31** (1990), 265–275.
2. M. Anderson, P. Kronheimer, and C. LeBrun, *Complete Ricci-flat Kähler manifolds of infinite topological type*, Comm. Math. Phys. **125** (1989), 637–642.
3. A. Besse, *Einstein manifolds*, Springer-Verlag, 1987.
4. L. Z. Gao, *The construction of negatively Ricci curved manifolds*, Math. Ann. **271** (1985), 185–208.
5. G. Gibbons and S. Hawking, *Gravitational multi-instantons*, Phys. Lett. B **78** (1978), 430–432.
6. D. Gromoll and W. T. Meyer, *Examples of complete manifolds with positive Ricci curvature*, J. Differential Geom. **21** (1985), 195–211.

7. M. Gromov and B. Lawson, *The classification of simply connected manifolds of positive scalar curvature*, Ann. of Math. (2) **111** (1980), 423–434.

8. R. Hamilton, *Three manifolds with positive Ricci curvature*, J. Differential Geom. **17** (1982), 255–306.

9. N. Hitchin, *Polygons and gravitons*, Math. Proc. Cambridge Philos. Soc. **85** (1979), 465–476.

10. B. Lawson, *The theory of gauge fields in four dimensions*, Amer. Math. Soc., Providence, RI, 1985.

11. A. Lichnerowicz, *Spineurs harmoniques*, C. R. Acad. Sci. Paris **257** (1963), 7–9.

12. S. B. Myers, *Riemannian manifold with positive mean curvature*, Duke Math. J. **8** (1941), 401–404.

13. J.-P. Sha and D. G. Yang, *Examples of manifolds of positive Ricci curvature*, J. Differential Geom. **29** (1989), 95–103.

14. ____, *Positive Ricci curvature on the connected sum of $S^n \times S^m$* , J. Differential Geom. **33** (1991), 127–137.

INDIANA UNIVERSITY

TULANE UNIVERSITY

Proceedings of Symposia in Pure Mathematics
Volume **54** (1993), Part 3

Volume Growth and Finite Topological Type

ZHONGMIN SHEN AND GUOFANG WEI

1. Introduction

The structure of an open manifold with nonnegative Ricci curvature has received much attention recently. On one hand, Sha-Yang's examples [**SY1**, **SY2**] demonstrate that such a manifold could have infinite topological type. On the other hand, the beautiful result of U. Abresch and D. Gromoll [**AG**] shows that if the manifold M^n is not too "large" at infinity and the sectional curvature is bounded from below, then it must be of finite topological type. Here the "largeness" at infinity is measured by the notion of diameter growth introduced in [**AG**]. Precisely they require that the diameter growth $\mathscr{D}(p, r) = o(r^{1/n})$ for some point p. Following this line, the first author also proved variants of this result [**S**].

A natural generalization of Abresch-Gromoll's result would be to place conditions on the volume growth instead of diameter growth, as the fact that the diameter growth is not more than r^α implies that the volume growth is not more than $r^{1+(n-1)\alpha}$. In fact it is generally believed that Abresch-Gromoll's result would continue to hold for a nonnegatively Ricci curved manifold whose volume growth is not more than r^2. The purpose of this paper is to present the following result.

We assume that Riemmannian manifolds under consideration will always have "weak bounded geometry." We say a complete manifold M has weak bounded geometry, if it satisfies the bounds

$$(1) \qquad K = \inf K_M > -\infty, \qquad v = \inf \operatorname{vol}(B(x, 1)) > 0.$$

THEOREM 1.1. *Let M^n be complete with weak bounded geometry* (1) *and* $\operatorname{Ric}_{(k)} \geq 0$ *outside a geodesic ball* $B(p, D)$ *for some* $2 \leq k \leq n - 1$. *There*

1991 *Mathematics Subject Classification.* Primary 53C20.
This paper is in final form and no version of it will be submitted for publication elsewhere.
Both authors supported in part by grants from the National Science Foundation.

is a constant $c = c(n, k, K, v, D)$ *such that if*

$$\overline{\lim}_{r \to +\infty} \frac{\mathrm{vol}(B(p, r))}{r^{1+1/(k+1)}} < c,$$

then there exists a compact set, C, *such that* $M^n \backslash C$ *contains no critical points of* p. *In particular,* M^n *has finite topological type.*

The proof of Theorem 1.1 will be given in §3. Here for some $1 \le k \le n - 1$, we say that the kth Ricci curvature is nonnegative, $\mathrm{Ric}_{(k)} \ge 0$, at some point x, if for every $(k + 1)$-dimensional subspace $V \subset T_x M$,

$$\sum_{i=1}^{k+1} \langle R(e_i, v)v, e_i \rangle \ge 0 \quad \text{for all } v \in V,$$

where $\{e_1, \dots, e_{k+1}\}$ is any orthonormal basis for V. Thus for example $\mathrm{Ric}_{(1)} \ge 0$ means exactly the same as $K_M \ge 0$ while $\mathrm{Ric}_{(n-1)} \ge 0$ is the same as $\mathrm{Ric} \ge 0$.

As a special case of Theorem 1.1, we have

COROLLARY 1.2. *Let* M^n *be a complete manifold with weak bounded geometry,* $\mathrm{Ric} \ge 0$ *outside a compact set and* $\mathrm{vol}(B(p, r)) = o(r^{1+1/n})$ *for some* $p \in M^n$. *Then* M^n *has finite topological type.*

REMARK. By the Bishop-Gromov comparison theorem, one can show that for any complete open manifold M^n with $\mathrm{Ric}(M) \ge 0$,

$$\mathrm{vol}(B(p, r)) \ge c(n)\mathrm{vol}(B(p, 1)) \cdot r, \qquad p \in M, r \ge 1;$$

see [CGT], Calabi and Yau [Y]. Thus our condition on the volume growth requires it to be close to the minimum growth. On the other hand there are manifolds with nonnegative Ricci curvature that do not have weak bounded geometry; see [CK].

2. Critical points of distance functions

The fundamental notion involved in such finite topological type result is that of the critical point of a distance function, first introduced by Grove and Shiohama. For this and the following fundamental lemma, the reader is referred to [G, C].

ISOTOPY LEMMA. *If* $r_1 \le r_2 \le +\infty$ *and if a connected component* C *of* $\overline{B(p, r_2)} \backslash B(p, r_1)$ *is free of critical points of* p, *then* C *is homeomorphic to* $C_1 \times [r_1, r_2]$, *where* C_1 *is a topological submanifold without boundary.*

We now look for conditions that will enable us to tell whether a point is a critical point of p or not. For $r > 0$ and a point p on a complete manifold

M, let
$$R(p, r) = \{\gamma(r), \gamma \text{ a ray from } p\}.$$

$R(p, r)$ consists of points of intersections of the geodesic sphere of radius r with all the rays emanating from p. Let $R_p(x) = d(x, R(p, r))$, where $r = d(p, x)$. The excess function is defined as

$$e_p(x) = \lim_{t \to \infty}(d(p, x) + d(x, S(p, t)) - t),$$

where $S(p, t) := \{x \in M ; d(p, x) = t\}$. Clearly,

$$(2) \qquad e_p(x) \leq \lim_{t \to +\infty}(d(p, x) + d(x, \gamma_p(t)) - t)$$

for any ray γ_p from p. The following inequality is then trivial.

$$(3) \qquad\qquad e_p(x) \leq R_p(x).$$

The significance of the excess function lies in the following

LEMMA 2.1 [S, Lemma 10]. *Suppose that M has sectional curvature $K_M \geq -K$ $(K > 0)$. Then for any critical point x of p,*

$$e_p(x) \geq \frac{1}{\sqrt{K}} \log \frac{e^{\sqrt{K} d(p, x)}}{\cosh \sqrt{K} d(p, x)}.$$

For our purpose, we need to obtain a better estimate for $e_p(x)$ than (3) in the case that $\mathrm{Ric}_{(k)} \geq 0$.

PROPOSITION 2.2. *Suppose $\mathrm{Ric}_{(k)} \geq 0$ on $M \backslash B(p, D)$ for some $2 \leq k \leq n - 1$. Then for all $x \in M \backslash B(p, 2D)$*

$$(4) \qquad\qquad e_p(x) \leq 8R_p(x) \left(\frac{R_p(x)}{d(p, x)}\right)^{1/k}.$$

PROOF. If $\mathrm{Ric}_{(k)} \geq 0$ on all of M, then the estimate (4) is known; see [S, Lemma 12], also compare [AG, C] in the case $k = n - 1$. Now if $\mathrm{Ric}_{(k)} \geq 0$ only on $M \backslash B(p, D)$, the estimate (4) can be obtained by a modification of the argument in [S]. We need the following

LEMMA 2.3 [S, Lemma 11]. *Let M^n be complete and $p, q \in M^n$. Suppose that q is not on the cut-locus of p (hence $d_p(x) := d(p, x)$ is smooth near q) and $\mathrm{Ric}_{(k)} \geq 0$ along the minimal geodesic σ from q to p. Then for any orthonormal set $\{e_1, \ldots, e_{k+1}\}$ in $T_q M$ with $\dot{\sigma}(0) \in \mathrm{span}\{e_i\}$,*

$$\sum_{i=1}^{k+1} D^2 d_p(e_i, e_i) \leq \frac{k}{d(p, q)}.$$

Fix any point $x \in M \setminus B(p, 2D)$. Let γ_p be the ray emanating from p such that for $z = \gamma_p(r) \in R(p, r)$, where $r = d(p, x)$,

$$(5) \qquad\qquad R_p(x) = d(x, z).$$

We may assume that $R_p(x) > 0$, otherwise it is done. Since $d(x, \gamma_p(t)) - t$ is nonincreasing in t, it follows from (2) that for all $0 \leq t_1 < t_2$,

$$(6) \qquad e_p(x) \leq d(\gamma_p(t_1), x) + d(\gamma_p(t_2), x) - (t_2 - t_1).$$

Let $\eta > 0$ be a small number ($\eta \leq 2R_p(x)$) and $R = R_p(x) + \eta$. It is easy to see that if a minimal geodesic σ from $\gamma_p(t_o)$ to a point y in $B(p, R)$ intersects with $B(p, D)$, then $t_o < D + 2R_p(x) + \eta$. Take $t_1 = D + 4R_p(x)$ and $t_2 = 2d(p, x) - t_1$. Let

$$e_{p_1 p_2}(y) = d(p_1, y) + d(p_2, y) - d(p_1, p_2),$$

where $p_i = \gamma_p(t_i)$. Then it follows from (6) that

$$(7) \qquad\qquad e_p(x) \leq e_{p_1 p_2}(x).$$

Following the line in [AG], we will show that

$$(8) \qquad e_{p_1 p_2}(x) \leq \frac{2k}{k-1} \left(\frac{C}{2(k+1)} R_p(x)^{k+1} \right)^{1/k},$$

where $C = \frac{k}{d(p_1, x) - R_p(x)} + \frac{k}{d(p_2, x) - R_p(x)}$.

Now if $R_p(x) > \frac{1}{20} d(p, x)$, then (4) follows from (3). Thus from now on we will always assume that $R_p(x) \leq \frac{1}{20} d(p, x)$. In this case

$$(9) \qquad\qquad C \leq \frac{8k}{d(p, x)}.$$

Assuming (8) and making use of (7) and (9), one can easily obtain (4).

The outline of the proof of (8) is given as follows. Let

$$\phi_R(t) = \frac{1}{(k-1)(k+1)} (t^{k-1} - R^{1-k})R^{k+1} + \frac{1}{2(k+1)} (t^2 - R^2).$$

Let

$$f(y) = C'\phi_R(d(x, y)) - e_{p_1 p_2}(y), \qquad y \in B(p, R),$$

where $C' > \frac{k}{d(p_1, x) - R} + \frac{k}{d(p_2, x) - R}$. By the choices of t_i, one see that for any $y \in B(p, R_p + \eta)$, the minimal geodesics from y to p_i and x, respectively, do not intersect with the bad set $B(p, D)$. Following the proof of Lemma 12 in [S] and using Lemma 2.3, one can show that for any $y \in B(p, R) \setminus \{x\}$, there is an orthonormal set $\{e_1, \ldots, e_{k+1}\}$ in $T_y M$ such that the following inequality holds in a generalized sense (see [S]):

$$\sum_{i=1}^{k+1} D^2 f(e_i, e_i) \geq C' - \frac{k}{d(p_1, y)} - \frac{k}{d(p_2, y)}$$

$$\geq C' - \frac{k}{d(p_1, x) - R} - \frac{k}{d(p_2, x) - R} > 0.$$

Thus f has no locally maximal point in $B(x, R)$. Notice that $f|_{S(x,R)} \leq 0$ and $f(z) > 0$, where z is defined in (5). One has that for any $0 < \rho < R$,

$$0 < f(z) \leq C' \phi_R(\rho) - \min_{y \in S(x,\rho)} e_{p_1 p_2}(y),$$

which implies

$$e_{p_1 p_2}(x) \leq \min_{y \in S(x,\rho)} e_{p_1 p_2}(y) + 2\rho \leq C' \phi_R(\rho) + 2\rho,$$

where we have made use of $|e_{p_1 p_2}(x) - e_{p_1 p_2}(y)| \leq 2 d(x, y)$. One obtains

$$e_{p_1 p_2}(x) \leq \min_{0 < \rho < R} \{ C' \phi_R(\rho) + 2\rho \}$$

$$\leq \frac{2k}{k-1} \left(\frac{C'}{2(k+1)} R^{k+1} \right)^{1/k}.$$

Letting $\eta \to 0$ (hence $R \to R_p(x)$) and $C' \to C$, one obtains (8). $\square$

Note that $\frac{1}{2} \leq \log(e^r / \cosh r)$ for $r \geq 10$. Hence combining Proposition 2.2 and Lemma 2.1, we have

LEMMA 2.4. *Let* M^n *be complete with* $\mathrm{Ric}_{(k)} \geq 0$ *on* $M \backslash B(p, D)$ *and* $K_M \geq -K$ $(K > 0)$. *If* $d(p, x) \geq \max\{10/\sqrt{K}, 2D\}$ *and*

$$R_p(x) \leq \frac{1}{16} K^{-k/(2(k+1))} d(p, x)^{1/(k+1)},$$

then x *is not a critical point of* p.

Now we shall introduce a notion of essential diameter of ends (compare [C]). Let M be a complete manifold and $p \in M$. For any $r > 0$, the essential diameter of ends at distance r from p is defined by

$$\mathscr{D}(p, r) = \sup_{\Sigma} \mathrm{diam}(\Sigma),$$

where the supremum is taken over all boundary components Σ of $M \backslash \overline{B(p, r)}$, with $\Sigma \cap R(p, r) \neq \emptyset$. Let Σ_r be a boundary component of $M \backslash \overline{B(p, r)}$ with $\Sigma_r \cap R(p, r) \neq \emptyset$. By the definition of $R_p(x)$, one has that for any $x \in \Sigma_r$, $R_p(x) \leq \mathscr{D}(p, r)$. Thus one has the following theorem which is proved in [S] in the case $\mathrm{Ric}_{(k)} \geq 0$ on all of M^n.

THEOREM 2.5. *Let M^n be complete with $\mathrm{Ric}_{(k)} \geq 0$ on $M \setminus B(p, D)$ and $K_M \geq -K \ (K > 0)$. Suppose that*

$$\overline{\lim}_{r \to +\infty} \frac{\mathscr{D}(p, r)}{r^{1/(k+1)}} < \frac{1}{16} K^{-k/(2(k+1))}.$$

Then there is a compact subset C in M^n such that $M \setminus C$ contains no critical point of p.

PROOF. Applying Lemma 2.4, one concludes that there is a large number r_0 such that for any $r \geq r_0$, if Σ_r is a boundary component of $M \setminus \overline{B(p, r)}$ with $\Sigma_r \cap R(p, r) \neq \emptyset$, then Σ_r is free of critical points. Now fixing Σ_{r_0}, let γ_p be a ray from p with $\gamma(r_0) \in \Sigma_{r_0}$. Denote by Σ_t the boundary component of $M \setminus \overline{B(p, t)}$ with $\gamma(t) \in \Sigma_t$. Thus the set

$$U = \bigcup_{t \geq r_0} \Sigma_t$$

is free of critical point of p. By using the Isotopy Lemma one can see that U coincides with an unbounded connected component of $M \setminus B(p, r_0)$, and U is homeomorphic to $\Sigma_{r_0} \times [0, +\infty)$. It is clear that there are only finitely many (bounded and unbounded) connected components of $M \setminus \overline{B(p, r_0)}$, which contain points with distance from p at least $r_0 + 1$. Thus there is $r_1 \geq r_0$ such that $M \setminus \overline{B(p, r_1)}$ is contained in the union of the unbounded components of $M \setminus \overline{B(p, r_0)}$. Therefore $M \setminus \overline{B(p, r_1)}$ is free of critical points. $\square$

3. Volume growth and diameter growth

The purpose of this section is to provide some relations between the volume growth and diameter growth for complete manifolds with nonnegative Ricci curvature.

LEMMA 3.1. *Let M^n be complete with $v = \inf \mathrm{vol}(B(x, 1)) > 0$. Then for any pont $p \in M^n$ and $r > 2$,*

$$\mathscr{D}(p, r) \leq \frac{8}{v} \mathrm{vol}(B(p, r+2) \setminus B(p, r-2)).$$

PROOF. The observation here is that under our assumption the volume of annuli can be used to estimate the essential diameter of ends. Let Σ_r be a boundary component of $M \setminus \overline{B(p, r)}$ such that $\Sigma_r \cap R(p, r) \neq \emptyset$. Then there is a ray γ_p such that $\gamma_p(r) \in \Sigma_r$.

Let $\{B(p_j, 1)\}_{j=1}^N$ be a maximal set of disjoint balls with radius 1 and center $p_j \in \sum_r$. Then

$$\bigcup_{j=1}^N B(p_j, 2) \supset \sum_r,$$

and

$$N \leq \frac{1}{v} \text{vol}(B(p, r+2) \backslash B(p, r-2)).$$

By the connectedness of $\sum_r$, one can show that for any point $x \in \sum_r$, there is a subset of $p_i's$, say, $q_1, \ldots, q_k$, $k \leq N$, such that $x \in B(q_1, 2)$, $\gamma_p(r) \in B(q_k, 2)$, and

$$B(q_t, 2) \cap B(q_{t+1}, 2) \neq \emptyset, \qquad 1 \leq t \leq k-1.$$

Now one can easily construct a piecewise smooth geodesic c joining x and $\gamma_p(r)$ through q_i's. Thus

$$d(x, \gamma_p(r)) \leq \text{length}(c) \leq 4N \leq \frac{4}{v} \text{vol}(B(p, r+2) \backslash B(p, r-2)).$$

Therefore

$$\mathscr{D}(p, r) \leq \frac{8}{v} \text{vol}(B(p, r+2) \backslash B(p, r-2)). \quad \square$$

PROPOSITION 3.2. *Let M^n be complete with $\text{Ric} \geq 0$ on $M \backslash B(p, D)$ and $\text{Ric} \geq -(n-1)H^2$ on $B(p, D)$. Then there is a constant $c = c(n, H, D)$ such that for all $R \geq r \geq 2D$,*

$$\text{vol}(B(p, R) \backslash B(p, r)) \leq c \int_r^R \frac{1}{t} \text{vol}(B(p, t)) \, dt,$$

where $c = c(n, H, D)$.

PROOF. Let $U_p M$ be the unit sphere in $T_p M$. For each $v \in U_p M$, let t_v denote the cut value of v. Thus if $U_p = \{tv \in T_p M | t < t_v\}$, then $\exp_p |_{U_p}$ is an imbedding onto $M \backslash C_p$, where C_p denotes the cut-locus, which is of measure zero.

For $0 < r < R$,

$$\text{vol}(B(p, R) \backslash B(p, r)) = \int_{U_p \cap \{tv|r<t<R\}} \det(\exp_{p*}) \, d\lambda$$

$$= \int_r^R \left(\int_{U_p \cap S_t^{n-1}} \det(\exp_{p*}) \, dv \right) dt.$$

Now fix a vector $v \in U_p M$. Let $\gamma_v(t)$ be the geodesic from p with $\dot{\gamma}(0) = v$. Denote $f_v(t) = \det(\exp_{p*} |_{tv})$. By the Basic Index Comparison [CE], we have for $r \leq t_v$,

$$\frac{f_v'(r)}{f_v(r)} \leq \int_0^r (n-1)g'(t)^2 - \text{Ric}(\dot{\gamma}_v, \dot{\gamma}_v)g(t)^2 \, dt - (n-1)r^{-1}$$

for all smooth functions $g(t)$ with $g(0) = 0$, $g(r) = 1$. By taking $g(t) = tr^{-1}(r \geq D)$, we have

$$\frac{f_v'(r)}{f_v(r)} \leq \int_0^D (n-1)H^2 r^{-2} t^2 \, dt.$$

Thus

$$(\log f_v(r))\big|_r^R \leq \frac{1}{3}(n-1)D^3 H^2 r^{-1} - \frac{1}{3}(n-1)D^3 H^2 R^{-1}$$

$$\leq \frac{1}{3}(n-1)D^2 H^2 \qquad (t_v \geq R \geq r \geq D).$$

Therefore $f_v(R) \leq c f_v(r)$, where we have put $c = \exp(\frac{1}{3}(n-1)D^2 H^2)$. Let $\bar{f}_v(t) = 0$ for $t \geq t_v$ and $\bar{f}_v(t) = f_v(t)$ for $t < t_v$. Then $\bar{f}_v(R) \leq c\bar{f}_v(r)$ $(R \geq r \geq D)$. Now

$$\frac{r^{n-1}}{R^{n-1}} \int_{S_R^{n-1}} \bar{f}_v(R) \, dv = \int_{S_r^{n-1}} \bar{f}_v(R) \, dv \leq c \int_{S_r^{n-1}} \bar{f}_v(r) \, dv.$$

Assuming $R \geq 2D$ and integrating with respect to r from D to R, we have

$$\int_{S_R^{n-1}} \bar{f}_v(R) \, dv \leq \frac{nR^{n-1}}{R^n - D^n} c \int_D^R \int_{S_r^{n-1}} \bar{f}_v(r) \, dv \, dr \leq \frac{2n}{R} c \text{vol} B(p, R).$$

Hence for $R \geq r \geq 2D$,

$$\text{vol}(B(p, R) \backslash B(p, r)) = \int_r^R \left(\int_{S_t^{n-1}} \bar{f}_v(t) \, dv \right) dt$$

$$\leq 2nc \int_r^R \frac{1}{t} \text{vol}(B(p, t)) \, dt. \qquad \square$$

COROLLARY 3.3. *Let M^n be complete with $\text{Ric} \geq 0$ on $M \backslash B(p, D)$ and $\text{Ric} \geq -(n-1)H^2$ on $B(p, D)$. Suppose $v = \inf \text{vol}(B(x, 1)) > 0$. Then there is a constant $c = c(n, H, D)$ such that for $r \geq 2D + 10$,*

$$\mathscr{D}(p, r) \leq c(n, H, D)v^{-1} \frac{\text{vol}(B(p, r+2))}{r+2}.$$

Now Theorem 1.1 follows from Theorem 2.5 and Corollary 3.3.

4. Sha-Yang's examples

Sha-Yang's examples are the first kind of the nonnegatively Ricci curved complete manifolds having infinite topological type. It is thus of much interest to compute its various geometric quantities such as the diameter growth and the volume growth. In fact in [SY1] it is pointed out that for the 7-dimensional example constructed there, the degree of diameter growth is $\frac{2}{3}$. What we presented here is a detailed computation for this and other examples in [SY2]. We acknowledge gratefully here the helpful conversations with D. Yang.

Sha-Yang's examples are built upon manifolds obtained by rotating a curve. We first establish a simple lemma about such manifolds.

LEMMA 4.1. *Let* $f : \mathbb{R}^+ \to \mathbb{R}^+$ *be a smooth function. The equation* $\rho = f(x)$ *defines a surface of revolution in* $\mathbb{R}^{n+1}$, *where* $\rho = \sqrt{y_1^2 + \cdots + y_n^2}$ *and* $(x, y_1, \ldots, y_n)$ *are the coordinates of* $\mathbb{R}^{n+1}$. *Assume for* x *large,* $f(x) = O(x^\alpha)$, $\alpha \leq 1$. *Then the diameter growth of such a surface is at most* r^α *and hence the volume growth is at most* $r^{1+(n-1)\alpha}$.

PROOF. Let $r = d(0, (x, y))$. Then $r = \int_0^x \sqrt{1 + (f'(x))^2} \, dx$. Our assumption implies that $f'(x) = O(x^{\alpha-1}) \leq O(1)$. Hence

$$(10) \qquad\qquad x \leq r \leq Cx,$$

for some positive constant C.

Now clearly the essential diameter $\mathcal{D}(0, r) \leq 2\pi f(x) = O(x^\alpha)$. Hence by (10) $\mathcal{D}(0, r) \leq O(r^\alpha)$. $\square$

Recall that Sha-Yang's examples $\mathbb{R}^{n,m}$ ($m, n \geq 2$) are obtained by performing surgery infinitely many times on $\mathbb{R}^{m+1} \times S^{n-1}$,

$$\mathbb{R}^{n,m} = S^{n-1} \times \left(\mathbb{R}^{m+1} \backslash \coprod_{k=0}^{\infty} D_k^{m+1} \right) \cup_{\mathrm{Id}} D^n \times \coprod_{k=0}^{\infty} S_k^m.$$

Here the metric on $\mathbb{R}^{m+1}$ is obtained by realizing it as a surface of revolution $\rho = f(x)$ of parabolic type in $\mathbb{R}^{m+2}$, and D_k^{m+1}'s are a sequence of disjoint geodesic balls having constant sectional curvature. In order for the surgery to preserve nonnegative Ricci curvature the geodesic balls D_k^{m+1} must be chosen to have larger and larger radius; see [SY2, (16)] and (11) below. This can be achieved by constructing f so that the surface $\mathbb{R}^{m+1}$ contains larger and larger spherical shells. One first chooses an infinite sequence $a_k > 0$ such that $\sum_{k=0}^{\infty} a_k = \frac{\pi}{2}$. See Figure 1 on next page.

The first step of construction is to enlarge the circular arc with opening angle a_2 by going out radially so that the arclength is a prescribed number R_2. Clearly the radius N_2 is related to a_2 and R_2 by the relation $N_2 = a_2^{-1} R_2$. Now one slides the circular sector thus obtained parallel down the x-axis a distance s_2 so that the circular arc with opening angle a_0 can be connected with this slide-down circular arc by a smooth curve of parabolic

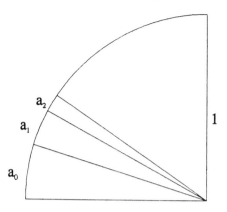

FIGURE 1

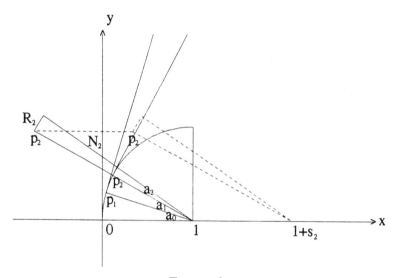

FIGURE 2

type. Elementary trigonometry shows that this is possible when

$$\frac{N_2 \cos a_1 - N_0}{\cos a_0} < s_2 < \frac{(N_2/N_0 - 1)N_2}{\cos(a_0 + a_1)},$$

i.e. p_2'' should lie in the region enclosed by the tangent lines at p_1 and p_2; see Figure 2.

Inductively, at the kth step, one enlarges the circular arc with the opening angle a_{2k} by going out radially so that its arclength is a prescribed number R_{2k}. Thus the radius $N_{2k} = a_{2k}^{-1} R_{2k}$. Now in order that the enlarged circular arc can be connected with the previous one already in position one has to slide it down a distance $s_0 + s_1 + \cdots + s_k$ where

$$\frac{N_{2k} \cos a_{2k-1} - N_{2k-2}}{\cos \sum_{j=0}^{2k-2} a_j} < s_k < \frac{(N_{2k}/N_{2k-2} - 1)N_{2k}}{\cos \sum_{j=0}^{2k-1} a_j}.$$

Now as is shown in [SY2], $\mathbb{R}^{n,m}$ will have nonnegative Ricci curvature if

$$(11) \qquad\qquad R_k \sim a_k^{-2/\alpha},$$

where $\alpha = 2(m-1)/n$.

Since the surgery reduces diameter, the diameter growth is controlled by the diameter growth of the hypersurface $\mathbb{R}^{m+1}$. By Lemma 4.1 we only need to find out the growth of the function that generates $\mathbb{R}^{m+1}$. From the above discussion, at $x = s_0 + s_1 + \cdots + s_k$, $f(x) = N_{2k} \sin \sum_{j=0}^{2k-1} a_j$. By taking $a_k = \frac{\pi}{4} 2^{-k}$, one easily finds $f(x) = O(x^\beta)$, where $\beta = \frac{\alpha+2}{2\alpha+2}$.

Thus for example, $\mathbb{R}^{2,2}$ ($m = 2$, $n = 2$) will have diameter growth of degree $\frac{3}{4}$, hence volume growth of degree $\frac{5}{2}$. We remark that as $m \to \infty$, $\beta \to \frac{1}{2}$. Therefore the degree of diameter growth of such manifold could be arbitrarily close to $\frac{1}{2}$.

References

[AG] U. Abresch and D. Gromoll, *On complete manifolds with nonnegative Ricci curvature*, J. Amer. Math. Soc. **3** (1990), 355–374.

[C] J. Cheeger, *Critical points of distance functions and applications to geometry*, Preprint, 1990.

[CE] J. Cheeger and D. Ebin, *Comparison theorems in Riemannian geometry*, North-Holland, Amsterdam, 1975.

[CGT] J. Cheeger, M. Gromov, and M. Taylor, *Finite propagation speed, kernel estimates for functions of the Laplace operator and the geometry of complete Riemannian manifolds*, J. Differential Geom. **17** (1982), 15–55.

[CK] C. Croke and H. Karcher, *Volume of small balls on open manifolds: lower bounds and examples*, Trans. Amer. Math. Soc. **309** (1988), 753–762.

[G] M. Gromov, *Curvature, diameter and Betti numbers*, Comm. Math. Helv. **56** (1981), 179–195.

[SY1] J. P. Sha and D. G. Yang, *Examples of metrics of positive Ricci curvature*, J. Differential Geom. **29** (1989), 95–103.

[SY2] ——*Positive Ricci curvature on the connected sums of $S^n \times S^m$*, J. Differential Geom. **33** (1991), 127–138.

[S] Z. Shen, *On complete manifolds of nonnegative kth Ricci curvature*, Trans. Amer. Math. Soc. (to appear).

[Y] S. T. Yau, *Some function-theoretic properties of complete riemannian manifolds and their applications to geometry*, Indiana Univ. Math. J. **25** (1976), 659–670.

Mathematical Sciences Research Institute

Massachusetts Institute of Technology

Proceedings of Symposia in Pure Mathematics
Volume **54** (1993), Part 3

Recent Developments in Sphere Theorems

KATSUHIRO SHIOHAMA

Dedicated to Professor T. Otsuki on his 75th Birthday

1. Introduction

An important problem in differential geometry is to investigate the relation between curvature and topology of complete Riemannian manifolds. A traditional topic is the sphere theorem which we want to discuss in the present article. Throughout this article let M be a connected and compact Riemannian n-manifold with $n \geq 2$. The classical sphere theorem initiated by Rauch and developed by Berger and Klingenberg may be summarized as follows.

THE CLASSICAL SPHERE THEOREM. *If M is simply connected and if the sectional curvature K_M of M satisfies $1/4 \leq K_M \leq 1$, then M is either homeomorphic to S^n (when the diameter $d(M)$ of M is greater than π), or else isometric to a symmetric space of the compact type of rank one (when $d(M) = \pi$).*

A generalized version of the rigidity case was obtained by Gromoll and Grove (see [**G-G1, G-G2**]). The δ-pinching condition with $\delta > 1/4$ for the sectional curvature is optimal for obtaining the topological sphere. It turns out that this condition provides a uniqueness of topological types of this class of manifolds. A more general class of Riemannian manifolds with certain restrictions for pinching sectional curvature will contain finitely many distinct topological types, as obtained by Cheeger [**C**], Weinstein [**W.A**] and others.

The discovery of the exotic sphere by Milnor [**M.J**] gives rise to the problem of finding a pinching condition to single out the differentiable structure of S^n. The classical differentiable sphere theorems were first obtained by Gromoll [**G.D**], Shikata [**S.Y**] and Calabi by different ideas and the dimension

1991 *Mathematics Subject Classification*. Primary 53C20.
This paper is in final form and no version of it will be submitted for publication elsewhere.

independency was first proved by Sugimoto-Shiohama [S-S]. Diffeomorphism theorems between nonsimply connected δ-pinched manifolds and spherical space forms were obtained by Ruh, Karcher, Grove, Shiohama, Im Hof and others. The center of mass technique plays an essential role to establish equivariant diffeomorphism between the universal Riemannian covering of M and S^n [G-K2]. Recently Katsuda [K.A], Durumeric [D.O], Yamaguchi [Y3], Anderson [A2] and others obtained generalizations of the classical differentiable sphere theorems in connection with the Gromov convergence theorem. For convergence theorems see [F1].

All of the above results are obtained under pinching conditions for sectional curvature. The estimate of the injectivity radius of the exponential map plays an important role.

In this article we discuss sphere theorems without pinching sectional curvature. The sectional curvature of M is not bounded above but bounded below by a constant. Under our situation we cannot expect to estimate the injectivity radius of the exponential map. Although the Gromov convergence theorem does not apply to have a smooth manifold as the Hausdorff limits of our manifolds, it is useful to discuss the Hausdorff distance between M and S^n. The Hausdorff closeness property is important because it affects the proofs of differentiable sphere theorems as well as finiteness theorems. It is very interesting to study such length spaces that are obtained as the Hausdorff limits of manifolds belonging to a certain class defined by geometry.

The organization of this article is as follows. In §2, we shall prepare basic tools. It seems that in spite of its importance the Alexandrov convexity theorem is not very familiar, and hence we shall give a sketch of the proof of it. In §3, we state isometric conditions known as classical results, by relaxing which sphere theorems will be derived. Topological and differentiable sphere theorems are stated in §§4 and 5. Some open problems are discussed in §6. Finally we shall state recent results obtained by pinching sectional curvature in the Appendix. Proofs are all short sketches.

2. Basic tools

We prepare the basic tools. They are referred to in [B-C, C-E, G-K-M, K4, S.T]. The following Theorems 2.1 and 2.2 are equivalent and frequently used in this article.

THE ALEXANDROV CONVEXITY THEOREM 2.1. *Let $K_M \geq \kappa$. Let γ and σ be minimizing geodesics parametrized proportionally to $[0, 1]$ with the same starting point and $\tau_{s,t}$ a minimizing geodesic joining $\gamma(s)$ to $\sigma(t)$ for $s, t \in [0, 1]$. Let $\tilde{\Delta}_{s,t} := (\tilde{\gamma}|[0, s], \tilde{\sigma}|[0, t], \tilde{\tau}_{s,t})$ be the corresponding triangle sketched on the complete simply connected surface $S^2(\kappa)$ of constant curvature κ such that it has the same edge lengths as the triangle $\Delta_{s,t} := (\gamma|[0, s], \sigma|[0, t], \tau_{s,t})$. If $\tilde{\theta}(s, t)$ is the angle of $\tilde{\Delta}_{s,t}$ opposite $\tilde{\tau}_{s,t}$, then*

$\tilde{\theta}(s, t)$ *is monotone nonincreasing in the following sense: if* $t_1 \geq t_2$ *and if* $s_1 \geq s_2$, *then*

$$\tilde{\theta}(s_1, t_1) \leq \tilde{\theta}(s_2, t_2).$$

The following well-known theorem due to Toponogov is a direct consequence of the Alexandrov convexity theorem.

THE TOPONOGOV COMPARISON THEOREM 2.2. *Let* $K_M \geq \kappa$ *and* $\Delta :=$ $(\gamma_0, \gamma_1, \gamma_2)$ *be an arbitrary geodesic triangle consisting of all minimizing geodesics whose angle opposite* γ_i *is* θ_i. *Then there exists a corresponding geodesic triangle* $\tilde{\Delta} := (\tilde{\gamma}_0, \tilde{\gamma}_1, \tilde{\gamma}_2)$ *sketched on the complete simply connected surface* $S^2(\kappa)$ *of constant curvature* κ *whose length* $L(\tilde{\gamma}_i)$ *is equal to that of* γ_i *in such a way that every angle* $\tilde{\theta}_i$ *opposite* $\tilde{\gamma}_i$ *for* $i = 0, 1, 2$ *satisfies*

$$\theta_i \geq \tilde{\theta}_i.$$

PROOF OF THEOREM 2.2 (by Assuming the Alexandrov Convexity Theorem 2.1). Let $\gamma_1, \gamma_2 : [0, 1] \to M$ be parametrized proportionally to arc length and $\gamma_1(0) = \gamma_2(0)$, and $\tilde{\theta}_0(s, t)$ for $0 \leq s, t \leq 1$ be the corresponding angle opposite $\tilde{\tau}_{s,t}$ of the triangle $\tilde{\Delta}_{s,t}$ sketched on $S^2(\kappa)$. Then $\theta_0 = \lim_{s,t \to 0} \tilde{\theta}_0(s, t)$ and $\tilde{\theta}_0(1, 1) = \tilde{\theta}_0$. The monotone nonincreasing property of $\tilde{\theta}_0$ proves that $\theta_0 \geq \tilde{\theta}_0$.

SKETCH OF THE PROOF OF THEOREM 2.1. We only consider the case where $\kappa > 0$ since other cases are easier. The proof is divided into two steps. In the first step Theorem 2.1 is proved for all geodesic triangles with circumferences less than $2\pi/\sqrt{\kappa}$. In the second step it is shown that every geodesic triangle has circumference not greater than $2\pi/\sqrt{\kappa}$. Moreover if there is a geodesic triangle with circumference $2\pi/\sqrt{\kappa}$, then the diameter of M is $\pi/\sqrt{\kappa}$. If the diameter of M is $\pi/\sqrt{\kappa}$, then M is isometric to the standard n-sphere of constant curvature κ, and nothing is left to prove in this case. For the proof of maximal diameter theorem, see Remark 3.a in §3.

Let $\Delta_{s,t} := (\gamma|[0, s], \sigma|[0, t], \tau_{s,t})$, $\Delta := \Delta_{1,1}$ and $p_0 := \gamma(0) = \sigma(0)$. Since the circumference $L(\Delta)$ of Δ satisfies $L(\Delta) < 2\pi/\sqrt{\kappa}$, all the corresponding triangles $\tilde{\Delta}_{s,t}$ and the angles are uniquely determined. If $L(\Delta)$ is small enough such that Δ is contained in the intersection of injectivity radius balls around its vertices, then the Rauch comparison theorem implies the conclusion. Let s_0, t_0 be the supremum of $s, t \in (0, 1]$ such that the conclusion holds for all $\Delta_{s,t}$ and for every minimizing geodesic $\tau_{s,t}$ joining $\gamma(s)$ and $\sigma(t)$. Suppose that $s_0 < 1$ and $t_0 < 1$. Then, for an $s_1 \in (s_0, 1]$ sufficiently close to s_0 and for a triangle $\overline{\Delta}' := (\tau_{s_0,t_0}, \tau_{s_1,t_0}, \gamma|[s_0, s_1])$ the Berger comparison theorem (see [B2]) implies that the angles of $\overline{\Delta}'$ at $\gamma(s_0)$ and $\gamma(s_1)$ are not less than the corresponding angles of the triangle $\tilde{\Delta}'$ sketched on $S^2(\kappa)$ with the same edge lengths as $\overline{\Delta}'$. Then the nonoverlapping union $\tilde{\Delta}_{s_0,t_0} \cup \tilde{\Delta}'$ with common edge in the middle forms a convex quadrangle on

$S^2(\kappa)$, and $\tilde{\theta}(s, t_0)$ is monotone nonincreasing for $s \in [s_0, s_1]$. Similarly one has the same property for $\tilde{\theta}(s_0, t)$ for $t \in [t_0, t_1]$ where $1 > t_1 > t_0$ is taken sufficiently close to t_0. Combining these properties for $\tilde{\theta}(s, t)$, the monotone property is extended beyond s_0 and t_0, a contradiction.

For the proof of second step we suppose that there exists a triangle Δ with $L(\Delta) > 2\pi/\sqrt{\kappa}$. In view of the maximal diameter theorem, we may assume that $d(M) < \pi/\sqrt{\kappa}$ and hence every edge has length less than $\pi/\sqrt{\kappa}$. There exist $s, t \in (0, 1)$ such that $L(\Delta_{s,t}) = 2\pi/\sqrt{\kappa}$. The convex quadrangle argument for this $\Delta_{s,t}$ then implies that $\Delta_{s,t}$ forms a simply closed geodesic of length $2\pi/\sqrt{\kappa}$. Thus Δ is contained entirely in $\Delta_{s,t}$, and hence it has a circumference $2\pi/\sqrt{\kappa}$, a contradiction.

GENERALIZED TOPONOGOV THEOREM 2.3. *Let* $K_M \geq \kappa > 0$. *Let* $\Delta := (\gamma_0, \gamma_1, \gamma_2)$ *be a generalized geodesic triangle such that* γ_0 *is not neccessarily minimizing but* γ_1, γ_2 *are minimizing. Assume that*

$$L(\gamma_0) \leq \pi/\sqrt{\kappa}, \qquad L(\gamma_0) \leq L(\gamma_1) + L(\gamma_2).$$

Then there exists a corresponding geodesic triangle $\tilde{\Delta} := (\tilde{\gamma}_0, \tilde{\gamma}_1, \tilde{\gamma}_2)$ *on* $S^2(\kappa)$ *such that* $L(\tilde{\gamma}_i) = L(\gamma_i)$ *for* $i = 0, 1, 2$ *with the properties that if* θ_i *and* $\tilde{\theta}_i$ *are angles of* Δ *and* $\tilde{\Delta}$ *opposite* γ_i *and* $\tilde{\gamma}_i$ *respectively, then*

$$\tilde{\theta}_1 \leq \theta_1, \quad and \quad \tilde{\theta}_2 \leq \theta_2.$$

Here we note that the circumference of Δ in Theorem 2.3 does not exceed $2\pi/\sqrt{\kappa}$. This fact is seen in the second step of the proof of Theorem 2.1. However we do not know whether or not the existence of a triangle with circumference $2\pi/\sqrt{\kappa}$ implies that M is isometric to the standard n-sphere of constant curvature κ.

The above theorem is equivalent to the following

THE HINGE THEOREM 2.4. *Let* M *be as in the above theorem. Let* $\gamma_0, \gamma_1 : [0, 1] \to M$ *be geodesics with* $\gamma_0(0) = \gamma_1(0) = p$ *such that* γ_0 *is not necessarily minimizing but* γ_1 *is minimizing and such that*

$$L(\gamma_0) \leq \pi/\sqrt{\kappa}, \qquad L(\gamma_0) \leq L(\gamma_1) + d(\gamma_0(1), \gamma_1(1)).$$

If $\hat{\gamma}_0, \hat{\gamma}_1 : [0, 1] \to S^2(\kappa)$ *are geodesics with* $\hat{\gamma}_0(0) = \hat{\gamma}_1(0)$ *and* $L(\hat{\gamma}_i) = L(\gamma_i)$ *for* $i = 0, 1$ *and if* $\langle \gamma_0'(0), \gamma_1'(0) \rangle = \langle \hat{\gamma}_0'(0), \hat{\gamma}_1'(0) \rangle$, *then*

$$d(\hat{\gamma}_0(1), \hat{\gamma}_1(1)) \geq d(\gamma_0(1), \gamma_1(1)).$$

REMARK 2.A. With the same notations as in Theorem 2.4 we have for every $s \in [0, 1]$,

$$d(\hat{\gamma}_0(s), \hat{\gamma}_1(1)) \geq d(\gamma_0(s), \gamma_1(1)).$$

However we cannot expect, for $t \in [0, 1)$,

$$d(\hat{\gamma}_0(s), \hat{\gamma}_1(t)) \geq d(\gamma_0(s), \gamma_1(t)),$$

since the triangle inequality for the lengths of this generalized triangle may fail for $t \in [0, 1)$.

REMARK 2.B. Let $\kappa > 0$. For a generalized triangle $\Delta := (\gamma_0, \gamma_1, \gamma_2)$ such that γ_0 is not necessarily minimizing and satisfies $L(\gamma_0) \leq \pi/\sqrt{\kappa}$ and $L(\gamma_0) \leq L(\gamma_1) + L(\gamma_2)$, let $\tilde{\Delta} := (\tilde{\gamma}_0, \tilde{\gamma}_1, \tilde{\gamma}_2)$ be the corresponding triangle with the parametrization $\gamma_0(0) = \gamma_1(0)$ and $\tilde{\gamma}_0(0) = \tilde{\gamma}_1(0)$. Let $\hat{\gamma}_0$ and $\hat{\gamma}_1$ be as in Theorem 2.4. Let $\tau_s : [0, 1] \to M$ and $\hat{\tau}_s, \tilde{\tau}_{s,1} : [0, 1] \to S^2(\kappa)$ be minimizing geodesics with $\tau_s(0) = \gamma_1(1), \tau_s(1) = \gamma_0(s)$ and $\hat{\tau}_s(0) = \hat{\gamma}_1(1)$, $\hat{\tau}_s(1) = \hat{\gamma}_0(s)$, and $\tilde{\tau}_{s,1}(0) = \tilde{\gamma}_1(1)$, $\tilde{\tau}_{s,1}(1) = \tilde{\gamma}_0(s)$. Then we have, for $s \in [0, 1]$,

$$L(\tilde{\tau}_{s,1}) \leq L(\tau_s) \leq L(\hat{\tau}_s).$$

From this fact we see that the Alexandrov convexity theorem provides us more information than the Toponogov comparison theorem.

A more general triangle comparison theorem has been discussed by Abresch [A.U] which was initiated by Elerath [E.D.]. This idea leads to obtain a Toponogov type comparison theorem for radial sectional curvature as stated in Remark 3.d.

A Morse-theoretic approach to the distance functions on M is employed. This idea was first introduced in [G-S] and later systematically used by Gromov to obtain many fruitful results on the topology of compact manifolds whose curvature is bounded below [G.M1]. A critical point of a Lipschitz continuous function $f : M \to R$ is defined as follows. For a point $x \in M$ let $\lim \nabla f(x)$ be the set of all limits of gradient vectors $\nabla f(x_j)$ at differentiable points $x_j \in M$ of f such that $\lim x_j = x$. A point $p \in M$ is by definition a noncritical point (critical point, respectively) of f iff $\lim \nabla f(p)$ is contained (not contained, respectively) in an open half-space of the tangent space $T_p M$ of M at p. If $0 \in \lim \nabla f(p)$, then p is a critical point of f. If q is a point at which the distance function $d(p, \cdot)$ to p takes a local maximum, then q is a critical point of $d(p, \cdot)$. A metric r-ball $B(p, r)$ around p is contractible if $d(p, \cdot)$ has no critical point in $B(p, r) \backslash \{p\}$. The contractibility radius of distance function to points on M will be employed instead of the injectivity radius of the exponential map of M. The idea of contractibility radius on M is combined with the following Generalized Scheonflies Theorem (see [R.B]) to obtain the topological sphere theorem. The contractibility radius estimate was used by Yamaguchi (see [Y1]) to obtain a finiteness theorem for the homotopy types of a certain class of manifolds, as stated in Remark 4.a in §4.

GENERALIZED SCHEONFLIES THEOREM 2.5. *If a compact manifold M is covered by two nonoverlapping closed disks, then it is homeomorphic to S^n.*

The Bishop-Gromov volume comparison theorem for concentric metric balls is used for the construction of Hausdorff approximation maps between a certain M and S^n. This and the next theorems play an important role throughout this article. Let $b_\kappa(t)$ for $t > 0$ be the volume of t-ball on $S^n(\kappa)$.

THE BISHOP-GROMOV VOLUME COMPARISON THEOREM 2.6. *Let M be a complete Riemannian manifold whose Ricci curvature Ric_M satisfies $\mathrm{Ric}_M \geq (n-1)\kappa$. For a fixed point p on M the function*

$$\mathrm{Vol}(B(p, t))/b_\kappa(t)$$

is monotone nonincreasing.

The Hausdorff distance and pointed Hausdorff distance between two manifolds will be used for the proofs of differentiable sphere theorems. Let (X_i, d_i), $i = 1, 2$, be metric spaces. A map $\phi : X_1 \to X_2$ is by definition an ε-Hausdorff approximation map iff

$$|d_2(\phi(x_1), \phi(x_2)) - d_1(x_1, x_2)| < \varepsilon,$$

for every $x_1, x_2 \in X_1$, and

$$B(\phi(X_1), \varepsilon) \supset X_2.$$

The Hausdorff distance $d_H(X_1, X_2)$ between X_1 and X_2 is defined by

$$d_H(X_1, X_2) := \inf\{\varepsilon ;\ \phi : X_1 \to X_2,\ \psi : X_2 \to X_1$$
$$\text{are } \varepsilon\text{-Hausdorff approximation maps}\}.$$

A map $\phi : X_1 \to X_2$ is by definition an (ε, L)-Hausdorff approximation map iff

$$|d_2(\phi(x_1), \phi(x_2)) - d_1(x_1, x_2)| < \varepsilon,$$

for every $x_1, x_2 \in X_1$ with $d(x_1, x_2) \leq L$, and

$$B(\phi(X_1), \varepsilon) \supset X_2.$$

If X_1 and X_2 are compact and if $d(X_1) \leq L$, $d(X_2) \leq L$, then an (ε, L)-Hausdorff approximation map is an ε-Hausdorff approximation map.

Let $p_1 \in X_1$ and $p_2 \in X_2$ be fixed points. A map $\phi : X_1 \to X_2$ is by definition an ε-pointed Hausdorff approximation map iff

$$B(\phi(X_1), \varepsilon) \supset B(p_2, 1/\varepsilon),$$

and for every $x_1, x_2 \in B(p_1, 1/\varepsilon)$

$$|d(\phi(x_1), \phi(x_2)) - d(x_1, x_2)| < \varepsilon.$$

The pointed Hausdorff distance $d_{p,H}((X_1, p_1), (X_2, p_2))$ is defined by

$$d_{p,H}((X_1, p_1), (X_2, p_2)) := \inf\{\varepsilon ;\ \phi : X_1 \to X_2 \text{ and } \psi : X_2 \to X_1$$
$$\text{are pointed } \varepsilon\text{-Hausdorff approximation maps}\}.$$

For the proofs of Theorems 4.3, 4.4 and for later use in §5, we shall introduce some notions of excess. For points p, q on M the excess function $e_{p,q} : M \to R$ of p, q is given by

$$e_{p,q}(x) := d(p, x) + d(q, x) - d(p, q), \qquad x \in M.$$

The restricted excess function $e^d : M \to \mathbf{R}$ for $d > 0$ is given by

$$e^d(p) := \sup_{x \in B(p,d)} \inf_{q \in S(p,d)} e_{p,q}(x),$$

where $S(p,d) := \{y \in M : d(p,y) = d\}$. The excess $e(M)$ and restricted excess $e^d(M)$ for $d > 0$ of M are defined as

$$e(M) := \inf_{p,q \in M} \sup_{x \in M} e_{p,q}(x),$$

and

$$e^d(M) := \sup_{p \in M} e^d(p).$$

The radius $\mathrm{rad}(M)$ of M is defined by

$$\mathrm{rad}(M) := \inf_{p \in M} \sup_{q \in M} d(p,q).$$

Every rank-one symmetric space of the compact type as in the classical sphere theorem has radius π and zero restricted excess for $d \leq \pi$. Among them only the sphere has zero excess, while all the others have excess π. It follows from definition that $e^d(M) = 0$ for every $d \in (0, \mathrm{inj}(M)]$. If $p, q \in M$ are chosen so as to satisfy $e(M) = \max e_{p,q}$, then

$$d(M) \leq d(p,q) + e(M).$$

REMARK 2.C. It folows from the Bishop-Gromov volume comparison Theorem 2.6 that (1) if $\mathrm{Ric}_M \geq n-1$ and if $\mathrm{Vol}(M) \geq \omega_n - \varepsilon$, then $\mathrm{rad}(M)$ and $d(M)$ are close to π and $e(M)$ is close to 0, and (2) if $\mathrm{Ric}_M \geq n-1$ and if $d(M) \geq \pi - \varepsilon$, then $e(M)$ is close to 0.

We emphasize that the Hausdorff limit of a sequence of manifolds is not discussed except in the proof of Theorem 5.5. In most cases the Hausdorff limit of a certain class of manifolds $\{M_i\}$ is taken under such circumstances that a manifold N exists as the Hausdorff limit and that each manifold M_i of the sequence is supposed not to have the same topological properties. It is quite natural to anticipate under certain conditions that if the Hausdorff distance between two manifolds of the same dimension is sufficiently close to zero, then they will have the same topological type. Thus a contradiction will be derived by some arguments on N. It will then turn out that there exists some constant ε such that $d_H(M_i, M_j) < \varepsilon$ implies that they are of the same topological type. The point in this argument is that we do not know the dependence of ε upon the a priori given geometric data which define the class of manifolds. To show the existence of such an ε as mentioned above is much easier than to express it in terms of a priori given geometric data which define the class of Riemannian manifolds. Since we do not take Hausdorff limits, the constants stated in theorems in §§4 and 5 are all (at least in principle) estimable explicitly in terms of given data.

3. Isometric conditions

We shall begin by stating classical well-known results due to Myers, Toponogov and others which provide sufficient conditions for M whose curvature is bounded below by a constant to be isometric to S^n. We denote by $\mathrm{Vol}(M)$ the volume of M and ω_n the volume of the standard unit n-sphere.

In the first place we shall state the results concerning the volume of M.

THEOREM 3.1. *If* $K_M \geq 1$, *then* $\mathrm{Vol}(M) \leq \omega_n$. *Here equality holds if and only if* M *is isometric to* S^n.

THEOREM 3.2. *If* $\mathrm{Ric}_M \geq n - 1$, *then* $\mathrm{Vol}(M) \leq \omega_n$. *Here equality holds if and only if* M *is isometric to* S^n.

THEOREM 3.3. *If* $K_M \leq 1$ *and if* M *is not simply connected and the order of the fundamental group* $\pi_1(M)$ *is* k, *then* $\mathrm{Vol}(M) \leq \omega_n/k$. *Here equality holds if and only if* M *is isometric to a spherical space form* S^n/Γ *and* Γ *is a subgroup of* $O(n+1)$ *with order* k.

THEOREM 3.4. *If* $\mathrm{Ric}_M \geq n - 1$ *and if* M *is not simply connected and the order of* π_1 *is* k, *then* $\mathrm{Vol}(M) \leq \omega_n/k$. *Here equality holds if and only if* M *is isometric to a spherical space form* S^n/Γ.

In the next place we shall state the results concerning with the diameter of M.

THE TOPONOGOV MAXIMAL DIAMETER THEOREM 3.5. *If* $K_M \geq 1$, *then* $d(M) \leq \pi$. *Here equality holds if and only if* M *is isometric to* S^n.

THE CHENG MAXIMAL DIAMETER THEOREM 3.6. *If* $\mathrm{Ric}_M \geq n - 1$, *then* $d(M) \leq \pi$. *Here equality holds if and only if* M *is isometric to* S^n.

TOPONOGOV'S THEOREM 3.7. *If* $K_M \geq 1$, *then any triple of points* p, q *and* r *on* M *has the property* $d(p, q) + d(q, r) + d(r, p) \leq 2\pi$. *Equality holds if and only if* M *is isometric to* S^n.

It should be noted that the injectivity radius $\mathrm{inj}(M)$ of M furnishes a lower bound for the volume of M, known as the isoembolic inequality (see [B3, K.J]) as follows.

THEOREM 3.8. *If the injectivity radius of a compact Riemannian* n-*manifold* N *satisfies* $\mathrm{inj}(N) \geq \rho$, *then* $\mathrm{Vol}(N) \geq (\frac{\rho}{\pi})^n \cdot \omega_n$. *Here equality holds if and only if* N *is isometric to the standard* n-*sphere of constant curvature* $(\frac{\pi}{\rho})^2$.

REMARK 3.A. An elementary proof of the Cheng maximal diameter theorem is obtained by using the Bishop-Gromov volume comparison theorem for concentric metric balls. The proof proceeds as follows. Let $p, q \in M$ be chosen as $d(p, q) = d(M)$. If we set $v_p(t) := \mathrm{Vol}\,B(p, t)/b_1(t)$ and $v_q(t) := \mathrm{Vol}\,B(q, t)/b_1(t)$, then both v_p and v_q are monotone and nonincreasing. Therefore $v_p(\pi) = \mathrm{Vol}(M)/\omega_n \leq v_p(\pi/2) = 2\,\mathrm{Vol}(B(p, \pi/2))/\omega_n$ implies $\mathrm{Vol}(B(p, \pi/2)) \geq \mathrm{Vol}(M)/2$ and similarly $\mathrm{Vol}(B(q, \pi/2)) \geq \mathrm{Vol}(M)/2$. Since $B(p, t) \cap B(q, \pi - t) = \varnothing$ for all $t \in (0, \pi)$, we observe that $B(p, \pi/2)$ is the interior of $M \backslash B(q, \pi/2)$. In particular every geodesic emanating from

p reaches to q at length π, and the cut locus to p (q, respectively) is $\{q\}$ ($\{p\}$, respectively). The index form comparison implies that the sectional curvature is constant 1 at all plane sections containing vectors tangent to geodesics emanating from p. This proves the isomety.

REMARK 3.B. The isometries in the above results occur when the volume, diameter or the circumference of triangles of M take the maximal values. One might anticipate that the diffeomorphism (or homeomorphism) between M and S^n may be obtained near the maximal values.

REMARK 3.C. In the case of Riemannian orbifolds we have the same estimate for diameter as the classical Myers theorem which has recently been proved by Hodgson and Tysk (see [H-T]) as stated.

THEOREM 3.9. *Let M be a cone manifold which is a simplicial complex with a rational homology n-manifold and admits a smooth Riemannian metric on the complement of the codimension two skeleton of M. If $\mathrm{Ric}_M \geq n-1$, then $d(M) \leq \pi$. Here equality holds if and only if $K_M = 1$ and M is the suspension of a low dimensional sphere to a cone manifold.*

REMARK 3.D. Some of the above results may be extended to radial sectional and Ricci curvature. The notion of radial sectional curvature was first introduced by Klingenberg [K3], in which he proved a homotopy sphere theorem under certain conditions for radial sectional curvature pinching. Let o be an arbitrary fixed point on a complete Riemannian n-manifold N and $\gamma : [0, \infty) \to N$ a unit speed geodesic with $\gamma(0) = o$. We say that the radial sectional curvature of N with base point at o is bounded below by $\kappa \geq 0$ iff

$$\inf\{K_N(\Pi)\,;\, \gamma'(t) \in \Pi,\ t \in [0, \pi/\sqrt{\kappa}),\ \gamma(0) = o\} \geq \kappa.$$

Also we say that the radial Ricci curvature of N with base point at o is bounded below by $(n-1)\kappa \geq 0$ iff

$$\inf\{\mathrm{Ric}_N(\gamma'(t))\,;\, \gamma(0) = o,\ t \in [0, \pi/\sqrt{\kappa})\} \geq (n-1)\kappa.$$

If the radial Ricci curvature of N with base point at o is bounded below by $(n-1)\kappa$, then the Bishop-Gromov volume comparison Theorem 2.6 for concentric metric balls around o is valid. If the radial sectional curvature of N with base point at o is bounded below by $\kappa \geq 0$, then the Toponogov Theorems 2.2 and 2.3 hold for triangles with edges at o, q_1 and q_2. If θ_1 and θ_2 are the angles at q_1 and q_2 of this triangle and if $\tilde{\theta}_1$ and $\tilde{\theta}_2$ are the angles of the corresponding triangle sketched on $S^2(\kappa)$, then we have

$$\theta_1 \geq \tilde{\theta}_1, \qquad \theta_2 \geq \tilde{\theta}_2.$$

In particular, if the radial sectional curvature of N with base point at o is bounded below by 1, we then have for every q_1 and q_2 in N

$$d(o, q_1) + d(o, q_2) + d(q_1, q_2) \leq \pi.$$

With these facts we obtain the following

THEOREM 3.10. *If the radial sectional curvature of M with base point at o is bounded below by 1, then*

$$\mathrm{Vol}(M) \leq \omega_n, \quad \text{and} \quad d(o, \cdot) \leq \pi,$$

where each of the equalities holds if and only if M is isometric to S^n. Moreover we have

$$d(M) \leq \pi.$$

THEOREM 3.11. *If the radial Ricci curvature of M with base point at o is bounded below by $n - 1$, then*

$$\mathrm{Vol}(M) \leq \omega_n, \quad \text{and} \quad d(o, \cdot) \leq \pi.$$

Here $\mathrm{Vol}(M) = \omega_n$ holds if and only if M is isometric to S^n.

I do not know if $d(M) = \pi$ in Theorem 3.10 and also $\max d(o, \cdot) = \pi$ in Theorem 3.11 will imply that M is isometric to S^n.

4. Topological sphere theorems

The following diameter sphere theorem (see [G-S]) is the starting point of our discussion in which we succeeded in taking off the upper bound for the sectional curvature and in which noncritical point of distance functions was first discussed.

THE DIAMETER SPHERE THEOREM 4.1. *If the sectional curvature and diameter of M satisfy*

$$d(M) > \pi/2, \qquad K_M \geq 1$$

then M is homeomorphic to S^n.

SKETCH OF THE PROOF. Let $p, q \in M$ be taken such that $d(p, q) = d(M)$. The Toponogov comparison theorem implies that the distance functions $d(p, \cdot)$ and $d(q, \cdot)$ to p and q have critical points only at p and q. M is exhibited as a union of two sufficiently small r-balls $B(p, r), B(q, r)$ around p and q and a cylinder $S^{n-1} \times [0, 1]$ joined along with their common boundaries, $S^{n-1} \times \{0\} = \partial B(p, r)$ and $S^{n-1} \times \{1\} = \partial B(q, r)$. Thus M is a topological sphere.

Making use of the generalized Scheoflies theorem we obtain the following (see [S.K]) which was recently extended to give diffeomorphism by Yamaguchi [Y2].

THEOREM 4.2. *Given an integer $n \geq 2$ and $\kappa > 0$, there exists a number $\varepsilon = \varepsilon(n, \kappa) > 0$ such that if M has the properties*

$$K_M \geq -\kappa^2, \quad \mathrm{Ric}_M \geq n - 1, \quad \mathrm{Vol}(M) \geq \omega_n - \varepsilon$$

then M is homeomorphic to S^n.

The following lemma presents how to estimate contractibility radius at a point; it is used for the proof of Theorem 4.2 and later in §5.

LEMMA. *Assume that M has the properties*:

$$K_M \geq -\kappa^2, \qquad \mathrm{Ric}_M \geq n - 1,$$

where κ is a real or pure imaginary number. For given constants $\ell \in (0, \pi)$, $\lambda \in (0, 1)$ and for a sufficiently small positive ε there exists a constant $c = c(n, \kappa, \lambda, \ell, \varepsilon) > 0$ with $\lim_{\varepsilon \to 0} c(n, \kappa, \lambda, \ell, \varepsilon) = \lambda\ell$ such that if

$$\frac{\mathrm{Vol}(B(p, \ell))}{b_1(\ell)} \geq 1 - \varepsilon$$

is fulfilled for a point $p \in M$, then the distance function to p has no critical point in $B(p, c)\backslash\{p\}$.

PROOF OF LEMMA. Let $U_p \subset M_p$ be the disk domain star-shaped with respect to the origin whose boundary is the tangent cut locus to p. For a nonzero vector $v \in R^n$ and for $\theta \in (0, \pi)$ let

$$\Theta_{\lambda, \ell}(v, \theta) := \{w \in R^n : \langle v, w\rangle/\|v\| \cdot \|w\| \geq \cos \theta, \ \lambda\ell \leq \|w\| \leq \ell\}.$$

We then define $\theta(\lambda, \ell, \varepsilon) > 0$ by the relation

$$\mathrm{Vol}[\exp_{\tilde{p}} \Theta_{\lambda, \ell}(v, \theta(\lambda, \ell, \varepsilon))] = \varepsilon b_1(\ell),$$

where $\tilde{p} \in S^n$. From Theorem 2.6 we observe that ∂U_p intersects $\Theta_{\lambda, \ell}(v, \theta)$ for every $v \in M_p$ (because if otherwise is supposed, then $\mathrm{Vol}(B(p, \ell)) < b_1(\ell)(1 - \varepsilon)$). Namely, the set of all unit vectors at p tangent to minimizing geodesics emanating from p having lengths no less than $\lambda\ell$ forms a $\theta(\lambda, \ell, \varepsilon)$-dense set in the unit sphere in $T_p M$. Setting

$$c(n, \kappa, \lambda, \ell, \varepsilon) := \begin{cases} \kappa^{-1} \tanh^{-1}\{\tanh(\kappa\lambda\ell) \cdot \cos \theta\}, & \text{for } \kappa^2 > 0, \\ \kappa^{-1} \tan^{-1}\{\tan(\kappa\lambda\ell) \cdot \cos \theta\}, & \text{for } \kappa^2 < 0, \end{cases}$$

we see from Theorem 2.2 that the distance function to p has no critical point in the ball $B(p, c(n, \kappa, \lambda, \ell, \varepsilon))\backslash\{p\}$. This proves the Lemma.

OUTLINE OF THE PROOF OF THEOREM 4.2. By a slight modification of the above lemma it is possible to estimate the uniform lower bound $c(n, \kappa, \varepsilon)$ of the minimal critical values of the distance function to every point x on M, and hence the contractibility radius of M is estimated. Here we have $\lim_{\varepsilon \to 0} c(n, \kappa, \varepsilon) = \pi$. Thus M can be covered by two nonoverlapping closed n-disks for a suitable choice of ε.

REMARK 4.A. It was proved in [Y1] that the precompact class of all complete n-manifolds M whose contractibility radius is bounded below by a constant $R > 0$ has finite homotopy types. As a cororally, it is shown that for given positive constants κ, D and R the class of all n-manifolds M with the properties

$$\mathrm{Ric}_M \geq -(n - 1)\kappa^2, \quad d(M) \leq D, \quad c(M) \geq R$$

has finitely many homotopy types, where $c(M)$ is by definition the contractibility radius of M. A more general notion of R-contractibility was introduced by Grove-Petersen as is seen in the proof of Theorem 4.4.

Note that if $\mathrm{Ric}_M \geq n-1$ and if the volume of M is close to ω_n, then $d(M)$ is also close to π. However the converse does not necessarily hold. This fact will mean that under the restriction of the diameter of M some additional condition may be needed to prove the topological sphere theorem. The following result was independently obtained by Eschenburg [E1] and Nakamura [N]. This will be extended to give diffeomorphism in Theorem 5.6 (see [O3]).

THEOREM 4.3. *Given an integer* $n \geq 2$ *and* $\rho, \kappa > 0$, *there exists a constant* $\varepsilon = \varepsilon(n, \rho, \kappa) > 0$ *such that if* M *has the properties*

$$\mathrm{Ric}_M \geq n-1, \quad K_M \geq -\kappa^2, \quad \mathrm{inj}(M) \geq \rho, \quad d(M) \geq \pi - \varepsilon$$

then M *is homeomorphic to* S^n.

As is stated in the isoembolic inequality, the volume of M in Theorem 4.3 has a lower bound depending only on n and ρ. It turns out that the above results are essentially contained in the following striking result by Grove-Petersen [G-P3].

THEOREM 4.4. *Given* $\kappa, D, v > 0$ *there exists an* $\varepsilon = \varepsilon(n, \kappa, D, v) > 0$ *such that if* M *has the properties:*

$$K_M \geq \kappa, \quad d(M) \leq D, \quad \mathrm{Vol}(M) \geq v, \quad and \quad e(M) \leq \varepsilon,$$

then M *is a homotopy sphere.*

As a consequence of the above theorem, we have the following (see [G-P3])

THEOREM 4.5. *Given an integer* $n \geq 2$ *and* $\kappa, v > 0$, *there exists a constant* $\varepsilon = \varepsilon(n, \kappa, v) > 0$ *such that if* M *has the properties*

$$K_M \geq \kappa, \quad \mathrm{Ric}_M \geq n-1, \quad \mathrm{Vol}(M) \geq v, \quad d(M) \geq \pi - \varepsilon,$$

then M *is homeomorphic to* S^n.

In view of the above results it is natural to ask whether or not the assumption for K_M is essential. The following example of metrics $\{g_i\}$ on $M = S^k \times S^{n-k}$ for $k \geq 2$, $n-k \geq 3$ constructed by Otsu (see [O1]) will show that the lower bound for K_M in Theorems 4.2 and 4.4 is indispensable. Making use of the Riemannian warped product, Otsu constructed $\{g_i\}$ with the following properties:

$$\mathrm{Ric}_{(M, g_i)} \geq n-1, \quad d(M, g_i) \geq \pi - \frac{1}{i}, \quad \mathrm{Vol}(M, g_i) \geq C(n, k) \cdot \omega_n,$$

and

$$-\min K_{(M, g_i)} = \max K_{(M, g_i)} =: a_i$$

where $\lim a_i = \infty$ and $\lim_{i \to \infty} \mathrm{inj}(M, g_i) = 0$, and $C(n, k) > 0$ a constant depending on k and n. From these examples of metrics we know that for every fixed small positive ε, the class of all Riemannian n-manifolds with $n \geq 6$ characterized by $\operatorname{Ric}_M \geq n - 1$ and $d(M) \geq \pi - \varepsilon$ contains infinitely many distinct homotopy types (see [O1]). Anderson also constructed complete metrics on $M = CP^2$ and $CP^2 \sharp CP^2$ with the properties that

$$\operatorname{Ric}_M \geq n - 1, \quad d(M) \geq \pi - \varepsilon, \quad \operatorname{Vol}(M) \geq v$$

for an arbitrary small $\varepsilon > 0$ and for a constant $v > 0$ (see [A1]). However we do not know if Theorem 4.4 holds without assuming the lower bound for the volume of M.

OUTLINE OF THE PROOFS OF THEOREMS 4.3, 4.4 AND 4.5. Let $p, q \in M$ be chosen so as to satisfy $d(M) = d(p, q)$, and for $x \in M$ set $d(p, q) = a + e_{p,q}(x)/2$, $d(q, x) = b + e_{p,q}(x)/2$. Then the balls $B(p, a)$, $B(q, b)$ and $B(x, e_{p,q}(x)/2)$ are nonoverlapping. If $\operatorname{Ric}_M \geq n - 1$ and $d(M) \geq \pi - \varepsilon$ for small ε, then Theorem 2.6 implies that

$$\omega_n \geq b_1(a) + b_1(b) + b_1(e_{p,q}(x)/2) \geq \omega_n - \omega_{n-1} \cdot \varepsilon + b_1(e_{p,q}(x)/2).$$

Therefore we have $e(M) \leq b_1^{-1}(\omega_{n-1} \cdot \varepsilon)$. This shows that M in Theorems 4.3 and 4.5 has a small excess, and hence Theorem 4.5 is derived from Theorem 4.4. Moreover under the assumptions in Theorem 4.3, we apply Theorem 2.2 to observe that for a fixed $r_0 \in (0, \rho)$ and for a suitable choice of small $\varepsilon > 0$ the distance functions to p and to q have no critical point in $M \backslash B(p, r_0) \cup B(q, r_0)$. This proves Theorem 4.3.

However we cannot control the topology of small metric balls on M under the assumptions in Theorems 4.4 and 4.5. The folowing fact was established in [G-P1]. Let $K_M \geq -\kappa^2$, $d(M) \leq D$ and $\operatorname{Vol}(M) \geq v$ for given constants $\kappa, D, v > 0$. Let $\Delta \subset M \times M$ be the diagonal of $M \times M$ and for an $r > 0$ let $\Delta(r) := \{(x, y) \in M \times M; d(x, y) < r\}$. Then there exist positive constants r, R depending only on $n \geq 2$, D, v and κ with the property that there exists a differentiable strong deformation retract $H : \Delta(r) \times [0, 1] \to \Delta(r)$ onto Δ in such a way that every retraction curve $t \mapsto H(x, y, t)$ for $(x, y) \in \Delta$ has length not greater than $R \cdot d(x, y)$ and that the curve near the diagonal is a minimizing geodesic. Thus $B(p, r_0)$ for a fixed $r_0 \in (0, r)$ is contractible in $B(p, R \cdot r_0)$ (which is called the R-contractibility). This shows that $M \backslash \{p\}$ is contractible in itself to $\{q\}$, and similarly $M \backslash \{q\}$ is contractible in itself to $\{p\}$, and hence M is homotopic to S^n. The proof of Theorem 4.4 is obtained by the solution of Poincaré conjecture for $n \geq 4$ and by a well-known result due to Hamilton for $n = 3$ (see [H.R]). This completes the proofs.

It should be noted that the class of all n-manifolds with the properties

$$K_M \geq \kappa, \quad d(M) \leq D, \quad \text{and} \quad \operatorname{Vol}(M) \geq v$$

has at most finitely many distinct homotopy types (see [G-P1]).

By using a similar method it has been proved in [G-P2] that there exists for $n \geq 2$ an $\varepsilon = \varepsilon(n) > 0$ such that if the volume and sectional curvature of M satisfy $K_M \geq 1$ and $\operatorname{Vol}(M) \geq \omega_n/2 - \varepsilon$, then M has the same homotopy type of either S^n or the real projective space RP^n. This result will be strengthened in §5 to give diffeomorphism between M and RP^n.

In connection with the isoembolic inequality it is quite natural to expect that if $\mathrm{Vol}(M)/\mathrm{inj}(M)^n$ is sufficiently close to ω_n/π^n, then M will be homeomorphic to S^n. It was proved (see [Y1]) that for given n and $L > 0$ the number of homotopy types of the class of all n-manifolds with the property

$$\mathrm{Vol}(M)/\mathrm{inj}(M)^n \leq L$$

is finite. This result was strengthend by Croke (see [C.C1]) to provide a topological sphere theorem without assuming any restriction for curvature as follows.

THEOREM 4.6. *If the volume and injectivity radius of M satisfy*

$$\omega_n/\pi^n \leq \mathrm{Vol}(M)/\mathrm{inj}(M)^n \leq \omega_n/\pi^n + c(n),$$

then M is homeomorphic to S^n. Here $c(n)$ is defined by

$$c(n) := \omega_{n-1}^n/2n^n(n+1)\omega_n^{n-1}.$$

5. Differentiable sphere theorems

The first result on the differentiable sphere theorem without assuming an upper bound for the sectional curvture of M was obtained in [O-S-Y], as stated.

THEOREM 5.1. *For given integer $n \geq 2$ there exists an $\varepsilon = \varepsilon(n) > 0$ such that if the sectional curvature and volume of M satisfy*

$$K_M \geq 1, \qquad \mathrm{Vol}(M) \geq \omega_n - \varepsilon,$$

then M is diffeomorphic to S^n. Moreover there exists a constant $L = L(n, \varepsilon) > 1$ with $\lim_{\varepsilon \to 0} L(n, \varepsilon) = 1$ such that the diffeomorphism $F : M \to S^n$ satisfies

$$L^{-1} \leq \|dF\| \leq L.$$

Instead of the assumption for the volume of M the radius of M is discussed to obtain the following (see [S-Y])

THEOREM 5.2. *There exists for $n \geq 2$ an $\varepsilon = \varepsilon(n)$ such that if M has the properties*

$$K_M \geq 1, \qquad \mathrm{rad}(M) \geq \pi - \varepsilon,$$

then M is diffeomorphic to S^n. Moreover there exists a constant $L = L(n, \varepsilon) \geq 1$ with $\lim_{\varepsilon \to 0} L(n, \varepsilon) = 1$ such that the diffeomorphism $F : M \to S^n$ satisfies

$$L^{-1} \leq \|dF\| \leq L.$$

Recently a differentiable sphere theorem for the Ricci curvature was established by Yamaguchi (see [Y2]) as follows.

THEOREM 5.3. *Given an integer $n \geq 2$ and $\kappa \in \mathbf{R}$ there exists a positive $\varepsilon = \varepsilon(n, \kappa)$ such that if M has the properties*

$$\mathrm{Ric}_M \geq n - 1, \quad K_M \geq -\kappa^2, \quad \mathrm{Vol}(M) \geq \omega_n - \varepsilon,$$

then M is diffeomorphic to S^n. Moreover there exists a constant $L = L(n, \kappa, \varepsilon) \geq 1$ with $\lim_{\varepsilon \to 0} L(n, \kappa, \varepsilon) = 1$ such that the diffeomorphism $F : M \to S^n$ has the property

$$L^{-1} \leq \|dF\| \leq L.$$

Under the assumption for the Ricci curvature and volume of M we observe that the radius of M in Theorem 5.3 is close to π. In this case the assumption for the volume of M is relaxed as follows (see [O3]).

THEOREM 5.4. *There exists for given $n \geq 2$, $v > 0$ and $\kappa > 0$ an $\varepsilon = \varepsilon(n, \kappa, v) > 0$ such that if M has the properties*

$$\mathrm{Ric}_M \geq n - 1, \quad K_M \geq -\kappa^2, \quad \mathrm{Vol}(M) \geq v, \quad \mathrm{rad}(M) \geq \pi - \varepsilon,$$

then M is diffeomorphic to S^n.

The most recent result by Anderson provides a differentiable sphere theorem without bounding sectional curvature but pinching Ricci curvature (see [A2]). This will be the first result without bounding sectional curvature from below.

THEOREM 5.5. *If the Ricci curvature and volume of M satisfy*

$$C \geq \mathrm{Ric}_M \geq n - 1, \quad \mathrm{Vol}(M) \geq \omega_n - \varepsilon$$

for a constant C and for a sufficiently small $\varepsilon > 0$, then M is diffeomorphic to S^n.

SKETCH OF THE PROOFS OF THEOREMS 5.1–5.3. The basic idea of the proofs of Theorems 5.1–5.3 is as follows. First of all we prove that the Hausdorff distance $d_H(M, S^n)$ between M and S^n is sufficiently small, depending on ε. To see this fix points $p \in M$ and $p^* \in S^n$ and let $U \subset S^n \backslash \{-p^*\}$ be the disk domain which is star-shaped with respect to p^* such that for a fixed linear isometry $I : T_{p^*} \to T_p M$, the map $f : S^n \backslash \{-p^*\} \to M$ defined by

$$f := \exp_p \circ I \circ (\exp_{p^*} |B(\pi))^{-1}$$

satisfies $f(U) = M \backslash C(p)$, where $C(p) \subset M$ is the cut locus of M to p. Thus $f|U : U \to M$ is an embedding.

Let M fulfill the assumptions in Theorem 5.3 for a sufficiently small $\varepsilon > 0$. Then we observe that $\mathrm{Vol}(S^n \backslash U) \leq \varepsilon$ and that $\|\det df\| \leq 1$ on U and that for a given small positive constant δ there exists an $L = L(\kappa, \delta) > 0$ such that $\|df\| \leq L$ on $U \backslash B(p^*, \pi - \delta)$. Setting

$$A[\delta] := \{q^* \in U : \|df_{q^*}\| > 1 + \delta\}$$

and

$$B[\delta] := \{ q^* \in U : \| df_{q^*} \| < 1 - \delta \},$$

we see from Theorem 2.6 that there exists a $\tau = \tau(\delta|\varepsilon) > 0$ with $\lim_{\varepsilon \to 0} \tau(\delta|\varepsilon)$ $= 0$ such that $\text{Vol}(B[\delta]) < \tau(\delta|\varepsilon)$. Computations will show that $\text{Vol}(A[\delta]) <$ $\tau(\delta|\varepsilon)$. This is achieved by using index form comparison where we only need the assumption $\text{Ric}_M \geq n - 1$. A metric discussion on S^n applies to obtain that for $\delta \gg \sigma \gg \varepsilon > 0$ and for every $x^*, y^* \in S^n \setminus \{-p^*\}$ we have

$$|d(f(x^*), f(y^*)) - d(x^*, y^*)| < \tau(\sigma|\varepsilon).$$

The map f may be extended to a τ-Hausdorff approximation map from S^n to M, whose inverse is also a τ-Hausdorff approximation map from M to S^n.

Since δ may be chosen arbitrary small, we have proved the following

PROPOSITION. *If* $\text{Ric}_M \geq n - 1$, $K_M \geq -\kappa^2$ *and if* $\text{Vol}(M) \geq \omega_n - \varepsilon$, *then* $d_H(M, S^n) \leq \tau(\varepsilon)$.

Secondly, we shall construct an immersion of M into R^{n+1} by means of the Hausdorff closeness. The standard unit sphere S^n carries $n + 1$ natural Morse functions by which it is written as

$$S^n = \left\{ x = (x_1, \ldots, x_{n+1}) \in R^{n+1} | \sum (x_i)^2 = 1 \right\}.$$

These functions can be expressed in terms of distance functions $\varphi_1^*, \varphi_2^*, \ldots,$ φ_{n+1}^* from points $p_1^*, p_2^*, \ldots, p_{n+1}^*$ such that $p_i^* := (0, \ldots, 0, 1, 0, \ldots, 0)$ (the ith component is 1 and others zero). The natural identification Φ^* : $S^n \to S^n \subset R^{n+1}$ is given by

$$\Phi^*(x^*) = (\cos \varphi_1^*(x^*), \ldots, \cos \varphi_{n+1}^*(x^*)), \qquad x^* \in S^n.$$

If $d_H(M, S^n) < \varepsilon$ for a sufficiently small $\varepsilon > 0$ and if $f : M \to S^n$ is a Hausdorff approximation map between M and S^n, then the distance function φ_i from the point $p_i := f^{-1}(p_i^*)$ will have the same behavior as φ_i^* for all $i = 1, \ldots, n + 1$. In fact if the sectional curvature is bounded below, then the Toponogov Theorem 2.2 implies that for a fixed small number $r_0 > 0$ and for every $i = 1, \ldots, n + 1$, $d(p_i, \cdot)$ has no critical point on $M \setminus B(p_i, r_0) \cup B(p_{n+1+i}, r_0)$. This r_0 will be chosen later so as to depend only on n. If φ_i is a smooth approximation of $d(p_i, \cdot)$, then it has no critical point on $M \setminus B(p_i, r_0) \cup B(p_{n+1+i}, r_0)$. Thus a smooth map $\Phi : M \to R^{n+1}$ is defined by

$$\Phi(x) := (\cos \varphi_1(x), \ldots, \cos \varphi_n(x)), \qquad x \in M.$$

Because $|\varphi_i^* \circ f - \varphi_i| < \varepsilon$ for all $i = 1, \ldots, n + 1$, we see that $d_H(\Phi(M), S^n)$ is small, and hence $\Phi(M)$ is C^0-close to S^n. For the proof of C^1-closeness between $\Phi(M)$ and S^n we consider R^{n+1}-bundle $T_0(M) = T(M) \oplus e(M)$ over M, where $e(M)$ is the trivial line bundle over M. $T_0(M)$ is equipped with the inner product $\langle\ ,\ \rangle_0$ such that $\langle u, v\rangle_0 := \langle u, v\rangle$, $\langle u, e\rangle_0 := 0$, and $\langle e, e\rangle_0 := 1$ for $u, v \in T(M)$ and $e \in e(M)$. Let $X_i := \nabla\varphi_i$ be the gradient vector field to φ_i and $E_i : M \to T_0(M)$ the cross section defined by

$$E_i := -\sin\varphi_i \cdot X_i + \cos\varphi_i \cdot e.$$

Then $\{E_1, \dots, E_{n+1}\}$ forms a basis of $T_0(M)$ if and only if Φ is an immersion. If $X_i^* := \nabla\varphi_i^*$ and $E_i^* := -\sin\varphi_i^* \cdot X_i^* + \cos\varphi_i^* \cdot e_i^*$ is the corresponding cross section of $S^n \to R^{n+1}$, then $\{E_1^*, \dots, E_{n+1}^*\}$ is nothing but the canonical orthonormal frame field along Φ^*. Setting

$$\theta_{i,j} := \langle X_i, X_j\rangle / \|X_i\| \cdot \|X_j\|, \qquad \theta_{i,j}^* := \langle X_i^*, X_j^*\rangle$$

we observe that $\{E_1, \dots, E_{n+1}\}$ is close to an orthonromal frame field along Φ if and only if $|\theta_{i,j} - \theta_{i,j}^* \circ f|$ is close to zero for all $i, j = 1, \dots, n+1$.

The crucial point is to estimate $|\theta_{i,j} - \theta_{i,j}^* \circ f|$. The Toponogov Theorem 2.2 is employed for this estimate in the proof of Theorem 5.1 and also the generalized Toponogov Theorem 2.3 for this estimate in the proof of Theorem 5.2. However we cannot use the same idea for the proof of Theorem 5.3. In the case of Theorem 5.3 we employ the property that the composed exponential map $f_p := \exp_p \circ I \circ (\exp_{p^*} |B(\pi))^{-1}$ for every $p \in M$ is a Hausdorff approximation map, and the composed map $f_p^{-1} \circ f_x$ for every $x \in M$ is C^0-close to an isometry on S^n. By choosing for every $x \in M$ a suitable linear isometry $I_x : T_{x^*}S^n \to T_xM$ in such a way that $f_p^{-1} \circ f_x$ is C^0-close to the identity on S^n, we obtain the desired estimate.

Once the estimate of $|\theta_{i,j} - \theta_{i,j}^* \circ f|$ has been established, it then turns out that $\{E_1, \dots, E_{n+1}\}$ tends to an orthonormal frame field as $\varepsilon \to 0$, and Φ is an immersion. This Φ itself may be viewed as a cross section of M into R^{n+1}, and it is an embedding if it is transversal to $\Phi(M)$. To see the transversality we define new sections $\widetilde{E}_i : \to R^{n+1}$ for $i = 1, \dots, n+1$ as

$$\widetilde{E}_i := -\sin\varphi_i \cdot d\Phi(X_i) + \cos\varphi_i \cdot \Phi.$$

A similar computation shows $\{\widetilde{E}_1, \dots, \widetilde{E}_{n+1}\}$ is close to $\{E_1^*, \dots, E_{n+1}^*\}$, and hence forms a basis field for R^{n+1} for a sufficiently small ε. Therefore $F := \Phi/\|\Phi\|$ is an embedding and the norm of its differential can be estimated in terms of given constants. This completes the proofs.

The following theorem due to Otsu (see [O3]) plays an important role for the proof of Theorem 5.4. In the study of collapsing phenomena of the Hausdorff limit for a certain class of Riemannian manifolds it will be (in most cases) assumed that the Hausdorff limit of a sequence of manifolds belonging to this class is a smooth Riemannian manifold. This result first establishes the existence of a smooth Riemannian manifold with controlled sectional curvature and injectivity radius as the Hausdorff limit of the sequence, and also provides an affirmative answer to a problem proposed by Fukaya in his invited talk at ICM-Kyoto (see [F1]).

THEOREM. *For given $n \geq 2$, $\kappa > 0$, $d > 0$ and $\tau \in (0, 1)$ there exist positive constants $\delta = \delta(n, \kappa, d, \tau)$, $i_0 = i_0(n, \kappa, d, \tau)$ and $\Delta = \Delta(n, \kappa, d, \tau)$ with $\lim_{\tau \to 0} \delta = \lim_{\tau \to 0} i_0 = 0$ and $\lim_{\tau \to 0} \Delta = \infty$ such that if M is a complete Riemannian n-manifold (not necessarily compact) with the properties*

$$K_M \geq -\kappa^2, \quad \mathrm{rad}(M) \geq d, \quad e^d(M) \leq \delta,$$

then there exists a complete Riemannian m-manifold N with $1 \leq m \leq n$ and a fibration $F : M \to N$ with the following properties:

(1) $|K_N| \leq \Delta$, $\mathrm{inj}(N) \geq i_0$,
(2) F *is a $(\tau L, L)$-Hausdorff approximation map for every $L > i_0$,*
(3) F *is a τ-almost Riemannian submersion.*

Here F is by definition a τ-almost Riemannian submersion iff

$$\left| \frac{\|dF(\xi)\|}{\|\xi\|} - 1 \right| \leq \tau$$

for every tangent vector ξ orthogonal to the fibre of F.

OUTLINE OF THE PROOF OF THEOREM 5.4. If $\mathrm{Ric}_M \geq n-1$ and $\mathrm{rad}(M) \geq \pi - \epsilon$, then $e(M)$ (and hence $e^d(M)$ for $0 < d \leq \pi - \epsilon$) is small. The above theorem then applies to obtain a smooth Riemannian m-manifold N and a fibration $F : M \to N$ with the properties (1)–(3). It turns out that the diameter of the fibres is uniformly bounded by a small number $\delta' = \delta'(n, \kappa, d, \epsilon)$ with $\lim_{\epsilon \to 0} \delta'(n, \kappa, d, \epsilon) = 0$. Suppose that $m < n$. Then the volume of M tends to zero as $\epsilon \to 0$, a contradiction. By means of $\dim M = \dim N = n$ (namely the fibre is trivial) and $d_H(M, N)$ being small, we observe that $e(N)$ is small. A similar discussion as in the proof of Theorem 4.3 applies to see that N (and hence M) is homeomorphic to S^n. It follows from the construction of F that for small positive $r \gg \epsilon$ and for every point $p \in M$, we have for a constant $C = C(n, \kappa) > 0$ with the property

$$|\mathrm{Vol}(B(p, r)) - b_0(r)| \leq C \cdot r^{1/4} \cdot b_0(r).$$

It follows from $\mathrm{Vol}(M) \geq \mathrm{Vol}(B(p, r)) + \mathrm{Vol}(B(p', \pi - r - \epsilon))$ for $p' \in M$ with $d(p, p') \geq \pi - \epsilon$ that

$$\mathrm{Vol}(M) \geq \frac{\omega_n}{b_1(r+\epsilon)} \cdot \mathrm{Vol}(B(p, r)) \geq \frac{b_0(r)}{b_1(r+\epsilon)} \cdot (1 - Cr^{1/4}) \cdot \omega_n.$$

Thus $\mathrm{Vol}(M)$ tends to ω_n for sufficiently small choice of r. The rest is contained in the proof of Theorem 5.3.

OUTLINE OF THE PROOF OF THEOREM 5.5. First of all a uniform estimate of the radius of harmonic balls on M is established as follows. Suppose it were false. Then there exists a sequence $\{M_i\}$ of manifolds satisfying that $C \geq \mathrm{Ric}(g_i) \geq n - 1$ and $\mathrm{Vol}(g_i) \geq \omega_n - \epsilon$ for a decreasing sequence $\{\epsilon_i\}$ with $\lim \epsilon_i = 0$ such that if $r_h(M_i)$ is the radius of harmonic balls on M_i, then $\{r_h(M_i)\}$ converges to zero. Let $p_i \in M_i$ be chosen such that $r_h(p_i) =$

$r_h(M_i)$. By rescaling the metric g_i of M_i we have $r_h(M_i, h_i) = 1$ for the metric $h_i := C_i \cdot g_i$ for every i, where $C_i > 0$ is a constant. Thus every (M_i, h_i) has the uniform estimate for the radius of harmonic balls not less than 1. The pointed Hausdorff convergence of the sequence $\{(M_i, p_i, h_i)\}$ has been established in [A2], and the limit $(M_\infty, p_\infty, g_\infty)$ of this sequence is a $C^{1,\alpha}$-Riemannian manifold of the same dimension, where $0 < \alpha < 0$. The limit (M_∞, h_∞) has the properties that $r_h(M_\infty, h_\infty) = 1$.

With respect to a harmonic coordinate system $\{u^1, \ldots, u^n\}$ for a metric g, the operator $g \to \mathrm{Ric}(g)$ is the second-order elliptic partial differential equation, as stated

$$L(g) := g^{ij} \frac{\partial^2 g_{rs}}{\partial u^i \partial u^j} + Q_{rs}(g, \partial g) = \mathrm{Ric}(g)_{rs},$$

where Q is a quadratic form in the first derivatives of g. The uniform bound for the harmonic radius together with the Sobolev embedding theorem yields uniform bounds on the coefficients in the above equation for the metrics h_i. Since $|\mathrm{Ric}(h_i)| \to 0$ as $i \to \infty$, h_∞ is a weak $C^{1,\alpha}$-solution of the Einstein equation

$$h^{ij} \frac{\partial h_{rs}}{\partial u^i \partial u^j} + Q_{rs}(h, \partial h) = 0,$$

and the regularity theory implies that h is a smooth metric with respect to which the Ricci curvature is flat.

On the other hand it follows from what is supposed that, for every fixed $t > 0$,

$$\frac{\mathrm{Vol}(B(p_i, t))}{b_0(t)} \geq 1 - \varepsilon_i$$

holds on (M_i, h_i). In particular we have $\mathrm{Vol}(B(p_\infty, t))/b_0(t) = 1$ for all $t > 0$. This fact means that every geodesic emanating from p_∞ is a ray. The Ricci-flatness implies that $\lim_{t \to \infty} \mathrm{Vol}(B(q_\infty, t))/b_0(t)$ is independent of the choice of point q_∞, and hence $\mathrm{Vol}(B(q_\infty, t))/b_0(t) = 1$ holds for every $q_\infty \in M_\infty$. Therefore all geodesics in M_∞ are straight lines. The Cheeger-Gromoll splitting theorem (see [C-G]) for complete open manifolds of nonnegative Ricci curvature implies that (M_∞, h_∞) is isometric to R^n. This is ridiculous since $r_h(M_\infty, h_\infty) = 1$.

Once a uniform bound for the radius of harmonic balls has been established, the Hausdorff limit of converging sequence $\{M_i\}$ is a $C^{1,\alpha}$-Riemannian manifold, which is nothing but S^n. This completes the proof of Theorem 5.5.

REMARK 5.A. In the above theorem we do not know the dependence of ε upon the geometric data. It will be very nice to give an a priori estimate of ε in terms of C and n.

In view of the discussion developed in the proof of Theorem 5.4 together with the isoembolic inequality we can prove the following Theorem 5.6 (see

[O3]). The proof will be omitted. We only remark that the radius on M in Theorem 5.6 need not be close to π but $e^d(M) = 0$ for $0 < d \leq \rho$, and this property ensures that the volume of M is close to ω_n. We also note that a Hausdorff converging sequence $\{M_i\}$ satisfying the assumptions in Theorem 5.6 has "nonnegative Ricci curvature" in a weak sense. It is not certain whether or not the Cheeger-Gromoll splitting theorem might apply to this case.

THEOREM 5.6. *Assume that* $\mathrm{Ric}_M \geq n-1$ *and* $K_M \geq -\kappa^2$ *and* $\mathrm{inj}(M) \geq \rho$ *for given positive constants* κ *and* ρ. *Then there exists an* $\varepsilon = \varepsilon(n, \kappa, \rho) > 0$ *such that if the diameter of* M *satisfies* $d(M) \geq \pi - \varepsilon$, *then* M *is diffeomorphic to* S^n.

Theorems 5.1 and 5.3 have a real projective analogue.

THEOREM 5.7. *For given integer* $n \geq 2$, *there exists an* $\varepsilon' = \varepsilon'(n)$ *such that if the sectional curvature, diameter and volume of* M *satisfy*

$$K_M \geq 1, \quad d(M) \leq \pi/2, \quad and \quad \mathrm{Vol}(M) \geq \omega_n/2 - \varepsilon,$$

then M *is diffeomorphic to the real projective space* RP^n.

OUTLINE OF THE PROOF OF THEOREM 5.7. We use the same notations as before. Let M satisfy the assumptions in Theorem 5.7. We denote by $(\gamma_1, \gamma_2; \theta, \theta')$ a geodesic biangle consisting of minimizing geodesics γ_1, γ_2 : $[0, \ell] \to M$ with $\ell = d(p, q)$, $p = \gamma_1(0) = \gamma_2(0)$, $q = \gamma_1(\ell) = \gamma_2(\ell)$ and $\cos \theta = \langle \gamma_1'(0), \gamma_2'(0) \rangle$, $\cos \theta' = \langle \gamma_1'(1), \gamma_2'(1) \rangle$. For constants $\ell > d > 0$, $D > 0$ and $r := \ell - d$ and for points $p^*, x_1^*, x_2^* \in S^n$ we define f_p : $B(\pi/2) \to M$ by $f_p := \exp_p \circ I \circ (\exp_{p^*} |B(\pi/2))^{-1}$ and set $f_p(x_i^*) = \gamma_i(d)$, $i = 1, 2$. Then we have

$$\mathrm{Vol}[B(p, D) \backslash B(\gamma_1(d), r) \cup B(\gamma_2(d), r)] \leq \mathrm{Vol}[B(p^*, D) \backslash B(x_1^*, r) \cup B(x_2^*, r)].$$

We first prove the following

ASSERTION. *Let* $(\gamma_1, \gamma_2; \theta, \theta')$ *be a geodesic biangle in* M *of length* ℓ *with* $\ell > \pi/4$. *Then for a sufficiently small* $\varepsilon > 0$ *we have*

$$\theta < \pi/4 \quad or \quad \theta > 3\pi/4, \qquad |\theta - \theta'| \text{ is sufficiently small.}$$

PROOF OF ASSERTION. For simlplicity we set $D := \pi/2$, $r := \pi/20$. Assume that $\theta \geq \pi/4$. The above inequality implies that

$$\mathrm{Vol}[B(p, \pi/2) \backslash B(\gamma_1(d), r) \cup B(\gamma_2(d), r)] \leq \frac{1}{2}\omega_n - 2b_1(r).$$

Here we used the relation $B(x_1^*, r) \cap B(x_2^*, r) = \varnothing$, which is guaranteed by $\theta \geq \pi/4$. Theorem 2.2 implies that if $r' \in (0, \pi)$ is given by

$$\cos 2r' = \cos^2 r + \sin^2 r \cdot \cos \theta',$$

then $d(\gamma_1(d), \gamma_2(d)) \le 2r'$. If $z \in M$ is the midpoint of a minimizing geodesic joining $\gamma_1(d)$ to $\gamma_2(d)$, then $B(z, r - r') \subset B(\gamma_1(d), r) \cup B(\gamma_2(d), r)$. Thus we get

$$\text{Vol}[B(\gamma_1(d), r) \cup B(\gamma_2(d), r)] \le 2b_1(r) - \text{Vol}(B(z, r - r')).$$

It follows from $\text{Vol}(M) = \text{Vol}[B(p, \pi/2)\backslash B(\gamma_1(d), r) \cup B(\gamma_2(d), r)] + \text{Vol}[B(\gamma_1(d), r) \cup B(\gamma_2(d), r)] \le \frac{1}{2}\omega_n - \text{Vol}(B(z, r-r'))$ that $r-r' \le b_1^{-1}(\varepsilon) =: \varepsilon_1$. This fact already means that θ' is close to π. Notice that both $f_p|B(\pi/4)$ and $f_q|B(\pi/4)$ are ε_1-Hausdorff approximation maps of euclidean $\pi/4$-ball onto $B(p, \pi/4)$ and $B(q, \pi/4)$ respectively. Therefore we get

$$|d(\gamma_1(\ell/2), \gamma_2(\ell/2)) - d(f_p^{-1}(\gamma_1(\ell/2)), f_p^{-1}(\gamma_2(\ell/2)))| < \varepsilon_1$$

and

$$|d(\gamma_1(\ell/2), \gamma_2(\ell/2)) - d(f_q^{-1}(\gamma_1(\ell/2)), f_q^{-1}(\gamma_2(\ell/2)))| < \varepsilon_1.$$

In particular, applying the cosine rule of spherical trigonometry to isosceles triangles, we have $|\cos\theta - \cos\theta'| \le 2\varepsilon_1/\sin^2(\ell/2)$. This proves that $\theta \ge 3\pi/4$ for sufficiently small $\varepsilon > 0$, and the proof of the assertion is complete.

Let $\mathscr{V}(p) \subset M$ be a disk domain containing p such that its boundary $\partial\mathscr{V}(p) = \mathscr{S}(p)$ is sufficiently close to the geodesic $\pi/4$-sphere around p and diffeomorphic to S^{n-1}. Here we note that the contractibility radius of M is not less than $\pi/4$. By means of the assertion there exists a smooth unit vector field X_p defined on an open dense subset M' of $M\backslash\mathscr{V}(p)$ such that the distance function to p is strictly increasing along every flow curve to X_p which is almost minimizing. Let $\beta_x : [\pi/4, \ell(x)) \to M'$ for every $x \in \mathscr{S}(p)$ be the maximal integral curve along X_p with the initial condition $\beta_x(\pi/4) = x$. Let $\Sigma(p) := \{\lim_{s\to\ell(x)}\beta_x(s) : x \in \mathscr{S}(p)\}$ be the set of all limit points of the flow curves. Then $\Sigma(p)$ carries the structure of $(n-1)$-dimensional topological manifold. The map $\pi : \mathscr{S}(p) \to \Sigma(p)$ defined by $\pi(x) := \lim_{s\to\ell(x)}\beta_x(s)$ is locally homeomorphic, and hence a covering map. From the assertion we see that π is double covering, and clearly $\pi_1(M) \approx \mathbf{Z}_2$.

Finally the universal Riemannian covering manifold $\widetilde{M}$ of M admits an embedding $\Phi : \widetilde{M} \to R^{n+1}$ with the property that $\Phi(\widetilde{M})$ is C^1-close to S^n. If $\delta : \widetilde{M} \to \widetilde{M}$ is the deck transformation, then we have $d(x, \delta(x)) \ge \pi(1 - \varepsilon/\omega_n)$ for every point $x \in \widetilde{M}$. By setting $F := \Phi/\|\Phi\|$, we observe that $F \circ \delta \circ F^{-1} : S^n \to S^n$ is C^1-close to the antipodal map. This proves Theorem 5.7.

At the end of this section we state a result concerning with the spherical space forms proved by Yamaguchi in [**Y2**].

THEOREM 5.8. *For given integers* $n, k \ge 2$ *and* $\kappa > 0$ *there exists an* $\varepsilon = \varepsilon(n, k, \kappa) > 0$ *such that if the order of the fundamental group* $\pi_1(M)$ *of* M *is* k *and if the volume and curvature of* M *satisfy*

$$\text{Ric}_M \ge n - 1, \quad K_M \ge -\kappa^2, \quad \text{Vol}(M) \ge \omega_n/k - \varepsilon,$$

then there exists a finite subgroup Γ *of* $O(n+1)$ *such that* M *is diffeomorphic to a spherical space form* S^n/Γ.

6. Problems

We would like to state problems concerning with sphere theorems which came out through the communications with Grove, Croke, Petersen, Yamaguchi, Anderson, Shioya and Otsu to whom I like to express my thanks.

First of all Problems related to the diameter of M will be stated.

PROBLEM 1. Assume that $K_M \geq 1$. Does there exist an $\varepsilon = \varepsilon(n, \kappa) > 0$ such that if $d(M) \geq \pi - \varepsilon$, then M is diffeomorphic to S^n?

We do not even know the following easier case:

PROBLEM 2. For given positive constant v does there exist an $\varepsilon = \varepsilon(n, v)$ such that if M has the properties

$$K_M \geq 1, \quad \mathrm{Vol}(M) \geq v, \quad d(M) \geq \pi - \varepsilon,$$

then M is diffeomorphic to S^n?

PROBLEM 3. Assume that $\mathrm{Ric}_M \geq n-1$ and $K_M \geq -\kappa^2$. Does there exist an $\varepsilon = \varepsilon(n, \kappa) > 0$ such that if $d(M) \geq \pi - \varepsilon$ (or if $\mathrm{rad}(M) \geq \pi - \varepsilon$), then M is homeomorphic to S^n?

In Problems 1 and 3 collapsing phenomena might occur for a Hausdorff converging sequence of manifolds. It seems that the Hausdorff limit of manifolds in Problem 2 may have singularities at points with maximal distance.

PROBLEM 4. Assume that $K_M \geq 1$ and that there exists a geodesic triangle $\Delta = (\gamma_0, \gamma_1, \gamma_2)$ such that the edge lengths of Δ satisfy $\delta \leq L(\gamma_i) \leq \pi - \delta$ for some constant $\delta > 0$ and $i = 0, 1, 2$. Does there exist an $\varepsilon = \varepsilon(n, \delta) > 0$ such that if $L(\Delta) \geq \pi - \varepsilon$, then M is diffeomorphic to S^n?

Now we shall state problems concerning with the volume of M.

PROBLEM 5. Assume that $\mathrm{Ric}_M \geq n-1$. Does there exist an $\varepsilon = \varepsilon(n) > 0$ such that if $\mathrm{Vol}(M) \geq \omega_n - \varepsilon$, then M has the same topological type as S^n?

A result due to Croke gives raise to the following

PROBLEM 6. Assume that $\mathrm{inj}(M) \geq \rho > 0$. Does there exist an $\varepsilon = \varepsilon(n, \rho) > 0$ such that if

$$\frac{\mathrm{Vol}(M)}{\mathrm{inj}(M)^n} \leq \frac{\omega_n}{\pi^n} + \varepsilon,$$

then M is diffeomorphic to S^n?

PROBLEM 7. In the Anderson Theorem 5.5, is it possible to obtain a precise estimate of ε in terms of C and n?

PROBLEM 8. Is it possible to construct a complete metric on an open manifold M having two ends with respect to which the radial Ricci curvature with base point at $o \in M$ is positive?

Appendix: Pinching cases

We shall state several results on sphere theorems which have recently been investigated by pinching curvature of M.

The notion of pointwise-pinching sectional curvature was first introduced by Ruh (see [R.E2]) in which he proved that there exists for $n \geq 2$ a $\delta = \delta(n) \in (1/4, 1)$ such that every pointwise δ-pinched compact and connected M is diffeomorphic to a spherical space form. A similar result was recently obtained by Margerin to give a diffeomorphism between a $\delta(n)$-pinched complete M and a spherical space form (see [M.C1]). For other information see [C.C-2, D.O, M.C-2].

In the proofs of classical differentiable sphere theorem two ideas were employed. Namely, one is the diffeotopy theorem and the other is the construction of an almost flat linear connection on the R^{n+1}-bundle $T_0(M)$ over M. These two ideas were combined by Suyama in [S.Y] to discover a new pinching constant. In [S.Y] he discovered a sharp estimate for the twist of Jacobi fields by using an iteration method for the norm of curvature tensor along geodesics.

In the classical sphere theorem the estimate of the injectivity radius was proved by using Morse theory on path space. Eschenburg obtained a simple proof of it by using the property of a filtration of locally δ-convex sets shrinking to a point (see [E2]). A similar idea was discussed by Wu in [W.H] under a more general situation of k-convex sets on manifolds of positive kth Ricci curvature. A result due to Hartman in [H.P] states that if a complete and simply connected M has the properties : $n \geq 3$, $K_M \leq 1$ and the $(n-2)$th Ricci curvature of M is not less than $1/4$, then $\mathrm{inj}(M) \geq \pi$. These facts were used by Shen (see [S.Z]) to prove that if M is simply connected and if the sectional and Ricci curvature of M satisfy $K_M \leq 1$, $\mathrm{Ric}_M > \delta(n)$, then M is homeomorphic to S^n.

A sphere theorem for pinching radial sectional curvature has recently been proved by Machigashira (see [M.Y]) as follows. If the radial sectional curvature of a simply connected M with base point at $o \in M$ lies in $(1/4, 1]$, then M is homeomorphic to S^n. Moreover there exists for given $n \geq 2$ and $\kappa \leq 0$ a constant $\delta = \delta(n, \kappa) \in (1/4, 1)$ such that if $K_M \geq -\kappa^2$ and if the radial sectional curvature of a simply connected M with base point at $o \in M$ lies in $(\delta, 1)$, then M is diffeomorphic to S^n.

A classical result due to Klingenberg states that an even-dimensional compact M has injectivity radius not less than π if $0 < K_M \leq 1$. It is proved by Coghlan and Itokawa (see [C-I]) that if such an M has volume not greater than $3\omega_n/2$, then M is homeomorphic to S^n. It will very soon been shown by Itokawa that such an M is diffeomorphic to S^n if the volume of M is sufficiently close to ω_n.

References

[A.U] U. Abresch, *Lower curvature bounds, Toponogov's theorem and bounded topology*, Ann. Sci. École Norm. Sup. (4) **18** (1985), 651–670 .

[A1] M. Anderson, *Metrics of positive Ricci curvature with large diameter*, Manuscripta Math. **68** (1990), 405–415.

[A2] ――――, *Convergence and rigidity of manifolds under Ricci curvature bound*, Invent. Math. **102** (1990), 429–445.

[B1] M. Berger, *Les variétés riemanniennes 1/4-pincées*, Ann. Scuola Norm. Sup. Pisa **14** (1960), 161–170.

[B2] ――――, *An extension of Rauch's metric comparison theorem and some applications*, Illinois J. Math. **6** (1962), 700–712.

[B3] ――――, *Une borne inferieure pour le volume d'une variété riemannienne en function du rayon d'injectivité*, Ann. Inst. Fourier (Grenoble) **30** (1980), 259–265.

[B-C] R. Bishop and R. Chrittenden, *Geometry of manifolds*, Academic Press, 1964.

[C] J. Cheeger, *Finiteness theorems for Riemannian manifolds*, Amer. J. Math. **92** (1970), 61–74.

[C-E] J. Cheeger and D. Ebin, *Comparison theorems in Riemannian geometry*, Math. Library 9, North-Holland, 1975.

[C-G] J. Cheeger and D. Gromoll, *On the structure of complete manifolds of nonnegative Ricci curvature*, Ann. of Math. (2) **96** (1972), 413–443.

[C.S] S. Y. Cheng, *Eigenvalue comparison theorems and its geometric application*, Math. Z. **143** (1975), 289–297.

[C-I] L. Coghlan and Y. Itokawa, *A sphere theorem for reverse volume pinching on even-dimensional manifolds*, Proc. Amer. Math. Soc. **111** (1991), 815–819.

[C.C1] C. Croke, *An isoembolic pinching theorem*, Invent. Math. **92** (1988), 385–387.

[C.C2] ――――, *An eigenvalue pinching theorem*, Invent. Math. **68** (1982), 253–256.

[D.O] O. Durumeric, *Manifolds of almost half of the maximal volume*, Proc. Amer. Math. Soc. **104** (1988), 277–283.

[E.D] D. Elerath, *An improved Toponogov comparison theorem for non-negatively curved manifolds*, J. Differential Geom. **15** (1980), 187–216.

[E1] J. H. Eschenburg, *Diameter, volume and topology for positive Ricci curvature*, (to appear).

[E2] ――――, *Local convexity and nonnegative curvature-Gromov's proof of the sphere theorem*, Invent. Math. **84** (1986), 507–522.

[F1] K. Fukaya, *Hausdorff convergence of Riemannian manifolds and its applications*, Adv. Studies in Pure Math. **18** (1990), 143–238, North-Holland.

[F2] ――――, *Collapsing Riemannian manifolds and its application*, Proc. Intern. Congress. Math. Kyoto 1990, vol. 1, 491–500,, Springer-Verlag.

[D.G] D. Gromoll, *Differenzierbare Strukturen und Metriken positiver Krümmung auf Sphären*, Math. Ann. **164** (1966), 353–371.

[G-G1] D. Gromoll and K. Grove, *A generalization of Berger's rigidity theorem for positively curved manifolds*, Ann. Sci. École Norm. Sup. **20** (1987), 227–239.

[G-G2] ――――, *The low dimensional metric foliations of euclidean spheres*, J. Differential Geom. **28** (1988), 143–156.

[G-K-M] D. Gromoll, W. Klingenberg, and W. Meyer, *Riemannsche Geometrie im Grossen*, Lecture Notes in Math., vol. 55, Springer-Verlag, 1968.

[G.M1] M. Gromov, *Curvature, diameter and Betti numbers*, Comment. Math. Helv. **56** (1981), 179–195.

[G.M2] ――――, *Structures métriques pour les variétés riemanniennes*, rédige par Lafontaine and Pansu, Cedic/Fernand Nathan, Paris, 1980.

[G-K1] K. Grove and H. Karcher, *On pinched manifolds with fundamental group Z_2*, Compositio Math. **27** (1973), 49–61.

[G-K2] ――――, *How to conjugate C^1-close group actions*, Math. Z **132** (1973), 11–20.

[G-K-R] K. Grove, H. Karcher, and E. Ruh, *Group actions and curvature*, Invent. Math. **23** (1974), 31–48.

[G-P1] K. Grove and P. Petersen, *Bounding homotopy types by geometry*, Ann. of Math. (2) **128** (1988), 195–206.

[G-P2] ――――, *Homotopy types of positively curved manifolds with large volume*, Amer. J. Math. **110** (1988), 1183–1188.

[G-P3] ――――, *On the excess of metric spaces and manifolds* (to appear).

[G-S] K. Grove and K. Shiohama, *A generalized sphere theorem*, Ann. of Math. **106** (1977), 201–211.

[H.P] P. Hartman, *Oscillation criteria for self-adjoint second-order differential systems and "principal sectional curvature"*, J. Differential Equations **34** (1979), 326–338.

[H.R] R. Hamilton, *Three manifolds with positive Ricci curvature*, J. Differential Geom. **17** (1982), 255–306.

[H-T] C. Hodgson and J. Tysk, *Eigenvalue estimates and isoperimetric inequalities for cone manifolds* (to appear).

[IH-R] H. C. Im Hof and E. Ruh, *An equivariant pinching theorem*, Comment. Math. Helv. **50** (1975), 389–401.

[K.J] J. Katzdan, *An isoperimetric inequality and Wiedersehen manifolds*, Seminar on Differential Geometry, Ann. of Math. Stud., vol. 102, Princeton Univ. Press, Princeton, NJ, 1982, pp. 143–157.

[K.A] A. Katsuda, *Gromov's convergence theorem and its application*, Nagoya Math. J. **100** (1985), 223–230.

[K1] W. Klingenberg, *Contributions to Riemannian geometry in the large*, Ann. of Math. (2) **69** (1959), 654–666.

[K2] ____, *Über Riemannsche Mannigfaltigkeiten mit positiver Krümmung*, Comment. Math. Helv. **35** (1961), 47–54.

[K3] ____, *Manifolds with restricted conjugate locus*, Ann. of Math. (2) **78** (1963), 527–547.

[K4] ____, *Riemannian geometry*, de Gruyter Studies in Math., vol. 1, de Gruyter, 1982.

[K-S] W. Klingenberg and T. Sakai, *Injectivity radius estimate for 1/4-pinched manifolds*, Arch. Math. (Basel) **34** (1980), 371–376.

[M.Y] Y. Machigashira, *Manifolds with pinched radial curvature*, Proc. Amer. Math. Soc. (to appear).

[M.C1] C. Margerin, *Pointwise pinched manifolds are space forms*, Geometric Measure Theory-Arcata 1984 (B. Allard and F. Almgren eds.), Proc. Sympos. Pure Math., vol. 44, Amer. Math. Soc., Providence, RI, 1986, pp. 307–328.

[M.C2] ____, *Un théoréme optimal pour le pincement faible en dimension 4*, C. R. Acad. Sci. Paris Sér. I Math. **303** (1986), 887–890.

[M.J] J. Milnor, *On manifolds homeomorphic to 7-spheres*, Ann. of Math. (2) **64** (1956), 399–405.

[M1] S. B. Myers, *Riemannian manifolds in the large*, Duke Math. J. **1** (1935), 39–49.

[M2] ____, *Riemannian manifolds with positive mean curvature*, Duke Math. J. **8** (1941), 401–404.

[N] G. Nakamura, *Diameter sphere theorems for manifolds of positive Ricci curvature*, Master Thesis, Nagoya Univ (Japanese).

[O1] Y. Otsu, *On manifolds of positive Ricci curvature with large diameter*, Math. Z. **206** (1991), 255–264.

[O2] ____, *Topology of complete open manifolds with non-negative Ricci curvature*, Geometry of Manifolds (K. Shiohama, ed.), Perspect. Math., vol. 8, Academic Press, 1990, pp. 295–302.

[O3] ____, *On manifolds of small excess* (to appear in Amer. J. Math.).

[O-S-Y] Y. Otsu, K. Shiohama and T. Yamaguchi, *A new version of differentiable sphere theorem*, Invent. Math. **98** (1989), 219–228.

[R.E1] E. Ruh, *Curvature and differential structures on spheres*, Comment. Math. Helv. **46** (1971), 161–167.

[R.E2] ____, *Riemannian manifolds with bounded curvature ratios*, J. Differential Geom. **17** (1982), 643–653.

[R.B] T. B. Rushing, *Topological embeddings*, Pure Appl. Math., vol. 52, Academic Press, 1973.

[S.T] T. Sakai, *Comparison and finiteness theorems in Riemannian geometry*, Geometry of Geodesics and Related Topics (K.Shiohama, ed.), Adv. Stud. Pure Math., vol. 3, North-Holland, 1984, pp. 125–181.

[S.Z] Z. Shen, *A sphere theorem for manifolds of positive Ricci curvature*, Indiana Univ. Math. J. **38** (1989), 229–233.

[S.Y] Y. Shikata, *On the differential pinching problem*, Osaka Math. J. **4** (1967), 279–287.

[S.K] K. Shiohama, *A sphere theorem for manifolds of positive Ricci curvature*, Trans. Amer. Math. Soc. **275** (1983), 811–819 .

[S-Y] K. Shiohama and T. Yamaguchi, *Positively curved manifolds with restricted diameters*, Geometry of Manifolds (K.Shiohama, ed.), Perspect. Math., vol. 8, Academic Press, 1989, pp. 345–350.

[S-S] M. Sugimoto and K. Shiohama, *On the differentiable pinching problem*, Math. Ann. **195** (1971), 1–16.

[S.Y] Y. Suyama, *Differentiable sphere theorem by curvature pinching*, J. Math. Soc. Japan **43** (1991), 527–553.

[T] V. A. Toponogov, *Riemannian spaces having their curvature bounded below by a positive number*, Amer. Math. Soc. Transl. vol. 37, Amer. Math. Soc., Providence, RI, 1964, 291–336.

[W.A] A. Weinstein, *On the homotopy type of positively pinched manifolds*, Arch. Math. (Basel) **18** (1967), 299–301.

[W.H] H. Wu, *Manifolds of partially positive curvature*, Indiana Univ. Math. J. (1987), 528–548.

[Y1] T. Yamaguchi, *Homotopy type finiteness theorems for certain precompact family of Riemannian manifolds*, Proc. Amer. Math. Soc. **102** (1988), 660–666.

[Y2] _____, *Lipschitz convergence of manifolds of positive Ricci curvature with large volume*, Math. Ann. **284** (1989), 423–436.

[Y3] _____, *A differentiable sphere theorem for volume-pinched manifolds*, Geometry of Geodesics and Related Topics (K.Shiohama, ed.), Adv. Stud. Pure Math., vol. 3, North-Holland, 1984, pp. 183–192.

KYUSHU UNIVERSITY, JAPAN

Proceedings of Symposia in Pure Mathematics
Volume **54** (1993), Part 3

On the Excess of Open Manifolds

TAKASHI SHIOYA

The excess of a metric space can be thought of as an indicator, which uses only the distance function on the metric space, of how close the given space is to some class of model spaces. Grove and Petersen [**GP1**] studied some relations between excess and the structures of manifolds (more generally metric spaces). They treated mainly compact Riemannian manifolds and proved a form of sphere theorem. Later Otsu [**O**] studied the differentiable structure of a (not necessarily compact) Riemannian manifold having small excess on each metric ball with a fixed radius and proved that if the sectional curvature of such a manifold is bounded from below by a constant, it is diffeomorphic to a fibre bundle with compact fibre.

In this note, we introduce a new variation of excess suitable for complete open manifolds and report our own results of [**Sy**], in which we proved some splitting theorems under certain restrictions of the ideal boundaries of nonnegatively curved complete open manifolds. The excess which will be first introduced in the present note relates the ideal boundary and, by applying the results of [**Sy**], any nonnegatively curved complete open manifold of small excess turns out to be split isometrically.

1. Excess of open manifolds

Let M be a complete open Riemannian manifold and d the distance function on M induced from the metric. The *generalized Busemann function* $b_p : M \to \mathbf{R}$ *with respect to a point* $p \in M$ (see [**Wu**]) is defined by

$$b_p(x) := \limsup_{q \to \infty} (d(p, q) - d(q, x)) \quad \text{for any } x \in M,$$

where $q \to \infty$ means the distance between $q \in M$ and a fixed point in M tending to infinity. It is easy to check that

$$b_p(x) = \lim_{t \to +\infty} (t - d(x, S_t(p))),$$

1991 *Mathematics Subject Classification.* Primary 53C20.
This paper is in final form and no version of it will be submitted for publication elsewhere.

where $S_t(p)$ is the metric sphere of radius $t \geq 0$ centered at p. Here, note that the existence of this limit is implied by the monotonicity of the function $t \mapsto t - d(x, S_t(p))$. Note also that b_p is a Lipschitz continuous function with Lipschitz constant 1. According to [**Wu**, Theorem A (a)], if the sectional curvature of M is nonnegative everywhere, then b_p is convex. The *excess of M at $p \in M$* is defined to be

$$e(p) := \sup_{x \in M}(d(p, x) - b_p(x)).$$

By the triangle inequality, this is nonnegative. Define two notions of excess of M (cf. [**Sn**, §1.4; **GP1**, §5]) by

$$e(M) := \sup_{p \in M} e(p) \quad \text{and} \quad E(M) := \inf_{p \in M} e(p).$$

Note that $e(M)$ and $E(M)$ may be $+\infty$. We introduce two more versions of excess of M which are finer than the above in the sense of the ensuing remark. Define

$$e_\infty(M) := \limsup_{p \to \infty} \frac{e(p)}{d(o, p)} \quad \text{and} \quad E_\infty(M) := \liminf_{p \to \infty} \frac{e(p)}{d(o, p)}$$

for a fixed point $o \in M$. The triangle inequality implies that $e_\infty(M)$ and $E_\infty(M)$ are independent of o.

REMARK 1.1. (1) $E_\infty(M) \leq e_\infty(M) \leq e(M)$ and $E(M) \leq e(M)$.

(2) If $e(M) < +\infty$, then $e_\infty(M) = E_\infty(M) = 0$.

(3) If a point $p \in M$ satisfies $e(p) = 0$, then the exponential map $\exp_p : T_p M \to M$ at p is a diffeomorphism.

2. The topology of open manifolds with restricted excess

Let M be a complete open Riemannian manifold. Cheeger and Gromoll's theorem (see [**CG2**] and [**Sr1**]) states that if the sectional curvature K_M of M is nonnegative everywhere, then M contains a compact totally convex submanifold S over which the normal bundle is diffeomorphic to M. Such an S is called a *soul* of M.

PROPOSITION 2.1. *If $K_M \geq 0$, then the diameter of a soul of M is less than $2E(M)$. In particular, if $K_M \geq 0$ and $E(M) = 0$, then M is diffeomorphic to $\mathbf{R}^n$.*

PROOF. Let $p \in M$ be any point and let $C_{p,t} := \{x \in M ; b_p(x) \leq t\}$ for $t \geq 0$. Then, $\{C_{p,t}\}_{t \geq 0}$ is a family of compact totally convex sets. It is easy to show that

(1) if $t_1 \leq t_2$, then $C_{p,t_1} \subset C_{p,t_2}$ and

$$C_{p,t_1} = \{x \in C_{p,t_2} ; d(x, \partial C_{p,t_2}) \geq t_2 - t_1\},$$

(2) $\bigcup_{t \geq 0} C_{p,t} = M$,

(3) $p \in \partial C_{p,0}$.

According to [**CG2**], there exists a soul S_p of M entirely contained in $C_{p,0}$.

We will prove that $C_{p,0}$ is contained in the closed metric ball $\overline{B}_{e(p)}(p)$ of radius $e(p)$ centered at p. In fact, taking any $x \in C_{p,0}$, since $b_p(x) \leq 0$ we have

$$d(p, x) \leq d(p, x) - b_p(x) \leq e(p),$$

which means that $x \in \overline{B}_{e(p)}(p)$. Thus, for any point $p \in M$, the soul S_p is entirely contained in $\overline{B}_{e(p)}(p)$, so that

$$\text{diam}\, S_p \leq 2e(p),$$

where diam means the diameter. This completes the proof. $\square$

Combining the results in [O] and [Sn], we have the following

PROPOSITION 2.2. *For any integer* $n \geq 2$ *and number* $\kappa > 0$ *there exists an* $\varepsilon(n, \kappa) > 0$ *such that if a complete open* n-*dimensional Riemannian manifold* M *satisfies*

$$K_M \geq -\kappa \quad \text{and} \quad e(M) < \varepsilon(n, \kappa),$$

then M *is homeomorphic to* $S \times \mathbf{R}^k$, *where* S *is a compact manifold.*

PROOF. By [O, Lemma 4(ii)], there exist $r = r(n, \kappa) > 0$ and $\varepsilon'(n, \kappa) > 0$ such that if a complete open n-dimensional Riemannian manifold M satisfies that $K_M \geq -\kappa$ and $e(M) < \varepsilon'(n, \kappa)$, then for any point $p \in M$ the open metric ball $B_r(p)$ is diffeomorphic to $S \times B_r(o, \mathbf{R}^k)$, where S is a compact manifold independent of p and $B_r(o, \mathbf{R}^k)$ the open metric ball in $\mathbf{R}^k$ of radius r centered at the origin o.

On the other hand, [Sn, Lemma 1.2] states that if $K_M \geq -\kappa$ and $e(p) < f(d(p, q), \kappa)$ for different points $p, q \in M$, then q is not a critical point of the distance function $d(p, \cdot)$ from p, where

$$f(d, \kappa) := \frac{1}{\sqrt{\kappa}} \ln \frac{\exp \sqrt{\kappa} d}{\cosh \sqrt{\kappa} d}.$$

Hence, if $K_M \geq -\kappa$ and $e(p) < f(d, \kappa)$ for $\kappa, d > 0$, then $M - B_d(p)$ is homeomorphic to $S_d(p) \times [0, +\infty)$ (see for instance [C, Isotopy Lemma 1.4]).

Now, define the desired constant

$$\varepsilon(n, \kappa) := \min\{\varepsilon'(n, \kappa), f(r(n, \kappa), \kappa)\}$$

and assume that $K_M \geq -\kappa$ and $e(M) < \varepsilon(n, \kappa)$. Then, for any point $p \in M$, we obtain as above that $B_{r(n,\kappa)}(p)$ is diffeomorphic to $S \times B_{r(n,\kappa)}(o, \mathbf{R}^k)$ and $M - B_{r(n,\kappa)}(p)$ homeomorphic to $S_{r(n,\kappa)}(p) \times [0, +\infty)$. This completes the proof. $\square$

3. The relation between the excess and the ideal boundary

Throughout this section, let M be an asymptotically nonnegatively curved complete open Riemannian manifold with base point $o \in M$ (see [A]); i.e. there exists a continuous function $\kappa \colon [0, +\infty) \to [0, +\infty)$ such that

(1) $\int_0^{+\infty} r\kappa(r)\,dr < +\infty$,

(2) $K_M|_p \geq -\kappa(d(o, p))$ for any $p \in M$.

Kasue [K] constructed the ideal boundary $M(\infty)$ with Tits metric d_∞ of M as the set of equivalence classes of rays. Here, $(M(\infty), d_\infty)$ is independent of the base point o. To describe the properties of the ideal boundary, we need some definitions. Let (X, ρ) be a metric space. We define the *length* $L(c)$ of a continuous curve $c: [a, b] \to X$ by

$$L(c) := \sup_{a = t_0 < \cdots < t_k = b} \sum_{i=0}^{k-1} \rho(c(t_i), c(t_{i+1}))$$

and define a new metric $\rho_i: X \times X \to [0, +\infty]$ by the following. For any $x, y \in X$, if x and y are contained in a common arcwise connected component, then $\rho_i(x, y)$ is the infimum of the lengths of all continuous curves $c: [a, b] \to X$ joining x and y; otherwise $\rho_i(x, y) := +\infty$. We call ρ_i the *interior metric* of (X, ρ) (see [G, BGS]). It is easily checked that $\rho_i \geq \rho$ and $(\rho_i)_i = \rho_i$. A metric space (X, ρ) is called a *length space* if $\rho_i = \rho$.

Let d_t be the interior metric of $(S_t(o), d)$ for $t \geq 0$. Then we have

$$\lim_{\substack{H \\ t \to +\infty}} (S_t(o), d_t/t) = (M(\infty), d_\infty)$$

(see [K, Proposition 2.3 (iv)]), where $\lim_{H\,t \to +\infty}$ means the Gromov-Hausdorff limit. It is known that $(M(\infty), d_\infty)$ is a compact length space. Defining a new metric l on $M(\infty)$ by

$$l(x, y) := 2 \sin \frac{\min\{d_\infty(x, y), \pi\}}{2} \quad \text{for any } x, y \in M(\infty),$$

we have $l_i = d_\infty$ and

$$\lim_{\substack{H \\ t \to +\infty}} (S_t(o), d/t) = (M(\infty), l)$$

(see [K, Proposition 2.4]).

The relation between the ideal boundary and the excess is seen in the following proposition. The *radius* $\mathrm{rad}(X, \rho)$ of a metric space (X, ρ) is defined by

$$\mathrm{rad}(X, \rho) := \inf_{y \in X} \sup_{x \in X} \rho(x, y).$$

Obviously, $\mathrm{diam}(X, \rho)/2 \leq \mathrm{rad}(X, \rho) \leq \mathrm{diam}(X, \rho)$.

PROPOSITION 3.1. *Setting* $\varphi(x) := 2 - 2\sin(\min\{x, \pi\}/2)$, *we have*
(1) $\varphi(\mathrm{rad}\,M(\infty)) \leq e_\infty(M)$,
(2) $\varphi(\mathrm{diam}\,M(\infty)) \leq E_\infty(M)$.

PROOF. For any $p \in M$ we have

$$(*) \qquad e(p) \geq d(o, p) - b_p(o) = \liminf_{q \to \infty}(d(o, p) + d(o, q) - d(p, q)).$$

For any $q \in M$ with $d(o, q) > d(o, p)$, we take a minimal normal geodesic segment $\sigma: [0, d(o, q)] \to M$ such that $\sigma(0) = o$ and $\sigma(d(o, q)) = q$ and set $q' := \sigma(d(o, p))$. The triangle inequality implies that

$$d(p, q) \leq d(p, q') + d(q, q') = d(p, q') + d(o, q) - d(o, p).$$

By this and $(*)$ we have

$$e(p) \geq \liminf_{q \to \infty}(2d(o, p) - d(p, q')) \geq 2d(o, p) - \sup_{x \in S_{d(o, p)}(o)} d(p, x)$$

for any $p \in M$, which yields that

$$e_\infty(M) \geq 2 - \lim_{t \to +\infty} \mathrm{rad}(S_t(o), d/t) = 2 - \mathrm{rad}(M(\infty), l) = \varphi(\mathrm{rad}(M(\infty), d_\infty))$$

and

$$\begin{aligned} E_\infty(M) &\geq 2 - \lim_{t \to +\infty} \mathrm{diam}(S_t(o), d/t) \\ &= 2 - \mathrm{diam}(M(\infty), l) = \varphi(\mathrm{diam}(M(\infty), d_\infty)). \quad \Box \end{aligned}$$

4. Splitting theorems for nonnegatively curved open manifolds

Throughout this section, let M be a complete open Riemannian manifold with nonnegative sectional curvature. Let us first consider the relation between the ideal boundary of M and the structure of M. The nonnegativity of the sectional curvature of M yields that $\mathrm{diam}(M(\infty), d_\infty)$ ranges over $[0, \pi] \cup \{+\infty\}$ and that

$$\mathrm{diam}(M(\infty), d_\infty) = +\infty \Leftrightarrow M(\infty) \text{ consists of exactly two points}$$
$$\Leftrightarrow M \text{ has exactly two ends}$$
$$\Leftrightarrow M \text{ is isometric to } S \times \mathbf{R},$$

where S is a compact manifold.

PROPOSITION 4.1 [BGS, K, Sy]). (1) *If* $\mathrm{diam}(M(\infty), d_\infty) = \pi$, *then* M *is isometric to* $N \times \mathbf{R}$.

(2) *If* $\mathrm{diam}(M(\infty), d_\infty) > \pi/2$, *then* M *is diffeomorphic to* $N \times \mathbf{R}$.

(3) *If* $\mathrm{rad}(M(\infty), d_\infty) = \pi$, *then* M *is isometric to* $S \times \mathbf{R}^k$, $k \geq 2$.

Here, N *is a complete Riemannian manifold and* S *a compact Riemannian manifold.*

(1) was first stated in [BGS, §4] as an exercise and proved by Kasue [K, Proposition 4.2]. (2) follows from Greene and Shiohama's theorem. (3) is proved by using (1) repeatedly. The precise proofs of (2) and (3) were given in [Sy].

If the radius of the ideal boundary is close enough to π, then M splits isometrically as follows.

THEOREM 4.2 [Sy]. *For any integer $n \geq 2$ there exists an $\varepsilon(n) > 0$ depending only on n such that if*

$$\dim M = n \quad and \quad \mathrm{rad}(M(\infty), d_\infty) > \pi - \varepsilon(n),$$

then M is isometric to $S \times N$, where S is a compact Riemannian manifold and N a complete Riemannian manifold diffeomorphic to $\mathbf{R}^k$.

Note that N as above is not necessarily flat, but has nonnegative sectional curvature. Proposition 4.1 and Theorem 4.2 lead us to the following

CONJECTURE 4.3. *If $\mathrm{rad}(M(\infty), d_\infty) > \pi/2$, then M is isometric to $S \times N$ as in Theorem 4.2 (or diffeomorphic to $S \times \mathbf{R}^k$ for a compact manifold S).*

Concerning the excess, we have the following corollary as a direct consequence of Propositions 3.1, 4.1 and Theorem 4.2.

COROLLARY 4.4. (1) *If $E_\infty(M) = 0$, then M is isometric to $N \times \mathbf{R}$.*
(2) *If $E_\infty(M) < 2 - \sqrt{2}$, then M is diffeomorphic to $N \times \mathbf{R}$.*
(3) *If $e_\infty(M) = 0$, then M is isometric to $S \times \mathbf{R}^k$.*
(4) *For any integer $n \geq 2$ there exists an $\varepsilon(n) > 0$ depending only on n such that if*

$$\dim M = n \quad and \quad e_\infty(M) < \varepsilon(n),$$

then M is isometric to $S \times N$.

Here, N is a complete Riemannian manifold and S a compact Riemannian manifold. Moreover, in (4), N is diffeomorphic to $\mathbf{R}^k$.

Finally we state the outline of the proof of Theorem 4.2. The proof is divided into the following three steps.

STEP 1 (Construction of coordinate functions on M). Take a maximal system of points $x_1, \ldots, x_k \in M(\infty)$ such that $d_\infty(x_i, x_j) = \pi/2$ for any $i \neq j$ and take a ray γ_i corresponding to x_i for each i. Let F_i be the Busemann function with respect to γ_i; i.e.,

$$F_i(p) := \lim_{t \to +\infty} (t - d(p, \gamma_i(t))) \quad \text{for any } p \in M.$$

Each F_i is a convex and Lipschitz continuous function with Lipschitz constant 1. By the assumption that $\mathrm{rad}(M(\infty), d_\infty)$ is close enough to π, every F_i has no critical points. According to [Sr2], the notion of the gradient vector field ∇F_i of F_i is generalized at all the points where F_i is not differentiable. The assumption yields that $\{\nabla F_i\}_{i=1, \ldots, k}$ is linearly independent everywhere. By the definition of $\{x_i\}$, the map $F := (F_1, \ldots, F_k): M \to \mathbf{R}^k$ turns out to be a proper map.

STEP 2 (The diffeomorphic splitting of M). By the Riemannian convolution smoothing process, we obtain the smooth approximation f_i of F_i such that $f := (f_1, \ldots, f_k): M \to \mathbf{R}^k$ is a proper submersion. Ehresmann's fibration theorem implies that f becomes a fibration whose fibre is a compact manifold S. Thus, M is diffeomorphic to $S \times \mathbf{R}^k$.

STEP 3 (The isometric splitting of M). To prove that M splits isometrically, we need the result due to Yim [Y2]. He proved that M contains a totally geodesic submanifold isometric to $S_0 \times N$, where S_0 is a soul of M and N a complete Riemannian manifold diffeomorphic to $\mathbf{R}^{k_0}$. Here, k_0 is the dimension of the space of parallel normal vector fields along S_0. By the discussion as in [Sr3], each ∇F_i is a parallel normal vector field. Hence we have

$$k_0 \geq k.$$

On the other hand, since both of the soul S_0 and the compact manifold S as in Step 2 are homotopy equivalent to M, we have

$$\dim S_0 = \dim S.$$

Moreover, it follows that

$$\dim S + k = n \geq \dim S_0 + k_0.$$

Combining these formulas, all the inequalities become equalities and thus M turns out to be isometric to $S_0 \times N$.

REFERENCES

[A] U. Abresch, *Lower curvature bounds, Toponogov's theorem, and bounded topology.* I/II, Ann. Sci. École Norm. Sup. (4) **18** (1985), 651–670; **20** (1987), 475–502.

[AG] U. Abresch and D. Gromoll, *On complete manifolds with nonnegative Ricci curvature,* J. Amer. Math. Soc. **3** (1990), 355–374.

[BGS] W. Ballmann, M. Gromov, and V. Schroeder, *Manifolds of nonpositive curvature,* Prog. Math., vol. 61, Birkhäuser, 1985.

[C] J. Cheeger, *Critical points of distance functions and applications to geometry,* Preprint.

[CG1] J. Cheeger and D. Gromoll, *The splitting theorem for manifolds of nonnegative Ricci curvature,* J. Differential Geom. **6** (1971), 119–128.

[CG2] ____, *On the structure of complete manifolds of nonnegative curvature,* Ann. of Math. (2) **96** (1972), 413–443.

[GS1] R. Greene and K. Shiohama, *Convex functions on complete noncompact manifolds: topological structure,* Invent. Math. **63** (1981), 129–157.

[GS2] ____, *Convex functions on complete noncompact manifolds: differentiable structure,* Ann. Sci. École Norm. Sup. (4) **14** (1981), 357–367.

[GW] R. Greene and H. Wu, *On the subharmonicity and plurisubharmonicity of geodesically convex functions,* Indiana Univ. Math. J. **22** (1973), 641–653.

[G] M. Gromov, *Structures métriques pour les variétés riemanniennes,* (J. Lafontaine and P. Pansu, eds.), Textes Math. 1, Cedic/Fernand Nathan, Paris, 1981.

[GS] K. Grove and K. Shiohama, *A generalized sphere theorem,* Ann. of Math. (2) **106** (1977), 201–211.

[GP1] K. Grove and P. Petersen, *On the excess of metric spaces and manifolds,* Preprint.

[GP2] ____, *Volume comparison à la Aleksandrov,* to appear in Acta. Math.

[K] A. Kasue, *A compactification of a manifold with asymptotically nonnegative curvature,* Ann. Sci. École Norm. Sup. (4) **21** (1988), 593–622.

[O] Y. Otsu, *On manifolds of small excess,* to appear in Amer. J. Math.

[Sr1] V. A. Sharafutdinov, *Complete open manifolds of nonnegative curvature,* Sibirsk Mat. Zh. **15** (1973), 177–191.

[Sr2] ____, *The Pogorelov-Klingenberg theorem for manifold homeomorphic to $\mathbf{R}^n$,* Sibirsk Mat. Zh. **18** (1977), 915–925.

[Sr3] ____, *Convex sets in a manifold of nonnegative curvature,* Mat. Zametki **26** (1979), 129–136.

[Sn] Z. Shen, *Finite topological type and vanishing theorems for Riemannian manifolds*, Ph.D. Thesis, SUNY, Stony Brook, 1990.

[S] K. Shiohama, *Topology of complete noncompact manifolds*, Geometry of Geodesics and Related Topics (K. Shiohama, ed.), Adv. Stud. Pure Math., vol. 3, Kinokuniya, Tokyo, 1984, pp. 423–450.

[SY] K. Shiohama and T. Yamaguchi, *Positively curved manifolds with restricted diameters*, Geometry of Manifolds (K. Shiohama, ed.), Perspec. Math., vol. 8, Academic Press, 1989, pp. 345–350.

[Sy] T. Shioya, *Splitting theorems for nonnegatively curved open manifolds with large ideal boundary*, to appear in Math. Z.

[T] V. A. Toponogov, *Spaces with straight lines*, Amer. Math. Soc. Transl., vol. 37, Amer. Math. Soc., Providence, RI, 1964, pp. 287–290.

[W1] G. Walschap, *Nonnnegatively curved manifolds with souls of codimension 2*, J. Differential Geom. **27** (1988), 525–537.

[W2] ____, *A splitting theorem for 4-dimensional manifolds of nonnegative curvature*, Proc. Amer. Math. Soc. **104** (1988), 265–268.

[Wu] H. Wu, *An elementary method in the study of nonnegative curvature*, Acta Math. **142** (1979), 57–78.

[Y1] J. Yim, *Distance nonincreasing retraction on a complete open manifold of nonnegative sectional curvature*, Ann. Global Anal. Geom. **6** no. 2, (1988), 191–206.

[Y2] ____, *Space of souls in a complete open manifold of nonnegative curvature*, J. Differential Geom. **32** no. 2, (1990), 429–455.

KYUSHU UNIVERSITY, JAPAN

Proceedings of Symposia in Pure Mathematics
Volume **54** (1993), Part 3

Local Theory of Affine 2-Spheres

U. SIMON AND C. P. WANG

0. Introduction

The purpose of this paper is to study the local theory of affine 2-spheres in A_3.

For any affine 2-sphere $f : M \to A_3$ of nondegenerate Blaschke metric, there exists a 1-parameter family of affine spheres $\{f_t | t \in \mathbb{R}\}$ containing $f = f_0$ with the property that they have the same Blaschke metric and the same constant mean curvature.

Next we present an existence result: we find intrinsic conditions for a nondegenerate metric G on a simply connected surface M such that G can be realized as the Blaschke metric of an affine sphere. Additionally we show that this 1-parameter family of affine spheres is completely determined by its Blaschke metric G provided that the scalar curvature of G is negative at some point.

For any definite (i.e., locally strongly convex) affine sphere with nonzero Pick invariant, we can choose a local isothermal coordinate system $\Omega \subset \mathbb{C}$ such that the Blaschke metric is given by $G = e^{2w} |dz|^2$ and $w : \Omega \to \mathbb{R}$ satisfies the differential equation $w_{z\bar{z}} + \frac{1}{2} e^{-4w} + \frac{1}{4} H e^{2w} = 0$ for some constant H. Conversely, given any solution of this equation we can construct an affine sphere in A_3 with $G = e^{2w} |dz|^2$ as its Blaschke metric and H its mean curvature. Similarly, for any indefinite affine sphere with nonzero Pick invariant we can choose an asymptotic parameter system (u^1, u^2) such that the metric can be expressed by $G = 2e^w du^1 du^2$ and w satisfies the differential equation $\partial_2 \partial_1 w + e^{-2w} + H e^w = 0$ for some constant H, where $\partial_i = \partial / \partial u^i$. Conversely, given any solution of this equation on a simply connected domain we can construct an affine sphere in A_3 with $G = 2e^w du^1 du^2$ as its

1991 *Mathematics Subject Classification.* Primary 53A15; Secondary 53A10, 53B30.

The first author was partially supported by the GADGET program of the European Community and by the Japan Society for the Promotion of Science, and the second author by a grant from TU Berlin.

This paper is in final form and no version of it will be submitted for publication elsewhere.

Blaschke metric and H its mean curvature. Note that in case $H = 0$ one can get the general solution of the PDE's mentioned above explicitly using the Bäcklund transformations (cf. [19, p. 13–14]).

This paper is organized as follows. In §1 we summarize the notations used in this paper. In §2 we study definite affine spheres, and in §3 the indefinite case. §4 is an appendix; we give a short survey of recent results about affine spheres.

1. Notations and equiaffine invariants

Let A_3 be a real affine space of dimension 3 and $\overline{\nabla}$ its standard flat connection. We fix a determinant Det with $\overline{\nabla}(\mathrm{Det}) = 0$; then (A_3, Det) forms an equiaffine 3-space.

Let $f: M \to A_3$ be a regular immersion of a simply connected surface. f induces a pseudo-Riemannian metric G on M, called the Blaschke metric. We denote by y the equiaffine normal for M in A_3. The structure equations for M in A_3 read:

(1.1)   Gauss. $\overline{\nabla}_v df(w) = df(^1\nabla_v w) + G(v, w)y, \qquad v, w \in TM$.

(1.2)   Weingarten. $dy(v) = df(-B(v)), \qquad v \in TM$.

The affine connection $^1\nabla$ defined by (1.1) is a symmetric connection for TM. The equiaffine Weingarten operator $B: TM \to TM$ given by (1.2) defines the symmetric Weingarten form $\widehat{B}$,

$$\widehat{B}(v, w) = G(B(v), w), \qquad v, w \in TM,$$

and the mean curvature $H = \frac{1}{2}\operatorname{tr} B$.

We denote by ∇ the Levi-Civita connection of G and define the difference tensor

(1.3) $C = {}^1\nabla - \nabla$.

It is well known that C is a symmetric (2,1)-type tensor on M and $\widehat{C}$ defined by

$$\widehat{C}(u, v, w) = G(C(u, v), w)$$

is a totally symmetric tensor, called the cubic form. The apolarity condition gives

(1.4) $\operatorname{tr} C(v) = 0$

for all $v \in TM$. The Pick invariant J is defined by

(1.5) $J = \frac{1}{2}\|C\|^2$,

where the norm on tensor spaces is induced by G.

All the tensors and the functions defined above are equiaffine invariants. The fundamental uniqueness theorem states that G and C (or G and $\widehat{C}$, resp.) determine the surface uniquely modulo equiaffine equivalences of A_3.

As a consequence of the fundamental theorem we get the following uniqueness results for proper affine spheres in arbitrary dimension $n \geq 2$:

1.6. PROPOSITION. *Let* f, f' : $M_n \to A_{n+1}$ *be proper affine hyperspheres satisfying*

$$^1\nabla = {}^1\nabla'.$$

Then f, f' *are affinely equivalent.*

PROOF. The Ricci tensor ^1Ric of the first connection $^1\nabla$ satisfies $^1\text{Ric} = (n-1)HG$ on affine spheres (see [23, (2.5)]). As $H = \text{const} \neq 0$ for proper affine spheres, both metrics satisfy $G = aG'$, where $a := H'H^{-1}$. Then the Levi-Civita connections coincide: $\nabla(G) = \nabla(G')$ and thus by (1.3) $C = {}^1\nabla - \nabla(G) = {}^1\nabla' - \nabla(G') = C'$ and $\hat{C} = a\hat{C}'$.   Q.E.D.

The second induced connection $^2\nabla$ (see [23, §1]) satisfies $C = {}^2\nabla + \nabla(G)$ and also $^2\text{Ric} = (n-1)HG$ on affine spheres. This gives the analogous result:

1.7. PROPOSITION. *Let* f, f' : $M_n \to A_{n+1}$ *be proper affine hyperspheres satisfying*

$$^2\nabla = {}^2\nabla'.$$

Then f, f' *are affinely equivalent.*

2. Locally strongly convex affine spheres in A_3

Let $f : M \to A_3$ be a locally strongly convex surface in A_3. The Blaschke metric G of M is positive definite. We introduce a complex coordinate system $\Omega \subset \mathbb{C} = \{z = u^1 + u^2 i | (u^1, u^2) \in \mathbb{R}^2\}$ for M with respect to G. In Ω we can write

$$(2.1) \qquad G = e^{2w}|dz|^2 = e^{2w}((du^1)^2 + (du^2)^2)$$

for some smooth function $w : \Omega \to \mathbb{R}$. Then we have an orthonormal basis $\{e_1, e_2\}$ for TM,

$$(2.2) \qquad e_1 = e^{-w}\partial_1 f, \qquad e_2 = e^{-w}\partial_2 f.$$

The dual basis $\{\theta^1, \theta^2\}$ for $\{e_1, e_2\}$ is given by

$$\theta^1 = e^w \, du^1, \qquad \theta^2 = e^w \, du^2.$$

Let $\{C_{ij}^k\}$ be the components of the tensor C defined by (1.3) with respect to the basis $\{e_1, e_2\}$. Since $\hat{C} = \{C_{ijk}\}$ is a symmetric tensor, the apolarity condition (1.4) implies

$$(2.3) \quad C_{111} = -C_{221} = -C_{212} = -C_{122}, \qquad C_{112} = -C_{222} = C_{121} = C_{211}.$$

Thus C is completely determined by $C_{111} + iC_{112}$. The Pick invariant is given by

$$(2.4) \qquad J = 2(C_{111}^2 + C_{112}^2).$$

We introduce the Cauchy-Riemann operators

$$\frac{\partial}{\partial z} = \frac{1}{2}\left(\frac{\partial}{\partial u^1} - i\frac{\partial}{\partial u^2}\right), \qquad \frac{\partial}{\partial \bar{z}} = \frac{1}{2}\left(\frac{\partial}{\partial u^1} + i\frac{\partial}{\partial u^2}\right),$$

and define two complex functions U and E on Ω by

(2.5) $\qquad U = (C_{111} - iC_{112})e^{3w}, \; E = (B_1^1 - B_2^2 - 2iB_1^2)e^{2w},$

where $\{B_i^j\}$ are the components of the Weingarten operator B. Then we have the following structure equations in complex coordinates Ω (cf. E. Calabi [5] and C. P. Wang [28]):

(2.6) $\qquad \begin{cases} f_{zz} = 2w_z f_z + Ue^{-2w} f_{\bar{z}}, \\ f_{z\bar{z}} = \frac{1}{2}e^{2w} y, \\ f_{\bar{z}\bar{z}} = \overline{U}e^{-2w} f_z + 2w_{\bar{z}} f_{\bar{z}}, \\ y_z = -Hf_z - \frac{1}{2}Ee^{-2w} f_{\bar{z}}, \\ y_{\bar{z}} = -\frac{1}{2}\overline{E}e^{-2w} f_z - Hf_{\bar{z}}, \end{cases}$

where y is the equiaffine normal for M in A_3. The integrability conditions for (2.6) are given by

(2.7) $\qquad \begin{cases} w_{z\bar{z}} + \frac{1}{2}|U|^2 e^{-4w} + \frac{1}{4}He^{2w} = 0, \\ U_{\bar{z}} = -\frac{1}{4}e^{2w} E, \\ E_{\bar{z}} = 2H_z e^{2w} + \overline{E}Ue^{-2w}. \end{cases}$

Conversely, let (w, H, U, E) be given functions satisfying (2.7) on a simply connected domain $\Omega \subset \mathbb{C}$, where w, H are real and U, E are complex, then we can solve the linear system (2.6) with the initial condition $(f(z_0), f_z(z_0), f_{\bar{z}}(z_0), y(z_0))$ satisfying

$\qquad \begin{cases} \overline{f(z_0)} = f(z_0), \quad f_{\bar{z}}(z_0) = \overline{f_z(z_0)}, \quad \overline{y(z_0)} = y(z_0), \\ \text{Det}(f_z(z_0), f_{\bar{z}}(z_0), y(z_0)) = \frac{1}{2}ie^{2w(z_0)}, \end{cases}$

and get a locally strongly convex surface $f: \Omega \to \mathbb{R}^3$ with the Blaschke metric $G = e^{2w}|dz|^2$ and the mean curvature H. Two surfaces obtained by two such initial conditions are equiaffinely equivalent.

Using (2.7) we can easily show that

2.8. THEOREM. *Let $f, f': M \to A_3$ be two locally strongly convex surfaces with the same Blaschke metric $G = G'$ and the same Weingarten operator $B = B'$. If the umbilics of f (or f') are nowhere dense on M, then f and f' are equiaffinely equivalent.*

PROOF. We need only to show that if the umbilics on M are nowhere dense, then the tensor C is completely determined by (G, B). Given (G, B) we can choose isothermal coordinates for G. Thus the function w in (2.1), the function E defined by (2.5) and the mean curvature H are known. Since $E \neq 0$ almost everywhere on M, we can determine the function U from the last formula of (2.7). Thus C is completely determined by (2.5).   Q.E.D.

In the following we assume that M is an affine sphere. Since M is totally umbilic, we have $E \equiv 0$ on M. Thus (2.7) takes the form

(2.9)
$$\begin{cases} w_{z\bar{z}} + \frac{1}{2}|U|^2 e^{-4w} + \frac{1}{4}He^{2w} = 0, \\ U_{\bar{z}} = 0, \qquad H_z = 0. \end{cases}$$

We note that if the triple (w, U, H) satisfies (2.9), then for any constant $t \in \mathbb{R}$ the triple $(w, e^{it}U, H)$ also satisfies (2.9), which determines a 1-parameter family of affine spheres $\{f_t\}$ in A_3. Each affine sphere of this family has the same Blaschke metric G and the same mean curvature H. We call $\{f_t | t \in \mathbb{R}\}$ the 1-parameter family of affine spheres associated with the affine sphere $f : M \to A_3$. Note that if $U \equiv 0$ on M then f is a quadratic surface and $f_t = f$.

2.10. PROPOSITION. *Two locally strongly convex affine spheres* f, f' : $M \to A_3$ *are in the same family if and only if they have the same Blaschke metric and the same mean curvature.*

PROOF. We need only to show that if $G = G'$, $H = H'$, then $U = e^{it}U'$ for some constant $t \in \mathbb{R}$. We take isothermal coordinates for $G = G'$. From (2.9) we know that if $G = G'$, $H = H'$, then $U = e^{it}U'$ for some real function t. We may assume $U \not\equiv 0$. As U and U' are holomorphic, we obtain $t_{\bar{z}} = 0$. Thus t is a constant.   Q.E.D.

Now we come to study the following problems: When can a Riemannian metric G on a noncompact, simply connected surface M be realized as the Blaschke metric of an affine sphere $f : M \to A_3$? Under what conditions is the 1-parameter family of affine spheres $\{f_t\}$ associated with f completely determined by its Blaschke metric G?

Suppose that G is the Blaschke metric of an affine sphere $f : M \to A_3$, where M is noncompact and simply connected. We introduce a global isothermal coordinate system $\Omega \subset \mathbb{C}$ for M with respect to G. Then we have

(2.11)
$$G = e^{2w}|dz|^2, \qquad R = -4e^{-2w}w_{z\bar{z}},$$

where R is the Gauss curvature of G. By (2.9) we get

(2.12)
$$(R - H)e^{6w} = 2|U|^2.$$

Thus R is bounded below by H. We assume that $R \neq \text{const}$. Since U is holomorphic, the zeros of $R - H$ form a discrete set S. As $\partial_z \partial_{\bar{z}} \ln|U|^2 = 0$, by (2.11) and (2.12) we have

(2.13)
$$\Delta \ln(R - H) = 6R \quad \text{on } M - S,$$

where Δ denotes the Laplacian of G (cf. Blaschke [1, p. 211]). Now let $r : S^1 \to M - S$ be a closed curve in $M - S$. Since

$$\frac{\partial}{\partial z}\ln((R - H)e^{6w}) = \frac{\partial}{\partial z}\ln U$$

is holomorphic on $M - S$, we know that

$$(2.14) \qquad \frac{1}{2\pi i} \oint_r \frac{\partial}{\partial z} \ln((R - H)e^{6w}) \, dz = \frac{1}{2\pi i} \oint_r d(\arg U) = n(r)$$

is an integer which depends only on the homology class of r in $M - S$. Thus for the Riemannian metric G with nonconstant curvature, (2.13) and (2.14) are necessary conditions for G to be the Blaschke metric for an affine sphere. We can prove that

2.15. THEOREM. *A Riemannian metric G with nonconstant curvature R on a noncompact simply connected surface M is the Blaschke metric for an affine sphere $f : M \to A_3$ if and only if the following intrinsic conditions on G are satisfied:*

(i) *There exists a constant H with $H \le \inf_M R$ such that the zeros of $R - H$ form a discrete set S, and in $M - S$ the formula $\Delta \ln(R - H) = 6R$ holds.*

(ii) *For any simple closed curve $r : S^1 \to M - S$ the value*

$$\frac{1}{2\pi i} \oint_r \frac{\partial}{\partial z} \ln((R - H)e^{6w}) \, dz$$

is an integer; here z is an isothermal coordinate for G such that $G = e^{2w} |dz|^2$.

2.16. REMARK. The condition (i) implies that $\frac{\partial}{\partial z} \ln((R - H)e^{6w})$ is holomorphic on $M - S$, thus the value $\frac{1}{2\pi i} \oint_r \frac{\partial}{\partial z} \ln((R - H)e^{6w}) \, dz$ depends only on the homology class of r in $M - S$.

2.17. COROLLARY. *A Riemannian metric G on a noncompact simply connected surface M is the Blaschke metric for an affine sphere $f : M \to A_3$ with nonzero Pick invariant if and only if there exists a constant H such that $R - H > 0$ and $\Delta \ln(R - H) = 6R$.*

PROOF OF THEOREM 2.15. Let G be a Riemannian metric on a noncompact simply connected surface M. Suppose that G satisfies the conditions (i) and (ii) in Theorem 2.15. We introduce a global isothermal coordinate system such that $G = e^{2w} |dz|^2$. Then the formula $\Delta \ln(R - H) = 6R$ on $M - S$ is equivalent to the fact that $\frac{\partial}{\partial z} \ln((R - H)e^{6w})$ is holomorphic on $M - S$. We define a holomorphic function U on $M - S$ by

$$U = \exp\left(\int_{z_0}^z \frac{\partial}{\partial z} \ln((R - H)e^{6w}) \, dz \right), \qquad z_0 \in M - S.$$

U is well defined on $M - S$ because of the condition (ii). Furthermore, from the identity

$$\frac{\partial}{\partial z} \ln |U|^2 = \frac{\partial}{\partial z} \ln U = \frac{\partial}{\partial z} \ln((R - H)e^{6w})$$

we know that $|U|^2 = c(R - H)e^{6w}$ on $M - S$ for some constant $c > 0$. Thus U is bounded around any point in S. We can extend U to a holomorphic function on M by defining $U = 0$ on S. Since $R = -4e^{-2w}w_{z\bar{z}}$, from the identity $|U|^2 = c(R - H)e^{6w}$ we get

$$w_{z\bar{z}} + \frac{1}{2}|U'|^2 e^{-4w} + \frac{1}{4}He^{2w} = 0,$$

where $U' = (\frac{c}{2})^{1/2}U$ is a holomorphic function on M. Therefore there exists an affine sphere $f: M \to A_3$ with $G = e^{2w}|dz|^2$ as its Blaschke metric and H as its mean curvature.   Q.E.D.

Now let G be a Riemannian metric which can be realized as the Blaschke metric for some affine spheres. We denote by O_G the set of affine spheres with G as their Blaschke metric. From Proposition 2.10 we know that O_G contains at least a 1-parameter family of affine spheres, all of which have the same constant mean curvature.

2.18. PROPOSITION. *If $R \not\equiv 0$, then O_G contains at most two 1-parameter families of affine spheres.*

PROOF. By Proposition 2.10 we need only to show that the (constant) mean curvature of the affine spheres in O_G can take at most two values. If $R \not\equiv 0$, then (2.13) is equivalent to the quadratic equation

(2.19) $$6R(R - H)^2 - \Delta R(R - H) + |\nabla R|^2 = 0.$$

Thus H can take at most two values.   Q.E.D.

2.20. REMARK. If $R \equiv 0$, it is well known that O_G consists of the affine spheres $\{x_1 x_2 x_3 = c | (x_1, x_2, x_3) \in A_3, \ c > 0\}$ and the affine spheres $\{x_3 = c(x_1^2 + x_2^2) | (x_1, x_2, x_3) \in A_3, \ c > 0\}$.

2.21. PROPOSITION. *Assume that $R \not\equiv 0$ and O_G consist of exactly two 1-parameter families of affine spheres $\{f_t\}$ and $\{f_t'\}$ with constant mean curvature H and H', respectively. Then we have*

(2.22)
$$\Delta R - 12R^2 = -6(H + H')R,$$
$$(\Delta R)^2 - 24R|\nabla R|^2 = 36(H - H')^2 R^2.$$

Moreover, $R \geq 0$.

PROOF. (2.22) follows immediately from (2.19). Now we prove $R \geq 0$. Again from (2.19) we get

$$(R - H)(R - H') = \frac{|\nabla R|^2}{6R}.$$

Since $R - H = J \geq 0$ and $R - H' = J' \geq 0$ we conclude that $R \geq 0$. Q.E.D.

2.23. COROLLARY. *Let G be the positive definite Blaschke metric of an affine sphere. If $R < 0$ at some point, then G determines uniquely a 1-parameter family of affine spheres.*

2.24. THEOREM. *Let* f, $f': M \to A_3$ *be two locally strongly convex affine spheres which are not quadrics. If* $C = C'$, *then* f *and* f' *are affinely equivalent.*

PROOF. From (2.3) we can calculate directly that

$$(2.25) \qquad \sum_{ijkl} C^l_{ik} C^k_{jl} \theta^i \theta^j = JG.$$

Since the left-hand side is completely determined by the cubic form C and $C = C'$, we can write $G' = e^{2v} G$ for some real function v. Now let $\{e_1, e_2\}$ be an orthonormal basis for G with dual basis $\{\theta^1, \theta^2\}$; then $\{e'_1, e'_2\} = \{e^{-v} e_1, e^{-v} e_2\}$ is an orthonormal basis for G' with dual basis $\{e^v \theta^1, e^v \theta^2\}$. From the identity

$$C = \sum_{ijk} C^k_{ij} \theta^i \theta^j e_k = \sum_{ijk} C_{ijk} \theta^i \theta^j e_k = C' = \sum_{ijk} C'_{ijk} \theta'^i \theta'^j e'_k$$

we obtain $C'_{ijk} = e^{-v} C_{ijk}$. Thus by (2.5) we have

$$U' = (C'_{111} - iC'_{112}) e^{3(w+v)} = e^{2v}(C_{111} - iC_{112}) e^{3w} = e^{2v} U.$$

As f and f' are affine spheres, the functions U and U' ($\not\equiv 0$) are holomorphic, so v is a constant. It is easy to see that $e^{4v} f$ and f' are equiaffinely equivalent. Q.E.D.

At any point p with $J(p) \neq 0$ we have $U(p) \neq 0$. Since U is holomorphic for an affine sphere, we can find a complex coordinate system z around p such that $U \equiv 1$ (note that $U \, dz^3$ is an invariant cubic form). Thus we have

2.26. THEOREM. (i) *Let* M *be an affine sphere in* A_3 *with Blaschke metric* G *and mean curvature* H. *Then around any point with nonzero Pick invariant we can find complex coordinates* z *with respect to* G *such that* $G = e^{2w}|dz|^2$, *and the function* w *is a solution of the differential equation*

$$(2.27) \qquad w_{z\bar{z}} + \frac{1}{2} e^{-4w} + \frac{1}{4} H e^{2w} = 0.$$

(ii) *Let* $w: \Omega \to \mathbb{R}$ *be a solution of the above equation with some constant* H *on a simply connected domain* $\Omega \subset \mathbb{C}$. *Then there exists an affine sphere* $f: \Omega \to A_3$ *with nonzero Pick invariant such that* $G = e^{2w}|dz|^2$ *is its Blaschke metric and* H *its mean curvature.*

3. Indefinite affine spheres in A_3

In this section we study affine spheres in A_3 with indefinite Blaschke metric. We call such affine spheres indefinite affine spheres.

Let $f: M \to A_3$ be an affine sphere. For G we introduce now asymptotic parameters $(u^1, u^2) \in \Omega \subset \mathbb{R}$:

$$(3.1) \qquad G = 2F \, du^1 \, du^2,$$

where $F : \Omega \to \mathbb{R}$ is a positive function.

Let $\{C_{ijk}\}$ be the components of the tensor $\widehat{C}$ with respect to the Gauss basis $\partial_1 f$, $\partial_2 f$. Since $\widehat{C}$ is totally symmetric, the apolarity condition (1.4) implies

$$C_{12k} = 0, \qquad k = 1, 2.$$

So there exist two real functions φ, $\psi : \Omega \to \mathbb{R}$ such that

(3.2) $\qquad C_{111} = \varphi, \quad C_{222} = \psi \quad \text{and} \quad C_{ijk} = 0 \text{ otherwise,}$

and

(3.3) $\qquad C_{11}^2 = F^{-1}\varphi, \quad C_{22}^1 = F^{-1}\psi \quad \text{and} \quad C_{ij}^k = 0 \text{ otherwise.}$

We calculate the components Γ_{ij}^k of the Levi-Civita connection ∇ of G and the components ${}^1\Gamma_{ij}^k$ of the connection ${}^1\nabla = \nabla + C$:

(3.4) $\quad \Gamma_{11}^1 = F^{-1}\partial_1 F, \quad \Gamma_{22}^2 = F^{-1}\partial_2 F \quad \text{and} \quad \Gamma_{ij}^k = 0 \text{ otherwise,}$

(3.5) $\quad \begin{cases} {}^1\Gamma_{11}^1 = F^{-1}\partial_1 F, \quad {}^1\Gamma_{11}^2 = F^{-1}\varphi, \quad {}^1\Gamma_{22}^1 = F^{-1}\psi, \\ {}^1\Gamma_{22}^2 = F^{-1}\partial_2 F \quad \text{and} \quad {}^1\Gamma_{ij}^k = 0 \text{ otherwise.} \end{cases}$

The structure equations for an affine sphere read

(3.6) $\quad \begin{cases} \partial_i \partial_j f = {}^1\Gamma_{ij}^k \partial_k f + G_{ij}y, \\ \partial_i y = -H\partial_i f. \end{cases}$

This system has the following integrability conditions:

(3.7) $\quad \begin{cases} \partial_1 \partial_2 \ln F + \varphi\psi F^{-2} + HF = 0, \\ \partial_2 \varphi = 0, \qquad \partial_1 \psi = 0. \end{cases}$

Since the scalar curvature R of G satisfies $R = -F^{-1}\partial_1\partial_2 F$ and $R = J + H$, (3.7) implies

(3.8) $\qquad\qquad\qquad J = \varphi\psi F^{-3}.$

We interpret now φ, ψ as equiaffine invariants; for this purpose we define two differential forms

(3.9) $\quad \begin{aligned} U_1 &= \mathrm{Det}(\partial_1\partial_1 f, \partial_1 f, y)(du^1)^3, \\ U_2 &= \mathrm{Det}(\partial_2\partial_2 f, \partial_2 f, y)(du^2)^3. \end{aligned}$

It is easy to see that U_1, U_2 are independent of the choice of the asymptotic parameters (u^1, u^2). Both forms are globally defined on M. From the structure equation (3.6) we get

$$U_1 = {}^1\Gamma_{11}^2 \mathrm{Det}(\partial_2 f, \partial_1 f, y)(du^1)^3 = 2\varphi \mathrm{Det}(e_1, e_2, y)(du^1)^3,$$

where $e_1 = \frac{1}{2}F^{-1/2}(\partial_1 f + \partial_2 f)$, $e_2 = \frac{1}{2}F^{-1/2}(\partial_1 f - \partial_2 f)$. Since $G(e_1, e_1) = -G(e_2, e_2) = 1$ and $G(e_1, e_2) = 0$ we have $\mathrm{Det}(e_1, e_2, y) = 1$. Thus we obtain

$$(3.10) \qquad U_1 = 2\varphi(du^1)^3, \qquad U_2 = 2\psi(du^2)^3,$$

both forms are equiaffine invariants.

From (3.1) and (3.2) we can easily see

3.11. THEOREM. *Two indefinite affine spheres* f, $f' : M \to A_3$ *are equiaffinely equivalent if and only if there exists a diffeomorphism* $\sigma : M \to M$ *such that*

$$G' = \sigma^* G, \quad U_1' = \sigma^* U_1, \quad U_2' = \sigma^* U_2.$$

Given an indefinite metric G on a noncompact simply connected surface M, we obtain two families of asymptotic curves on M by $G(v, w) = 0$. We denote by (u^1, u^2) the asymptotic parameters for M with respect to G and call two real cubic forms (U_1, U_2) a *pair of asymptotic forms* with respect to G if

$$U_1 = \varphi(du^1)^3, \qquad U_2 = \psi(du^2)^3$$

for some functions φ, ψ on M such that

$$\partial_2 \varphi = \partial_1 \psi = 0.$$

For any indefinite affine sphere $f : M \to A_3$, we can define a triple (G, U_1, U_2) such that G is its Blaschke metric and (U_1, U_2) defined by (3.10) is a pair of asymptotic forms with respect to G. Moreover, we have

$$(3.12) \qquad R - \varphi\psi F^{-3} = H = \mathrm{const}.$$

Conversely, one can easily show the following result:

3.13. THEOREM. *Given a triple* (G, U_1, U_2) *satisfying* (3.12) *for some constant* H, *where* (U_1, U_2) *is a pair of asymptotic forms with respect to the indefinite metric* G, *we can construct an affine sphere* $f : M \to A_3$ *with* G *as its Blaschke metric and* H *its mean curvature.*

We note that if (G, U_1, U_2) is a triple satisfying (3.12), then, for any number $t \in \mathbb{R}$ $(t \neq 0)$, $(G, tU_1, t^{-1}U_2)$ also satisfies (3.12). By Theorem 3.13 each triple determines an indefinite affine sphere $f_t : M \to A_3$. We will call $\{f_t | t \in \mathbb{R}\}$ the 1-parameter family of affine spheres associated with f.

3.14. REMARK. The affine spheres in such a 1-parameter family have the same Blaschke metric and mean curvature.

3.15. PROPOSITION. *Let* f, $f' : M \to A_3$ *be two affine spheres with the same indefinite Blaschke metric and mean curvature. If* $J \neq 0$, *then* $f' = f_t$ *for some* $t \in \mathbb{R}$.

PROOF. Assume $G = G'$, $H = H'$. We choose the same asymptotic parameters (u^1, u^2) for $G = G'$. From (3.12) we get

$$\varphi\psi = \varphi'\psi' \neq 0.$$

Since $\varphi = \varphi(u^1)$, $\varphi' = \varphi'(u^1)$, $\psi = \psi(u^2)$ and $\psi' = \psi'(u^2)$, we obtain $\varphi' = t\varphi$ and $\psi' = t^{-1}\psi$ for some constant $t \in \mathbb{R}$, $t \neq 0$. Thus $(G', U_1', U_2') = (G, tU_1, t^{-1}U_2)$, and $f' = f_t$.   Q.E.D.

3.16. THEOREM. *Let G be an indefinite metric on a noncompact simply connected surface M. Then G is the Blaschke metric for some affine sphere $f: M \to A_3$ with $J \neq 0$ if and only if there exists a constant H such that $R - H \neq 0$ on M and*

$$(3.17) \qquad \Delta \ln(R-H)^2 = 12R.$$

PROOF. It was proved by Blaschke (see [1, p. 211]) that the Blaschke metric G for an affine sphere satisfies (3.17). Conversely, suppose that G is an indefinite metric on a noncompact simply connected surface M satisfying (3.17) for some constant H with $R - H \neq 0$ on M; we choose asymptotic coordinates (u^1, u^2) for G. From $\Delta = 2F^{-1}\partial_1\partial_2$ and $R = -F^{-1}\partial_2\partial_1 F$ we have

$$\partial_1\partial_2 \ln((R-H)F^3)^2 = 0.$$

Then there exist functions $\varphi = \varphi(u^1) \neq 0$, $\psi = \psi(u^2) \neq 0$ such that

$$R - H = F^{-3}\varphi(u^1)\psi(u^2).$$

We define $U_1 = 2\varphi(u^1)(du^1)^3$ and $U_2 = 2\psi(u^2)(du^2)^3$, then (G, U_1, U_2) is a triple satisfying the condition of Theorem 3.13. Therefore we can construct an affine sphere $f: M \to A_3$ with G as its Blaschke metric.   Q.E.D.

As in the definite case we can prove

3.18. PROPOSITION. *Let G be a nonflat indefinite metric on M. Then*

(i) *there exist at most two families of affine spheres with nonzero Pick invariant and with G as its Blaschke metric;*

(ii) *if there exist exactly two different families of affine spheres with nonzero Pick invariant and with G as their Blaschke metric, then necessarily*

$$\Delta R - 12R^2 = cR, \qquad (\Delta R)^2 - 24R|\nabla R|^2 = c'R^2$$

for some constants c and c'.

3.19. THEOREM. *Two indefinite affine spheres f, $f' : M \to A_3$ with nonzero Pick invariants are affinely equivalent if and only if $C = C'$, where C is the tensor defined by* (1.3).

PROOF. We choose an asymptotic coordinate system (u^1, u^2) for the Blaschke metric G of f. Then by (3.3) we get

$$\sum_{ijkl} C_{il}^k C_{kj}^l du^i du^j = JG.$$

Since the left-hand side is completely determined by the cubic form C, we conclude that $G' = e^v G$ for some function v on M. Thus (u^1, u^2) are also asymptotic parameters for G'. Therefore we can write $G = 2F\,du^1\,du^2$, $G' = 2F'\,du^1\,du^2$ and $F' = e^v F$. But

$$F^{-1}\varphi = C_{11}^2 = C_{11}'^2 = F'^{-1}\varphi' = e^{-v}F^{-1}\varphi',$$

so we have $\varphi' = e^v\varphi$. From the fact $\partial_2\varphi = \partial_2\varphi' = 0$ we get $\partial_2(e^v) = 0$. Similarly from $C_{22}^1 = C_{22}'^1$ we can get $\partial_1(e^v) = 0$. Thus $v = \text{const}$. It is easy to see that f' is equiaffinely equivalent to $e^{2v}f$.   Q.E.D.

At the end we are going to prove an analogue to Theorem 2.26. Let $f: M \to A_3$ be an indefinite affine sphere with $J \neq 0$. Let (u^1, u^2) be asymptotic parameters for the Blaschke metric G of f. As $U_1 = 2\varphi(du^1)^3$ and $U_2 = 2\psi(du^2)^3$ are global forms on M such that $\partial_2\varphi = \partial_1\psi = 0$ and $J = F^{-3}\varphi\psi \neq 0$, we can choose new asymptotic parameters $(\bar{u}^1, \bar{u}^2)$ such that $U_1 = 2(d\bar{u}^1)^3$ and $U_2 = 2(d\bar{u}^2)^3$. Thus $\bar{\varphi} = \bar{\psi} = 1$ and we have

3.20. THEOREM. (i) *Let* $f: M \to A_3$ *be an indefinite affine sphere with* $J \neq 0$. *Then we can choose asymptotic coordinates* (u^1, u^2) *with respect to the Blaschke metric* G *of* f *such that* $G = 2e^w\,du^1\,du^2$, $U_1 = 2(du^1)^3$, $U_2 = 2(du^2)^3$ *and the function* w *satisfies*

$$(3.21) \qquad \partial_1\partial_2 w + e^{-2w} + He^w = 0.$$

(ii) *Given any solution* w *of* (3.21) *for some constant* H *on a simply connected domain* $\Omega \subset \mathbb{R}^2$, *there exists an indefinite affine sphere* $f: \Omega \to A_3$ *such that* $G = 2e^w\,du^1\,du^2$ *is the Blaschke metric of* f, H *its constant mean curvature and* $J = e^{-3w}$.

3.22. REMARK. (i) If $H = 0$, the general solution w of (3.21) can be represented by two arbitrary 1-variable functions $g(u^1)$, $h(u^2)$ as follows:

$$w = \ln\left(\frac{\frac{c}{2}\int_{2u_0^1}^{2u^1}\exp(g(u))du + \frac{1}{c}\int_{u_0^2}^{u^2}\exp(-h(u))du}{\exp(\frac{1}{2}(g(2u^1) - h(u^2)))}\right).$$

where $c > 0$, u_0^1, u_0^2 are constants (cf. [19, p. 12–14]; we thank B. Palmer for pointing out this reference to us). From this integral we get the following consequence: any improper affine sphere with $J \neq 0$ with indefinite metric can be uniquely described in terms of two functions $g(u^1)$, $h(u^2)$ in asymptotic coordinates (u^1, u^2) for the metric.

(ii) The foregoing remark is related to the representation of indefinite affine surfaces using the Lelieuvre formulas (see Blaschke [1, §52]).

4. Appendix: Recent results about affine spheres

The commentary [26] to Blaschke's collected works [2] gives in §§III.4.2 and III.5.5 a survey about affine spheres up to 1984. The survey is continued in §4 of [25]. At the Leeds Conference K. Nomizu gave the report [17].

The §4 of it concerns affine spheres. In the following, we give additional references for certain topics:

(1) **Compact affine spheres with boundary.** Results are reviewed in [24].

(2) **Locally strongly convex complete affine spheres.** In addition to the survey given in §III.5.5 of the commentary to Blaschke's collected works there are two papers of A. M. Li (see [10, 11]) who found a gap in the proof for the global classification given so far. Li finally could give a correct proof.

(3) **Affine spheres with constant sectional curvature.** In dimension 2 this class is classified due to recent work of Li-Penn [14], Magid-Ryan [15] and Simon [21]. In dimension 3 the classification was partly given by Yu [29] and Magid-Ryan [16]. The last paper investigates the case of indefinite metrics and nonzero Pick invariant J (note that $J = 0$ in this case does not imply that the affine sphere is a quadric).

In arbitrary dimension the case of definite metric was first solved by L. Vrancken (see [27]).

The indefinite case is an open problem for dimension $n > 3$.

(4) **Examples of improper affine spheres in dimension 2.** M. Kozlowski gives in four papers [7, 8, 9, 30] special solutions of Monge-Ampère equations. These solutions describe improper affine spheres.

REFERENCES

1. W. Blaschke, *Vorlesungen über Differentialgeometrie*. II, Affine Differentialgeometrie, Springer, 1923.
2. ____, *Gesammelte Werke*, Thales Verlag, Essen, 6 vols., 1982–1986.
3. E. Calabi, *Improper affine hyperspheres of convex type and a generalization of a theorem by K. Joergens*, Michigan Math. J. **5** (1985), 105–126.
4. ____, *Complete affine hypersurfaces*. I. Symposia Math. **10** (1972), 19–38.
5. ____, *Convex affine maximal surfaces*, Results Math. **13** (1988), 199–223.
6. S. Y. Cheng and S. T. Yau, *Complete affine hypersurfaces*. I: *The completeness of affine metrics*, Pure Appl. Math. **39** (1986), 839–866.
7. M. Kozlowski, *The Monge-Ampère equation in affine differential geometry*. Anz. Österreich. Akad. Wiss. Math.-Natur. Kl. **126** (1988), 95–96.
8. ____, *One parameter families of improper affine spheres*, Anz. Österreich. Akad. Wiss. Math.-Natur. Kl. **126** (1989), 81–82.
9. ____, *Improper affine spheres*, Anz. Österreich. Akad. Wiss. Math.-Natur. Kl. **125** (1988), 95–96.
10. A. M. Li, *Calabi conjecture on hyperbolic affine hyperspheres*, Math. Z. **203** (1990), 483–491.
11. ____, *Calabi conjecture on hyperbolic affine hyperspheres*. (2). Math. Ann. **293** (1992), 485–493.
12. ____, *Affine completeness and Euclidean completeness*, Lecture Notes in Math., vol. 1481, Springer-Verlag, 1991, pp. 115–125.
13. ____, *Some theorems in affine differential geometry*, Acta Math. Sinica **5** (1989), 345–354.
14. A. M. Li and G. Penn, *Uniqueness theorems in affine differential geometry*. II, Results Math. **13** (1988), 308–317.
15. A. Magid and P. J. Ryan, *Flat affine spheres in R^3*, Geom. Dedicata **33** (1990), 277–288.
16. ____, *Affine 3-spheres with constant affine curvature*, Trans. Amer. Math. Soc. (to appear)

17. K. Nomizu, *A survey of recent results in affine differential geometry*, Geometry and Topology of Submanifolds III (L. Verstraelen and A. West, eds.), (Leeds Conf. 1990), World Scientific, Singapore 1991, pp. 227–256.

18. K. Nomizu and U. Pinkall, *Cayley surfaces in affine differential geometry*, Tôhoku Math. J. **41** (1989), 589–596.

19. C. Rogers and W. F. Shadwick, *Bäcklund transformations and their applications*, Mathematics in Science and Engineering, vol 161, Academic Press, 1982.

20. R. Schneider, *Zur affinen Differentialgeometrie im Großen. II: Über eine Abschätzung der Pickschen Invariante auf Affinesphären*, Math. Z. **102** (1969), 1–8.

21. U. Simon, *Local classification of two dimensional affine spheres with constant curvature metric*, J. Differential Geom. Appl. (Brno) **1**, (1991), 123–132.

22. ____, *Hypersurfaces in affine differential geometry*, Geom. Dedicata 17 (1984) 157–168.

23. ____, *The fundamental theorem in affine hypersurface theory*, Geom. Dedicata **26** (1988), 125–137.

24. ____, *Dirichlet problems and the Laplacian in affine differential geometry*, Lecture Notes in Math. vol. 1369, Springer-Verlag, 1989 pp. 243–260.

25. ____, *Recent developments in affine differential geometry*, Proc. Conf. Differential Geom. and its Appl. (Dubrovnik, Yugoslavie), June 26–July 3, 1988, (N. Bokan, I. Comic, J. Nikic and M. Prvanovic, eds.), Novi Sad, 1988, pp. 327–247.

26. ____, *Zur Entwicklung der affinen Differentialgeometrie nach Blaschke*, in W. Blaschke's Gesammelte Werke, Bd. IV, Thales Verlag, 1985, pp. 35–88.

27. L. Vrancken, A. M. Li, and U. Simon, *Affine spheres with constant sectional curvature*, Math. Z. **206** (1991), 651–658.

28. C. P. Wang, *Some examples of complete hyperbolic affine 2-spheres in R^3*, Lecture Notes in Math. vol. 1481, Springer-Verlag, 1991, pp. 271–280.

29. Yu Jianhui, *Affine spheres with constant sectional curvature in A^4*, Sichuan Univ., 1989.

30. M. Kozlowski, *Some improper affine spheres in A_3*, Proc. Conf. Global Analysis Global Diff. Geometry (TU Berlin, 1990), Lecture Notes in Math. vol. 1481, Springer-Verlag, 1991, pp. 104–107.

TU BERLIN, MA 8-3, GERMANY

NANKAI UNIVERSITY, PEOPLE'S REPUBLIC OF CHINA

Proceedings of Symposia in Pure Mathematics
Volume **54** (1993), Part 3

Riemannian Manifolds
with Completely Integrable Geodesic Flows

R. J. SPATZIER

1. Introduction

Last century many mathematicians hoped that most Hamiltonian systems were completely integrable. That complete integrability is an exceptional phenomenon was first realized by Poincaré. After this turnabout little work was done until the discovery of KAM-theory which describes the dynamics of perturbations of completely integrable systems. The subject has since enjoyed a renaissance. A large number of completely integrable systems were discovered in the last three decades. For geodesic flows, however, the situation is more restrictive. Hardly any new examples had been exhibited in the time between the classical systems and a handful of geodesic flows on homogeneous spaces found in the late 70s and early 80s. Recently some more general examples have been found by G. P. Paternain and the author.

The paper begins by reviewing the basic theory and the classical examples. We then explain how symmetry gives rise to completely integrable geodesic flows on certain homogeneous spaces. Finally we describe how we used Riemannian submersions and glueing in our recent constructions. We close with some remarks about the topology of spaces with completely integrable geodesic flows.

2. Complete integrability

Let us first review the basic theory of completely integrable flows.

Let X^{2n} be a symplectic manifold and ω its symplectic form. This means

1991 *Mathematics Subject Classification*. Primary 28D10, 28D20.

Partially supported by the National Science Foundation and an American Mathematical Society Centennial Fellowship.

This paper is in final form and no version of it will be submitted for publication elsewhere.

ω is a nondegenerate closed 2-form on X. For example, the form

$$dx_1 \wedge dy_1 + \cdots + dx_n \wedge dy_n$$

is a symplectic form on $\mathbf{R}^{2n}$ with coordinates $(x_1, \ldots, x_n, y_1, \ldots, y_n)$. This is the prototype of any symplectic form ω since Darboux's theorem asserts the existence of local coordinates in which ω takes the above form.

Any C^∞-function f defines a vector field ξ_f by duality: for $v \in TX$ we define

$$\omega(\xi_f, v) = df(v).$$

One calls ξ_f a *Hamiltonian vector field* and f its *Hamiltonian*. Now define the *Poisson bracket* of two C^∞-functions f and g by

$$\{f, g\} \stackrel{\text{def}}{=} \xi_f(g).$$

Note that $\xi_{\{f,g\}} = [\xi_f, \xi_g]$. On $\mathbf{R}^{2n}$ with symplectic form as above the Poisson bracket is given by

$$\{f, g\} = \sum_{i=1}^{n} \frac{\partial f}{\partial x_i} \frac{\partial g}{\partial y_i} - \frac{\partial f}{\partial y_i} \frac{\partial g}{\partial x_i}.$$

The Poisson bracket furnishes the space of smooth functions with a Lie bracket. In particular, we say that f and g *Poisson commute* if $\{f, g\} = 0$. Recall that a *first integral* for a flow ϕ_t is a ϕ_t-invariant function.

DEFINITION 2.1. A flow ϕ_t on X^{2n} is called *completely integrable* if there are n Poisson commuting C^∞ first integrals $f_1, \ldots, f_n$ that are independent a.e. (i.e., the 1-forms $df_1, \ldots, df_n$ are independent a.e.). One calls such functions *first integrals in involution*.

As an example on $\mathbf{R}^{2n}$ consider the flow ϕ_t given by

$$\phi_t(x, y) = (x + ty, y).$$

Clearly the n coordinate functions y_i are first integrals in involution, and ϕ_t is completely integrable.

On $\mathbf{R}^{2n}$ completely integrable flows can be integrated by quadratures; i.e., one can find its solutions a.e. by a finite number of algebraic operations (including inversion of functions) and by integrals of known functions.

The dynamics of completely integrable flows is quite simple. For example, the existence of first integrals alone precludes ergodicity of ϕ_t w.r.t. a volume form. The existence of n first integrals in involution however gives a much finer structure. In fact, the Hamiltonian flows of the f_i generate an action of $\mathbf{R}^n$ that commutes with ϕ_t and one obtains the following theorem of Arnold [2].

THEOREM 2.2. *Set* $f = (f_1, \ldots, f_n)$. *Assume that the Hamiltonian flows* ξ_{f_i} *are complete on the level surfaces of* f *(e.g., assume the level surfaces are compact). For a.e.* $a \in \mathbf{R}^n$, *the connected components* C *of the level surface* $f^{-1}(a)$ *are diffeomorphic to* $\mathbf{R}^k \times T^{n-k}$ *where* T^{n-k} *is the* $(n-k)$-*torus. Moreover,* ϕ_t *is linear on* C; *i.e.,* ϕ_t *is given by* $x \mapsto x + t\nu$ *where* ν *depends on* C.

If the level surfaces are compact then their connected components are tori, and ϕ_t is almost periodic on a.e. level surface. On the singular level surfaces, however, the dynamics of ϕ_t might be very complicated.

KAM-theory shows that many of these invariant tori persist for small perturbations of completely integrable systems. Thus while the latter systems are the exception, there are many systems with similar dynamics. Of course, the dynamics will in general be chaotic on the complement of the invariant tori.

3. Geodesic flows

Let M be a Riemannian manifold with inner product $\langle , \rangle$ and TM its tangent bundle. Define the *geodesic flow* $g_t : TM \to TM$ by $g_t(v) = \dot{\gamma}(t)$ where γ is the geodesic tangent to $v \in TM$ at $t = 0$.

Let $\pi : TM \to M$ denote the map that projects a tangent vector to its footpoint. Then define a 1-form Ξ on TM by

$$\Xi_x(v) = \langle x, d\pi_x(v) \rangle$$

where $x \in TM$ and $v \in T_x TM$. The *canonical symplectic form* on TM then is defined by $\omega = -d\Xi$. Note that ω is dual to the canonical symplectic form on T^*M by the Riemannian metric. The geodesic flow preserves ω. In fact, g_t is the Hamiltonian flow for the function $E(v) = \frac{1}{2}\|v\|^2$ where $\|v\|$ denotes the Riemannian length of a tangent vector v of M.

4. Examples

4.1. The classical examples. We will give a brief overview of the classical examples. More details are readily available in the literature.

Flat metrics. Identify $T\mathbf{R}^n$ with $\mathbf{R}^{2n}$. Let $(x, y) = (x_1, \ldots, x_n, y_1, \ldots, x_n)$ be coordinates as above. Then the geodesic flow g_t is given by

$$g_t(x, y) = (x + ty, y).$$

As seen above, this flow is integrable. Note that the integrals descend to the tangent bundle of the n-torus.

Surfaces of revolution. Let M be a surface of revolution, i.e., a 2-dimensional Riemannian manifold that admits a faithful S^1-action by isometries. Let X be the corresponding Killing field. Denote by $\pi : TM \to M$ the canonical projection. By Clairaut's theorem, the function $F : TM \to \mathbf{R}$ given by

$$v \mapsto \langle v, X(\pi(v)) \rangle$$

is invariant under the geodesic flow g_t. Thus the Hamiltonian E for g_t and F Poisson-commute. Since they are also independent, g_t is completely integrable.

The ellipsoid. Let A be a positive definite symmetric matrix. Then the equation $\langle A^{-1}x, x \rangle = 1$ defines an n-dimensional ellipsoid. We furnish it with the induced metric from $\mathbf{R}^n$. In 1838, Jacobi established

THEOREM 4.1. *The geodesic flow for the triaxial ellipsoid is completely integrable.*

This generalizes to ellipsoids in any dimension. While Jacobi's work dates from the last century there is still considerable current interest in this example. We will briefly describe three methods of proof for this theorem (cf. [13] for a more detailed account).

PROOF BY SEPARATION OF VARIABLES. Jacobi finds a symplectic coordinate transform (X, Y) of the original coordinates (x, y) such that in the new variables (X, Y) the Hamiltonian is independent of Y, and thus completely integrable. The change of coordinates is defined by a "generating function" $S(Y, y)$ by $S_y = x$ and $S_Y = -X$ where $H(S_y, y) = K(Y)$ is the Hamiltonian in the coordinates (X, Y). Thus solving the system of ordinary differential equations is "reduced" to solving a partial differential equation. $\square$

PROOF BY CONFOCAL QUADRICS. Moser considers the confocal quadrics with defining functions $Q_z(x, y) = \langle (zI - A)^{-1}x, y \rangle$. Note that the ellipsoid is a level surface of Q_0. Let α_k denote the eigenvalues of A. Then one can expand

$$Q_z(x, y) = \sum_{k=1}^n \frac{F_k(x, y)}{z - \alpha_k}.$$

Remarkably, the F_k Poisson commute and yield complete integrability for the geodesic flow of the ellipsoid when restricted to its tangent bundle. $\square$

PROOF BY ISOSPECTRAL DEFORMATION. Suppose the Hamiltonian system $\dot{x} = f(x)$ can be represented as a matrix equation $\dot{L} = [L, M]$ where L and

M are functions of x. Then

$$L(t) = u(t)L(0)u(t)^{-1}$$

where $M = \dot{u}\, u^{-1}$. Thus

$$I_k = \operatorname{tr}(L^k)$$

are integrals of motion.

Moser and also Adler and van Moerbeke established the complete integrability of the ellipsoid using isospectral deformations [1, 14]. □

Other metrics on S^2. One can describe the metrics on S^2 with completely integrable geodesic flow quite explicitly in terms of isothermal coordinates [7, Theorem 5.4.4]. Particular examples of these are the *Zoll metrics* which are metrics all of whose geodesics are closed with the same period. The geodesic flows of the latter actually are symplectomorphic with that of a round sphere [20].

Left-invariant metrics on SO(3). The geodesic flows of such metrics describe the motion of a rigid body about a fixed point under its own inertia. This mechanical problem was completely solved by Euler in 1765. It is also the classical example of a Hamiltonian system with a large symmetry group, and thus is a special case of what follows.

4.2. Systems with symmetry. Here we will describe the complete integrability of geodesic flows on symmetric spaces of the compact type and certain other homogeneous spaces. Two methods, the Thimm method and the translation method of the Fomenko school, prove useful here. Since both use the moment map in an essential way, we will start with it.

Let a Lie group G act on a symplectic manifold N by symplectomorphisms. Let g denote the Lie algebra of G. We call the action *Hamiltonian* if there is a homomorphism $\theta : g \to \{C^\infty\text{-functions on } N\}$ such that $\xi_{\theta(X)}$ is the vector field of the action of $X \in g$ on N. Then we can define the *moment map* $\Phi : N \to g^*$ by

$$[\Phi(n)](X) = [\theta(X)](n).$$

The moment map intertwines the action of G on N with the coadjoint action.

Next we need a Poisson structure on the set of smooth functions on g^*. For two functions f and g on g^*, define the Poisson bracket at $\alpha \in g^*$ by

$$\{f, g\}(\alpha) \stackrel{\text{def}}{=} \alpha([df_\alpha, dg_\alpha]).$$

Alternatively, one can restrict the functions to coadjoint orbits and use the canonical symplectic structure on these to get a Poisson bracket. Two facts are essential.

(i) The pullback $\Phi^*(f) = f \circ \Phi$ commutes with any G-invariant Hamiltonian on N.

(ii) The pullback Φ^* is a homomorphism of Poisson algebras.

The idea now is to pull back sufficiently many Poisson commuting functions on g^* to get first integrals in involution for any G-invariant Hamiltonian. This is where the Thimm method and the translation method differ.

The Thimm method. Thimm's original method was a rather explicit case-by-case analysis and established the complete integrability of the geodesic flow on real and complex Grassmannians and a few other examples [18]. Guillemin and Sternberg conceptualized his method later and obtained further examples [8, 9]. We will follow Guillemin and Sternberg in our presentation.

Let N be a symplectic space with a Hamiltonian action of a Lie group G. Such an action is called *multiplicity-free* if the algebra of the G-invariant functions on N is commutative under the Poisson bracket [8, p. 361]. Let $\Phi : N \to l^*$ denote the moment map of the action. Let $\{1\} = G_l \subset G_{l-1} \subset \cdots \subset G_1 = G$ be an ascending chain of Lie subgroups of G, and denote their Lie algebras by l_i. Furnish each coadjoint orbit of G_i in l_i^* with the Kostant-Kirillov symplectic structure. Then each subgroup G_{i+1} acts on each orbit in l_i^* in a Hamiltonian way. The moment maps are just the restrictions of the dual maps $j_i : l_i^* \to l_{i+1}^*$ to the coadjoint orbits. We will call the chain G_i *multiplicity-free* if the actions of the G_{i+1} on the coadjoint orbits of G_i in l_i^* are multiplicity-free. This is quite a strong condition on the chain G_i as it forces the G_i to be locally isomorphic to $SO(n)$ or $SU(n)$ or products of these [10, 11].

If the G_i are a multiplicity-free chain and the action of G on N is multiplicity-free, then any G-invariant Hamiltonian on N is completely integrable [8, p. 366]. This is the essence of the Thimm method. It applies in particular when N is the tangent bundle of a Riemannian manifold M and G acts on TM by derivatives of an isometric action on M. If the action on TM is multiplicity-free then M is a homogeneous space G/K. The pairs (G, K), called *Gelfand pairs*, are quite special and have been classified by Krämer in [12]. For $G = SU(n)$ for example, the possible K are $SO(n)$, $Sp([n/2])$, $U(1) \cdot Sp([n/2])$, $S(U(k) \times U(l))$ and $SU(k) \times SU(l)$ where $k + l = n$. Also any Riemannian symmetric pair (G, K) is Gelfand, but not vice versa.

Sometimes a variation of the Thimm method yields new examples [16]. Suppose G acts on M by isometries and on TM by their derivatives. Let $\mathbf{R}$ act on TM via the geodesic flow. Then $G \times \mathbf{R}$ acts on TM. If this action is multiplicity-free then the geodesic flow is again completely integrable. We use this argument to show the complete integrability of the geodesic flow of certain left-invariant metrics on the Wallach manifold $SU(3)/T^2$ where T^2 is the maximal torus in $SU(3)$ [19]. The condition on the metric is simply that some geodesic is not the orbit of a 1-parameter subgroup of G. Similar remarks apply to $Sp(2)/Sp(1)$ and other homogeneous spaces.

The translation method. Consider Ad^*-invariant functions f on g^*. They are all polynomials. Fix a regular semisimple element $a \in g^*$. Then one can expand

$$f(X + ta) = \sum P_i^f(x, a)t^i$$

as a polynomial in t. Then the family of functions $P_i^f(x, a)t^i$ with f varying over all Ad^*-invariant functions Poisson commutes. Call a family of functions on a symplectic space X^{2n} *maximal commutative linear* if one can choose a basis $f_1, \ldots, f_n$ of a.e. independent functions.

THEOREM 4.2. *If G is semisimple of the compact type then the family of functions P_i^f are a maximal linear commutative algebra on every coadjoint orbit in g^*.*

This theorem was first suggested by Manakov for $G = SO(n)$. Fomenko and Mishchenko proved it for generic coadjoint orbits for general G while Brailov established it in full generality [7, Theorem 4.1.2].

Using the moment map as above one obtains the following

COROLLARY 4.3. *Let (G, K) be a Gelfand pair. Then the geodesic flow for any left invariant metric on G/K is completely integrable. In particular, all globally symmetric spaces of the compact type are completely integrable.*

4.3. Systems with hidden symmetries. In this section we will describe joint work with Gabriel Paternain [16]. Two new techniques are introduced.

The submersion method. Let M be a Riemannian manifold. Suppose H acts on M by isometries and without fixed points. Then $B \stackrel{\text{def}}{=} M/H$ is a manifold which can be endowed with the submersion metric inherited from M. Moreover, the geodesic flow on B is covered by the geodesic flow on M restricted to the horizontal vectors [3]. We show that the tangent bundle of B is symplectomorphic with the symplectic reduction of TM by H. It follows that any collection of H-invariant functions in involution descends to a set of Poisson-commuting functions on TB. Thus sometimes the basespace of a Riemannian submersion has a completely integrable geodesic flow. Of course, the main problem is to show the independence of the first integrals thus obtained.

In particular, let S^1 act on a manifold M with a completely integrable geodesic flow such that the first integrals are invariant under the S^1-action and let N be a surface of revolution. Then S^1 acts on $M \times N$ diagonally and $M \times_{S^1} N$ has a completely integrable geodesic flow.

As a specific example, $CP^n \# -CP^n$ is diffeomorphic with $S^{2n-1} \times_{S^1} S^2$. One observes from Thimm's construction of the first integrals on S^{2n-1} that they are invariant under the S^1-action. Thus $CP^n \# -CP^n$ carries a metric with completely integrable geodesic flow.

Similarly one obtains completely integrable geodesic flows on $SU(3) \times_{S^1} S^2$ where S^1 acts on $SU(3)$ by isometries. In particular, S^1 may act on $SU(3)$ by left and right translations simultaneously. Thus one obtains such geodesic flows on S^2- bundles over the Eschenburg manifolds $SU(3)/S^1$ [4]. Some of these are known not to have the homotopy type of any compact homogeneous space [17].

The Eschenburg manifolds $SU(3)/S^1$ themselves are a different kind of submersion example, one not of product type. We show complete integrability of a large number of these. Again quite a few of these examples are not homotopic to a homogeneous space [4].

Finally, we exhibit an exotic sphere Σ^7 with completely integrable geodesic flow. This sphere is a biquotient of $Sp(2)$ by $Sp(1)$, and had been used before by Gromoll and Meyer as an example of an exotic sphere with nonnegative sectional curvature [6]. The integrals in this example arise both from the submersion method combined with a Thimm construction as well as from the isometry group of this exotic sphere.

Gluing. We have mainly one example of this type, namely $CP^n \# CP^n$ for n odd. For simplicity suppose $n = 3$. Topologically this space is obtained from two copies of $S^5 \times_{S^1} D^2$ where D^2 is the 2-disk and S^1 acts diagonally, glued along their boundary $S^5 \times_{S^1} S^1 = S^5$ by an orientation reversing map [5]. If we start with a Berger sphere and a D^2 that looks cylindrical near the boundary circle, then the boundary S^5 is a totally geodesic round sphere [5]. We find first integrals on the Berger sphere such that the induced integrals on $S^5 \times_{S^1} D^2$ are invariant under complex conjugation on the boundary S^5, and fit together smoothly with the integrals on the second $S^5 \times_{S^1} D^2$. It follows that $CP^3 \# CP^3$ carries a completely integrable geodesic flow.

5. Topology and complete integrability

One would love to have a deeper understanding of topological obstructions to complete integrability of the geodesic flow. So far one knows that the geodesic flow of closed surfaces of genus bigger than 1 with a real analytic metric does not have real analytic first integral independent of the Hamiltonian. Taymanov, in fact, has found a remarkable extension of this result.

THEOREM 5.1. *Let M be a closed real analytic Riemannian manifold. Suppose the geodesic flow is completely integrable with real analytic first integrals. Then the fundamental group of M has an abelian subgroup of finite* *index. Furthermore, the first Betti number is bounded by the dimension of* M .

For simply connected manifolds however no topological obstruction has been found. Recall that a manifold is rationally elliptic if the total rational homotopy is finite dimensional. At least for the real analytic category, Paternain conjectures

CONJECTURE 5.2. *Let* M *be a simply connected closed Riemannian manifold with completely integrable geodesic flow* g_t. *Then* M *is rationally elliptic and the topological entropy of* g_t *is* 0 .

Remarkably, the two conclusions are related [15]. Indeed, Paternain shows that 0 topological entropy implies rational ellipticity. He also proved the claims of the conjecture if g_t is completely integrable with periodic integrals. This means that g_t has $n - 1$ independent first integrals independent of the Hamiltonian of g_t all of which generate circle actions.

REFERENCES

1. M. Adler and P. van Moerbeke, *Completely integrable systems, Euclidean Lie algebras and curves*, Adv. in Math. **38** (1980), 267–317.
2. V. I. Arnold, *Mathematical methods of classical mechanics*, Springer-Verlag, 1984.
3. A. L. Besse, *Einstein manifolds*, Ergeb. Math. Grenzgeb. (3) vol. 10, Springer-Verlag, 1987.
4. J.-H. Eschenburg, *New examples of manifolds with strictly positive curvature*, Invent. Math. **66** (1982), 469–480.
5. J. Cheeger, *Some examples of manifolds of nonnegative curvature*, J. Differential Geom. **8** (1972), 623–628.
6. D. Gromoll and W. Meyer, *An exotic sphere with nonnegative sectional curvature*, Ann. of Math. (2) **100** (1974), 401–406.
7. A. T. Fomenko, *Integrability and nonintegrability in geometry and dynamics*, Kluwer, Dordrecht, 1988.
8. V. Guillemin and S. Sternberg, *Symplectic techniques in physics*, Cambridge Univ. Press, 1984.
9. _____, *On collective complete integrability according to the method of Thimm*, Ergodic Theory and Dynamical Systems 3 (1983), 219–230.
10. G. Heckman, *Projections of orbits and asymptotic behavior of multiplicities for compact Lie groups*, Invent. Math. **67** (1982), 333–356.
11. M. Krämer, *Multiplicity free subgroups of compact connected Lie groups*, Arch. Math. (Basel) **27** (1976), 28–36.
12. _____, *Sphärische Untergruppen in kompakten zusammenhängenden Liegruppen*, Compositio Math. **38** (1979), 129– 153.
13. J. Moser, *Integrable Hamiltonian systems and spectral theory*, Lezioni Fermiane, Scuola Norm. Sup., Pisa, 1981.
14. _____, *Geometry of quadrics and spectral theory*, The Chern Symposium 1979, Springer-Verlag, 1980, pp. 147–188.
15. G. P. Paternain, *On the topology of manifolds with completely integrable flows*, Ergod. Th. & Dynam. Syst. **12** (1992), 109–121.
16. G. P. Paternain and R. J. Spatzier, *New examples of manifolds with completely integrable geodesic flows*, to appear in Advances of Math.
17. R. J. Spatzier and M. Strake, *Some examples of higher rank manifolds of nonnegative curvature*, Comment. Math. Helv. **65** (1990), 299–317.

18. A. Thimm, *Integrable geodesic flows on homogeneous spaces*, Ergodic Theory and Dynamical Systems **1** (1981), 495–517.

19. N. R. Wallach, *Compact homogeneous Riemannian manifolds with strictly positive curvature*, Ann. of Math. (2) **96** (1972), 277–295.

20. A. Weinstein, *Fourier integral operators, quantization and the spectra of Riemannian manifolds*, Gèomètrie Symplectique et Physique Mathèmatique, Colloq. Internat. CNRS No. 237 (Aix-en-Provence, 1974), Editions Centre National Recherche Sci. Paris 1975, 289–298.

UNIVERSITY OF MICHIGAN

Proceedings of Symposia in Pure Mathematics
Volume **54** (1993), Part 3

Differentiable Structure on Spheres
and Curvature

YOSHIHIKO SUYAMA

1. Introduction

Let (M, g) be a complete, simply connected riemannian manifold of dimension n. The purposes of this note are to present a new pinching constant in the differentiable pinching problem, and to indicate the idea of its proof in comparison with related results. A detailed account will be published elsewhere [9].

Our theorem is as follows.

THEOREM 1. *Let (M, g) be a 0.681-pinched riemannian manifold. Then M is diffeomorphic to the standard sphere S^n.*

We say that (M, g) is a δ-pinched riemannian manifold, if the sectional curvature K satisfies $\delta \le K \le 1$.

The sphere theorem implies that a riemannian manifold with $\frac{1}{4} < K \le 1$ is homeomorphic to the standard sphere [**1, 5**]. On the other hand, there exist many exotic spheres, which are homeomorphic but not diffeomorphic to the standard spheres. These facts gave rise to the following question : When can the conclusion in the sphere theorem be replaced by diffeomorphism? We call it the differentiable pinching problem. For the first time, Gromoll [**2**], Calabi and Shikata [**7**] gave some results on the problem. Later on, their results were improved as follows:

THEOREM (Sugimoto-Shiohama [**8**]). *Let (M, g) be a 0.87-pinched riemannian manifold. Then M is diffeomorphic to the standard sphere S^n.*

THEOREM (Im Hof–Ruh [**4**]). *There exists a decreasing sequence δ_n with limit $\delta_n \to 0.68$ as n tends to infinity such that the following assertion holds: If (M, g) is a δ_n-pinched riemannian manifold of dimension n*

1991 *Mathematics Subject Classification.* Primary 53C20; Secondary 53C22, 58B20.
The final version of this paper has been submitted for publication elsewhere.

609

and $\mu : G \times M \to M$ is an isometric action of the Lie group G on M, then

(1) there exists a diffeomorphism $F : M \to S^n$,
(2) there exists a homomorphism $\psi : G \to O(n+1, \mathbf{R})$ such that
(3) $\psi(g) = F \circ \mu(g, \cdot) \circ F^{-1}$ for all $g \in G$.

We are interested in a pinching constant independent of the dimension of M. In Im Hof–Ruh's results, if we take the number δ independent of the dimension of M, then δ becomes considerably large, i.e., $\delta = 0.98$ for $n > 5$. Our pinching constant 0.681 is almost same as the number $\lim \delta_n = 0.68$ given by Im Hof–Ruh. But their numbers are determined by different equations from each other.

2. Ideas of proof

Sugimoto-Shiohama's idea was as follows : A complete, simply connected and δ-pinched riemannian manifold M^n is diffeomorphic to the standard sphere S^n if a diffeomorphism $f : S^{n-1} \to S^{n-1}$, which is naturally defined for δ-pinched riemannian manifold M [see §4], is diffeotopic to the identity map. We shall call this the diffeotopy idea. So, the problems in their case were how to construct a diffeotopy, and how to find an explicit estimate of δ to guarantee such a diffeotopy. On the other hand, the main idea in a series of papers Ruh [6], Grove-Karcher-Ruh [3] and Im Hof-Ruh, was to lead from a connection with small curvature on the stabilized tangent bundle E of M to a flat connection on the bundle. This first connection with small curvature on the bundle was defined as an anology with the restriction to $T(\mathbf{R}^{n+1})|_{S^n}$ of the canonical flat connection of tangent bundle $T(\mathbf{R}^{n+1})$ over $\mathbf{R}^{n+1}$ [see §4]. We shall call this the flat connection idea. Using the resulting flat connection, they defined a generalized Gauss map $F : M \to S^n$, which gave a diffeomorphism. So, the problems in this case were how to construct a flat connection from the connection with small curvature on E, and how to find an explicit estimate of δ in order that the Gauss map could be a diffeomorphism.

The emphasis of this note is to combine these different ideas from our viewpoint to obtain a new pinching constant.

We use the diffeotopy idea in the proof of Theorem 1; that is, we find a sufficient condition that the diffeomorphism $f : S^{n-1} \to S^{n-1}$ is diffeotopic to the identity map. But our diffeotopy is constructed in a quite different way from Sugimoto-Shiohama's. Sugimoto-Shiohama made a diffeotopy by joining a point x of S^{n-1} to its image $f(x)$ by geodesic in S^{n-1}. To do this, they determined an arcwise connected neighborhood of the identity map in the group of diffeomorphisms of S^{n-1}. On the other hand, our construction is based on the following consideration: The group $SO(n, \mathbf{R})$ of special orthogonal matrices belongs to an arcwise connected component of the group of diffeomorphisms of S^{n-1}. Therefore we should determine a connected neighborhood of $SO(n, \mathbf{R})$ in some way. Let put $T(x) = Ux$ for $U \in SO(n, \mathbf{R})$ and $x \in \mathbf{R}^n$. The differential $(dT)_x = U$ for $x \in \mathbf{R}^n$. Therefore, a differential of map f is useful to determine a connected neighborhood of $SO(n, \mathbf{R})$.

Our idea is as follows : $f : S^{n-1} \to S^{n-1}$ is homothetically extended to a diffeomorphism $F : \mathbf{R}^n - \{0\} \to \mathbf{R}^n - \{0\}$. The restriction of the differential dF to S^{n-1} becomes a map of S^{n-1} into the space $M(n, \mathbf{R})$ of $n \times n$-matrices. We approximate $dF : S^{n-1} \to M(n, \mathbf{R})$ by a map $\alpha : S^{n-1} \to SO(n, \mathbf{R})$. For a differentiable map $\alpha : S^{n-1} \to SO(n, \mathbf{R})$, we denote by α_x the matrix corresponding to $x \in S^{n-1}$. Then, we construct the diffeotopy by joining α_x to a constant matrix by geodesic in $SO(n, \mathbf{R})$ for each $x \in S^{n-1}$. We note that the above construction of diffeotopy is almost similar to a way of the construction of flat connection of the stabilized tangent bundle [see §4].

3. Diffeotopy theorem

In this section, we state precisely our diffeotopy theorem. Let S^{n-1} be the standard sphere with curvature 1. Let $f : S^{n-1} \to S^{n-1}$ be a diffeomorphism. We put $F(tx) = tf(x)$ for $t > 0$. We define a norm of differential of $\alpha : S^{n-1} \to SO(n, \mathbf{R})$ by

$$\|d\alpha\| = \max \{\|(d_X\alpha.)U\| \mid X \in T_x(S^{n-1}) \text{ and } U \in \mathbf{R}^n$$
$$\text{with } \|X\| = \|U\| = 1\},$$

where $\| X \|$ denotes the euclidean norm of X. Similarly, the norm of a map $A : S^{n-1} \to M(n, \mathbf{R})$ is defined by

$$\|A\| = \max \{ \|A_xU\| \mid x \in S^{n-1} \text{ and } U \in \mathbf{R}^n \text{ with } \|U\| = 1 \}.$$

We say that f is diffeotopic to the identity map of S^{n-1}, if there exists a differentiable map $H : [0, 1] \times S^{n-1} \to S^{n-1}$ satisfying the following:

(1) $H(1, x) = f(x)$ and $H(0, x) = x$.

(2) The map $H_t = H(t, \cdot)$ is a diffeomorphism of S^{n-1} for each t.

We say that $\alpha : S^{n-1} \to SO(n, \mathbf{R})$ is an approximation of df on S^{n-1}, if there exist real numbers C and N and they satisfy the following:

(1) $N < 1$.   (2) $\alpha_x(x) = (d_xF)(x)$ for $x \in S^{n-1}$.

(3) $\| \alpha - dF \| \leq C$.   (4) $\| d\alpha \| \leq N$.

Note that the inequality $C \leq N$ always holds. Furthermore, the inequality $\| d\alpha \| < 1$ implies that all image α_x for $x \in S^{n-1}$ belongs to a normal coordinate neighborhood of some α_{x_0} in $SO(n, \mathbf{R})$. For the approximation α of df, we define a positive function $P(t)$ for $t \in [0, \pi]$: Put $C_2 = (N - C)/2$, $C_3 = (N + C)/2$ and take t_0 and t_1 such that

$$\cos \left(\frac{3}{2}N(\pi - t_0)\right) = -1 \quad \text{and} \quad \cos \left(\frac{3}{2}N(\pi - t_1)\right) = 0.$$

Then, we put

$$P(t)^2 = C_2^2 \left(\frac{\sin(Nt/2)}{\sin(N\pi/2)} \right)^2 + C_3^2 \left(\frac{\sin(Nt)}{\sin(N\pi)} \right)^2 + 2C_2 C_3 \frac{\sin(Nt)}{\sin(N\pi)} \psi(t),$$

where $\psi(t)$ is defined by

$$\psi(t) = \begin{cases} \dfrac{\sin(Nt/2)}{\sin(N\pi/2)} & (0 \le t \le t_0), \\[2mm] -\dfrac{\sin(Nt/2)}{\sin(N\pi/2)} \cos(\tfrac{3}{2}N(\pi - t)) & (t_0 \le t \le t_1), \\[2mm] -\dfrac{t}{\pi} \cos(\tfrac{3}{2}N(\pi - t)) & (t_1 \le t \le \pi). \end{cases}$$

THEOREM 2. *Let* $f : S^{n-1} \to S^{n-1}$ *be a diffeomorphism. Suppose there exists an approximation* α *of* df *such that* $P(t) < 1$ *for* $t \in [0, \pi]$. *Then* f *is diffeotopic to the identity map.*

4. Construction of α and estimate of norms

Let $\delta > \frac{1}{4}$. The manifold M is homeomorphic to the standard sphere by the sphere theorem. In particular, we use the following properties of M. Let q_0 and q_1 be a pair of points with maximal distance $d(q_0, q_1)$ on M, where d denotes the distance function induced by the riemannian metric g. Set $X(p) = d(q_0, p) - d(q_1, p)$ for $p \in M$, and define $C = X^{-1}(0)$, $M_0 = X^{-1}((-\infty, 0])$, $M_1 = X^{-1}([0, \infty))$. The exponential maps $\exp_0$ and $\exp_1$ with centers at q_0 and q_1 respectively are bijective maps if they are restricted to an open ball of radius π. The set C is diffeomorphic to S^{n-1}. Let S_0 and S_1 denote unit spheres in tangent spaces at points q_0 and q_1. Then the diffeomorphism $f : S_0 \to S_1$ is defined by requiring $\exp_0(tx)$ and $\exp_1(tf(x))$ to coincide for some $t = t(x)$ satisfying $\pi/2 \le t(x) \le \pi/2\sqrt{\delta}$. Note that the point of intersection lies on C. We denote by $q(x)$ the point $\exp_0(t(x)x) = \exp_1(t(x)f(x)) \in C$.

We fix suitable orthonormal bases of tangent spaces $T_0(M)$ at q_0 and $T_1(M)$ at q_1, respectively. Put $\tau^0(x, t) = \exp_0(tx)$ and $\tau^1(f(x), t) = \exp_1(tf(x))$ for $x \in S_0$. For a vector $X \in T_{q(x)}(C)$, we denote by X_0 (resp. X_1) the vector of $T_0(M)$ (resp. $T_1(M)$) obtained by parallel translation of $X - g(X, \dot{\tau}^0(x))\dot{\tau}^0(x) \in T_{q(x)}(M)$ (resp. $X - g(X, \dot{\tau}^1(f(x)))\dot{\tau}^1(f(x)) \in T_{q(x)}(M)$) along a geodesic $\tau^0(x, t)$ (resp. $\tau^1(f(x), t)$). Then our approximation $\alpha_x \in SO(n, \mathbf{R})$ of df_x for $x \in S_0$ is defined as follows:

(1) $\alpha_x(X_0) = X_1$ for $X \in T_{q(x)}(C)$,
(2) $\alpha_x(x) = f(x)$,

where X_i $(i = 0, 1)$, x and $f(x)$ denote the component vectors with respect to the bases. Then we have the estimate of $\| \alpha - dF \|$ by using Levi-Civita connection D of M. A kind of estimate of $\| \alpha - dF \|$ was given by Sugimoto-Shiohama. But, we can give the estimate sharper than it. Furthermore, on the above construction of diffeotopy we use the estimate in a quite different way from that of Sugimoto-Shiohama.

PROPOSITION 1. *We put* $c = \sqrt{(1+\delta)/2}$. *Then we have*

$$\| \alpha - dF \| \leq \frac{1}{2}\frac{1-\delta}{1+c^2}\left[\frac{c\{\exp(\frac{\pi}{2\sqrt{\delta}}) - \exp(-\frac{\pi}{2\sqrt{\delta}})\}}{2\sin(\frac{c\pi}{2\sqrt{\delta}})} - 1\right]\left[1 + \frac{1}{\sqrt{\delta}\sin(\frac{\pi}{2\sqrt{\delta}})}\right].$$

On the other hand, to estimate $\| d\alpha \|$ we use the stabilized tangent bundle E and the connection ∇ with small curvature of E due to Ruh: Let $1(M)$ be a trivial line bundle $M \times \mathbf{R}$. Let $E = T(M) \oplus 1(M)$. A cross-section $\mathbf{e} : M \to E$ is defined by $\mathbf{e}_p = (0, 1)_p$, where 0 is the zero vector of $T_p(M)$. We take a fibre metric h of E by

$$h_p(X, Y) = g_p(X, Y), \quad h_p(X, \mathbf{e}) = 0 \quad \text{and} \quad h_p(\mathbf{e}, \mathbf{e}) = 1$$

for $X, Y \in T_p(M)$. The metric connection ∇ of E is defined by

$$\nabla_X Y = D_X Y - cg(X, Y)\mathbf{e} \quad \text{and} \quad \nabla_X \mathbf{e} = cX$$

for $X, Y \in T(M)$. A norm $\| R^\nabla \|$ of curvature R^∇ of ∇ is defined by

$$\| R^\nabla \| = \max \{\| R^\nabla(X, Y)Z \| \mid X, Y \text{ and } Z \in T_p(M) \text{ with}$$
$$\|X\| = \|Y\| = \|Z\| = 1\}.$$

Then we have $\| R^\nabla \| \leq \frac{2}{3}(1 - \delta)$.

Let P be a principal bundle over M of orthonormal frames of E. The structure group of P is $\mathrm{O}(n + 1, \mathbf{R})$. We take a horizontal lift $u^0(x, t)$ (resp. $u^1(f(x), t))$ in P of $\tau^0(x, t)$ (resp. $\tau^1(f(x), t))$ under a fixed initial condition $u^0(x, 0)$ (resp. $u^1(f(x), 0))$. Then there exists $b_x \in \mathrm{O}(n + 1, \mathbf{R})$ such that $u^0(x, t(x))b_x = u^1(f(x), t(x))$. The scale $\| db \|$ is determined by the norm $\| R^\nabla \|$. Furthermore, by investigating the difference between α_x and b_x we have the estimate of $\| d\alpha \|$.

PROPOSITION 2. *We have*

$$\| db \| \leq \frac{2}{3}\frac{1-\delta}{\delta}\left(1 + \frac{1}{\sqrt{\delta}\sin(\pi/2\sqrt{\delta})}\right).$$

If $\| db \| < 1$, then the stabilized tangent bundle E becomes a flat bundle $M \times \mathbf{R}^{n+1}$. Originally, Ruh [6] constructed a flat connection of E in this way. (But our norm $\| db \|$ defined at §3 was later introduced in [3].) The norm $\| d\alpha \|$ can be estimated by using the norm $\| db \|$ as follows.

PROPOSITION 3. *We have*

$$\| d\alpha \| \leq \frac{2}{\sin^2(c\pi/2\sqrt{\delta})(1 - \cos(\pi\sqrt{\delta}))} \| db \|.$$

References

1. M. Berger, *Les variétés riemanniennes* (1/4)-*pincées*, Ann. Scuola Norm. Sup. Pisa **14** (1960), 161–170.

2. D. Gromoll, *Differenzierbare Strukturen unt Metriken positiver Krümmung auf Sphären*, Math. Ann. **164** (1966), 351–371.

3. K. Grove, H. Karcher, and E. A. Ruh, *Group actions and curvature*, Invent. Math. **23** (1974), 31–48.

4. H. C. Im Hof and E. A. Ruh, *An equivariant pinching theorem*, Comment. Math. Helv. **50** (1975), 389–401.

5. W. Klingenberg, *Über Riemannsche Mannigfaltigkeiten mit positiver Krümmung*, Comment. Math. Helv. **35** (1961), 47–54.

6. E. A. Ruh, *Curvature and differentiable structure on spheres*, Comment Math. Helv. **46** (1971), 127–136.

7. Y. Shikata, *On the differentiable pinching problem*, Osaka Math. J. **4** (1967), 279–287.

8. M. Sugimoto and K. Shiohama, *On the differentiable pinching problem*, Math. Ann. **195** (1971), 1–16.

9. Y. Suyama, *Differentiable sphere theorem by curvature pinching*, J. Math. Soc. Japan **43** (1991), 527–553.

FUKUOKA UNIVERSITY, JAPAN

Proceedings of Symposia in Pure Mathematics
Volume 54 (1993), Part 3

Spectral Theory for Operator Families
on Riemannian Manifolds

Z. I. SZABO

Introduction

Spectral geometry today means the geometric investigation of the Laplacian spectrum on Riemannian spaces. It is well known that this spectrum does not determine the metric in general. So far, only higher-dimensional examples were known for this phenomenon. Just this year (1991), Gordon, Webb and Wolpert [GWW1] announced the beautiful result saying that the Dirichlet-Laplace spectrum does not determine even the shape of 2-dimensional membranes. This result answers negatively the long-standing and very intensively investigated question of Kac [Ka1]: Can one hear the shape of a drum?

In this paper, we involve many operators into the investigation, in order to have a better chance for the determination of the metric by the "joint spectrum" of these operators. This sort of investigation was motivated for the present author by the Lichnerowicz Conjecture (1944) concerning harmonic manifolds.

A Riemannian manifold is said to be harmonic if the harmonic functions have the classical Mean Value Property. The Lichnerowicz Conjecture says that the harmonic spaces are exactly the 2-point homogeneous spaces.

This conjecture has been established on simply connected and compact manifolds or on complete spaces with nonnegative scalar curvature, by the present author [Sz1]. This year, Damek and Ricci [DR1; DR2] announced the surprising result saying that the Lichnerowicz Conjecture is not true on noncompact manifolds. What is more, they constructed infinitely many nonsymmetric negatively curved harmonic spaces, each of them diffeomorphic

1991 *Mathematics Subject Classification.* Primary 58G25, 53C25.

Research partially supported by the NSF grant DMS-9213805, by The Max-Planck Institut für Mathematik Bonn, and by IHES, Bures Sur Yvette, France.

This paper is in final form and no version of it will be submitted for publication elsewhere.

to $\mathbf{R}^n$. The least dimension for these counterexamples is 7.

In Chapter 1, we describe and compare the compact and the noncompact cases in detail. Now we only indicate how the above-mentioned "many operators" emerge in the investigations.

$\theta_p(q)$ stands for the polar density function in the polar coordinate system around a point p. It turns out that a space is harmonic if and only if $\theta_p(q)$ is radial of the form $\theta_p(q) = \theta(r(p,q))$, where r is the geodesic distance of points p and q.

Introduce also the averaging operator $E_{p;r}(u)$ on the geodesic sphere with centre p and radius r, and produce the following so-called density Laplacians $\Delta_\theta^{(k)}$, defined by

$$\Delta_\theta^{(k)} u(p) = \frac{\partial^{2k} E_{p;r}(u)}{\partial r^{2k}} \bigg/ r = 0; \quad \forall k \in \mathbf{N}.$$

Important points of [Sz1] are the following characterizations for harmonicity:

(1) A space is harmonic if and only if the radial kernels of the form $H(p,q) = h(r(p,q))$ are closed with respect to convolution; i.e. they form a closed commutative convolution algebra.

(2) A space is harmonic if and only if the density Laplacians $\Delta_\theta^{(k)}$ are polynomials of the Laplacian Δ.

In the compact case, it turns out that the radial density function $\theta(r)$ on a harmonic space is the same as on one of the compact 2-point homogeneous spaces. Moreover, also the spectrum of the operators $\Delta_\theta^{(k)}$ is the same as the corresponding spectrum on that 2-point homogeneous space. Some further considerations lead to the establishing of the conjecture on simply connected and compact manifolds.

In the noncompact case, there emerge harmonic spaces whose density function is different from the density functions of 2-point homogeneous spaces but—in a very surprising manner—there are also such nonsymmetric harmonic spaces whose density function is the same as that of the hyperbolic quaternionic spaces. This last example shows that the density function $\theta_p(q)$ or the convolution algebra of radial kernels or the spectrum of the density Laplacians $\Delta_\theta^{(k)}$ does not determine the metric. This phenomenon looks like uncertainty for the metrics representing the same spectral geometric situation for the operators $\Delta_\theta^{(k)}$.

In order to get a spectrum notion which determines also the metric, we have to involve further operators. In §2 of Chapter 1, we introduce the general Jacobi kernels $H(x,y)$ (one of them is the density kernel $\theta(x,y)$) which generate the Jacobi Schrodinger operators $\square_H^{(k)}$ and the Jacobi Laplace operators $\Delta_H^{(k)}$ similarly as the density function $\theta(x,y)$ generates the density operators $\Delta_\theta^{(k)}$. This larger class of operators determines the metric much more strongly than the density operators. For instance, we have

(1) A space is 2-point homogeneous if and only if all Jacobi kernels are radial functions—or equivalently, the Jacobian convolution algebra consists of radial kernels.

(2) A space is 2-point homogeneous if and only if all the Jacobi operators $\square_H^{(k)}$, $\Delta_H^{(k)}$ are polynomials of the Laplacians.

The first statement is the weak form of the Lichnerowicz conjecture, which is true in any case. The second statement is the converse of one of Helgason's theorems.

This paper is devoted to the spectral investigation of the above Jacobian differential operator family.

CHAPTER 1
TWO-POINT HOMOGENEOUS SPACES
AND THE LICHNEROWICZ CONJECTURE

I. An overview of problems concerning 2-point homogeneous spaces

1. Symmetry of 2-point homogeneous spaces. A surprisingly large number of problems deal with 2-point homogeneous spaces. These spaces were introduced by Birkhoff [**Bi1**] and Busemann [**Bu1**] defined as locally compact metric spaces (M^n, d) such that for any two pairs of points (x, y) and (x', y') satisfying $d(x, y) = d(x', y')$ there exists an isometry T on M^n so that $T(x) = x'$ and $T(y) = y'$. Since the transitive isometry group acting on such spaces (M^n, d) is a Lie group, furthermore the isotropy group of isometries I_p inscribes a sphere as indicatrix at any tangent space $T_p(M^n)$, all these spaces are Riemannian manifolds.

The 2-point homogeneous spaces have been classified by Tits [**T1**] and Wang [**Wan1**] and it turned out just from this list that all these spaces were symmetric. It was a longstanding problem to find a direct proof for this symmetry [**Hel**, **Na1**, **Ma1**, **Wo1**]. Very recently the present author [**Sz2**] found a short topological proof for this fact. Since this topological trick is used in this paper as well, we give some details here.

The local symmetry of a locally 2-point homogeneous space directly follows from the following

FUNDAMENTAL LEMMA 1.1 [Sz2]. *Let $S_0^{n-1} \subset \mathbf{R}^n$ be the Euclidean unit sphere around the origin 0 in $\mathbf{R}^n$ and let*

$$A_{\mathbf{m}}(\cdot)\colon T_{\mathbf{m}}(S_0^{n-1}) \to T_{\mathbf{m}}(S_0^{n-1}), \qquad \mathbf{m} \in S_0^{n-1},$$

be a continuous endomorphism field on S_0^{n-1} such that

(1) *it is selfadjoint ($A_{\mathbf{m}}^* = A_{\mathbf{m}}$) and "skew" in the sense $A_{-\mathbf{m}} = -A_{\mathbf{m}}$; furthermore*

(2) *the eigenvalues $\lambda_1 \leq \lambda_2 \leq \cdots \leq \lambda_{n-1}$ of $A_{\mathbf{m}}$ are constant on S_0^{n-1}.*
Then $A_{\mathbf{m}} = 0$.

Apply this lemma to the field $A_{\mathbf{m}}(\cdot) = (\nabla_{\mathbf{m}}R)(\cdot, \mathbf{m})\mathbf{m}$ of a locally 2-point homogeneous space M^n at any tangent plane $T_p(M^n)$, $p \in M^h$, where $R(X, Y)Z$ is the curvature and $\nabla_{\mathbf{m}}$ is the covariant derivative with respect to the unit vector $\mathbf{m}$. The properties $A_{\mathbf{m}}^* = A_{\mathbf{m}}$ and $A_{-\mathbf{m}} = -A_{\mathbf{m}}$ are trivially satisfied and, since $A_{\mathbf{m}}$ is invariant under the transitive action of the isotropy group I_p on $S_p^{n-1} \subset T_p(M^n)$, property (2) is satisfied as well.

Therefore, $(\nabla_{\mathbf{m}}R)(\cdot, \mathbf{m})\mathbf{m} = 0$ everywhere, which implies the local symmetry $\nabla R = 0$ easily (see [**Bes1**, Proposition 2.35]).

The universal covering space $\widetilde{M}^n$ of a globally 2-point homogeneous space is globally symmetric by a theorem of Cartan. The global symmetry of the base space M^n as well as the rank-one-property is also proved in [**Sz2**]. Short direct proof of the classification of rank-one-symmetric spaces (i.e., not using Cartan's root-system-method) has been given by Cowling, Dooley, Koranyi, and Ricci in [**CDKR1**] and by Karcher in [**Kar1**].

The proof of the fundamental lemma is based on the following purely topological statement.

Consider the Euclidean unit sphere $S_0^{n-1} \subset \mathbf{R}^n$ again. The tangent plane $T_{\mathbf{m}}(S_0^{n-1})$ at a point $\mathbf{m} \in S_0^{n-1}$ is identified with the vectors $\mathbf{v} \perp \mathbf{m}$ to provide a natural identification between the tangent spaces at $\mathbf{m}$ and the antipodal point $-\mathbf{m}$.

BORSUK-ULAM-TYPE THEOREM 1.2 [**Sz2**]. (A) *Let* $X(\mathbf{m})$ *be a continuous tangent vector field on* S_0^{n-1}. *There exists an antipodal point pair* $\pm\mathbf{m}_0$ *such that* $X(\mathbf{m}_0) = -X(-\mathbf{m}_0)$.

(B) *In general, let* $X(\mathbf{m}) \subseteq T_{\mathbf{m}}(S_0^{n-1})$ *be a continuous nontrivial distribution of linear subspaces on* S_0^{n-1}. *There exists an antipodal point pair* $\pm\mathbf{m}_0$ *such that the subspaces* $X(\mathbf{m}_0)$ *and* $X(-\mathbf{m}_0)$ *are not independent; i.e.,* $\dim(X(\mathbf{m}_0) \cap X(-\mathbf{m}_0)) > 0$.

This statement is only similar but actually independent of the Borsuk-Ulam Theorem. Actually the proof of our statement is easier:

For the proof of (A), we have to show that the symmetric part

(1.1) $X_{\text{sym}}(\mathbf{m}) = \frac{1}{2}(X(\mathbf{m}) + X(-\mathbf{m}))$

vanishes at some point. Assume the contrary and consider the normalized vector field

(1.2) $f(\mathbf{m}) = X_{\text{sym}}(\mathbf{m})/|X_{\text{sym}}(\mathbf{m})|: S_0^{n-1} \to S_0^{n-1}$.

Since $f(\mathbf{m}) \perp \mathbf{m}$, so $f_\varepsilon(\mathbf{m}) := \cos\varepsilon\mathbf{m} + \sin\varepsilon f(\mathbf{m})$ is a smooth homotopy from $f(\mathbf{m})$ to the identity map. Therefore, $\deg f = 1$. On the other hand, from $f(\mathbf{m}) = f(-\mathbf{m})$ we easily get $\deg f = $ even and this contradiction completes the proof of (A).

For the proof of (B) consider a continuous vector field ν tangent to X which vanishes only at one point, say at $\mathbf{m}_1$. Since $S_0^{n-1} \setminus \{\mathbf{m}_1\}$ is contractable, such a vector field obviously exists. By (A), there exists $\mathbf{m}_0 \neq \pm\mathbf{m}_1$

such that $v(\mathbf{m}_0) = -v(-\mathbf{m}_0) \neq 0$; i.e., $X(\mathbf{m}_0)$ and $X(-\mathbf{m}_0)$ are not independent.

In case of the Fundamental Lemma suppose that there is a nonzero eigenvalue $\lambda \neq 0$ and let $E(\lambda, \mathbf{m})$ be the eigensubspace distribution. Since $E(\lambda, \mathbf{m}) = E(-\lambda, -\mathbf{m})$, by $A_{\mathbf{m}} = -A_{-\mathbf{m}}$, so $E(\lambda, \mathbf{m}) \perp E(\lambda, -\mathbf{m})$ at any point $\mathbf{m}$. This is impossible because of (B) and the proof of the Fundamental Lemma is finished.

Notice that our approach is purely topological rather than group theoretical as in the case of the previous authors. We keep this approach throughout this paper investigating general Riemannian manifolds without any homogeneity assumption.

It should be mentioned that this avoiding of the group theory is the main point also of Chavel's booklet [**Cha1**] written about rank-one-symmetric spaces. Chavel uses Jacobian field theory investigating several so-called pinching problems of Riemannian geometry. This tool will be involved also in this paper.

2. The Blaschke conjecture. The 2-point homogeneous spaces have very nice geometric properties. Most of the problems in this field ask for converse statements: Are these "nice properties" characteristic for these spaces?

For instance, the cut value is constant finite number with respect to any point and any direction on a compact 2-point homogeneous space. The complete Riemannian manifolds satisfying this property are said to be Blaschke manifolds and the famous Blaschke conjecture asserts the converse statement: The Blaschke manifolds are compact 2-point homogeneous spaces.

This question is still open; however there are some very nice contributions to this field. Besse [**Bes1**] devoted a whole book to this question; we concentrate only on the most important points.

The first important contribution was made by Green [**Gre1**] confirming the conjecture on 2-dimensional manifolds. The strongest result has been achieved by Berger [**Ber1**] so far, proving the conjecture on such Blaschke manifolds M^n where M^n is homeomorphic to the sphere S^n or to the real projective space $\mathbf{R}P^n$.

Berger's proof rests on a tricky geometric inequality of Kazdan [**Ka1**] and on other basic results which describe the topology of Blaschke manifolds. The starting point of this topological investigation is the *Allamigeon-Warner Theorem* [**All1, War1, Bes1**] asserting the following facts:

THEOREM 1.3 [**All1, War1**]. (A) *The Blaschke manifolds are compact spaces having simply closed geodesics with the same length, say L.*

(B) *The cut locus $C(m)$ with respect to any point m is the greatest geodesic sphere around m with radius $L/2$. $C(m)$ is a differentiable submanifold and the geodesics starting from m intersect it orthogonally. More precisely, for any unit vector $\mathbf{v} \in T_q(M^n)$ with $q \in C(m)$ and $\mathbf{v} \in T_q(C(m))^\perp$ there exists a unique geodesic $\gamma(s)$ satisfying $\gamma(0) = m$, $\gamma(L/2) = q$ and $\dot{\gamma}(L/2) = \mathbf{v}$.*

The above periodicity of the geodesic determines the topology (the type of the homotopy and the possible dimensions of the cut locus) on Blaschke manifolds. This is described in the following *Bott-Samelson Theorem* [**BS1, Bes1**].

THEOREM 1.4 [**BS1**]. *Let M^d be a Blaschke manifold of dimension d and let $d - k$ be the dimension of the cut locus $C(m)$. The homotopy type of M is denoted by $HT(M)$. The possibilities for d, k and $HT(M)$ are as follows*:

(1) $k = 1$; $d = 2n$; $HT(M) = HT(CP^n)$,
(2) $k = 3$; $d = 4n$; $HT(M) = HT(HP^n)$,
(3) $k = 7$; $d = 16$; $HT(M) = HT(CaP^2)$,
(4) $k = d - 1$; $HT(M) = HT(S^d)$,
(5) $k = 0$; $\pi_1(M) = Z_2$; $HT(M) = HT(RP^d)$

In spite of the above-mentioned very attractive results, the Blaschke conjecture has remained one of the most exciting unsolved questions in the field.

3. The Lichnerowicz conjecture. The long history of harmonic manifolds started with the work of Ruse who attempted to introduce radial harmonic functions on Riemannian manifolds for developing harmonic analysis similar to the Euclidean case. It turned out that this was possible only in very special cases, namely if the density function $\omega_p(q) = \sqrt{|\det g_{ij}|}$ with respect to the geodesic normal coordinate neighborhood around an arbitrary point p is a radial function around p.

Using the well-known symmetry $\omega_p(q) = \omega_q(p)$, this property means that there exists a function $\phi: R_+ \to R_+$ such that $\omega_p(q) = \phi(r(p, q))$ where $r(p, q)$ is the geodesic distance between p and q.

The Riemannian spaces having this radial property on the density function $\omega(p, q)$ are said to be *harmonic manifolds*. The origin of this name is a theorem of Willmore [**Wi1**] asserting that a Riemannian manifold is harmonic if and only if the classical Mean Value Theorem of harmonic functions holds on the space.

The 2-point homogeneous spaces are obviously harmonic manifolds. The Lichnerowicz Conjecture [**Li1**] says the converse: The harmonic manifolds are just the 2-point homogeneous spaces. Lichnerowicz [**Li1**] and Walker [**Wal1**] proved that this statement is true in case dim ≤ 4 and that among the symmetric spaces exactly the 2-point homogeneous spaces are harmonic manifolds.

It should be mentioned that the cases dim ≤ 3 are trivial since, by the well-known formula

(1.3) $\nabla_m \nabla_m \omega_p = -\frac{1}{3} \text{Ricci}(m, m)$, $m \in T_p(M)$,

all *the harmonic manifolds are Einstein spaces*.

Notice that from this property also the *real analyticity of harmonic spaces follows*, by the DeTurck-Kazdan theorem [**DeTK1**].

In spite of much work only little progress was made for a long time. Most of the authors used local methods for the investigations. Among the few global results we mention Allamigeon's Observation [**All1**] asserting that a simply connected and complete harmonic manifold is a Blaschke manifold in the compact case and it is diffeomorphic to $\mathbf{R}^n$ in the noncompact case. This observation can be easily checked, since the cut point with respect to a geodesic $\gamma(s)$ and the point $p = \gamma(0)$ is where the density function $\omega_p(s)$ vanishes.

By the Allamigeon Observation we get that the *simply connected compact harmonic manifolds have simply closed geodesics with the same length.*

Another global result is Besse's Nice Imbedding Theorem [**Bes1**]. In this theorem, an isometric imbedding $\phi: M^n \to \mathbf{R}^d$ is constructed for compact simply connected harmonic spaces such that $\phi(M^n)$ is minimal on a certain sphere, and, furthermore, all the geodesics are congruent screw lines in $\mathbf{R}^d$.

The Lichnerowicz Conjecture is completely solved by the following theorems.

First, the present author proved the following theorem in 1989.

THEOREM 1.5 [**Sz1**]. *The Lichnerowicz conjecture is true on simply connected and compact manifolds, or more generally (use universal covering spaces!), the conjecture is established on compact harmonic manifolds which have finite fundamental group.*

Using Myers' Theorem [**KN1**] (asserting that a complete and positively curved Einstein manifold is compact and has finite fundamental group) or, more generally, by the Cheeger-Gromoll Theorem [**CG1**] (asserting that the universal covering space $\widetilde{M}^n$ of a complete Riemannian manifold with nonnegative Ricci curvature can be decomposed into the Riemannian product $\widetilde{M}^n = \mathbf{R}^k \times M^{n-k}$, where M^{n-k} is a compact space), we immediately get

THEOREM 1.6 [**Sz1**]. *The Lichnerowicz conjecture is true on complete harmonic manifolds with nonnegative scalar curvature.*

These theorems left the conjecture for the noncompact case (or for the complete spaces with negative scalar curvature) still open. Very recently Damek and Ricci [**DR1, DR2**] announced the following beautiful result saying that the Lichnerowicz Conjecture is not true in the noncompact case.

THEOREM 1.7 [**DR1, DR2**]. *The left invariant metrics on the solvable extensions of the Heisenberg type groups define harmonic manifolds. Since this family of spaces contains beside the negatively curved rank-1 symmetric spaces a number of nonsymmetric negtively curved spaces as well, the Lichnerowicz Conjecture is not true for noncompact manifolds.*

Next we describe both cases in detail in order to find the reason why the conjecture fails in the noncompact case. We will see that the procedure developed for the compact case stops working in the noncompact case only at the very last step. First we describe

3.1. *The proof in the compact case.*

The considerations of [Sz1] start by several characterizations of harmonic spaces. These characterizations concern general spaces and not only the compact ones. Using universal covering spaces *we suppose that the space* (M^n, g) *is simply connected and complete.*

First we introduce *averaging operator* $E_p(u)$ where $p \in M^n$ and u is a continuous function on M^n. $E_p(u)$ is a radial function around p which is defined to have the constant value

$$(1.4) \qquad E_{p;r}(u) := \int u \cdot \theta(\exp_p r\mathbf{e}_p)\, d\mathbf{e}_p \Big/ \int \theta(\exp_p r\mathbf{e}_p)\, d\mathbf{e}_p$$

on the geodesic sphere with radius r and centre p, where $d\mathbf{e}_p$ is the normalized Euclidean measure of the unit vectors $\mathbf{e}_p \in T_p(M^n)$.

By the Allamigeon-Warner Theorem 1.3, this averaging operator E_p is globally defined on simply connected and complete harmonic manifolds and it preserves also the smoothness of the functions.

STEP 1 [Sz1]. *A space is harmonic if and only if the Laplacian* Δ *commutes with the averaging operators* E_p *; i.e.,*

$$(1.5) \qquad E_p(\Delta u) = \Delta E_p(u)$$

for every point $p \in M^n$ *and smooth function* u.

By this statement, if u is an eigenfunction of Δ with the eigenvalue λ, so is the radial function

$$(1.6) \qquad \phi(r(p, x)) = E_p(u)(x)$$

on a harmonic manifold. Furthermore the radial eigenfunction $\phi(r)$ satisfies the differential equation

$$(1.7) \qquad \phi'' + \frac{\theta'}{\theta}\phi' + \lambda\phi = 0,$$

so by the analyticity of a harmonic manifold we have

STEP 2 [Sz1]. *For any eigenvalue* λ *of* Δ *on a harmonic manifold there exists a uniquely determined radial eigenfunction* $\phi_\lambda(r)$ *with eigenvalue* λ *such that* $\phi_\lambda(0) = 1$ *and* $\phi_\lambda'(0) = 0$. *Furthermore for any eigenfunction* u *with eigenvalue* λ *the following Mean Value Property holds:*

$$(1.8) \qquad E_{p,r}(u) = u(p)\phi_\lambda(r).$$

Notice that for harmonic functions (i.e., when $\lambda = 0$) $\phi_0(r) = 1$. So the above statement is the generalization of the Willmore Theorem [Wi1].

Next we establish further characterization of harmonic spaces. The first one concerns the *heat kernel* $H_t(x, y)$ *of Riemannian manifolds.*

On compact manifolds this smooth kernel is uniquely determined by the following properties:

(1.9)
$$(1) \quad \left(\Delta_y + \frac{\partial}{\partial t} \right) H_t(x, y) = 0 \quad \forall x \in M^n,$$
$$(2) \quad \lim_{t \to +0} H_t(x, \cdot) = \delta_x \ (\text{Dirac } \delta\text{-functions}).$$

On noncompact manifolds, the uniqueness requires further restrictions. In this case the Ricci curvature is assumed to be bounded below [Y1]; furthermore, by a theorem of Dodziuk [Dod1], there exists a uniquely determined minimal positive heat kernel $H_t(x, y)$ (i.e., $0 < H_t(x, y) \le q_t(x, y)$ for any positive fundamental solution). Dodziuk shows also that this kernel generates the semigroup.

(1.10)
$$T_t = e^{-t\Delta} = \int_0^\infty e^{-t\lambda} \, dE_\lambda$$

on $L^2(M)$ which may serve for an equivalent definition of the heat kernel $H_t(x, y)$ on open Riemannian manifolds.

STEP 3 [Mi1, Sz1]. *A simply connected and complete Riemannian manifold is harmonic if and only if the heat kernel $H_t(x, y)$ is radial of the form $H_t(x, y) = h_t(r(x, y))$.*

The next characterization of harmonicity concerns the *convolution algebra of radial kernel functions.*

Consider a function $f(r): \mathbf{R}^+ \to \mathbf{R}$. The supporting radius of f is defined as the supremum of support of f. If this supporting radius is less than the injectivity radius of a manifold (M^n, g), then f defines a *radial kernel function*

(1.11)
$$F(x, y) := f(r(x, y))$$

on $M^n \times M^n$. The convolution of the kernel functions $G(x, y)$, $F(x, y)$ (provided $G(x, \cdot), F(\cdot, x) \in L^2(M^n)$, $\forall x \in M^n$) is defined as usual by

(1.12)
$$G * F(x, y) := \int_{M^n} G(x, z) F(z, y) \, dz.$$

The convolution of two radial kernel functions is not a radial kernel on a general Riemannian manifold. More precisely we have

STEP 4 [Sz1]. *A space is harmonic if and only if the radial kernels are closed with respect to convolution (i.e., the convolution of any two radial kernels is a radial kernel again).*

From this statement we get, that the *radial kernels form a commutative convolution algebra on harmonic manifolds.*

The proof of the above statement is based upon the Mean Value Property described in Step 2. In fact, if $F(x, y) = f(r(x, y))$ is a radial kernel and furthermore $\phi_\lambda(x, y) = \phi_\lambda(r(x, y))$ is a radial eigenfunction of Δ on a harmonic manifold, by (1.8) we have

$$(1.13) \qquad F * \phi_\lambda(x, y) = \left(\int_0^\infty f(r)\phi_\lambda(r)\theta(r)\, dr \right) \phi_\lambda(r(x, y));$$

i.e., $F * \phi_\lambda$ is a radial kernel indeed. For general convolutions $F_1 * F_2$, the statement follows by L^2-expansions. Notice, too, that the convolution of radial kernels depends only on the generating functions f_1 and f_2 and on the density function θ on a harmonic manifold. So we have

COROLLARY 1.8. *If two harmonic manifolds have the same radial density function $\theta(r)$ then the corresponding convolution algebras of the radial kernels are isomorphic.*

A similar coincidence can be proved with respect to the heat kernel $H_t(x, y)$.

COROLLARY 1.9. *Simply connected and complete harmonic manifolds with the same radial density function $\theta(r)$ have the same radial heat kernel $H_t(x, y) = h_t(r(x, y))$.*

A simple proof of this corollary will be given in §4.2.

In the following theorem we establish isometric immersions of a harmonic manifold M^n into its L^2-function space, using smooth radial kernels $H(x, y) = h(r(x, y))$ such that $\|h\|_\theta < \infty$ and $0 < \|h'\|_\theta < \infty$; i.e., $h(r(x, \cdot))$, $h'(r(x, \cdot)) \in L^2(M^n)$; $\forall x \in M^h$. We need also the constant $Q_h = \sqrt{n/\Omega_{n-1}}\|h'\|_\theta^2$ for defining the radial kernel $\mathbf{H}(x, y) := Q_h H(x, y)$, where Ω_{n-1} is the volume of the Euclidean unit sphere S^{n-1}. The proof of the following statement is based upon the theorem described the preceding step.

STEP 5 (The Nice Imbedding of Harmonic Manifolds [Sz1]). *For any radial function h as above, the map*

$$(1.14) \qquad \mathbf{r}_h: M^n \to L^2(M^n), \, \mathbf{r}_h: x \to \mathbf{H}(x, \cdot) = Q_h h(r(x, \cdot))$$

has the following properties on a harmonic manifold M^n.

(1) *It is an isometric immersion.*

(2) *The geodesics of the image manifold $\mathbf{r}_h(M^n)$ are congruent screw lines in the space $L^2(M^n)$ (i.e., the geodesics have the same constant Frenet curvature series $\kappa_1, \kappa_2, \kappa_3, \ldots$).*

(3) *Since $\|\mathbf{r}_h(x)\| = Q_h \Omega_{n-1} \int h(r)\theta(r)\, dr$ is constant, the image $\mathbf{r}_h(M^n)$ is lying on a sphere S of the space $L^2(M^n)$. This image is a minimal surface on this sphere S if and only if the radial function $h(r)$ is an eigenfunction of the Laplacian.*

This statement still concerns the general harmonic manifolds and not only the compact ones. A general radial function $h(r(x, y))$ leads to a "very screwed" image $\mathbf{r}_h(M^n)$; i.e., it may span the whole function space $L^2(M^n)$ and the geodesics have an infinite nonzero Frenet curvature series $\kappa_1, \kappa_2, \kappa_3, \dots$.

The consideration of simply connected compact harmonic manifolds starts now. In this case, we establish the Nice Imbedding by the globally defined radial eigenfunctions $\phi_\lambda(x, y) = \phi_\lambda(r(x, y))$ of the Laplacian, whose existence is insured by Steps 1 and 2. More precisely, let $0 < \lambda_1 < \lambda_2 < \cdots$ be the spectrum (without multiplicity!) of Δ on a simply connected compact harmonic manifold and let

$$(1.15) \qquad L^2(M^n) = V^0 + V^1 + \cdots + V^i + \cdots$$

be the corresponding spectral decomposition of $L^2(M^n)$ into the finite-dimensional eigensubspaces V^i. If $\phi_{i1}; \phi_{i2}; \dots; \phi_{ik_i}$ is an arbitrary orthnormal basis in V^i, the radial kernel $\phi_{\lambda_i}(x, y) = \phi_{\lambda_i}(r(x, y))$ is just the kernel

$$(1.16) \qquad \phi_{\lambda_i}(x, y) = \frac{k_i}{\text{vol}(M^n)} \sum_{l=1}^{k_i} \phi_{il}(x)\phi_{il}(y).$$

The Nice Imbedding

$$(1.17) \qquad \mathbf{r}_{\phi_{\lambda_i}} : M^n \to V^i$$

maps the compact harmonic manifold into the finite-dimensional eigensubspace V^i isometrically such that the geodesics are closed congruent screw lines with the same length, say L (see the Allamigeon Observation). Furthermore the image $\mathbf{r}_{\phi_{\lambda_i}}(M^n)$ is a minimal surface on a sphere S of V^i.

These imbeddings of compact harmonic manifolds have been originally constructed by A. Besse in [**Bes1**] and our statement given in Step 5 is a generalization to the general (compact or noncompact) case and to arbitrary radial kernels.

Next we assume that the compact harmonic manifold (M^n, g) is normalized such that the closed geodesics have the length $L = 2\pi$. In this case the radial eigenfunctions $\phi_{\lambda_i}(r)$ are 2π-periodic functions and the functions $\phi_{\lambda_i}(x, \cdot) = \phi_{\lambda_i}(r(x, \cdot))$ $\forall x \in M^n$ span the finite-dimensional function space V^i. From the finiteness of $\dim V^i$ we get

STEP 6 [**Sz1**]. *The radial eigenfunctions $\phi_{\lambda_i}(r)$ of a normalized compact harmonic manifolds are of the form $\phi_{\lambda_i}(r) = P_i(\cos r)$ where the P_i's are polynomials.*

STEP 7 [**Sz1**]. *The squared density function $\theta^2(r)$ is also a trigonometric polynomial of the form $\theta^2(r) = T(\cos r)$ for any compact normalized harmonic manifold.*

The functions $\phi_{\lambda_i}(r)$ are eigenfunctions of the following Schrodinger-type operator

$$(1.18) \qquad \frac{d^2}{dr^2} + \frac{\theta'}{\theta}\frac{d}{dr} = \frac{d^2}{dr^2} + \frac{1}{2}\frac{(\theta^2)'}{\theta^2}\frac{d}{dr}.$$

The polynomial property $\phi_{\lambda_i}(r) = P_i(\cos r)$ of the eigenfunctions means strong restrictions on the polynomial $\theta^2(r) = T(\cos r)$ which can occur only for special polynomials T. The following statement is the heart of our procedure. It is proved in several steps using more lemmas.

MAIN STEP 8 [Sz1]. *The polynomial $T(x)$ has only the roots $+1$, -1, so the density function $\theta(r)$ of a compact normalized harmonic manifold is of the form*

$$(1.19) \qquad \theta(r) = D(1 - \cos r)^A(1 + \cos r)^B = D\sin^p r(1 - \cos r)^q.$$

It should be mentioned that for the density function $\theta(r)$ of a 2-point homogeneous space we have

$$S^n: \theta = D\sin^{n-1} r, \qquad \mathbf{P}^n(\mathbf{C}): \theta(r) = D\sin r(1 - \cos r)^{(n-2)/2},$$

$$(1.20)$$

$$\mathbf{P}^n(\mathbf{H}): \theta = D\sin^3 r(1 - \cos r)^{(n-4)/2}, \qquad \mathbf{P}^{16}(\mathrm{Cay}): \theta = D\sin^7 r(1 - \cos r)^4.$$

One can see that these are the only possibilities for simply connected compact harmonic manifolds. In fact, from $\theta(r) = r^{n-1} + \cdots$ we have $p + 2q = n - 1$, $D = 2^{2q} = 4^q$; furthermore from the Bott-Samelson Theorem 1.4 (which describes the cut locus of a Blaschke manifold) we get $2q = \dim$ (cut locus) $= 0$, $n - 2$, $n - 4$ or 8 (and in the last case $n = 16$). These prove the above statement.

We do not use this remark in the following considerations and we choose a much more elementary way for completing of our proof.

STEP 9 [Sz1]. *Any normalized simply connected compact harmonic manifold has a Laplacian eigenfunction of the form*

$$(1.21) \qquad u(r) = \cos r + (q/(p + q + 1)) = \cos r + c.$$

The full spectrum of the space is

$$(1.22) \qquad \{\lambda_k = k(k + p + q)\}_{k\in\mathbf{N}}$$

and $u(r)$ belongs to the first nontrivial eigenvalue $\lambda_1 = p + q + 1$.

In the following very last step we establish the Lichnerowicz conjecture on simply connected compact harmonic manifolds (or more generally on compact manifolds which have finite fundamental group).

STEP 10 [Sz1]. *Choose the first nontrivial radial eigenfunction $u(r) = \cos r + c$ for the Nice Imbedding*

$$(1.23) \qquad \mathbf{r}_u: M^n \to V^1$$

of the simply connected compact harmonic manifold M^n. Then the geodesics on $\mathbf{r}_u(M^n)$ are unit circles (i.e., plane-curves) in V^1. Consequently, the space is symmetric, which establishes the conjecture.

The above symmetry can be proved as follows: The reflexion of V^1 with respect to the subspace $N_p \perp T_p(\mathbf{r}_u(M^n))$ at an arbitrary point $p \in \mathbf{r}_u(M^n)$ leaves the surface $\mathbf{r}_u(M^n) \subset V^1$ invariant and it induces the geodesic involution as an isometry of the space at any point p.

3.2. The noncompact counterexamples.

THE HEISENBERG-TYPE ALGEBRAS OF KAPLAN.

The Heisenberg-type (or shortly H-type) algebras were introduced by A. Kaplan [Ka1] as follows.

Let $\mathbf{n}$ be a real Lie algebra equipped with a scalar product, which can be written as orthogonal direct sum

$$(1.24) \qquad \mathbf{n} = \mathbf{v} \oplus \mathbf{z},$$

where $[\mathbf{v}, \mathbf{v}] = \mathbf{z}$ and $[\mathbf{v}, \mathbf{z}] = [\mathbf{z}, \mathbf{z}] = 0$. The linear mapping $\mathscr{I} : \mathbf{z} \to \mathrm{End}(\mathbf{v})$ is defined by the formula

$$(1.25) \qquad \langle \mathscr{I}_Z X, X' \rangle = \langle Z, [X, X'] \rangle \quad \forall X, X' \in \mathbf{v}, \forall Z \in \mathbf{z};$$

therefore

$$\mathscr{I}_Z^t = -\mathscr{I}_Z \quad \forall Z \in \mathbf{z}.$$

The $\mathbf{n}$ is said to be H-type algebra if for all Z in $\mathbf{z}$

$$(1.26) \qquad \mathscr{I}_Z^2 = -\langle Z, Z \rangle I,$$

where I denotes the identity map.

These algebras can be attached to Clifford modules. The Clifford algebra over the quadratic space $\{\mathbf{z}, \langle\ \rangle\}$, $C(z)$, is the free associative generated by 1 and $\mathbf{z}$, modulo the relations

$$(1.27) \qquad Z^2 + \langle Z, Z \rangle 1 = 0, \qquad Z \in \mathbf{z}.$$

So the $\mathscr{I}$ extends to a unitary representation of $C(\mathbf{z})$ on $\mathbf{v}$. One can get every such Clifford module in this way. Since the Clifford modules are completely classified [La1], there can be given a classification also for H-type algebras [Ri1].

The H-type algebras are strongly related to the negatively curved rank 1 symmetric spaces. In fact, let $\mathbf{k} \oplus \mathbf{a} \oplus \mathbf{n}$ be the Iwasawa decomposition of a semisimple Lie algebra g belonging to these spaces. Then, equipped with the inner product $-(p + 4q)^{-1} B(\cdot, \theta)$, $\mathbf{n}$ is a H-type algebra [Kor1]. These special Iwasawa H-type algebras have been characterized and classified by Cowling, Dooley, Koranyi, and Ricci [CDKR1] as follows:

We say that a H-type algebra $\mathbf{n}$ satisfies the $\mathscr{I}^2$-condition if whenever $X \in \mathbf{v}$ and $Z, Z' \in \mathbf{z}$ with $\langle Z, Z' \rangle = 0$, then there exists Z'' in z such that

$$(1.28) \qquad \mathscr{I}_Z \mathscr{I}_{Z'} X = \mathscr{I}_{Z''} X.$$

THEOREM 1.8 [CDKR1]. *An H-type algebra* **n** *satisfies the $\mathscr{J}^2$-condition if and only if it is isometrically isomorphic to one of the Iwasawa H-type algebras* $\mathbf{R}^a$, $\mathbf{n}_1^a$, $\mathbf{n}_3^{a,0}$ *or to* $\mathbf{n}_7^1$ *where these algebras can be introduced as follows.*

First, by C_1^a is denoted the Clifford module $\mathbf{C}^a$ (identified with $\mathbf{R}^{2a}$) over $\mathbf{R}$, where

$$(1.29) \qquad \mathscr{J}_Z v = vZ\mathbf{i} \quad \forall Z \in \mathbf{R}, \forall v \in \mathbf{C}^a.$$

Next, denote by $\mathbf{H}$ the space of quaternions, identified with $\mathbf{R}^4$, and denote by $\phi: \mathbf{R}^3 \to \mathbf{H}$ the mapping $(r, s, t) \to r\mathbf{i} + s\mathbf{j} + t\mathbf{k}$. By $C_3^{a,b}$ is denoted the Clifford module $\mathbf{H}^{a+b}$ over $\mathbf{R}^3$ where

$$
\begin{aligned}
&\mathscr{J}_Z(v_1, \ldots, v_a, v_{a+1}, \ldots, v_{a+b}) \\
(1.30) \qquad &= (v_1\phi(Z), \ldots, v_a\phi(Z), \phi(Z)v_{a+1}, \ldots, \phi(Z)v_{a+b}).
\end{aligned}
$$

In the end, denote by $\phi: \mathbf{R}^7 \to \mathbf{H}$ and $\psi: \mathbf{R}^7 \to \mathbf{H}$ the mappings $(Z_1, \ldots, Z_7) \to Z_1\mathbf{i} + Z_2\mathbf{j} + Z_3\mathbf{k}$ and $(Z_1, \ldots, Z_7) \to Z_4 + Z_5\mathbf{i} + Z_6\mathbf{j} + Z_7\mathbf{k}$ respectively. Let C_7^1 be the Clifford module $\mathbf{H}^2$ over $\mathbf{R}^7$, where

$$(1.31) \qquad \mathscr{J}_z(\nu_1, \nu_2) = (\nu_1\phi(Z), \Phi(Z)v_2) + (\Psi(Z)v_2 - \overline{v}_1\Psi(Z)).$$

The H-type algebras $\mathbf{n}_1^a$, $\mathbf{n}_3^{a,b}$, $\mathbf{n}_7^1$ are attached to the above modules. The H-type algebra $\mathbf{R}^a$ occurs when $\mathbf{v} = 0$ or $\mathbf{z} = 0$. It is easy to see that $\mathbf{n}_3^{a,b}$ is isometrically isomorphic to $\mathbf{n}_3^{b,a}$ but otherwise these algebras are nonisomorphic. Furthermore the algebras $\mathbf{n}_3^{a,b}$ with $a \cdot b \neq 0$ and $\mathbf{n}_7^a$ with $a > 1$ do not satisfy the $\mathscr{J}^2$-condition.

THE SOLVABLE EXTENSION NA OF A H-TYPE GROUP N. This extension was introduced by Damek [Da1, Da2] and was investigated in more detail by Cowling, Dooley, Koranyi, and Ricci in [CDKR1, CDKR2]. The group NA is defined by the semidirect product $(\mathbf{v} \oplus \mathbf{z}) \times \mathbf{R}^+$ with multiplication given by the formula

$$(1.32) \quad (X, Z, t)(X', Z', t') = (X + t^{1/2}X', Z + tZ' + \tfrac{1}{2}t^{1/2}[X, X'], tt').$$

The group N is represented in NA as a subgroup given by the elements of the form $(X, Z, 1)$. The Lie algebra

$$(1.33) \qquad\qquad \mathbf{s} = \mathbf{n} \oplus \mathbf{a}$$

of NA is described by the following formulas:

$$\left[\frac{\partial}{\partial t}, X\right] = \frac{1}{2}X, \quad \left[\frac{\partial}{\partial t}, Z\right] = Z,$$

$$(1.34) \qquad [\mathbf{n}, \mathbf{n}]_{NA} = [\mathbf{n}, \mathbf{n}]_N \subset \mathbf{n}, \quad \forall X \in \mathbf{v}, Z \in \mathbf{z}.$$

This Lie algebra (identified with the tangent space at the unit element $e = (0, 0, 1)$) can be endowed by the inner product described by the rule

$$(1.35) \qquad \langle (X, Z, s), (X', Z', s') \rangle = \langle X, X' \rangle + \langle Z, Z' \rangle + ss',$$

which can be extended into a left-invariant Riemannian metric (NA, can) on the group NA. The Riemannian spaces $\mathbf{R}^a A$, $N_1^a A$, $N_3^{a,0} A$, $N_7^1 A$ are the negatively curved 2-point homogeneous spaces. The space $\mathbf{R}^a A$ is the real hyperbolic space of scalar curvature $-(a+1)$, resp. $-(a+1)/4$, corresponding to the cases $\mathbf{v} = 0$, resp. $\mathbf{z} = 0$. The other NA spaces, for instance $N_3^{a,b} A$ with $a \cdot b \neq 0$ and $N_7^a A$ with $a > 1$, are nonsymmetric spaces. An accurate description of nonsymmetric NA-spaces can be found in [DR1]. The smallest dimension for these spaces is 7.

Damek and Ricci proved the harmonicity on all NA-spaces. Their proof rests on some results of [CDKR1]. One of them describes the totally geodesics subgroups in the spaces (N, can) resp. (NA, can) as follows:

THEOREM 1.9 [CDKR1]. (A) *Suppose* $\dim \mathbf{v} > 0$ *and* $\dim \mathbf{z} > 0$ *and consider an element* $(X, Z) \in N$ *such that* $X \neq 0$ *and* $Z \neq 0$. *Then the smallest closed totally geodesics subgroup of* N *containing* (X, Z) *is given by the formula*

$$(1.36) \qquad N_{(X,Z)} = \{(t_1 X + t_2 \mathcal{J}_Z(X), t_3 Z) : \mathbf{t} \in \mathbf{R}^3\}.$$

(B) *If* $(X, Z, t) \in NA$ *is such that* $X \neq 0$ *and* $Z \neq 0$ *then the smallest closed totally geodesic subgroup containing* (X, Z, t) *is the extension* $N_{(X,Z)} A$ *of* $N_{(X,Z)}$.

This theorem says that if a space NA is not the real hyperbolic space then it is covered by isometrically equivalent 4-dimensional complex hyperbolic planes of scalar curvature $-(1 + 3/4)$.

The other tool of [DR1, DR2] is the generalized Cayley transform introduced in [CDKR1]. This transformation maps the unit ball

$$(1.37) \qquad B = \{(X, Z, t) \in \mathbf{s} = \mathbf{n} \oplus \mathbf{a} | \, |X|^2 + |Z|^2 + t^2 = \tau^2 < 1\}$$

onto NA, by the formula
(1.38)
$$C(X, Z, t) = \frac{1}{(1-t)^2 + |Z|^2} (2(1 - t + \mathcal{J}_Z)(X), 2Z, 1 - t^2 - |X|^2 - |Z|^2).$$

By the pull-back of the metric, we represent (NA, can) on B. The restriction of C onto the total geodesic subgroups $N_{(X,Z)} A$ is just the usual Cayley transform of the complex hyperbolic planes. So it can be easily seen that the geodesics through the origin $0 \in B$ are just the straight lines such that the geodesic distance, r, from the origin depends only on the Euclidean distance $\tau = \sqrt{|X|^2 + |Z|^2 + t^2}$. The exact formulation of the Damek-Ricci Theorem is as follows:

THEOREM 1.10 [DR1, DR2]. *The Riemannian density on* (NA, can) *is the left-invariant Haar measure*

(1.39) $dm_L = a^{-Q-1} dX \, dZ \, da \qquad Q = \frac{1}{2} \dim v + \dim z.$

Let τ, ω be radial coordinates on B; furthermore $m = \dim v$, $k := \dim z$.
The measure

$$
\begin{aligned}
d\gamma &= 2^{m+k+1}(1 - \tau^2)^{-Q-1} dX \, dZ \, dt \\
&= C_{m,k} \tau^{m+k} (1 - \tau^2)^{-Q-1} d\tau \, d\sigma(\omega) \\
&= C_{m,k} (\cosh r + 1)^{k/2+1} (\cosh r - 1)^{(m+k)/2} dr \, d\sigma(\omega) \\
&= C_{m,k} (\sinh r)^{k+2} (\cosh r - 1)^{m/2-1} dr \, d\sigma(\omega)
\end{aligned}
$$

(1.40)

is the image of the Haar measure dm_L via the map C^{-1}.

 The first statement can be easily checked. For the proof of the second statement, the authors computed the Jacobi map dC and its determinant $\det dC$.

The radial property of $d\gamma$ proves the harmonicity of the NA-spaces, providing a very wide class of nonsymmetric harmonic spaces.

For comparison with the compact case notice that the squared density function of an NA-space is the polynomial of the $\cosh r$ function of the form $\theta^2(r) = T(\cosh r)$ such that the T has two roots namely $+1$ and -1. A similar statement is the main point in the compact case. There we proved this polynomial property with respect to the function $\cos r$. Also on the NA-spaces, a simple linear Laplacian eigenfunction of the form $\phi_\lambda(r) = A \cosh r + B$ exists but this radial eigenfunction does not belong to the L^2 function space and so the Nice Imbedding of NA cannot be established by $\phi_\lambda(r)$. So the method developed for the compact case stops working on noncompact manifolds only at the very last step. Just this suggests that this method may be useful in the classification of harmonic spaces.

4. Further observations on NA-spaces.

4.1. *Curvature; The Osserman conjecture.* By harmonicity, the NA-spaces are Einstein manifolds. Next we compute the Riemannian curvature in an accurate form.

The Riemannian connection for any left-invariant metric on a Lie group is easily seen to satisfy (cf. [Wo2])

(1.41) $\langle \nabla_X Y, W \rangle = \frac{1}{2}(\langle [X, Y], W \rangle - \langle [Y, W], X \rangle + \langle [W, X], Y \rangle),$

where X, Y, W are arbitrary elements of the corresponding Lie algebra. So (1.34) shows that this connection of (NA, can) is given by

$$\nabla_X Y = \frac{1}{2}[X, Y] + \frac{1}{2}\langle X, Y \rangle \frac{\partial}{\partial t},$$

$$\nabla_Z X = \nabla_X Z = -\frac{1}{2} \mathscr{J}_Z(X),$$

(1.42)
$$\nabla_{Z_1} Z_2 = \langle Z_1; Z_2 \rangle \frac{\partial}{\partial t},$$

$$\nabla_{\frac{\partial}{\partial t}} X = \nabla_{\frac{\partial}{\partial t}} Z = \nabla_{\frac{\partial}{\partial t}} \frac{\partial}{\partial t} = 0,$$

$$\nabla_X \frac{\partial}{\partial t} = -\frac{1}{2} X, \qquad \nabla_Z \frac{\partial}{\partial t} = -Z,$$

$\forall X, Y \in \mathbf{v}$ and $Z, Z_1, Z_2 \in \mathbf{z}$.

Damek [**Da2**] has proved that the sectional curvature of a space (NA, can) is nonpositive. Next we give a more accurate description of this curvature by computing the eigenvalues of the Jacobian curvature

(1.43)
$$R(\cdot, y)y = \nabla \cdot \nabla_y y - \nabla_y \nabla \cdot y - \nabla_{[\cdot, y]} y$$

for an arbitrary unit vector

(1.44)
$$y = \left(X, Z, c\frac{\partial}{\partial t} \right) \in \mathbf{s} = \mathbf{n} \oplus \mathbf{a} \quad \text{with } X \neq 0, Z \neq 0.$$

The selfadjoint operator $R(\cdot, y)y$ acts on the Jacobi hyperplane $y^\perp \perp y$ and its eigenvalues are just the principal sectional curvatures with respect to the planes lying on y.

First we decompose the Jacobi hyperplane into the orthogonal direct sum

(1.45)
$$y^\perp = Q_y \oplus P_y^0 \oplus P_y^1,$$

where $Q_y = y^\perp \cap T_e(N_{(X,Z)}A)$ is the Jacobi hyperplane on the totally geodesic subgroup $N_{(X,Z)}A$, the subspace $P_y^0 \subset \mathbf{v}$, $P_y^0 \perp Q_y$ is defined by the relations

(1.46)
$$[P_y^0, X] = [P_y^0, \mathscr{J}_Z(X)] = 0,$$

and furthermore $P_y^1 \subset \mathbf{v} \oplus \mathbf{z}$ is the orthogonal complement to $Q_y \oplus P_y^0$. It is easy to see that P_y^1 is spanned just by the subspaces

(1.47)
$$Z^\perp; \mathscr{J}_{Z^\perp}(X); \mathscr{J}_{Z^\perp}\mathscr{J}_Z(X),$$

where $Z^\perp$ is the orthogonal hyperplane to Z in $\mathbf{z}$. The $\mathscr{J}^2$-property can be checked on the subspace P_y^1 as follows:

Introduce the unit vectors $X_e := X/|X|$, resp. $Z_e = Z/|Z|$, and define the map $K_y : Z^\perp \to Z^\perp$ by

(1.48)
$$K_y(z^\perp) := [X_e, \mathscr{J}_{z^\perp}\mathscr{J}_{z_e}(X_e)].$$

It can be easily seen that $K_y^t = -K_y$ and for the imaginary eigenvalues μ of K_y we have $|\mu| \leq 1$. Furthermore the space satisfies the $\mathscr{J}^2$-condition if and only if $|\mu| = 1$ for any eigenvalue of K_y at every unit vector y with $X \neq 0$, $Z \neq 0$.

Let $V + z^\perp \in P_y := P_y^0 \oplus P_y^1$ be an arbitrary vector. By a lengthy but straightforward calculation we get

$$
(1.49) \quad
\begin{aligned}
R(V + z^\perp, y)y = {}& -\frac{1}{4}V - \frac{3}{4}\mathcal{J}_{[X,V]}(X) + \mathcal{J}_{z^\perp}(\mathbf{q}) \\
& + \left(\frac{3}{4}|X|^2 - 1\right)z^\perp + [\mathbf{q}, V],
\end{aligned}
$$

where $\mathbf{q} = \frac{3}{4}\mathcal{J}_Z(X) - \frac{3}{4}cX$. From these formulas we have

THEOREM 1.11. (A) *If* $y \in \mathbf{v} \cup \mathbf{z} \cup \frac{\partial}{\partial t}$, *the Jacobi curvature* $R(\cdot, y)y$ *has the eigenvalues* -1, *resp.* $-1/4$, *with multiplicity* $k = \dim \mathbf{z}$, *resp.* $m = \dim \mathbf{v}$. *Consequently an NA-space is an Einstein space with constant scalar curvature* $-(k + m + 1)(k + (m/4))$.

(B) *If* $X \neq 0$, $Z \neq 0$ *for the vector* $y = (X, Z, c\frac{\partial}{\partial t})$, *all the subspaces* Q_y, P_y^0, P_y^1 *are invariant under the action of* $R(\cdot, y)y$ *and there we get the following eigenvalues.*

(1) *On* Q_y, *the eigenvalues are* -1, *resp.* $-1/4$, *with multiplicity* 1, *resp.* 2.

(2) *On* P_y^0, $-1/4$ *is the only eigenvalue with multiplicity* $m_0 = \dim P_y^0$.

(3) *On* P_y^1, *the eigenvalues* λ *of* $R(\cdot, y)y$ *depend on the eigenvalues*

$$
(1.50) \quad 0 \leq |\mu_1| < |\mu_2| < \cdots < |\mu_l| \leq 1
$$

of K_y *by the following third-order equation:*

$$
(1.51) \quad (\lambda + 1)(\lambda + \tfrac{1}{4})^2 + (\tfrac{3}{4})^3 |X|^4 |Z|^2 (|\mu_i|^2 - 1) = 0.
$$

More precisely, let

$$
(1.52) \quad Z^\perp = L_1 \oplus L_2 \oplus \cdots \oplus L_l
$$

be the eigensubspace decomposition with respect to the operator K_y; *i.e.,* $K_y^2(L_i) = -|\mu_i|^2 L_i$. *If* $|\mu_l| = 1$, *the* $R(\cdot, y)y$ *has the eigenvalues* -1, *resp.* $-1/4$, *with the same multiplicity equal to* $\dim L_l$. *For an eigenvalue* $0 \leq |\mu_i| < 1$, *the* $R(\cdot, y)y$ *has three eigenvalues of the form*

$$
(1.53) \quad
\begin{aligned}
-1 < \lambda_{i1} &\leq -\frac{3}{4} \leq \lambda_{i2} < -\frac{1}{4} < \lambda_{i3} \leq 0, \\
\lambda_{i1} + \lambda_{i2} &+ \lambda_{i3} = -\frac{3}{2}
\end{aligned}
$$

which have the same multiplicity equal to $\dim L_i$.

A Riemannian space is said to satisfy the *Osserman-property* if the eigenvalues $\lambda_1 \leq \lambda_2 \cdots \leq \lambda_{n-1}$ of the Jacobian curvature $R(\cdot, y)y$ are constant; i.e. they are independent on the unit vector y. The *Osserman conjecture* says that these spaces are exactly the 2-point homogeneous spaces. This conjecture has been proved by Chi [Ch1] except in the cases dim $= 4k$. The spaces (NA, can) do not give counterexample for this conjecture, since by the above theorem we have

COROLLARY. *A space* (NA, can) *satisfies the Osserman property if and only if it satisfies the* $\mathscr{J}^2$-*condition; i.e., the space is* 2-*point homogeneous.*

4.2. *Spectral geometric coincidences on the spaces* $N_3^{(a,b)}A$. The radial density function $\theta(r)$ of an NA space depends only on $k = \dim z$ and $m = \dim v$; therefore the nonisometric spaces $N_3^{(a,b)}A$, where $k = 3$ and $m = 4(a + b)$, have the same density function

(1.54) $$\theta(r) = C_{m,3}(\sinh r)^5(\cosh r - 1)^{m/2-1}$$

as the $(m + 4)$-dimensional quaternionic hyperbolic space $N_3^{(a+b,0)}A$. The spaces $N_3^{(a,b)}A$ with $ab \neq 0$ are not 2-point homogeneous since, by the noncommutativity of quaternions, the $\mathscr{J}^2$-condition and consequently the Osserman property do not hold on these spaces. Yet they have the same density function $\theta(r)$ as the 2-point homogeneous space $N_3^{(a+b,0)}A$.

This coincidence of the density function $\theta(r)$ illuminates the most difficult point of the classification problem of harmonic spaces. Namely, at present it is not clear whether there exist other spaces (for instance nonhomogeneous ones) having the same density function (1.54). It seems that the classification—program of harmonic spaces can be achieved only up to the determination of all possible density functions $\theta(r)$.

For a general treatment, one can introduce the notion of *density equivalent spaces*:

Let g_1 and g_2 be Riemannian metrics defined on the same manifold M^n. These spaces are said to be density equivalent, if at every point $p \in M^n$ there exists an endomorphism $A_p : T_p(M^n) \to T_p(M^n)$ such that

(1.55) $$g_1(X, Y) = g_2(A_p(X), A_p(Y)), \quad \forall X, Y \in T_p(M^n),$$

(1.56) $$\omega_{1p}(q) = \omega_{2p}(\exp_{2p} \circ A \circ \exp_{1p}^{-1}(q)).$$

The *totally symmetric density tensors* of a space (M^n, q) are defined by the derivatives:

$$\omega_p^{(l)}(\mathbf{m}_p, \mathbf{m}_p; \dots; \mathbf{m}_p) := \nabla_{\mathbf{m}_p}^{(l)} \omega_p, \qquad \mathbf{m}_p \in T_p(M^n), |\mathbf{m}_p| = 1$$

and the *scalar density invariants* are defined by the following trace of the density tensors:

(1.57) $$\mathrm{Tr}\, \omega_p^{(l)} := \int \omega_p^{(l)}(\mathbf{m}_p, \dots, \mathbf{m}_p)\, d\mathbf{m}_p.$$

Density equivalent spaces have equivalent density tensors (for instance, equivalent Ricci curvature) and pointwise identical scalar density invariants (for instance, they have the same scalar curvature).

The spaces $N_3^{(a,b)}A$ with $a + b = $ constant are obviously density equivalent. This equivalence implies further interesting spectral geometric coincidences.

(1) On the nonisometric spaces $N_3^{(a,b)}A$ with $a + b = $ constant, *the geodesic balls with the same radius have the same volume.*

At this point the work of Gray and Vanhecke [GV1] should be mentioned, where they investigated those Riemannian spaces on which the volume of a geodesic ball with radius R is equal to the volume of the Euclidean ball with the radius R. The authors conjectured that such spaces were Euclidean and they proved this statement for low-dimensional cases $(\dim \leq 5)$. The question is still open in higher-dimensional cases but the above examples show that such a statement might be true only on Euclidean spaces or, perhaps, on spaces of constant (real or holomorphic) sectional curvature.

(2) From the equation

$$(1.58) \qquad\qquad \phi_\lambda'' + \frac{\theta'}{\theta}\phi_\lambda' + \lambda\phi_\lambda = 0$$

of the radial eigenfunctions we get, that *any eigenvalue λ of Δ is represented by the same radial eigenfunction $\phi_\lambda(r)$ on the nonisometric spaces $N_3^{(a,b)}A$* with $a + b = $ constant. It would be interesting to know whether the geodesic balls with the same radius have the same Dirichlet, resp. Neumann, spectrum on these spaces. By the Mean Value Property (1.8), this coincidence can be directly proved for such spectrum elements λ_i which can be represented by eigenfunctions which do not vanish at the centre of the ball. Some further coincidences can be proved too, but we do not go into detail here.

(3) By Corollary 1.9, *the nonisometric spaces $N_3^{(a,b)}A$ with $a+b = $ constant have the same radial heat kernel $H_t(x, y) = h_t(r(x, y))$.* More precisely, the generating function $h_t(r)$ is common in this case and after representing these spaces on the unit ball $B \subset \mathbf{s} = \mathbf{n} \oplus \mathbf{a}$ by the Cayley transform (1.38), also the function $H_t(0, y) = h_t(r(0, y))$ around the origin $0 \in B$ is the same for these spaces $N_3^{(a,b)}A$. For a general point $p \in B$ there can be stated a *heat kernel equivalence* similar to the density equivalence given in (1.55) and (1.56).

Notice that *also the Brownian motion is the same on the nonisometric spaces $N_3^{(a,b)}A$ with $a + b = $ constant.*

In fact, let $\beta \subset B$ be an arbitrary Borel set. The probability that the process started at $0 \in B$ is at the time t in β is

$$(1.59) \qquad\qquad rw(0, t, \beta) := \int_\beta H_t(0, y)\,dy.$$

This probability is obviously the same on the spaces $N_3^{(a,b)}$ with $a + b = $ constant.

An easy proof of Corollary 1.9 is as follows: Let $B_{p,R}$ be a ball with centre p and radius R on a harmonic space and let $0 < \lambda_1 < \lambda_2 < \cdots$ be its

Dirichlet-Laplace spectrum without multiplicity. Furthermore let V^i be the Dirichlet-Laplace eigensubspace concerning λ_i. If there is an eigenfunction $u \in V^i$ not vanishing at the centre p then also the radial eigenfunction $\phi_{\lambda_i}(r(p, \cdot))$ around the centre p belongs to V^i, by the Mean Value Property (1.8). Such eigenvalues

(1.60) $$0 < \lambda_{w1} < \lambda_{w2} < \lambda_{w3} < \cdots$$

form the so-called *spectrum with warm centre* and the rest of the spectrum

(1.61) $$\lambda_{c1} < \lambda_{c2} < \lambda_{c3} < \cdots$$

is said to be the *spectrum with cold centre*. The spectrum with warm centre is uniquely determined by the radius R and the density $\theta(r)$ on a harmonic manifold.

Choose an orthonormal basis $\phi_{i1}, \phi_{i2}, \ldots, \phi_{il_i}$ in V^i such that

(1.62) $$\phi_{i1}(\cdot) = \phi_{\lambda_{wi}}(r(p, \cdot))/\|\phi_{\lambda_{wi}}\|.$$

All the other eigenfunctions ϕ_{ik} vanish at p, by the Mean Value Property (1.8). So for the Dirichlet heat kernel $H_{B_{p;R}}(t, x, y)$ we have

(1.63) $$H_{B_{p;R}}(t, p, y) = \sum_{i,k} e^{-\lambda_i t} \phi_{ik}(p)\phi_{ik}(y) = \sum_i e^{-\lambda_{wi} t} \phi_{wi1}(p)\phi_{wi1}(y).$$

Therefore the function $H_{B_{p;R}}(t, p, \cdot)$ is radial around p and it is uniquely determined by R and $\theta(r)$. On the other hand, by a theorem of Dodziuk [D1], we have

(1.64) $$H(t, p, q) = \sup_{R \to \infty} H_{B_{p;R}}(t, p, q)$$

on open manifolds; therefore the heat kernel $H(t, p, q)$ is radial one of the form $H(t, p, q) = h_t(r(p, q))$ and it is uniquely determined by the density $\theta(r)$ on simply connected and complete harmonic manifolds.

(4) The strongest coincidence on the nonisometric spaces $N_3^{(a,b)}A$ with $a + b = $ constant is stated in Corollary 1.8. It says that if we identify the radial kernels $H(x, y) = h(r(x, y))$ generated by the same function $h(r)$, this map is an isomorphism which preserves the linear, multiplicative and also the convolution structure of the corresponding radial kernel algebras.

For a more general treatment, consider a Riemannian manifold (M^n, g) whose injectivity radius δ is positive. Consider the radial kernels $H(x, y) = h(r(x, y))$ on M^n, where $h(r)$ is a continuous function with compact support such that its supporting radius is less than δ. By linear combinations, pointwise multiplications: $H_1 \cdot H_2(x, y) = H_1(x, y)H_2(x, y)$ and by convolution $H_1 * H_2$ we generate a larger kernel algebra which is said to be the *radial kernel algebra* or *density algebra*. The density algebras of the Riemannian manifolds (M_1, g_1) and (M_2, g_2) are said to be *naturally isomorphic* if the identification of the radial kernels $H_1(\cdot, \cdot) = h(r_1(\cdot, \cdot))$ and

$H_2(\cdot,\cdot) = h(r_2(\cdot,\cdot))$, generated by the same function $h(r)$, extends to an isomorphism of the corresponding density algebras.

By the above examples, *nonisometric spaces can have naturally isomorphic density algebras*. On the other hand, the density function $\omega_p(q)$, the Laplace spectrum of the manifold as well as the heat kernel and the Brownian motion can be described by the density algebra. Since these notions describe the space at "the level of quantum mechanics," the above indeterminacy of the metric looks like *uncertainty for the metric representing "the same quantum mechanical systems."* Because of this reason, the development of a theory of spaces having equivalent density algebras is desirable. This may also illuminate the puzzle: *How is God playing dice with the space-time metrics of Einstein?*

In this paper, we follow the opposite way. Roughly speaking, we get rid of the above uncertainty seeking certainty by involving other kernel functions into the investigations. We introduce these kernels by the invariants of the Jacobian endomorphism field $A_{c(r)}(\cdot)$ which is defined along a geodesics $c(r)$ by the equation

$$(1.65) \qquad A''_{c(r)} + R_{c(r)}A_{c(r)0} = 0, \qquad A_{c(0)} = 0, \qquad A'_{c(0)} = \mathrm{id},$$

where $R_{\dot{c}}(\cdot) := R(\cdot, \dot{c})\dot{c}$ is the Jacobian curvature along $c(r)$; furthermore $A_{c(r)}$ acts on the Jacobi subspace $V_{c(r)} \perp \dot{c}(r)$. One of these invariants is the density

$$(1.66) \qquad \theta_{c(0)}(c(r)) = \det A_{c(r)}.$$

It turns out, for instance, that the radial property of all these so-called Jacobi kernels is equivalent to the 2-point homogeneous property of the space. In general, these kernels describe the metric much more precisely than the single density kernel $\theta_p(q)$. These kernels define several differential operators on the manifold. The main point of this paper is to give a spectral theory for all these operators. The exact details are as follows.

II. The ring of the Jacobi kernels and the Jacobi differential operators $\square_H^{(k)}$, $\Delta_H^{(k)}$

Let p be a point of a Riemannian manifold (M^n, g) and, for a unit vector $\mathbf{e}_p \in T_p(M^n)$, $e_p(r)$ stands for the geodesic with $e_p(0) = p$ and $\dot{e}_p(0) = \mathbf{e}_p$. In order to get kernels $H_p(q) = H(p, q) = H^q(p)$ which are differentiable also at the diagonal points (p, p), we introduce the Jacobian field

$$(2.1) \qquad \mathbf{A}_{e_p(r)} = \frac{1}{r}A_{e_p(r)}$$

which will be denoted also by $\mathbf{A}_{p,q}$ or by $\mathbf{A}(p, q)$, where $q = e_p(r)$. If the tangent spaces along $e_p(r)$ are identified by parallel displacement, the well-known symmetry

(2.2) $$\mathbf{A}^*(p,q) = \mathbf{A}(q,p)$$

can be proved [Bes1].

The normal density kernel $\omega_p(q)$ is one of the invariants of $\mathbf{A}_p(q)$, namely

(2.3) $$\omega_p(q) = \det \mathbf{A}_p(q)$$

[Bes1]; consequently $\omega_p(q) = \omega_q(p)$. We introduce also the other invariants as follows.

For an arbitrary polynomial

(2.4) $$\mathrm{pol}(\mathbf{A}_{p,q}, \mathbf{A}^*_{p,q})$$

of both fields $\mathbf{A}$ and $\mathbf{A}^*$, consider the kernel

(2.5) $$H_p(q) := \mathrm{Tr}\,\mathrm{pol}(\mathbf{A}_{p,q}, \mathbf{A}^*_{p,q}).$$

The set of kernels we get in this way are not closed with respect to the multiplication $(H_1 \cdot H_2)(p,q) = H_1(p,q)H_2(p,q)$. Therefore by taking all the polynomial expressions

(2.6) $$\mathrm{pol}(H_1, H_2, \ldots, H_k)$$

of the above kernels we generate the so-called Jacobian ring, $\mathcal{J}R$, of the Jacobian kernels.

All these Jacobian kernels are locally defined differentiable functions. Furthermore this ring is closed with respect to the transposition

(2.7) $$H^t(p,q) = H(q,p)$$

and it contains the unit element $H_p(q) = 1$ and the density kernel $\omega_p(q) = \det \mathbf{A}_p(q)$ as well.

If we use only the field $\mathbf{A}_{pq}$ in the above procedure, we get the so-called pure Jacobian ring, $P\mathcal{J}R$, of the pure Jacobian kernels which consists of symmetric kernels $(H(p,q) = H(q,p))$. By the Fundamental Theorem of the elementary symmetric polynomials, this pure ring is finitely generated (namely by $\alpha^{(k)} := \mathrm{Tr}\,\mathbf{A}^k$, $k = 0, 1, \ldots, n-1$) and it contains the kernels 1 and $\omega_p(q)$ as well. The ring $\mathcal{J}R$ is also called total Jacobian ring. The elements $H \in \mathcal{J}R$ are invariant w.r.t. the isometries of the space; i.e. $H(\phi(p), \phi(q)) = H(p,q)$ for any isometry ϕ.

Now we define several differential operators with the help of the Jacobi kernels $H_p(q)$. First, for a function $H \in \mathcal{J}R$, we define the unnormed averaging $UE_{H;p,r}(u)$ of a continuous function u along the sphere $S_{p,r}$ with the centre p and radius r by

(2.8) $$UE_{H;p;r}(u) = \int_{S^{n-1}} u(e_p(r))H_p(e_p(r))\,de_p,$$

where de_p means the normalized Euclidean measure of the unit vectors $\mathbf{e}_p \in S^{n-1} \subset T_p(M^n)$. If $\int H_p(e_p(r))\,de_p = UE_{H;p;r}(1) \neq 0$ (which condition is not satisfied in general) we can define the proper averaging $E_{H;p;r}$ by

(2.9) $$E_{H;p;r}u = UE_{H;p;r}(u)/UE_{H;p;r}(1).$$

The kth order Jacobi-Schrodinger operators $\square_H^{(k)}$ are defined by the following $2k$th order derivative at the origin $r = 0$:

$$(2.10) \qquad \square_H^{(k)} u(p) = \frac{\partial^{2k} U E_{H;p;r}(u)}{\partial r^{2k}} \Big/ r = 0.$$

The reason for this name is that the operators $\square_H^{(1)}$ are of the form

$$(2.11) \qquad \square_H^{(1)} = C\Delta + DR,$$

where C and D are constant and R is the curvature scalar.

If $\int H_p(e_p(r)) de_p \neq 0$ for small values $0 < r < \varepsilon$, then the kth order Jacobi-Laplace operator $\Delta_H^{(k)}$ is defined by

$$(2.12) \qquad \Delta_H^{(k)} u(p) = \frac{\partial^{2k} E_{H;p;r}(u)}{\partial r^{2k}} \Big/ r = 0.$$

The first-order operators $\Delta_H^{(1)}$ are of the form $\Delta_H^{(1)} = C\Delta$, $C = $ constant.

The operators are clearly invariant w.r.t. the isometries of the space.

A detailed investigation of these operators is missing from the literature and only the operators $\Delta_1^{(k)} = \Delta^{(k)} = \square_1^{(k)}$ were considered to some extent by Willmore [Wi3], Gray [GW1] and Gilkey [Gi1] so far. These operators are said to be *unweighted Laplacians* and the operators $\square_\omega^{(k)}$; $\Delta_\omega^{(k)}$ are called *density operators*.

Explicit expressions for the operators of lower order can be computed from the power series

$$
\begin{aligned}
(2.13) \qquad \mathbf{A}_{e_p} &= \mathrm{id} + \frac{1}{6} r^2 R_{ep} + \frac{1}{12} r^3 R'_{e_p} + \frac{r^4}{5!}(R^2_{e_p} + 3R''_{e_p}) \\
&\quad + \frac{r^5}{6!}(4R'_{e_p} R_{e_p} + 2R_{e_p} R'_{e_p} + 4R'''_{e_p}) + O(r^6)
\end{aligned}
$$

of the Jacobian field, where $R'_{e_p}(\cdot) = (\nabla_{e_p} R)(\cdot \ \mathbf{e}_p)\mathbf{e}_p$, etc. For instance, for the fields

$$(2.14) \qquad S_{pq} := \frac{1}{2}(\mathbf{A}_{pq} + \mathbf{A}^*_{pq}), \qquad L_{pq} := \frac{1}{2}(\mathbf{A}_{pq} - \mathbf{A}^*_{pq}),$$

$$(2.15) \qquad Q_{pq} := \frac{1}{2}[\mathbf{A}^*_{pq}, \mathbf{A}_{pq}] = [S_{pq}, L_{pq}] := S_{pq} L_{pq} - L_{pq} S_{pq}$$

we have

$$(2.16) \qquad S_{pq} = \mathrm{id} + \frac{1}{6} r^2 R_{\mathbf{e}_p} + O(r^3),$$

$$(2.17) \qquad L_{pq} = \frac{r^5}{6!}[R'_{\mathbf{e}_p}, R_{\mathbf{e}_p}] + O(r^6),$$

$$(2.18) \qquad Q_{pq} = \frac{r^7}{6 \cdot 6!}[[R'_{\mathbf{e}_p}, R_{\mathbf{e}_p}], R_{\mathbf{e}_p}] + O(r^8).$$

Furthermore for the operators $\Delta^{(2)}$ and $\Delta_\omega^{(2)}$ we get

$$(2.19) \qquad n(n+2)\Delta^{(2)} = 3\Delta^2 + 2\rho^{ij}\nabla_i\nabla_j + 2(\nabla_j\rho^{ij})\nabla_i,$$

$$(2.20) \qquad n(n+2)\Delta_\omega^{(2)} = n(n+2)\Delta^{(2)} - 2R\Delta - 4\rho^{ij}\nabla_i\nabla_j - (\nabla^i R)\nabla_i$$

where R is the scalar curvature and ρ_{ij} is the Ricci curvature.

On Kahler manifolds (M, g, J) we introduce a new kernel K_{pq} defined by

$$(2.22) \qquad K_{p,q} = g(A_{pq}(J(\dot{e}_p(r))), J(\dot{e}_p(r))), \qquad q = e_p(r).$$

The symmetry $K_{pq} = K_{qp}$ easily follows from $A_{pq}^* = A_{qp}$. From the power series

$$K_{p;e_p(r)} = r - \frac{r^3}{6}R(\mathbf{e}_p, J(\mathbf{e}_p), \mathbf{e}_p, J(\mathbf{e}_p))$$

$$(2.22) \qquad\qquad - \frac{r^4}{12}(\nabla_{\mathbf{e}_p}R)(\mathbf{e}_p, J(\mathbf{e}_p), \mathbf{e}_p, J(\mathbf{e}_p)) + O(r^5),$$

we see that this kernel is not differentiable at the diagonal points (p, p). Therefore we introduce

$$(2.23) \qquad \mathbf{K}_{p,q} := r(p, q)K_{p,q}$$

which is called *Kahler-Jacobi kernel*. We will see that this single kernel strongly determines the structure of Kahler manifolds.

III. The weak form of the Lichnerowicz conjecture

1. The conjecture for Jacobi harmonic spaces. A Riemannian space is said to be Jacobi-harmonic if any Jacobi kernel $H_p(q) \in \mathscr{J}R$ is a radial function around any point $p \in M^n$. The 2-point homogeneous spaces are obviously Jacobi-harmonic spaces. The converse statement is also true:

THEOREM 3.1. *A Riemannian space is 2-point homogeneous if and only if it is Jacobi-harmonic.*

The proof is based on the Fundamental Lemma 1.1.

On Jacobi-harmonic manifold the pure kernels

$$(3.1) \qquad \mathrm{Tr}(\mathbf{A}_{e_p(r)} - \mathrm{id})^k = \frac{r^{2k}}{6^k}\mathrm{Tr}\, R_{\mathbf{e}_p}^k + O(r^{2k+1})$$

are of the form $H(p, q) = h(r(p, q))$ by the symmetry $H(p, q) = H(q, p)$ of these kernels. Therefore, $\mathrm{Tr}\, R_{\mathbf{e}_p}^k = $ constant for any k and so the Jacobi curvature $R_{\mathbf{e}_p}(\cdot) = R(\cdot, \mathbf{e}_p)\mathbf{e}_p$ has constant eigenvalues over the whole manifold. I.e. the space satisfies the Osserman property on the Jacobian curvature tensor.

From (2.18) we get similarly that the field

$$(3.2) \qquad \Lambda_{\mathbf{e}_p} := [[R'_{\mathbf{e}_p}, R_{\mathbf{e}_p}], R_{\mathbf{e}_p}]$$

has constant eigenvalues at each point p on the unit sphere $S \subset T_p(M^n)$. Since $\Lambda^*_{e_p} = \Lambda_{e_p}$ and $-\Lambda_{-e_p} = \Lambda_{e_p}$,

$$(3.3) \qquad \Lambda_{e_p} = R'_{e_p} R^2_{e_p} - 2R_{e_p} R'_{e_p} + R^2_{e_p} R'_{e_p} = 0$$

by the Fundamental Lemma.

Now we show that from this equation,

$$(3.4) \qquad [R'_{e_p}, R_{e_p}] = 0$$

follows.

Let $\mathbf{v}$ be an eigenvector of R_{e_p} with eigenvalue λ and let

$$(3.5) \qquad R'_{e_p}(\mathbf{v}) = \sum \mathbf{v}_i$$

be the decomposition following the eigenspaces of R_{e_p}; i.e. $R_{e_p}(\mathbf{v}_i) = \lambda_i \mathbf{v}_i$, where $\lambda_1 < \lambda_2 \cdots < \lambda_k$ are different eigenvalues of R_{e_p}. From

$$(3.6) \qquad \Lambda_{e_p}(\mathbf{v}) = \sum_{i=1}^{k} (\lambda^2 - 2\lambda\lambda_i + \lambda_i^2)\mathbf{v}_i = \sum(\lambda - \lambda_i)\mathbf{v}_i = 0$$

we have $\mathbf{v}_i = 0$ if $\lambda_i \neq \lambda$; consequently the eigenspaces of R_{e_p} are invariant under the action of R'_{e_p}. This completes the proof of (3.4). At last we show that from (3.4), $R'_{e_p} = 0$ follows. In fact, take a geodesic $\gamma(r)$ and let

$$(3.7) \qquad T_{\gamma(r)}(M^n) = \sum V^i_{\gamma(r)}$$

be the common eigensubspace decomposition with respect to the commuting selfadjoint operators $R_{\dot\gamma(r)}$ and $R'_{\dot\gamma(r)}$; i.e., for any $\mathbf{v}_i \in V^i_{\gamma(r)}$ we have

$$(3.8) \qquad R_{\dot\gamma(r)}(\mathbf{v}_i) = \lambda_i \mathbf{v}_i, \qquad R'_{\dot\gamma(r)}(\mathbf{v}_i) = \mu_i(r)\mathbf{v}_i$$

and for any $i \neq j$ we have $(\lambda_i, \mu_i(r)) \neq (\lambda_j, \mu_j(r))$. Even though the distribution $V^i_{\gamma(r)}$ might be nondifferentiable where the multiplicity of the eigenvalue $\mu_i(r)$ is changing, these distributions are differentiable on an everywhere dense open subset [Sz6]. Take a differentiable unit vector field $\mathbf{v}(\gamma(r)) \in V^i_{\gamma(r)}$ on a connected component (i.e., on an interval) of this subset. Then we have

$$\mu^i = \langle R'_{\dot\gamma}(\mathbf{v}), v \rangle = \langle (\lambda_i \mathbf{v})' - R_{\dot\gamma}(\mathbf{v}'), \mathbf{v} \rangle = \lambda_i \langle \mathbf{v}', \mathbf{v} \rangle - \langle \mathbf{v}', R_{\dot\gamma}(\mathbf{v}) \rangle = 0.$$

Therefore $\mu^i = 0$, $R'_{\mathbf{m}} = 0$ and $\nabla R = 0$ everywhere. Since the space is harmonic as well, it is 2-point homogeneous by the Lichnerowicz theorem [Li1].

By tracing this proof we see that the radial property of the pure kernels $\mathrm{Tr}\, \mathbf{A}^k$, $k = 1, 2, \ldots, n-1$, implies the Osserman property on a manifold (M^n, g) so, by Chis theorem [Chi1], we have

PROPOSITION 3.2. *Suppose* $n \neq 4k$. *Then* (M^n, g) *is locally* 2-*point homogeneous if and only if the pure kernels* $\mathrm{Tr}\, \mathbf{A}^k$, $k = 1, 2, \ldots, n-1$, *are radial kernels.*

For the general case we have involved also the nonpure kernels $\mathrm{Tr}[\mathbf{A}, \mathbf{A}^*]^k$, $k = 1, 2, \ldots, (n-1)$; so our proof uses altogether $2(n-1)$ number of Jacobi kernels for deciding the 2-point homogeneous property on a manifold.

On Kahler manifolds, this property can be decided by the single Kahler-Jacobi kernel $K_{p;q}$.

PROPOSITION 3.3. *A Kahler manifold is locally Hermitian* 2-*point homogeneous (i.e., it is a space of constant holomorphic sectional curvature) if and only if the Kahler-Jacobi kernel* $K_{p;q}$ *is radial around any point* p.

The statement immediately follows from (2.22), since from the radial property of $K_{p;q}$ we get

$$(3.10) \qquad R(\mathbf{e}_p, \mathscr{F}(\mathbf{e}_p), \mathbf{e}_p, \mathscr{F}(\mathbf{e}_p)) = \text{constant}$$

and so the space has constant holomorphic sectional curvature [**KN2**].

We finish this paragraph by a Lichnerowicz type characterization of the symmetric spaces.

The *Jacobi symmetries* $\mathscr{F}$ Symm(p) at a point p are defined by the orthogonal transformations $A \in O(n)$ of the Euclidean tangent space $T_p(M^n)$ such that the map

$$(3.11) \qquad \widetilde{A} = \exp_p \circ A \circ \exp_p^{-1}$$

leaves each Jacobi kernel $H_p(q)$ invariant; i.e.,

$$(3.12) \qquad H_p(\widetilde{A}(q)) = H_p(q).$$

The following theorems immediately follow from the DeRham decomposition from Theorems 1.5, 1.6, 3.1 and from the *Berger-Simons theorem* [**Ss1**]. The last theorem asserts that the holonomy group, $\mathrm{Hol}(p)$, of an irreducible Riemannian manifold transitively acts on the unit sphere $S \subset T_p(M^n)$ except the case when the space is a symmetric space of rank ≥ 2.

COROLLARY 3.4. *A Riemannian space is locally symmetric if and only if*

$$\mathrm{Hol}(p) \subseteq \mathscr{F}\, \mathrm{Symm}(p), \quad \forall p;$$

i.e., if the holonomy group is a subgroup of the Jacobi symmetries at each point p.

COROLLARY 3.5. *Let* (M^n, g) *be a complete Riemannian manifold with nonnegative scalar curvature* R. *Such a space is locally symmetric if and only if the maps*

$$(3.13) \qquad \widetilde{A} = \exp_p \circ A \circ \exp_p^{-1} \quad \forall A \in \mathrm{Hol}(p)$$

are volume preserving at any point p.

The strength of the previous theorems becomes apparent in comparing with the following classical theorem

THEOREM [KN2]. *A Riemannian space is symmetric if and only if the maps* $\tilde{A} = \exp_p A \exp_p^{-1}$ $\forall A \in \text{Hol}(p)$ *are isometries.*

2. Mean Value Theorems; the converse of a theorem of Helgason. The Mean Value Theorems on analytic Riemannian spaces can be established by the power series

$$(3.14) \qquad E_{H;p;r}(u) = u(p) + \sum_{k=1}^{\infty} \frac{\Delta_H^{(k)} u(p)}{(2k)!},$$

where u is an analytic function and $H \in \mathscr{J}R$ with $H_p(p) \neq 0$. A function u is called common harmonic with respect to the operators $\Delta_H^{(k)}$, $k = 1, 2, \ldots$ if $\Delta_H^{(k)} u = 0$.

Using the fact that analytic Riemannian manifolds are analytic with respect to the normal coordinate systems as well, we have

PROPOSITION 3.6. *Suppose that the space is analytic. Then a function* u *is common harmonic with respect to the operators* $\Delta_H^{(k)}$, $k = 1, 2, \ldots, \infty$, *if and only if the local Mean Value Property*

$$(3.15) \qquad E_{H;p;r}(u) = u(p)$$

holds at each point of the definition domain of the function u.

In the following theorem we characterize the harmonic, resp. 2-point, homogeneous spaces by the polynomial property of the Jacobi Laplacians $\Delta_H^{(k)}$. This statement can be considered as the *converse of a theorem of Helgason* which asserts that the invariant differential operators on a 2-point homogeneous space are the polynomials of the Laplacian.

THEOREM 3.7. (A) *A Riemannian space is harmonic if and only if the density operators* $\Delta_\theta^{(k)} = \Delta_\omega^{(k)}$ *are polynomials of the Laplacian* Δ.

(B) *A Riemannian space is* 2-point homogeneous *if and only if all Jacobi Laplace operators* $\Delta_H^{(k)}$ *are polynomials of the Laplacian.*

The proof of both statements is based on elementary arguments.
From the classical formula

$$(3.16) \qquad \Delta = \Delta_S - \frac{\partial^2}{\partial r^2} - \frac{\theta'}{\theta} \frac{\partial}{\partial r}$$

and from the Stokes Theorem we get

$$
\begin{aligned}
(3.17) \qquad E_{\theta;p;r}(\Delta u) &= E_{\theta;p;r}\left(-u'' - \frac{\theta'}{\theta} u'\right) \\
&= E_{\theta;p;r}\left(-u'' - \frac{n-1}{r} - \frac{\omega'}{\omega} u'\right).
\end{aligned}
$$

When the space is harmonic the function ω'/ω is radial. So by taking the $2k$th derivature at $r = 0$, a recursion formula of the form

$$(3.18) \qquad \Delta_\theta^{(k+1)} = \Delta_\theta^{(k)}\Delta + P_k(\Delta, \Delta_\theta^{(2)}, \ldots, \Delta_\theta^{(k-1)})$$

follows where P_k is polynomial of the arguments. The polynomial property of $\Delta_\theta^{(k)}$ can be proved by induction.

Conversely, if the operators $\Delta_\theta^{(k)}$ are polynomials of the Laplacian then from (2.20) we get that the space is an Einstein space and therefore it is analytic by the De Turck-Kazdan Theorem. The harmonic functions are also analytic by the Bernstein Theorem. Since the Laplace harmonic functions are common harmonics, with respect to the operators $\Delta_\theta^{(k)} = \mathrm{Pol}_{(k)}(\Delta)$,

$$(3.19) \qquad E_{\theta;p;r}(u) = u(p) + \sum \frac{\Delta_\theta^{(k)}(u)/p}{(2k)!} = u(p);$$

i.e., the space satisfies the Mean Value Property on harmonic functions. Therefore the space is harmonic by the Willmore Theorem [**Wi1**].

When all Jacobi Laplace operators are polynomials of the Laplacian then

$$(3.20) \qquad E_{H;p;r}(u) = u(p) = E_{\theta;p;r}(u)$$

for any harmonic function u. So, by solving the Dirichlet problem, we have

$$(3.21) \qquad \frac{H_p(e_p(r))}{\int H_p(e_p(r))\,de_p} = \frac{\theta_p(e_p(r))}{\int \theta_p(e_p(r))\,de_p};$$

i.e., each Jacobi kernel $H_p(q)$ is radial. The proof of the 2-point homogeneous property can be completed by Theorem 3.1. These considerations directly imply the following *Mean Value Theorems*.

THEOREM 3.8. (A) (*Willmore* [**Wi1**]) *A Riemannian space is harmonic if and only if for any Laplace harmonic function u, the local Mean Value Property*

$$(3.22) \qquad E_{\theta;p;r}(u) = u(p) = E_{\omega;p;y}(u)$$

holds.

(B) *A Riemannian space is 2-point homogeneous if and only if for any Laplace harmonic function u and for any Jacobi kernel $H(p, q)$ with $H(p, p) \neq 0$, the Mean Value Property*

$$(3.23) \qquad E_{H;p;r}(u) = u(p)$$

holds.

REMARK (FURTHER LICHNEROWICZ-TYPE CHARACTERIZATIONS). On simply connected compact Riemannian manifolds, the Mean Value Theorem characterizes the 2-point homogeneous spaces; i.e., this theory holds exactly on these spaces only. This suggested that it would be possible to find such characteristic theorems for other spaces, for instance for symmetric spaces.

Next we describe such kinds of statements. The following theorem had been proved by Funk as early as 1913.

Let $D \subset \mathbf{R}^n$ be a star-domain with respect to a point $p \in D$. Then D is central symmetric with respect to p if and only if any hyperplane lying on p divides D into two parts which have identical volume.

Grunbaum conjectured that the same statement should be true concerning the hypersurface area of the boundary ∂D (which is supposed to be smooth). This conjecture has been proved by Schneider [Sch1]. Schneider actually proved a stronger statement asserting the theorem concerning the principal curvature measures of ∂D as well.

For the extension of these theorems onto Riemannian manifolds we define the hyperplanes V_p at a point $p \in M^n$ by the hypersurfaces which are described by the geodesics tangent to the hyperplanes $v_p \subset T_p(M^n)$ (i.e., $v_p = T_p(V_p)$) of the tangent space $T_p(M^n)$.

THEOREM 3.9. (A) *The Funk theorem, concerning the volume of small star-domains, is true exactly on volume symmetric spaces, i.e., where the geodesics involutions are volume preserving.*

(B) *The Schneider Theorem, concerning the area measure of the boundary ∂D of small star-domains, is true exactly on locally symmetric space.*

The theorem can be proved by using of the Fourier transform. We do not go into detail here.

CHAPTER 2
SPACES WITH COMMUTING
JACOBI DIFFERENTIAL OPERATORS

I. Spaces with cyclically parallel Ricci tensor

The curvature condition

$$(1.1) \qquad L_{ijk} := \nabla_i \rho_{jk} + \nabla_j \rho_{ki} + \nabla_k \rho_{ij} = 0,$$

for the Ricci curvature ρ_{ij}, often emerges in Riemannian geometry. For instance, all the Einstein spaces and all the volume symmetric spaces have this property. Further investigations of this condition can be found in [Gra3, Sn1].

Now we add two new statements to this field. First of them is the generalization of a theorem of DeTurck and Kazdan [DeTK1]. This theorem states the real analyticity of the (positive definite!) Einstein manifolds.

THEOREM 1.1 [Sz4]. *The Riemannian spaces with cyclically parallel Ricci tensor (i.e., satisfying (1.1)) are real analytic spaces.*

This theorem can be proved similarly as the DeTurck-Kazdan Theorem. In harmonic coordinates, the equation (1.1) has the symbol

$$(1.2) \qquad h_{ij} \to \xi_i |\xi|^2 h_{jk} + \xi_j |\xi|^2 h_{ki} + \xi_k |\xi|^2 h_{ij}.$$

This symbol is obviously injective, since for all $\xi \neq 0$, from

$$(1.3) \qquad \xi_i h_{jk} + \xi_j h_{ki} + \xi_k h_{ij} = 0,$$

we have

$$|\xi|^2 \xi^j \xi^k h_{jk} = 0, \qquad 2|\xi|^2 \xi^j h_{jk} + \xi^i \xi^j h_{ij} \xi_k = 0,$$

$$(1.4) \qquad |\xi|^2 h_{jk} + \xi^i \xi_j h_{ik} + \xi^i \xi_k h_{ij} = 0,$$

and so $h_{jk} = 0$. This injectivity proves the theorem completely.

THEOREM 1.2 [Sz4]. *A Riemannian space has cyclically parallel Ricci tensor if and only if the Laplacian Δ commutes with the second unweighted operator*

$$(1.5) \qquad \Delta^{(2)} = \frac{1}{n(n+2)}(3\Delta^2 + 2\rho^{ij}\nabla_i\nabla_j + 2(\nabla_j\rho^{ij})\nabla_i).$$

The proof of this statement is a lengthly technical computation using the Ricci and the Bianchi identities. From

$$(1.6) \qquad \Delta\Delta^{(2)} - \Delta^{(2)}\Delta = \frac{2}{3n(n+1)}L^{sli}\nabla_s\nabla_l\nabla_i + \{\text{lower-order terms}\}$$

we get that in the case $\Delta\Delta^{(2)} = \Delta^{(2)}\Delta$, $L_{ijk} = 0$ follows. Conversely, if $L_{ijk} = 0$ then

$$(1.7) \qquad 0 = L_{sl}^l = \nabla_s R + 2\nabla^l\rho_{sl} = 2\nabla_s R = 4\nabla_l\rho_s^l,$$

where R is the scalar curvature. So Δ and $\Delta^{(2)}$ commute if and only if Δ commutes with the operator $\rho^{ij}\nabla_i\nabla_j$. This last commutativity is equivalent to

$$L_{ijk} = 0,$$

$$\nabla_i\nabla^i\rho^{ls} + 2\rho^{ij}R_{ij}^{ls} + 2\rho_i^l\rho^{is} = \nabla_i L^{ils} = 0,$$

$$(1.8) \qquad \frac{4}{3}(\nabla^s\rho^{ij})R_{lijs} - 2\rho^{ij}\nabla_l\rho_{ij} = \frac{1}{3}\nabla^i\nabla^s L_{sli} = 0.$$

So from $L_{ijk} = 0$, $\Delta\Delta^{(2)} = \Delta^{(2)}\Delta$ follows, which completes the proof of the above theorem.

Kowalski [Kow1] has proved that the commutativity $\Delta\Delta^{(2)} = \Delta^{(2)}\Delta$ is equivalent to

$$(1.9) \qquad L_{ijk} = 0, \qquad \nabla_l\|\rho\|^2 = \frac{4}{3}R_{ijkl}\nabla^i\rho^{jk}.$$

But he did not notice that his second equation is equivalent to $\nabla^i\nabla^j L_{ilj} = 0$ and so it follows from the first one.

It should be mentioned another observation of Kowalski, namely

THEOREM 1.3 [Kow2]. *A Riemannian space is an Einstein space if and only if the operator $\Delta^{(2)}$ is a polynomial of the Laplacian Δ.*

This statement is the reformulation of the following theorem of Gray and Willmore:

THEOREM 1.4 [GW1]. *A Riemannian space is an Einstein space if and only if the Mean Value Theorem for harmonic functions holds up to the order* 8.

II. The Laplacian commuting with the operators $\square_H^{(k)}$

Next we establish necessary and sufficient condition for the Laplacian Δ to be commuting with each of the operators $\square_H^{(k)}$, defined by a kernel $H(x, y)$. We give two versions of this statement; the first of them concerns the commutativity of Δ with convolution operators.

Let $H(x, y)$ be a kernel of class C^2 and let $f(r)$ be an arbitrary C^2-function with compact support such that its supporting radius is less than the injectivity radius of the manifold. The kernels of the form $fH(x, y) := f(r(x, y))H(x, y)$ are said to be *radialized kernels*. The convolution

$$(2.1) \qquad fH * u(x) = \int fH(x, y)u(y)\,dy$$

is now considered on the C^2-functions $u(y)$ with compact support.

We need also the following notions for the next theorem. A kernel $K(x, y)$ is said to be *parallel on geodesics* if for any geodesic $\gamma(r)$, there exists a function $k_y(r)$ such that the restriction $K_y(x, y)$ of $K(x, y)$ onto γ is of the form $K_y(x, y) = k_y(r(x) - r(y))$.

We introduce also the kernel $\mathbf{H}(x, y) := H(x, y)/\omega(x, y)$, where $\omega(x, y)$ is the density kernel. In the analytic case, the connection between the convolution $f\mathbf{H}*$ and the operators $\square_H^{(k)}$ is described by the following *pre-Pizzetti formula*

$$(2.2) \qquad f\mathbf{H} * u = \Omega_{n-1} \sum_k \frac{1}{(2k)!} \left(\int f(r)r^{n+2k+1}\,dr \right) \square_H^{(k)} u.$$

BASIC THEOREM 2.1 [Sz4]. *The Laplacian Δ commutes with the convolution operators $f\mathbf{H}*$, $\forall f$, if and only if each of the following conditions holds.*

(1) *The kernel $H^2(x, y)/\omega(x, y)$ is parallel on the geodesics.*

(2) *The kernel $\mathbf{H} = H/\omega$ satisfies the ultra hyperbolic equation*

$$(2.3) \qquad \Delta_x \mathbf{H}(x, y) = \Delta_y \mathbf{H}(x, y).$$

The proof rests on elementary arguments. From the Stokes Theorem,

$$(2.4) \qquad \int K(x, y)\Delta_y u(y)\,dy = \int (\Delta_y K(x, y))u(y)\,dy$$

we get that $f\mathbf{H} * \Delta u = \Delta f\mathbf{H} * u$, $\forall u$, if and only if $\Delta_x f\mathbf{H}(x, y) = \Delta_y f\mathbf{H}(x, y)$, i.e., if

$$(2.5) \qquad \begin{aligned} &f'(r(x, y))\mathbf{H}(x, y) \left(\ln \frac{(H^y)^2}{\omega^y} \right)'(x) + f(r(x, y))(\Delta_x \mathbf{H}^y)(x) \\ &= f'(r(x, y))\mathbf{H}(x, y) \left(\ln \frac{H_x^2}{\omega_x} \right)'(y) + f(r(x, y))(\Delta_y \mathbf{H}_x)(y), \end{aligned}$$

where $K(x, y) = K_x(y) = K^y(x)$ and the prime means radial derivative.

Introducing the kernel

(2.6)
$$Z_\gamma(t, s) = \ln \frac{H_\gamma^2(t, s)}{\omega_\gamma(t, s)}$$

on an arbitrary geodesic $\gamma(r)$, we get that $\Delta(f\mathbf{H}*) = (f\mathbf{H}*)\Delta$, $\forall f$, is satisfied if and only if

(2.7)
$$\frac{\partial Z_\gamma}{\partial s}(t, s) = -\frac{\partial Z_\gamma}{\partial t}(t, s)$$

holds on each geodesic γ; furthermore the kernel H satisfies the equation

(2.8)
$$\Delta_x \mathbf{H}(x, y) = \Delta_y \mathbf{H}(x, y).$$

The proof can be completed by the following observation.

LEMMA 2.2. *The general C^2-solutions of (2.7) are just the functions of the form*

(2.9)
$$Z_\gamma(t, s) = \varphi_\gamma(t - s),$$

where $\varphi_\gamma: \mathbf{R} \to \mathbf{R}$ is a C^2-function.

The functions (2.9) obviously satisfy (2.7). Conversely, if $Z_\gamma(t, s)$ is a solution of (2.7), then

(2.10)
$$\frac{\partial^2 Z_\gamma}{\partial s^2} = -\frac{\partial^2 Z_\gamma}{\partial s \partial t} = -\frac{\partial^2 Z_\gamma}{\partial t \partial s} = \frac{\partial^2 Z_\gamma}{\partial t^2}$$

and therefore

(2.11)
$$Z_\gamma(t, s) = \tilde{\varphi}(t - s) + \psi(t + s).$$

By (2.7) we get

(2.12)
$$-\tilde{\varphi}'(t - s) + \psi'(t + s) = -\tilde{\varphi}'(t - s) - \psi'(t + s)$$

and so $\psi' = 0$, $\psi = \text{constant}$ and $Z_\gamma(t, s) = \varphi(t - s)$. This completes the proof both of Lemma and of Theorem 2.1.

By the pre-Pizzetti formula (2.2), we get.

BASIC THEOREM 2.3 [Sz4]. *Let $H(x, y)$ be an analytic kernel on an analytic manifold. The Laplacian commutes with the operator $\square_H^{(k)}$, $\forall k \in \mathbf{N}$, if and only if*

(1) *the kernel H^2/ω is parallel on the geodesics,*
(2) *the kernel $\mathbf{H} = H/\omega$ satisfies the ultra hyperbolic equation*

(2.13)
$$\Delta_x H(x, y) = \Delta_y H(x, y).$$

III. Volume symmetric spaces

A Riemannian space is said to be volume symmetric if the geodesic involutions are volume preserving, i.e., if the density function $\omega_p(q)$ is central symmetric on each normal coordinate system.

These spaces have been introduced by D'Atri and Nickerson [D'AN1] and have been studied by many authors.

The symmetric spaces as well as the harmonic spaces are obviously volume symmetric spaces. Also a large class of nonsymmetric, naturally reductive spaces consists of volume symmetric spaces [DA1].

All of the odd-order derivatives $\nabla_{\mathbf{e}_p}^{(2k+1)}\omega_p$ vanish on these spaces. These identities are said to be the odd conditions. The first nontrivial odd condition says:

$$(3.1) \qquad \nabla_{\mathbf{e}_p}^{(3)}\omega_p = \frac{1}{3}(\nabla_{\dot{\mathbf{e}}_p}\rho)(\dot{\mathbf{e}}_p,\dot{\mathbf{e}}_p) + 0; \quad \text{i.e., } L_{ijk} = 0,$$

which is just the curvature condition (1.1). Therefore all the (positive definite!) volume symmetric spaces are real analytic manifolds, by Theorem 1.1.

By a theorem of Kowalski and Vankecke [KV2, Theorem (2.5)] any analytic, symmetric $(K(p,q) = K(q,p))$ and central symmetric $(K_p(q) = K_p(-q))$ kernel $K(p,q)$ is parallel of the form $K_\gamma(p,q) = k_\gamma(r(p,q))$ on the geodesics γ of an analytic space. So we get

THEOREM 3.1 [Sz4]. *The volume symmetric Riemannian spaces are real analytic spaces. Furthermore these are just the spaces which have parallel density function of the form* $\omega_\gamma(p,q) = \phi_\gamma(r(p,q))$ *on each geodesic* γ.

The following characterization of volume symmetric spaces immediately follows from Theorem 2.3.

THEOREM 3.2 [Sz4]. *A Riemannian space is volume symmetric if and only if the Laplacian* Δ *commutes with the density operators* $\square_\omega^{(k)}$; $\forall k \in \mathbf{N}$.

In fact, in case $H = \omega$ we have $H^2/\omega = \omega$ and $\mathbf{H} = H/\omega = 1$. Therefore Δ commutes with the operators $\square_\omega^{(k)}$ on volume symmetric spaces, by the Theorems 3.1 and 2.3. Conversely, in case of the above commutativity, we have the following. From $\Delta\square_\omega^{(2)} = \square_\omega^{(2)}\Delta$ we get $L_{ijk} = 0$; i.e., the space is analytic by Theorem 1.1. Since $\omega(p,q) = \omega(q,p)$, $\omega_\gamma(p,q) = \phi_\gamma(r(p) - r(q)) = \phi_\gamma(r(q) - r(p)) = \phi_\gamma(r(p,q))$, by Theorem 2.3. This completes the proof of Theorem 3.2.

Now consider an arbitrary pure Jacobi kernel

$$(3.2) \qquad \alpha^{(i)}(p,q) = \operatorname{Tr}\mathbf{A}_{p;q}^i.$$

These kernels are symmetric $(\alpha^{(i)}(p,q) = \alpha^{(i)}(q,p))$; futhermore $\nabla_{\mathbf{e}_p}^{(3)}\alpha_p^{(i)} = q_i(\nabla_{\dot{\mathbf{e}}_p}\rho)(\dot{\mathbf{e}}_p,\dot{\mathbf{e}}_p)$ holds for a constant q_i. Also, from the commutativity

$$(3.3) \qquad \Delta\square_{\alpha_i}^{(2)} = \square_{\alpha_i}^{(2)}\Delta$$

we get $L_{ijk} = 0$. A Riemannian space is said to be $\alpha^{(i)}$-symmetric if the kernel $S^{(i)} = (\alpha^{(i)})^2/\omega$ is central symmetric $(S_p^{(i)}(q) = S_p^{(i)}(-q))$ on each normal coordinate system around p.

By the same arguments as before, we get

THEOREM 3.3 [Sz4]. (A) *The $\alpha^{(i)}$-symmetric spaces are real analytic spaces and they are characterized by the property that the kernel $S^{(i)}$ is parallel of the form*

$$(3.4) \qquad S_\gamma^{(i)}(p\,,q) = s_\gamma^{(i)}(r(p\,,q))$$

on the geodesic γ.

(B) *The Laplacian Δ commutes with the operators $\square_{\alpha^{(i)}}^{(k)}$, $\forall k \in \mathbf{N}$, if and only if the space is $\alpha^{(i)}$-symmetric and satisfies the ultra hyperbolic equation*

$$(3.5) \qquad \Delta_x \frac{\alpha^{(i)}}{\omega}(x\,,y) = \Delta_y \frac{\alpha^{(i)}}{\omega}(x\,,y).$$

For the special case $i = 0$ (i.e., when $\alpha^{(i)} = 1$, $S^{(0)} = 1/\omega$ and $\square_1^{(k)} = \Delta^{(k)}$) we have.

The Laplacian Δ commutes with the unweighted operators $\Delta^{(k)}$, $\forall k \in \mathbf{N}$, if and only if the space is volume symmetric and satisfies the equation

$$(3.6) \qquad \Delta_x \frac{1}{\omega}(x\,,y) = \Delta_y \frac{1}{\omega}(x\,,y).$$

A space is called *probabilistic commutative* if any two operators $\Delta^{(k)}$ and $\Delta^{(l)}$ commute. These are just the spaces on which the random walks commute [RU1]. Because of this relationship to probability, these spaces have been studied by many authors.

Notice that the probabilistic commutative spaces are volume symmetric by the preceding statement. However, the still-open conjecture for the converse statement (i.e., the probabilistic commutative spaces are exactly the volume symmetric spaces) does not seem to be true.

Closing this section we mention the results of Prufer [Pr1] and of Kowalski-Vanhecke [KV1] relating to the above statements.

THEOREM [Pr1]. *Let M^n be an analytic volume symmetric space and consider the integral operator*

$$(3.7) \qquad I_R(u)(p) := E_{\theta\,,p\,,R}(u)$$

for a radius R. Then the Laplacian commutes with I_R, $\forall R$.

THEOREM [KV1]. *Let G/K be a homogeneous Riemannian manifold such that the invariant differential operators commute. Then the space is volume symmetric. In addition, the characteristic polynomial of the metric tensor g_{ij} defines central symmetric kernels on any normal coordinate neighborhood.*

These results immediately follow from the above, more general results.

The proof of the preceding theorem is methodically different from our approach. There the authors use a theorem of Sumimoto [Su1].

IV. Symmetric spaces

Gelfand [Ge1] has observed that the invariant differential operators on a symmetric space commute. This statement can be checked easily as follows:

Consider the smooth invariant kernels $H(x, y)$ on a symmetric space G/K such that $H(0, \cdot)$ has compact support. This space of kernels is obviously closed with respect to the linear combinations, multiplications and convolutions. All these kernels are symmetric. In fact, let $m(x, y)$ be the midpoint on the geodesic segment $\overline{xy}$ and let σ_m be the geodesic involution at the point $m(x, y)$. Then we get

$$(4.1) \qquad H(x, y) = H(\sigma_m(x), \sigma_m(y)) = H(y, x).$$

Therefore we have

$$(4.2) \qquad H_1 * H_2 = (H_1 * H_2)^t = H_2^t * H_1^t = H_2 * H_1;$$

i.e., the convolution algebra of the invariant kernels is commutative on a symmetric space. Produce the invariant differential operators $\square_H^{(k)}$. From the pre-Pizzetti formula (2.2) we get that all these operators commute as well. More than that, all these operators are selfadjoint on a symmetric space.

Notice that the above procedure is working under a weaker symmetry hypothesis. Namely, only assume that for any geodesic γ and for any point $p \in \gamma$ there exists an isometry ϕ with $\phi(p) = p$ and $\phi(\gamma) = \gamma$ so that ϕ induces a nontrivial involution on γ. The other geodesics through p may be noninvariant or fixed at ϕ. We call these spaces ray-symmetric spaces. The ray-symmetric spaces are nonsymmetric homogeneous spaces in general (the homogeneity can be proved in the same way as on symmetric spaces). Yet they have commutative (even selfadjoint) invariant differential operator algebra.

Also a large class of nonsymmetric naturally reductive spaces consists of those spaces, which have commutative invariant differential operator algebra [TV1]. Therefore the Gelfand Theorem cannot be inverted directly.

It would be interesting to know whether the spaces having commuting Jacobi-Schrodinger operators $\square_H^{(k)}$ are homogeneous spaces? Furthermore, are the ray-symmetric spaces the spaces on which the Jacobi-Schrodinger operators generate a selfadjoint operator algebra?

In the following theorem we characterize the symmetric spaces by two commutativity properties.

THEOREM 4.1 [Sz4]. *A Riemannian space is locally symmetric if and only if*

(1) *The Laplacian Δ commutes with the operators $\square_{\alpha^{(i)}}^{(k)}$, $\forall k \in \mathbf{N}$ and $i = 1, \ldots, n-1$, where $\alpha^{(i)} = \mathrm{Tr}\, \mathbf{A}^i$*

(2) *The Jacobi kernel*

$$\beta(p, q) := \mathrm{Tr}[\mathbf{A}_{p;q}, \mathbf{A}^*_{p;q}]^2 \tag{4.3}$$

vanishes. This assumption can be replaced by the weaker property $\square_\beta^{(7)} = 0$.

On symmetric spaces, $\mathbf{A} = \mathbf{A}^*$. Notice that the endomorphism $[\mathbf{A}, \mathbf{A}^*]$ is selfadjoint and therefore from $\beta = 0$, $[\mathbf{A}, \mathbf{A}^*] = 0$ follows; i.e., the Jacobi endomorphism $\mathbf{A}$ is normal.

The properties (1) and (2) trivially hold on symmetric spaces.

Conversely, if (1) holds, then by Theorem 3.3,

$$\alpha_\gamma^{(i)}(p, q) = a_\gamma^{(i)}(r(p, q)), \qquad \mathrm{Tr}\, R_\gamma^i = \text{constant} \tag{4.4}$$

on each geodesic γ. I.e., the Jacobi curvature operator $R_{\dot\gamma}(\cdot) = R(\cdot, \dot\gamma)\dot\gamma$ has constant eigenvalues on γ. By the formula (2.18) of Chapter 1, from $[\mathbf{A}, \mathbf{A}^*] = 0$, $[[R'_{\dot\gamma}, R_{\dot\gamma}], R_{\dot\gamma}] = 0$ follows. The same identity follows from $\square_\beta^{(7)} = 0$ since this operator is nothing but multiplication by the scalar field $\int \mathrm{Tr}[[R'_{e_p}, R_{e_p}], R_{e_p}]^2\, d\mathbf{e}_p$. The proof of the theorem can be completed similarly as the proof of Theorem 3.1 in Chapter 1.

Finishing this section, we mention a theorem of Kowalski and Vanhecke which asserts that the Gelfand theorem can be inverted on Kahler manifolds.

THEOREM [KV3]. *A Kahler manifold is Hermitian symmetric if and only if the invariant differential operators commute.*

CHAPTER 3
SPECTRAL THEOREMS

In this closing chapter we establish a "joint spectrum notion" for the infinitely many Jacobi-Schrodinger operators $\square_H^{(k)}$, respectively Jacobi-Laplace operators $\Delta_H^{(k)}$. Since these operators are noncommuting, this joint spectrum will be "matrix valued" in general, which will describe also the algebraic structure of the operators considered. Instead of $\square_H^{(k)}$, $\Delta_H^{(k)}$ we consider the convolution operators with the radialized Jacobi kernels. This leads to more general theory, since it can be established on a Riemannian manifold of class C^4. On analytic manifolds, the connections between the operators $\square_H^{(k)}$ and the above convolution operators are described by the pre-Pizzetti formula (2.2) of Chapter 2. I.e., the operators $\square_H^{(k)}$ are the infinitesimal generators of the convolution algebra of the radialized Jacobi kernels.

The exact details are as follows. Consider a complete Riemannian manifold M^n whose injectivity radius δ is positive. For instance, all the compact spaces or homogeneous spaces have this property. The *radialized Jacobi kernels* are the kernels of the form

$$H(x, y) = \sum_{i=1}^k \frac{F_i(x, y)H_i(x, y)}{\omega(x, y)}, \tag{1.1}$$

where $k < \infty$, $H_i(x, y)$ are Jacobi kernels, $\omega(x, y)$ is the density kernel and $F_i(x, y) = f_i(r(x, y))$ are smooth radial kernels whose supporting radius is finite and less than δ.

The space of radialized Jacobi kernels is obviously closed with respect to finite *linear combinations* and with respect the transposition t defined by

$$(1.2) \qquad\qquad H^t(x, y) = H(y, x).$$

The radialized density kernels $(H_i = \omega)$ are exactly the radial kernels on M^n. Using also *point-wise multiplication*

$$(1.3) \qquad\qquad H_1 \cdot H_2(x, y) = H_1(x, y)H_2(x, y)$$

and *convolutions*

$$(1.4) \qquad\qquad H_1 * H_2(x, y) = \int H_1(x, z)H_2(z, y)\, dz,$$

we generate a "bigger space of kernels" in general. Notice that this space is *closed also with respect to radialization*, since it corresponds to multiplication by the radialized density kernels (=radial kernels).

In the compact case, one can introduce a pre-Hilbert norm on this "bigger kernel space," defined by

$$(1.5) \qquad\qquad \langle H_1, H_2 \rangle = \int H_1 * H_2^t(x, x)\, dx = \operatorname{Tr} H_1 * H_2^t,$$

where Tr means the trace-operator

$$(1.6) \qquad\qquad \operatorname{Tr} H = \int H(x, x)\, dx.$$

In this case, the completion leads to the Hilbert space $\{\mathscr{J}S, \langle\ ,\ \rangle\}$ equipped with further structures

$$(1.7) \qquad\qquad \{\mathscr{J}S, \langle\ ,\ \rangle, t, \cdot, *\}$$

which is called *compact Jacobi spectral algebra*. The spectral algebra generated by the radial kernels only is called *density spectral algebra*. In the noncompact case, the situation is more complicated. One can introduce a pre-Banach norm on the "bigger space," if for any kernel H the value

$$(1.8) \qquad\qquad \|H\|_B := \sup_{x \in M^n} \sqrt{H * H^t(x, x)}$$

is finite. In this case, the completion leads to the *noncompact Jacobi spectral algebra*

$$(1.9) \qquad\qquad \{\mathscr{J}S, \|\ \|_B, t, \cdot, *\}.$$

Notice that this space is a Hilbert space, if for the kernels H the value $H * H^t(x, x)$ is constant. This property obviously holds on homogeneous spaces and also on those spaces, where the Laplacian commutes with the convolution operators

$$(1.10) \qquad H*: L^2(M^n) \to L^2(M^n), \qquad H * u(x) = \int H(x, y)u(y)\, dy,$$

for any kernel H of the "bigger kernel space." In fact, in the last case, all kernels H are parallel along the geodesics (see Basic Theorem 2.1 in Chapter 2) and therefore $H * H^t(x, x)$ is constant on M^n.

If the norm $\| \ \|_B$ cannot be introduced on a noncompact case, the *Jacobi spectral algebra* is the "bigger kernel space" $\{\mathscr{J}S, t, \cdot, *\}$.

On two Riemannian spaces M_1 and M_2, these spectral algebras can be compared as follows. First identify the generator kernels; i.e., consider the same Jacobi kernels and the same radial kernels on both manifolds.

The manifolds M_1 and M_2 are said to be *Jacobi isospectral* if the above identification extends to an isomorphism of the corresponding spectral algebras. M_1 and M_2 are said to be *density-isospectral* if the identification of the same radial kernels extends to an isomorphism of the corresponding density spectral algebras.

The spaces $N_3^{(a,b)}A$ show that the density spectral algebras do not determine the metric; i.e., density isospectral spaces may be nonisometric. The larger Jacobi spectral algebra determines the metric much more strongly. Namely, with the aid of $\mathscr{J}S$ one can decide when the space is 2-point homogeneous, symmetric, volume symmetric, Einstein or when it has cyclically parallel Ricci tensor.

For deepening the knowledge about spectral algebras, we prove a *Spectral Theorem for compact spectral algebras*.

The following considerations concern general real convolution algebras $\{V; \langle \ , \ \rangle; *; t\}$ of kernels $H(x, y)$ on a compact Riemannian manifold, which form a Hilbert space and which are closed with respect to the transposition t as well. I.e., V is a real Hilbert algebra. The spectral decomposition of complex Hilbert algebras has been given by Ambrose [**Am1**] which is the generalization of the classical Wedderburn Theorem concerning finite-dimensional cases. Since in most of the standard books only this version is given [**Loo1**], we describe a simple elementary proof for the real Hilbert algebras. The most sophisticated point of our procedure will be the classical Frobenius Theorem.

I. The selfadjoint convolution algebras

A convolution algebra is said to be selfadjoint if it consists of symmetric kernels only. This is a stronger property than commutativity, since for selfadjoint algebras we have

$$(1.11) \qquad H_1 * H_2 = (H_1 * H_2)^t = H_2^t * H_1^t = H_2 * H_1.$$

The following facts are very well known for a selfadjoint convolution algebra V of a compact space M^n. In this case, the convolutions

$$(1.12) \qquad H * \phi(x) = \int H(x, y)\phi(y)\, dy, \qquad H \in V, \phi \in L^2(M^n),$$

acting on the $L^2(M^n)$ function space, split the space into the orthogonal direct sum

(1.13) $$L^2(M^n) = L_0 \oplus L_1 \oplus L_2 \oplus \cdots$$

of the joint invariant subspaces L_i such that L_0 is the zero-space ($H * L_0 = 0$, $\forall H \in V$) and the actions

(1.14) $$H*: L_i \to L_i$$

on L_i are multiplications by the real numbers $\lambda_i(H)$. More than that, the function $\lambda_i(H)$ is a continuous linear functional on V satisfying the character property

(1.15) $$\lambda_i(H_1 * H_2) = \lambda_i(H_1)\lambda_i(H_2)$$

as well. If the subspaces L_i are maximal with the above property, the decomposition (1.13) is uniquely determined up to order. We consider this maximal decomposition. Each subspace L_i, $i > 0$, is of finite dimension, say d_i. For an orthonormal basis $\{\phi_{i1}, \phi_{i2}, \ldots, \phi_{id_i}\}$ of L_i, consider the kernel

(1.16) $$\mathbf{e}_i(x, y) = \sum_{k=1}^{d_i} \phi_{ik}(x)\phi_{ik}(y).$$

This kernel is uniquely determined, since $\mathbf{e}_i*$ acts on L_i as the identity map and it is the zero map on L_j where $j \neq i$. Furthermore any kernel function $H \in V$ can be written as an l^2-series in the following form

(1.17) $$H = \sum \lambda_i(H)e_i;$$

and therefore

(1.18) $$\|H\|^2 = \sum d_i \lambda_i^2(H) < \infty.$$

Another easy consideration shows that the characters $\lambda_i(H)$ give the complete system of all continuous real characters for a selfadjoint convolution algebra V. In fact, for any such character $\lambda(H)$, the kernel $\mathbf{e}_\lambda$, defined by

(1.19) $$\lambda(H) = \langle H, \mathbf{e}_\lambda \rangle,$$

is equal to one of the above defined kernels $\mathbf{e}_i$. So we have

PROPOSITION 1.1. *For any selfadjoint convolution algebra V of a compact space M^n, there exists a uniquely determined orthonormal basis $\mathbf{v}_1, \mathbf{v}_2, \ldots$ (called character basis) such that*
    (1) $H * \mathbf{v}_i = \lambda_i(H)\mathbf{v}_i$; *i.e. the basis elements are joint eigenvectors for the convolutions $H*$,*
    (2) $\mathrm{Tr}\,\mathbf{v}_i = \int \mathbf{v}_i(x, x)\,dx > 0$.
The kernels $\mathbf{v}_i$ are of the following form:

(1.20) $$\mathbf{v}_i(x, y) = \frac{1}{\sqrt{d_i}}\mathbf{e}_i(x, y).$$

Therefore $\operatorname{Tr} \mathbf{v}_i = \sqrt{d_i}$ *and*

(1.21)
$$\mathbf{v}_i * \mathbf{v}_j = \frac{\delta_{ij}}{\sqrt{d_i}} \mathbf{v}_j.$$

II. The maximal torus algebras

Now consider a general convolution algebra V and decompose it into the orthogonal direct sum

(1.22)
$$V = S_v \oplus \mathbf{A}_v,$$

where S_v, resp. $\mathbf{A}_v$, is the space of symmetric $(H^t = H)$, resp. of skew symmetric, $(H^t = -H)$ kernels. For the Lie bracket

(1.23)
$$[H_1, H_2] := H_1 * H_2 - H_2 * H_1$$

we obviously get

(1.24)
$$[\mathbf{A}_v, \mathbf{A}_v] \subseteq \mathbf{A}_v, \quad [S_v, S_v] \subseteq \mathbf{A}_v, \quad [S_v, \mathbf{A}_v] \subseteq S_v.$$

A closed subspace $T \subset S_v$ of the symmetric kernels is said to be a *maximal torus algebra* if it is a maximal Abelian subalgebra with respect to the convolution $*$. Through each element $H \in S_v$ there exist maximal torus algebras which contain the powers $H * H * \cdots * H = H^k$ of H.

Fix an arbitrary maximal torus algebra T and let $\mathbf{t}_1, \mathbf{t}_2, \mathbf{t}_3, \ldots$ be the uniquely determined character basis of T, described in Proposition 1.1. The $L^2(M^n)$ function space can be decomposed as

(1.25)
$$L^{(2)}(M^n) = L_v^0 \oplus L_v^1,$$

where L_v^0 is the maximal 0-space of the convolution algebra V (i.e., $V * L_v^0 = 0$) and V acts nontrivially on L_v^1.

Let

(1.26)
$$L_v^1 = L_T^0 \oplus L_T^1 \oplus L_T^2 \oplus \cdots$$

be the joint eigensubspace decomposition of L_v^1 with respect to T, where L_T^0 is the 0-space of T on L_v^1 and the spaces L_T^i with $i > 0$ are of finite dimension, say d_i.

PROPOSITION 1.2. $L_T^0 = 0$ *and* $T * V * T = V$.

In fact, choose an orthonormal basis $\{\phi_{01}, \phi_{02}, \ldots\}$ in L_T^0 and $\{\phi_{i1}, \phi_{i2}, \ldots, \phi_{i2}, \ldots, \phi_{id_i}\}$ in each L_T^i where $i > 0$ and let $\Phi_1, \Phi_2, \Phi_3, \ldots$ be a unified orthonormal basis in L_v^1; i.e., $\Phi_i = \varphi_{jk}$. Then

(1.27)
$$\mathbf{t}_i(x, y) = \frac{1}{\sqrt{d_i}} \sum_{k=1}^{di} \phi_{ik}(x) \phi_{ik}(y);$$

furthermore for any kernel $H \in V$ we have

(1.28)
$$H(x, y) = \sum_{j,k} a^{jk} \Phi_j(x) \Phi_k(y),$$

where

(1.29)
$$\|H\|^2 = \sum_{j,k} (a^{jk})^2 < \infty.$$

Therefore, any kernel $H \in V$ can be decomposed in the form

(1.30)
$$H = H_{11} + H_{01} + H_{10} + H_{00},$$

where the kernels H_{ab} are as follows:
(1.31)
$$H_{11} = \sum_{i,j} \sqrt{d_i}\sqrt{d_j} t_i * H * t_j, \qquad H_{01} = \sum_i \sqrt{d_i}(H - H_{11}) * t_i,$$

$$H_{10} = \sum_i \sqrt{d_i} t_i * (H - H_{11}), \qquad H_{00} = H - H_{11} - H_{01} - H_{01} - H_{10}.$$

By the last formulas, the kernels H_{ab} belong to V. If $L_T^0 \neq 0$, there exists a nonvanishing kernel of the form H_{01} or H_{10} or H_{00}. In the first case, the kernel $H = H_{10}^t * H_{10}$, in the second case $H = H_{01} * H_{01}^t$ and in the third case $H = H_{00} * H_{00}^t$ is a nonzero symmetric kernel satisfying $H \in T$ and $H * T = T * H = 0$. Since T is maximal, this contradiction proves the proposition completely.

III. The boxes and their structure; Frobenius theorem emerging

The kernel spaces

(1.32)
$$\tau_{ij} := t_i * V * t_j$$

are said to be *boxes* and the diagonal boxes

(1.33)
$$\tau_{ii} = t_i * V * t_i$$

are said to be *main boxes*. Consider the decomposition

(1.34)
$$L_v^1 = L_T^1 + L_T^2 + L_T^3 \cdots$$

constructed in the preceding step and choose an orthonormal basis $\{\phi_{i1}, \phi_{i2}, \cdots, \phi_{id_i}\}$ in each L_T^i. The kernels from a box τ_{ij} are of the form

(1.35)
$$H(x, y) = \sum_{k=1}^{di} \sum_{l=1}^{dj} a_{ij}^{kl} \phi_{ik}(x) \phi_{jl}(y)$$

and in particular

(1.36)
$$t_i(x, y) = \frac{1}{\sqrt{d_i}} \sum_k \phi_{ik}(x) \phi_{ik}(y).$$

So *each box* τ_{ij} *is of finite dimension.* By Proposition 1.2, $T * V * T = V$, and therefore the decomposition

$$(1.37) \qquad V = \sum_{i,j} \tau_{ij}$$

splits the whole V into an orthogonal direct sum of finite dimensional subspaces, which is called box-decomposition.

PROPOSITION 1.3. (A) *For the structure of the boxes we have*

$$(1.38) \qquad \tau_{ij}^t = \tau_{ji}, \qquad \tau_{ik} * \tau_{lj} \subset \delta_{kl}\tau_{ij}.$$

Consequently, the main boxes τ_{ii} *are closed with respect to transposition and convolution as well.*

(B) *The elements of any main box* τ_{ii} *commute with the elements of* T; *more precisely*

$$(1.39) \qquad \mathbf{t}_j * \tau_{ii} = \tau_{ii} * \mathbf{t}_j = (\delta_{ji}/\sqrt{d_i})\tau_{ii}.$$

The kernels $q\mathbf{t}_i$, $q \in \mathbf{R}$, *are the only symmetric kernels in a main box* τ_{ii}.

(C) *For any element* $A \in \tau_{ij}$ *we have*

$$(1.40) \qquad A * A^t = \frac{\|A\|^2}{\sqrt{d_i}}\mathbf{t}_i, \qquad A^t * A = \frac{\|A\|^2}{\sqrt{d_j}}\mathbf{t}_j.$$

Formulas (1.38) and (1.39) obviously hold, by (1.21). If $H \in \tau_{ii}$ is symmetric kernel, so are the kernels $T * H = H * T$, by (1.39). Since T is maximal, $H = q\mathbf{t}_i$ follows. This proves the second part of (B). For the symmetric kernels $A * A^t$ and $A^t * A$ of (C) we have

$$A * A^t \in \tau_{ii}, \qquad A^t * A \in \tau_{jj}.$$

Therefore, they are of the form (1.40), by (B), and the proof of the proposition is completed.

The Frobenius Theorem emerges in the following:

PROPOSITION 1.4. (A) *For a main box* τ_{ii}, *the element*

$$(1.41) \qquad \mathbf{e}_i = \sqrt{d_i}\mathbf{t}_i$$

is the multiplicative unit; furthermore any nonzero element $A \in \tau_{ii}$ *has the multiplicative inverse*

$$(1.42) \qquad A^{-1} = \frac{d_i}{\|A\|^2}A^t \in \tau_{ii}.$$

I.e., any τ_{ii} *is an associative division algebra and is isomorphic to* $\mathbf{R}, \mathbf{C}$ *or* $\mathbf{H}$, *by the classical Frobenius Theorem.*

(B) *If* $\tau_{ij} \neq 0$, *then*

$$d_i = \dim L_T^i = \dim L_T^j = d_j, \qquad \dim \tau_{ii} = \dim \tau_{ij} = \dim \tau_{jj}$$

and the main boxes τ_{ii} and τ_{jj} are isometrically isomorphic. More precisely, for any nonzero element $B \in \tau_{ij}$, the maps

$$(1.43) \qquad \begin{aligned} \Psi_B &: \tau_{ij} \to \tau_{ii}, & \Psi_B(\beta) &:= B * \beta^t, \\ \Psi_B^t &: \tau_{ij} \to \tau_{jj}, & \Psi_B^t(\beta) &:= \beta * B^t \end{aligned}$$

are linear isomorphisms and the map

$$(1.44) \qquad \Phi_B : \tau_{ii} \to \tau_{jj}, \qquad \Phi_B(A) := \frac{d_i}{\|B\|^2} B^t * A * B$$

is isometric isomorphism between the (real, complex or quaternion) algebras τ_{ii} and τ_{jj}.

Statement (A) immediately follows from (1.40). For the proof of (B), consider the representation

$$(1.45) \qquad B(x, y) = \sum b^{kl} \phi_{ik}(x) \phi_{jl}(y)$$

for $B \in \tau_{ij}$. By (1.40), the row vectors as well as the column vectors of (b^{kl}) form an orthogonal system of independent vectors. Therefore $d_i = d_j$ (in fact, in case $d_i < d_j$, the number d_j of d_i-dimensional column vectors cannot be independent), and the matrix (b^{kl}) is nondegenerate. The proof can be completed by a simple application of (1.40).

IV. The total decomposition of V

The indices i and j are said to be *connected* if $\tau_{ij} \neq 0$. This relation is obviously reflexive and symmetric. Furthermore it is transitive, since $\tau_{ik} = \tau_{ij} * \tau_{jk} \neq 0$ if $\tau_{ij} \neq 0$ and $\tau_{jk} \neq 0$. Let I_k, $k = 1, 2, 3, \ldots$, be the classes of this equivalence and for a class I_k we define the subspaces $Q_k \subset V$ and $L_k \subset L_v^1$ by

$$(1.46) \qquad Q_k := \sum_{i_k ; j_k \in I_k} \tau_{i_k j_k}, \qquad L_k = \sum_{i_k \in I_k} L_T^{i_k}.$$

Then the decompositions

$$(1.47) \qquad V = \sum_k Q_k, \qquad L_v^1 = \sum L_k$$

are orthogonal direct decompositions such that

$$(1.48) \quad V * Q_k, \quad Q_k * V \subset Q_k, \quad Q_k * Q_j = 0 \text{ if } k \neq j, \quad V * L_k \subset L_k.$$

$|I_k|$ stands for the order ($=$ the number of elements) of the index class I_k. This can be finite as well as infinite.

Next we show that the subalgebras Q_k are minimal two-sided ideals so that each of them is isomorphic to one of the following full matrix algebras:

$\mathrm{gl}(|I_k|, \mathbf{R})$, $\mathrm{gl}(|I_k|, \mathbf{C})$, $\mathrm{gl}(|I_k|, \mathbf{H})$. If $|I_k| = \infty$, the corresponding matrix algebra consists of such matrices which have the Hilbert-Schmidt norm

$$(1.49) \qquad \|a_{ij}\|^2 = \sum_{i,j=1}^{|I_k|} a_{ij}\bar{a}_{ij} < \infty.$$

For showing this, we have to establish a matrix parametrization on each component Q_k. It is enough to construct such parametrization on the boxes $\tau_{i_k j_k}$, i_k, $j_k \in I_k$, which provide a basis for Q_k. The algebras $\mathbf{C}$, resp. $\mathbf{H}$, are considered to be identified with $\mathbf{R}^2$, resp. $\mathbf{R}^4$.

By Proposition 1.4 (B), the algebraic structure of the main boxes $\tau_{i_k i_k}$ in Q_k is uniquely determined, say it is S_k ($= \mathbf{R}$ or $\mathbf{C}$ or $\mathbf{H}$). First parametrize each main box $\tau_{i_k i_k}$ by S_k isomorphically. Then identify an arbitrary box $\tau_{i_k j_k}$, $i_k \neq j_k$, with S_k:

$$(1.50) \qquad B \in \tau_{i_k j_k} \leftrightarrow b \in S_k$$

such that

$$(1.51) \qquad \|B\| = |b|, \qquad B^t \leftrightarrow \bar{b};$$

furthermore the isomorphism $\Phi_B \colon \tau_{i_k i_k} \to \tau_{j_k j_k}$, defined in (1.44), is described in the above parametrization as

$$(1.52) \qquad \Phi_B \colon a \to b^{-1}ab.$$

There exist exactly two identifications for each box $\tau_{i_k j_k}$, $i_k \neq j_k$, satisfying these properties. Any such matrix parametrization defines the isomorphism

$$(1.53) \qquad Q_k \leftrightarrow \mathrm{gl}(|I_k|, S_k),$$

so we have

SPECTRAL THEOREM 1.5. *Any compact spectral algebra V of compact Riemannian manifolds is a direct sum of algebras of the form* $\mathrm{gl}(|I_k|, \mathbf{R})$, $\mathrm{gl}(|I_k|, \mathbf{C})$, $\mathrm{gl}(|I_k|, \mathbf{H})$.

Notice that on compact homogeneous spaces, each component Q_k is final dimensional by the Peter-Weyl theorem.

A component Q_k has a nontrivial *centre* Z_k if and only if Q_k is finite dimensional, i.e., $|I_k| < \infty$. In the real and quaternonic cases, this centre is one dimensional and it is 2 dimensional in the complex case.

The centre Z_v of V is spanned by the centre of the finite-dimensional components Q_k. The algebra V is selfadjoint if and only if each spectral component Q_k is equal to $\mathbf{R}$. The algebra V is commutative if and only if its spectral components Q_k are $\mathbf{R}$ and $\mathbf{C}$ only. The symmetric spaces, or more generally the ray-symmetric spaces, have selfadjoint spectral algebras. In this case, the unit kernels $\mathbf{e}_i(x, \cdot)$ around each point x are joint eigenfunctions of the invariant differential operators $\square_H^{(k)}$, $\Delta_H^{(k)}$ and additionally they are invariant under the action of the isotropy group I_x. I.e., these functions are exactly the spherical functions on a symmetric space, whose theory was developed by Harish-Chandra. Just by this reason, we call the unit kernels $e_i(x, y)$ of the finite-dimensional spectral components Q_i *spherical kernels* as well.

For a kernel $H \in V$, consider the spectral decomposition $H = H_1 + H_2 + \cdots$, $H_i \in Q_i$. By the matrix parametrization $\mathrm{mat}(H_i)$, $i = 1, 2, 3, \ldots$, of the components, we get the *matrix-valued spectrum of H*, which describes the convolution action $H* : V \to V$ on the invariant subspaces Q_i by means of matrix product. This matrix-valued spectrum obviously determines the proper spectrum of the convolution operator $H* : L^2(M^n) \to L^2(M^n)$. For instance, one can use the formulas

$$(1.54) \qquad \sum_{i=1}^{\infty} \mathrm{Tr}\,\mathrm{mat}(H_i)^k = \mathrm{Tr}\,H^k = \sum d_i \lambda_i^k(H)$$

for this computation. One can introduce a matrix-valued spectrum for the differential operators $\square_H^{(k)}$, $\Delta_H^{(k)}$ as well. On smooth manifolds, the spectral components Q_i are spanned by smooth kernels and, using approximations for the averaging operators $UE_{H;p;r}$, $E_{H;p;r}$ we get that the operations $\square_{Hx}^{(k)} H(x, y)$, $\Delta_{Hx}^{(k)} H(x, y)$ leave the subspaces Q_i invariant. The corresponding matrices define the matrix-valued spectrum of these differential operators.

On selfadjoint spaces (for instance, on symmetric spaces or on ray-symmetric spaces) this matrix-valued spectrum is real-number valued. On commutative spaces, this spectrum is complex-number valued.

This spectrum notion suggests a whole list of *fundamental problems* to be solved. At the very end of this paper, we only indicate these problems and we will return to these questions in a subsequent work.

The *Main Question* of this field is as follows:

(1) Are the Jacobi isospectral spaces isometric?

We have seen that many geometric properties can be decided with the help of the Jacobi spectral algebra. It should be mentioned that in the case of symmetric spaces, the isometry can be stated in the following form.

PROPOSITION 1.6. *Let M_1 and M_2 be complete Jacobi isospectral manifolds. If one of them is symmetric then the universal covering spaces $\widetilde{M}_1$ and $\widetilde{M}_2$ are isometric.*

I.e., the Jacobi spectrum does not determine the fundamental group in general.

It seems to be reasonable to make use of the following classical theorem for the solution of the Main Question.

Let M_1 and M_2 be Riemannian manifolds such that there exists an L^2-norm-preserving and multiplication-preserving linear isomorphism Φ between *the spaces $C(M_1)$ and $C(M_2)$ of continuous functions with compact support. Then there exists an isometry $\phi\colon M_1 \to M_2$ such that $\Phi(f) = f \circ \phi^{-1}$, $\forall f \in C(M_1)$.*

The main point of the proof of this theorem is that the functions from $C(M)$ separate the points and the maximal ideals can be identified with points, since any maximal ideal X consists exactly of such functions which vanish on a uniquely determined point $x \in M$.

For the application of this theorem, we have to produce functions on the manifold M itself with the help of the kernels $H(x, y) \in \mathscr{J}S$. One of the possibilities is to produce the functions

$$(1.55) \qquad r_H(x) = \int H(x, y)\, dy$$

which are called *Jacobian root functions*. If this Jacobian root function space is "thick" enough so that these functions separate the points of the manifold, the Main Question can be positively answered immediately.

The problem is that the root function space may be very "thin." For instance, these functions are constant on homogeneous spaces. However just this suggests the following ideas:

A space is said to be *Jacobi homogeneous* if its root functions are constant. Notice that these spaces have the following remarkable property.

For arbitrary points x_1 and x_2, consider the map

$$(1.56) \qquad \Phi\colon H(x_1, \cdot) \to H(x_2, \cdot), \qquad \forall H \in \mathscr{J}S$$

between the function spaces $\mathscr{J}S_{x_1}$ and $\mathscr{J}S_{x_2}$ on a Jacobi homogeneous manifold. Then the Φ is an L^2-norm-preserving and multiplication-preserving linear isomorphism. This suggests an affirmative answer to the following question:

(2) Is it true that the Jacobi homogeneous spaces are homogeneous?

We have seen that those spaces are Jacobi homogeneous where the Laplacian commutes with the operators $\square_H^{(k)}$. Therefore a positive answer can be expected for the following question as well.

(3) Is it true that the Jacobi commutative spaces, or more generally the spaces where the Laplacian commutes with the operators $\square_H^{(k)}$, are homogeneous?

We also repeat here an earlier question, discussed in §4 of Chapter 2.

(4) Are the Jacobi selfadjoint spaces exactly the ray-symmetric spaces?

If the space is not Jacobi homogeneous, one can follow the following track. Consider the space of maximal ideals in the Jacobian root function space.

(5) Is it true that the space of maximal ideals is exactly the orbit space of isometries acting on the space?

The answering of these questions would lead to a complete answer to the Main Question. More than that, the above-sketched procedure would clarify how the Jacobi spectral algebra determines the isometries of the space. In particular, it would answer clearly how it is possible to decide by the aid of this spectrum whether the space is homogeneous or not.

A large number of further natural questions can be formulated for the decision of other properties (for instance, real analyticity or Kahler-property) with the help of this spectrum. For the present author, one of the most attractive questions is

(6) How can it be seen from the Jacobi spectrum whether a space can be represented as a minimal surface in a euclidean sphere?

The nice imbedding of harmonic manifolds suggests that the answering of this question requires an extended study of the spherical kernels $e_i(x, y)$.

Acknowledgments

Most of the results of this article were found while the author was staying at the Max-Planck-Institute für Mathematik in Bonn in the academic years 1987–88 and 1989–90. I am indebted for the excellent working conditions there. I would like to express many thanks to Professors U. Abresch, W. Ballmann, H. Karcher, J. Kazdan, O. Kobayashi, N. Koiso and A. Koranyi for valuable discussions.

References

[Al1] A. Allamigeon, *Properties globales es espaces de Riemann harmoniques*, Ann. Inst. Fourier (Grenoble) **15** (1965), 91–132.

[Am1] W. Ambrose, *Structure theorems for a special class of Banach algebras*, Trans. Amer. Math. Soc. **57** (1945), 364–386.

[AS1] W. Ambrose and I. M. Singer, *On homogeneous Riemannian manifolds*, Duke Math. J. **25** (1958), 647–669.

[Ber1] M. Berger, *Blaschke manifolds on S^n and $\mathbb{P}^n$*, published in [Bes1], 351–364.

[Ber2] ___, *Eigenvalues of the Laplacian*, Proc. Sympos. Pure Math., vol. 16, Amer. Math. Soc., Providence, RI, 1970, pp. 121–125.

[Ber3] ___, *Geometry of the spectrum*. I, Proc. Sympos. Pure Math., vol. 27, part II, Amer. Math. Soc., Providence, RI, 1975, pp. 129–152.

[BGM1] M. Berger, P. Gauduchon and E. Mazet, *Le spectre d'une variete' Riemannienes*, Lecture Notes in Math., vol. 194, Springer-Verlag, 1974.

[BK1] M. Berger and J. L. Kazdan, *A Sturm-Liouville inequality with applications to an isoperimetric inequality for volume, injectivity radius and to wiedersehn manifolds*, General Inequalities (E. F. Beckenbach, ed.). Vol. 2, Birkhauser, 1980, pp. 367–377.

[Bes1] A. L. Besse, *Manifold all of whose geodesics are closed*, Springer-Verlag, 1978.

[Bes2] ___, *Einstein manifolds*, Springer-Verlag, 1986.

[Bi1] G. Birkhoff, *Metric foundation of geometry*. I, Trans. Amer. Math. Soc. **55** (1944), 465–492.

[Bo1] R. Bott, *On manifolds all of whose geodesics are closed*, Ann. of Math. (2) **60** (1954), 375–387.

[BS1] R. Bott and H. Samelson, *Applications of the theory of Morse to symmetric space*, Amer. J. Math. **80** (1958), 964–1029; corrections; ibid. **83** (1961), 207–208; published also in [Bes1].

[Bu1] H. Busemann, *Metrics methods in Finsler geometry and in the foundations of geometry*, Ann. of Math. Stud., vol. 8, Princeton Univ. Press, Princeton, NJ, 1942.

[Cha1] I. Chavel, *Riemannian symmetric spaces of rank one*, Marcel Dekker, New York, 1972.

[Cha2] ___, *Eigenvalues in Riemannian geometry*, Academic Press, 1984.

[CG1] J. Cheeger and D. Gromoll, *The splitting theorem for manifolds of non-negative Ricci curvature*, J. Differential Geom. **6** (1971), 119–128.

[CG2] ____, *On the structure of complete manifold of non negative curvature*, Ann. of Math. **96** (3) (1972), 413–443.

[Chi1] Q. S. Chi, *A curvature characterization of certain locally rank-one symmetric spaces*, J. Differential Geom. **28** (1988), 187–202.

[CDKR1] M. Cowling, A. H. Dooley, A. Koranyi and F. Ricci, *H-type groups and Iwasawa decompositions*, Adv. in Math. **87** (1991), 1–41.

[CDKR2] ____, *H-type groups and Iwasawa decompositions*, II, preprint.

[Da1] E. Damek, *Geometry of a semidirect extension of a Heisenberg type nilpotent group*, Colloq. Math. **53** (1987), 255–268.

[Da2] ____, *Curvature of a semidirect extension of a Heisenberg type nilpotent group*, Colloq. Math. **53** (1987), 249–253.

[DR1] E. Damek and F. Ricci, *A class of non-symmetric harmonic Riemannian spaces*, Rapporto Interno (12), 1991, Bull. Amer. Math. Soc. **27** (1992), 139–142 .

[DR2] ____, *Harmonic analysis on solvable extensions of H-type groups*, Preprint (1991).

[D'A1] J. E. D'Atri, *Geodesic spheres and symmetries in naturally reductive homogeneous spaces*, Michigan Math. J. **22** (1975), 71–76.

[D'AN1] J. E. D'Atri and H. K. Nickerson, *Divergence preserving geodesic symmetries*, J. Differential Geom. **3** (1969), 467–476.

[D'AN2] ____, *Geodesic symmetries in spaces with special curvature tensors*, J. Differential Geom. **9** (1974) 251–262.

[D'AZ1] J. E. D'Atri and W. Ziller, *Naturally reductive metrics and Einstein metrics on compact Lie groups*, Mem. Amer. Math. Soc. No. 215 (1979).

[DeTK1] J. DeTurck and J. L. Kazdan, *Analyticity of Einstein manifolds*, published in [**Bes2**], 283–287.

[Do1] J. Dodziuk, *Maximum principle for parabolic inequalities and the heat flow on open manifold*, Indiana Univ. Math. J. **32** (1983) 703–716.

[Ge1] I. M. Gelfand, *Spherical functions on Riemannian symmetric spaces*, Dokl. Akad. Nauk SSSR **70** (1950), 5–8.

[Gi1] P. B. Gilkey, *Spectral geometry of the higher order Laplacian*, Duke Math. J. **47** (1980), 511–528.

[Gi2] ____, *The index theorem and the heat equation*, Publish or Perish, Boston, MA, 1974.

[GWW1] C. Gordon, D. Webb and S. Wolpert, *One cannot hear the shape of a drum*, Preprint (1991). Bull. Amer. Math. Soc. **27** (1992), 134–138.

[Gra1] A. Gray, *Riemannian manifolds with geodesic symmetries of order* 3 , J. Differential Geom. **7** (1972), 343–369.

[Gra2] ____, *The volume of a small geodesic ball in a Riemannian manifold*, Michigan Math. J. **20** (1973), 329–344.

[Gra3] ____, *Einstein-like manifolds which are not Einstein*, Geom. Dedicata **7** (1978), 259–280.

[GraV1] A. Gray and L. Vanhecke, *Riemannian geometry as determined by the volume of small geodesic balls*, Acta Math. **142** (1979), 157–198.

[GraW1] A. Gray and T. J. Willmore, *Mean-value theorems for Riemannian manifolds*, Proc. Roy. Soc. Edinburgh Sec. A **92** (1982), 343–364.

[Gre1] L. W. Green, *Aufwiderschusflachen*, Ann. of Math. (2) **78** (1968), 289–299.

[H1] S. Helgason, *Differential operators on homogeneous spaces*, Acta Math. **102** (1959), 239–299.

[H2] ____, *Differential geometry and symmetric spaces*, Academic Press, 1962.

[H3] ____, *Groups and geometric analysis*, Academic Press, 1984.

[Kc1] M. Kac, *Can one hear the shape of a drum?* Amer. Math. Monthly **73** (4) (1966), 1–23.

[Kp1] A. Kaplan, *Fundamental solutions for a class of hypoelliptic PDE generated by composition of quadratic forms*, Trans. Amer. Math. Soc. **258** (1980), 147–153.

[Kp2] ____, *Riemannian nilmanifolds attached to Clifford modules*, Geom. Dedicata **11** (1981), 127–136.

[Kr1] H. Karcher, *A geometric classification of positively curved symmetric spaces and the isoparametric construction of the Cayley plane*, Asterisque, no. 163–164, Soc. Math. France, Paris, 1988, pp. 6, 111–135; 1989, p. 282.

[Kz1] J. L. Kazdan, *Geometric inequalities*, published in [**Bes1**] 361–364, and [**BerK1**].

[KN1] S. Kobayashi and K. Nomizu, *Foundations of differential geometry*, Interscience, vol. 1, 1963; vol. 2, 1969.

[Kor1] A. Koranyi, *Geometric properties of Heisenberg-type groups*, Adv. in Math. **56** (1985), 28–38.

[Kow1] O. Kowalski, *Some curvature identities for commutative spaces*, Czechoslovak Math. J. **32 (107)** 1982, 389–397.

[Kow2] ____, *Normal forms of the Laplacian and its iterations in the symmetric spaces of rank one*, Quart. J. Pure Appl. Math. **57** (1983), 215–223.

[KP1] O. Kowalski and F. Prufer, *On parobabilistic commutative spaces*, Monatsh. Math. **107** (1989), 57–68.

[KV1] O. Kowalski and L. Vanhecke, *Opereteours differentiels invariants et symmetries geodesiques preservant le volume*, C.R. Acad. Sci. Paris Sér. I 296 (1983) 1001–1003.

[KV2] ____, *Two point functions on Riemannian manifolds*, Ann. Global Anal. Geom. **3** (1985), 95–119.

[KV3] ____, *The Gelfand theorem and its converse for Kahler manifolds*, Proc. Amer. Math. Soc. **102** (1988), 150–152.

[La1] T. Y. Lam, *The algebraic theory of quadratic forms*, Benjamin, 1973.

[Li1] A. Lichnerowicz, *Sur les espaces riemanniens complement harmoniques*, Bull. Soc. Math. France **72** (1944), 146–168.

[LO1] L. H. Loomis, *An introduction to abstract harmonic analysis*, Van Nostrand, Toronto, New York, and London, 1953.

[Ma1] H. Matsumoto, *Quelques remarques sur les espaces Riemanniens isotropes*, C. R. Acad. Sci. Paris Sér. I. **272** (1971), 316–319.

[Mi1] D. Michel, *Comparison des notions de varietes Riemanniens globalement harmoniques et fortement harmoniques*, C. R. Acad. Sci. Paris Sér. A **282** (1976), 1007–1010.

[N1] T. Nagano, *Homogeneous sphere bundles*, Nagoya Math. J. **15** (1959), 29–55.

[O1] R. Osserman, *Curvature in the eighties*, Amer. Math. Monthly, **97** (1990), No. 8.

[OS1] R. Osserman and P. Sarnak, *A new curvature invariant and entropy of geodesic flows*, Invent. Math. **77** (1984), 455–462.

[P1] F. Prufer, *On compact Riemannian manifolds with volume preserving symmetries*, Ann. Global Anal. Geom. **7** (1989), 133–140.

[Ri1] C. Riehm, *The automorphism group of a composition of quadratic forms*, Trans. Amer. Math. Soc. **269** (1982), 403–414.

[RO1] P. H. Roberts and H. D. Ursell, *Random walk on a sphere and on Riemannian manifolds*, Phil. Trans. Roy. Soc. London Ser. A **252** (1960), 317–386.

[Ru1] H. S. Ruse, *On commutative Riemannian manifolds*, Tensor **26** (1972), 180–184.

[RWW1] H. S. Ruse, A. G. Walker and T. J. Willmore, *Harmonic spaces*, Cremonese, Rome, 1961.

[Scg1] R. Schiming, *Harmonic differential operators*, Forum Math. **3** (1991), 177–203.

[Sch1] R. Schneider, *Curvature measures and integral geometry of convex bodies*, Rend. Sem. Mat. Univ. Politec. Torino **38** (1980), 79–98 (1981).

[Sn1] U. Simon, *On differential operators of second order on Riemannian manifolds*, Colloq. Math. **31** (1974), 223–229.

[Ss1] J. Simons, *On transitivity of holonomy systems*, Ann. of Math. (2) **76** (1962), 213–234.

[Su1] T. Sumimoto, *On the commutator of differential operators*, Hokkaido Math. J. **1** (1972), 30–42.

[Sz1] A. I. Szabo, *Lichnerowicz Conjecture on harmonic manifolds*, J. Differential Geom. **31** (1990), 1–28.

[Sz2] ____, *Short topological proof for the symmetry of 2 point homogeneous spaces*, Invent. Math. **106** (1991), 61–64, preprint no. 64, Max Planck Inst. für Math., Bonn., 1989.

[Sz3] ____, *Higher order Laplacians. I, Harmonic and 2 point homogeneous spaces*, Preprint no. 65, Max Planck Inst. für Math., Bonn, 1989.

[Sz4] ____, *Higher order Laplacians. II, Laplacian commuting with the higher orders*, Spectral Theorems, Preprint no. 66, Max Planck Inst. für Math., Bonn, 1989.

[Sz5] ____, *Structure theorems on Riemannian spaces satisfying* $R(X, Y) \cdot R = 0$, J. Differential Geom. **17**, 531–582.

[T1] J. Tits, *Sur certaines classes d'espaces homogenes de groupes de Lie*, Mem. Cl. Sci, Collect. Octaro, II. Ser., Acad. R. Belg. (1955).

[TV1] F. Tricerri and L. Vanhecke, *Homogeneous structures on Riemannian manifolds*, London Math. Soc. Lecture Note Ser., no. 83, Cambridge Univ Press, 1983.

[Y1] S. T. Yau, *On the heat kernel of a complete Riemannian manifold*, J. Math. Pures Appl. **57** (1978), 191–201.

[Wal1] A. C. Walker, *On Lichnerowicz's conjecture for harmonic 4-spaces*, J. London Math. Soc. **24** (1948–49), 317–329.

[Wan1] H. C. Wang, *Two point homogeneous spaces*, Ann. of Math. **55** (1952), 177–191.

[War1] F. W. Warner, *Conjugate loci of constant order*, Ann. of Math. **86** (1967), 192–212.

[Wi2] T. J. Willmore, *Mean value theorems in harmonic spaces*, J. London Math. Soc. **25** (1950), 54–57.

[Wi2] ____, *2-point invariant functions and k-harmonic manifolds*, Rev. Roumaine Math. Pures Appl. **13** (1968), 1051–1057.

[Wi3] ____, *An extension of Pizzetti's formula to Riemannian manifolds*, Anal. on Manifolds (Conf. Univ. Metz, 1979), Asterisque, no. 80, Soc. Math. France, Paris, 1980.

[WO1] J. A. Wolf, *Spaces of constant curvature*, McGraw-Hill, 1961.

[WO2] ____, *Curvature in nilpotent Lie groups*, Proc. Amer. Mat. Soc. **15** (1964), 271–274.

CITY UNIVERSITY OF NEW YORK, LEHMAN COLLEGE

AND

RESEARCH INSTITUTE OF MATHEMATICS, BUDAPEST, HUNGARY

Proceedings of Symposia in Pure Mathematics
Volume **54** (1993), Part 3

Riemannian Foliations and Tautness

PHILIPPE TONDEUR

1. Introduction

A foliation $\mathscr{F}$ on a Riemannian manifold (M, g) is Riemannian, and g a bundle-like metric for $\mathscr{F}$, if locally the foliation is given by the level sets of a Riemannian submersion. This class of foliations has been introduced by Reinhart in [**RE1**], and studied extensively in the eighties. The accounts in book form [**MO2, RE3, T**] present developments from this period. The purpose of this paper is to describe some recent results related to the concept of tautness. Everything written here is in the context of smooth foliations on smooth manifolds.

2. Tautness criterion

A foliation $\mathscr{F}$ on M is said to be taut, if there exists a Riemannian metric g on M, such that all leaves of $\mathscr{F}$ are minimal submanifolds of (M, g). The idea is then to characterize the tautness of $\mathscr{F}$ by cohomological properties of $\mathscr{F}$. These characterizations are based on the following criterion. For simplicity, we assume throughout that the tangent bundle L and the normal bundle $Q = TM/L$ of $\mathscr{F}$ are oriented (and hence also M oriented). The dimension of (the leaves of) $\mathscr{F}$ is denoted p, $0 < p < n = \dim M$.

THEOREM (Rummler-Sullivan [**RU, SU**]). *Let g_L be a Riemannian metric on L, with volume form ω_L on the leaves. Then g_L is induced by a Riemannian metric g on M for which all leaves of $\mathscr{F}$ are minimal, if and only if ω_L is the restriction of a p-form χ on M which satisfies the condition*

$$d\chi(X_1, \ldots, X_{p+1}) = 0,$$

if p of the vector fields $X_1, \ldots, X_{p+1}$ are section of $L \subset TM$.

For $p = n - 1$, this condition simply states that χ is a closed form. To explain this criterion, we need the following metric concepts for a foliation

---

1991 *Mathematics Subject Classification.* Primary 57R30; Secondary 58A14, 58G11.
This paper is in final form and no version of it will be submitted for publication elsewhere.

$\mathcal{F}$ on a Riemannian manifold (M, g). Let ∇^M be the Levi-Civita connection of g. We identify the orthogonal complement $L^\perp$ of L with respect to g with the normal bundle $Q = TM/L$. Then for Z a section of $L^\perp$ the formula

$$W(Z)X = -\pi^\perp(\nabla_X^M Z),$$

for $X \in \Gamma L$, defines a linear bundle map $W(Z): L \to L$. Here $\pi^\perp$ denotes the orthogonal projection of TM to L (with kernel $L^\perp$). $W(Z)$ is the Weingarten map of $\mathcal{F}$, a selfadjoint operator. Its trace

$$\kappa(Z) = \text{trace } W(Z)$$

defines the mean curvature one-form on normal vector fields $Z \in \Gamma L^\perp$. On tangent vector fields $X \in \Gamma L$ one sets $\kappa(X) = 0$. Thus κ is a one-form on M.

The characteristic form $\chi_\mathcal{F}$ of $\mathcal{F}$ on (M, g) is a p-form on M, which evaluated on a local oriented orthonormal frame E_i $(i = 1, \ldots, p)$ of L gives the value 1 (i.e., is the canonical volume associated to g_L), and for arbitrary $Y_1, \ldots, Y_p \in \Gamma TM$ is given by

$$\chi_\mathcal{F}(Y_1, \ldots, Y_p) = \det(g_M(Y_i, E_j)_{ij}).$$

Note that $i(Z)\chi_\mathcal{F} = 0$ for $Z \in \Gamma L^\perp$. The following formula is due to Rummler [**RU**]:

$$\Theta(Z)\chi_\mathcal{F}|L = -\kappa(Z) \cdot \chi_\mathcal{F}|L \quad \text{for } Z \in \Gamma L^\perp.$$

Since $i(Z)\chi_\mathcal{F} = 0$, this says that the p-form $i(Z)d\chi_\mathcal{F} + \kappa(Z) \cdot \chi_\mathcal{F}$ evaluates to 0 along L. The Rummler-Sullivan criterion is an easy consequence of this formula.

The Rummler-Sullivan criterion can be conveniently interpreted in terms of a spectral sequence associated to the foliation. This is the spectral sequence determined by the following multiplicative filtration of the DeRham complex $\Omega^\cdot = \Omega^\cdot(M)$

$$F^r\Omega^m = \{\omega \in \Omega^m | i(X_1) \cdots i(X_{m-r+1})\omega = 0 \quad \text{for } X_1, \ldots, X_{m-r+1} \in \Gamma L\}.$$

It is a decreasing filtration by differential ideals. Clearly

$$F^0\Omega^m = \Omega^m \quad \text{and} \quad F^{m+1}\Omega^m = 0.$$

Further

$$F^r\Omega^{p+r} = \Omega^{p+r} \quad (p = \dim \mathcal{F}),$$

since $(p+r) - r + 1 = p + 1$, and every $(p+r)$-form evaluated on $p+1$ vector fields tangent to $\mathcal{F}$ vanishes. Note that $F^{r+1}\Omega^{p+r}$ consists of all $(p + r)$-forms evaluating to zero on $(p + r) - (r + 1) + 1 = p$ vector fields tangent to $\mathcal{F}$. Thus by definition

$$F^{r+1}\Omega^{p+r} \equiv \text{``}\mathcal{F}\text{-trivial''}(p + r)\text{-forms}.$$

Rummler's formula can be restated as

$$\Theta(Z)\chi_{\mathscr{F}} + \kappa(Z) \cdot \chi_{\mathscr{F}} \in F^1\Omega^p,$$

or equivalently

$$d\chi_{\mathscr{F}} + \kappa \wedge \chi_{\mathscr{F}} \equiv \varphi_0 \in F^2\Omega^{p+1}.$$

Note that $F^{q+1}\Omega^m = 0$, so that for $q = 1$ Rummler's formula simply states

$$d\chi_{\mathscr{F}} + \kappa \wedge \chi_{\mathscr{F}} = 0.$$

In terms of these data, the Rummler-Sullivan criterion takes the following form [HA1, KT2].

THEOREM. *A foliation $\mathscr{F}$ is taut if and only if some element in $E_2^{0,p}$ can be realized by a p-form which is positive along the leaves.*

For a closed manifold M, Haefliger proved in [HA1] that the Rummler-Sullivan criterion depends only on the transversal structure of the foliation. For the special class of Riemannian foliations, this criterion takes an even simpler form, to be explained below.

3. Basic cohomology

Let $\mathscr{F}$ be a foliation on M. A differential form $\omega \in \Omega^r(M)$ is basic, if

$$\iota_X\omega = 0, \quad \Theta_X\omega = 0 \quad \text{for all } X \in \Gamma L.$$

The set $\Omega_B(\mathscr{F}) \subset \Omega(M)$ is a subcomplex, since $d/\Omega_B(\mathscr{F}): \Omega_B(\mathscr{F}) \to \Omega_B(\mathscr{F})$. We denote $d/\Omega_B(\mathscr{F}) = d_B$. The basic cohomology is then by definition

$$H_B^{\cdot}(\mathscr{F}) = H^{\cdot}(\Omega_B(\mathscr{F}), d_B).$$

It plays the role of the DeRham cohomology of the leaf space of $\mathscr{F}$. In terms of the spectral sequence associated to $\mathscr{F}$, it is easy to see that

$$\Omega_B^r(\mathscr{F}) \cong E_1^{r,0}.$$

Moreover, d_1 induces on $\Omega_B^{\cdot}(\mathscr{F})$ precisely d_B. Thus

$$H_B^r(\mathscr{F}) \cong E_2^{r,0}.$$

Note that the range of these groups is by very definition $0 \le r \le q$, $q = \text{codim}\,\mathscr{F}$. Thus the top basic cohomology group is

$$H_B^q(\mathscr{F}) \cong E_2^{q,0}.$$

The idea is to compare this top term in the basis with the top term $E_2^{0,p}$ in the formal fiber which occurred before. This works out pefectly in the case of Riemannian foliations on closed oriented manifolds. In this case the basic cohomology is finite-dimensional [EHS].

4. The Riemannian case

We consider an oriented and transversely oriented Riemannian foliation $\mathscr{F}$ of codimension q on a closed oriented manifold M. After partial results by several authors, the following result was finally proved by Masa.

THEOREM [**MA**]. $\mathscr{F}$ *is taut if and only if* $H_B^q(\mathscr{F}) \neq 0$.

For a compact or nilpotent structural Lie algebra in the sense of Molino, this was proved by Haefliger [**HA1, HA2**]. If the leaves have polynomial growth, this was proved by Carrière [**C**]. For the case of a basic mean curvature form, this was proved by Kamber and Tondeur [**KT1, KT2**]. For the case $q = n - 1$ this was proved by Molino and Sergiescu [**MS**]. For the case $q = 1$ or 2 this was proved by Alvarez López [**A3**].

The idea is to deduce this result from the tautness criterion of §2 by establishing $E_2^{q,0} \cong E_2^{0,p}$. The technical difficulty is as follows. The algebra $(\Omega(M), d)$ carries the C^∞-topology, in terms of which the successive terms (E_r, d_r) of the spectral sequence are topological differential algebras. But E_1 is not Hausdorff, as observed already in [**HA1**]. The closure of the trivial subspace 0_1 of E_1 leads to $\mathscr{E}_1 = E_1/\overline{0}_1$, and $\mathscr{E}_2 = H(\mathscr{E}_1)$. If $\mathscr{F}$ is Riemannian, then $E_2, \mathscr{E}_2$ are finite dimensional by [**A1, SA**], and $\mathscr{E}_2^{q,0} \cong \mathscr{E}_2^{0,p}$ by [**A2**]. The fact established by Masa [**MA**] is that for Riemannian $\mathscr{F}$ one has $E_2 \cong \mathscr{E}_2$, and from this the tautness criterion $H_B^q(\mathscr{F}) \neq 0$ follows.

5. Transversal curvature and tautness

In this section we discuss conditions sufficient to imply the tautness of a Riemannian foliation $\mathscr{F}$. The normal bundle Q of $\mathscr{F}$ carries a canonical metric and torsion-free connection ∇. Associated to this connection are the usual curvature data: curvature tensor, Ricci curvature, and sectional curvature. These are the transversal curvature data of the foliation. Heuristically they correspond to the curvature data of the (local) Riemannian model space, or of the leaf space of $\mathscr{F}$.

In particular consider the transversal Ricci operator $\rho \colon Q \to Q$, and the transversal curvature operator $\mathscr{R} \colon \Lambda^2 Q^* \to \Lambda^2 Q^*$. Then the following result was established in [**A4, MRT2**].

THEOREM. *Let $\mathscr{F}$ be a transversely oriented Riemannian foliation of codimension $q \geq 2$ on a closed oriented Riemannian manifold (M, g). Assume either $\rho > 0$ or $\mathscr{R} > 0$. Then $\mathscr{F}$ is taut.*

The idea is to use the following cohomology vanishing results of [**HB, MRT1**].

THEOREM. *Let $\mathscr{F}$ be a transversely oriented Riemannian foliation of codimension $q \geq 2$ on a closed oriented Riemannian manifold. Then the following hold:*

   (i) *if $\rho > 0$, then $H_B^1(\mathscr{F}) = 0$;*
   (ii) *if $\mathscr{R} > 0$, then $H_B^r(\mathscr{F}) = 0$ for $0 < r < q$.*

While Hebda's result (i) generalizes a classical result of Bochner, property (ii) of [**MRT1**] generalizes a result of Gallot-Meyer [**GM**]. The preceding theorem can now be obtained from this vanishing result as follows. According to [**A4**], for every Riemannian foliation $\mathscr{F}$ there is a well-defined cohomology class $\zeta(\mathscr{F}) \in H_B^1(\mathscr{F})$, whose vanishing characterizes the tautness of $\mathscr{F}$. The vanishing conditions above imply the nullity of this obstruction class.

It is worthwhile to note that this result holds in particular on a Riemannian manifold (M, g) with positive sectional curvature $K > 0$. This follows from O'Neill's formula [**ON**, p. 465]

$$K_\nabla(e_\alpha, e_\beta) = K(e_\alpha, e_\beta) + 3|A_{e_\alpha} e_\beta|^2,$$

relating the transversal sectional curvature K_∇ and ambient sectional curvature K. Here e_α, e_β denotes a pair of orthogonal unit vectors, both orthogonal to the foliation, and A the integrability tensor of $L^\perp$.

The proof of the cohomology vanishing theorem given in [**MRT1**] is as follows. The general idea is to deal with the basic cohomology by an appropriate generalization of the Laplacian and the Bochner method. Reinhart's definition in [**RE2**] is a straightforward generalization of the Laplace-Beltrami operator. However it is not presented as a differential operator on a differentiable manifold, and it is therefore not clear if the usual theorems apply. The definition by El Kacimi and Hector in [**EH**] is well suited to prove finite-dimensionality results. However, because the definition does not allow for an explicit local formula, the Bochner technique and the Weitzenböck formula do not apply. The operator used by Kamber and Tondeur in [**KT2, KT3, KT4**] is well adapted to the Bochner technique. However, the assumption of a basic mean curvature form restricts the scope of its application. In [**MRT1**] an elliptic operator on the DeRham complex of the total space of a Riemannian foliation is introduced, such that its restriction to basic differential forms coincides with the Laplace operator introduced by Reinhart. The advantage of this set-up is that on the one hand the standard theory of partial differential equations applies, and on the other hand the local computations on basic forms in distinguished coordinate neighborhoods are identical with the computations in the usual Riemannian case. In particular the vanishing theorems based on the positivity of certain curvature expressions can be proved essentially as in the special case of Riemannian manifolds.

REFERENCES

[**A1**] J. A. Alvarez López, *A finiteness theorem for the spectral sequence of a Riemannian foliation*, Illinois J. Math. **33** (1989), 79–92.

[**A2**] ____, *Duality in the spectral sequence of Riemannian foliations*, Amer. J. Math. **111** (1989), 905–926.

[**A3**] ____, *On Riemannian foliations with minimal leaves*, Ann. Inst. Fourier (Grenoble) **40** (1990), 163–176.

[**A4**] ____, *The basic component of the mean curvature of Riemannian foliations*, Ann. Global Anal. Geom. (to appear).

[**C**] Y. Carrière, *Feuilletages riemanniens a croissance polynomiale*, Comment. Math. Helv. **63** (1988), 1–20.

[**EH**] A. El Kacimi and G. Hector, *Décomposition de Hodge basique pour un feuilletage riemannien*, Ann. Inst. Fourier (Grenoble) **36** (1986), 207–227.

[EHS] A. El Kacimi, G. Hector, and V. Sergiescu, *La cohomologie basique d'un feuilletage riemannien est de dimension finie*, Math. Z. **188** (1985), 593–599.

[GM] S. Gallot and D. Meyer, *Opérateurs de courbure et Laplacien des formes différentielles d'une variété riemannienne*, J. Math. Pures Appl. **54** (1975), 259–284.

[HA1] A. Haefliger, *Some remarks on foliations with minimal leaves*, J. Differential Geom. **15** (1980), 269–284.

[HA2] ——, *Pseudogroups of local isometries*, Res. Notes in Math., vol. 131, Pitman, Boston, 1985, pp. 174–197.

[HA3] ——, *Feuilletages riemanniens*, Séminaire Bourbaki, 41 ème année, 1988–89, no. 707 (March 1989).

[HB] J. Hebda, *Curvature and focal points in Riemannian foliations*, Indiana Univ. Math. J. **35** (1986), 321–331.

[KT1] F. Kamber and Ph. Tondeur, *Duality for riemannian foliations*, Proc. Sympos. Pure Math. vol. 40, part 1, Amer. Math. Soc., Providence, RI, 1983, pp. 609–618.

[KT2] ——, *Foliations and metrics*, Proc. of a year in Differential Geometry, University of Maryland, Progr. in Math., vol. 32, Birkhäuser, 1983, pp. 103–152.

[KT3] ——, *Duality theorems for foliations*, Astérisque, no. 116, Soc. Math. France, Paris, 1984, pp. 108–116.

[KT4] ——, *De Rham-Hodge theory for riemannian foliations*, Math. Ann. **277** (1987), 415–431.

[MA] X. Masa, *Duality and minimality in Riemannian foliations*, Comm. Math. Helv. **67** (1992), 17–27.

[MO1] P. Molino, *Géometrie globale des feuilletages riemanniens*, Nederl. Akad. Wetensch. Proc. Ser. A **85** (1982), 45–76.

[MO2] ——, *Riemannian foliations*, Progr. in Math. vol. 73, Birkhäuser, 1988.

[MRT1] M. Min-oo, E. Ruh and Ph. Tondeur, *Vanishing theorems for the basic cohomology of Riemannian foliations*, J. Reine Angew. Math. **415** (1991), 167–174.

[MRT2] ——, *Transversal curvature and tautness for Riemannian foliations*, Proc. Conf. Global Anal. and Global Differential Geom. (Berlin, 1990), Lecture Notes in Math., Springer-Verlag, **1481** (1991), 145–146.

[MS] P. Molino and V. Sergiescu, *Deux remarques sur les flots riemanniens*, Manuscripta Math. **51** (1985), 145–161.

[ON] B. O'Neill, *The fundamental equations of a submersion*, Michigan Math. J. **13** (1966), 459–469.

[RE1] B. Reinhart, *Foliated manifolds with bundle-like metrics*, Ann. of Math. (2) **69** (1959), 119–132.

[RE2] ——, *Harmonic integrals on foliated manifolds*, Amer. J. Math. **81** (1959), 529–536.

[RE3] ——, *Differential geometry of foliations*, Ergeb. Math. Grenzgeb. (3), vol. 99, Springer-Verlag, 1983.

[RU] H. Rummler, *Quelques notions simples en géométrie riemannienne et leurs applications aux feuilletages compacts*, Comment. Math. Helv. **54** (1979), 224–239.

[SA] K. S. Sarkaria, *A finiteness theorem for foliated manifolds*, J. Math. Soc. Japan **30** (1978), 687–696.

[SE] V. Sergiescu, *Cohomologie basique et dualité des feuilletages riemanniens*, Ann. Inst. Fourier (Grenoble) **35** (1985), 137–158.

[SU] D. Sullivan, *A homological characterization of foliations consisting of minimal surfaces*, Comment. Math. Helv. **54** (1979), 218–223.

[T] Ph. Tondeur, *Foliations on Riemannian manifolds*, Universitext, Springer-Verlag, New York, 1988.

UNIVERSITY OF ILLINOIS

Proceedings of Symposia in Pure Mathematics
Volume **54** (1993), Part 3

Eigenvalue Problems for Manifolds with Singularities

JOHAN TYSK

We will discuss some methods for estimating the eigenvalues of manifolds with singularities of codimension at least two. More specifically, we will estimate the eigenvalues of some

1. branched coverings, cf. [**T**];
2. branched minimal immersions, cf. [**C-T**];
3. cone-manifolds with small cone-angles (orbifolds fall in this category), cf. [**H-T**].

Let us now describe these problems in some greater detail.

1. Branched coverings

Let $\phi\colon M^n \to N^n$ be a k-sheeted branched Riemannian covering of compact manifolds where the metric $\phi^*(ds_N^2)$ is smooth outside a set of codimension at least two. Let E be a small open set with smooth boundary containing the singular set. Restrict ϕ to obtain $\phi_-\colon M_- \to N_-$, where $M_- = M - E$ and N_- is the image of M_- under ϕ_-. Let H_{M_-} and H_{N_-} be the Dirichlet heat kernels of the respective manifolds with boundary. Since we endow M with the pull-back metric of N under ϕ, $H_{M_-}(x, y, t)$ and $H_{N_-}(\phi(x), \phi(y), t)$ both solve the heat equation on M_-. By comparing the values of these kernels at $t = 0$, one can show, using the maximum principle for the heat equation, that for all $t > 0$

$$H_{M_-}(x, y, t) \le H_{N_-}(\phi(x), \phi(y), t).$$

Integrating over M_- and shrinking the set E one sees that

$$(1) \qquad \sum_{i=0}^{\infty} e^{-\mu_i t} \le k \sum_{i=0}^{\infty} e^{-\lambda_i t},$$

1991 *Mathematics Subject Classification*. Primary 58G25, 53C42.

Partially supported by National Science Foundation grant DMS 90-04062 and by a grant from the Swedish Natural Science Research Council (NFR).

This paper is in final form and no version will be submitted for publication elsewhere.

where $\{\mu_i\}$ and $\{\lambda_i\}$ are the eigenvalues of M and N respectively.

Applications. Let M be a complete oriented minimal surface of finite total curvature in $\mathbb{R}^3$. Then the Gauss map has a holomorphic extension $\mathscr{G}: \overline{M} \to \mathbb{S}^2$, where $\overline{M}$ is a Riemann surface compactification of M. The second variation operator takes the form $\Delta + 2$ on $(\overline{M}, \mathscr{G}^*(ds_{\mathbb{S}^2}^2))$. The number of negative eigenvalues of this operator, or the index of M, therefore satisfies

$$\text{index}(M) \cdot e^{-2t} \le \sum_{\mu_i < 2} e^{-\mu_i t} \le (\deg \mathscr{G}) \sum_{i=0}^{\infty} e^{-\lambda_i t}$$

by estimate (1), where $\{\mu_i\}$ and $\{\lambda_i\}$ are the eigenvalues of M and $\mathbb{S}^2$ respectively. Since the λ_i's are explicitly known, this shows that

$$\text{index}(M) \le C \int_M (-K),$$

where C is some explicit constant, and K is the curvature of M (cf. [**T**]).

2. Branched minimal immersions

We illustrate the method used here by considering one particular example. Let M^2 be an oriented minimal surface in $\mathbb{R}^n$, $n \ge 3$. Then the Gauss map $\mathscr{G}: M \to G_{2,n}$ is antiholomorphic, where $G_{2,n}$ is the Grassmannian of oriented two-planes in $\mathbb{R}^n$. Let us now consider the eigenvalues of $(M^2, \mathscr{G}^*(ds_{G_{2,n}}^2))$. To estimate the number of these eigenvalues that are less than some given constant, we would like to find upper bounds for the Dirichlet heat kernels of domains in $\mathscr{G}(M)$ with compact closure that avoid the singular set. To establish such bonds we embed $G_{2,n}$ into some Euclidean space $\mathbb{R}^N$ and view $\mathscr{G}(M)$ as a submanifold of this Euclidean space. According to the Sobolev inequality in [**M-S**] we then have, for C^1 functions on $\mathscr{G}(M)$,

$$(2) \qquad \left(\int_{\mathscr{G}(M)} f^2 \right)^{1/2} \le C \int_{\mathscr{G}(M)} (|\nabla f| + |f| |\mathscr{H}|),$$

where $|\mathscr{H}|$ is the length of the mean curvature vector of $\mathscr{G}(M)$ as a submanifold of $\mathbb{R}^N$. Since $\mathscr{G}(M)$ is a minimal variety in $G_{2,n}$, $|\mathscr{H}|$ is less than or equal to the length $|B|$ of the second fundamental form of $G_{2,n}$ as a submanifold of $\mathbb{R}^N$. Hence, in inequality (2) we can replace $|\mathscr{H}|$ by $\max_{G_{2,n}} |B|$. We can then use (2) to derive estimates of the form

$$\text{Tr} H_D(t) \le \phi(t) \cdot \text{Area } D,$$

where $\text{Tr} H_D$ denotes the trace of the Dirichlet heat kernel of a domain D in $\mathscr{G}(M)$, and ϕ is an explicitly given function depending only on n. By exhausting $\mathscr{G}(M)$ with such domains we find, for the heat kernel $H_{\mathscr{G}(M)}$ of $\mathscr{G}(M)$,

(3) $$\operatorname{Tr} H_{\mathscr{G}(M)}(t) \leq \phi(t) \cdot \operatorname{Area} \mathscr{G}(M) = \phi(t) \int_M (-K).$$

Applications. For complete minimal surfaces M^2 in $\mathbb{R}^n$ one obtains from inequality (3)

$$\operatorname{index}(M) \leq C(n) \int_M (-K);$$

compare the applications in §1. Since the second variation operator of M as a submanifold of $\mathbb{R}^n$ is an operator on the normal bundle of M, one must in the case of higher codimension use a theorem in [H-S-U] relating the trace of heat kernels on vector bundles to the trace of heat kernels on functions to obtain the estimate above. In a similar way one obtains for minimal surfaces M^2 in $\mathbb{S}^n$ that

$$\operatorname{index}(M) \leq C(n)(2 \operatorname{Area} M - 2\pi\chi(M)),$$

where $\chi(M)$ denotes the Euler characteristic of M. The argument above also applies if M is a minimal surface with boundary Γ in $\mathbb{R}^n$ and one obtains

$$\operatorname{index}(M) \leq C(n) \left(\int_\Gamma \kappa(s)\, ds - 2\pi\chi(M) \right),$$

where $\kappa(s)$ is the geodesic curvature of Γ (cf. [C-T]).

3. Cone-manifolds

An n-dimensional cone-manifold is a simplicial complex M which is a rational homology n-manifold; i.e., the link of each vertex has the rational homology of an $(n-1)$-dimensional sphere. We also require M to be a complete metric space with a smooth Riemannian metric defined on the complement of the codimension two skeleton of M and on each closed simplex. The singular locus Σ consists of points in M with no neighborhood isometric to a ball in a Riemannian manifold. At each point of Σ in the interior of an $(n-2)$-simplex there is a cone angle which is the sum of the dihedral angles of the n-simplices containing the point. We can define sectional and Ricci curvatures as usual in the smooth part of M. On the singular locus the curvature should be regarded as a measure. There is concentrated positive curvature at points where the cone angle is less than 2π and concentrated negative curvature where the cone-angle is greater than 2π.

The singular manifolds we studied above are cone-manifolds, where in §1 the cone-angles are multiples of 2π. Here we will instead assume that the cone-angles are strictly less than 2π (see [H-T]). Note, for instance, that orbifolds fall into this category of cone-manifolds. One can show that, under this assumption on the cone-angles, the length-minimizing geodesics do *not* pass through the singular locus (in contrast to the case of cone-angles larger than 2π where an entire pencil of length-minimizing geodesics pass through the singular locus). Thus at any point of a complete connected cone-manifold with cone-angles less than 2π the exponential map is well defined and surjective. One can use this fact to generalize eigenvalue estimates (compare [C, L]) and isoperimetric inequalities (see [G]) for smooth manifolds to such cone-manifolds. Before listing some of these results let us introduce the notation $\mathrm{Cone}_K(S; R)$ for a cone of constant curvature K with base S and radius R, where S is a spherical cone-manifold of constant curvature one, and let $B(p; \delta)$ denote a ball of radius δ centered at p in our cone-manifold. Topologically, $\mathrm{Cone}_K(S; R)$ is obtained from the space $S \times [0, R)$ by identifying $S \times 0$ to a point. The metric has the form $ds^2 = dr^2 + s_K(r)^2 d\theta^2$ where $d\theta^2$ denotes the metric on S, $r \in [0, R)$ and

$$s_K(r) = \begin{cases} \frac{1}{\sqrt{K}} \sin\left(\sqrt{K} r\right) & \text{if } K > 0; \\ r & \text{if } K = 0; \\ \frac{1}{\sqrt{|K|}} \sinh\left(\sqrt{|K|} r\right) & \text{if } K < 0; \end{cases}$$

If $K \leq 0$ then $\mathrm{Cone}_K(S; R)$ is defined for all $0 < R \leq \infty$. If $K > 0$ then $\mathrm{Cone}_K(S; R)$ is defined for $0 < R \leq \pi/\sqrt{K}$ and we define the suspension of S to be the completion of $\mathrm{Cone}_K(S; \pi/\sqrt{K})$. This suspension is thus obtained by gluing together two closed cones of radius $\pi/2\sqrt{K}$. These cones and suspensions are the analogues of standard balls and spheres for constant curvature cone-manifolds. We also note that at each point p on a cone-manifold M there is a tangent cone TM_p isometric to a Euclidean cone $\mathrm{Cone}_0(S_p; \infty)$ where S_p is the spherical cone-manifold of constant curvature one consisting of all unit tangent vectors at p.

THEOREM 1. *Let M be an n-dimensional cone-manifold with Ricci curvature bounded below by $K(n-1)$, for some real number K. Then for any $\delta > 0$, $p \in M$, we have*

$$\lambda(B(p; \delta)) \leq \lambda(B_K(\delta)),$$

where $\lambda(B_K(\delta))$ is the first Dirichlet eigenvalue of a disk of radius δ in a simply connected space form of curvature K. Equality occurs if and only if $B(p; \delta)$ has constant curvature K and is isometric to the cone $\mathrm{Cone}_K(S_p; \delta)$.

THEOREM 2. *Let M be as above with $K > 0$. Let*

$$\beta = \frac{V(M)}{V(M_K)},$$

where $V(\cdot)$ denotes volume and M_K denotes the simply-connected space form of constant curvature K. Given any $\Omega \subset M$ which is a finite disjoint union of domains with piecewise smooth boundary, let D be the disk in M_K for which

$$V(\Omega) = \beta V(D).$$

Then, with $A(\cdot)$ *denoting* $(n-1)$-*dimensional area,*

$$A(\partial\Omega) \geq \beta A(\partial D),$$

with equality if and only if Ω *is isometric to a constant curvature cone* $\text{Cone}_K(S;\delta)$.

THEOREM 3. *Let* M *be as Theorem* 1. *Then*

$$\lambda(M) \geq nK$$

and $\lambda(M) = nK$ *if and only if* M *has constant curvature* K *and is the suspension of an* $(n-1)$-*dimensional spherical cone-manifold.*

Applications. From the theorems above one obtains the following characterization of cone-manifolds of maximal diameter (see [H-T]; compare also [C]).

THEOREM 4. *Let* M *be as above with* $K > 0$. *Then the diameter of* M *satisfies*

$$d(M) \leq \frac{\pi}{\sqrt{K}}$$

with equality holding if and only if M *has constant curvature* K *and is the suspension of an* $(n-1)$-*dimensional spherical cone-manifold.*

REFERENCES

[C] S.-Y. Cheng, *Eigenvalue comparison theorems and its geometric applications*, Math. Z. **143** (1975), 289–297.

[C-T] S.-Y. Cheng and J. Tysk, *Schrödinger operators and index bounds for minimal surfaces*, Preprint.

[G] M. Gromov, *Paul Levy's isoperimetric inequality*, Preprint.

[H-S-U] H. Hess, R. Schrader, and D. A. Uhlenbrock, *Kato's inequality and the spectral distribution of Laplacians on compact Riemannian manifolds*, J. Differential Geom. **15** (1980), 27–37.

[H-T] C. Hodgson and J. Tysk, *Eigenvalue estimates and isoperimetric inequalities for cone-manifolds*, to appear in the Bulletin of the Australian Mathematical Society.

[L] A. Lichnerowicz, *Géométrie des groupes des transformations*, Dunod, Paris, 1958.

[M-S] J. H. Michael and L. M. Simon, *Sobolev and mean value inequalities on generalized submanifolds of* $\mathbb{R}^n$, Comm. Pure Appl. Math. **26** (1973), 361–379.

[T] J. Tysk, *Eigenvalue estimates with applications to minimal surfaces*, Pacific J. Math. **128** (1987), 361–366.

UPPSALA UNIVERSITY

Proceedings of Symposia in Pure Mathematics
Volume **54** (1993), Part 3

Some Rigidity Aspects of Riemannian Fibrations

GERARD WALSCHAP

The theory of foliations has traditionally played a major role in differential geometry. Although they were primarily studied from a topological point of view, it was realized early on that new insight could be gained in the presence of a Riemannian metric on the ambient space (see, e.g., [**Re**]). Specifically, a foliation of a Riemannian manifold is said to be *metric* or *Riemannian* if its leaves are locally everywhere equidistant. These foliations have been extensively studied over the years. A detailed exposition of the basic results can be found in [**Mo**]. In this note, we shall focus instead on foliated spaces with curvature bounds, and see that, in many cases, curvature conditions on the ambient Riemannian manifold together with completeness of the metric sharply restrict the amount of allowable foliations. Consider for example a submersion $\pi: M \to B$ between Riemannian manifolds. The "vertical" distribution $\mathscr{V} = \ker \pi_*$ induces a foliation of M. π is said to be *Riemannian* if its derivative is isometric on the "horizontal" distribution $\mathscr{H} = \mathscr{V}^\perp$, i.e., if $|\pi_* E| = |E^h|$ for $E = E^h + E^v \in \mathscr{H} \oplus \mathscr{V}$. In this case, the foliation induced by π is metric. Moreover, every metric foliation is locally determined by Riemannian submersions, so that the latter play a crucial role in understanding the former. They were first extensively studied by O'Neill [**O'N**], who expressed the curvatures in terms of the following two tensor fields characterizing the submersion:

(1) the integrability or O'Neill tensor $A: \mathscr{H} \times \mathscr{H} \to \mathscr{V}$ given by

$$A_X Y = (\nabla_X Y)^v = \tfrac{1}{2}[X, Y]^v;$$

(2) the second fundamental tensor $S: \mathscr{H} \times \mathscr{V} \to \mathscr{V}$ of the fibers,

$$S_X T = (\nabla_T X)^v.$$

(The formulas here differ from those in [**O'N**].) One relevant feature is that Riemannian submersions are curvature nondecreasing: if $X, Y \in TM$ are

1991 *Mathematics Subject Classification.* Primary 53C20, 53C12.
This paper is in final form and no final version will be submitted for publication elsewhere.

horizontal, then

$$(*) \qquad\qquad K_{\pi_* X, \pi_* Y} = K_{X,Y} + 3|A_X Y|^2.$$

Another geometrically appealing fact is that geodesics which start out perpendicular to some fiber stay orthogonal to the fibers for all time (this is actually a local property which carries over from Riemannian foliations; see [Re]). One can then horizontally lift a geodesic $\gamma: [0, a] \to B$ to M and obtain a family $\{h_\gamma^t\}$ of diffeomorphisms between the fibers over $\gamma(0)$ and $\gamma(t)$ for $t \in (0, a)$. If u is a vector tangent to the fiber, then $J(t) = h_{\gamma_*}^t u$ is a nowhere vanishing vertical Jacobi field along the corresponding horizontal lift c, with derivative

$$J' = -A_{\dot{c}}^* J + S_{\dot{c}} J,$$

where $A_x^*: \mathscr{V} \to \mathscr{H}$ denotes the pointwise adjoint of $A_x: \mathscr{H} \to \mathscr{V}$. Since the behavior of Jacobi fields is controlled by the curvature, the same can be said of the tensors A and S. This imposes strong restrictions on the submersion when the metric of M is complete (see [GG2, W3]).

Among the many examples of Riemannian submersions, the following three deserve special attention:

(a) The closest thing to a metric product $B \times F \to B$ is a warped product $B \times_h F$ of two Riemannian manifolds (B, g_B) and (F, g_F): the underlying space is topologically $B \times F$, and the metric is $\pi_B^* g_B + (h \circ \pi_B) \pi_F^* g_F$, where π_B, π_F are the respective projections, and h is a smooth positive function on B. π_B is easily seen to be a Riemannian submersion, and the horizontal distribution is integrable or, equivalently, $A \equiv 0$. All space forms have (local) representations as warped products (cf. [B]).

(b) Let G be a group of isometries acting freely on a Riemannian manifold M, with closed orbits. Then there exists a (unique) Riemannian metric on the quotient M/G such that the projection $M \to M/G$ becomes a Riemannian submersion. The Hopf fibrations $S^1 \to S^{2n+1} \to CP^n$ fall into this category (cf. [GHL]).

(c) The procedure in (b) by means of which a metric on M projects down to B can be reversed if M is a fiber bundle. Suppose for simplicity that $\mathbf{R}^n \to M \xrightarrow{\pi} B$ is a Euclidean vector bundle over a Riemannian manifold B. Any $O(n)$-invariant metric on $\mathbf{R}^n$ induces an inner product on $\mathscr{V} = \ker \pi_*$. On the other hand, given a Riemannian connection on the bundle, the associated distribution $\mathscr{H}$ inherits an inner product from TB via $\pi_{*|\mathscr{H}}^{-1}$. Defining the inner product on $TM = \mathscr{H} \oplus \mathscr{V}$ as a direct sum, we obtain a metric on M for which π becomes a Riemannian submersion with totally geodesic fibers (see [V]). A well-known example is the "connection metric" on the tangent bundle of a Riemannian manifold (M, g). Here, the fibers are flat, and the horizontal distribution is the one associated with the Levi-Civita connection of g.

Let us next examine how these constructions interact with curvature. According to (∗), if M has nonnegative curvature, then so does B. Procedure (b) has thus been used by many authors to obtain metrics of nonnegative curvature on such manifolds as $CP^2 \# \pm CP^2$ [C], some exotic spheres [GM, Ri], and the tangent bundle of the n-sphere [CG].

In the last example, it turns out that the vector bundle projection $TS^n \to S^n$ is itself a Riemannian submersion, and S^n is the soul of TS^n in the terminology of [CG]. It is still unknown whether any complete noncompact manifold M^n of nonnegative curvature admits a Riemannian submersion onto a soul S. When the metric is analytic, however, a straightforward argument using a result of Marenich [M] shows that, on some neighborhood of S, the metric projection onto S is a Riemannian submersion. Moreover, in some special cases such as $n \leq 4$, one has a global submersion without requiring analyticity [W1]. In fact, if the fibers of $M^4 \to S$ are totally geodesic, then the metric on M arises from example (b). More precisely, assume for simplicity that M^4 is simply connected. Then M is either a metric product $N \times P^k$ (with N compact and P^k diffeomorphic to $\mathbf{R}^k$), or the quotient of a metric product by an isometric S^1-action [W2]. Surprisingly, a similar result holds for compact manifolds: if M^4 is a compact, simply connected manifold with nonnegative curvature which admits a totally geodesic metric foliation, then M is either an isometric product of 2-spheres, or the quotient of an isometric product $B \times S^2$—here B is a lens space—by a diagonal S^1-action [W3].

One last remark before leaving the noncompact case: if E is a vector bundle over a compact manifold of positive curvature, then one might try procedure (c) to obtain a metric with nonnegative curvature on E. This works provided there exists a vector bundle connection on E with "almost" parallel curvature tensor [SW]. This is the case, of course, for the tangent bundle of a symmetric space. No such requirement is necessary, however, for Ricci curvature: every vector bundle over a compact manifold of positive Ricci curvature admits a metric with positive Ricci curvature (see [BB, N, Po]).

Insofar as the variety of metric foliations of a given manifold is concerned, one expects space forms to be more accessible. In [GG1], it is shown that one-dimensional metric foliations of constant curvature spaces are either flat (the tensor $A \equiv 0$) or homogeneous (the leaves are locally spanned by a Killing field). A more involved argument actually yields a classification of all metric foliations with dimension ≤ 3 on Euclidean spheres [GG2]. These are always homogeneous; in fact they are orbit foliations of a Lie subgroup of isometries. As a consequence, all Riemannian fibrations of these spheres are congruent to Hopf fibrations, with the possible exception of $S^{15} \to M^8$.

On flat or negatively curved manifolds, the strongest restrictions require compactness. For example, any Riemannian fibration of a flat manifold M

over a compact base is at least locally the projection of a metric product onto one of the factors [Pe, W3]. When the ambient space M itself is compact, the result generalizes to foliations [W4]. Without compactness, it is already false in $\mathbf{R}^3$: consider for example the one-dimensional foliation generated by glide rotations. In constant negative curvature, there are no odd-dimensional metric foliations (provided M is compact), and even-dimensional ones must have minimal leaves (see [KW, W4]). The proof in the general case uses ergodic theory, but for one-dimensional leaves, this also follows from the fact that any compact Riemannian manifold which admits a one-dimensional Riemannian foliation has vanishing Gromov invariant and Pontrjagin numbers [G]. Here again the result fails if one drops the compactness assumption: hyperbolic space admits an abundance of Riemannian foliations, many of which are not even rigid (in the sense that they can be deformed; cf. [GG1]). One last comment about the compact case: the author knows of no even-dimensional metric foliation on a compact hyperbolic manifold. One might conjecture that there are none,[1] and more generally, that a compact manifold of negative (but not necessarily constant) curvature does not admit any Riemannian foliation.

Much less is known when the curvature is nonconstant and not strictly negative. The simplest case is when the submersions defining the foliation are locally a projection $M_1 \times M_2 \to M_1$ of a Riemannian product. We then say the foliation *splits*. This amounts to the vanishing of the two tensors A and S in (1) and (2). Under appropriate curvature assumptions, however, if one vanishes, then so does the other. For example, the curvature identities in [O'N] imply that any totally geodesic—i.e., $S \equiv 0$—metric foliation of a manifold with nonpositive curvature splits. For nonnegative curvature, one can show that any flat—i.e., $A \equiv 0$—metric foliation splits, provided the manifold is complete [A, W3]: to see this intuitively, assume for simplicity that the leaves have codimension 1. Since the curvature of the ambient space is nonnegative, the principal curvatures of a family of parallel hypersurfaces along a horizontal geodesic cannot be larger than those (with same initial condition) in Euclidean space (in fact in [E] it is shown that classical comparison theory—such as the Rauch theorem—can be derived from principal curvature comparison). It follows that the foliation must be totally geodesic (in Euclidean space, a negative principal curvature goes to $-\infty$ in finite time), and hence splits.

Thus, for example, the only complete warped products with nonnegative curvature are isometric products.

Finally, there is another way of describing the splitting situation: notice that if a Riemannian submersion $M \to B$ splits, then every plane spanned by a horizontal and a vertical vector must have zero curvature. This condition is

---

[1] In the latest version of [W4], it is shown that compact hyperbolic manifolds admit no metric foliations.

actually sufficient if M is compact and has either nonpositive or nonnegative curvature everywhere [W3]. Observe that it is trivially satisfied in case M is flat, thereby generalizing a result mentioned above.

REFERENCES

[A] K. Abe, *Applications of a Riccati type differential equation to Riemannian manifolds with totally geodesic distributions*, Tôhoku Math. J. **25** (1973), 425–444.

[BB] L. Bérard-Bergery, *Certains fibrés à courbure de Ricci positive*, C. R. Acad. Sci. Paris Sér. I Math. **286** (1978), 929–931.

[B] A. Besse, *Einstein manifolds*, Springer-Verlag, Berlin and Heidelberg, 1987.

[C] J. Cheeger, *Some examples of manifolds of nonnegative curvature*, J. Differential Geom. **8** (1973), 623–628.

[CG] J. Cheeger and D. Gromoll, *On the structure of complete manifolds of nonnegative curvature*, Ann. of Math. (2) **96** (1972), 413–443.

[E] J. Eschenburg, *Comparison theorems and hypersurfaces*, Manuscripta Math. **59** (1987), 295–323.

[GHL] S. Gallot, D. Hulin, and J. Lafontaine, *Riemannian geometry*, Springer-Verlag, Berlin and Heidelberg, 1987.

[GG1] D. Gromoll and K. Grove, *One-dimensional metric foliations in constant curvature spaces*, Differential Geometry and Complex Analysis, Springer-Verlag, Berlin and Heidelberg, 1985, pp. 165–168.

[GG2] ____, *The low dimensional metric foliations of Euclidean spheres*, J. Differential Geom. **28** (1988), 143–156.

[GM] D. Gromoll and W. Meyer, *An exotic sphere with nonnegative sectional curvature*, Ann. of Math. (2) **100** (1974), 401–406.

[G] M. Gromov, *Volume and bounded cohomology*, Inst. Hautes Études Sci. Publ. Math. **56** (1982), 213–307.

[KW] H. Kim and G. Walschap, *Riemannian foliations on compact hyperbolic manifolds*, Preprint. Indiana U. Math. J. **41** (1992), pp. 37–41.

[M] V. B. Marenich, *The structure of open manifolds of nonnegative curvature*, Soviet Math. Dokl. **39** (1989), 404–407.

[Mo] P. Molino, *Riemannian foliations*, Birkhauser, Boston, 1988.

[N] J. Nash, *Positive Ricci curvature on fibre bundles*, J. Differential Geom. **14** (1979), 241–254.

[O'N] B. O'Neill, *The fundamental equations of a submersion*, Michigan Math. J. **13** (1966), 459–469.

[Pe] P. Petersen V, *Rigidity of fibrations in nonnegative curvature*, Preprint, UCLA, 1989.

[Po] W. A. Poor, *Some exotic spheres with positive Ricci curvature*, Math. Ann. **216** (1975), 245–252.

[Re] B. Reinhart, *Foliated manifolds with bundle-like metrics*, Ann. of Math. (2) **69** (1959), 119–132.

[Ri] A. Rigas, *Some bundles of nonnegative curvature*, Math. Ann. **232** (1978), 187–193.

[SW] M. Strake and G. Walschap, *Connection metrics of nonnegative curvature on vector bundles*, Manuscripta Math. **66** (1990), 309–318.

[V] J. Vilms, *Totally geodesic maps*, J. Differential Geom. **4** (1970), 73–79.

[W1] G. Walschap, *Nonnegatively curved manifolds with souls of codimension 2*, J. Differential Geom. **27** (1988), 525–537.

[W2] ____, *The soul at infinity in dimension 4*, Proc. Amer. Math. Soc. **112** (1991), pp. 563–567.

[W3] ____, *Measure-invariant flows in constant curvature*, J. Geom. Anal. (to appear).

[W4] ____, *Foliations of symmetric spaces*, Amer. J. Math. (to appear).

UNIVERSITY OF OKLAHOMA

Proceedings of Symposia in Pure Mathematics
Volume **54** (1993), Part 3

Hausdorff Convergence and Sphere Theorems

JYH-YANG WU

Around 1980, Gromov [**G1, G2**] gave an abstract definition of Hausdorff distance between two compact metric spaces. It can be described as follows.

Consider two compact metric spaces X and Y. If there is a metric on the disjoint union $X \amalg Y$, extending the metrics on X and Y in which $B(X, \varepsilon) = \{z \in X \amalg Y : d(z, X) < \varepsilon\} \supset Y$ and $B(Y, \varepsilon) \supset X$, i.e., the classical Hausdorff distance between X and Y in $X \amalg Y$ is less than ε, then we shall say that the Gromov-Hausdorff distance, $d_H(X, Y)$, between X and Y is less than ε.

It is natural to ask if two closed Riemannian manifolds are topologically similar to each other, provided that the Gromov-Hausdorff distance between them is small? In general, without any further assumption, the Gromov-Hausdorff distance does not give us much information. This can be seen from the following examples.

EXAMPLE 1. Let (M, g) be a closed Riemannian n-manifold. Then

$$\lim_{\varepsilon \to 0} d_H((M, \varepsilon^2 g), \text{point}) = 0.$$

EXAMPLE 2. Let (M, g) and (N, h) be two closed Riemannian n-manifolds. Consider the small manifold $(N, \varepsilon^2 h)$ for small positive number ε. Let $M \# N_\varepsilon$ be the connected sum of (M, g) and $(N, \varepsilon^2 h)$ by gluing them along a small disk and rounding off the corner. Then

$$\lim_{\varepsilon \to 0} d_H((M \# N_\varepsilon), M) = 0.$$

A general simple characterization of precompact classes of compact metric spaces is the following [**G1**]:

LEMMA. *A class $\mathscr{M}$ of compact metric spaces X is precompact in the Gromov-Hausdorff topology if and only if there is a function $h : \mathbb{R}^+ \to \mathbb{N}$ such that for any $\varepsilon > 0$, any $X \in \mathscr{M}$ can be covered with $h(\varepsilon)$ ε-balls.*

1991 *Mathematics Subject Classification.* Primary 53C20.

Partially supported by a National Science Foundation grant.

This paper is in final form and no version of it will be submitted for publication elsewhere.

According to the relative volume comparison theorem, it is now well known that the class $\mathrm{M}_{k,\cdot}^{\cdot,D}(n)$ of closed Riemannian n-manifolds M with the Ricci curvature $\mathrm{Ric}_M \geq (n-1)k$ and the diameter $d(M) \leq D$ is precompact in the Gromov-Hausdorff topology. For the class $\mathcal{M}_{k,\cdot}^{\cdot,D}(n)$ of closed Riemannian manifolds M with the sectional curvature $K_M \geq k$ and the diameter $d(M) \leq D$, Gromov [G3] proved that the sum of the Betti numbers of any element in $\mathcal{M}_{k,\cdot}^{\cdot,D}(n)$ is uniformly bounded in terms of the given constants. These results suggest that the class $\mathcal{M}_{k,\cdot}^{\cdot,D}(n)$ is a good object to relate the topology and the geometry.

In general, the topological and/or geometrical properties of the boundary metric spaces of certain precompact classes of closed Riemannian n-manifolds help us to understand the relation between the topology and the geometry of this class.

Let $\mathcal{M}_{k,d,v}^{K,D,V}(n)$ denote the class of closed Riemannian n-manifolds M with $k \leq K_M \leq K$, $d \leq d(M) \leq D$, and the volume $v \leq v(M) \leq V$. Here we list some properties of the boundary metric spaces of these classes.

(I) [GP, Y1]). Let $X = \lim M_i$, $M_i \in \mathcal{M}_{k,\cdot}^{\cdot,D}(n)$.

(1) X is a length space, i.e., any two points in X can be joined by a minimal geodesic.

(2) The Toponogov curvature of $X \geq k$; i.e., the Toponogov comparison theorem holds on X.

(3) $\dim X \leq n$.

(4) If X is a smooth Riemannian n-manifold, then M_i and X are diffeomorphic for large i. If $\dim X < n$, then M_i fiber over X with fiber almost nonnegatively curved manifold for large i.

(II) [GPW]. Let $X = \lim M_i$, $M_i \in \mathcal{M}_{k,\cdot,v}^{\cdot,D,\cdot}(n)$.

(1) X is a homology n-manifold; i.e., $H_*(X, X-x) = H_*(\mathbb{R}^n, \mathbb{R}^n - 0)$ for all x in X.

(2) The product space $X \times N$ is a topological manifold for any closed topological manifold N with dimension ≥ 2.

(3) For $n \neq 3$, M_i and M_j are homeomorphic for large i and j. In particular, the class $\mathcal{M}_{k,\cdot,v}^{\cdot,D,\cdot}(n)$ contains only finitely many homeomorphism types.

(III) [P]. Let $X = \lim M_i$, $M_i \in \mathcal{M}_{k,\cdot,v}^{K,D,\cdot}(n)$.

(1) X is a smooth n-manifold.

(2) There exist diffeomorphisms $f_i: X \to M_i$ such that the sequence of pull-back metrics $f_i^* g_i$ converges, in C^1, to a $C^{1,\alpha}$ $(0 < \alpha < 1)$ metric g_∞ on X. This is the so-called Gromov compactness theorem. In particular, we have the Cheeger finiteness theorem: the class $\mathcal{M}_{k,\cdot,v}^{K,D,\cdot}(n)$ contains only finitely many diffeomorphism types.

(3) There is a well-defined exponential map $\exp: TX \to X$ and the injectivity radius $i(X) \geq \limsup i(M_i)$.

These results are very helpful for us to investigate some perturbation phenomena. Here we also list some applications by using these techniques.

1. The Berger almost $1/4$-pinching theorem [B]

There is a positive number $\delta(n)$ depending only on the dimension $2n$ such that if M is a compact simply connected Riemannian $(2n)$-manifold with $1/4 - \delta(n) \le K_M \le 1$, then

(1) M is homeomorphic to the $(2n)$-sphere, S^{2n}, or

(2) M is diffeomorphic to a CROSS (compact rank one symmetric space).

Very roughly speaking, to prove this result, one considers a convergent sequence of Riemannian $(2n)$-manifolds $M_i \to X$ satisfying $1/4 - \delta_i \le K_{M_i} \le 1$ with $\delta_i \to 0$. If $d(X) > \pi$, the diameter sphere theorem [GS] implies that X is homeomorphic to S^{2n}. Hence M_i is homeomorphic to S^{2n} for large i. If $d(X) = \pi$, then apply (III) to show that X is actually a smooth Riemannian $(2n)$-manifold with $1/4 \le K_X \le 1$, and then appeal to the Berger $1/4$-pinching theorem.

This approach is now quite typical. The general strategy of the Hausdorff convergence technique is to consider a sequence of Riemannian n-manifolds M_i in the precompact class we want to study, which converge to a metric space X. Then we should ask ourselves the following three questions:

(1) What common topological and/or geometric properties of the M_i's can be carried over to the limit space X?

(2) Does the limit space X have the special properties we expected?

(3) To what extent can these special properties of X be pulled back to the manifolds M_i for large i?

In what follows, we shall give two sphere theorems to illustrate this technique.

2. A reversed volume-pinching sphere theorem

THEOREM 1. *There exists a positive number $\delta(n) > 3/2$ depending only on n such that if M is a compact simply connected Riemannian $(2n)$-manifold with $0 < K_M \le 1$ and $v(M) \le \delta(n)\omega_{2n}$, where ω_{2n} is the volume of the standard unit $(2n)$-sphere, S^{2n}, then M is homeomorphic to S^{2n}.*

REMARK. L. Coghlan and Y. Itokawa [CI] proved this for $\delta(n) = 3/2$.

PROOF. Suppose that this theorem is not true. Then there exists a convergent sequence of Riemannian manifolds M_j with $0 < K_{M_j} \le 1$ and $v(M_j) \le \delta_j \omega_{2n}$ where $3/2 < \delta_j < 2$ and $\delta_j \to 3/2$ such that M_j are not homeomorphic to S^{2n}.

According to the Klingenberg estimate on the injectivity radius, we have $i(M_j) \ge \pi$. On the other hand, the Rauch comparison theorem implies that the volume of the r-ball around any point p in M_j, $v(B(p,r)) \ge v(r)$ for all $r \le \pi$ where $v(r)$ denotes the volume of the r-ball in S^{2n}. Hence

$d(M_j) \leq 2\pi$ for all j. By (III), M_j converges to a smooth manifold X with a $C^{1,\alpha}$ metric. Moreover, $i(X) \geq \pi$ and $v(X) \leq \frac{3}{2}\omega_{2n}$.

Now choose two points p, q in X with $d(p, q) = d(X) \geq \pi$ and define

$$l = \max\{\min\{d(p, x), d(q, x)\} \,|\, x \in X - (B(p, \pi/2) \cup B(q, \pi/2))\}.$$

Hence, $l \geq \pi/2$. To obtain a contradiction, we consider three cases.

CASE 1. $l > \pi$. That is, there is a point x in $X - (B(p, \pi/2) \cup B(q, \pi/2))$ with $d(p, x) \geq l$ and $d(q, x) \geq l$. Thus, the balls $B(p, \pi/2)$, $B(q, \pi/2)$ and $B(x, l - \pi/2)$ are disjoint. Hence

$$v(X) \geq v(B(p, \pi/2)) + v(B(q, \pi/2)) + v(B(x, l - \pi/2)) > \frac{3}{2}\omega_{2n}.$$

This is impossible.

CASE 2. $l < \pi$. Choose η so small that $l + \eta < \pi$. Then every point $x \in X$ is either in $B(p, l+\eta)$ or in $B(q, l+\eta)$. Namely, X is covered with two $(2n)$-disks $B(p, l+\eta)$ and $B(q, l+\eta)$. Hence, X is homeomorphic to S^{2n}. In turn, M_j is, by (III), homeomorphic to S^{2n}. This is a contradiction to our assumption.

CASE 3. $l = \pi$. Then there is a point $x \in X$ such that $\min\{d(p, x), d(q, x)\} = \pi$. Thus the balls $B(p, \pi/2)$, $B(q, \pi/2)$ and $B(x, \pi/2)$ are disjoint. Again we have

$$\frac{3}{2}\omega_{2n} \geq v(X) \geq v(B(p, \pi/2)) + v(B(q, \pi/2)) + v(B(x, \pi/2)) \geq \frac{3}{2}\omega_{2n}.$$

Hence, all equalities hold and $\overline{B}(p, \pi/2) \cup \overline{B}(q, \pi/2) \cup \overline{B}(x, \pi/2) = X$. In particular, $d(p, q) = d(p, x) = d(q, x) = \pi$.

If X is a smooth Riemannian manifold, then the Bishop volume comparison theorem will imply that $K_X = 1$ everywhere. Then the Cartan-Ambrose-Hicks theorem gives that X is isometric to S^{2n} and $v(X) = \omega_{2n}$. This will provide a contradiction. However, since X is just a $C^{1,\alpha}$ Riemannian manifold, we do not have the notion of the sectional curvature for X. To overcome the difficulty, we shall use the Toponogov curvature for X as mentioned in (I).

Note that $\partial B(x, \pi/2) \cap \partial B(p, \pi/2) \neq \emptyset$ and $\partial B(x, \pi/2) \cap \partial B(q, \pi/2) \neq \emptyset$. Since $\partial B(x, \pi/2)$ is connected and

$$\partial B(x, \pi/2) = (\partial B(x, \pi/2) \cap \partial B(p, \pi/2))$$
$$\cup (\partial B(x, \pi/2) \cap \partial B(q, \pi/2)),$$

$(\partial B(x, \pi/2) \cap \partial B(p, \pi/2))$ and $(\partial B(x, \pi/2) \cap \partial B(q, \pi/2))$ cannot be disjoint. Hence, there is a point y in $\partial B(x, \pi/2) \cap \partial B(p, \pi/2) \cap \partial B(q, \pi/2)$. Consider a minimal geodesic σ from x to y, and the minimal geodesics γ_1 (resp. γ_2) from y to p (resp. q). Then the minimal geodesics $(\sigma \cup \gamma_1)$ and $(\sigma \cup \gamma_2)$ start from x to y and then branch to the points p and q. This is impossible because (I)-(2) says that the Toponogov comparison theorem still holds on X.

The only way out is that our assumption is wrong. Therefore, the theorem holds. □

3. The diameter and volume pinching sphere theorems

Let M be a compact Riemannian n-manifold with the Ricci curvature, $\mathrm{Ric}_M \geq n - 1$. The Myers theorem implies that $d(M) \leq \pi$, and the fundamental group, $\pi_1(M)$, of M is finite. The maximal diameter theorem [C] states that if $d(M) = \pi$, then M is isometric to the unit n-sphere, S^n, in $\mathbb{R}^{n+1}$. On the other hand, the Bishop volume comparison theorem implies that $v(M) \leq \omega_n$ and the equality holds if and only if M is isometric to S^n. These two results suggest the following diameter and volume pinching questions.

(+) $\mathrm{Ric}_M \geq n - 1$.

(Q1) If $d(M)$ is close to π, will M be homeomorphic to S^n?

The answer is NO, for $n > 3$. Recently, Otsu [O] has constructed a family of metrics g_i on $M = S^{n-m} \times S^m$ ($n \geq 5$, $m = 2, 3, \ldots, n - 2$) with the properties

$$(*) \qquad \mathrm{Ric}_{(M, g_i)} \geq n - 1, \quad v(M, g_i) \geq v, \quad d(M, g_i) \to \pi$$

where v is a positive constant depending only on n and m. On the other hand, Anderson [A] also found metrics g_i on $M = CP^n$ for all n with the same property $(*)$.

For $n = 3$, it is still open. However, if a sequence of 3-manifolds M_i satisfy the property $(*)$, then M_i is diffeomorphic to S^3 (see [W] for the details). To prove this result, one can follow the ideas in the proof of (II) and show that for large i, M_i is simply connected. Then appeal to the Hamilton's result on the 3-manifolds with positive Ricci curvature [H].

(Q2) If $v(M)$ is close to ω_n, will M be homeomorphic to S^n?

This is still open. However, under an extra assumption on the upper bound of the Ricci curvature, $n-1 \leq \mathrm{Ric}_M \leq C$, Anderson [A] showed that if $v(M)$ is close to ω_n, then M is diffeomorphic to S^n.

(++) $K_M \geq 1$.

The diameter sphere theorem implies that M is homeomorphic to S^n, provided that $d(M) > \pi/2$.

(Q3) If $d(M)$ is close to π, will M be diffeomorphic to S^n?

This is still open, too. Here we represent a partial answer to this question.

THEOREM 2. *Given n and $i_0 > 0$, there is a positive number $d^* < \pi$ depending only on these constants such that if M is a compact Riemannian n-manifold with $K_M \geq 1$ and $i(M) \geq i_0$, then M is diffeomorphic to S^n, provided that $d(M) \geq d^*$.*

REMARK. This theorem can also be viewed as a generalization of the classical differentiable sphere theorem which states that there is a positive number

$\delta_n > 1$ such that if M is a complete simply connected Riemannian n manifolds with $\delta_n \geq K_M \geq 1$, then M is diffeomorphic to S^n, since in this case the injectivity radius of M is greater than $\pi/2$.

PROOF. Suppose that the theorem is not true; then one can find a sequence of Riemannian n-manifolds M_j which are not diffeomorphic to S^n and satisfy

$$K_{M_j} \geq 1, \quad i(M_j) \geq i_0, \quad d(M_j) \geq \pi - \varepsilon_j$$

where $\varepsilon_j \to 0$ as $j \to \infty$. We can assume that M_j converges to a compact metric space X which satisfies

(1) the Toponogov curvature $K_X \geq 1$ and
(2) $d(X) = \pi$.

We claim that X is isometric to S^n. To verify this, we need some lemmas.

LEMMA 1. *If x_j, $y_j \in M_j$ realize the diameter of M_j, then every geodesic $\gamma: [0, \pi] \to M_j$ with $\gamma(0) = x_j$ satisfies $d(x_j, \gamma(\pi)) \geq \pi - 2\varepsilon_j$; that is, γ is almost minimizing.*

PROOF. Take a minimal geodesic σ from x_j to y_j and apply the Toponogov comparison theorem to the hinge (γ, σ, x_j). One obtains $d(x_j, \gamma(\pi)) \geq d(x_j, y_j) - d(y_j, \gamma(\pi)) \geq \pi - 2\varepsilon_j$. □

Let $x = \lim x_j$ and $y = \lim y_j$. By Lemma 1, every geodesic $\gamma: [0, \pi] \to X$ with $\gamma(0) = x$ is minimizing and $\gamma(\pi) = y$.

Given any point p in S^n, denote its antipodal point by $\overline{p}$. Now fix a point p in S^n, and choose an isometry $I_j: T_p S^n \to T_{x_j} M_j$. Define a map $f_j: B(p, \pi) \to M_j$ by $f_j = \exp_{x_j} \circ I_j \circ \exp_p^{-1}$. The Toponogov comparison theorem implies that f_j is a distance-nonincreasing map. Hence, according to the Arzela-Ascoli theorem, f_j converges to a map $f: B(p, \pi) \to X$. By the above argument, f can be extended to a map $f: S^n \to X$.

LEMMA 2. *Given any point q in S^n, $d(f(q), f(\overline{q})) = \pi$.*

PROOF. For $q = p$ or $\overline{p}$, Lemma 2 obviously holds. Hence, we can assume that q is neither p nor $\overline{p}$. The minimal geodesics γ and $\overline{\gamma}$ from p to $\overline{p}$, passing through q and $\overline{q}$, respectively, form a great circle in S^n. Let $\sigma = f(\gamma)$ and $\overline{\sigma} = f(\overline{\gamma})$. The fact that $i(M_j) \geq i$ implies that $d(\sigma(t), \overline{\sigma}(t)) = 2t$ for all $t \in [0, i/2]$. Choose $t > 0$ so small that $\min\{d(p, q), d(p, \overline{q})\} \geq t$. The Toponogov comparison theorem applies for the triangle $(\sigma(t), \overline{\sigma}(t), y)$ to yield $d(f(q), f(\overline{q})) \geq d(q, \overline{q}) = \pi$. Thus, $d(f(q), f(\overline{q})) = \pi$. □

LEMMA 3. *Given any y_1, y_2 in S^n, $d(y_1, y_2) = d(f(y_1), f(y_2))$. That is, f is an isometry from S^n to X.*

PROOF. By the Toponogov comparison theorem, we have $d(y_1, y_2) \geq d(f(y_1), f(y_2))$ and $d(\overline{y}_1, y_2) \geq d(f(\overline{y}_1), f(y_2))$. Thus, $\pi \geq d(f(y_1), f(y_2))$

$+ d(f(\overline{y}_1), f(y_2)) \geq d(f(y_1), f(\overline{y}_1)) = \pi$. Hence, all equalities hold and this gives $d(y_1, y_2) = d(f(y_1), f(y_2))$. $\square$

Lemma 3 shows that X is isometric to S^n. By (I)-(4), M_j is diffeomorphic to S^n. But this, of course, contradicts the fact that M_j is not diffeomorphic to S^n and completes the proof of this theorem. $\square$

Here we also state a Ricci curvature version of this theorem:

THEOREM. *Given any $i > 0$, k and $n \geq 2$, there exists a positive number $\varepsilon^* = \varepsilon^*(i, k, n)$ such that if M is a complete Riemannian n-manifold satisfying*

$$\mathrm{Ric}_M \geq n - 1, \quad i(M) \geq i, \quad K_M \geq k, \quad d(M) \geq \pi - \varepsilon^*,$$

then M is diffeomorphic to S^n.

SKETCH OF THE PROOF. Once again consider a sequence of Riemannian n-manifolds M_j satisfying

$$\mathrm{Ric}_{M_j} \geq n - 1, \quad i(M_j) \geq i, \quad K_{M_j} \geq k, \quad d(M_j) \geq \pi - \varepsilon_j$$

where $\varepsilon_j \to 0$ as $j \to \infty$. Recently, Anderson and Cheeger [AC] showed that the class of closed Riemannian n-manifolds M with $\mathrm{Ric}_M \geq k$, $d(m) \leq D$ and $i(M) \geq i > 0$ is C^α precompact. Namely, the properties (III)-(1), (2) still hold if one replaces the $C^{1,\alpha}$ by C^α in (III)-(2). This result allows us to assume that M_j converges to a C^α Riemannian n-manifold X. It is easy to see that the relative volume comparison theorem still holds for the limit manifold X. Then use the argument of Shiohama's proof [S] of the maximal diameter sphere theorem to conclude that $v(X) = v(S^n)$. This theorem follows now from (I)-(4) and a result of Yamaguchi:

THEOREM (Yamaguchi [Y2]). *Let M_j be a sequence of Riemannian n-manifolds satisfying*

$$\mathrm{Ric}_{M_j} \geq n - 1, \quad K_{M_j} \geq k, \quad v(M_j) \to v(S^n);$$

then the limit space $X = \lim M_j$ is isometric to S^n. $\square$

(Q4) If $v(M)$ is close to ω_n, will M be diffeomorphic to S^n?

Recently, Otsu, Shiohama and Yamaguchi [OSY] have given an affirmative answer to this question.

Let us conclude this note by proposing the following question:

QUESTION. What can we say about the limit space X in (I), (II), and (III) if we replace the sectional curvature by the Ricci curvature?

REFERENCES

[A] M. Anderson, *Metrics of positive Ricci curvature with large diameter*, Preprint (1990).

[AC] M. Anderson and J. Cheeger, *C^α-compactness for manifolds with Ricci curvature and injectivity radius bounded below*, J. Differential Geom. **35** (1992), 265–281.

[B] M. Berger, *Sur les varietes riemanniennes princees juste au-dessaus de* 1/4, Ann. Inst. Fourier (Grenoble) **33** (1983), 135–150.

[C] S. Y. Cheng, *Eigenvalue comparison theorem and its geometric applications*, Math. Z. **143** (1975), 289–297.

[CI] L. Coghlan and Y. Itokawa, *A sphere theorem for reverse volume pinching on even-dimensional manifolds*, Proc. Amer. Math. Soc. (to appear).

[G1] M. Gromov, *Groups of polynomial growth and expanding maps*, Inst. Hautes Études Sci. Publ. Math. **53** (1981), 183–215.

[G2] ――, *Structure metrique pour les varites riemanniennes*, Cedic/Fernand Nathan, Paris, 1981.

[G3] ――, *Curvature, diameter and Betti numbers*, Comment. Math. Helv. **56** (1981), 179–195.

[GP] K. Grove and P. Petersen V, *Manifolds near the boundary of existence*, J. Differential Geom. **33** (1991), 379–394.

[GPW] K. Grove, P. Petersen, V, and J.-Y. Wu, *Geometric finiteness theorems via controlled topology*, Invent. Math. **99** (1990), 205–211.

[GS] K. Grove and K. Shiohama, *A generalized sphere theorem*, Ann. of Math. (2) **106** (1977), 210–211.

[H] R. Hamilton, *Three manifolds with positive Ricci curvature*, J. Differential Geom. **17** (1982), 255–306.

[O] Y. Otsu, *On manifolds of positive Ricci curvature with large diameter*, Math. Zeit (to appear).

[OSY] Y. Otsu, K. Shiohama, and T. Yamaguchi, *A new version of differential sphere theorem*, Invent. Math. **98** (1989), 219–228.

[P] S. Peters, *Convergence of Riemannian manifolds*, Compositio Math. **62** (1987), 3–16.

[S] K. Shiohama, *A sphere theorem for manifolds of positive Ricci curvature*, Trans. Amer. Math. Soc. **275** (1983), 811–819.

[W] J.-Y. Wu, *Convergence of Riemannian 3-manifolds under Ricci curvature bounds*, Amer. J. Math. (to appear).

[Y1] T. Yamaguchi, *Collapsing and pinching under a lower curvature bound*, Ann. of Math. **133** (1991), 317–357.

[Y2] ――, *Lipschitz convergence of manifolds of positive Ricci curvature with large volume*, Math. Ann. **284** (1989), 423–436.

UNIVERSITY OF MARYLAND, COLLEGE PARK

Proceedings of Symposia in Pure Mathematics
Volume **54** (1993), Part 3

Automorphism Groups and Fundamental Groups
of Geometric Manifolds

ROBERT J. ZIMMER

1. Statement of main results

The aim of this paper is to survey some of the recent results concerning the relationship of the structure of the fundamental group of a manifold and that of the automorphism group of a geometric structure on the manifold. We shall focus our attention here on rigid geometric structures in the sense of Gromov [**G2**] (defined below), which is a generalization of the classical notion of structure of finite type in the sense of Cartan [**C, GS**]. The most natural relevant case of such a structure is that of an affine connection, and if the reader is so inclined he may restrict attention to that case without losing, for the most part, the essence of the nature of the results or of the techniques of the proofs. There is also considerable information available in the case of general nonrigid structures, but for this we shall refer the reader to [**Z6, Z4**]. We shall also not discuss the special (nonrigid) situation of complex structures, on which of course there is a large literature.

Let M be a smooth n-manifold. We shall often need M to be real analytic, but shall be explicit about when this assumption is being made. We let $G^{(r)}$ be the group of r-jets of diffeomorphisms of $\mathbb{R}^n$ fixing the origin, and $P^{(r)}(M)$ the principal $G^{(r)}$-bundle of r-frames on M. We recall that an r-frame at $m \in M$ is simply an r-jet of a local diffeomorphism $(\mathbb{R}^n, 0) \to (M, m)$. Thus for $r = 1$, $G^{(r)}$ is naturally isomorphic to $\mathrm{GL}(n, \mathbb{R})$ and $P^{(1)}(M)$ is the usual frame bundle. Suppose V is a manifold on which $G^{(r)}$ acts smoothly. We let $E_V \to M$ be the associated fiber bundle with fiber V. By a structure of order r and type V on M we mean a section of the bundle E_V. Such a structure will be called of algebraic type if V is a real algebraic variety on which $G^{(r)}$ (which we recall is a real algebraic group)

1991 *Mathematics Subject Classification.* Primary 58D19.
Research partially supported by the National Science Foundation.
This paper is in final form and no version will be submitted for publication elsewhere.

acts regularly. By a G-structure of order r we mean a structure of type $G^{(r)}/G$, where $G \subset G^{(r)}$ is a closed subgroup. This is of course equivalent to the assertion that one has a reduction of the bundle $P^{(r)}(M)$ to G. A G-structure will be of algebraic type if G is an algebraic subgroup.

We call a G-structure unimodular if the image of G in $\mathrm{GL}(n, \mathbb{R}) = G^{(1)}$ under the natural projection of jets $G^{(r)} \to G^{(1)}$ is contained in the group of matrices whose determinant is ± 1. (In other words the G-structure defines a volume density.) More generally, a structure ω of type V will be called unimodular if for each $m \in M$, the $G^{(r)}$-orbit in V determined by $\omega(m)$ has stabilizers whose image in $\mathrm{GL}(n, \mathbb{R}) = G^{(1)}$ under the natural projection of jets $G^{(r)} \to G^{(1)}$ is contained in the group of matrices whose determinant is ± 1. Thus, any unimodular structure of type V determines a volume density on M.

The group $\mathrm{Diff}(M)$ acts naturally on $P^{(r)}(M)$, hence on E_V for any $G^{(r)}$-space V, and hence on the space of structures of type V. If ω is such a structure, we let $\mathrm{Aut}(M, \omega)$ be the stabilizer of ω in the group $\mathrm{Diff}(M)$. In the case of rigid structures, $\mathrm{Aut}(M, \omega)$ will be a Lie group (but possibly infinite and of dimension 0), and it is of course a basic classical question to understand the relationship between ω, M, and $\mathrm{Aut}(M, \omega)$. While our main interest here will be specifically in the relationship of $\mathrm{Aut}(M, \omega)$ and $\pi_1(M)$, we mention one other basic result, first proved in [Z3], that we will need to establish along the way. We emphasize that rigidity of the structure is not required in this theorem.

THEOREM 1.1 [Z3]. *Suppose H is a noncompact connected simple Lie group, M is compact, and ω is a unimodular G-structure of algebraic type. If there is a nontrivial action $H \to \mathrm{Aut}(M, \omega)$, then there is an embedding of Lie algebras $\mathfrak{h} \to \mathfrak{g}$. Furthermore, under this inclusion the corresponding representation $\mathfrak{h} \to \mathfrak{gl}(n, \mathbb{R})$ (obtained by composition with the derivative of projection of jets $G \to \mathrm{GL}(n, \mathbb{R})$) contains $\mathrm{ad}_{\mathfrak{h}}$ as a subrepresentation.*

By taking $G = \mathrm{SO}(p, q)$, for example, we obtain considerable information on the possible automorphism groups of pseudo-Riemannian manifolds. This becomes particularly striking in the case of Lorentz manifolds, where one deduces

COROLLARY 1.2 [Z3]. *Let M be a compact Lorentz manifold. Then the connected component of the identity of $\mathrm{Aut}(M)$ is either:*

(a) *locally isomorphic to $\mathrm{SL}(2, \mathbb{R}) \times K$ where K is compact; or*

(b) *a compact extension of a 2-step nilpotent group.*

We shall indicate the proof of Theorem 1.1 in §3 below. For the proof of the corollary, see [Z3].

Let ω be a structure of order r and type V. For $k \geq r$ and $x, y \in M$, we let $\mathrm{Aut}^{(k)}(\omega, x, y)$ be the k-jets of local diffeomorphisms of M taking x to y and $\omega(x)$ to $\omega(y)$ up to order k. We set $\mathrm{Aut}^{(k)}(\omega, m, m) =$

$\mathrm{Aut}^{(k)}(\omega, m)$, the k-jets at m of local diffeomorphisms of M fixing m and ω at m up to order k.

DEFINITION 1.3 [G2]. The structure ω is k-rigid if for each $m \in M$, and $j \geq k$, the natural projection map $\mathrm{Aut}^{(j)}(\omega, m) \to \mathrm{Aut}^{(k)}(\omega, m)$ is injective. A structure is called rigid if it is k-rigid for some k.

Thus, k-rigidity entails the assertion that the k-jet of an infinitesimal automorphism at a point determines the infinitesimal automorphism at that point.

EXAMPLE 1.4. Pseudo-Riemannian structures are rigid structures of order 1. Affine connections are rigid structures of order 2. More generally, any G-structure of finite type in the sense of Cartan is rigid.

We can now state some of the main results concerning the relation with fundamental groups. The remaining sections of the paper will essentially be devoted to developing the ideas necessary to explain the proofs of these results.

For Theorems 1.5–1.10, we let M be a compact real analytic manifold, and ω a real analytic, rigid, unimodular structure of algebraic type on M. We let $\mathrm{Aut}^{\mathrm{an}}(M, \omega)$ be the real analytic automorphisms. (For example, we can take ω to be an analytic affine connection and volume density, in which case $\mathrm{Aut}^{\mathrm{an}}(M, \omega)$ is simply the group of affine, volume-preserving analytic diffeomorphisms.) Let H be a connected noncompact simple Lie group with finite center. We suppose there is a nontrivial action $H \to \mathrm{Aut}^{\mathrm{an}}(M, \omega)$, which without loss of generality we may assume is faithful.

THEOREM 1.5 (Gromov [G2]). *There exists a representation $\sigma: \pi_1(M) \to \mathrm{GL}(q, \mathbb{R})$ for some q such that the algebraic hull (i.e., Zariski closure) of $\sigma(\pi_1(M))$ contains a group locally isomorphic to H.*

For example, by a theorem of C. C. Moore [M], the algebraic hull of an amenable group is amenable. Thus, we deduce from Theorem 1.5 that no compact analytic manifold with amenable fundamental group admits an H-action preserving an analytic connection and volume. In particular, of course, M cannot be simply connected. The proof of Theorem 1.5 will be discussed in §4. In the simply connected case, Corollary 1.2 can be strengthened as follows.

THEOREM 1.6 (d'Ambra [d'A]). *Let M be a simply connected real analytic Lorentz manifold. Then the group of automorphisms is compact.*

Returning now to the hypotheses described before Theorem 1.5, we have

THEOREM 1.7 [Z7]. *Assume $\mathbb{R}$-rank$(H) \geq 2$ and that $\pi_1(H)$ is finite. Suppose that $\pi_1(M)$ admits a faithful linear representation $\sigma: \pi_1(M) \to \mathrm{GL}(q, \mathbb{C})$ for some q such that either:*
  (i) *$\sigma(\pi_1(M))$ is discrete; or*
  (ii) *$\sigma(\pi_1(M)) \subset \mathrm{GL}(q, \overline{\mathbb{Q}})$.*

Then $\pi_1(M)$ *contains a lattice in a linear Lie group* L, *where* L *contains a group locally isomorphic to* H.

The proof of Theorem 1.7 will be discussed §5. If $\pi_1(H)$ is not finite, one can often obtain strong information directly from 1.7 by passing to a subgroup $H_1 \subset H$ for which $\pi_1(H_1)$ is finite and which still satisfies the $\mathbb{R}$-rank condition.

COROLLARY 1.8. *With the hypotheses of Theorem 1.7,* $\pi_1(M)$ *is not isomorphic to a discrete subgroup of a Lie group of real rank strictly smaller than that of* H. *In particular, it is not isomorphic to a discrete subgroup of a Lie group of real rank one.*

THEOREM 1.9 [SZ]. *Assume* $\mathbb{R}$-rank$(H) \geq 2$, *and* $\pi_1(H)$ *is finite. Then* $\pi_1(M)$ *is not isomorphic to the fundamental group of a complete Riemannian manifold of negative sectional curvature bounded away from* 0 *and* $-\infty$.

While the above results give obstructions to the appearance of a group Λ as the fundamental group of a manifold admitting a suitable H-action, there are groups which may appear as such a fundamental group, but only in manifolds of large dimension. Namely:

THEOREM 1.10 [Z7]. *Fix* H *as above with* $\mathbb{R}$-rank$(H) \geq 2$, *and fix a positive integer* m. *Then there is a finitely generated group* Λ *with the properties*:

(1) *If* dim $M \leq m$ (*and* M *is as above*), *there is no isogeny* (*i.e., surjection with finite kernel*) $\pi_1(M) \to \Lambda$; *and*

(2) *There exists a compact analytic manifold* X (*necessarily of dimension greater than* m) *with an isogeny* $\pi_1(X) \to \Lambda$ *such that there is a real analytic action of* H *on* X *preserving an analytic pseudo-Riemannian metric.*

The group Λ may in fact be taken to be a suitable cocompact lattice in $\mathrm{SL}(p, \mathbb{R})$ for large enough p, where p is a prime. We sketch the proof in §5.

We can obtain additional information on $\pi_1(M)$ if we make further dynamical assumptions on the H-action. See for example Theorem 5.3 below.

These results all concern connected subgroups of $\mathrm{Aut}^{\mathrm{an}}(M, \omega)$. For actions of discrete groups, one sees immediately that the situation is more complex. For example, the natural action of $\mathrm{SL}(n, \mathbb{Z})$ on T^n shows that Theorem 1.5 is no longer true, since one has actions on manifolds with abelian fundamental group. In this case, the action of $\mathrm{SL}(n, \mathbb{Z})$ on the fundamental group of T^n is of course nontrivial, which one can never have for actions of connected groups. The relevance of this remark is highlighted by the following (new) result whose proof we give in §6.

THEOREM 1.11. *Let* (M, ω) *be a compact analytic manifold with a real analytic, rigid, unimodular structure of algebraic type, and suppose* $\pi_1(M)$ *is isomorphic to an infinite finitely generated nilpotent group. Let* $n \geq 6$. *Then there is no analytic ergodic action of* $\Gamma = \mathrm{SL}(n, \mathbb{Z})$ *on* M *preserving* ω *such that the natural homomorphism* $\Gamma \to \mathrm{Out}(\pi_1(M))$ *is trivial.*

We remark that the same is true if $\mathrm{SL}(n, \mathbb{Z})$ is replaced by any finite index subgroup of $\mathrm{SL}(n, \mathbb{Z})$, or more generally by a finite index subgroup Γ of the group of integer points of a simple algebraic group over $\mathbb{Q}$ with $\mathbb{Q}$-rank at least 2, and which satisfies $H^2(\Gamma, \mathbb{R}) = 0$. This latter condition holds for most groups for which the associated symmetric space is non-Hermitian [**B**].

In the simply connected case, one can obtain stronger results. See Theorem 6.1 below.

2. Infinitesimal, local, and global automorphisms

In this section we summarize (mostly without proof) some basic results relating general properties of infinitesimal, local, and global automorphisms.

For any structure ω, and $x, y \in M$, by $\mathrm{Aut}^{\mathrm{loc}}(\omega, x, y)$ we mean the group of germs of automorphisms of ω which are defined in a neighborhood of x and take x to y. We set $\mathrm{Aut}^{\mathrm{loc}}(\omega, m, m) = \mathrm{Aut}^{\mathrm{loc}}(\omega, m)$.

Throughout this section we take M *to be a manifold with a rigid structure* ω.

THEOREM 2.1 (Gromov). (a) *There is a positive integer* k, *and an open dense set* $U \subset M$ *such that for* $x, y \in U$, *every element of* $\mathrm{Aut}^{(k)}(\omega, x, y)$ *extends uniquely to an element of* $\mathrm{Aut}^{\mathrm{loc}}(\omega, x, y)$. *In particular, for* $m \in U$, *every element of* $\mathrm{Aut}^{(k)}(\omega, m)$ *extends uniquely to an element of* $\mathrm{Aut}^{\mathrm{loc}}(\omega, m)$.

(b) *If in addition* (M, ω) *is compact and analytic, we may take* $U = M$.

A proof of this result is given in [**G1**] or [**G2**]. Theorem 2.1 is similar in spirit and closely related to earlier basic work of Singer [**S**], in which he proves that a Riemannian manifold which is homogeneous up to some large (but finite) order is actually locally homogeneous.

If X is a vector field defined on an open subset of M, then X is called a Killing field or a symmetry field of ω if $L_X\omega = 0$. We let $\mathrm{Kill}(M, \omega)$ be the space of globally defined Killing fields, and for $m \in M$, we let $\mathrm{Kill}^{\mathrm{loc}}(\omega, m)$ be the space of germs at m of Killing fields defined in a neighborhood of m.

THEOREM 2.2. *Suppose further that* M *is analytic and simply connected, and that* ω *is analytic. Then for every* $m \in M$, *every element of* $\mathrm{Kill}^{\mathrm{loc}}(\omega, m)$ *extends uniquely to an element of* $\mathrm{Kill}(M, \omega)$.

This theorem is due to Nomizu [**N**] for the case in which ω is a complete parallelism. It was observed that this is still true by basically the same argument for structures of finite type by Amores [**A**], and for rigid structures by Gromov [**G2**]. The next result is a straightforward generalization of classical assertions about structures of finite type.

THEOREM 2.3. *If ω is a rigid structure, then $\mathrm{Aut}(M, \omega)$ is a Lie group in such a way that the action on M is smooth.*

One difficulty in applying Theorem 2.3 is the fact that in general $\mathrm{Aut}(M, \omega)$ may be an infinite discrete group. It is obviously of interest to have natural conditions under which this group is connected, or at least has finitely many components.

THEOREM 2.4 (Gromov [**G2**]). *If ω is a rigid analytic structure of algebraic type, and M is simply connected and compact, then $\mathrm{Aut}(M, \omega)$ has finitely many components, as does the stabilizer of each point in m.*

PROOF (sketch). Suppose ω is of order r and type V. Consider the kth jet bundle $E_V^{(k)} \to M$ of the basic associated bundle $E_V \to M$. We can naturally view this as a bundle having as fiber a naturally associated algebraic variety $V^{(k)}$ and with structure group $G^{(r+k)}$ which acts on $V^{(k)}$ regularly. The k-jet extension of the structure ω, say $j^k\omega$, defines continuous map $\varphi: M \to V^{(k)}/G^{(r+k)}$ which will be $\mathrm{Aut}(M, \omega)$-invariant. The assertion $\varphi(x) = \varphi(y)$ is equivalent to the assertion that $\mathrm{Aut}^{(k)}(\omega, x, y)$ is nonempty. It follows from Theorem 2.1 that for k sufficiently large the fibers of φ are precisely the orbits of the pseudogroup $\mathrm{Aut}^{\mathrm{loc}}(M, \omega)$. One can also show that these are submanifolds. The points in $V^{(k)}/G^{(r+k)}$ are not necessarily closed, but by basic results about algebraic group actions each point is open in its closure, and every closed set contains a closed point. It follows that there is a closed fiber of φ, i.e., a closed submanifold $N \subset M$ which is an orbit of the pseudogroup $\mathrm{Aut}^{\mathrm{loc}}(M, \omega)$. Therefore each $m \in N$ has an open neighborhood in N contained in an orbit of the local flow generated by a local Killing field near m. By Theorem 2.2 and the completeness of the Killing fields (since M is compact), we deduce that each $m \in N$ has an open neighborhood in N contained in the $\mathrm{Aut}(M, \omega)$-orbit of m. Therefore, the $\mathrm{Aut}(M, \omega)$ orbit of m has only finitely many connected components, and so to prove the theorem it suffices to see that all stabilizers in $\mathrm{Aut}(M, \omega)$ of points in M have finitely many components. However by completeness of Killing fields and Theorem 2.2, every element of $\mathrm{Aut}^{\mathrm{loc}}(\omega, m)^0$, the connected component of the identity in $\mathrm{Aut}^{\mathrm{loc}}(\omega, m)$, extends to a unique element of the global stabilizer $\mathrm{Aut}(M, \omega)_m$. Thus, we have inclusions $\mathrm{Aut}^{\mathrm{loc}}(\omega, m)^0 \subset \mathrm{Aut}(M, \omega)_m \subset \mathrm{Aut}^{\mathrm{loc}}(\omega, m)$. However by Theorem 2.1, we can identify $\mathrm{Aut}^{\mathrm{loc}}(\omega, m)$ with $\mathrm{Aut}^{(k)}(\omega, m)$ for k sufficiently large. This is an algebraic group, and hence has only finitely many components.

3. Actions on principal bundles, the algebraic hull, and the generalized Borel density theorem

In this section we present some basic information on actions of groups on principal bundles, and in particular discuss the notion of the algebraic hull of such a group action. We use the Borel density theorem to show how to obtain information about this invariant.

Let $P \to M$ be a principal G-bundle and suppose a group H acts on P by principal bundle automorphisms, i.e., it acts on P commuting with the action of G. If G is an algebraic group, the algebraic hull of the action of H on P is (the conjugacy class of) an algebraic subgroup that describes the smallest algebraic subgroup for which P has a measurable H-invariant reduction. This in turn controls the nature of the invariant sections of any associated bundle with an algebraic variety as the fiber. To define the algebraic hull, it is easier to first consider the case in which we have an ergodic measure μ on M for the action of H. The general case can easily be reduced to this case via the ergodic decomposition. (See [**Z5**] for a discussion in this context.) We also remark that the algebraic hull can be defined for actions on principal bundles where the structure group is (the group of k-points of) an algebraic group over any local field of characteristic 0. While this is useful even for considerations of a differential geometric nature, we shall for simplicity restrict our attention here to real groups.

THEOREM 3.1. *If* G *is real algebraic and* H *acts by principal bundle automorphisms of* $P \to M$ *with a quasi-invariant ergodic measure* μ *on* M, *then there is an algebraic subgroup* $L \subset G$, *unique up to conjugacy, with the following property*:

(i) *There is a measurable* H-*invariant section of the associated bundle* $P/L \to M$, *but there is no such invariant section for any proper algebraic subgroup of* L.

Furthermore, L *satisfies*:

(ii) *If* V *is any real algebraic variety on which* G *acts regularly, and there is a measurable* H-*invariant section* φ *of the associated bundle* $E_V \to M$, *then for* μ-*almost every* $m \in M$, *the* G-*orbit in* V *corresponding to* $\varphi(m)$ *has each of its stabilizers containing a conjugate of* L.

In particular, if there is an (even measurable) H-*invariant reduction of* P *to an algebraic subgroup* G_1, *then* G_1 *contains a conjugate of* L.

DEFINITION 3.2. (i) Under the hypotheses of Theorem 3.1, the group L (or more precisely its conjugacy class in G) is called the algebraic hull of the action of H on P.

(ii) If we remove the ergodicity hypothesis, but assume we still have a given quasi-invariant measure (e.g., any smooth measure for a smooth action on a manifold), then by the algebraic hull we will mean the measurable map $M \to \{$Conjugacy classes of algebraic subgroups of $G\}$ obtained from (i) and the ergodic decomposition. (Thus, this map is constant on the ergodic components on M.)

We remark that in (ii), the algebraic hull is well defined as such a measurable map up to equality on conull sets.

For a proof and discussion of Theorem 3.1, see [**Z1**]. The discussion there also shows that the notion of algebraic hull is more generally defined for actions of groups on measurable principal bundles, and that the algebraic hull is invariant under measurable isomorphism of principal G-bundles on which H acts by principal bundle automorphisms. The following two propositions are clear, but we record them for ease of future use.

PROPOSITION 3.3. *If $P \to M$ is a measurable principal G_1-bundle, $Q \to M$ is a measurable principal G_2-bundle, $p: G_1 \to G_2$ is a surjection of algebraic groups, $f: P \to Q$ is a measurable p-compatible bundle map (covering the identity), H acts on P, Q by bundle automorphisms commuting with f, and $L_i \subset G_i$ are the respective algebraic hulls, then we can choose L_i (i.e., fix an element in the conjugacy class) such that $p(L_1) \subset L_2$ and is a Zariski dense subgroup of finite index. (Over $\mathbb{C}$, it would be a surjection, but here one needs to account for the fact that images of real algebraic groups need only be of finite index in their Zariski closure.)*

PROPOSITION 3.4. *Let $P \to M$ be a principal Γ-bundle where Γ is not necessarily algebraic. (E.g., we shall later need the case of Γ discrete.) Assume H acts by principal bundle automorphisms. Let $\sigma: \Gamma \to G$ be a homomorphism into an algebraic group, and let L be the Zariski closure of the image. Let $Q \to M$ be the associated principal G-bundle, on which we then have a natural induced action of H. Then the algebraic hull of the action of H on Q is contained in L (for almost every point m).*

We now recall the Borel density theorem. We restrict our attention to simple Lie groups, although it can be formulated more generally. Suppose H is a connected simple noncompact real algebraic group. If $L \subset H$ is a closed subgroup and H/L has a finite H-invariant measure, then Borel's theorem asserts that L is Zariski dense in H. (See [Z1] for a proof.) Equivalently, if L is an algebraic subgroup and H/L has a finite invariant measure, then $L = H$. Yet another equivalent formulation is the assertion that if H acts regularly on a variety, any H-invariant probability measure is supported on the fixed points of H. This has the following consequence.

PROPOSITION 3.5. *Let H be as above, and $\rho: H \to G$ a nontrivial rational homomorphism of real algebraic groups. Let H act on M with a finite invariant measure, and let $P = M \times G$ be the product principal G-bundle, on which H acts by the product action. Then the algebraic hull is the Zariski closure of $\rho(H)$, and in particular it contains a group locally isomorphic to H.*

PROOF. An H-invariant section of P/L for an algebraic subgroup $L \subset G$ is simply an H-map $M \to G/L$. Since the invariant measure on M pushes forward to one on G/L, Borel density implies that almost every point in the image is contained in the H-fixed points. Thus, L contains a conjugate of $\rho(H)$.

COROLLARY 3.6. *Suppose H is a simple noncompact Lie group with finite center acting (nontrivially) on a manifold M with a finite invariant smooth measure. Then there is an open dense conull set for which the stabilizers are discrete.*

PROOF. It suffices to see this set is conull, since it is clearly open. If $\mathfrak{h}_m$ is the Lie algebra of the stabilizer of m, then $m \to \mathfrak{h}_m$ is an H-map where H acts on the latter via Ad. Thus, for almost all m, $\mathfrak{h}_m$ is invariant under $\mathrm{Ad}(H)$, and hence by simplicity is either 0 or $\mathfrak{h}$. However, if H fixes a set of positive measure, so does the maximal compact subgroup, and hence by local linearizability for compact groups, the maximal compact subgroup, and hence H, acts trivially.

We now prove Theorem 1.1. A generalization of this argument, which we discuss in the next section, will be a basic step in the proofs of the other main results of §1.

PROOF OF THEOREM 1.1. For each $m \in M$, let $V_m \subset TM_m$ be the tangent space to the H-orbit. We have a natural isomorphism (for a.e. m) $\mathfrak{h} \cong V_m$ in such a way that for $h \in H$, $dh_m|V_m \colon V_m \to V_{hm}$ corresponds to $\mathrm{Ad}(h)$. Thus, by Propositions 3.3, 3.5, the algebraic hull of the action on the first-order frame bundle, say L, contains a group isomorphic to $\mathrm{Ad}(H)$. However, since there is an invariant G-structure, we have $L \subset G$, completing the proof.

4. Application to global Killing fields

In this section we use the techniques of the preceding section, combined with the results of §2, to obtain results on the space of globally defined Killing vector fields on the universal cover of M. As we shall see, this is a key point in relating the group H to $\pi_1(M)$. The development in this section is basically due to Gromov [G2].

There will be two classes of actions on principal bundles that will be most relevant for our considerations. If H acts smoothly on a manifold M, we always have such an action on $P^{(r)}(M) \to M$, and if H is simply connected we always have such an action on $M^\sim \to M$ viewed as a principal $\pi_1(M)$-bundle. In the latter case the structure group is of course not algebraic in general; however we obtain bundles with an algebraic structure group by forming the associated principal bundle for any representation $\pi_1(M) \to G$ where G is algebraic.

For the first class of bundles, we immediately see the relevance of the algebraic hull for obtaining further information on the automorphism group.

PROPOSITION 4.1. *Suppose H is any group acting on M preserving a structure ω of algebraic type on M. For every $k \geq 1$, let $L_k \subset G^{(k)}$ be the algebraic hull for the action on $P^{(k)}(M)$. Then for almost every $m \in M$, and every $k \geq 1$, $\mathrm{Aut}^{(k)}(\omega, m)$ contains a group isomorphic to L_k. Therefore in* *the rigid, analytic case, for k sufficiently large* $\mathrm{Aut}^{\mathrm{loc}}(\omega, m)$ *contains a group isomorphic to* L_k. *Thus if* $\tilde{m} \in \tilde{M}$ *projects to* $m \in M$, $\mathrm{Kill}(\tilde{M}, \tilde{\omega})_{\tilde{m}}$, *the globally defined Killing fields on* $\tilde{M}$ *for the lifted structure* $\tilde{\omega}$ *which vanish at* $\tilde{m}$, *contains a Lie algebra isomorphic to* $\mathfrak{L}_k$, *the Lie algebra of* L_k.

PROOF. By definition, the algebraic hull fixes the prolongation of ω to order k, and the corresponding subgroup of $\mathrm{Aut}^{(k)}(\omega, m)$ is simply obtained by conjugating from $(\mathbb{R}^n, 0)$ to (M, m). The last assertions then follow from the results of §2.

We now assume H to be a noncompact simple Lie group with finite center, and show how the results of §3 can be used to improve Proposition 4.1 for actions of these groups. There will be two essential improvements. First, we will be able to assert that the algebraic hulls L_k must each contain a group locally isomorphic to H. This of course immediately gives more information in 4.1. However, as in the proof of Theorem 1.1 in §3, we shall also see that the algebraic hull must be well behaved with respect to the tangent spaces (and higher order tangent spaces) to the orbits of the action of H.

THEOREM 4.2. *Suppose H is a noncompact simple Lie group with finite center acting (nontrivially) on M preserving a unimodular structure ω of algebraic type on M, where M is compact. Then*:

(1) *For almost every $m \in M$, and every $k \geq 1$, the algebraic hull L_k as well as* $\mathrm{Aut}^{(k)}(\omega, m)$ *contains a group locally isomorphic to H.*

(2) *Suppose further that ω is also analytic and rigid. Then*:

(i) *for almost every m and for k sufficiently large* $\mathrm{Aut}^{\mathrm{loc}}(\omega, m)$ *contains a group locally isomorphic to H.*

(ii) *Identify the Lie algebra $\mathfrak{h}$ of H with a Lie algebra of globally defined Killing fields on $\tilde{M}$ via the H action on M. For k sufficiently large, and almost every $m \in M$, let L be the algebraic hull at m (i.e., of the ergodic component containing m) of the action on $P^{(k)}(M)$. Suppose $\tilde{m} \in \tilde{M}$ projects to m. Then there is a Lie algebra $\mathfrak{L}$ of globally defined Killing fields on $\tilde{M}$ (depending on $\tilde{m}$) with the following properties*:

(a) $\mathfrak{L}$ *is isomorphic with the Lie algebra of L.*

(b) *All elements of $\mathfrak{L}$ vanish at $\tilde{m}$.*

(c) $\mathfrak{L}$ *normalizes $\mathfrak{h}$, i.e., $[\mathfrak{L}, \mathfrak{h}] \subset \mathfrak{h}$. Furthermore, the associated map* $\mathfrak{L} \to \mathrm{Der}(\mathfrak{h})$ *is in fact a surjection onto* $\mathrm{ad}(\mathfrak{h})$.

PROOF. (1) follows from 4.1 and the proof of Theorem 1.1 in §3. (2.i) then follows from Theorem 2.1. We take $\mathfrak{L}$ to be the local Lie algebra corresponding to $L \subset \mathrm{Aut}^{\mathrm{loc}}(\omega, m)$, each element of which extends uniquely to a globally defined Killing field by Theorem 2.2. To see (2.ii.c), we argue as in the proof of Theorem 1.1 in §3. Namely, let $T^{(k)}(M)$ be the bundle of k-jets of vector fields. This is an associated bundle to the principal bundle $P^{(k)}(M) \to M$. For each $m \in M$, we let $V_m \subset T^{(k)}(M)$ be the space of

k-jets of vector fields corresponding to the action of $\mathfrak{h}$ on M. We have for almost every m a natural identification $\mathfrak{h} \cong V_m$ in such a way that for $h \in H$, $d^{(k)}h_m|V_m: V_m \to V_{hm}$ corresponds to $\mathrm{Ad}(h)$ It follows that the algebraic hull L bay be chosen so that it also leaves V_m invariant and (by Proposition 3.4), if L^0 is the connected component of L, then $L^0|V_m$ will act by $\mathrm{Ad}(H)$. Since local or global Killing fields are uniquely determined by their k-jets at a point for k sufficiently large, it is clear that this implies assertion (2.ii.c).

Theorem 4.2 shows that the existence of a significant collection of global Killing fields on $M^\sim$ that do not come directly from the original action on M. As a result, we have

COROLLARY 4.3. *Suppose H is a noncompact simple Lie group with finite center acting (nontrivially) on a compact manifold M preserving a unimodular, rigid, analytic, structure ω of algebraic type. Identify the Lie algebra $\mathfrak{h}$ of H with a Lie algebra of globally defined Killing fields on $M^\sim$ via the action of H on M. Let $\mathfrak{z}$ be the centralizer of $\mathfrak{h}$ in the Lie algebra of all globally defined Killing fields on $M^\sim$. For $x \in M^\sim$, let $\mathfrak{z}(x)$ and $\mathfrak{h}(x)$ be the image of $\mathfrak{z}$ and $\mathfrak{h}$ respectively under the evaluation map at x. Then for almost all x, $\mathfrak{z}(x) \supset \mathfrak{h}(x)$.*

PROOF. Let $x \in M^\sim$ satisfy the conclusions of Theorem 4.2. Let L be the local Lie group of diffeomorphisms fixing x corresponding to $\mathfrak{L}$. For $h \in H$ sufficiently small, we can choose a small $g \in L$ such that g acts on $\mathfrak{h}$ by $\mathrm{Ad}(h^{-1})$. It follows that $h \circ g$ is a locally defined diffeomorphism near x that acts trivially on $\mathfrak{h}$, and hence commutes with H. Furthermore, $h \circ g(x) = h(x)$. It follows that if we let Z be the local group at x corresponding to $\mathfrak{z}$, then the local Z-orbit at x contains an open set in the local H-orbit of x. Hence, $\mathfrak{z}(x) \supset \mathfrak{h}(x)$.

While both 4.2 and 4.3 are of interest in terms of establishing the existence of new automorphisms and Killing fields, we now show how 4.3 can be used to obtain information on $\pi_1(M)$, and in particular, how it easily implies Theorem 1.5.

We first make the following simple (essentially tautological) remark which allows the connection of the above results to the fundamental group.

PROPOSITION 4.4. *Let M be a smooth manifold. Suppose V is a finite-dimensional vector space consisting of vector fields on $M^\sim$ and suppose that V is invariant under the natural $\pi_1(M)$-action on $M^\sim$. Let $E_V \to M$ be the associated bundle with fiber V. Then the evaluation map $M^\sim \times V \to TM^\sim$ factors to a vector bundle map $E_V = (M^\sim \times V)/\pi_1(M) \to (TM^\sim)/\pi_1(M) = TM$. If H is a connected, simply connected group acting smoothly on M, and each element of V is fixed under the induced action on $M^\sim$, then this map $E_V \to TM$ also commutes with the H-actions.*

We now prove Theorem 1.5.

PROOF OF THEOREM 1.5. Choose a point $x \in M^{\sim}$ satisfying the conclusions of 4.3. Choose V in Proposition 4.4 to be the Lie algebra $\mathfrak{z}$ in 4.3. Since H commutes with $\pi_1(M)$ on $M^{\sim}$, and $\mathfrak{z}$ is the centralizer of $\mathfrak{h}$, it follows that $\mathfrak{z}$ is $\pi_1(M)$-invariant. By Corollary 4.3, the image of $E_{\mathfrak{z}} \to TM$ contains the (measurable) subbundle consisting of the tangent space to the H-orbits. The algebraic hull for the action on this subbundle contains $\mathrm{Ad}(H)$ by Proposition 3.5, and hence so does the algebraic hull of the action on $E_{\mathfrak{z}}$, by Proposition 3.3. However by Proposition 3.4, this algebraic hull in turn is contained in the Zariski closure of the image of $\pi_1(M)$ under the representation $\pi_1(M) \to \mathrm{GL}(\mathfrak{z})$. Thus $\pi_1(M) \to \mathrm{GL}(\mathfrak{z})$ is the sought after representation.

The above argument also establishes the following result we shall need later.

COROLLARY 4.5. *Suppose H is a noncompact simple Lie group with finite center acting on a compact manifold M preserving a unimodular, rigid, analytic structure ω of algebraic type on M. Assume further that $\pi_1(H)$ is finite. Then for almost every $x \in M^{\sim}$, the $H^{\sim}$ orbit of x is proper; i.e., the stabilizer $H_x^{\sim}$ is compact and the map $H^{\sim} \to M^{\sim}$, $h \to hx$, induces a homeomorphism $H^{\sim}/H_X^{\sim} \to H^{\sim}x$.*

PROOF. If we let U be the open conull set given by Theorem 3.6, then it is clear that every orbit in $P(M)$ that projects to a point in U satisfies this properness condition, since H acts on the tangent space to the orbit foliation by Ad, and the latter is a surjection with finite kernel onto a closed subgroup. By 4.3 and 4.4, the same will be true for the frame bundle of the vector bundle $E_{\mathfrak{z}} \to M$. Since the latter is an associated bundle to the universal cover via a representation $\pi_1(M) \to \mathrm{GL}(\mathfrak{z})$, is is easy to check that the required properness condition holds on $M^{\sim}$.

5. Applications of superrigidity

The results of §§3 and 4 apply to any noncompact, connected, simple Lie group H of finite center. Under the additional assumption that $\mathbb{R}\text{-}\mathrm{rank}(H) \geq 2$, we shall obtain further results. The key additional fact that we obtain from this assumption is superrigidity, which yields a nearly complete description of the measure-theoretic structure of H-actions on principal bundles with algebraic structure group. This result, first proved in [Z2], is a generalization of Margulis's superrigidity theorem, which is essentially equivalent to the special case in which the base manifold M is H/Γ, where Γ is a lattice subgroup of H. See [Z1] for a discussion of both of these versions of superrigidity and [Z4] for some other geometric applications.

If $P \to M$ is a principal G-bundle on which H acts by automorphisms, a measurable section of P (which always exists) yields a measurable trivialization $P \cong M \times G$. The action of H on P is then given by $h \cdot (m, g) = (hm, \alpha(h, m)g)$, where $\alpha: H \times M \to G$ satisfies the cocycle identity. The superrigidity theorem asserts that, under suitable hypotheses, one can choose the measurable trivialization such that the cocycle α is of a particularly transparent form.

THEOREM 5.1 (Superrigidity [Z1, Z8]). *Let* $P \to M$ *be a principal G-bundle where* G *is a real algebraic group and* M *is compact. Let* H *be a connected, simple Lie group with* $\mathbb{R}$-$\mathrm{rank}(H) \geq 2$, *and suppose* H *acts by principal bundle automorphisms of* P, *preserving a finite measure on* M. *Lifting the action to* $H^{\sim}$, *we may assume* H *is also simply connected.*

(a) *Assume the* H-*action on* M *is ergodic and the algebraic hull of the action on* P *is (Zariski) connected. Then one may choose the measurable trivialization* $P \cong M \times G$ *so that the corresponding cocycle* α *has the following description. There is a continuous homomorphism* $\theta: H \to G$ *and a compact subgroup* $K \subset G$ *commuting with* $\theta(H)$ *such that* $\alpha(h, m) = \theta(h)c(h, m)$ *where* $c(h, m) \in K$.

(b) *Without the hypothesis of ergodicity, one may apply* (a) *to each ergodic component. Without the hypothesis that the algebraic hull is connected, one may obtain the same conclusion by pulling back* P *over a finite (measurable) extension of* M (*for each ergodic component*).

Assertion (a) is of course the central one, and (b) is basically one of technical adjustments that rarely cause a serious problem. We shall basically ignore these technicalities in the sequel. For a proof in the case the algebraic hull is semisimple, see [Z1]. For the extension to the form with the above hypotheses, see [Z8].

We shall also need a basic theorem of Ratner, which verified a conjecture of Raghunathan.

THEOREM 5.2 (Ratner [R]). *Let* G *be a connected Lie group,* $\Gamma \subset G$ *a discrete subgroup, and* $H \subset G$ *a connected simple noncompact Lie group. Let* ν *be a finite* H-*invariant ergodic measure on* G/Γ, *where the action of* H *is given by the embedding in* G. *Then there is a closed connected subgroup* L *with* $H \subset L \subset G$ *and a point* $x \in G/\Gamma$, *say with stabilizer in* G *being* $g\Gamma g^{-1}$, *such that:*

(i) $L \cap g\Gamma g^{-1}$ *is a lattice subgroup in* L; *and*

(ii) ν *is the measure on* G/Γ *corresponding to the* L-*invariant volume on* $L/L \cap g\Gamma g^{-1}$ *under the natural bijection* $L/L \cap g\Gamma g^{-1} \cong Lx \subset G/\Gamma$.

We now indicate the proof of Theorem 1.7.

PROOF OF THEOREM 1.7 (sketch). We consider first the easier case in which there is some faithful representation $\sigma: \pi_1(M) \to \mathrm{GL}(q, \mathbb{C})$ with discrete image. We let $\mathrm{GL}(q, \mathbb{C}) = G$, and $\Gamma = \pi_1(M)$ which we also identify with its image in G. Fixing a measurable trivialization of the Γ-bundle $M^{\sim} \to M$ yields a measurable trivialization of the associated principal G-bundle for which the associated cocycle $\beta: H \times M \to G$ actually takes values in Γ. Applying Theorem 5.1 (and ignoring the technicalities of 5.1.(b)), we can choose another measurable trivialization so that the corresponding cocycle α

satisfies the conclusions of 5.1. Since any two measurable trivializations of a principal bundle differ by a measurable map from the base to the structure group, α and β will differ by such a function. I.e., there is a measurable $\varphi \colon M \to G$ such that $\varphi(hm)\beta(h, m)\varphi(m)^{-1} = \alpha(h, m) = c(h, m)\theta(h)$, where c and θ are as in 5.1. There are now two cases to consider.

If θ is trivial, then we can write $\beta(h, m) = \varphi(hm)^{-1}c(h, m)\varphi(m)$. This implies that there is some set $X \subset M$ of positive measure and a compact set $A \subset G$ such that whenever $m, hm \in X$ we have $\beta(h, m) \in A$. Since there is a finite H-invariant measure, Poincaré recurrence implies that for almost every $m \in X$, we can find a sequence $h_n \in H$, $h_n \to \infty$, so that $h_n m \in X$. However, since the image of Γ in G is discrete, this contradicts the properness assertion of Corollary 4.5.

Suppose now that θ is not trivial. Then we have $c(h, m)^{-1}\varphi(hm)\beta(h, m) = \theta(h)\varphi(m)$. Letting $\lambda \colon M \to K\backslash G/\Gamma$ be the composition of φ with the natural projection onto the double coset space, we see that λ is an H-map. Thus, pushing the measure on M forward by λ, there is a $\theta(H)$-invariant finite measure on $K\backslash G/\Gamma$. Lifting by Haar measure of K, there is a $\theta(H)$-invariant finite measure on G/Γ, and the conclusion of theorem now follows by applying Theorem 5.2.

The proof in case (ii) in the statement of Theorem 1.7, i.e., when we assume there is a faithful (but not necessarily discrete) homomorphism into $\mathrm{GL}(q, \mathbb{Q}^-)$, is similar but a bit more involved. The main point is that any finitely generated subgroup of $\mathrm{GL}(q, \mathbb{Q}^-)$ has a natural realization as a discrete subgroup of a product of finitely many Lie groups with finitely many algebraic groups over p-adic fields. One can prove Theorem 5.1 for bundles with p-adic structure group and then use the same sort of argument. See [Z7] for details.

We now consider the proof of Theorem 1.10. Once again, we provide only an indication of the proof.

PROOF OF THEOREM 1.10 (sketch). Consider the proof of Theorem 1.7 above where we consider the special case in which $\Gamma = \pi_1(M)$ is a lattice in a Lie group G (which we now take only to be a subgroup of some $\mathrm{GL}(q)$). Suppose Γ has the following property, to which we will return in moment:

(∗)   If $L \subset G$ is a proper connected subgroup which intersects Γ in a lattice in L, then L is solvable.

If this is the case, the measure ν in the proof of 1.7 must be the standard G-invariant measure on G/Γ. This implies that we have a measure-preserving H-map $M \to K\backslash G/\Gamma$. Therefore, for each $h \in H$, the (measure-theoretic) entropy of h acting on M is at least that of h acting on $K\backslash G/\Gamma$. The latter is explicitly computable in terms of eigenvalues of $\mathrm{Ad}_G(h)$. In other words, if Γ satisfies property (∗) above, we obtain a lower bound on possible values of entropy for H acting on M with this given fundamental group. On the other hand, the values of entropy on M can be read off from the exponents of the action, which by superrigidity can be described in terms of the eigenvalues of $\theta(h)$, where θ is as in 5.1. If the dimension of M is bounded above, this explicitly bounds the entropy from above. Comparing these bounds we see that if the dimension of M is too small, the bounds are incompatible, and hence there is no such action. Finally, that there exist cocompact lattices satisfying $(*)$ in $\mathrm{SL}(p, \mathbb{R})$ for all primes p is an arithmetic theorem of Kottwitz. See [Z7] for a proof of Kottwitz's theorem and further details of the above proof.

With further dynamical assumptions on the H-action, we obtain further information on $\pi_1(M)$. Namely, let us call an H-action an Anosov action if there is some $\mathbb{R}$-split semisimple element $h \in H$ (i.e., $\mathrm{Ad}(h)$ is diagonalizable over $\mathbb{R}$) whose action on M is hyperbolic normal to the H-orbits. With the assumption of a smooth invariant measure, any locally free Anosov action will be ergodic, as will its lift to any finite cover of M. This condition, which is called "engaging" in [Z6], enables one to prove

THEOREM 5.3 [Z6]. *Let H be a connected simple Lie group with $\mathbb{R}$-rank(H) ≥ 2. Suppose there is an engaging (e.g. Anosov) action of H on a compact M preserving a unimodular structure. Let d be the smallest dimension of a nontrivial representation of the Lie algebra $\mathfrak{h} \to \mathfrak{gl}(d, \mathbb{C})$. Then any representation $\pi_1(M) \to \mathrm{GL}(r, \mathbb{C})$ where $r < d$ has finite image.*

The proof uses superrigidity in a spirit similar to the proof of Theorem 1.7 above. See [Z6] for details.

6. Actions of discrete groups

If M is a simply connected compact manifold with a rigid, unimodular, analytic structure ω of algebraic type, and we have an action $\Gamma \to \mathrm{Aut}^{\mathrm{an}}(M, \omega)$ where Γ is a discrete group, we may sometimes apply Theorem 2.4 and results about the finite-dimensional representation theory of Γ to obtain information about the action.

THEOREM 6.1. *With the assumptions as above, assume $\Gamma \subset H$ is a lattice subgroup where H is a connected, simple Lie group of finite center, finite fundamental group, and $\mathbb{R}$-rank$(H) \geq 2$. Then Γ preserves a Riemannian metric on M, or equivalently, the image of Γ in $\mathrm{Aut}^{\mathrm{an}}(M, \omega)$ is contained in a compact subgroup. In particular, if $\Gamma = \mathrm{SL}(n, \mathbb{Z})$, $n \geq 3$, then the action factors through a finite quotient of Γ. (This latter assertion is more generally true if H is the real points of a simple $\mathbb{Q}$-group with $\mathbb{Q}$-rank$(H) \geq 2$, and Γ is a corresponding arithmetic subgroup of integral points.)*

PROOF. By Theorem 2.4 $\mathrm{Aut}^{\mathrm{an}}(M, \omega)$ is almost connected. By Margulis's superrigidity theorem [M, Z1], if the image of Γ is not compact, $\mathrm{Aut}^{\mathrm{an}}(M, \omega)$ contains a group locally isomorphic to H. But this is impossible by Theorem 1.5 since M is simply connected.

We now consider the nonsimply connected case, where one can no longer appeal to Theorem 2.4. If Γ acts on M, we have a natural induced homomorphism $\Gamma \to \mathrm{Out}(\pi_1(M))$. In certain cases, we can generalize the results for connected group actions to the discrete case if we assume triviality of this homomorphism. An example is Theorem 1.11 and we now turn to the proof of this result. We need the following general fact.

PROPOSITION 6.2. *Suppose a group* Γ *acts on a compact manifold* M *preserving a rigid structure* ω *and a finite ergodic measure. Let* L_k *be the algebraic hull for the action on the bundle* $P^{(k)}(M) \to M$. *If there is a sufficiently large* k *for which* L_k *is compact, then the image of* Γ *in* $\mathrm{Aut}(M, \omega)$ *has compact closure.*

PROOF. Since there is a measurable Γ-invariant reduction of $P^{(k)}(M)$ to L_k, there is a finite Γ-invariant measure on $P^{(k)}(M)$. However for k sufficiently large, since ω is rigid, $\mathrm{Aut}(M, \omega)$ (and hence any closed subgroup) acts properly on $P^{(k)}(M)$. However, it is easy to see that a group that acts properly on a locally compact space preserving a finite measure must be compact.

PROOF OF THEOREM 1.11. Since $H^2(\Gamma, \mathbb{R}) = 0$ for all lattices in $\mathrm{SL}(n, \mathbb{R})$ for $n \geq 6$, [B], and every homomorphism from Γ into an abelian group has finite image [Z1], every element of $H^2(\Gamma, \mathbb{Z})$ is trivial on a subgroup of finite index. If we let Λ be the group of diffeomorphisms of $M^\sim$ covering elements of Γ, we have an exact sequence $0 \to \pi_1(M) \to \Lambda \to \Gamma \to 0$. A straightforward inductive argument then shows that by passing to a subgroup of Γ of finite index, this is a split extension. Therefore, replacing Γ by a subgroup of finite index, we can assume the action lifts to $M^\sim$ in such a way as to commute with $\pi_1(M)$. This in turn defines a cocycle $\Gamma \times M \to \pi_1(M)$, which since Γ has Kazhdan's property T and $\pi_1(M)$ is nilpotent must be equivalent to a cocycle taking values in a finite subgroup of $\pi_1(M)$. (See [Z1, Theorem 9.1.1] for a proof.) This implies that the ergodic components of the Γ action on $M^\sim$ are actually finite Γ-invariant measures, and in fact each one is simply the smooth measure on $M^\sim$ restricted to a Γ-invariant set of positive smooth measure.

Since every homomorphism of Γ into a compact Lie group has finite image (and finite groups cannot act ergodically on a manifold of positive dimension), we deduce from 6.2 that for large k we must have the algebraic hull L_k being nonfinite for every ergodic component. It follows from Proposition 4.1 that the space of global Killing fields on $M^\sim$, say W, is nonzero. In fact, if for $x \in M^\sim$ we let $W_x = \{X \in W | X(x) = 0\}$, then 4.1 shows that for almost all x we have $W_x \neq 0$. We have a natural representation $\Gamma \to \mathrm{GL}(W)$ and by Margulis's superrigidity theorem the algebraic hull of the image will be locally isomorphic to $\mathrm{SL}(n, \mathbb{R})$. The map $x \to W_x$ is a Γ-map $\varphi: M^\sim \to \mathrm{Grass}(W)$. Since each ergodic component of Γ on $M^\sim$ has finite measure, by the Borel density theorem the image of φ must be supported on $\mathrm{SL}(n, \mathbb{R})$ fixed points in $\mathrm{Grass}(W)$. Hence, φ is essentially constant on ergodic components. I.e., for almost all x, any $X \in W_x$ must vanish on the ergodic component through x. However, each ergodic component has positive measure with respect to the smooth measure, and this implies that $X = 0$, providing a contradiction.

Proposition 6.2 can also be used to prove the following result. See [Z4] for a complete proof and discussion of related results.

THEOREM 6.3 [Z4]. *Let H be a connected simple Lie group with $\mathbb{R}$-rank(H) ≥ 2, and let $\Gamma \subset H$ be a lattice subgroup. Let M be a compact manifold and ω a rigid unimodular G-structure of algebraic type on M. Suppose there is an action $\Gamma \to \mathrm{Aut}(M, \omega)$. Then either:*

(i) *There is an embedding of Lie algebras $\mathfrak{h} \to \mathfrak{g}$; or*

(ii) *The image of Γ is contained in a compact subgroup of $\mathrm{Aut}(M, \omega)$.*

COROLLARY 6.4 [Z4]. *Let $\Gamma \subset \mathrm{SL}(n, \mathbb{R})$ be a lattice, $n \geq 3$. Let M be a compact manifold with $\dim(M) < n$, and ω a rigid unimodular G-structure of algebraic type on M. Then any action $\Gamma \to \mathrm{Aut}(M, \omega)$ factors through a finite quotient of Γ.*

REFERENCES

[A] A. M. Amores, *Vector fields of a finite type G-structure*, J. Differential Geom. **14** (1979), 1–6.

[B] A. Borel, *Stable real cohomology of arithmetic groups*, Ann. Sci. École Norm. Sup. **7** (1974), 235–272.

[C] E. Cartan, *Oevres completes*, Gauthier-Villars, Paris, 1952–1955.

[D'A] G. D'Ambra, *Isometry groups of Lorentz manifolds*, Invent. Math. **92** (1988), 555–565.

[G1] M. Gromov, *Partial differential relations*, Springer, New York, 1986.

[G2] _____, *Rigid transformation groups*, Geometrie Differentielle (D. Bernard and Y. Choquet-Bruhat, editors), Hermann, Paris, 1988.

[GS] V. Guillemin and S. Sternberg, *Deformation theory of pseudogroup structures*, Mem. Amer. Math. Soc. No. 64, 1966.

[M] G. A. Margulis, *Discrete subgroups of Lie groups*, Springer, New York, 1991.

[Mo] C. C. Moore, *Amenable subgroups of semisimple groups an proximal flows*, Israel J. Math. **34** (1979), 121–138.

[N] K. Nomizu, *On local and global existence of Killing fields*, Ann. of Math. (2) **72** (1970), 105–112.

[R] M. Ratner, *On Raghunathan's measure conjecture*, Ann. of Math. **134** (1991), 545–607.

[S] I. M. Singer, *Infinitesimally homogeneous spaces*, Comm. Pure Appl. Math. **13** (1960), 685–697.

[SZ] R. Spatzier and R. J. Zimmer, *Fundamental groups of negatively curved manifolds and actions of semisimple groups*, Topology **30** (1991), 591–601.

[Z1] R. J. Zimmer, *Ergodic theory and semisimple groups*, Birkhäuser, Boston, 1984.

[Z2] _____, *Strong rigidity for ergodic actions of semisimple Lie groups*. Ann. of Math. (2) **112** (1980), 511–529.

[Z3] _____, *On the automorphism group of a compact Lorentz manifold and other geometric manifolds*, Invent. Math **83** (1986), 411–426.

[Z4] _____, *Lattices in semisimple groups and invariant geometric structures on compact manifolds*, Discrete Groups in Geometry and Analysis (R. Howe, editor), Birkhäuser, Boston, 1987, pp. 152–210.

[Z5] _____, *Ergodic theory and the automorphism group of a G-structure*, Group Representations, Ergodic Theory, Operator Algebras, and Mathematical Physics (C. C. Moore, editor), Springer, New York, 1987, 247–278.

[Z6] ____, *Representations of fundamental groups of manifolds with a semisimple transformation group*, J. Amer. Math. Soc. **2** (1989), 201–213.

[Z7] ____, *Superrigidity, Ratner's theorem, and fundamental groups*, Israel J. Math. **74** (1991), 199–207.

[Z8] ____, *On the algebraic hull of the automorphism group of a principal bundle*, Comment. Math. Helv. **65** (1990), 375–387.

UNIVERSITY OF CHICAGO

Recent Titles in This Series

(*Continued from the front of this publication*)

30 **R. O. Wells, Jr., editor,** Several complex variables (Williams College, Williamstown, Massachusetts, July/August 1975)

29 **Robin Hartshorne, editor,** Algebraic geometry – Arcata 1974 (Humboldt State University, Arcata, California, July/August 1974)

28 **Felix E. Browder, editor,** Mathematical developments arising from Hilbert problems (Northern Illinois University, Dekalb, May 1974)

27 **S. S. Chern and R. Osserman, editors,** Differential geometry (Stanford University, Stanford, California, July/August 1973)

26 **Calvin C. Moore, editor,** Harmonic analysis on homogeneous spaces (Williams College, Williamstown, Massachusetts, July/August 1972)

25 **Leon Henkin, John Addison, C. C. Chang, William Craig, Dana Scott, and Robert Vaught, editors,** Proceedings of the Tarski symposium (University of California, Berkeley, June 1971)

24 **Harold G. Diamond, editor,** Analytic number theory (St. Louis University, St. Louis, Missouri, March 1972)

23 **D. C. Spencer, editor,** Partial differential equations (University of California, Berkeley, August 1971)

22 **Arunas Liulevicius, editor,** Algebraic topology (University of Wisconsin, Madison, June/July 1970)

21 **Irving Reiner, editor,** Representation theory of finite groups and related topics (University of Wisconsin, Madison, April 1970)

20 **Donald J. Lewis, editor,** 1969 Number theory institute (State University of New York at Stony Brook, Stony Brook, July 1969)

19 **Theodore S. Motzkin, editor,** Combinatorics (University of California, Los Angeles, March 1968)

18 **Felix Browder, editor,** Nonlinear operators and nonlinear equations of evolution in Banach spaces (Chicago, April 1968)

17 **Alex Heller, editor,** Applications of categorical algebra (New York City, April 1968)

16 **Shiing-Shen Chern and Stephen Smale, editors,** Global analysis, Part III (University of California, Berkeley, July 1968)

15 **Shiing-Shen Chern and Stephen Smale, editors,** Global analysis, Part II (University of California, Berkeley, July 1968)

14 **Shiing-Shen Chern and Stephen Smale, editors,** Global analysis, Part I (University of California, Berkeley, July 1968)

13 **Dana S. Scott (Part 1) and Thomas J. Jech (Part 2), editors,** Axiomatic set theory (University of California, Los Angeles, July/August 1967)

12 **William J. LeVeque and Ernst G. Straus, editors,** Number theory (Houston, Texas, January 1967)

11 **S. S. Chern, L. Ehrenpreis, J. Korevaar, W. H. J. Fuchs, and L. A. Rubel, editors,** Entire functions and related parts of analysis (University of California, San Diego, July 1966)

10 **Alberto P. Calderón, editor,** Singular integrals (University of Chicago, April 1966)

9 **Armand Borel and George D. Mostow, editors,** Algebraic groups and discontinuous subgroups (University of Colorado, Boulder, July 1965)

8 **Albert Leon Whiteman, editor,** Theory of numbers (California Institute of Technology, Pasadena, November 1963)

7 **Victor L. Klee, editor,** Convexity (University of Washington, Seattle, June 1961)

(See the AMS catalog for earlier titles)

ISBN 0-8218-1496-6